AF478310

Theory and Phenomenology of Sparticles

An account of four-dimensional $N=1$ supersymmetry in High Energy Physics

THEORY AND PHENOMENOLOGY of
Sparticles

An account of four-dimensional $N=1$
supersymmetry in High Energy Physics

Manuel Drees
University of Bonn

Rohini Godbole
Indian Institute of Science, Bangalore

Probir Roy
Tata Institute of Fundamental Research, Mumbai

World Scientific

NEW JERSEY · LONDON · SINGAPORE · BEIJING · SHANGHAI · HONG KONG · TAIPEI · CHENNAI

Published by

World Scientific Publishing Co. Pte. Ltd.

5 Toh Tuck Link, Singapore 596224

USA office: 27 Warren Street, Suite 401-402, Hackensack, NJ 07601

UK office: 57 Shelton Street, Covent Garden, London WC2H 9HE

British Library Cataloguing-in-Publication Data
A catalogue record for this book is available from the British Library.

First published 2004
Reprinted 2005

THEORY AND PHENOMENOLOGY OF SPARTICLES

ISBN 981-02-3739-1
ISBN 981-256-531-0 (pbk)

Printed by FuIsland Offset Printing (S) Pte Ltd, Singapore

*To those of our parents who are living
and the memory of those who are not.*

"तत्त्वम् भावेन"

'वैशेषिकसूत्र' - कणाद

"Truth via existence"

'The Atomic Code' - Kaṅāda

"Die Theorie ist ein Werkzeug
das wir durch Anwendung erproben"

'Die Logik der Forschung' - Karl Popper

"Theory is a tool which we test
through applications"

'The logic of scientific research' - Karl Popper

PREFACE

Supersymmetry (or SUSY in short) is a proposed invariance under generalized spacetime transformations linking fermions and bosons. One can say without exaggeration that it is one of the most strikingly beautiful recent ideas in Physics. Ever since the codification of the spin-statistics theorem in the nineteen thirties, most physicists had developed an attitude towards fermions and bosons that can be summed up in a Kiplingesque adage: a fermion is a fermion, a boson is a boson and ne'er the twain shall meet. Supersymmetry breaks this attitude and makes them meet. It enables a fermion to transform into a boson and vice versa. It admits supermultiplets with fermionic and bosonic members. The couplings of those members get related and their masses are split by supersymmetry breaking effects.

Historically, supersymmetry originated at the beginning of the nineteen seventies through aesthetic reasoning rather than from factual evidence. (Earlier formulations, motivated by hadronic resonances and strings, had not been complete). The greater ultraviolet convergence of supersymmetric field theories and the nonrenormalization of some of their interactions immediately generated much formal interest. The close link between local supersymmetry transformations and gravity also attracted a lot of attention. However, the obviously different physical properties of the known fermions and bosons made it appear initially that any connection of supersymmetry with phenomena in particle physics might be remote. Yet astonishingly, within just about a decade of its first mathematical formulation, the situation in this respect changed dramatically. A very plausible reason was found, namely the stability of the observed weak interaction scale $\sim 10^2$ GeV vis-à-vis much higher scales, for the existence of broken supersymmetry in the real world. This led to the anticipation of new **sparticles**, superpartners of the old particles with an intra-supermultiplet mass splitting $M_s \sim 10^2$–10^3 GeV, possibly within the study reach of current or foreseeable high energy accelerators. Soon afterwards, with the birth of superstring theory, came additional theoretical motivation for supersymmetry as a low energy legacy of Planck scale physics.

Three decades after the inception of supersymmetry, the literature today is flush with phenomenological papers anticipating the experimental discovery of softly broken $N=1$ supersymmetry in the near future. The key new feature of this phenomenology is the predicted occurrence of sparticles. For each known elementary particle (including the confined quarks and gluons) there is supposedly at least one sparticle. It differs in spin from the particle by 1/2 and is endowed with opposite statistics, a mass difference of the order of M_s and couplings strictly related to the known particle couplings. When produced in the laboratory, sparticles are expected to exhibit unusual and unique kinematic behavior and decay patterns enabling the unambiguous confirmation of supersymmetry. Many calculations have been performed on sparticle production and major experiments are planned for detecting such signals. One of the major goals of the Large Hadron Collider, now being constructed in Geneva at CERN, is to discover these sparticles. Furthermore, precision study of the properties of the latter has been stated as a reason to propose the construction of high luminosity, high energy lepton colliders.

Supersymmetry also enables one to connect the unification of fundamental forces at high energies with phenomena studied in the laboratory. This comes about through the extrapolation of scale dependent 'running' coupling strengths of strong, weak and electromagnetic interactions, obtained from experiments performed at presently accessible energies, to much higher energy scales. The construction of supersymmetric grand unified theories has been an ambitious step towards the above unification. The relation between such theories and a higher dimensional superstring theory, which supposedly describes nature at and above Planckian energies ($\sim 10^{18}$ GeV), has been the subject of much study too. A further possible role for supersymmetry would be in linking particle physics with large scale structure in cosmology. This is envisaged with the lightest sparticle, expected in the minimal scenario to be stable, being the leading candidate for the cold component of cosmological nonbaryonic dark matter which is supposed to constitute nearly 86% of the total mass of matter in the Universe.

Supersymmetric field theories are most easily formulated in terms of spinorial supercharges. This needs an extension of our usual spacetime to superspace. In addition to the four spacetime coordinates, a pair of conjugate spinorial coordinates are included in the minimal version of superspace. The densities of the spinorial supercharge and its conjugate, which take fermions into bosons and vice versa, are generalized local functions (distributions) of superspace coordinates. Together with the generators of the Poincaré group, these charges form a graded Lie algebra: the super-Poincaré algebra. A supersymmetric system is one whose action is invariant under infinitesimal transformations of this algebra. Just as quantum fields can be defined in our usual spacetime, so can superfields be defined in superspace as polynomials in the spinorial coordinates with ordinary bosonic and fermionic fields as coefficients. The beauty and power of supersymmetry in a field theoretic context are best appreciated through the superfield formalism set up in superspace. Furthermore, the link between theoretical and phenomenological aspects of supersymmetry appears clearest when expressed in this language. It is this link that the present work aspires to bring into a clear view.

The aim of this book is to provide the reader with an introductory, pedagogical and user-friendly treatment of those observational and unifying aspects of softly broken $N{=}1$ supersymmetry in (3+1) dimensions that fall within a phenomenological ambit. However, we have tried to make it self-contained by including the requisite superfield formalism in the beginning (Part One). There are several excellent texts in the market (cited in the *Bibliography*) with expositions of these aspects. Our effort in this direction is not supposed to be in competition with them, but is only meant to help the uninitiated reader learn enough formalism to be able to understand the phenomenological consequences of supersymmetry which are dealt with in Part Two. We provide an introduction to the subject in Ch.1 by sketching the historical development of supersymmetry ideas in quantum field theory leading to the supersymmetric resolution of the naturalness (or, more specifically in terms of an overlying grand unified theory, the gauge hierarchy) problem which plagues the Standard Model of particle interactions. We then attempt to impart – in terms of one loop corrections in a toy model – a bird's eye view of naturalness and nonrenormalization in supersymmetric field theory.

The pedagogical part of the book, written in the style of a textbook with more or less complete derivations, starts with Part One entitled "Supersymmetry Formalism". It is structured as follows. The preliminary ideas and tools of the subject are introduced in a quantum mechanical setting in Ch.2. The $N{=}1$ supersymmetry algebra in (3+1) dimensions is derived and the contents of simple supermultiplets are discussed in Ch.3. In Ch.4 we start considering supersymmetry in a field theoretic context by introducing free superfields. Interacting superfields are treated in Ch.5 with discussions of the Wess-Zumino model, supersymmetric quantum electrodynamics (SQED), supersymmetric quantum chromodynamics (SQCD) and supersymmetric chiral gauge theory (SχGT). In Ch.6, supersymmetric perturbation theory is treated in superspace by means of supergraphs and the nonrenormalization theorem is explained with this methodology; one loop β- and γ-functions are also derived. Ch.7 contains a treatment of supersymmetry breaking, both of the explicit and of the spontaneous kinds, with special reference to the goldstino in the latter case.

We change gears after this. Part Two, entitled "Supersymmetry Phenomenology" and written more like a research monograph, does not always include derivations in which case suitable references are given. This part is devoted to a discussion of the supersymmetrization of the Standard Model as well as of the new phenomena that are consequently predicted to occur above the electroweak energy scale. The basic structure of the simplest extension, known as the Minimal Supersymmetric Standard Model (MSSM), is outlined in Ch.8 which introduces the various new sparticle fields and their couplings that become necessary on account of exact supersymmetry. We discuss the soft explicit breaking of supersymmetry and the mass mixing among various sparticles with the same charge in Ch.9; physical mass eigenstate sparticles as well as the interaction vertices in which they participate are also enumerated. Ch.10 describes the Higgs sector of the MSSM with Higgs mass spectra plus couplings including radiative corrections and highlights relations that follow from the supersymmetric identification of Higgs quartic self couplings with gauge couplings. Possible constraints on the soft supersymmetry breaking parameters in the MSSM from high scale physics are discussed in Ch.11. The origin of the said parameters from gravity mediation between observable and hidden sector superfields is discussed in detail in Ch.12; the latter also contains an annex presenting a brief introduction to the essential features of N=1 supergravity that are used in this chapter. The gauge mediation mechanism of supersymmetry breaking and related models are discussed in Ch.13. A critique of the MSSM is given in Ch.14 along with a survey of some proposed alternative scenarios. Issues related to present and future collider searches for sparticles and supersymmetric Higgs bosons, including limits already obtained, are covered in Ch.15. Ch.16 is devoted to a discussion of the cosmological consequences of supersymmetry with a focus on supersymmetric Dark Matter, gravitino cosmology and baryogenesis. The concluding Ch.17 discusses the future prospects of the entire supersymmetric scenario, in particular its viability in relation to forthcoming experiments by quantitatively specifying the amounts of fine tuning that get introduced as lower bounds on the masses of the yet unobserved sparticles keep getting pushed upwards.

We would like to make some disclaimers by stating what this book is not. First, we have not gone into nonperturbative aspects of supersymmetry such as the vacuum structure or phases of a supersymmetric gauge theory. Specifically, we have chosen to omit any serious treatment of dynamical supersymmetry breaking though we refer to and discuss it occasionally. Second, though this book has been written with a phenomenological ethos, it is not a comprehensive account of *all* phenomenological aspects of four dimensional supersymmetry. For the sake of keeping the length reasonable, we have not given any significant coverage to indirect phenomenological probes of supersymmetry, referring to them only sketchily. These include additional MSSM contributions to CP violation, possible supersymmetric implications for heavy flavor physics as well as detailed discussions of supersymmetric loop corrections in the MSSM interrelated with constraints from precision electroweak data. An informative review of these topics may be found in the article by Eigen, Gaitskell, Kribs and Matchev cited in the *Bibliography* at the end. We have also avoided any significant discussion of the interface between unified supersymmetric theories and String Theory. Finally, though we mention some braneworld scenarios, we have not gone in detail into supersymmetric models with extra dimensions.

Coming to the Chapter References, our selection has been illustrative (and hence subjective) rather than comprehensive. While these references are by no means complete, we have tried to list most relevant review article collections and books in the *Bibliography* at the end. Apologies are made for any omissions that may have occured here. A word also about the motivation behind writing this book. Our intent has not been to 'sell' supersymmetry at laboratory energies. We have chosen to present a specific minimal model (and its extensions) for the supersymmetric stabilization of the electroweak scale, after setting up the requisite formalism. We discuss its strengths and weaknesses as well as those of its features that open it to experimental confirmation or demolition in forthcoming colliders. One can then judge for oneself how desirable it all is. We shall deem our effort as worthwhile if the practitioners of phenomenlogical as well as theoretical supersymmetry find this work useful and if a beginner is motivated by reading it towards the pursuit of research in this direction.

It has taken several years to write this book. All of us, in the interim, have utilized visits to various institutes, departments and laboratories for the purpose of writing it. We have also benefited immensely from interactions with many colleagues: theorists and experimentalists alike. We acknowledge fruitful discussions and/or research collaborations on the subject with H. Baer, S. Banerjee, G. Bélanger, T. Bhattacharya, G. Bhattacharyya, F. Borzumati, F. Boudjema, B. Brahmachari, U. Chattopadhyay, D. Choudhury, S.R. Das, A. Datta, D. Denegri, M. Dittmar, D.A. Dicus, A. Djouadi, D.K. Ghosh, G.F. Giudice, M. Glück, M. Guchait, K. Hamaguchi, S. Heinemeyer, J. Kalinowski, R. Kaul, A. Kundu, E. Ma, N.K. Mondal, P. Majumdar, S.P. Martin, S. Mukhi, B. Mukhopadhyaya, P. Nath, M.M. Nojiri, G. Polesello, S. Raychaudhuri, E. Reya, D.P. Roy, S. Roy, U. Sarkar, S. Sen Gupta, M. Sher, K. Sridhar, X. Tata, S.P. Trivedi, T. Vachaspati, D. Wyler and Y. Yamada. We are grateful to high energy theory groups at the Tata Institute of Fundamental Research (Mumbai), the Centre for High Energy Physics of the Indian Institute of Science (Bangalore), Technical University (Munich), Laboratoire d'Annecy de Physique des Particules, Oxford University, University of California (Riverside), University of Hawaii (Manoa), University of Wisconsin (Madison), the Kavli Institute of Theoretical Physics (Santa Barbara), UNESP (Sao Paolo), Korea Institute of Advanced Studies (Seoul), College of William and Mary (Williamsburg) as well as at the CERN, DESY, INFN (Frascati) and Thomas Jefferson Laboratories for their hospitality. We are indebted to G.V. Ogale for preparing the latexscript and to R.S. Pawar for drawing the line diagrams with xfig. Additional secretarial assistance from R.N. Bathija and M.R. Shinde is acknowledged. We thank S. Raychaudhuri for designing the cover and A. Mukherjee for help in adjusting figure postscript files as well as in preparing the index.

M.D.
R.M.G.
P.R.
Spring 2004

Contents

INTRODUCTION AND OVERVIEW

PART ONE

SUPERSYMMETRY FORMALISM

Chapter 3. Algebraic Aspects   29

Chapter 4. Free Superfields in Superspace   49

Chapter 5. Interacting Superfields   71

Chapter 6. Superspace Perturbation Theory and Supergraphs   97

PART TWO

SUPERSYMMETRY PHENOMENOLOGY

Contents

Chapter 12. Gravity Mediated Supersymmetry Breaking

Chapter 13. Gauge Mediated Supersymmetry Breaking

Chapter 14. Beyond the MSSM

Some tips on using this book

The reader's familiarity with basic quantum field theory and the Standard Model of particle interactions at a textbook level (e.g. Peskin and Schroeder, *opere citato, Bibliography*), has been assumed. The introductory Ch.1 is largely for motivation and can be more easily followed by those already with some exposure to supersymmetry. The uninitiated reader, who may find it difficult to follow this chapter fully, is advised to browse through it first and come back for serious reading after finishing the more formal and pedagogical Part One. The latter, comprising Chs. 2–7, is a must for beginners. However, Ch.6 may be omitted by the less theoretically inclined reader who will then need to accept the nonrenormalization theorem and the Renormalization Group Evolution equations on faith. A cautionary note is that our conventions in describing supersymmetry transformations as well as a four component spinor in terms of its two component parts are different from those of Wess and Bagger, *op. cit., Bibl.* However, our Feynman rules are those of Haber and Kane, *loco citato, Bibl.*

Those, who already know supersymmetric field theory but would like to apprise themselves of the phenomenological consequences of softly broken $N{=}1$ supersymmetry in four dimensions, need only skim through Part One. That would be mainly to familiarize themselves with our notation and the Feynman rules of SQED, SQCD and SχGT. They can then directly go to Part Two, consisting of Chs. 8–16. Chs. 8–10 describe the MSSM largely in the "low" energy context and form the backbone of our phenomenological discussions. Again, those – who are already familiar with "low" energy MSSM but would like to see the links between high scale physics and laboratory signatures, cosmological aspects and non-minimal scenarios – may browse though these chapters and go directly to Ch.11. Ch.15 may be omitted without loss of continuity by readers uninterested in collider signals of supersymmetry. A similar statement can be made for the cosmological discussions of Ch.16. A final tip on references: using those given at the end of each chapter and the review articles/books cited in the *Bibliography*, a more complete set can be obtained by literature survey through the SPIRES database http://www.slac.stanford.edu/spires/hep/.

Corrections

A list of corrections (including typographical errors) will be posted on the website with the url 'http://www.th.physik.uni-bonn.de/Groups/Drees/book' and will be updated from time to time. The cooperation of readers in pointing out additional errors in the book to drees@th.physik.uni-bonn.de will be much appreciated.

MSSM sparticles and lower mass bounds[1]

Sfermions[2]		Lower mass bound (GeV)	Bosinos[2]		Lower mass bound (GeV)
$\tilde{u}$	u-squark	250	$\tilde{g}$	gluino	300 $(m_{\tilde{q}} = M_{\tilde{g}})$
$\tilde{d}$	d-squark	–do–			195 (otherwise)
$\tilde{c}$	c-squark	–do–			
$\tilde{s}$	s-squark	–do–	$\tilde{\chi}_1^0$	lightest neutralino	59 (mSUGRA) 40 (otherwise)
$\tilde{t}$	stop	135 $(\tilde{t}_1)$			
$\tilde{b}$	sbottom	91 $(\tilde{b}_1)$	$\tilde{\chi}_2^0$	next lightest neutralino	62.4
$\tilde{e}$	selectron[3]	99 $(\tilde{e}_R)$	$\tilde{\chi}_3^0$	second heaviest neutralino	99.9
$\tilde{\nu}_e$	e-sneutrino[3]	45	$\tilde{\chi}_4^0$	heaviest neutralino	–do–
$\tilde{\mu}$	smuon[3]	95 $(\tilde{\mu}_R)$	$\tilde{\chi}_1^\pm$	lighter chargino	103 (gauginolike) 99 (higgsinolike)
$\tilde{\nu}_\mu$	mu-sneutrino[3]	45			
$\tilde{\tau}$	stau[3]	80 $(\tilde{\tau}_1)$	$\tilde{\chi}_2^\pm$	heavier chargino	–do–
$\tilde{\nu}_\tau$	tau-sneutrino[3]	45	$\tilde{G}$	gravitino	1.0×10^{-14}

---

[1] M. Schmitt in *Supersymmetric Particle Searches*, *Review of Particle Physics*, Particle Data Group, Phys. Rev. **D66** (2002) 010001. Other assumptions, which go into the extraction of these bounds, are detailed in this review which also contains some recent improvements in older bounds plus stronger constraints in the plane of two sparticle masses (e.g. $M_{\tilde{\chi}_1^0}$ vs $m_{\tilde{\ell}}$, cf. Ch.15).

[2] All are expected to have masses in the sub-TeV to a few TeV range, though the gravitino $\tilde{G}$ could be much lighter or somewhat heavier. Two mass diagonal states $\tilde{f}_{1,2}$ exist for each charged sfermion $\tilde{f}$, being orthogonal combinations of mixed left $(\tilde{f}_L)$ and right $(\tilde{f}_R)$ chiral sfermion fields. However, the mixing is expected to be significant only for third generation sfermions. For any sfermion $\tilde{f}$, the corresponding antisfermion will be denoted by $\overline{\tilde{f}}$. Uncharged sfermions (sneutrinos) are purely left chiral and uncharged bosinos are Majorana fermions. An axino $\tilde{a}$ will also be called for as the superpartner of the axion a in case the latter is included in the Standard Model.

[3] We shall sometimes use the symbol $\tilde{e}_i$ to denote a charged slepton, i being a generation index and equal to $1, 2, 3$ for the selectron, smuon, stau respectively. Also, $\tilde{q}$ and $\tilde{\ell}$ will be used generically to stand for an electroweak doublet squark and a similar slepton respectively.

Acronyms and Abbreviations

AD	Affleck-Dine (Mechanism)
ALEPH	Apparatus (detector) for LEP Physics
AMSB	Anomaly Mediated Supersymmetry Breaking
ATLAS	A Toroidal LHC ApparatuS (Detector)
BB	Big Bang
BBN	Big Bang Nucleosynthesis
BR	Branching Ratio
BSM	Beyond Standard Model
CDF	Collider Detector at Fermilab
CDM	Cold Dark Matter
CERN	Organization of European Nuclear Research (Geneva)
CKM	Cabibbo-Kobayashi-Maskawa (Matrix)
c.l.	Confidence Level
CM	Center of Mass
CMS	Compact Muon Solenoid (LHC Detector)
CMB	Cosmic Microwave Background
CMSSM	Constrained Minimal Supersymmetric Model
$\tilde{\text{C}}$MSSM	Constrained Minimal Supersymmetric Model (Modified)
D$\emptyset$	A detector at the TEVATRON, Fermilab
DR	Dimensional Reduction (Regularization procedure)
$\overline{\text{DR}}$	Modified Dimensional Reduction
DELPHI	DEtector with Lepton, Photon & Hadron Identification (at LEP)
DESY	Deutsches Elektronen Synchrotron (Hamburg)
DY	Drell-Yan (Process)
EDM	Electric Dipole Moment
$\not{E}$	Missing Energy
E_T	Transverse Energy
$\not{E}_T$	Missing Transverse Energy
em	Electromagnetic
EW	Electroweak
FCNC	Flavor Changing Neutral Current
FN	Frogatt-Nielsen (Fields, Superfields)
FRW	Friedmann-Robertson-Walker (Metric)
FP	Faddeev-Popov (Ghost)
FSR	Final State Radiation
GF	Gauge Fixed
GIM	Glashow-Iliopoulos-Maiani (Mechanism)
GMSB	Gauge Mediated Supersymmetry Breaking
GRS	Grisaru-Roček-Siegel (Rules)
GUT	Grand Unified Theory
HDM	Hot Dark Matter
HERA	Hadron Elektron Ring Anlage (Device), Hamburg
ISR	Initial State Radiation
JLC	Joint Linear Collider (Proposed)
L3	A Detector at LEP

LC	Linear Collider
LEP	Large Electron Positron (Ring)
LHC	Large Hadron Collider (CERN)
LSP	Lightest Supersymmetric Particle
LO	Leading Order
L-R	Left-Right
Λ_{CDM}	Cosmological Constant plus CDM (Model)
MACHO	MAssive Compact Halo Object
$\overline{\mathrm{MS}}$	Modified Minimal Subtraction (Renormalization scheme)
MSSM	Minimal Supersymmetric Standard Model
mGMSB	Minimal gauge mediated supersymmetry breaking (Model)
mSUGRA	Minimal Supergravity (Model)
MSW	Mikhayev-Smirnov-Wolfenstein (Effect)
NLSP	Next Lightest Supersymmetric Particle
NLC	Next Linear Collider (Proposed)
NLO	Next-to-the Leading Order
NMSSM	Next-to-the Minimal Supersymmetic Standard Model
OPAL	Omni Purpose Apparatus (Detector) for LEP
OS, SF	Opposite Sign, Same Flavor
$P_1 \cdot P_2$ c.o.m.	Center of Mass Frame of the Partons P_1 and P_2
QCD	Quantum Chromodynamics
QED	Quantum Electrodynamics
QFT	Quantum Field Theory
R_p	R-parity
$\not{R}_p$	R-parity Violating
$\not{R}_p$MSSM	Extended MSSM with $\not{R}_p$ Interactions
RGE	Renormalization Group Evolution
RPC	R-parity Conserving
SχGT	Supersymmetric Chiral Gauge Theory
sFG	Super-Feynman Gauge
SLC	Stanford Linear Collider
SM	Standard Model
SQED	Supersymmetric Quantum Electrodynamics
SQCD	Supersymmetric Quantum Chromodynamics
SS	Salam-Strathdee (Propagators)
SSB	Spontaneous Supersymmetry Breaking
STr	Supertrace
SUGRA	Supergravity
SUSY	Supersymmetry
SYM	Super-Yang-Mills (theory)
TESLA	TeV Energy Superconducting Linear Accelerator (Proposed)
TEV-I (II)	Run I (II) at the TEVATRON
UA1	Underground Area 1 (CERN Experiment)
VEV	Vacuum Expectation Value
WIMP	Weakly Interacting Massive Particle
WZ	Wess-Zumino (Supermultiplet, Model, Gauge)
YM	Yang-Mills (Theory)

Notation, Conventions and Basic Superfield Definitions

Natural units $c = \hbar = 1$.

x^μ = spacetime coordinate $\equiv (t, \vec{x})$.

$\eta^{\mu\nu} \equiv$ flat Minkowski metric $(1, -1, -1, -1) = \eta_{\mu\nu}$.

$\eta^{\mu\rho}\eta_{\rho\nu} = \delta^\mu_\nu = \mathrm{diag}(1,1,1,1)$.

$\epsilon_{\mu\nu\alpha\beta}$ is the Levi-Civita tensor density with $\epsilon_{0123} = 1$.

$$\partial_\mu \equiv \frac{\partial}{\partial x^\mu}, \ \partial^\mu = \frac{\partial}{\partial x_\mu}.$$

$X[\partial_\mu]Y \equiv \frac{1}{2}X\partial_\mu Y - \frac{1}{2}(\partial_\mu X)Y$.

$M_{Pl} \equiv (8\pi G_{\mathrm{Newton}})^{-1/2} \simeq 2.43 \times 10^{18}$ GeV.

$\kappa \equiv \sqrt{8\pi G_{\mathrm{Newton}}} = M_{Pl}^{-1}$.

$[X, Y] \equiv XY - YX$.

$[X, Y]_+ \equiv XY + YX$.

P_μ = generators of four translations = $i\partial_\mu$ on incoming waves.

$M_{\mu\nu}$ = generators of homogenous Lorentz transformations.

ϕ = complex scalar field.

A_μ or A_μ^a = real gauge field, abelian or nonabelian.

T^a = internal symmetry generator.

g = gauge coupling.

ψ = Dirac four spinor.

ψ^C = charge conjugate of $\psi \equiv C\bar{\psi}^T$.

λ_M = Majorana four spinor = λ_M^C.

a, b = four component spinor sub-indices in the $(\frac{1}{2}, 0) \oplus (0, \frac{1}{2})$ representation.

ψ_a = spinor component of ψ.

$P_L = \dfrac{1}{2}(1 - \gamma_5)$: left chiral projector.

$P_R = \dfrac{1}{2}(1 + \gamma_5)$: right chiral projector.

A, B = two spinor indices in the $(\frac{1}{2}, 0)$ representation.

$\dot{A}, \dot{B}$ = two spinor indices in the $(0, \frac{1}{2})$ representation.

$\epsilon^{AB} \equiv \begin{pmatrix} 0 & 1 \\ -1 & 0 \end{pmatrix}, \ \epsilon_{AB} \equiv \begin{pmatrix} 0 & -1 \\ 1 & 0 \end{pmatrix}$: metric in the $(\frac{1}{2}, 0)$ representation space.

$\epsilon^{\dot{A}\dot{B}} \equiv \begin{pmatrix} 0 & 1 \\ -1 & 0 \end{pmatrix}, \ \epsilon_{\dot{A}\dot{B}} \equiv \begin{pmatrix} 0 & -1 \\ 1 & 0 \end{pmatrix}$: metric in the $(0, \frac{1}{2})$ representation space.

θ, ϵ = Grassmann coordinate two component spinor in the $(\frac{1}{2}, 0)$ representation.

$\bar{\theta}, \bar{\epsilon}$ = Grassmann coordinate two component spinor in the $(0, \frac{1}{2})$ representation.

$\theta_A, \ \bar{\theta}^{\dot{B}}$ = component of $\theta, \bar{\theta}$.

$\partial_A \equiv \dfrac{\partial}{\partial \theta^A}, \ \partial^A = \dfrac{\partial}{\partial \theta_A}, \ \bar{\partial}^{\dot{A}} \equiv \dfrac{\partial}{\partial \bar{\theta}_{\dot{A}}}, \ \bar{\partial}_{\dot{A}} = \dfrac{\partial}{\partial \bar{\theta}^{\dot{A}}}.$

$Q_A = -i(\partial_A + i\sigma^\mu_{A\dot{B}}\bar{\theta}^{\dot{B}}\partial_\mu)$: spinorial supercharge in the $(\frac{1}{2}, 0)$ representation.

$\bar{Q}^{\dot{A}} = -i(\bar{\partial}^{\dot{A}} + i\bar{\sigma}^{\mu\dot{A}B}\theta_B\partial_\mu)$: spinorial supercharge in the $(0, \frac{1}{2})$ representation.

$\xi, \chi, \lambda, \zeta$ = two component spinors in the $(\frac{1}{2}, 0)$ representation.

$\bar{\xi}, \bar{\chi}, \bar{\lambda}, \bar{\zeta}$ = two component spinors in the $(0, \frac{1}{2})$ representation.

$\zeta\chi \equiv \zeta^A\chi_A, \ \bar{\zeta}\bar{\chi} \equiv \bar{\zeta}_{\dot{A}}\bar{\chi}^{\dot{A}}.$

$\sigma^\mu_{A\dot{B}} \equiv (I, \vec{\sigma}), \ \bar{\sigma}^{\mu\dot{A}B} \equiv (I, -\vec{\sigma})$, where $\sigma_{1,2,3}$ are the Pauli matrices.

$\sigma^{\mu\nu} \equiv \dfrac{i}{4}[\sigma^\mu, \sigma^\nu], \ \bar{\sigma}^{\mu\nu} \equiv \dfrac{i}{4}[\bar{\sigma}^\mu, \bar{\sigma}^\nu].$

$\psi = \begin{pmatrix} \xi \\ \bar{\chi}^T \end{pmatrix}, \ \lambda_M = \begin{pmatrix} \lambda \\ \bar{\lambda}^T \end{pmatrix}$

z = superspace coordinate $\equiv (x, \theta, \bar{\theta}).$

$\mathcal{D}$ = chirally covariant derivative in the $(\frac{1}{2}, 0)$ representation.

$\bar{\mathcal{D}}$ = chirally covariant derivative in the $(0, \frac{1}{2})$ representation.

$\mathcal{D}_A \equiv \partial_A - i\sigma^\mu_{A\dot{B}}\bar{\theta}^{\dot{B}}\partial_\mu.$

$$\mathcal{D}^A = -\partial^A + i\bar{\theta}_{\dot{B}}\bar{\sigma}^{\mu\dot{B}A}\partial_\mu.$$

$$\bar{\mathcal{D}}^{\dot{A}} \equiv \bar{\partial}^{\dot{A}} - i\bar{\sigma}^{\mu\dot{A}B}\theta_B\partial_\mu.$$

$$\bar{\mathcal{D}}_{\dot{A}} = -\bar{\partial}_{\dot{A}} + i\theta^B\sigma^\mu_{B\dot{A}}\partial_\mu.$$

$$y^\mu \equiv x^\mu - i\theta\sigma^\mu\bar{\theta}.$$

$$\bar{y}^\mu = x^\mu + i\theta\sigma^\mu\bar{\theta}.$$

$$d^4\theta = d^2\bar{\theta}d^2\theta.$$

$$d^8z \equiv d^4x\,d^4\theta.$$

$$d^6z \equiv d^4x\,d^2\theta.$$

$$d^6\bar{z} \equiv d^4x\,d^2\bar{\theta}.$$

Φ = left chiral superfield, $\overline{\mathcal{D}}\Phi = 0$.

$\Phi^\dagger$ = right-chiral superfield, $\mathcal{D}\Phi^\dagger = 0$.

$$\Phi_1\cdot\Phi_2 \equiv \int d^6z\,\Phi_1(z)\Phi_2(z).$$

$$\Phi_1^\dagger\cdot\Phi_2^\dagger \equiv \int d^6\bar{z}\,\Phi_1^\dagger(\bar{z})\Phi_2^\dagger(\bar{z}).$$

$$\Phi(y,\theta) = \phi(y) + \sqrt{2}\theta\xi(y) + \theta\theta F(y).$$

$$\Phi^\dagger(\bar{y},\bar{\theta}) = \phi^\star(\bar{y}) + \sqrt{2}\bar{\theta}\bar{\xi}(\bar{y}) + \bar{\theta}\bar{\theta}F^\star(\bar{y}).$$

The symbol | to the right of a superfield means that it is evaluated at $\theta = 0 = \bar{\theta}$.

$\mathcal{W}(\Phi)$ = superpotential.

$$F_i^\star = -\mathcal{W}_i(\phi) = -\partial\mathcal{W}/\partial\Phi_i|.$$

$$F_i = -\overline{\mathcal{W}}_i(\phi) = -\partial\mathcal{W}^\dagger/\partial\Phi_i^\dagger|.$$

$V \equiv V^\dagger$ = vector superfield.

$V \equiv 2gV^aT^a$ in the nonabelian case.

$$W_A \equiv \text{left chiral field-strength spinorial superfield} \equiv -\frac{1}{4}\bar{\mathcal{D}}\bar{\mathcal{D}}\mathcal{D}_A V.$$

$$\overline{W}^{\dot{A}} \equiv \text{right chiral field-strength spinorial superfield} \equiv -\frac{1}{4}\mathcal{D}\mathcal{D}\bar{\mathcal{D}}^{\dot{A}} V.$$

$$D^a \equiv g\phi^\star T^a\phi.$$

$R_p \equiv R_{\text{parity}}$, even (odd) for particles (sparticles), is $(-1)^{3(B-L)+2S}$ in the Minimal Supersymmetric Standard Model and its simple extensions.

INTRODUCTION AND OVERVIEW

Chapter 1

SUPERSYMMETRY: WHY AND HOW

1.1   History and Motivation

We first give a brief factual account of the history [1.1] of supersymmetry, leaving a more pedagogical development to later sections and chapters. This has evidently been a history of experimenters chasing a theoretically driven idea. The notion of a symmetry transformation between fermionic and bosonic modes emerged [1.2] in connection with string theory. However, this was constructed on a two dimensional world sheet rather than in the real world of $3+1$ dimensions. N=1 supersymmetry in the latter (in terms of supercharges that take fermions into bosons and vice versa) was first proposed and formulated as a graded Lie algebra by Golfand and Likhtman [1.3] in 1971. Akulov and Volkov [1.4] later gave a nonlinear realization of it together with the idea of spontaneous breakdown. Finally, in 1974, Wess and Zumino [1.5] as well as Salam and Strathdee [1.5] constructed field theories with supersymmetry (cf. Ch.5) and the subject immediately attracted attention on a large scale. Supersymmetry was shown by Haag, Łopuzański and Sohnius (cf. Ch.3) to be the only possible extension of the known spacetime symmetries of particle interactions. Several important results were derived on the more convergent ultraviolet behavior of supersymmetric field theories, exploiting the cancellation between fermionic and bosonic loops. In particular, a theorem on the nonrenormalization of superpotential terms (cf. Ch.6) was proved [1.6]. Scalar field theories are generically not natural in the sense of Weinberg, Susskind and 't Hooft [1.7]. But it became clear after a while that supersymmetric field theories, despite containing scalar fields, are natural [1.8].

The four momentum operator P^μ is essential to an algebraic formulation of supersymmetry. As elaborated in Ch.3, if a supercharge, carrying spin 1/2, takes a boson to a fermion or vice versa, the anticommutator between two supercharges with arbitrary spinorial components must be proportional to P^μ. If the vacuum is annihilated by a supercharge, a vanishing energy for it is then ensured. In exact supersymmetry, P^μ and $P^2 \equiv P^\mu P_\mu$ commute with the supercharge and one is led to mass degenerate supermultiplets of states differing in spin by 1/2. Since no such mass degeneracy has been seen[1] among particles occurring in Nature,

---

[1]There were early attempts to put a photon and a neutrino together in a supermultiplet. It soon became

supersymmetry must be a badly broken symmetry. The intra-supermultiplet mass splitting, characteristically denoted as M_s, then becomes a scale of some significance. A question of immediate interest arises in consequence: what is the order of magnitude of M_s? A related issue concerns the existence of **sparticles**, i.e. superpartners of the known particles. The latter cannot make up complete supermultiplets by themselves. Therefore each particle in a broken supersymmetric world must have a new superpartner which we call a sparticle. A sparticle is typically heavier than the corresponding particle by a mass difference $\mathcal{O}(M_s)$ and is lying yet undiscovered. Furthermore, the application of naturalness arguments [1.8] to the weak scale ($\sim$ 100 GeV), generated in the SM by an unnatural scalar field sector, has suggested [1.9] that[2] $M_s \lesssim \mathcal{O}(\text{TeV})$ and that sparticles should be discovered in forthcoming high energy accelerator experiments probing these energies. The highly successful Standard Model (**SM**) of particle interactions has been minimally extended [1.10] to include all these sparticles and is now called the Minimal Supersymmetric Standard Model (**MSSM**). Our aim in this book is to develop this theme concretely to the extent that its links with experiments, now being conducted or planned, become clear.

There have already been major experimental efforts to search for sparticles, undertaken all through the 1980's and 1990's. The production and decays of sparticles are uniquely characterized by large (more than tens of GeV) missing transverse energy at least in R-parity conserving supersymmetric scenarios where the undetected lightest supersymmetric particle (LSP) carries it away. Early hints at the beginning of the eighties in the UA1 experiment, performed at the $S\bar{P}PS$ machine at CERN, did not materialize into believable signals but were later identified with more mundane processes of the Standard Model. Afterwards, the LEP e^+e^- storage ring at CERN and the Tevatron $\bar{p}p$ collider at Fermilab have been heavily deployed in searches for sparticles, but without any success so far. The same can be said for searches in ep collision experiments performed at HERA. The four LEP experimental groups, ALEPH, DELPHI, L3 and OPAL, as well as the two major experimental collaborations at the TEVATRON, CDF and D$\emptyset$, have published the strongest experimental lower bounds on the masses of numerous sparticles; they have also established exclusion zones in parameter spaces of various supersymmetric extensions of the Standard Model (Ch.15). The TEVATRON experiments are being extended to RUN II with higher integrated luminosity. Two major collaborations in the Large Hadron Collider (LHC), being built at CERN, ATLAS and CMS, are preparing dedicated experiments which will probe sparticles in the TeV mass range. The exploration of sparticles has been stated as a major goal in proposals for e^+e^- linear colliders with CM energies in the range 500 GeV–1.5 TeV, now being pursued vigorously. There are also nonaccelerator experiments trying to detect the very weakly interacting LSP pervading the universe as cold dark matter (Ch.16). Thus we are in for another decade of intense experimental activity full of exciting possibilities.

One criticism, frequently levelled against the supersymmetry idea sketched above, is the

---

clear that the supercharge, being the generator of a spacetime symmetry, must commute with all generators of internal symmetries, e.g. electroweak symmetry. Thus all members of a supermultiplet must have identical internal symmetry properties. Such is not the case between the photon and any of the three known neutrinos.

[2]Such a statement is far from obvious. M_s could very well be of the order of other possible scales in Physics. These include the reduced Planck scale $M_{Pl} \equiv (8\pi G_N)^{-1/2} \simeq 2.4 \times 10^{18}$ GeV and the speculated grand unification scale $M_U \sim 2 \times 10^{16}$ GeV. As we shall see later, the above conclusion, reached on the basis of naturalness arguments, is quite a deep statement.

"inelegance" in postulating one new state for every known particle. But extended symmetry considerations did lead physicists in the past to postulate new particles which were subsequently discovered. An example, which we elaborate here, is the extension of nonrelativistic quantum electrodynamics of the electron to cover Lorentz invariance. Such an extension requires the existence of the positron, as can be understood from the standpoint of divergences. We know that the classical self energy of an electron of radius r_e, namely

$$E_{\text{self}}^{cl.} = \frac{3}{5}(e^2/4\pi r_e)$$

in rationalized units, is linearly divergent as $r_e \to 0$. One can guess [1.11] that this calculation becomes unreliable for radii less than the "classical electron radius" $R_0 \sim \frac{3}{5}(e^2/4\pi m_e) \simeq 1.7$ fm for which E_{self}^{cl} equals the rest energy of the electron. In the diagrammatic language of old fashioned perturbation theory [1.12], this contribution is given by Fig.1.1a — the solid, wiggly and dashed lines standing for the electron, the photon and a time slice respectively.

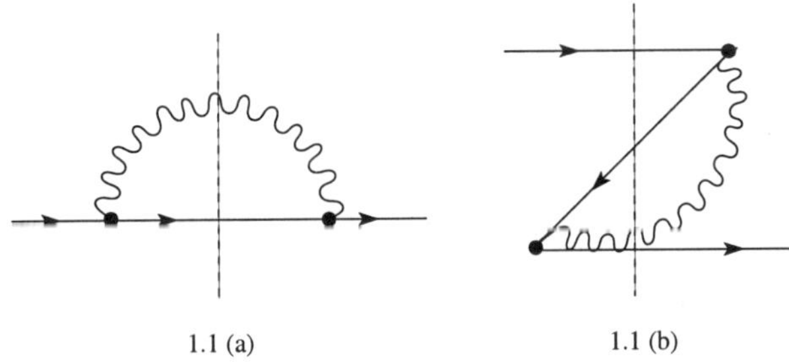

1.1 (a)   1.1 (b)

Fig.1.1. Electron self energy diagrams in old fashioned perturbation theory.

Of course, in a relativistic quantum description of the electron, it has been possible to probe r_e far below R_0 (indeed below 10^{-3} fm). This is because the linear divergence, mentioned earlier, gets cured here by the presence of the positron. Owing to the latter, there is also[3] Fig.1-1b now. In this contribution the electron annihilates the positron, created in a pair from a vacuum fluctuation, while the remaining electron goes out. As a result, there is an intrinsic uncertainty in the electron's position of the order of its Compton wavelength $r_e \sim 1/m_e$. The linear divergence cancels in the sum of these two contributions and the self energy becomes [1.13]

$$E_{\text{self}}^{quant.} = \frac{3e^2 m_e}{16\pi^2} \ln(m_e r_e) \ .$$

The expression for E_{self} is still logarithmically divergent as $r_e \to 0$, but this mild divergence[4] can be easily tackled within the renormalization program. One postulates a bare mass of the electron which is also logarithmically divergent, owing to a counterterm inserted in the Lagrangian, so that the renormalized mass $m = m^{\text{bare}} + E_{\text{self}}$ is finite.

---

[3]This reasoning has been highlighted by H. Murayama, hep-ph/9410285, *ibid*/0002242.

[4]The mildness of this divergence can be seen as follows. Even if r_e is replaced by the smallest length known in Physics, namely the Planck length $\lambda_{Pl} = M_{Pl}^{-1}$, the above expression becomes only 10% of the rest energy of the electron.

The cancellation of the linear divergence is really a consequence of chiral symmetry. The latter refers to an invariance under the transformation of the electron field $\psi_e \to e^{i\varphi\gamma_5}\psi_e$ (with φ being a real parameter), which becomes a symmetry of relativistic electrodynamics in the limit when $m_e \to 0$. Taking this limit in which $E_{\text{self}} \to 0$ enhances the symmetry of the theory which then includes chiral invariance. This makes the smallness of the mass of the electron *natural* in the sense of Weinberg, Susskind and 't Hooft [1.7] since the electron is protected by this symmetry from acquiring a huge mass due to self energy corrections. According to those authors, a small parameter in a theory is natural if, and only if, setting it to zero enhances the symmetry of the system. Being a symmetry breaking parameter, its smallness then gets protected against large radiative corrections by the concerned symmetry. Indeed, this criterion can be extended to the theory itself. In the modern Wilsonian view, every Lagrangian density $\mathcal{L}$ should be defined with a cutoff Λ and should be written as $\mathcal{L}(\Lambda)$, keeping renormalization in mind. Λ represents the highest energy scale upto which $\mathcal{L}(\Lambda)$ is the appropriate Lagrangian density and can be perceived as the energy scale where new physics comes into play. Now, a Lagrangian density $\mathcal{L}(\Lambda)$ is "natural" upto and below the energy scale Λ if any set of small parameters $\{\delta_n\}$, appearing in $\mathcal{L}(\Lambda)$, is associated with some approximate symmetry of $\mathcal{L}(\Lambda)$ which is exactly recovered in the $\delta_n \to 0$ limit. In this case quantum corrections – characterized by the scale Λ – will also vanish as $\{\delta_n\} \to 0$ and will remain small for nonvanishing but small $\{\delta_n\}$.

In order to make the above discussion more quantitative, let us write the tree level Lagrangian density of the low energy theory as $\mathcal{L}_{\text{tree}} = \sum_n \delta_n O_n$. Here the $\{O_n\}$ are a set of general operators, indexed by nonnegative integers n with δ_n as the corresponding coefficients, while the summation covers all such n that occur. The inclusion of quantum effects, characterized by the scale Λ, then leads to the following general form for the low energy effective Lagrangian density[5]:

$$\mathcal{L}_{\text{eff}} = \sum_{n,i} c_{n,i}(\delta_n)^i \Lambda^{[(i-1)(d_n-4)]} O_n \ . \tag{1.1}$$

In (1.1) the summation over nonnegative integers i can, in principle, go from zero to infinity; however, in practice, only a few terms matter. Moreover, d_n is the mass dimension of the operator O_n and $c_{n,i}$ are dimensionless coefficients which can depend only logarithmically on Λ. Since we take the low energy tree level Lagrangian density to be renormalizable, $\delta_n = 0$ whenever O_n has $d_n > 4$. For such operators, the sum over i in (1.1) collapses to the single term with $i = 0$ and (1.1) describes the usual expansion of the low energy effective Lagrangian density with higher dimensional operators suppressed by appropriate powers of Λ^{-1}. These latter terms are irrelevant to any discussion of naturalness, since they disappear when $\Lambda \to \infty$.

Turning to operators with $d_n \leq 4$, we can distinguish two cases. (1) A small coefficient δ_n is "naturally small" only if $c_{n,o} = 0$, since then the corresponding coefficient will remain small in the full effective low energy Lagrangian density as well. As already noted above, in all known examples, a symmetry is needed to ensure that $c_{n,o}$ vanishes to all orders in

---

[5]There could be additional terms in (1.1) involving powers of products of different δ_n's, but they do not change the discussion in substance.

perturbation theory. Illustrative examples are gauge symmetries "protecting" gauge couplings and chiral symmetries "protecting" fermion masses or Yukawa couplings. (2) On the other hand, if $c_{n,o} \neq 0$, there is no reason to assume δ_n to be small or zero in the tree level low energy Lagrangian density; such a choice would be "unnatural". This may not lead to serious problems for operators with $d_n = 4$. In this case our argument only shows that the natural scale for $\sum_i c_{n,i}(\delta_n)^i$ should be at least $\delta_n + O(\alpha/\pi)$, α being $(4\pi)^{-1}$ times the square of some (typically gauge) coupling strength. Thus even if, at the tree level, δ_n is chosen in magnitude to be much less than $O(\alpha/\pi)$, the coefficient of O_n in $\mathcal{L}_{\text{eff}}$ will naturally become of that order. An example is the quartic Higgs self coupling in the SM which is "naturally" at least $O(10^{-2})$ in $\mathcal{L}_{\text{eff}}$. (In this particular case, however, the experimental lower bound on the mass of the physical Higgs boson leads to a much stronger lower limit). The problem of "naturalness" becomes really severe only for operators with dimension $d_n < 4$. As shown in (1.1), the corresponding coefficients in the low energy effective Lagrangian density *diverge* like Λ^{4-d_n} if $c_{n,0} \neq 0$. The lowest dimensional relevant operator in the SM is the Higgs mass term, which has dimension two. Since the relevant coefficient is not protected by a symmetry, we expect it to receive *quadratically divergent* quantum corrections. In the next section we shall show explicitly that such divergences do indeed occur in the SM. An exorbitant degree of fine tuning between the bare mass and the radiative correction becomes necessary to keep the renormalized Higgs mass near the weak scale. We shall then show in §1.3 how supersymmetry removes this quadratic divergence (i.e. makes the corresponding $c_{n,o}$ vanish) and solves the problem by protecting the renormalized Higgs mass.

Let us return to the question of the mass of the electron. On dimensional grounds, one might naively expect m_e to grow like Λ after loop corrections. As already noted, it does not do so. The fact that it grows instead as $\ln(\Lambda/m_e)$ is because of chiral symmetry which makes m_e a "naturally small" parameter of the theory. Notice that chiral symmetry can be formulated only within a relativistic framework where a positron is obligatory. The existence of a new particle here is therefore linked to the greater convergence of the theory at short distances (or high energies) and is ultimately related to a symmetry. A similar motivation can be given for supersymmetry. Supersymmetry, or more specifically the existence of sparticle superpartners with masses near the weak scale, cures the problem of quadratic divergences through cancellations between fermionic and bosonic loops. This can be understood on the basis of symmetries as follows. Supersymmetry links boson masses to fermion masses, which are "protected" by chiral symmetry[6]. The weak scale M_W can then be naturally chosen to be many orders of magnitude below the Planck scale M_{Pl} or the hypothetical scale M_U of grand unification and kept protected. Operatively, the nonrenormalization theorem of supersymmetry (cf. Ch.6) provides this protection. Thus supersymmetry holds the key to the stability and naturalness of the weak scale vis-à-vis M_U or M_{Pl}. This really is the *raison d'être* for the extension of the phenomenologically successful Standard Model of particle interactions to the Minimal Supersymmetric Standard Model to which a large part of this book will be devoted. In the next sections we shall illustrate this main argument through explicit calculations at the one loop level.

---

[6]We note in passing that supersymmetry also allows one to "naturally" choose arbitrarily small, even vanishing, scalar self couplings, by relating them either to gauge or to Yukawa couplings.

1.2   Quadratic Divergence and Unnaturalness

We illustrate the problem of the quadratic divergence in the Higgs sector of the SM through an explicit calculation. The example studied is that of the two point function (inverse propagator) of the Higgs scalar at vanishing external momentum, computed at the one loop level. This quantity is roughly $-i$ times the squared scalar mass appearing in the Lagrangian. This particular object has been chosen since its calculation is simple and yet suffices to highlight the problem. Let ϕ be the SM neutral Higgs field with $v = \sqrt{(1/\sqrt{2}G_F)} \simeq 246$ GeV defined to be $\sqrt{2}\langle\phi\rangle$ so that the shifted physical field h is given by

$$\Re e\ \phi = \frac{1}{\sqrt{2}}(h + v) \ . \tag{1.2}$$

Take f to be a generic matter fermion field (of one species) with a Yukawa coupling to ϕ via the term (we follow the conventions of Bjorken and Drell [1.14])

$$\begin{aligned} \mathcal{L}_{\bar{f}f\phi} &= -\lambda_f \bar{f}_L f_R \phi + \text{h.c.} \\ &= -\frac{\lambda_f}{\sqrt{2}} h \bar{f} f - \frac{\lambda_f v}{\sqrt{2}} \bar{f} f, \end{aligned} \tag{1.3}$$

where $f_{L,R}$ are left, right chiral components of f. Thus, on account of spontaneous symmetry breaking, the fermion develops a tree level mass $m_f = \lambda_f v/\sqrt{2}$.

Let us now proceed to compute the one loop f-$\bar{f}$ contribution to the scalar two point function, as illustrated in Fig.1.2. We have

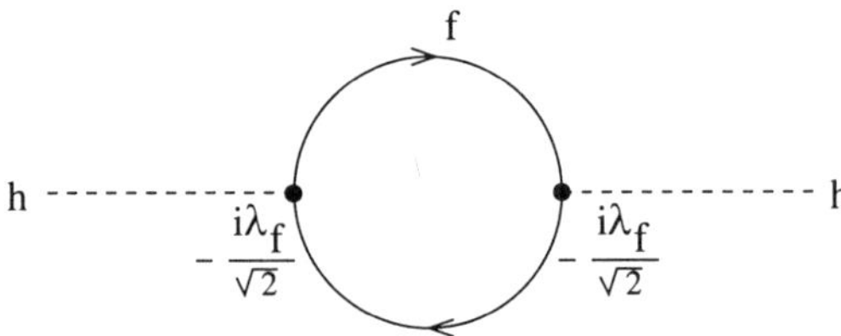

Fig.1.2. Fermionic loop contribution to the scalar two point function.

$$\begin{aligned} \Pi^f_{hh}(0) &= (-1) \int \frac{d^4k}{(2\pi)^4} \text{Tr}\left(-i\frac{\lambda_f}{\sqrt{2}}\right) \frac{i}{\not{k} - m_f} \left(-i\frac{\lambda_f}{\sqrt{2}}\right) \frac{i}{\not{k} - m_f} \\ &= -2\lambda_f^2 \int \frac{d^4k}{(2\pi)^4} \frac{k^2 + m_f^2}{(k^2 - m_f^2)^2} \\ &= -2\lambda_f^2 \int \frac{d^4k}{(2\pi)^4} \left[\frac{1}{k^2 - m_f^2} + \frac{2m_f^2}{(k^2 - m_f^2)^2} \right] . \end{aligned} \tag{1.4}$$

The first term in the final RHS of (1.4) is quadratically divergent and is moreover independent of the scalar mass m_S. First of all, this divergence is very severe. Suppose the integral is cut off by a Λ parameter which is then set equal to the Planck mass $M_{Pl} \simeq 2.4 \times 10^{18}$ GeV, the highest scale known in physics. Then the one loop correction to m_S^2 would be 30 orders of magnitude larger than m_S^2 itself since m_S is restricted [1.15], by the requirement of perturbative unitarity in the amplitude $W^+W^- \to W^+W^-$, to be $\leq \mathcal{O}\,(1\ \text{TeV})$. Furthermore, the correction (1.4) being independent of m_S is an indication of the fact that m_S is an *unnatural* parameter in the SM. Setting $m_S = 0$ does not increase the symmetry of that theory. That means that there exists no symmetry in the SM which protects the Higgs mass. For simplicity, we have dealt with only the fermion antifermion loop contribution to the Higgs self energy and ignored the gauge boson loop and (self-coupled) Higgs loop contributions. Each of the latter contains a quadratic divergence and has the same problem as above[7].

Of course, one could simply renormalize such quadratic divergences away in the same way that logarithmic divergences are disposed of. But the legacy of the severity of the quadratic divergence would still remain. Thus the residual finite correction in (1.4) would be or order $m_f^2 \lambda_f^2/(8\pi)$. Such a correction would be managably small for a standard model fermion like the top quark. However, the SM is expected to give way to a more fundamental theory, e.g. a Grand Unified one [1.16] unifying all forces in it, at a high energy scale $M_U \sim 10^{16}$ GeV. In this case the leading contribution will come from a fermion-antifermion pair which can couple to h and have the highest mass, with m_f expected to be $O(M_U)$, causing the loop correction to the scalar mass squared, i.e. δm_S^2, to be $O(M_U^2)$. One would have to do an unnatural amount of fine tuning (1 in 10^{20}) between the bare scalar mass squared $m_{S,0}^2$ and the renormalization δm_S^2 in order to keep the renormalized mass squared

$$m_S^2 = m_{S,0}^2 + \delta m_S^2 \tag{1.5}$$

to less than a $(\text{TeV})^2$. The argument can be amplified through the consideration of quartic scalar couplings.

To make the above discussion more concrete, let us take the Grand Unifying group to be [1.16] $SU(5)$ with $\Sigma\,(H, \bar{H})$ representing Higgs fields in the **24 (5,$\bar{5}$)** representation. While the mass of Σ is expected to be $O(M_U)$, that of the weak doublet parts of H and $\bar{H}$ should be of $O(M_W)$. At the tree level, the unifying scale M_U is generated via $M_U = g_U \langle \Sigma \rangle$, where g_U is the unified gauge coupling strength. The one loop effective action contains an interaction term $\lambda \bar{H} \Sigma^2 H$, from the graph of Fig.1.3, where the wiggly lines represent gauge bosons of the $SU(5)$ theory. If we take momentum scales at the two external H and Σ lines to be of order M_W and M_U respectively, we shall have

$$\lambda(M_W^2) \sim \lambda(M_U^2) + \frac{g_U^4}{16\pi^2} \ln \frac{M_U^2}{M_W^2}\,. \tag{1.6}$$

The induced mass of all components of H is $\lambda \langle \Sigma \rangle^2$. Even if λ is taken to be very small at the

---

[7]In principle, one can cancel the total one loop quadratic divergences by explicitly cancelling bosonic and fermionic contributions through some postulated relation between the boson and fermion masses. However, because such a cancellation is 'accidential', rather than being enforced by a symmetry, it will not work in higher loop order.

tree level, $\lambda(M_W^2)$ becomes $O(g_U^4)$ after the one loop correction. Without an extreme fine

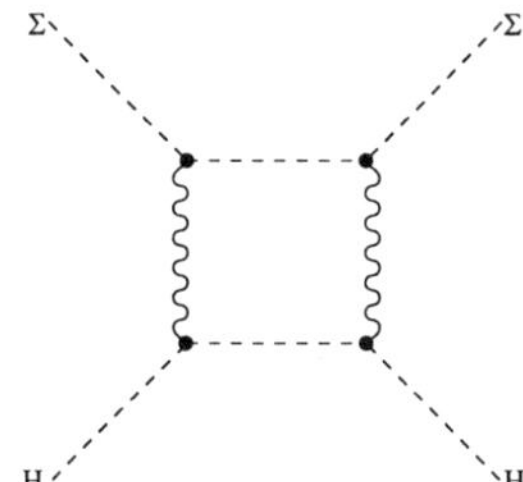

Fig.1.3. One loop graph for the $\bar{H}\Sigma^2 H$ vertex in an $SU(5)$ Grand Unified theory.

tuning of $\lambda(M_W^2)$, the induced mass of the weak scale Higgs doublet will then be at an unacceptably large level. Moreover, the fine tuning would be very different in different orders of perturbation theory. This, basically, is the gauge hierarchy problem arising out of the radiative instability of scalar masses: the latter like to be close to the highest mass scale in the theory.

1.3   Naturalness, Nonrenormalization, Supersymmetry

Loops induced by other scalar fields, contributing to the Higgs two point function, can also be considered. Let us construct a toy model [1.17] by introducing to the system of §1.2 two additional complex scalar ("sfermion") fields $\tilde{f}_L$, $\tilde{f}_R$ with the following coupling to the Higgs field:

$$
\begin{aligned}
\mathcal{L}_{\tilde{f}\tilde{f}\phi} &= \tilde{\lambda}_f |\phi|^2 (|\tilde{f}_L|^2 + |\tilde{f}_R|^2) + (\lambda_f A_f \phi \tilde{f}_L \tilde{f}_R^\star + \text{h.c.}) \\
&= \frac{1}{2}\tilde{\lambda}_f h^2 (|\tilde{f}_L|^2 + |\tilde{f}_R|^2) + v\tilde{\lambda}_f h(|\tilde{f}_L|^2 + |\tilde{f}_R|^2) \\
&\quad + \frac{h}{\sqrt{2}}(\lambda_f A_f \tilde{f}_L \tilde{f}_R^\star + \text{h.c.}) + \cdots .
\end{aligned}
\tag{1.7}
$$

In the second step of (1.7), we have rewritten, by means of (1.2), the interaction in terms of the h-field and have displayed only the h-dependent terms. The coefficient of the last RHS term is, in fact, arbitrary; the factor λ_f, multiplying the new unknown coupling strength A_f, has been put in only by convention. (1.7) makes the following additional contribution to the two point function via the loops of Fig.1.4:

$$
\begin{aligned}
\Pi_{hh}^{\tilde{f}}(0) &= -\tilde{\lambda}_f \int \frac{d^4k}{(2\pi)^4} \left(\frac{1}{k^2 - m_{\tilde{f}_L}^2} + \frac{1}{k^2 - m_{\tilde{f}_R}^2} \right) + \\
&\quad (\tilde{\lambda}_f v)^2 \int \frac{d^4k}{(2\pi)^4} \left[\frac{1}{(k^2 - m_{\tilde{f}_L}^2)^2} + \frac{1}{(k^2 - m_{\tilde{f}_R}^2)^2} \right] \\
&\quad + |\lambda_f A_f|^2 \int \frac{d^4k}{(2\pi)^4} \frac{1}{k^2 - m_{\tilde{f}_L}^2} \frac{1}{k^2 - m_{\tilde{f}_R}^2} .
\end{aligned}
\tag{1.8}
$$

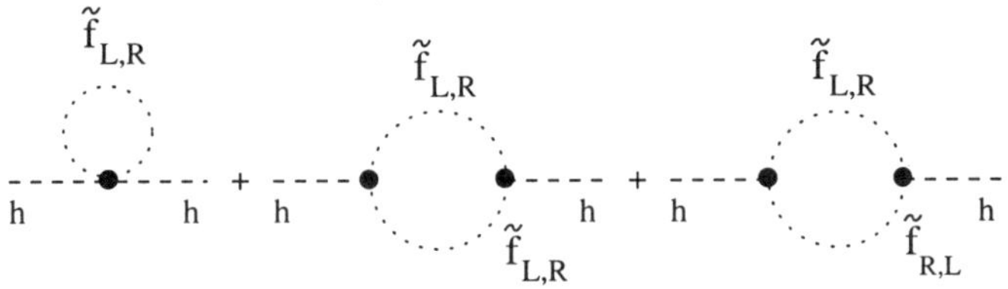

Fig.1.4. Sfermion loop contributions to Higgs self energy

Only the first line in (1.8), which comes from the leftmost diagram of Fig.1.4, contains a quadratic divergence. This, however, can be cancelled with that in the fermionic contribution (1.4), i.e. $\Pi_{hh}^{f}(0) + \Pi_{hh}^{\tilde{f}}(0)$ becomes free of any quadratic divergence, provided the following coupling constant equality is obeyed:

$$\tilde{\lambda}_f = -\lambda_f^2. \tag{1.9}$$

Note that the inequality $\tilde{\lambda}_f < 0$ is required in (1.7) to keep the Hamiltonian bounded from below. Another important point to note is that the above cancellation of the quadratic divergence is independent of the masses $m_{\tilde{f}_L}$, $m_{\tilde{f}_R}$ or the coupling strength A_f.

Now that the quadratic divergence has disappeared from $\Pi_{hh}^{f}(0) + \Pi_{hh}^{\tilde{f}}(0)$, the remaining logarithmic ones can be cancelled by contributions from logarithmically infinite counterterms introduced in the Lagrangian density as part of the renormalization procedure. In the $\overline{\text{MS}}$ renormalization scheme[8] [1.18], one can replace the logarithmic divergence in our loop integrals by the logarithm of the square of the **renormalization scale** μ. Utilizing the B_0-function of Passarino and Veltman, we can then make [1.19] the following types of replacements:

$$\int \frac{d^4k}{i\pi^2} \left(\frac{1}{k^2 - m_1^2} - \frac{1}{k^2 - m_2^2} \right) \;\to\; (m_1^2 - m_2^2)B_0(0, m_1^2, m_2^2)$$

$$\equiv\; m_1^2 \left(1 - \ln\frac{m_1^2}{\mu^2} \right) - m_2^2 \left(1 - \ln\frac{m_2^2}{\mu^2} \right), \tag{1.10a}$$

$$\int \frac{d^4k}{i\pi^2} \frac{1}{(k^2 - m^2)^2} \;\to\; -\ln\frac{m^2}{\mu^2}. \tag{1.10b}$$

---

[8]A caveat is in order here. The $\overline{\text{MS}}$ scheme was originally proposed with dimensional regularization which has a problem in supersymmetry since the numbers of bosons and fermions do not match as one goes off four dimensions. For supersymmetric loop computations, one needs to adopt the modified dimensional reduction or $\overline{\text{DR}}$ scheme where the momentum integrals are evaluated in continued dimensions and the subtraction is performed as in $\overline{\text{MS}}$, but the Dirac algebra in the numerator is done strictly in four dimensions. A more extensive discussion of the $\overline{\text{DR}}$ scheme will come in Ch.6.

The consequent expression for the sum of (1.4) and (1.8) can be simplified by choosing

$$m_{\tilde{f}_L} = m_{\tilde{f}_R} = m_{\tilde{f}} \ . \tag{1.11}$$

The choices (1.9) and (1.11) as well as the substitutions (1.10) lead to the result:

$$\Pi_{hh}^{f}(0) + \Pi_{hh}^{\tilde{f}}(0) = i\frac{\lambda_f^2}{16\pi^2}\left[-2m_f^2\left(1 - \ln\frac{m_f^2}{\mu^2}\right) + 4m_f^2\ln\frac{m_f^2}{\mu^2} \right.$$

$$\left. +2m_{\tilde{f}}^2\left(1 - \ln\frac{m_{\tilde{f}}^2}{\mu^2}\right) - 4m_{\tilde{f}}^2\ln\frac{m_{\tilde{f}}^2}{\mu^2} - |A_f|^2\ln\frac{m_{\tilde{f}}^2}{\mu^2} \right] . \tag{1.12}$$

Thus, if along with (1.11), we also require the relations

$$m_f = m_{\tilde{f}} \ , \tag{1.13a}$$

$$A_{\tilde{f}} = 0 \ , \tag{1.13b}$$

we will have

$$\Pi_{hh}^{f}(0) + \Pi_{hh}^{\tilde{f}}(0) = 0 \ . \tag{1.14}$$

Eq. (1.14) can be restated as follows. If the fermion Yukawa coupling strength squared equals the quartic coupling between the Higgs and the scalars $\tilde{f}_{L,R}$, if the masses of the fermion f and of the scalars $\tilde{f}_{L,R}$ are identical and if the A_f parameter is zero, the *entire* one loop renormalization of the Higgs self energy $\Pi_{hh}(0)$ vanishes.

We are now ready to give a supersymmetric interpretation of the above. In an exactly supersymmetric theory, the two scalars $\tilde{f}_{L,R}$ are the left and right superpartners (sfermions) of the fermion f. Moreover, the coupling strength equality (1.9), the mass equalities (1.11) and (1.13a) and the required null value of the (supersymmetry breaking) parameter A_f (1.13b) are all ensured by supersymmetry. Indeed, with these conditions, the vanishing of the renormalization of the Higgs self energy holds in all perturbation orders as a consequence of the **nonrenormalization theorem** (cf. §6.7) valid in supersymmetric theories. This is the essence of naturalness due to supersymmetry. The naturalness aspect is also made clear by the introduction of a certain kind of small supersymmetry breaking, namely that the breaking is confined to the masses m_f and $m_{\tilde{f}}$ being different and to A_f being nonzero but does *not* change the coupling equality (1.9). These are specific instances of parameters typical of softly broken supersymmetry, i.e. as coefficients of supersymmetry breaking operators of mass dimension less than four in the Hamiltonian. Suppose we characterize this supersymmetry breaking in terms of two small parameters A_f and δ, with

$$\delta^2 = m_{\tilde{f}}^2 - m_f^2 \ . \tag{1.15}$$

(Here we have chosen to maintain[9] (1.11) while relaxing (1.13) via a small mass splitting.) Thus δ characterizes the mass splitting within the f-$\tilde{f}$ supermultiplet. With the assumption

---

[9]In a more general discussion, one could introduce another supersymmetry breaking mass parameter, splitting $m_{\tilde{f}_L}$ and $m_{\tilde{f}_R}$, which would then enter the RHS of (1.16). But the basic conclusion would still be the same.

that $|\delta|$, $|A_f| \ll m_f$, we can approximate $\ln(m_{\tilde f}^2/\mu^2) \simeq \ln(m_f^2/\mu^2) + \delta^2/m_f^2$ and rewrite (1.12) as

$$\Pi_{hh}^f(0) + \Pi_{hh}^{\tilde f}(0) \simeq -i\frac{\lambda_f^2}{16\pi^2}\left[4\delta^2 + (2\delta^2 + |A_f|^2)\ln\frac{m_f^2}{\mu^2}\right] + \mathcal{O}(\delta^4, |A_f|^2\delta^2) . \tag{1.16}$$

Hence the one loop renormalization of the Higgs self energy is linearly proportional[10] to the small supersymmetry breaking parameters δ^2 and $|A_f|^2$, restricting the correction to one of modest magnitude, though m_f may be quite large.

Thus the introduction of the superpartners $\tilde f_{L,R}$ with the interactions of (1.7) has served two purposes: (1) the quadratic divergence in the scalar self energy is cancelled; (2) the scalar mass is shielded from large loop corrections involving heavy particles *so long as* the mass splitting between the heavy fermion and boson superpartners is itself of the order of the scalar mass. This then is a toy model example of how naturalness is restored by supersymmetry in the scalar sector of the SM. We have confined ourselves here to discussing fermion and sfermion loop contributions to the Higgs self energy. But the same conclusions follow *mutatis mutandis* if loop contributions from gauge bosons and their superpartners are combined or Higgs bosons and their superpartners are added together.

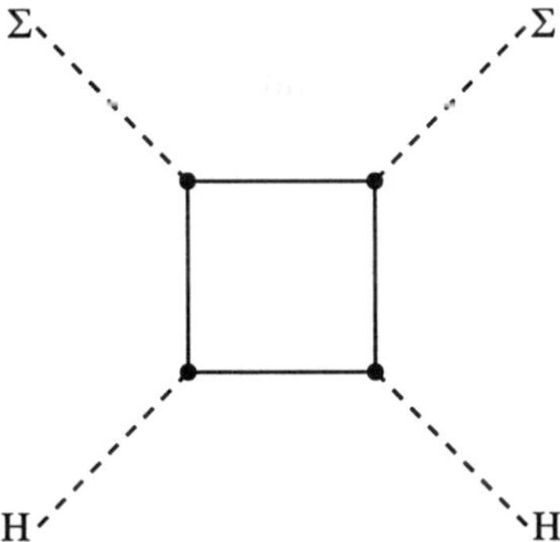

Fig.1.5. Additional one loop graph for the $\bar{H}\Sigma^2 H$ vertex in the supersymmetric $SU(5)$ theory.

Returning to the discussion, given at the end of §1.2, of the supersymmetric $SU(5)$ Grand Unified Theory, there will now be a new one loop diagram contributing to the $\bar{H}\Sigma^2 H$ vertex in addition to Fig.1.3. This is shown in Fig.1.5 where the solid lines represent appropriate fermionic superpartners of $SU(5)$ gauge bosons and Σ-fields. The two graphs cancel in the leading terms and the quartic coupling strengths λ at those two different scales are now related by

$$\lambda(M_W^2) \sim \lambda(M_U^2)\ln\frac{M_W^2}{M_U^2} .$$

---

[10]The persistence of the renormalization scale μ in (1.16) need not worry us since the LHS is not a physically measurable quantity.

Hence the previous hierarchical instability does not materialize and the problem of the radiative instability of the gauge hierarchy is solved[11] [1.20] as a consequence of the nonrenormalization theorem of supersymmetry (cf. Ch.6). The earlier additive term in the RHS of (1.6), proportional to g_U^4, has got cancelled. The multiplicative logarithmic factor comes in the following way. The scalar quartic coupling is a coupling in the superpotential (cf. Ch.5). Since the latter is not renormalized (cf. Ch.6), the renormalization of such a coupling has to be balanced by the wavefunction renormalizations of the multiplying superfields. Owing to dimensional reasons, the latter can at most have a logarithmic dependence on the two mass scales. The same must then be true of λ. These issues will become much clearer after the discussion in §6.6.

In our introduction and overview, as given in this chapter, we have tried to provide a motivation for softly broken supersymmetry other than just its mathematical beauty. It is needed as a stabilizer of the weak scale M_W. The latter is radiatively unstable in the Standard Model; the instability of the Higgs mass m_h accrues via the Higgs VEV to M_W. Stabilization within the Standard Model can be achieved only by fine tuning. As a result, despite its logical consistency and impressive experimental support, the Standard Model is an unnatural theory. Supersymmetry with soft breaking makes the theory radiatively stable and natural, provided the sparticles are not much heavier than a few TeV.

---

[11]This is a far cry, however, from explaining the *origin* of the hierarchy, namely the ratio of the magnitudes of the weak and the unification scales within a supersymmetric grand unified framework.

References

[1.1] G.L. Kane and M. Shifman (ed.), *op. cit.*, *Bibl.*

[1.2] P. Ramond, Phys. Rev. **D3** (1971) 2415. A. Neveu and J.H. Schwarz, Nucl. Phys. **B31** (1971) 86. J-L. Gervais and B. Sakita, Nucl. Phys. **B34** (1971) 632.

[1.3] Yu.A. Golfand and E.P. Likhtman, JETP Lett. **13** (1971) 3214. M. Shifman, *op. cit.*, *Bibl.*

[1.4] D.V. Volkov and V.P. Akulov, Phys. Lett. **B46** (1973) 109. J. Wess and V.P. Akulov, *op. cit.*, *Bibl.*

[1.5] J. Wess and B. Zumino, Nucl. Phys. **B70** (1974) 39. A. Salam and J. Strathdee, Nucl. Phys. **B76** (1974) 477; *loc. cit.*, *Bibl.*

[1.6] J. Wess and B. Zumino, Phys. Lett. **49B** (1974) 52. J. Iliopoulos and B. Zumino, Nucl. Phys. **B76** (1974) 310. S. Ferrara, J. Iliopoulos and B. Zumino, Nucl. Phys. **B77** (1974) 41. A. Salam and J. Strathdee, Nucl. Phys. **B76** (1974) 477; Phys. Rev. **D11** (1975) 1521. M.T. Grisaru, M. Roček and W. Siegel, Nucl. Phys. **B159** (1979) 429.

[1.7] S. Weinberg, Phys. Rev. **D13** (1976) 974; *ibid.* **D19** (1979) 1277. L. Susskind, Phys. Rev. **D20** (1980) 2619. G. 't Hooft, in *Recent Developments in Gauge Theories*, Proc. 1979 Cargése Summer Institute (ed. G. 't Hooft et al., Plenum, New York, 1980).

[1.8] W. Fischler, H.P. Nilles, J. Polchinski, S. Raby and L. Susskind, Phys. Rev. Lett. **47** (1981) 575. R. Kaul and P. Majumdar, Print 81-0373 unpublished and Nucl. Phys. **B199** (1982) 36.

[1.9] L. Maiani in *Proc. Gif-sur-Yvette Summer School* (Paris, 1980), p3. E. Witten, Nucl. Phys. **B88** (1981) 513. M. Veltman, Acta. Phys. Polon. **B12** (1981) 437.

[1.10] P. Fayet, Phys. Lett. **B64** (1976) 159; *ibid.* **B69** (1977) 489; *Recherche* **19** (1988) 304; Ecole Normale Superieure Report No. LPTENS-94-90 (1994). H.E. Haber and G.L. Kane, *loc. cit.*, *Bibl.*

[1.11] L.D. Landau and E.M. Lifshitz, #1, *op. cit.*, *Bibl.*

[1.12] J.J. Sakurai, *op. cit.*, *Bibl.* W. Heitler, *op. cit.*, *Bibl.*

[1.13] V.F. Weisskopf, Phys. Rev. **56** (1939) 72.

[1.14] J.D. Bjorken and S.D. Drell, *op. cit.*, *Bibl.*

[1.15] B. Lee, C. Quigg and H. Thacker, Phys. Rev. **D16** (1977) 1519. D.A. Dicus and V.S. Mathur, Phys. Rev. **D7** (1973) 3111.

[1.16] G.G. Ross, *op. cit.*, *Bibl.* R.N. Mohapatra, *op. cit.*, *Bibl.*

[1.17] M. Drees, *loc. cit., Bibl.*

[1.18] G. Passarino and M. Veltman, Nucl. Phys. **B160** (1979) 151.

[1.19] M. Drees, K. Hagiwara and A. Yamada, Phys. Rev. **D45** (1992) 1725, cf. Appendix.

[1.20] E. Witten, *loc. cit.* [1.9]. S. Dimopoulos and H. Georgi, Nucl. Phys. **B193** (1981) 150. N. Sakai, Z. Phys. **C11** (1981) 153. R. Kaul, Phys. Lett. **B109** (1982) 19.

PART ONE

SUPERSYMMETRY FORMALISM

Chapter 2

PRELIMINARIES

2.1 Grassmann Elements and Variables

To start with, we need to learn the basics of Grassmann elements and variables. The latter constitute the new coordinates in superspace which we need to formulate supersymmetry. Motivated by fermionic fields $\psi(x)$ which anticommute, we are first led to postulate [2.1] anticommuting Grassmann elements ϵ_i (i taking discrete values $1, 2, \cdots, n$) with a product law such that

$$\epsilon_i \epsilon_j = -\epsilon_j \epsilon_i \ .$$

In particular,

$$\epsilon_i^2 = 0 \ .$$

On the other hand, Grassmann elements have a commutative product law with any ordinary number r, i.e.

$$\epsilon_i r = r \epsilon_i \ .$$

More generally, one can include both kinds by defining a set of graded elements α_i such that

$$\alpha_i \alpha_j = (-1)^{n_i n_j} \alpha_j \alpha_i \ ,$$

where

$$n_i = \begin{array}{l} 0 \text{ for } \alpha_i \text{ bosonic,} \\ 1 \text{ for } \alpha_i \text{ fermionic.} \end{array}$$

One can extend the Grassmann concept to that of an anticommuting variable θ. For instance, consider [2.1] conjugate anticommuting variables θ, $\bar{\theta}$ with

$$\theta\bar{\theta} + \bar{\theta}\theta = 0 \ ; \quad \theta^2 = 0 = \bar{\theta}^2 \ ; \quad \bar{\bar{\theta}} = \theta \ .$$

These generate a Grassmann algebra. "Analytic" functions of the type $f(\theta) = f_0 + f_1\theta$, with $f_{0,1}$ being arbitrary complex numbers, constitute a two dimensional space. In this space the derivative w.r.t. θ can be defined as

$$\frac{d}{d\theta} f(\theta) = f_1$$

and f_0 can be reached by taking[1]

$$\theta f(\theta) = f_0 \theta \; .$$

Similarly, the functions $\bar{f}(\bar{\theta}) = \bar{f}_0 + \bar{f}_1 \bar{\theta}$ constitute another two dimensional space in which

$$\frac{d}{d\bar{\theta}} \bar{f}(\bar{\theta}) = \bar{f}_1$$

and

$$\bar{\theta} \bar{f}(\bar{\theta}) = \bar{f}_0 \bar{\theta} \; .$$

A general function $f(\theta, \bar{\theta})$ of these variables would be an element of the Grassmann algebra with the form

$$f(\theta, \bar{\theta}) = f_0 + f_1 \theta + \bar{f}_2 \bar{\theta} + f_3 \theta \bar{\theta} \; , \tag{2.1}$$

$f_0, f_1, \bar{f}_2$ and f_3 being complex numbers. In (2.1) f_0 can be reached by

$$\theta \bar{\theta} f(\theta, \bar{\theta}) = f_0 \theta \bar{\theta}$$

and $f_1, \bar{f}_2$ respectively by

$$\frac{\partial}{\partial \theta} \left\{ \bar{\theta} f(\theta, \bar{\theta}) \right\} = -f_1 \bar{\theta} \; ,$$

$$\frac{\partial}{\partial \bar{\theta}} \left\{ \theta f(\theta, \bar{\theta}) \right\} = -\bar{f}_2 \theta \; .$$

N.B. For consistency, each of the above partial derivatives has to anticommute with θ and $\bar{\theta}$.

In order to isolate f_3 in (2.1), we avoid the pitfalls of double differentiation with anti-commuting variables and – following [2.1] – choose to use Grassmann integration in a way that

$$\int d\bar{\theta} d\theta \, f(\theta, \bar{\theta}) = f_3 \; .$$

This is accomplished by considering $d\theta$ and $d\bar{\theta}$ as anticommuting variables and using the Berezin integration rules [2.1]

$$1 = \int d\theta \, \theta = \int d\bar{\theta} \, \bar{\theta} \; , \tag{2.2a}$$

$$0 = \int d\theta = \int d\bar{\theta} = \int d\theta \, \frac{\partial}{\partial \theta} f(\theta, \bar{\theta}) = \int d\bar{\theta} \, \frac{\partial}{\partial \bar{\theta}} f(\theta, \bar{\theta}) \; . \tag{2.2b}$$

These statements automatically fix the rules for partial integration.

Two other properties attributed to Grassmann integration are linearity and translational invariance. Linearity means that

$$\int d\theta \left\{ \alpha f(\theta) + \beta g(\theta) \right\} = \alpha \int d\theta f(\theta) + \beta \int d\theta g(\theta) \; ,$$

---

[1]If one insists on $f(\theta)$ itself being bosonic (as is the case with a bosonic superfield, cf. Ch.4), f_1 would have to be fermionic. Now $df(\theta)/d\theta$ would be $-f_1$, since one would need to anticommute the $d/d\theta$ through f_1. Similarly, if $f(\theta, \bar{\theta})$ is required to be bosonic, f_1 and $\bar{f}_2$ would need to be fermionic. Different requirements would be needed if the functions are fermionic.

where α and β are complex constants and similarly for the $\bar\theta$-integration. Translational invariance implies that

$$\int d\theta_i f(\theta_i + \theta_j) = \int d\theta_i f(\theta_i) \ , \tag{2.3}$$

where θ_i and θ_j are two anticommuting variables with

$$\int d\theta_i \theta_j = \delta_{ij} \ . \tag{2.4}$$

and ditto for $\bar\theta$. Additionally, Grassmann delta functions can be defined by

$$\int d\theta \delta(\theta) f(\theta) = f_0, \int d\bar\theta \delta(\bar\theta) \bar f(\bar\theta) = \bar f_0 \ , \tag{2.5a}$$

$$\int d\theta \delta(\theta - \theta') f(\theta) = f(\theta'), \int d\bar\theta \delta(\bar\theta - \bar\theta') \bar f(\bar\theta) = \bar f(\bar\theta') \tag{2.5b}$$

so that one can take the representations $\delta(\theta - \theta') = \theta - \theta', \delta(\bar\theta - \bar\theta') = \bar\theta - \bar\theta'$. Also,

$$\int d\bar\theta \delta(\bar\theta) \int d\theta \delta(\theta) f(\theta, \bar\theta) = f_0 \ ,$$

$$\int d\bar\theta \delta(\bar\theta - \bar\theta') \int d\theta \delta(\theta - \theta') f(\theta, \bar\theta) = f(\theta', \bar\theta') \ ,$$

$$\int d\bar\theta d\theta e^{-\bar\theta\theta} = 1 \ .$$

Before closing this section, we discuss the change of integration variables. Let $\{\theta_i\}$ be a denumerable set of independent Grassmann variables spanning some kind of a linear space. A linear transformation, admitted by the latter (with repeated indices being understood as summed), is $\theta_i \to \theta_i' = K_{ij}\theta_j$, where K_{ij} form a bosonic square matrix. Recall that – in a linear space spanned by a denumerable set $\{z_i\}$ of real variables z_i, $-\infty < z_i < \infty$, admitting the transformation with $z_i \to L_{ij}z_j$ (and hence $dz_i \to L_{ij}dz_j$) – one has

$$\prod_i \int dz_i f(Lz) = \frac{1}{\det L} \prod_i \int dz_i f(z) \ . \tag{2.6}$$

The Grassmann case is different since (2.2a) requires that $d\theta \to K^{-1}d\theta$ under $\theta \to K\theta$, where K is a bosonic constant, and more generally from (2.4), $d\theta_i \to (K^{-1})_{ji}d\theta_j$. So the fermionic version of (2.6) is

$$\prod_i \int d\theta_i f(K\theta) = \det K \prod_i \int d\theta_i f(\theta) \ . \tag{2.7}$$

These considerations readily extend to the complex case:

$$\prod_{i,j} \int d\bar z_i dz_j f(\bar z L z) = \frac{1}{\det L} \prod_{i,j} \int d\bar z_i dz_j f(\bar z z) \ , \tag{2.8a}$$

$$\prod_{i,j} \int d\bar\theta_i d\theta_j f(\bar\theta K \theta) = \det K \prod_{i,j} \int d\bar\theta_i d\theta_j f(\bar\theta\theta) \ . \tag{2.8b}$$

Specifically, for a common set of linear coefficients M_{ij}, it is easy to see that

$$\left[\prod_k \int dz_k \delta^{(k)}(M_{kj}z_j)\right]^{-1} = \left[\prod_k \int d\bar{z}_k \delta^{(k)}(\bar{z}_j M_{jk})\right]^{-1} = \det M = \prod_{k,\ell} \int d\bar{\theta}_k d\theta_\ell e^{-\bar{\theta}_i M_{ij}\theta_j}.$$

$$(2.9)$$

Eq.(2.9) finds application in the exponentiation of the Faddeev-Popov determinant and superdeterminant in quantized Yang-Mills and super-YM theories (cf. §6.5) respectively.

2.2 Supersymmetric Harmonic Oscillator

We shall illustrate how supersymmetry can arise in quantum mechanics by choosing the familiar problem of a simple harmonic oscillator.

Bose case

Normalizing the mass to unity, we write the Lagrangian for the one dimensional harmonic oscillator in terms of canonical variables (q, p) and the angular frequency ω as

$$L_B = \frac{1}{2}\dot{q}^2 - \frac{1}{2}\omega^2 q^2. \tag{2.10}$$

The (Bose) creation and annihilation operators $a^\dagger$ and a relate respectively to q and p via

$$q(t) = \frac{1}{\sqrt{2\omega}}(ae^{-i\omega t} + a^\dagger e^{i\omega t})$$

and

$$p(t) = i\sqrt{\frac{\omega}{2}}(-ae^{-i\omega t} + a^\dagger e^{i\omega t}) = \dot{q} \ .$$

They obey the commutation relation $[a, a^\dagger] = 1$ following the Heisenberg equation $[q, p] = i$. The Hamiltonian is

$$H_B = \dot{q}p - L_B = \frac{\omega}{2}(a^\dagger a + aa^\dagger) = \omega\left(a^\dagger a + \frac{1}{2}\right). \tag{2.11}$$

Fermi case

Here the canonical variables can be taken to be the complex fields ψ and $i\psi^\dagger$ and the Lagrangian as

$$L_F = \frac{i}{2}(\psi^\dagger\dot{\psi} - \dot{\psi}^\dagger\psi) - \frac{\omega}{2}(\psi^\dagger\psi - \psi\psi^\dagger) \ .$$

The (Fermi) creation and annihilation operators $b^\dagger$ and b relate to these via

$$\psi(t) = be^{-i\omega t}, \ i\psi^\dagger(t) = ib^\dagger e^{i\omega t}$$

and obey the anticommutation relation[2] $[b, b^\dagger]_+ = 1$ in consequence of the quantization rule $[\psi, i\psi^\dagger]_+ = i$. The canonical momenta corresponding to ψ and $i\psi^\dagger$ are[3] $-\frac{i}{2}\psi^\dagger$ and $-\frac{1}{2}\psi$ respectively. The Hamiltonian is

$$H_F = \dot\psi\left(-\frac{i}{2}\psi^\dagger\right) + i\dot\psi^\dagger\left(-\frac{1}{2}\psi\right) - L_F = \frac{\omega}{2}(b^\dagger b - bb^\dagger) = \omega\left(b^\dagger b - \frac{1}{2}\right).$$

The negative sign of the $bb^\dagger$ term corresponds to the negative energy contribution from fermionic antiparticles.

Superposed case

With both bosonic and fermionic harmonic oscillators, we have a Lagrangian

$$L_{B+F} = \frac{1}{2}\dot{q}^2 - \frac{1}{2}\omega^2 q^2 + \frac{i}{2}\left(\psi^\dagger\dot\psi - \dot\psi^\dagger\psi\right) - \frac{\omega}{2}\left(\psi^\dagger\psi - \psi\psi^\dagger\right). \tag{2.12}$$

and the Hamiltonian

$$H_{B+F} = \omega\left(a^\dagger a + \frac{1}{2} + b^\dagger b - \frac{1}{2}\right) = \omega\left(a^\dagger a + b^\dagger b\right). \tag{2.13}$$

The cancellation of the nullpoint energy in (2.13) assumes great significance when these considerations are extended to an uncountable infinity of harmonic oscillators as occur in field theory. An additional noteworthy feature is the invariance of the Lagrangian – modulo a total time derivative which does not affect the action $S = \int_{-\infty}^{\infty} dt\, L$ – under the following infinitesimal transformations defined by Grassmann parameters $\epsilon, \bar\epsilon$:

$$\delta q = \epsilon\psi + \psi^\dagger\bar\epsilon, \tag{2.14a}$$

$$\delta\psi = (-i\dot{q} - \omega q)\bar\epsilon, \tag{2.14b}$$

$$\delta\psi^\dagger = (i\dot{q} - \omega q)\epsilon. \tag{2.14c}$$

We can rewrite the transformations (2.14) in terms of a pair of conjugate fermionic generators Q and $\bar Q$ as

$$\delta q = [(\bar\epsilon\bar Q + \epsilon Q), q], \tag{2.15a}$$

$$\delta\psi = [\bar\epsilon\bar Q, \psi], \tag{2.15b}$$

$$\delta\psi^\dagger = [\epsilon Q, \psi^\dagger]. \tag{2.15c}$$

A comparison between (2.14) and (2.15) leads to the specification

$$\bar Q = (-i\dot{q} - \omega q)\psi^\dagger = -\sqrt{2\omega}ab^\dagger, \tag{2.16a}$$

---

[2]We use the notation $[A, B] \equiv AB - BA$, $[A, B]_+ \equiv AB + BA$.

[3]One needs to anticommute the partial derivative with respect to $\dot\psi$ through the $\psi^\dagger$ factor in the first term of L_F.

$$Q = (i\dot{q} - \omega q)\psi = -\sqrt{2\omega}a^{\dagger}b \ . \tag{2.16b}$$

In (2.14a) δq must be real which means that $\psi^{\dagger}\bar{\epsilon}$ has to be the hermitian conjugate of $\epsilon\psi$. It then follows from (2.16) that the hermitian conjugate of ϵQ is $-\bar{\epsilon}\bar{Q}$ here so that $\bar{\epsilon}\bar{Q} + \epsilon Q$ is antihermitian. The use of (2.13) and (2.16) further leads to

$$\dot{Q} = i[H, Q] = 0 \ . \tag{2.17}$$

This means that the fermionic generator Q can be viewed as a **conserved supercharge**. Moreover,

$$H = \frac{1}{2}[Q, \bar{Q}]_{+} \ . \tag{2.18}$$

Physical illustration

The supersymmetric harmonic oscillator is physically realizable. We give a simple example here. Consider a spin half particle of charge e and mass m (gyromagnetic ratio $|e|/m$) executing circular classical orbits with angular frequency $\omega = |e|\mathcal{B}/m$ in a plane perpendicular to a uniform magnetic field $\vec{\mathcal{B}}$ along the z (3rd) axis, say. The energy spectrum of such a particle can be written as [2.2]

$$E_n = \left(n + \frac{1}{2}\right)\frac{|e|\mathcal{B}}{m} + \frac{|e|\mathcal{B}}{2m}\langle\sigma^3\rangle \ .$$

Here n is a nonnegative integer and $\langle\sigma^3\rangle = +1, -1$ for the spin up ($\uparrow$) and spin down ($\downarrow$) cases. Suppose we identify all spin up (spin down) states as bosons (fermions)[4]. The absolute choice for this is arbitrary but the relative difference between these two sets of states is not. Now we have a supersymmetric system. The supersymmetric energy degeneracy shows up through the equality

$$E_{n-1,\uparrow} = E_{n,\downarrow} \ .$$

The bosonic creation, annihilation operators $a^{\dagger}, a$ are given as before as linear combinations of ωq and $\dot{q}$, where q is the coordinate in which the spin independent part of the Hamiltonian is manifestly [2.2] that of a harmonic oscillator. The fermionic creation, annihilation operators $b^{\dagger}, b$ respectively equal σ_{+}, σ_{-} with $\sigma_{\pm} = (\sigma^1 \pm i\sigma^2)/2$, σ^i being the Pauli matrices. The Hamiltonian is given by (2.13). The supercharge can be constructed using (2.16), while (2.17) and (2.18) can be shown to be satisfied.

General discussion

One can identify (2.14) as a set of infinitesimal supersymmetry transformations and Q as the supercharge. One can then postulate the validity of (2.17) and (2.18) for a general supersymmetric system in quantum mechanics. For such a general interacting system, though, the Hamiltonian may not decompose additively into separate fermionic and bosonic parts. However, considerable constraints ensue on the system because of supersymmetry by which we mean invariance under (2.14) or equivalently (2.15) and the corresponding validity of (2.17), (2.18). In particular, the energy spectrum *can contain no negative eigenvalues*. This

---

[4]Since we are talking about single particle states, the question of statistics is irrelevant here.

can be seen by the insertion of a complete set of states $|n\rangle\langle n|$ in the middle of the RHS of (2.18) and then taking a forward matrix element of either side with an energy eigenstate $|\alpha\rangle$ of energy eigenvalue E_α. Thus

$$E_\alpha = \langle\alpha|H|\alpha\rangle = \frac{1}{2}\sum_n \left\{ |\langle\alpha|Q|n\rangle|^2 + |\langle\alpha|\bar{Q}|n\rangle|^2 \right\} \geq 0 \ .$$

Moreover, the annihilation of the ground state $|\Omega\rangle$ by the supercharge is intrinsically related to the zero energy of $|\Omega\rangle$:

$$Q|\Omega\rangle = 0 \Leftrightarrow H|\Omega\rangle = 0 \ .$$

Contrariwise, if $H|\Omega\rangle \neq 0$, i.e. the energy of the groundstate is nonzero, $Q|\Omega\rangle \neq 0$ i.e. supersymmetry is spontaneously broken. The situation is illustrated in terms of a potential $V(x)$ in Fig.2.1.

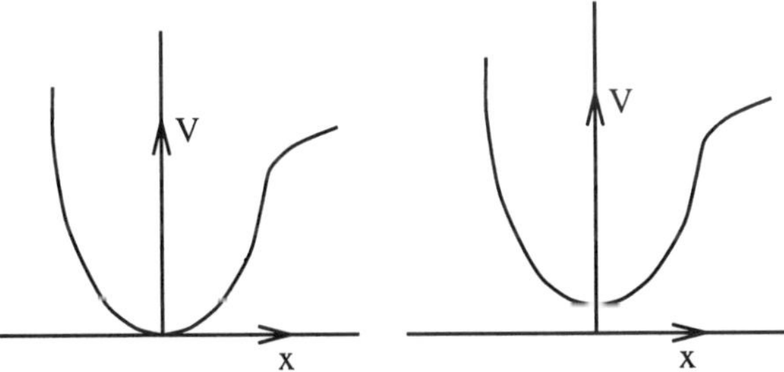

Fig.2.1. Potentials for (a) invariant or (b) spontaneously broken supersymmetric vacua.

We summarize the important features of the supersymmetric harmonic oscillator, learnt above.

- The null point energy cancels out between bosonic and fermionic contributions.

- There is a conserved supercharge.

- The Hamiltonian is half the anticommutator between the supercharge and its hermitian conjugate, implying the positivity of energy.

- The ground state energy is zero or else supersymmetry is spontaneously broken.

2.3 Glimpse of Superspace

We can introduce superspace in quantum mechanics [2.3, 2.4] by considering a hermitian supercoordinate X as a function of the three variables $(t, \theta, \bar{\theta})$. Superspace is the space spanned by X. Here t is time and $\theta, \bar{\theta}$ are complex conjugate Grassmann variables. X will have an expansion

$$X(t, \theta, \bar{\theta}) = q(t) + \theta\psi(t) + \psi^\dagger(t)\bar{\theta} + D(t)\theta\bar{\theta} \tag{2.19}$$

in analogy with (2.1). Of the four functions of time describing the supercoordinate in (2.19), $q(t)$ and $D(t)$ are real bosonic components while $\psi(t)$ and $\psi^\dagger(t)$ are complex fermionic ones. A general global supersymmetry transformation in superspace changes the supercoordinate variables as under

$$(t, \theta, \bar\theta) \rightarrow (t + i\bar\epsilon\theta + i\epsilon\bar\theta, \theta + \epsilon, \bar\theta + \bar\epsilon) \, , \tag{2.20}$$

t remaining real and $\epsilon, \bar\epsilon$ being t-independent conjugate Grassmann parameters with $(\bar\epsilon\theta)^\dagger = \bar\theta\epsilon = -\epsilon\bar\theta$.

A linear realization of supersymmetry requires the transformation $X \rightarrow X + \delta X$ with

$$\delta X(t, \theta, \bar\theta) = (\epsilon Q + \bar\epsilon\bar Q)X(t, \theta, \bar\theta) \, , \tag{2.21}$$

where the generators $Q, \bar Q$ are given by

$$Q = i\bar\theta\frac{\partial}{\partial t} + \frac{\partial}{\partial\theta} \, ,$$

$$\bar Q = i\theta\frac{\partial}{\partial t} + \frac{\partial}{\partial\bar\theta} \, .$$

It follows that the components of X must transform as

$$\delta q = \epsilon\psi + \psi^\dagger\bar\epsilon \, , \tag{2.22a}$$

$$\delta\psi = \bar\epsilon(-i\dot q + D) \, , \tag{2.22b}$$

$$\delta\psi^\dagger = \epsilon(i\dot q + D) \, , \tag{2.22c}$$

$$\delta D = i\frac{d}{dt}(\psi^\dagger\bar\epsilon - \epsilon\psi) \, . \tag{2.22d}$$

The identification $D = -\omega q$ (this value of D obtains from a constraint equation, as shown later) matches (2.22a-c) exactly with the corresponding transformations (2.14). Moreover, the coefficient of the $\theta\bar\theta$ term in X transforming in (2.22d) like a total time derivative means that such a term in a Lagrangian would yield a supersymmetric action. This fact is sometimes used in constructing the Lagrangians of supersymmetric theories.

One can recast the harmonic oscillator Lagrangian (2.12) in a superspace form which makes its supersymmetry invariance manifest. We introduce covariant derivatives $\mathcal{D}, \bar{\mathcal{D}}$ in superspace:

$$\mathcal{D} = \frac{\partial}{\partial\theta} - i\bar\theta\frac{\partial}{\partial t} \, ,$$

$$\bar{\mathcal{D}} = \frac{\partial}{\partial\bar\theta} - i\theta\frac{\partial}{\partial t} \, .$$

Their covariance can be explicitly checked by noting that $\delta\mathcal{D}X = \mathcal{D}\delta X$ and $\delta\bar{\mathcal{D}}X = \bar{\mathcal{D}}\delta X$. Now we write the Lagrangian as

$$L = \left[-\frac{1}{2}(\bar{\mathcal{D}}X)(\mathcal{D}X) + \frac{1}{2}\omega X^2 \right]_D \, . \tag{2.23}$$

In (2.23) the suffix D outside the square bracket means that only the coefficient of the $\theta\bar{\theta}$ term is to be taken. This then would lead to a supersymmetric action since δL is a total time derivative not contributing to the variation of the action. On working out the components, one finds that

$$L = \frac{1}{2}\dot{q}^2 + \frac{i}{2}(\psi^\dagger\dot{\psi} - \dot{\psi}^\dagger\psi) - \frac{\omega}{2}(\psi^\dagger\psi - \psi\psi^\dagger) + \frac{1}{2}D^2 + D\omega q \ .$$

Here D is an auxiliary variable which gets fixed from $\partial L/\partial D = 0$ to be $-\omega q$ so that one obtains (2.12).

2.4 Supersymmetry and Spacetime Transformations

As advertised in Ch.1, supersymmetry enables fermions to transform into bosons and vice versa. Thus generically the supercharge acts as

$$Q|\text{boson}, \text{fermion}\rangle = |\text{fermion}, \text{boson}\rangle \ .$$

Since the spins of the states before and after the transformation differ half-integrally, Q must be spinorial. In fact, it is easy to see [2.5] that Q (and hence any supersymmetry transformation) cannot commute with rotation, though (2.18) and its four dimensional generalization imply that it does commute with the four translation generators P_μ. Let $U_{2\pi}$ be a 2π rotation operator around some axis, i.e.

$$U_{2\pi}|\text{boson}\rangle = |\text{boson}\rangle$$
$$U_{2\pi}|\text{fermion}\rangle = -|\text{fermion}\rangle \ .$$

Now

$$\begin{aligned}
U_{2\pi}QU_{2\pi}^{-1}|\text{fermion}\rangle &= -U_{2\pi}QU_{2\pi}^{-1}U_{2\pi}|\text{fermion}\rangle \\
&= -U_{2\pi}Q|\text{fermion}\rangle \\
&= -U_{2\pi}|\text{boson}\rangle \\
&= -|\text{boson}\rangle \\
&= -Q|\text{fermion}\rangle \ .
\end{aligned}$$

Similarly,

$$U_{2\pi}QU_{2\pi}^{-1}|\text{boson}\rangle = -Q|\text{boson}\rangle.$$

Since all bosons and fermions form a complete basis,

$$U_{2\pi}QU_{2\pi}^{-1} = -Q \ .$$

More generally, Q turns out not to commute with homogeneous Lorentz transformations. Thus though $[Q, P_\mu] = 0$, as stated earlier, $[Q, M_{\mu\nu}] \neq 0$ where $M_{\mu\nu}$ are the generators of the homogeneous Lorentz group. Since the latter are defined in a $(3 + 1)$ dimensional spacetime, supersymmetry must necessarily be a spacetime symmetry. This fact, as well as the occurrence of the conserved supercharge Q in anticommutators such as (2.18), are two of the hallmarks of supersymmetry.

References

[2.1] F.A. Berezin, *op. cit.*, *Bibl.* L.D. Faddeev, *loc. cit.*, *Bibl.* B. DeWitt, *op. cit.*, *Bibl.*

[2.2] L.D. Landau and E.M. Lifshitz, #2, *op. cit.*, *Bibl.*.

[2.3] P. Salmonson and J.W. van Holten, Nucl. Phys. **B96** (1982) 509. E. Witten, Nucl. Phys. **B188** (1981) 533. F. Cooper, A. Khare and U. Sukhatme, Phys. Rep. **251** (1995) 267; *op. cit.*, *Bibl.*

[2.4] S.P. Misra, *op. cit.*, *Bibl.*

[2.5] M.F. Sohnius, *loc. cit.*, *Bibl.*

Chapter 3

ALGEBRAIC ASPECTS

3.1 Supersymmetry Algebra

Let us start with the fullest continuous spacetime symmetry of particle interactions observed so far, namely that of the Poincaré group. We work in a four dimensional Minkowski spacetime with a flat metric

$$\eta_{\mu\nu} = \eta^{\mu\nu} = \text{diag } (1, -1, -1, -1)$$

i.e. $\eta^{00} = \eta_{00} = 1$ and $\eta^{pr} = \eta_{pr} = -\delta_{pr}$ with p, r being spatial indices. The Lie algebra of the ten parameter Poincaré group is defined through inhomogenous Lorentz transformations

$$x'^{\mu} = (\delta^{\mu}_{\ \nu} + \omega^{\mu}_{\ \nu})x^{\nu} + a^{\mu}, \quad \omega_{\mu\nu} = -\omega_{\nu\mu} \ ,$$

in the neighborhood of the identity, where $\omega_{\mu\nu}$ is a second rank antisymmetric constant tensor and a^{μ} is a constant four vector. The unitary operators for four translations and homogeneous Lorentz transformations are $U(a) = e^{ia^{\mu}P_{\mu}}$ and $U(\Lambda) = e^{-\frac{i}{2}\omega^{\mu\nu}M_{\mu\nu}}$ respectively. The corresponding hermitian generators P_{μ} and $M_{\mu\nu}$ can be seen to satisfy the relations [3.1]

$$[P_{\mu}, P_{\nu}] = 0 \ , \tag{3.1a}$$

$$[M_{\mu\nu}, P_{\rho}] = i(\eta_{\nu\rho}P_{\mu} - \eta_{\mu\rho}P_{\nu}) \ , \tag{3.1b}$$

$$[M_{\mu\nu}, M_{\rho\sigma}] = -i(\eta_{\mu\rho}M_{\nu\sigma} - \eta_{\mu\sigma}M_{\nu\rho} - \eta_{\nu\rho}M_{\mu\sigma} + \eta_{\nu\sigma}M_{\mu\rho}) \ . \tag{3.1c}$$

These constitute the Poincaré algebra while the Lorentz algebra comprises just (3.1c).

The spinorial realization of the Lorentz algebra obtains in terms of the 4×4 matrices γ^{μ} ($\mu = 0, 1, 2, 3$) which form the elements of a Clifford algebra over Minkowski spacetime. These obey the anticommutation relations

$$[\gamma^{\mu}, \gamma^{\nu}]_{+} = 2\eta^{\mu\nu} I \ .$$

It may be verified that the representations

$$\Sigma_{\mu\nu} \equiv \frac{i}{4}[\gamma_{\mu}, \gamma_{\nu}] \ ,$$

$$M_{\mu\nu} = -x_\mu P_\nu + x_\nu P_\mu + \Sigma_{\mu\nu}$$

satisfy (3.1). These relations identify P_μ as the four momentum operator and M_{pr} as the total angular momentum tensor with Σ_{pr} as its spin part. M_{op} generate Lorentz boosts.

A commonly used representation for γ_μ and $\gamma_5 \equiv i\gamma^0\gamma^1\gamma^2\gamma^3$ is that of Ref. [3.2] which we call the **Dirac representation**. In this

$$\gamma^0 = \begin{pmatrix} I & 0 \\ 0 & -I \end{pmatrix}, \gamma^p = \begin{pmatrix} 0 & \sigma^p \\ -\sigma^p & 0 \end{pmatrix},$$

with σ^p $(p = 1, 2, 3)$ being the Pauli spin matrices and

$$\gamma_5 = i\gamma_0\gamma_1\gamma_2\gamma_3 = \begin{pmatrix} 0 & I \\ I & 0 \end{pmatrix}.$$

In theoretical discussions of supersymmetry it is convenient to employ the **Weyl representation**. Unless explicitly mentioned otherwise, we shall henceforth use the Weyl representation for γ-matrices. Here

$$\gamma^\mu = \begin{pmatrix} 0 & \sigma^\mu \\ \bar\sigma^\mu & 0 \end{pmatrix},$$

$$\gamma_5 = \gamma^5 = i\gamma^0\gamma^1\gamma^2\gamma^3 = \begin{pmatrix} -I & 0 \\ 0 & I \end{pmatrix},$$

$$\Sigma^{\mu\nu} = \frac{i}{4}\begin{pmatrix} \sigma^\mu\bar\sigma^\nu - \sigma^\nu\bar\sigma^\mu & 0 \\ 0 & \bar\sigma^\mu\sigma^\nu - \bar\sigma^\nu\sigma^\mu \end{pmatrix} \equiv \begin{pmatrix} \sigma^{\mu\nu} & 0 \\ 0 & \bar\sigma^{\mu\nu} \end{pmatrix},$$

with $\sigma^\mu = (I, \sigma^p)$, $\bar\sigma^\mu = (I, -\sigma^p)$, σ^p $(p = 1, 2, 3)$ being the Pauli spin matrices and

$$\sigma^{\mu\nu} \equiv \frac{i}{4}(\sigma^\mu\bar\sigma^\nu - \sigma^\nu\bar\sigma^\mu), \ \bar\sigma^{\mu\nu} \equiv \frac{i}{4}(\bar\sigma^\mu\sigma^\nu - \bar\sigma^\nu\sigma^\mu) = (\sigma^{\mu\nu})^\dagger,$$

the dagger meaning a hermitian conjugate. The matrices $\sigma_{\alpha\beta}$ and $\bar\sigma_{\alpha\beta}$ obey

$$\epsilon^{\mu\nu\alpha\beta}\sigma_{\alpha\beta} = 2i\sigma^{\mu\nu}, \epsilon^{\mu\nu\alpha\beta}\bar\sigma_{\alpha\beta} = -2i\bar\sigma^{\mu\nu},$$

where $\epsilon_{0123} = 1$. Note that $\sigma^\mu = \bar\sigma_\mu$ and $\bar\sigma^\mu = \sigma_\mu$. There are also the identities

$$\text{Tr}(\sigma^\mu\bar\sigma^\nu) = 2\eta^{\mu\nu},$$

$$\sigma^\mu\bar\sigma^\nu + \sigma^\nu\bar\sigma^\mu = 2\eta^{\mu\nu}I = \bar\sigma^\mu\sigma^\nu + \bar\sigma^\nu\sigma^\mu.$$

$$\text{Tr}(\sigma^{\mu\nu}\sigma^{\alpha\beta}) = \frac{1}{2}(\eta^{\mu\alpha}\eta^{\nu\beta} - \eta^{\mu\beta}\eta^{\nu\alpha}) + \frac{i}{2}\epsilon^{\mu\nu\alpha\beta}.$$

The **Coleman, Mandula theorem** [3.3] is a rigorous result on the interplay between continuous symmetries of the world other than Poincaré invariance and the latter. It has been proved on the basis of some very reasonable assumptions. These are: the existence of both a nontrivial analytic S-matrix and a nondegenerate vacuum plus the occurrence of massive particle states in finite dimensional positive energy representations of the Lorentz group. We shall not give the proof of the theorem here since it is somewhat involved. The interested reader should look up Ref. [3.3]. A more pedagogical account may be found in Ref. [3.4]. The theorem makes the following statement on the full Lie algebra pertaining to all the continuous symmetries of the S-matrix. *This Lie algebra, containing (as Lie subalgebras) both the above Poincaré algebra and any other Lie algebra that is defined by generators $\{T^a\}$ and structure constants t_c^{ab} such that*

$$[T^a, T^b] = it_c^{ab}T^c, \tag{3.2}$$

must be a direct sum of the two, i.e.

$$[T^a, P_\mu] = 0 = [T^a, M_{\mu\nu}] . \tag{3.3}$$

In other words, the spacetime symmetry of the Poincaré algebra cannot be conjoined with any other continuous symmetry of the S-matrix except in a trivial way, as is the case for internal symmetries.

It is evident from the example of §2.2 that supersymmetry evades the strait-jacket of the CM theorem. It does this by going beyond Lie algebras and embracing graded Lie algebras which involve anticommutators also. In particular, we take a Z_2-graded structure of the type

$$[\text{even}, \text{even}] = \text{even}, \tag{3.4a}$$

$$[\text{even}, \text{odd}] = \text{odd}, \tag{3.4b}$$

$$[\text{odd}, \text{odd}]_+ = \text{even} \tag{3.4c}$$

and identify bosonic (fermionic) generators as even (odd). We already have (3.2) and (3.3) since the bosonic subalgebras must still obey the *CM* theorem [3.3]. But now we include a **supercharge** Q_a, the subscript[1] a being a spinor index, as an extra spinorial fermionic generator of supersymmetry transformations. With Q_a present, the CM theorem does not apply and one can construct nontrivially combined spacetime and other continuous symmetries within the purview of a graded Lie algebra. In fact, one is now covered by the **Haag, Łopuszánski, Sohnius** [3.5] **theorem** which we shall again state without proof. That says: *the most general continuous symmetry of the S-matrix, consistent with the assumptions of the CM theorem, is that pertaining to a Z_2-graded Lie algebra where the odd generators belong to the representations[2] $(\frac{1}{2}, 0)$ and $(0, \frac{1}{2})$ of the homogenous Lorentz group and the even generators are a direct sum of the Poincaré and other symmetry generators (i.e. the latter two sets of generators mutually commute).*

Before taking up the implications of this HLS statement with regard to supersymmetry, we want to discuss the Lorentz group representation properties of the fermionic generators. Let us return to (3.1). The definitions

$$J_p = \frac{1}{2}\epsilon_{prs}M_{rs}, K_p = -M_{0p} ,$$

---

[1] This subscript should not be confused with the superscript a, which is an index of the adjoint representation of the Lie algebra.

[2] This notation is explained in the subsequent discussion.

with p, r, s as spatial indices, as well as the construction of the complex linear combinations

$$\mathcal{J}_p^{\pm} = \frac{1}{2}(J_p \pm iK_p) \ ,$$

enable us to rewrite [3.1] the algebra of the homogeneous Lorentz group $SO(1,3)$ in the last of the relations (3.1) as that of the homomorphic $SU(2)_+ \otimes SU(2)_-$:

$$
\begin{aligned}
[\mathcal{J}_p^+, \mathcal{J}_q^+] &= i\epsilon_{pqr}\mathcal{J}_r^+ \ , \\
[\mathcal{J}_p^-, \mathcal{J}_q^-] &= i\epsilon_{pqr}\mathcal{J}_r^- \ , \\
[\mathcal{J}_p^+, \mathcal{J}_q^-] &= 0 \ .
\end{aligned}
\tag{3.5}
$$

Consequently, each finite dimensional irreducible representation of the homogeneous Lorentz group can be characterized by integral or half-integral "angular momentum" index pairs (j_1, j_2), j_1 referring to the $SU(2)_+$ factor and j_2 to the $SU(2)_-$ one. Thus $(0,0)$ refers to a scalar, while $\left(\frac{1}{2},0\right)$ and $\left(0,\frac{1}{2}\right)$ specify the left chiral and right chiral parts of a Dirac spinor respectively. A massive four component Dirac spinor corresponds to the representation $\left(\frac{1}{2},0\right) \oplus \left(0,\frac{1}{2}\right)$. Fermionic generators, which have Lorentz group transformation properties of Dirac spinors, must also transform as those representations. Note furthermore that $\left(\frac{1}{2},\frac{1}{2}\right)$ describes a four vector (such as P_μ), while $(1,0)$ and $(0,1)$ pertain respectively to the selfdual and anti-selfdual parts of a second rank tensor (e.g. $M_{\mu\nu}$).

The HLS theorem allows nontrivial commutation relations between Q_a and $M_{\mu\nu}$, P_μ. Take first the commutator between Q_a and $M_{\mu\nu}$. Being of the form (3.4b) and with Q_a as the only odd operator present, the RHS of the said commutator will be linearly related to Q_a. By using the Jacobi identity[3] with two M's and one Q, one can show [3.6] that the corresponding linear coefficients form a representation of the Lorentz algebra. This means that Q_a carries a representation of the Lorentz group. The HLS theorem enables us to select that representation to be $\left(\frac{1}{2},0\right) \oplus \left(0,\frac{1}{2}\right)$, i.e. a to be a four component Dirac spinor index $(a = 1,2,3,4)$ and have[4]

$$[M_{\mu\nu}, Q_a] = -(\Sigma_{\mu\nu})_{ab}Q_b \ . \tag{3.6}$$

We can, as a minimal choice [3.6], restrict Q_a further to be a Majorana spinor which has fewer degrees of freedom than a Dirac spinor. Our definition of a Majorana spinorial field χ, equated to its charge conjugate χ^C, is

$$\chi = \chi^C \equiv C\bar{\chi}^T \ ,$$

where $\bar{\chi} \equiv \chi^\dagger \gamma^0$; the superscript T refers to the transpose and the charge conjugation matrix C obeys the properties

$$C^{-1}\gamma_5 C = \gamma_5, \ \ C^{-1}\gamma_\mu C = -\gamma_\mu^T, \ \ C^{-1}\gamma_\mu\gamma_5 C = (\gamma_\mu\gamma_5)^T, \ \ C = -C^T, \ \ C^{-1} = C^\dagger \ .$$

Thus we have

$$Q_a = C_{ab}\bar{Q}_b \ . \tag{3.7}$$

---

[3] This is $[M_{\mu\nu}, [M_{\rho\sigma}, Q_a]] + [M_{\rho\sigma}, [Q_a, M_{\mu\nu}]] + [Q_a, [M_{\mu\nu}, M_{\rho\sigma}]] = 0 \ .$

[4] In spinor space $U^{-1}(\Lambda)Q_a U(\Lambda) = (e^{-\frac{i}{2}\omega^{\mu\nu}\Sigma_{\mu\nu}})_{ab}Q_b \ .$

Next, consider the odd-even commutator between Q_a and P_μ. Once again, the RHS has to be linearly proportional to Q_a. Consistency with the requirements of the internal symmetry algebra and the Lorentz algebra constrains [3.6] the commutator to the form

$$[Q_a, P_\mu] = (x\gamma_\mu + y\gamma_\mu\gamma_5)_{ab}Q_b \ ,$$

x, y being constants. We have used here a consequence of the HLS theorem that there is no fermionic symmetry generator in the $\left(1, \frac{1}{2}\right) \oplus \left(\frac{1}{2}, 1\right)$ representation. Now the charge conjugation of both sides of the above equation and the requirement of consistency with (3.7) demands that x be real and y be imaginary. Moreover, the Jacobi identity[5] with two P's and one Q requires $x^2 = y^2$. Hence both x and y must vanish so that one has

$$[Q_a, P_\mu] = 0 \ . \tag{3.8}$$

An immediate consequence of (3.8) is that P^2 is a Casimir operator commuting with Q_a, i.e. for every supersymmetry multiplet all members are mass degenerate. In nature this is not even approximately true because of supersymmetry breaking. As we shall see in Ch.7, a challenge to any realistic supersymmetric theory is to have a credible mechanism which breaks supersymmetry without spoiling its desirable features.

We come next to the anticommutator between Q_a and Q_b which itself is even and symmetric under $a \leftrightarrow b$. The only even generators are those of the Poincaré group since others, by the CM theorem, would correspond to some internal symmetry absent in $Q_{a,b}$. The $a \leftrightarrow b$ symmetry can be ensured by utilizing the fact that the matrices $\gamma^\mu C$ and $\Sigma^{\mu\nu}C$ are even under transposition. Thus one can write

$$[Q_a, Q_b]_+ = r(\gamma^\mu C)_{ab}P_\mu + s(\Sigma^{\mu\nu}C)_{ab}M_{\mu\nu} \ ,$$

with r, s as constant parameters. Now the (Q, Q, P_μ) Jacobi identity[6] requires [3.6] that s vanish. On the other hand, the coefficient r in the first term can be scaled to any value simply by redefining Q_a. We choose $r = -2$ so as to have

$$[Q_a, Q_b]_+ \ = \ -2(\gamma^\mu C)_{ab}P_\mu \ , \tag{3.9a}$$

$$[Q_a, \bar{Q}_b]_+ \ = \ 2(\gamma^\mu)_{ab}P_\mu \ , \tag{3.9b}$$

$$[\bar{Q}_a, \bar{Q}_b]_+ \ = \ 2(C^{-1}\gamma^\mu)_{ab}P_\mu \ . \tag{3.9c}$$

(3.7) takes (3.9a) to (3.9b,c). The appearance of P_μ in the RHS above is instrumental in linking supersymmetry to gravity, as will become clear later. Moreover, note that the mass dimension of the spinorial charge Q is now fixed to be 1/2.

One can also see that the charge conjugation relation (3.7) plus the commutation and anticommutation relations (3.6), (3.8) and (3.9) are invariant under a chiral rotation

$$Q_a \to \exp(-i\varphi\gamma_5)_{ab}Q_b \ , \quad \bar{Q}_a \to \bar{Q}_b\exp(-i\varphi\gamma_5)_{ba} \ ,$$

---

[5] This is $[P_\mu, [P_\nu, Q_a]] + [P_\nu, [Q_a, P_\mu]] + [Q_a, [P_\mu, P_\nu]] = 0 \ .$
[6] This is $[Q_a, [Q_b, P_\mu]]_+ + [Q_b, [Q_a, P_\mu]]_+ + [P_\mu, [Q_a, Q_b]_+] = 0 \ .$

with φ being real. In fact, this invariance can be implemented unitarily through an axial $U(1)$ generator R by means of the unitary operator $e^{i\varphi R}$ as follows:

$$e^{i\varphi R} Q_a e^{-i\varphi R} = \exp(-i\varphi\gamma_5)_{ab} Q_b ,$$

$$e^{i\varphi R} \bar{Q}_a e^{-i\varphi R} = \bar{Q}_b \exp(-i\varphi\gamma_5)_{ba} ,$$

leading to the commutator

$$[Q_a, R] = (\gamma_5)_{ab} Q_b . \tag{3.10}$$

Thus we can enlarge the supersymmetry algebra by including this R via (3.10) together with

$$[R, P_\mu] = 0 = [R, M_{\mu\nu}] . \tag{3.11}$$

This global $U(1)$ symmetry of the algebra is called R-invariance.

Finally, combining (3.6 – 3.10) our enlarged supersymmetry algebra can be written as:

$$[Q_a, Q_b]_+ \;\; = \;\; -2(\gamma^\mu C)_{ab} P_\mu , \tag{3.12a}$$

$$[Q_a, P_\mu] \;\; = \;\; 0 , \tag{3.12b}$$

$$[M_{\mu\nu}, Q_a] \;\; = \;\; -(\Sigma_{\mu\nu})_{ab} Q_b , \tag{3.12c}$$

$$[Q_a, R] \;\; = \;\; (\gamma_5)_{ab} Q_b . \tag{3.12d}$$

(3.12) and (3.1) together comprise the super-Poincaré algebra.

A direct consequence [3.7] of (3.12) is the equality in the number of bosons and fermions in any supersymmetric representation acting on which P_μ conserves the number of particles. The latter is the case both for onshell states and, with $P_\mu = i\partial_\mu$, in the offshell case. The equality can be seen geometrically as follows. Suppose the number of concerned fermions is less than the number of bosons. This means that the map $Q_a|\text{boson}\rangle = |\text{fermion}\rangle$ is an into map $\forall a$. Similarly, $Q_a|\text{fermion}\rangle = |\text{boson}\rangle$ must also be an into map since the character of the map is controlled by the operators $\{Q_a\}$ and not by the representation space. Let us first map via Q_a the bosonic subset (say) of the representation space into the fermionic subset and then map the latter back into a bosonic subset by Q_b (Fig.3.1). Now we repeat

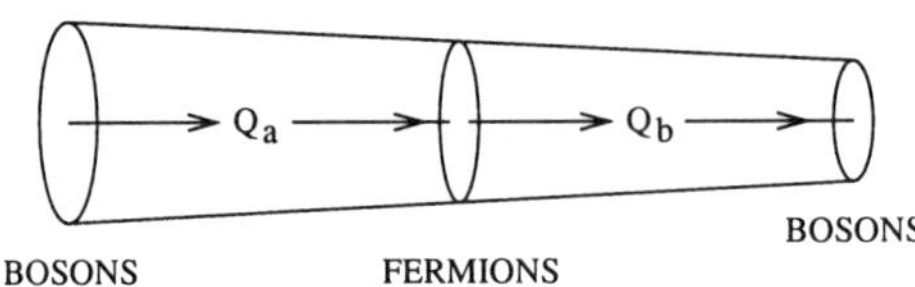

Fig.3.1. Into maps from bosonic to fermionic and back to bosonic subspaces.

the procedure with Q_a, Q_b interchanged and consider the union of the two resultant maps. We then have a bosonic subset that is reduced from the original one. This is inconsistent with the relation $[Q_a, Q_b]_+ \propto P_\mu$, which makes a one-to-one onto map. Thus each Q must also make a similar onto (rather than an into) map and there must be equal numbers of fermionic and bosonic degrees of freedom in any supermultiplet.

In order to derive the above result algebraically, we take the subspace of states $|i\rangle$ (in a supermultiplet) with the same eigenvalue p^μ of P^μ. Because of (3.12b), any product of Q_a and Q_b acting on $|i\rangle$ yields another state $|i'\rangle$ in the same subspace on account of closure. Now take the trace over all such states (including each helicity) of the operator $(-1)^{2S}[Q_a, Q_b]_+$, where S is the spin angular momentum. On using subspace completeness $\Sigma_i |i\rangle\langle i| = 1$ and the anticommutation[7] of $(-1)^{2S}$ with any fermionic operator Q, one gets

$$\sum_i \langle i|(-1)^{2S}Q_aQ_b|i\rangle \ + \ \sum_i \langle i|(-1)^{2S}Q_bQ_a|i\rangle$$

$$= \ \sum_i \langle i|(-1)^{2S}Q_aQ_b|i\rangle + \sum_i \sum_j \langle i|(-1)^{2S}Q_b|j\rangle\langle j|Q_a|i\rangle$$

$$= \ \sum_i \langle i|(-1)^{2S}Q_aQ_b|i\rangle + \sum_j \langle j|Q_a(-1)^{2S}Q_b|j\rangle = 0 \ .$$

Since, by (3.9), $[Q_a, \bar{Q}_b]_+$ is proportional to P^μ, it now follows that

$$0 = \sum_i \langle i|(-1)^{2S}P^\mu|i\rangle = p^\mu \sum_i \langle i|(-1)^{2S}|i\rangle \ .$$

Since the RHS sum just picks out the number of bosonic degrees of freedom n_B minus the number of fermionic degrees of freedom n_F in the supermultiplet,

$$n_B = n_F \ .$$

The above proof goes through only if $p^\mu \neq 0$, i.e. for nonzero momentum states. In particular, it does not apply to the ground state. In general, the difference $n_B - n_F$ is called the **Witten index** [2.3] of the system. If a system possesses a supersymmetric ground state (whose energy necessarily vanishes, cf. §2.2), the latter may or may not have a nonzero Witten index. But a system with a nonsupersymmetric vacuum, i.e. with spontaneously broken supersymmetry, will have a nonzero energy and hence a vanishing Witten index. Thus a vanishing Witten index is a necessary (though not sufficient) condition for the spontaneous breakdown of supersymmetry. However, we do not go into that aspect in this book. Interested readers are directed to Witten's article, Ref. [2.3].

Comments on extended supersymmetries

Before concluding this section, we wish to remark on extensions of the above supersymmetry algebra. Only *one* extra fermionic generator Q_a has been considered above. In fact, one could extend the consideration to a denumerably finite set $\{Q_a^i\}$ with $i\ (= 1, 2, \cdots, N)$ acting as an internal symmetry index referring to a global $SO(N)$ symmetry among them. This would lead to an N-extended supersymmetry [3.6]. Though that topic is outside the

---

[7]This can be seen from the vanishing of the anticommutator of $(-1)^{2S}$ and Q_a, as operated on all bosonic and fermionic states.

purview of this book, let us comment here on N-extended supersymmetry. The subject of extended supersymmetry is theoretically very interesting on several counts. First, among the commutation or anticommutation relations of the $N{=}1$ supersymmetry algebra, (3.12b), (3.12c) and (3.12d) get trivially extended to the $N > 1$ case, but (3.12a), namely the anticommutator $[Q_a^i, Q_b^j]_+$, does not. The latter develops additional additive pieces in the RHS known as **Central Charge** terms that are of considerable theoretical utility. Second, massive states in systems with extended supersymmetry are closely related to Bogomolny-Prasad-Sommerfield (BPS) monopoles in supersymmetric gauge theories. Third, $N{=}4$ supersymmetric Yang-Mills theory in $(3+1)$ dimensions has been claimed [3.7] to be finite. Moreover, this theory, if imbued with a nearly infinite number of colors, has been conjectured [3.8] to be equivalent to an $N{=}1$ supergravity theory (cf. Annex, Ch.12) defined in a particular ten dimensional curved spacetime, namely the topological product of a five dimensional Anti-de-Sitter space and a five sphere; this is supposed to be a consequence of an even deeper equivalence in superstring theory. Finally, Yang-Mills theories with extended supersymmetry in $(3+1)$ dimensions have lent themselves to exact nonperturbative treatments of their phase structures. Reviews of some of these issues may be found in the articles cited in Ref. [3.9]. The maximum permitted value of N, compatible with the tensor nature of gravity [3.10], is, in fact, 8. Note, however, that for any $N > 1$, there is a significant proliferation of additional matter fermions. With $N{=}2$, for instance, mirror partners [3.11] of the known fermions are predicted apart from all superpartners; moreover, fermion gauge group representations (with respect to a high scale grand unified theory, say) are necessarily vectorlike [3.7] rather than chiral, as required by the observed weak interactions. Not only is there a lack of evidence of such mirror partners, making them phenomenologically disfavored, such objects may be theoretically unwelcome if the observed chiral symmetric $V - A$ form is a deep feature of weak interactions upto a very high scale. It is, of course, possible that in nature a higher N supersymmetric field theory breaks down to one with $N{=}1$ at such a high scale. However, our aim is to discuss those aspects of supersymmetry which have lent themselves to serious phenomenological pursuit. So we restrict ourselves strictly to the $N{=}1$ case and choose not to cover the subject of extended supersymmetry in this book. In particular, we do not have any Central Charge.

3.2   Two Component Notation

We come to the two component formalism [3.12] which is particularly apposite for discussing supersymmetry. We have already introduced the algebra of $SU(2)_+ \otimes SU(2)_-$ which has a homomorphic relation with $SO(1,3)$. The massive Dirac four spinor transforms as the direct sum of the two fundamental representations $\left(\frac{1}{2},0\right) \oplus \left(0,\frac{1}{2}\right)$ of the former. We can construct a two component Weyl spinor ξ_A $(A = 1,2)$ which transforms as the representation $\left(\frac{1}{2},0\right)$. Mathematically, this means that the transformation equation for ξ_A is

$$\xi_A' = M_A{}^B \xi_B \ ,$$

where $M_A{}^B$, a complex $2{\times}2$ matrix, is an element of the group $SL(2,C)$ which is the universal covering group of $SO(1,3)$. Similarly, another Weyl spinor $\bar{\chi}_{\dot{A}}$ $(\dot{A} = 1,2)$, belonging to the complex conjugate representation $\left(0, \frac{1}{2}\right)$, can be introduced transforming as

$$\bar{\chi}'_{\dot{A}} = (M^\star)_{\dot{A}}^{\ \dot{B}} \bar{\chi}_{\dot{B}} \ .$$

The **purely antisymmetric** tensor ϵ^{AB} and its inverse ϵ_{AB}, i.e. with $\epsilon^{AB}\epsilon_{BC} = \delta^A_{\ C}$, $\epsilon^{12} = -\epsilon^{21} = 1$ and $\epsilon_{12} = -\epsilon_{21} = -1$, i.e.

$$\epsilon^{AB} = \begin{pmatrix} 0 & 1 \\ -1 & 0 \end{pmatrix}, \ \epsilon_{AB} = \begin{pmatrix} 0 & -1 \\ 1 & 0 \end{pmatrix},$$

which obey

$$\epsilon_{AB}\epsilon_{CD} = \epsilon_{AC}\epsilon_{BD} - \epsilon_{AD}\epsilon_{BC} \ , \tag{3.13a}$$

$$\epsilon^{AB}\epsilon_{CD} = -(\delta^A_{\ C}\delta^B_{\ D} - \delta^A_{\ D}\delta^B_{\ C}) \tag{3.13b}$$

and identical relations with the corresponding dotted indices, can be used for raising and lowering the spinorial indices of ψ or $\bar{\chi}$. Thus

$$\xi^A = \epsilon^{AB}\xi_B \ , \tag{3.14a}$$

$$\xi_A = \epsilon_{AB}\xi^B \ , \tag{3.14b}$$

$$\bar{\chi}^{\dot{A}} = \epsilon^{\dot{A}\dot{B}}\bar{\chi}_{\dot{B}} \ , \tag{3.14c}$$

$$\bar{\chi}_{\dot{A}} = \epsilon_{\dot{A}\dot{B}}\bar{\chi}^{\dot{B}} \ . \tag{3.14d}$$

ξ can be thought of as a two component column and $\bar{\chi}$ as a two component row. Also, $\bar{\xi}$ can be introduced like $\bar{\chi}$ and χ like ξ. The relation between ξ and $\bar{\xi}$ or χ and $\bar{\chi}$ is one of conjugation, i.e. $\xi^\dagger = \bar{\xi}$ and $\bar{\chi}^\dagger = \chi$, or in component form[8]

$$\xi_A = (\bar{\xi}_{\dot{A}})^\dagger, \ \xi^A = (\bar{\xi}^{\dot{A}})^\dagger \ , \tag{3.15a}$$

$$\bar{\chi}_{\dot{A}} = (\chi_A)^\dagger, \ \bar{\chi}^{\dot{A}} = (\chi^A)^\dagger \ . \tag{3.15b}$$

One can moreover verify [3.1] that, since $\det M = 1$, the transformation properties of ξ^A and $\bar{\chi}^{\dot{A}}$ are as under

$$\xi'^A = \epsilon^{AC} M_C^{\ D}\epsilon_{DB}\xi^B = (M^{-1^T})^A_{\ B}\xi^B, \tag{3.16a}$$

$$\bar{\chi}'^{\dot{A}} = \epsilon^{\dot{A}\dot{C}}(M^\star)_{\dot{C}}^{\ \dot{D}}\epsilon_{\dot{D}\dot{B}}\bar{\chi}^{\dot{B}} = (M^{\star-1^T})^{\dot{A}}_{\ \dot{B}}\bar{\chi}^{\dot{B}}. \tag{3.16b}$$

From the above relations, one can easily form a number of $SL(2,C)$ invariant and co-variant bilinear forms:[9]

$$\xi\chi \equiv \xi^A\chi_A \ , \tag{3.17a}$$

$$\bar{\chi}\bar{\xi} \equiv \bar{\chi}_{\dot{A}}\bar{\xi}^{\dot{A}} = (\xi\chi)^\dagger \ , \tag{3.17b}$$

---

[8]There is a difference between hermitian conjugation, denoted by a dagger, and complex conjugation, denoted by a star, for the components of (3.16). Thus $\xi_A \equiv (\bar{\xi}^{\dot{A}})^\star$ etc. One can formally write $\xi^A = (\bar{\xi}_{\dot{B}})^\star \bar{\sigma}_0^{\dot{B}A}$ and $\bar{\xi}^{\dot{A}} = (\xi_B)^\star \sigma_0^{B\dot{A}}$.

[9]In our notation we follow the convention that the contraction of undotted spinor indices is from the upper left to the lower right while that of dotted ones is from the lower left to the upper right.

$$\xi\sigma^\mu\bar\chi \equiv \xi^A\sigma^\mu_{A\dot B}\bar\chi^{\dot B} = (\chi\sigma^\mu\bar\xi)^\dagger \ , \tag{3.17c}$$

$$\bar\chi\bar\sigma^\mu\xi \equiv \bar\chi_{\dot A}\bar\sigma^{\mu\dot A B}\xi_B = (\bar\xi\bar\sigma^\mu\chi)^\dagger \ . \tag{3.17d}$$

The correctness of the spinor index structures of σ^μ and $\bar\sigma^\mu$, as displayed above, can be seen [3.1] from their $SL(2,C)$ properties. One may further note the identities

$$\bar\sigma^{\mu\dot A B} = \epsilon^{\dot A\dot C}\epsilon^{BD}\sigma^\mu_{D\dot C} \ , \tag{3.18a}$$

$$\sigma^\mu_{A\dot B} = \epsilon_{AC}\epsilon_{\dot B\dot D}\bar\sigma^{\mu\dot D C} \ , \tag{3.18b}$$

$$\sigma^\mu_{A\dot B}\sigma_{\mu C\dot D} = 2\epsilon_{AC}\epsilon_{\dot B\dot D} \ , \tag{3.18c}$$

$$\bar\sigma^{\mu\dot A B}\bar\sigma_\mu^{\dot C D} = 2\epsilon^{BD}\epsilon^{\dot A\dot C} \ , \tag{3.18d}$$

$$\sigma^\mu_{A\dot B}\bar\sigma_\mu^{\dot C D} = 2\delta_A^{\ D}\delta^{\dot C}_{\ \dot B} \ , \tag{3.18e}$$

$$(\sigma^\mu\bar\sigma^\nu + \sigma^\nu\bar\sigma^\mu)_A^{\ B} = 2\eta^{\mu\nu}\delta_A^{\ B} \ , \tag{3.18f}$$

$$(\bar\sigma^\mu\sigma^\nu + \bar\sigma^\nu\sigma^\mu)^{\dot A}_{\ \dot B} = 2\eta^{\mu\nu}\delta^{\dot A}_{\ \dot B} \ , \tag{3.18g}$$

$$\sigma^\mu\bar\sigma^\nu\sigma^\rho + \sigma^\rho\bar\sigma^\nu\sigma^\mu = 2(\eta^{\mu\nu}\sigma^\rho + \eta^{\nu\rho}\sigma^\mu - \eta^{\mu\rho}\sigma^\nu) \ , \tag{3.18h}$$

$$\bar\sigma^\mu\sigma^\nu\bar\sigma^\rho + \bar\sigma^\rho\sigma^\nu\bar\sigma^\mu = 2(\eta^{\mu\nu}\bar\sigma^\rho + \eta^{\nu\rho}\bar\sigma^\mu - \eta^{\mu\rho}\bar\sigma^\nu) \ , \tag{3.18i}$$

$$\mathrm{Tr}\ \sigma^\mu\bar\sigma^\nu\sigma^\rho\bar\sigma^\tau = 2(\eta^{\mu\nu}\eta^{\rho\tau} + \eta^{\mu\tau}\eta^{\nu\rho} - \eta^{\mu\rho}\eta^{\nu\tau} - i\epsilon^{\mu\nu\rho\tau}) \ . \tag{3.18j}$$

We will now *postulate* that these Weyl spinor components are of a Grassmann nature (cf §2.1), i.e. they all anticommute among themselves. In fact, we shall use $\xi, \bar\chi$ etc to denote two component spinors related to fermionic fields. Furthermore, we also explicitly introduce conjugate Weyl spinor doublets θ_A, $\bar\theta^{\dot A}$ as Grassmann variables which anticommute, not only among themselves, but also with those fermionic fields. It may then be seen that

$$\xi\chi = \chi\xi = (\bar\xi\bar\chi)^\dagger = (\bar\chi\bar\xi)^\dagger \ .$$

Thus $\theta\theta \neq 0$ but equals $-2\theta_1\theta_2$ and similarly $\bar\theta\bar\theta = 2\bar\theta^{\dot 1}\bar\theta^{\dot 2}$. Moreover, **any object, in which three or more factors of θ or $\bar\theta$ components occur, must vanish**. It requires now straightforward (though sometimes tedious) algebra to verify the following identities:

$$\xi\sigma^\mu\bar\chi = -\bar\chi\bar\sigma^\mu\xi \ , \tag{3.19a}$$

$$\xi\sigma^{\mu\nu}\chi = -\chi\sigma^{\mu\nu}\xi, \ \bar\xi\bar\sigma^{\mu\nu}\bar\chi = -\bar\chi\bar\sigma^{\mu\nu}\bar\xi \ , \tag{3.19b}$$

$$\theta^A\theta^B = -\frac{1}{2}\epsilon^{AB}\theta\theta \ , \tag{3.19c}$$

$$\theta_A\theta_B = \frac{1}{2}\epsilon_{AB}\theta\theta \ , \tag{3.19d}$$

$$\bar\theta_{\dot A}\bar\theta_{\dot B} = -\frac{1}{2}\epsilon_{\dot A\dot B}\bar\theta\bar\theta \ , \tag{3.19e}$$

$$\bar\theta^{\dot A}\bar\theta^{\dot B} = \frac{1}{2}\epsilon^{\dot A\dot B}\bar\theta\bar\theta \ , \tag{3.19f}$$

$$\theta\xi\,\theta\chi = -\frac{1}{2}\xi\chi\,\theta\theta \ , \tag{3.19g}$$

$$\bar\theta\bar\xi\,\bar\theta\bar\chi = -\frac{1}{2}\bar\xi\bar\chi\,\bar\theta\bar\theta \ , \tag{3.19h}$$

$$\xi\zeta\,\bar\chi\bar\tau = \frac{1}{2}\xi\sigma^\mu\bar\chi\,\zeta\sigma_\mu\bar\tau \ , \tag{3.19i}$$

$$\bar\xi\bar\zeta\,\chi\tau = \frac{1}{2}\bar\xi\bar\sigma^\mu\chi\,\bar\zeta\bar\sigma_\mu\tau \ , \tag{3.19j}$$

$$\theta\sigma^\mu\bar\theta\,\theta\sigma^\nu\bar\theta = \frac{1}{2}\eta^{\mu\nu}\theta\theta\,\bar\theta\bar\theta \ , \tag{3.19k}$$

$$\zeta\xi\,\chi\sigma^\mu\bar\tau = -\frac{1}{2}\zeta\chi\,\xi\sigma^\mu\bar\tau + \zeta\sigma^{\mu\nu}\chi\,\xi\sigma_\nu\bar\tau \ , \tag{3.19l}$$

$$\zeta\xi\,\bar\chi\bar\sigma^\mu\tau = -\frac{1}{2}\tau\xi\,\bar\chi\bar\sigma^\mu\zeta - \tau\sigma^{\mu\nu}\xi\,\bar\chi\bar\sigma_\nu\zeta \ , \tag{3.19m}$$

$$(\sigma^\mu\bar\theta)_A\theta\sigma^\nu\bar\theta = \bar\theta\bar\theta\left[\frac{1}{2}\eta^{\mu\nu}\theta_A - i(\sigma^{\mu\nu}\theta)_A\right] \ , \tag{3.19n}$$

$$(\theta\sigma^\mu)_{\dot A}\bar\theta\bar\sigma^\nu\theta = -\theta\theta\left[\frac{1}{2}\bar\theta_{\dot A}\eta^{\mu\nu} + i(\bar\theta\bar\sigma^{\mu\nu})_{\dot A}\right] \ . \tag{3.19o}$$

(3.19 g o) are Fierz identities, a more complete set of which may be found in Appendix A of Ref. [3.13]. Note also the consequence of (3.19b) that $\theta\sigma^{\mu\nu}\theta = 0 = \bar\theta\bar\sigma^{\mu\nu}\bar\theta$.

In this notation, a Dirac four spinor ψ with components ψ_a can be defined in the Weyl representation as follows:

$$\psi = \begin{pmatrix} \xi \\ \bar\chi^T \end{pmatrix} , \ \text{i.e.} \ \psi_a = \begin{pmatrix} \xi_A \\ \bar\chi^{\dot A} \end{pmatrix} . \tag{3.20}$$

In the LHS of (3.20) the index a runs over four components while in the RHS the A and $\dot A$ indices for the Weyl spinors ξ_A and $\bar\chi^{\dot A}$ respectively run over two components $(1,2)$ and $(\dot 1,\dot 2)$ each. It is trivial to see that $\frac{1}{2}(1-\gamma_5)\psi \equiv \psi_L$ is the chiral projection of ψ on ξ and similarly $\frac{1}{2}(1+\gamma_5)\psi \equiv \psi_R$ is the chiral projection of ψ on $\bar\chi$, consistent with ξ_A and $\bar\chi_{\dot A}$ being Weyl spinors belonging to the representations $\left(\frac{1}{2},0\right)$ and $\left(0,\frac{1}{2}\right)$ respectively. The conjugate of (3.20), as calculated in the Weyl representation, is

$$\bar\psi_a = (\psi^\dagger\gamma^0)_a = (\xi_A^\dagger \ \bar\chi^{\dot A\dagger})\begin{pmatrix} 0 & I \\ I & 0 \end{pmatrix} = (\bar\chi^{\dot A\dagger} \ \xi_A^\dagger) = (\chi^A \ \bar\xi_{\dot A}) \ , \tag{3.21}$$

the last step following by use of (3.15). Note also that

$$\bar\psi^T = \begin{pmatrix} \chi \\ \bar\xi^T \end{pmatrix} , \ \text{i.e.} \ \bar\psi_b^T = \begin{pmatrix} \chi^B \\ \bar\xi_{\dot B} \end{pmatrix} .$$

The relationship between the four spinor space and its twin two spinor subspaces is further amplified by the matrix structures

$$\delta_{ab} = \begin{pmatrix} \delta_A{}^B & 0 \\ 0 & \delta^{\dot A}{}_{\dot B} \end{pmatrix} , \ (\gamma^\mu A_\mu)_{ab} = \begin{pmatrix} 0 & \sigma^\mu_{A\dot B} \\ \bar\sigma^{\mu\dot A B} & 0 \end{pmatrix} A_\mu \ ,$$

$\forall$ four vectors A_μ. In particular, the Dirac equation $(i\gamma^\mu \partial_\mu - m)\psi = 0$ splits into two two spinor equations

$$i\sigma^\mu \partial_\mu \bar\chi = m\xi \ , \quad i\bar\sigma^\mu \partial_\mu \xi = m\bar\chi \ .$$

The charge conjugation matrix C is [3.13]

$$C = i\gamma^2\gamma^0 = \begin{pmatrix} -i\sigma^2 & 0 \\ 0 & i\sigma^2 \end{pmatrix} \tag{3.22}$$

in the Weyl representation. Now the charge conjugated Dirac spinor ψ^C obtains as follows.

$$\psi_a^C = C_{ab}\bar\psi_b^T = \begin{pmatrix} -i\sigma^2_{AB}\chi^B \\ i\sigma^{2\dot A\dot B}\bar\xi_{\dot B} \end{pmatrix} = \begin{pmatrix} \chi_A \\ \bar\xi^{\dot A} \end{pmatrix} , \text{ i.e. } \psi^C = \begin{pmatrix} \chi \\ \bar\xi^T \end{pmatrix} . \tag{3.23}$$

where one has used the results $-i\sigma^2_{AB} = \epsilon_{AB}$ and $i\sigma^{2\dot A\dot B} = i\epsilon^{\dot A\dot C}\epsilon^{\dot B\dot D}\sigma^2_{\dot C\dot D} = \epsilon^{\dot A\dot B}$. A self-conjugate Majorana four spinor λ^M can now be introduced as (we use the notation λ_A, $\bar\lambda^{\dot A}$ to mean two component spinors)

$$\lambda^M = \begin{pmatrix} \lambda \\ \bar\lambda^T \end{pmatrix} , \text{ i.e. } \lambda_a^M = \begin{pmatrix} \lambda_A \\ \bar\lambda^{\dot A} \end{pmatrix} , \tag{3.24}$$

so that $\lambda_a^M = (\lambda_a^M)^C$. In particular, since Q_α is a Majorana four spinor, we can employ the notation Q_A, $\bar Q^{\dot A}$ for the corresponding two spinors and write

$$Q_a = \begin{pmatrix} Q_A \\ \bar Q^{\dot A} \end{pmatrix} . \tag{3.25}$$

In two component notation the algebra (3.12) can be written as

$$[Q_A, \bar Q_{\dot B}]_+ = 2\sigma^\mu_{A\dot B}P_\mu, \ [\bar Q^{\dot A}, Q^B]_+ = 2\bar\sigma^{\mu\dot A B}P_\mu \ , \tag{3.26a}$$

$$[Q_A, Q_B]_+ = [\bar Q^{\dot A}, \bar Q^{\dot B}]_+ = 0 \ , \tag{3.26b}$$

$$[Q_A, P_\mu] = [\bar Q^{\dot A}, P_\mu] = 0 \ , \tag{3.26c}$$

$$[M_{\mu\nu}, Q_A] = -(\sigma_{\mu\nu})_A{}^B Q_B \ , \tag{3.26d}$$

$$[M_{\mu\nu}, \bar Q^{\dot A}] = -(\bar\sigma_{\mu\nu})^{\dot A}{}_{\dot B}\bar Q^{\dot B} \ , \tag{3.26e}$$

$$[Q_A, R] = Q_A \ , \tag{3.26f}$$

$$[\bar Q^{\dot A}, R] = -\bar Q^{\dot A} \ . \tag{3.26g}$$

Next, we come to the **parity operator** $\mathcal{P}$. Under a space reflection $M_{pr} \to M_{pr}$ but $M_{p0} \to -M_{p0}$, i.e. $J_p \to J_p$, but $K_p \to -K_p$ so that $\mathcal{J}_p^\pm \to \mathcal{J}_p^\mp$. Thus the representations $(\frac{1}{2}, 0)$ and $(0, \frac{1}{2})$ transform into each other under a space reflection and this means that under parity Q_A transforms into a linear combination of $\bar Q^{\dot A}$'s. To obtain the precise linear relation, note that, parity being commutative with rotations, it will take Q_A to an operator with the same rotation property. Now the rotation matrix for Q_A, corresponding to an arbitrary three dimensional rotation $\vec{\alpha}$, is $D(R,\vec{\alpha}) = \exp(i\vec{\alpha}\cdot\vec{\sigma}/2)$, i.e.

$$U^{-1}(R)Q_A U(R) = D(R,\vec{\alpha})_A{}^B Q_B \; .$$

From (3.16b) we see that the transformation for $\bar{Q}^{\dot{A}}$, with $D^\star(R,\vec{\alpha}) = \exp(-i\vec{\alpha}\cdot\vec{\sigma}^\star/2)$, is

$$U^{-1}(R)\bar{Q}^{\dot{A}}U(R) = \epsilon^{\dot{A}\dot{C}}D^\star(R,\vec{\alpha})_{\dot{C}}{}^{\dot{D}}\epsilon_{\dot{D}\dot{F}}\bar{Q}^{\dot{F}}.$$

With $\epsilon = \begin{pmatrix} 0 & 1 \\ -1 & 0 \end{pmatrix}$ and $\vec{\sigma}^\star = \epsilon\vec{\sigma}\epsilon$, one can derive that

$$U^{-1}(R)\epsilon_{\dot{A}\dot{B}}\bar{Q}^{\dot{B}}U(R) = D(R,\vec{\alpha})_{\dot{A}}{}^{\dot{B}}\epsilon_{\dot{B}\dot{C}}\bar{Q}^{\dot{C}}.$$

In other words, $\epsilon_{\dot{A}\dot{B}}\bar{Q}^{\dot{B}}$ has the same rotational properties as Q_A. Hence we can take

$$\mathcal{P}Q_A\mathcal{P}^{-1} = p\,\epsilon_{\dot{A}\dot{B}}\bar{Q}^{\dot{B}},$$

where p is a complex constant. By considering this operation on $[Q_A,\bar{Q}^{\dot{B}}]_+$, one can show that p must be a phase which can be absorbed in the definition of Q_A. Thus

$$\mathcal{P}Q_A\mathcal{P}^{-1} = \epsilon_{\dot{A}\dot{B}}\bar{Q}^{\dot{B}}, \quad \mathcal{P}Q^A Q_A\mathcal{P}^{-1} = -\bar{Q}_{\dot{A}}\bar{Q}^{\dot{A}}. \tag{3.27}$$

Before closing this section, we give a partial dictionary of covariant spinor bilinears written in four component and two component notations. Let us write the Dirac four spinor in the chiral notation

$$\psi = \begin{pmatrix} \xi_+ \\ \bar{\xi}_-^T \end{pmatrix}, \quad \bar{\psi} = \begin{pmatrix} \xi_-^T & \bar{\xi}_+ \end{pmatrix}.$$

Thus we have replaced the ξ and $\bar{\chi}$ of (3.20) by ξ_+ and $\bar{\xi}_-$ respectively. For a Majorana four spinor, we shall continue to write

$$\lambda_M = \begin{pmatrix} \lambda \\ \bar{\lambda}^T \end{pmatrix}, \quad \bar{\lambda}_M = \begin{pmatrix} \lambda^T & \bar{\lambda} \end{pmatrix}.$$

If ψ_1 and ψ_2 are two Dirac four spinors, the following identities hold:

$$\bar{\psi}_1\psi_2 = \xi_{1-}\xi_{2+} + \bar{\xi}_{1+}\bar{\xi}_{2-}\,, \tag{3.28a}$$

$$\bar{\psi}_1\gamma_5\psi_2 = -\xi_{1-}\xi_{2+} + \bar{\xi}_{1+}\bar{\xi}_{2-}\,, \tag{3.28b}$$

$$\bar{\psi}_1\gamma^\mu\psi_2 = \xi_{1-}\sigma^\mu\bar{\xi}_{2-} + \bar{\xi}_{1+}\bar{\sigma}^\mu\xi_{2+}\,, \tag{3.28c}$$

$$\bar{\psi}_1\gamma^\mu\gamma_5\psi_2 = \xi_{1-}\sigma^\mu\bar{\xi}_{2-} - \bar{\xi}_{1+}\bar{\sigma}^\mu\xi_{2+}\,, \tag{3.28d}$$

$$\bar{\psi}_1\Sigma^{\mu\nu}\psi_2 = \xi_{1-}\sigma^{\mu\nu}\xi_{2+} + \bar{\xi}_{1+}\bar{\sigma}^{\mu\nu}\bar{\xi}_{2-}\,, \tag{3.28e}$$

$$\bar{\psi}_1\psi_{2L} = \xi_{1-}\xi_{2+}\,, \tag{3.28f}$$

$$\bar{\psi}_1\psi_{2R} = \bar{\xi}_{1+}\bar{\xi}_{2-}\,, \tag{3.28g}$$

$$\bar{\psi}_1 \gamma^\mu \psi_{2L} = \bar{\xi}_{1+} \bar{\sigma}^\mu \xi_{2+} \ , \tag{3.28h}$$

$$\bar{\psi}_1 \gamma^\mu \psi_{2R} = \xi_{1-} \sigma^\mu \bar{\xi}_{2-} \ , \tag{3.28i}$$

$$\overline{\psi_1^C} \psi_2 = \xi_{1+} \xi_{2+} + \xi_{1-} \xi_{2-} \ , \tag{3.28j}$$

$$\overline{\psi_1^C} \gamma_5 \psi_2 = \xi_{1+} \xi_{2-} - \xi_{1-} \xi_{2+} \ . \tag{3.28k}$$

For two Majorana four spinors λ_{1M} and λ_{2M} or a single Majorana four spinor λ_M, we have

$$\bar{\lambda}_{1M} \lambda_{2M} = \lambda_1 \lambda_2 + \bar{\lambda}_1 \bar{\lambda}_2 = \bar{\lambda}_{2M} \lambda_{1M} \ , \tag{3.29a}$$

$$\bar{\lambda}_{1M} \gamma_5 \lambda_{2M} = -\lambda_1 \lambda_2 + \bar{\lambda}_1 \bar{\lambda}_2 = \bar{\lambda}_{2M} \gamma_5 \lambda_{1M} \ , \tag{3.29b}$$

$$\bar{\lambda}_{1M} \gamma^\mu \lambda_{2M} = \bar{\lambda}_1 \bar{\sigma}^\mu \lambda_2 + \lambda_1 \sigma^\mu \bar{\lambda}_2 = -\bar{\lambda}_{2M} \gamma^\mu \lambda_{1M} \ , \tag{3.29c}$$

$$\bar{\lambda}_{1M} \gamma^\mu \gamma_5 \lambda_{2M} = \lambda_1 \sigma^\mu \bar{\lambda}_2 - \bar{\lambda}_1 \bar{\sigma}^\mu \lambda_2 = \bar{\lambda}_{2M} \gamma_\mu \gamma_5 \lambda_{1M} \ , \tag{3.29d}$$

$$\bar{\lambda}_{1M} \Sigma^{\mu\nu} \lambda_{2M} = \lambda_1 \sigma^{\mu\nu} \lambda_2 + \bar{\lambda}_1 \bar{\sigma}^{\mu\nu} \bar{\lambda}_2 = -\bar{\lambda}_{2M} \Sigma^{\mu\nu} \lambda_{1M} \ . \tag{3.29e}$$

$$\bar{\lambda}_{1M} \lambda_{2M(L,R)} = \bar{\lambda}_{2M} \lambda_{1M(L,R)} \ , \tag{3.29f}$$

$$\bar{\lambda}_{1M} \gamma_\mu \lambda_{2M(L,R)} = -\bar{\lambda}_{2M} \gamma_\mu \lambda_{1M(L,R)} \ , \tag{3.29g}$$

$$\bar{\lambda}_M \gamma_\mu \lambda_M = 0 \ . \tag{3.29h}$$

There are additional relations, namely,

$$\bar{\lambda}_M \psi_L = \lambda \xi_+ \ , \tag{3.30a}$$

$$\bar{\lambda}_M \psi_R = \bar{\lambda} \bar{\xi}_- \ , \tag{3.30b}$$

$$\overline{\psi_L} \lambda_M = \bar{\lambda} \bar{\xi}_+ \ , \tag{3.30c}$$

$$\overline{\psi_R} \lambda_M = \lambda \xi_- \ , \tag{3.30d}$$

for a combination of a chiral Dirac and a Majorana spinor.

In (3.28), (3.29) and (3.30) the LHS has been written in four component notation while one has used the two component one in the RHS. Furthermore, (3.29) and (3.19) imply that

$$\bar{\lambda}_M \gamma^\mu \gamma_5 \lambda_M \ \bar{\lambda}_M \gamma^\nu \gamma_5 \lambda_M = \eta^{\mu\nu} (\bar{\lambda}_M \lambda_M)^2 = -\eta^{\mu\nu} (\bar{\lambda}_M \gamma_5 \lambda_M)^2 \ ,$$

where all the Majorana four spinors, in case they are fields, are necessarily taken at the same spacetime point.

3.3 Particle Supermultiplets

Since P^2 commutes with Q and $\bar{Q}$, the mass of a particle remains unchanged within a supermultiplet representation generated by the latter. The same holds for internal symmetry generators and therefore all internal quantum numbers are also invariant within such a supermultiplet. The former property is not expected to hold in broken supersymmetry, but the latter should remain valid if the supersymmery breaking terms maintain the internal symmetry. The angular momentum J^p, however, does not commute with the spinorial supercharge. From (3.6) or (3.26d,e), we can derive

$$[J^p, Q_A] = -\frac{1}{2}(\sigma^p)_A{}^B Q_B \,, \tag{3.31a}$$

$$[J^p, \bar{Q}^{\dot{A}}] = -\frac{1}{2}(\bar{\sigma}^p)^{\dot{A}}{}_{\dot{B}} \bar{Q}^{\dot{B}} \,. \tag{3.31b}$$

Let us consider massive and massless particles separately.

Massless case

Without loss of generality, we can Lorentz transform ourselves to the frame in which the four momentum of the particle is $P^\mu = \omega(1,0,0,1)$, where J^3 measures its helicity. Now $\sigma^\mu P_\mu = \omega(I - \sigma^3)$, so that the supersymmetry algebra in (3.26) can be rewritten in terms of spinor components as

$$[Q_1, \bar{Q}_{\dot{1}}]_+ = 0 \,, \tag{3.32a}$$

$$[Q_2, \bar{Q}_{\dot{2}}]_+ = 4\omega \,, \tag{3.32b}$$

$$[Q_1, \bar{Q}_{\dot{2}}]_+ = [Q_2, \bar{Q}_{\dot{1}}]_+ = 0 \,. \tag{3.32c}$$

Since $\bar{Q}_{\dot{1}}$ is the adjoint of the operator Q_1 and we work in a Hilbert space with positive norm, (3.32a) implies that $Q_1 = \bar{Q}_{\dot{1}} = 0$. Therefore, only Q_2 and $\bar{Q}_{\dot{2}}$ (N.B. $\bar{Q}_{\dot{2}} = \bar{Q}^{\dot{1}}$) are left and moreover we can redefine them as $Q = \bar{Q}_{\dot{2}}/(2\sqrt{\omega})$, $\bar{Q} = Q_2/(2\sqrt{\omega})$ so that we get a supersymmetry algebra

$$[Q, \bar{Q}]_+ = 1 \,, \tag{3.33a}$$

$$[Q, Q]_+ = [\bar{Q}, \bar{Q}]_+ = 0 \,. \tag{3.33b}$$

From (3.31) we have $[J^3, Q_2] = \frac{1}{2}Q_2$ and $[J^3, \bar{Q}_{\dot{2}}] = -\frac{1}{2}\bar{Q}_{\dot{2}}$, i.e.

$$J^3 Q = Q\left(J^3 - \frac{1}{2}\right), \tag{3.34a}$$

$$J^3 \bar{Q} = \bar{Q}\left(J^3 + \frac{1}{2}\right). \tag{3.34b}$$

Thus, applied to an eigenstate of J^3 with eigenvalue j_3, say, Q ($\bar{Q}$) decreases (increases) the latter by 1/2. Suppose $(j_3)_{\max} = j_0$ and let $|j_0\rangle$ be a normalized eigenstate of J^3 (i.e. $\langle j_0 | j_0 \rangle = 1$) with $J^3 |j_0\rangle = j_0 |j_0\rangle$. Then perforce one must have

$$\bar{Q}|j_0\rangle = 0 \,. \tag{3.35}$$

On the other hand, $Q|j_0\rangle$ is[10] an eigenstate of J^3 with eigenvalue $j_0 - \frac{1}{2}$. We can define a properly normalized state $|j_0 - \frac{1}{2}\rangle \equiv Q|j_0\rangle$, automatically having the properties

$$J^3|j_0 - \frac{1}{2}\rangle = \left(j_0 - \frac{1}{2}\right)|j_0 - \frac{1}{2}\rangle , \tag{3.36a}$$

$$Q|j_0 - \frac{1}{2}\rangle = 0 , \tag{3.36b}$$

$$\bar{Q}|j_0 - \frac{1}{2}\rangle = |j_0\rangle . \tag{3.36c}$$

j_0	Name	Helicities	States	Particles	Degrees of freedom
$\frac{1}{2}$	Chiral supermultiplet(s)	$-\frac{1}{2}$ or $\frac{1}{2}$	$1 + 1$	1 spin 1/2 complex chiral fermion +	2
		0	$1 + 1$	1 complex scalar	2
1	Gauge supermultiplet	-1	1	1 spin one gauge boson +	2
		1	1		
		$-\frac{1}{2}$	1	1 spin 1/2 Majorana gaugino	
		$\frac{1}{2}$	1		2
2	$N = 1$ supergravity multiplet	-2	1	1 spin two graviton +	2
		2	1		
		$-\frac{3}{2}$	1	1 spin 3/2 Majorana gravitino	
		$\frac{3}{2}$	1		2

Table 3.1. Different supermultiplets of massless particles.

It is clear from the above discussion that the algebra of the physically relevant generators $Q, \bar{Q}$ and J^3 admits two states – with helicities j_0 and $j_0 - \frac{1}{2}$. Now a local, causal and Lorentz-invariant field theory necessarily has CPT invariance which dictates that, for every state of helicity j, transforming under the representation R (say) of any existing internal symmetry group, there must be another state of helicity $-j$ transforming under the conjugate representation $\bar{R}$. This leads to two options for constructing irreducible representations of the superalgebra. (1) If $R \neq \bar{R}$, one can make a supermultiplet comprising a single helicity state described by a complex fermion field, which is a Weyl spinor, and a spin zero state described by a complex scalar field – as stated in the first row of Table 3.1. We shall later show matter and Higgs supermultiplets to be of this kind. (2) If R is self-conjugate, a single irreducible representation can be constructed by also including the conjugate states of opposite helicity. Row 2 (3) of Table 3.1 shows an example of such a self-conjugate supermultiplet comprising bosonic states described by a real vector (tensor) field plus a Majorana fermion of spin 1/2 (3/2). This gives rise to a gauge (gravity) supermultiplet containing a **gaugino** (**gravitino**) as its fermionic component.

---

[10]$Q|j_0\rangle$ has the total angular momentum quantum number $|j_0 - 1/2|$.

Massive case

Here we can go to the rest frame of the particle with $P^\mu = (m, 0, 0, 0)$, m being its mass, so that $\sigma^\mu P_\mu = mI$. Thus we have now

$$[Q_A, \bar{Q}_{\dot{B}}]_+ = 2m(I)_{A\dot{B}} \,, \tag{3.37a}$$

$$[Q_A, Q_B]_+ = [\bar{Q}_{\dot{A}}, \bar{Q}_{\dot{B}}]_+ = 0 \,. \tag{3.37b}$$

(3.37) can be rewritten in more detail as

$$[Q_1, \bar{Q}_{\dot{1}}]_+ = [Q_2, \bar{Q}_{\dot{2}}]_+ = 2m \,, \tag{3.38a}$$

$$[Q_1, \bar{Q}_{\dot{2}}]_+ = [Q_2, \bar{Q}_{\dot{1}}]_+ = 0 \,, \tag{3.38b}$$

$$[Q_1, Q_2]_+ = [\bar{Q}_{\dot{1}}, \bar{Q}_{\dot{2}}]_+ = 0 \,. \tag{3.38c}$$

Let us consider irreducible representations $|m, j, \lambda\rangle$ of the Poincaré algebra, characterized by spin j, its third component λ and a fixed mass m (corresponding to the Casimir operator P^2) with the properties

$$P^2|m, j, \lambda\rangle = m^2|m, j, \lambda\rangle \,, \tag{3.39a}$$

$$J^3|m, j, \lambda\rangle = \lambda|m, j, \lambda\rangle \,, \tag{3.39b}$$

$$\vec{J}^2|m, j, \lambda\rangle = j(j+1)|m, j, \lambda\rangle \,, \tag{3.39c}$$

$$\langle m, j', \lambda'|m, j, \lambda\rangle = \delta_{j'j}\delta_{\lambda'\lambda} \,. \tag{3.39d}$$

From among the above states we can always find one which is annihilated by Q_A. Thus if $|\psi\rangle$ is annihilated by neither Q_1 nor Q_2, $|\chi\rangle = Q_1 Q_2|\psi\rangle$ is such a state; if ψ is annihilated by $Q_{1,2}$ but not by $Q_{2,1}$, then such a state is $|\chi\rangle = Q_{2,1}|\psi\rangle$. By virtue of (3.31), $J^p|\chi\rangle$ is also annihilated by Q_A, i.e. the space of states $|\chi\rangle$ is rotationally invariant.

One can now rewrite $|\chi\rangle$ specifically as $|m, j_0, \lambda_0\rangle$ and moreover construct $\bar{Q}^{\dot{1}}|m, j_0, \lambda_0\rangle$ and $\bar{Q}^{\dot{2}}|m, j_0, \lambda_0\rangle$. The latter also are states in the representation of the super-Poincaré algebra. On account of (3.31b), we have $[J^3, \bar{Q}^{\dot{1}}] = \frac{1}{2}\bar{Q}^{\dot{1}}$ and $[J^3, \bar{Q}^{\dot{2}}] = -\frac{1}{2}\bar{Q}^{\dot{2}}$, i.e.

$$J^3\bar{Q}^{\dot{1}} = \bar{Q}^{\dot{1}}\left(J^3 + \frac{1}{2}\right) \,, \tag{3.40a}$$

$$J^3\bar{Q}^{\dot{2}} = \bar{Q}^{\dot{2}}\left(J^3 - \frac{1}{2}\right) \,. \tag{3.40b}$$

(3.40) imply that the two constructed states have their third spin components as $\lambda_0 + \frac{1}{2}$ and $\lambda_0 - \frac{1}{2}$ respectively whereas $\bar{Q}^{\dot{1}}\bar{Q}^{\dot{2}}|m, j_0, \lambda_0\rangle$ has λ_0. Any further application of $\bar{Q}^{\dot{1}}$ or $\bar{Q}^{\dot{2}}$ on $\bar{Q}^{\dot{1}}\bar{Q}^{\dot{2}}|m, j_0, \lambda_0\rangle$ annihilates it. Thus, for each fixed pair of values of m and j_0, we have a $4(2j_0+1)$-dimensional irreducible representation of the super-Poincaré algebra. The space of the corresponding states contains $2j_0 + 1$ subspaces ($-j_0 \leq \lambda_0 \leq j_0$), each of which contains four eigenstates of J^3 with values $\lambda_0, \lambda_0 + \frac{1}{2}, \lambda_0 - \frac{1}{2}, \lambda_0$ corresponding to the operators $I, \bar{Q}^{\dot{1}}$,

$\bar{Q}^{\dot{2}}$ and $\bar{Q}^{\dot{1}}\bar{Q}^{\dot{2}}$ respectively[11]. It may furthermore be noted from (3.27) that $\bar{Q}^{\dot{1}}\bar{Q}^{\dot{2}}|m, j_0, \lambda_0\rangle$ has a parity opposite to that of $|m, j_0, \lambda_0\rangle$. Schematically, the situation is as shown in Fig.3.2.

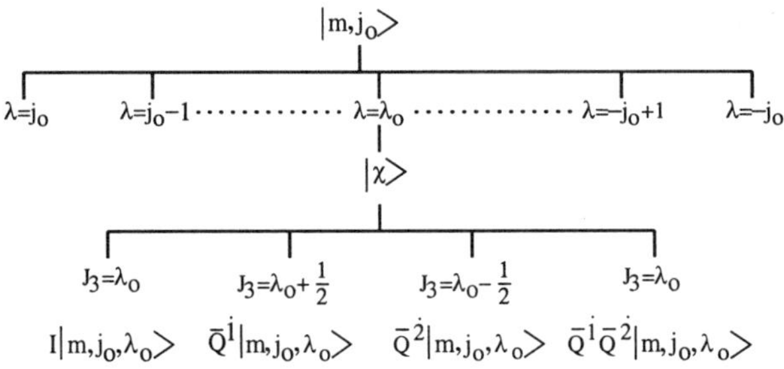

Fig.3.2. General schematic of massive supermultiplet states.

We shall give two examples to make the discussion more concrete. Let us start with $j_0 = 0$, i.e. $|\chi\rangle$ is a scalar state, cf. Fig.3.3. Now there are four states only with $J^3 = 0, +\frac{1}{2}, -\frac{1}{2}$ and 0. Again, the chiral fermion is described by a Weyl spinor with two components while

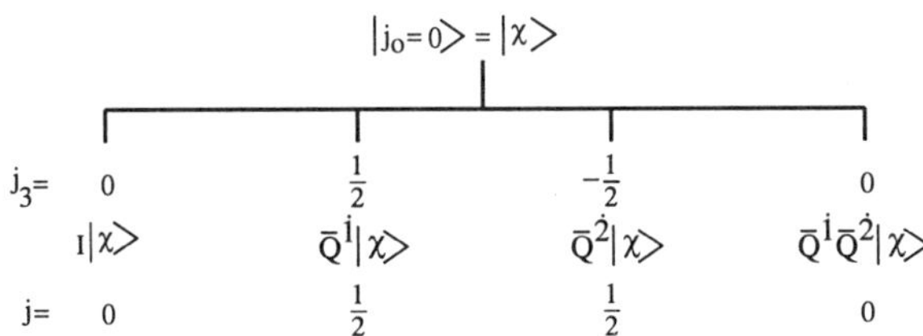

Fig.3.3. Schematic of the massive Wess-Zumino supermultiplet ($j_0 = 0$) states.

the rest are a scalar and a pseudoscalar. This, simplest of all supermultiplets, is known as the **Wess-Zumino supermultiplet**. Two such supermultiplets are needed to create a Dirac fermion out of two Weyl spinors. The corresponding four spin zero states can be described as two complex **sfermions** which can be interpreted as the superpartners of the left chiral and the right chiral components of the Dirac fermion. As a second example, take $j_0 = 1/2$, i.e. $|\chi\rangle$ is a fermionic state. There are eight states altogether. They correspond to one pseudoscalar, one vector (both have the same parity) and two Weyl spinors of opposite parity. The last two can combine into one Dirac fermion. The entire scheme is depicted in

---

[11]Since $\vec{J}^2$ does not commute with $\bar{Q}^{\dot{1}}, \bar{Q}^{\dot{2}}$, any of these states in general can (but need not) have an eigenvalue j different from j_0, when operated on by $\vec{J}^2$. For the $j_0 = 0$ (1/2) cases, depicted in Fig. 3.3 (3.4), the j values are shown. In particular, in Fig. 3.4, $|j = 1, j_3 = 0\rangle = (\bar{Q}^{\dot{1}}|1/2, 1/2\rangle + \bar{Q}^{\dot{2}}|1/2, -1/2\rangle)/\sqrt{2}$ and $|j = 0, j_3 = 0\rangle = (\bar{Q}^{\dot{1}}|1/2, 1/2\rangle - \bar{Q}^{\dot{2}}|1/2, -1/2\rangle)/\sqrt{2}$.

Fig.3.4.

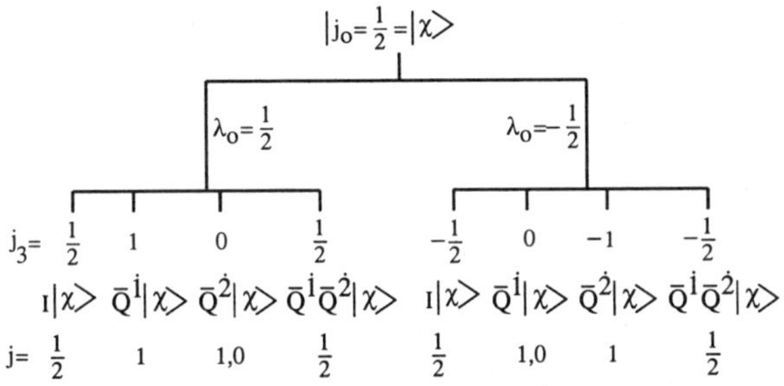

Fig.3.4. Schematic of massive supermultiplet states for $j_0 = 1/2$.

We can summarize the salient features of supermultiplet spectra in three points.

- There are four helicity states with helicities $j_0 + \dfrac{1}{2}, j_0, -j_0, -j_0 - \dfrac{1}{2}$, j_0 being zero, integral or half-integral. Two of them correspond to a fermion and the remaining two to either one spinning or a pair of spinless bosons. These massless states constitute either one self-conjugate supermultiplet or two complex chiral ones.

- There are $4(2j_0 + 1)$ helicity states in a massive supermultiplet (j_0 same as above) with helicities $\lambda_0, \lambda_0 + \dfrac{1}{2}, \lambda_0 - \dfrac{1}{2}, \lambda_0$ (where $\lambda_0 = -j_0, -j_0 + 1, \cdots, j_0$). The two states with helicity λ_0 have opposite parity, while those with $\lambda_0 \pm \dfrac{1}{2}$ have the same parity.

- The mass and internal quantum numbers of a particle remain unchanged within a supermultiplet representation, though the masses of different states will be split by supersymmetry breaking.

References

[3.1] H.J.W. Müller-Kirsten and A. Wiedemann, *op. cit.*, *Bibl.*

[3.2] J.D. Bjorken and S.D. Drell, *op. cit.*, *Bibl.*

[3.3] S. Coleman and J. Mandula, Phys. Rev. **159** (1967) 1251.

[3.4] S. Weinberg #3, *op. cit.*, *Bibl.*

[3.5] R. Haag, J. Łopuszánski and M.F. Sohnius, Nucl. Phys. **B188** (1975) 61. J. Łopuszánski, *op. cit.*, *Bibl.*

[3.6] P. West #2, *op. cit.*, *Bibl.* P. Fayet, Phys. Lett. **B142** (1984) 263; *ibid.* **B159** (1985) 121.

[3.7] M.F. Sohnius, *loc. cit.*, *Bibl.*

[3.8] J.M. Maldacena, Adv. Theor. Math. Phys. **2**, 231 (1998).

[3.9] M.E. Peskin #1, *loc. cit.*, *Bibl.* J.M. Maldacena, Int. J. Mod. Phys. **A1551** (2000) 840. A. Bilal, hep-th/0106246. A.S. Galperin, E.A. Ivanov, V.I. Ogievetsky and E.S. Sokatchev, *op. cit.*, *Bibl.*

[3.10] P. van Nieuwenhuizen, *loc. cit.*, *Bibl.*

[3.11] L.J. Hall in *Supersymmetry and supergravity/Nonperturbative QCD*, P. Roy and V. Singh (eds.), *op. cit.*, *Bibl.* P. Fayet, *loc. cit.*, Ref. [3.6].

[3.12] J. Wess and J. Bagger, *op. cit.*, *Bibl.*

[3.13] D. Bailin and A. Love, *op. cit.*, *Bibl.*

Chapter 4

FREE SUPERFIELDS IN SUPERSPACE

4.1 General Superfield in Superspace

In §2.3 we had a glimpse of superspace in relation to the supersymmetric harmonic oscillator problem where time t generalizes to $(t, \theta, \bar{\theta})$. In order to implement supersymmetry in a $(3+1)$-dimensional field theory, we need to take supercoordinates $(x^\mu, \theta^A, \bar{\theta}_{\dot{A}})$ which span our superspace wherein superfields will be functions of these. If we think of θ^A and $\bar{\theta}_{\dot{A}}$ as spinorial coordinate variables spanning the fermionic subspace of superspace, we can define differentiations $\partial_A \equiv \partial/\partial\theta^A$, $\partial^A \equiv \partial/\partial\theta_A$, $\bar{\partial}^{\dot{A}} \equiv \partial/\partial\bar{\theta}_{\dot{A}}$, $\bar{\partial}_{\dot{A}} \equiv \partial/\partial\bar{\theta}^{\dot{A}}$ with respect to them as generalizations of what was done in §2.1. By definition,

$$\partial_A \theta^B = \delta_A{}^B \,, \tag{4.1a}$$

$$\partial^A \theta_B = \delta^A{}_B \,, \tag{4.1b}$$

$$\bar{\partial}_{\dot{A}} \bar{\theta}^{\dot{B}} = \delta_{\dot{A}}{}^{\dot{B}} \,, \tag{4.1c}$$

$$\bar{\partial}^{\dot{A}} \bar{\theta}_{\dot{B}} = \delta^{\dot{A}}{}_{\dot{B}} \,, \tag{4.1d}$$

$$\partial_A \theta_B = -\epsilon_{AB} \,, \tag{4.1e}$$

$$\partial^A \theta^B = -\epsilon^{AB} \,, \tag{4.1f}$$

$$\bar{\partial}^{\dot{A}} \bar{\theta}^{\dot{B}} = -\epsilon^{\dot{A}\dot{B}} \,, \tag{4.1g}$$

$$\bar{\partial}_{\dot{A}} \bar{\theta}_{\dot{B}} = -\epsilon_{\dot{A}\dot{B}} \,. \tag{4.1h}$$

It also follows trivially that $0 = \partial_A \bar{\theta}_{\dot{B}} = \partial^A \bar{\theta}^{\dot{B}} = \partial_A \bar{\theta}^{\dot{B}} = \partial^A \bar{\theta}_{\dot{B}} = \bar{\partial}^{\dot{A}} \theta_B = \bar{\partial}^{\dot{A}} \theta^B = \bar{\partial}_{\dot{A}} \theta_B = \bar{\partial}_{\dot{A}} \theta^B$.

The following operator relations are also valid on any function of θ or $\bar{\theta}$:

$$\epsilon^{AB} \partial_B = -\partial^A \,, \tag{4.2a}$$

$$\epsilon_{AB} \partial^B = -\partial_A \,, \tag{4.2b}$$

49

$$\epsilon_{\dot{A}\dot{B}}\bar{\partial}^{\dot{B}} = -\bar{\partial}_{\dot{A}} \ , \tag{4.2c}$$

$$\epsilon^{\dot{A}\dot{B}}\bar{\partial}_{\dot{B}} = -\bar{\partial}^{\dot{A}} \ . \tag{4.2d}$$

All components of ∂, $\bar{\partial}$ anticommute with one another, i.e.

$$0 = [\partial_A, \partial_B]_+ = [\bar{\partial}_{\dot{A}}, \bar{\partial}_{\dot{B}}] = [\partial_A, \bar{\partial}_{\dot{B}}]$$

Operating on products of fermionic fields, ψ, χ etc. and/or Grassmann coordinates, they obey the Leibniz rule with a relative negative sign, e.g.

$$\partial(\psi\chi) = (\partial\psi)\chi - \psi(\partial\chi)$$

etc. Note, moreover, that

$$\begin{aligned}
[\partial_A, \theta^B]_+ &= \delta_A{}^B \ , \\
[\bar{\partial}_{\dot{A}}, \bar{\theta}^{\dot{B}}]_+ &= \delta_{\dot{A}}{}^{\dot{B}} \ , \\
[\partial_A, \bar{\theta}^{\dot{B}}]_+ &= 0 \ , \\
[\bar{\partial}_{\dot{A}}, \theta^B]_+ &= 0 \ .
\end{aligned}$$

Furthermore, two additional useful relations are obeyed by $\partial\partial \equiv \partial^A\partial_A$ and $\bar{\partial}\bar{\partial} = \bar{\partial}_{\dot{A}}\bar{\partial}^{\dot{A}}$:

$$\partial\partial(\theta\theta) = \bar{\partial}\bar{\partial}(\bar{\theta}\bar{\theta}) = 4 \ . \tag{4.3}$$

By definition, the product of two identical components of ∂ always vanishes and ditto for identical components of $\bar{\partial}$. It follows then that **the product of any three or more components of ∂ vanishes and the same holds for components of $\bar{\partial}$.**

Integration measures and rules in the θ, $\bar{\theta}$ subspace, cf. (2.2) and (2.5), are:

$$d^2\theta = -\frac{1}{4}d\theta^A d\theta_A \ , \tag{4.4a}$$

$$d^2\bar{\theta} = -\frac{1}{4}d\bar{\theta}_{\dot{A}}d\bar{\theta}^{\dot{A}} \ , \tag{4.4b}$$

$$d^4\theta = d^2\bar{\theta}d^2\theta \ , \tag{4.4c}$$

$$\int d^2\theta = \int d^2\bar{\theta} = \int d^2\theta\,\theta^A = \int d^2\bar{\theta}\,\bar{\theta}_{\dot{A}} = 0 \ , \tag{4.4d}$$

$$\int d^2\theta\,\theta^A\theta^B = -\frac{1}{2}\epsilon^{AB} \ , \tag{4.4e}$$

$$\int d^2\bar{\theta}\,\bar{\theta}_{\dot{A}}\bar{\theta}_{\dot{B}} = -\frac{1}{2}\epsilon_{\dot{A}\dot{B}} \ , \tag{4.4f}$$

$$\int d^2\theta\,\theta\theta = \int d^2\bar{\theta}\,\bar{\theta}\bar{\theta} = 1 \ , \tag{4.4g}$$

$$\int d^4\theta\theta\theta\ \bar\theta\bar\theta = 1 \ , \tag{4.4h}$$

$$\int d^2\theta\delta^{(2)}(\theta) = \int d^2\bar\theta\delta^{(2)}(\bar\theta) = 1 \ , \tag{4.4i}$$

$$\delta^{(2)}(\theta) = \theta\theta, \ \delta^{(2)}(\bar\theta) = \bar\theta\bar\theta \ , \tag{4.4j}$$

$$\int d^2\theta f(\theta,\bar\theta) = \frac{1}{4}\partial\partial f(\theta,\bar\theta) \ , \tag{4.4k}$$

$$\int d^2\bar\theta f(\theta,\bar\theta) = \frac{1}{4}\bar\partial\bar\partial f(\theta,\bar\theta) \ , \tag{4.4l}$$

$$\int d^2\theta\partial_A f(\theta,\bar\theta) = \int d^2\bar\theta\bar\partial^{\dot A} f(\theta,\bar\theta) = 0 \ . \tag{4.4m}$$

$$\int d^4\theta f(\theta,\bar\theta) = \frac{1}{16}\partial\partial\ \bar\partial\bar\partial f(\theta,\bar\theta) \ . \tag{4.4n}$$

The RHS of (4.4k) and (4.4l) are independent of θ and $\bar\theta$ respectively because any general function $f(\theta,\bar\theta)$ can have at most a quadratic dependence on these variables. The same argument shows why the RHS of (4.4n) is independent of $\theta,\bar\theta$. The function $f(\theta,\bar\theta)$ of Chapter 2 can be generalized now to

$$f(\theta,\bar\theta) = f_0 + f_1^A\theta_A + \bar f_{2\dot A}\bar\theta^{\dot A} + f_3\theta\theta + \bar f_4\bar\theta\bar\theta + f_5^A\theta_A\bar\theta\bar\theta + \bar f_{6\dot A}\bar\theta^{\dot A}\theta\theta + f_7\theta\theta\ \bar\theta\bar\theta \ ,$$

with

$$\int d^4\theta f(\theta,\bar\theta) = f_7 \ ,$$

$$\int d^4\theta\delta^{(2)}(\bar\theta)f(\theta,\bar\theta) = f_3 \ ,$$

$$\int d^4\theta\delta^{(2)}(\theta)f(\theta,\bar\theta) = \bar f_4 \ ,$$

$$\int d^4\theta\delta^{(2)}(\bar\theta)\theta^A f(\theta,\bar\theta) = \frac{1}{2}f_1^A \ ,$$

$$\int d^4\theta\delta^{(2)}(\theta)\bar\theta_{\dot A}f(\theta,\bar\theta) = \frac{1}{2}\bar f_{2\dot A} \ ,$$

$$\int d^4\theta\theta^A f(\theta,\bar\theta) = \frac{1}{2}f_5^A \ ,$$

$$\int d^4\theta\bar\theta_{\dot A}f(\theta,\bar\theta) = \frac{1}{2}\bar f_{6\dot A} \ ,$$

$$\int d^4\theta\delta^{(2)}(\theta)\delta^{(2)}(\bar\theta)f(\theta,\bar\theta) = f_0 \ .$$

The generalization of the infinitesimal transformations (2.20) to superspace can be given in terms of two component anticommuting spinor parameters ϵ^A, $\bar\epsilon_{\dot A}$ as

$$z \equiv (x^\mu,\theta,\bar\theta) \rightarrow (x^\mu - i\theta\sigma^\mu\bar\epsilon + i\epsilon\sigma^\mu\bar\theta, \theta+\epsilon, \bar\theta+\bar\epsilon) \ . \tag{4.5}$$

Dimensional consistency requires attributing a mass dimension $-1/2$ to θ and hence to ϵ. Also, ϵ^A and $\bar{\epsilon}_{\dot{A}}$, being constant parameters rather than local functions of x^μ, make (4.5) a **global** or **rigid supersymmetric transformation** rather than a local one. The relative sign between the $\theta\sigma^\mu\bar{\epsilon}$ and $\epsilon\sigma^\mu\bar{\theta}$ terms gets fixed[1] by the requirement that x^μ remain real since $(\theta\sigma^\mu\bar{\epsilon})^\dagger = \epsilon\sigma^\mu\bar{\theta}$. As in §2.3, we look for a linear realization of supersymmetry. (For alternative nonlinear realizations see Ref. [4.1]). Like (2.21), an infinitesimal transformation $\delta z \equiv \delta(x, \theta, \bar{\theta})$ on any function f of the supercoordinates should be expressible in terms of two component supersymmetry generators $Q, \bar{Q}$ as

$$\delta f(z) = \delta x^\mu \partial_\mu f + \delta\theta^A \partial_A f + \delta\bar{\theta}_{\dot{A}} \bar{\partial}^{\dot{A}} f = i(\epsilon Q + \bar{\epsilon}\bar{Q})f(x, \theta, \bar{\theta}) \ ,$$

With $\bar{Q}$ interpreted as $Q^\dagger$, the following explicit representations of $Q, \bar{Q}$ then obtain from (4.5):

$$Q_A = -i(\partial_A + i\sigma^\mu_{A\dot{B}}\bar{\theta}^{\dot{B}}\partial_\mu) \ , \tag{4.6a}$$

$$\bar{Q}^{\dot{A}} = -i(\bar{\partial}^{\dot{A}} + i\theta^B\sigma^\mu_{B\dot{B}}\epsilon^{\dot{B}\dot{A}}\partial_\mu) \ . \tag{4.6b}$$

The appearance of the factor i in the right hand side of $\delta f(z)$ above is significant. This is necessitated by the fact that (cf. 2.21) now $\bar{\epsilon}\bar{Q} = (\epsilon Q)^\dagger$, so that $\bar{\epsilon}\bar{Q} + \epsilon Q$ is hermitian, as contrasted with the situation in §2.2 and §2.3. Moreover, using (3.18a), (4.6b) can be rewritten as

$$\bar{Q}^{\dot{A}} = -i(\bar{\partial}^{\dot{A}} + i\bar{\sigma}^{\mu\dot{A}B}\theta_B\partial_\mu) \ ,$$

whereas, by virtue of (4.2c), one has

$$\bar{Q}_{\dot{A}} = i(\bar{\partial}_{\dot{A}} + i\theta^B\sigma^\mu_{B\dot{A}}\partial_\mu) \ .$$

Similar alternative expressions can also be derived for Q. It is straightforward to check from (4.6) that (3.26a) is indeed obeyed with $P_\mu = i\partial_\mu$, which is what the four momentum operator is, acting on an incoming state in the coordinate representation. Q has the mass dimension $\frac{1}{2}$.

Next, we try a field theoretic generalization of (2.19) and thereby come to superfields [4.2]. In this chapter we shall be concerned with classical superfields mostly. For the moment, these will be taken to commute with one another. That will ensure that the bosonic component fields will commute with all other fields and that all fermionic component fields will mutually anticommute. Let us consider a Lorentz invariant Z_2-even (cf.§3.1) function $\mathcal{F}(z)$ of the supercoordinate z defined on the superspace introduced above. This is our generalization of the decomposition (2.19), namely

$$\mathcal{F}(z) \equiv \mathcal{F}(x, \theta, \bar{\theta}) \ = \ f(x) + \sqrt{2}\theta\xi(x) + \sqrt{2}\bar{\theta}\bar{\chi}(x) + \theta\theta M(x) + \bar{\theta}\bar{\theta}N(x)$$

$$+\theta\sigma^\mu\bar{\theta}A_\mu(x) + \theta\theta\ \bar{\theta}\bar{\lambda}(x) + \bar{\theta}\bar{\theta}\ \theta\zeta(x) + \frac{1}{2}\theta\theta\ \bar{\theta}\bar{\theta}D(x) \ . \tag{4.7}$$

---

[1] One could have chosen $x^\mu \to x^\mu + i\theta\sigma^\mu\bar{\epsilon} - i\epsilon\sigma^\mu\bar{\theta}$ instead of (4.5), but then one would need to put an overall negative sign in the infinitesimal transformation of $f(z)$ above. Otherwise, there would be a flip in the signs of the space derivative terms in $Q, \bar{Q}$ (see 4.6 below) – eventually leading to a mismatch between the signs of the bosonic and fermionic kinetic energy terms in the supersymmetric Lagrangian densities of Ch.5.

This is the most general form since $\bar{\theta}\bar{\sigma}^{\mu}\theta = -\theta\sigma^{\mu}\bar{\theta}$ and $\theta\sigma^{\mu\nu}\theta = 0 = \bar{\theta}\bar{\sigma}^{\mu\nu}\bar{\theta}$. The $\sqrt{2}$ coefficients have been put in for convenience with respect to supersymmetry transformations. One can consider the hermitian conjugate $\mathcal{F}^{\dagger}(z)$ as an independent superfield. Note that $\mathcal{F}(z)$ need not have a well-defined parity but does have a precise mass dimension. In (4.7) we have introduced several component fields of the superfield $\mathcal{F}(x,\theta,\bar{\theta})$. They consist of four scalar fields $f(x)$, $M(x)$, $N(x)$ and $D(x)$, one vector field $A_{\mu}(x)$ plus two left handed Weyl spinor fields $\xi_{A}(x)$, $\zeta_{A}(x)$ and two right handed Weyl spinor fields $\bar{\chi}^{\dot{A}}(x)$ and $\bar{\lambda}^{\dot{A}}(x)$. All fields being complex, there are sixteen real bosonic and sixteen real fermionic component fields. These are, of course, off-shell degrees of freedom since we have not imposed the equations of motion. They do, nonetheless correspond (though not necessarily in a one to one way) to particles that neatly fall into complete supermultiplets.

The effect of the infinitesimal supersymmetry transformation (4.5) on the components of $\mathcal{F}$ can be worked out [4.2, 4.3] from the requirement that $\delta\mathcal{F}$ has the form (4.7) with all components replaced by their variations. Thus,

$$\delta\mathcal{F} \equiv \mathcal{F}(x^{\mu} - i\theta\sigma^{\mu}\bar{\epsilon} + i\epsilon\sigma^{\mu}\bar{\theta},\ \theta + \epsilon, \bar{\theta} + \bar{\epsilon}) - \mathcal{F}(x,\theta,\bar{\theta}) = i(\epsilon Q + \bar{\epsilon}\bar{Q})\mathcal{F},$$

implying

$$\delta f = \sqrt{2}\epsilon\xi + \sqrt{2}\bar{\epsilon}\bar{\chi}\,, \tag{4.8a}$$

$$\delta(\sqrt{2}\xi_{A}) = 2\epsilon_{A}M + (\sigma^{\mu}\bar{\epsilon})_{A}(-i\partial_{\mu}f + A_{\mu})\,, \tag{4.8b}$$

$$\delta(\sqrt{2}\bar{\chi}^{\dot{A}}) = 2\bar{\epsilon}^{\dot{A}}N - (\bar{\sigma}^{\mu}\epsilon)^{\dot{A}}(i\partial_{\mu}f + A_{\mu})\,, \tag{4.8c}$$

$$\delta M = \bar{\epsilon}\bar{\lambda} + \frac{i}{\sqrt{2}}\partial_{\mu}\xi\sigma^{\mu}\bar{\epsilon}\,, \tag{4.8d}$$

$$\delta N = \epsilon\zeta - \frac{i}{\sqrt{2}}\epsilon\sigma^{\mu}\partial_{\mu}\bar{\chi}\,, \tag{4.8e}$$

$$\delta A_{\mu} = \epsilon\sigma_{\mu}\bar{\lambda} + \zeta\sigma_{\mu}\bar{\epsilon} - \frac{i}{\sqrt{2}}\epsilon\partial_{\mu}\xi + \frac{i}{\sqrt{2}}\partial_{\mu}\bar{\chi}\bar{\epsilon}$$
$$+ i\sqrt{2}\epsilon\sigma_{\mu\nu}\partial^{\nu}\xi - i\sqrt{2}\bar{\epsilon}\bar{\sigma}_{\mu\nu}\partial^{\nu}\bar{\chi}\,, \tag{4.8f}$$

$$\delta\bar{\lambda}^{\dot{A}} = \bar{\epsilon}^{\dot{A}}D - \frac{i}{2}\bar{\epsilon}^{\dot{A}}\partial^{\mu}A_{\mu} - i(\bar{\sigma}^{\mu}\epsilon)^{\dot{A}}\partial_{\mu}M + (\bar{\sigma}^{\mu\nu}\bar{\epsilon})^{\dot{A}}\partial_{\mu}A_{\nu}\,, \tag{4.8g}$$

$$\delta\zeta_{A} = \epsilon_{A}D + \frac{i}{2}\epsilon_{A}\partial^{\mu}A_{\mu} - i(\sigma^{\mu}\bar{\epsilon})_{A}\partial_{\mu}N - (\sigma^{\mu\nu}\epsilon)_{A}\partial_{\mu}A_{\nu}\,, \tag{4.8h}$$

$$\delta D = i\partial_{\mu}(\zeta\sigma^{\mu}\bar{\epsilon} + \bar{\lambda}\bar{\sigma}^{\mu}\epsilon)\,. \tag{4.8i}$$

(4.8) describe the general infinitesimal globally supersymmetric transformations of the component fields mixing fermionic and bosonic ones.

The following three observations are in order.

- δD being a four divergence, with the usual assumption of discarding surface terms, any D-term (i.e. coefficient of $\theta\theta\,\bar{\theta}\bar{\theta}$ in a superfield) in the Lagrangian density would yield a supersymmetric action. The latter is denoted $\forall\mathcal{F}$ by $[\mathcal{F}]_{D} = \int d^{4}\theta\mathcal{F}$.

- Linear combinations of superfields are again superfields (since $Q, \bar{Q}$ are linear differential operators), i.e. superfields form linear representations of the supersymmetry algebra. Products of superfields will be general superfields.

- Superfield representations are, in general, highly reducible. Some of the component fields (called auxiliary fields) do not contribute to an on-shell description, i.e. are physically redundant. Specific types of irreducible superfields are : chiral, vector etc., as detailed in the following sections. However, products of irreducible superfields may or may not be irreducible superfields.

4.2 Chiral Covariant Derivatives

Before coming to specific types of superfields, it is important to strengthen our technical apparatus by defining chiral covariant derivatives since the covariant constraints are implemented through them. Though the operator $\partial_\mu = -iP_\mu$ commutes with the supersymmetry charges $Q, \bar{Q}$, the spinorial derivatives ∂_A and $\bar{\partial}_{\dot{A}}$ do not. As a result, they do not have covariant supersymmetric transformations. Let us take the transformation (4.5) and define $x'^\mu \equiv x^\mu - i\theta\sigma^\mu\bar{\epsilon} + i\epsilon\sigma^\mu\bar{\theta}$, $\theta' \equiv \theta + \epsilon$, $\bar{\theta}' \equiv \bar{\theta} + \bar{\epsilon}$. We then have

$$\partial_A = \frac{\partial\theta'^B}{\partial\theta^A}\frac{\partial}{\partial\theta'^B} + \frac{\partial x'^\mu}{\partial\theta^A}\frac{\partial}{\partial x'^\mu} = \frac{\partial}{\partial\theta'^A} - i(\sigma^\mu\bar{\epsilon})_A\frac{\partial}{\partial x'^\mu} \ .$$

Now the derivative

$$\mathcal{D}_A \equiv \partial_A - i\sigma^\mu_{A\dot{B}}\bar{\theta}^{\dot{B}}\partial_\mu \tag{4.9a}$$

does transform covariantly, as can be explicitly verified [4.2] by checking from (4.6) that $[\mathcal{D}_A, Q_B]_+ = 0 = [\mathcal{D}_A, \bar{Q}^{\dot{B}}]_+$. We can also define a contravariant version

$$\mathcal{D}^A \equiv \epsilon^{AB}\mathcal{D}_B = -\partial^A + i\bar{\theta}_{\dot{B}}\bar{\sigma}^{\mu\dot{B}A}\partial_\mu \ . \tag{4.9b}$$

Coming to $\bar{\partial}$, the corresponding derivatives are

$$\bar{\mathcal{D}}^{\dot{A}} = \bar{\partial}^{\dot{A}} - i\bar{\sigma}^{\mu\dot{A}B}\theta_B\partial_\mu \ , \tag{4.10a}$$

$$\bar{\mathcal{D}}_{\dot{A}} = -\bar{\partial}_{\dot{A}} + i\theta^B\sigma^\mu_{B\dot{A}}\partial_\mu \ , \tag{4.10b}$$

obeying $[\bar{\mathcal{D}}_{\dot{A}}, \bar{Q}^{\dot{B}}]_+ = 0 = [\bar{\mathcal{D}}_{\dot{A}}, Q_B]_+$ and $\mathcal{D}_A\mathcal{D}_B = \frac{1}{2}\epsilon_{AB}\mathcal{D}\mathcal{D}$, $\bar{\mathcal{D}}_{\dot{A}}\bar{\mathcal{D}}_{\dot{B}} = -\frac{1}{2}\epsilon_{\dot{A}\dot{B}}\bar{\mathcal{D}}\bar{\mathcal{D}}$ etc. Note that $\mathcal{D}\mathcal{D} \equiv \mathcal{D}^A\mathcal{D}_A$, $\bar{\mathcal{D}}\bar{\mathcal{D}} \equiv \bar{\mathcal{D}}_{\dot{A}}\bar{\mathcal{D}}^{\dot{A}}$.

The chiral covariant and contravariant derivatives, introduced above, obey the following relations which can be obtained by straightforward, though sometimes tedious, algebra.

$$[\mathcal{D}_A, \mathcal{D}_B]_+ = [\bar{\mathcal{D}}_{\dot{A}}, \bar{\mathcal{D}}_{\dot{B}}]_+ = 0 \ , \tag{4.11a}$$

$$[\mathcal{D}_A, \bar{\mathcal{D}}_{\dot{B}}]_+ = 2i\sigma^\mu_{A\dot{B}}\partial_\mu \ , \tag{4.11b}$$

$$[\mathcal{D}^A, \bar{\mathcal{D}}^{\dot{B}}]_+ = 2i\bar{\sigma}^{\mu\dot{B}A}\partial_\mu \ . \tag{4.11c}$$

$$\mathcal{D}\sigma^\mu\bar{\mathcal{D}} + \bar{\mathcal{D}}\bar{\sigma}^\mu\mathcal{D} = 4i\partial^\mu \ , \tag{4.11d}$$

$$[\mathcal{D}_A, \bar{\mathcal{D}}\bar{\mathcal{D}}] = 4i\sigma^\mu_{A\dot{B}}\bar{\mathcal{D}}^{\dot{B}}\partial_\mu \ , \tag{4.11e}$$

$$[\mathcal{D}^A, \bar{\mathcal{D}}\bar{\mathcal{D}}] = -4i\bar{\mathcal{D}}_{\dot{B}}\bar{\sigma}^{\mu\dot{B}A}\partial_\mu \ , \tag{4.11f}$$

$$[\bar{\mathcal{D}}_{\dot{A}}, \mathcal{D}\mathcal{D}] = -4i\mathcal{D}^B\sigma^\mu_{B\dot{A}}\partial_\mu \ , \tag{4.11g}$$

$$[\bar{\mathcal{D}}^{\dot{A}}, \mathcal{D}\mathcal{D}] = 4i\bar{\sigma}^{\mu\dot{A}B}\mathcal{D}_B\partial_\mu \ , \tag{4.11h}$$

$$[\mathcal{D}\mathcal{D}, \bar{\mathcal{D}}\bar{\mathcal{D}}] = 8i\mathcal{D}\sigma^\mu\bar{\mathcal{D}}\partial_\mu + 16\partial^2 \ , \tag{4.11i}$$

$$[\bar{\mathcal{D}}\bar{\mathcal{D}}, \mathcal{D}\mathcal{D}] = 8i\bar{\mathcal{D}}\bar{\sigma}^\mu\mathcal{D}\partial_\mu + 16\partial^2 \ , \tag{4.11j}$$

$$\mathcal{D}^A\bar{\mathcal{D}}\bar{\mathcal{D}}\mathcal{D}_A = \bar{\mathcal{D}}_{\dot{A}}\mathcal{D}\mathcal{D}\bar{\mathcal{D}}^{\dot{A}} = 8\partial^2 + \frac{1}{2}\left[\mathcal{D}\mathcal{D}, \bar{\mathcal{D}}\bar{\mathcal{D}}\right]_+ \ , \tag{4.11k}$$

$$\mathcal{D}\mathcal{D}\ \bar{\mathcal{D}}\bar{\mathcal{D}}\ \mathcal{D}\mathcal{D} = -16\partial^2\mathcal{D}\mathcal{D} \ , \tag{4.11l}$$

$$\bar{\mathcal{D}}\bar{\mathcal{D}}\ \mathcal{D}\mathcal{D}\ \bar{\mathcal{D}}\bar{\mathcal{D}} = -16\partial^2\bar{\mathcal{D}}\bar{\mathcal{D}} \ . \tag{4.11m}$$

Any product of three or more components of $\mathcal{D}$ vanishes and ditto for the components of $\bar{\mathcal{D}}$.

$$\mathcal{D}_A\mathcal{D}_B\mathcal{D}_C = 0 = \bar{\mathcal{D}}_{\dot{A}}\bar{\mathcal{D}}_{\dot{B}}\bar{\mathcal{D}}_{\dot{C}} \ .$$

Moreover, $\mathcal{D}$ and $\bar{\mathcal{D}}$ carry a mass dimension of $1/2$. Additionally, taking $\int d^4x\partial_\mu(\cdots)$, i.e. surface terms, to vanish, $\partial_A(\bar{\partial}^{\dot{A}})$ can be replaced by $\mathcal{D}_A(\bar{\mathcal{D}}^{\dot{A}})$ within a spacetime integral $\int d^4x$. Thus, $\forall \mathcal{F}(z) \equiv \mathcal{F}(x,\theta,\bar{\theta})$ superfields, covering all superfunctions $f(z) \equiv f(x,\theta,\bar{\theta})$, defined in superspace, the use of (4.4k-m) and of (4.3) leads one to

$$\int d^6z\mathcal{D}_A\mathcal{F} \equiv \int d^4x d^2\theta\partial_A\mathcal{F} = 0 \ , \tag{4.12a}$$

$$\int d^6\bar{z}\bar{\mathcal{D}}^{\dot{A}}\mathcal{F} \equiv \int d^4x d^2\bar{\theta}\bar{\partial}^{\dot{A}}\mathcal{F} = 0 \ , \tag{4.12b}$$

$$-\frac{1}{4}\int d^6z\bar{\mathcal{D}}\bar{\mathcal{D}}\mathcal{F} = \frac{1}{4}\int d^6z\bar{\partial}\bar{\partial}\mathcal{F} = \int d^8z\mathcal{F} \ , \tag{4.12c}$$

$$-\frac{1}{4}\int d^6\bar{z}\mathcal{D}\mathcal{D}\mathcal{F} = \frac{1}{4}\int d^6\bar{z}\partial\partial\mathcal{F} = \int d^8z\mathcal{F} \ , \tag{4.12d}$$

$$\frac{1}{16}\int d^4x\mathcal{D}\mathcal{D}\ \bar{\mathcal{D}}\bar{\mathcal{D}}\mathcal{F} = \int d^8z\mathcal{F} \ , \tag{4.12e}$$

$$\int d^6z\mathcal{F} = \frac{1}{4}\int d^4x\partial\partial\mathcal{F} = -\int d^4x\frac{\mathcal{D}\mathcal{D}}{4}\mathcal{F} \ , \tag{4.12f}$$

$$\int d^6\bar{z}\mathcal{F} = \frac{1}{4}\int d^4x\bar{\partial}\bar{\partial}\mathcal{F} = -\int d^4x\frac{\bar{\mathcal{D}}\bar{\mathcal{D}}}{4}\mathcal{F} \ . \tag{4.12g}$$

In (4.12) we have used the notations

$$d^8z \equiv d^4x d^4\theta \ , \ d^6z \equiv d^4x d^2\theta \ \text{and} \ d^6\bar{z} \equiv d^4x d^2\bar{\theta} \ .$$

For $\mathcal{F} = \mathcal{F}_1\mathcal{F}_2$, (4.12a) and (4.12b) respectively imply that

$$\int d^6z\,\mathcal{F}_1\mathcal{D}_A\mathcal{F}_2 = -\int d^6z\,(\mathcal{D}_A\mathcal{F}_1)\mathcal{F}_2\,,$$

$$\int d^6\bar{z}\,\mathcal{F}_1\bar{\mathcal{D}}^{\dot{A}}\mathcal{F}_2 = -\int d^6\bar{z}\,(\bar{\mathcal{D}}^{\dot{A}}\mathcal{F}_1)\mathcal{F}_2\,.$$

The signs in the RHS above would be opposite if $\mathcal{F}_1$ were odd under the Z_2 grading (Ch. 3). Note furthermore that these results, as well as (4.12a,b), remain valid if the partial superspace integrations $\int d^6z$, $\int d^6\bar{z}$ are extended to the full $\int d^8z$.

We now come to two specific constraints of the kind mentioned at the end of §4.1: **chirality** and **reality**.

(1) Chirality for a superfield means that it has a vanishing covariant derivative. Let $\mathcal{F} = \Phi$ obey

$$\bar{\mathcal{D}}_{\dot{A}}\Phi = 0\,, \tag{4.13a}$$

$$\mathcal{D}_A\Phi^\dagger = 0\,, \tag{4.13b}$$

where any one of the above two may be seen to be the consequence of the other. One can then derive that

$$\bar{\mathcal{D}}\bar{\mathcal{D}}\,\mathcal{D}\mathcal{D}\Phi = -16\partial^2\Phi\,, \tag{4.14a}$$

$$\mathcal{D}\mathcal{D}\,\bar{\mathcal{D}}\bar{\mathcal{D}}\Phi^\dagger = -16\partial^2\Phi^\dagger\,, \tag{4.14b}$$

$$\int d^6\bar{z}\,\Phi = -\frac{1}{4}\int d^4x\,\bar{\mathcal{D}}\bar{\mathcal{D}}\Phi = 0\,, \tag{4.14c}$$

$$\int d^6z\,\Phi^\dagger = -\frac{1}{4}\int d^4x\,\mathcal{D}\mathcal{D}\Phi^\dagger = 0\,, \tag{4.14d}$$

$$\int d^6z\,\Phi = -\frac{1}{4}\int d^4x\,\mathcal{D}\mathcal{D}\Phi = \int d^4x\,\frac{\mathcal{D}\mathcal{D}\,\bar{\mathcal{D}}\bar{\mathcal{D}}\,\mathcal{D}\mathcal{D}}{64\partial^2}\Phi = \int d^8z\,\frac{\mathcal{D}\mathcal{D}}{4\partial^2}\Phi\,, \tag{4.14e}$$

$$\int d^6\bar{z}\,\Phi^\dagger = -\frac{1}{4}\int d^4x\,\bar{\mathcal{D}}\bar{\mathcal{D}}\Phi^\dagger = \int d^4x\,\frac{\bar{\mathcal{D}}\bar{\mathcal{D}}\,\mathcal{D}\mathcal{D}\,\bar{\mathcal{D}}\bar{\mathcal{D}}}{64\partial^2}\Phi^\dagger = \int d^8z\,\frac{\bar{\mathcal{D}}\bar{\mathcal{D}}}{4\partial^2}\Phi^\dagger\,. \tag{4.14f}$$

In (4.14e,f) above we have inverted (4.14a,b) and have written them in the nonlocal form

$$\Phi = -\frac{\bar{\mathcal{D}}\bar{\mathcal{D}}\,\mathcal{D}\mathcal{D}}{16\partial^2}\Phi\,,$$

$$\Phi^\dagger = -\frac{\mathcal{D}\mathcal{D}\,\bar{\mathcal{D}}\bar{\mathcal{D}}}{16\partial^2}\Phi^\dagger\,.$$

The nonlocal operator[2] $\frac{1}{\partial^2}$ is defined in a way that $\frac{1}{\partial^2}g = f \Rightarrow g = \partial^2 f$. Moreover, (4.13a,b) imply that Φ, $\Phi^\dagger$ can be written equal to $-\frac{1}{4}\bar{\mathcal{D}}\bar{\mathcal{D}}U$, $-\frac{1}{4}\mathcal{D}\mathcal{D}U^\dagger$ respectively where U is a general superfield. Now (4.12c) and (4.12d) respectively imply that

$$\int d^6z\,\Phi = \int d^8z\,U\,,$$

---

[2]Partial superspace integrations, with $\mathcal{F} = \mathcal{F}_1\mathcal{F}_2$ and utilizing (4.12a,b), need not necessarily hold if the covariant derivative is multiplied by the nonlocal $1/\partial^2$, as in (4.14e,f).

$$\int d^6\bar{z}\,\Phi^\dagger = \int d^8 z\, U^\dagger$$

Generalized forms of (4.13) and (4.14) are valid when Φ ($\Phi^\dagger$) is replaced by any analytic function of it.

(2) Reality as a constraint implies, for $\mathcal{F} = V$ say,

$$V = V^\dagger. \tag{4.15}$$

Let us delineate the two types of superfields introduced above as follows.

- A superfield such as Φ, obeying (4.13a), is called left chiral while one such as $\Phi^\dagger$, obeying (4.13b), is called right chiral. Sometimes these are called chiral and antichiral superfields respectively.

- A superfield respecting (4.15) is called a vector superfield.

It is trivial to see that either of the chiral superfield constraints is incompatible with that of the vector superfield since the latter is hermitian and the hermiticity of Φ would imply $\Phi =$ constant because of (4.11b,c).

4.3 Left and right chiral superfields

We describe these by Φ and $\Phi^\dagger$ respectively obeying $\bar{\mathcal{D}}_{\dot{A}}\Phi = 0 = \mathcal{D}_A\Phi^\dagger$. As will be clear from what follows, these constraints enable each of the two superfields to describe a complete irreducible chiral supermultiplet. But first it is convenient to define left and right chiral superspace coordinates by formally shifting x^μ to $y^\mu \equiv x^\mu - i\theta\sigma^\mu\bar{\theta}$ and to $\bar{y}^\mu \equiv x^\mu + i\theta\sigma^\mu\bar{\theta}$ respectively. We now state two pairs of useful identities:

$$\bar{\mathcal{D}}_{\dot{A}}y^\mu = 0, \; \mathcal{D}_A\bar{y}^\mu = 0 \;, \tag{4.16a}$$

$$\bar{\mathcal{D}}_{\dot{A}}f(y,\theta) = 0, \; \mathcal{D}_A f^\star(\bar{y},\bar{\theta}) = 0 \;, \tag{4.16b}$$

where f is any function. Any of the spinorial indices in (4.16) can obviously be raised or lowered without changing the relations. By a straightforward use of the chain rule in (4.9a) and (4.10b) one can see that, under the transformation $(x,\theta,\bar{\theta}) \to (y,\theta,\bar{\theta})$, $\mathcal{D}_A$ and $\bar{\mathcal{D}}_{\dot{A}}$ change to

$$\mathcal{D}_A^{(y)} = \partial_A - 2i\sigma^\mu_{A\dot{B}}\bar{\theta}^{\dot{B}}\partial_\mu^{(y)}, \; \mathcal{D}^{A(y)} = -\partial^A + 2i\bar{\theta}_{\dot{B}}\bar{\sigma}^{\mu\dot{B}A}\partial_\mu^{(y)}, \tag{4.17a}$$

$$\bar{\mathcal{D}}_{\dot{A}}^{(y)} = -\bar{\partial}_{\dot{A}}, \; \bar{\mathcal{D}}^{\dot{A}(y)} = \bar{\partial}^{\dot{A}}. \tag{4.17b}$$

On the other hand, the transformation $(x,\theta,\bar{\theta}) \to (\bar{y},\theta,\bar{\theta})$ changes them to

$$\mathcal{D}_A^{(\bar{y})} = \partial_A, \; \mathcal{D}^{A(\bar{y})} = -\partial^A, \tag{4.18a}$$

$$\bar{\mathcal{D}}_{\dot{A}}^{(\bar{y})} = -\bar{\partial}_{\dot{A}} + 2i\theta^B\sigma^\mu_{B\dot{A}}\partial_\mu^{(\bar{y})}, \; \bar{\mathcal{D}}^{\dot{A}(\bar{y})} = \bar{\partial}^{\dot{A}} - 2i\bar{\sigma}^{\mu\dot{A}B}\theta_B\partial_\mu^{(\bar{y})}. \tag{4.18b}$$

Any superfield, which is a function of y and θ ($\bar{y}$ and $\bar{\theta}$) only, would automatically be left chiral (right chiral). The decomposition of these superfields in terms of their component fields proceeds analogously to that in (4.7) and can be given as

$$\Phi(y,\theta) = \phi(y) + \sqrt{2}\theta\xi(y) + \theta\theta F(y) \, , \tag{4.19a}$$

$$\Phi^\dagger(\bar{y},\bar{\theta}) = \phi^\star(\bar{y}) + \sqrt{2}\bar{\theta}\bar{\xi}(\bar{y}) + \bar{\theta}\bar{\theta} F^\star(\bar{y}) \, , \tag{4.19b}$$

i.e.

$$\int d^2\theta \Phi(y,\theta) = F(y) \, ,$$

$$\int d^2\bar{\theta} \Phi^\dagger(\bar{y},\bar{\theta}) = F^\star(\bar{y}) \, .$$

In (4.19) y ($\bar{y}$) can be thought of as a real spacetime coordinate in the left (right) chiral basis. Here the mass dimensions of ϕ, ξ and F are $1, 3/2$ and 2 respectively. Eqs. (4.19) have the following explicit meaning. For $\mathcal{F}(x,\theta,\bar{\theta})$ of (4.7), the constraint (4.13a) has the simple implication that

$$\mathcal{F}(z) \equiv \mathcal{F}(x,\theta,\bar{\theta}) \equiv \mathcal{F}(y + i\theta\sigma\bar{\theta},\theta,\bar{\theta}) = \Phi(y,\theta) \, ,$$

i.e. no dependence on $\bar{\theta}$ is present and therefore $\int d^8 z \Phi(z) = 0$. Similarly, the imposition of (4.13b) means that

$$\mathcal{F}(x,\theta,\bar{\theta}) \equiv \mathcal{F}(\bar{y} - i\theta\sigma\bar{\theta},\theta,\bar{\theta}) = \Phi^\dagger(\bar{y},\bar{\theta}) \, ,$$

with all θ-dependence absent, i.e. $\int d^8 z \Phi^\dagger(z) = 0$.

The effects of the above transformations on Q and $\bar{Q}$ also are simple. It follows directly from (4.6) that under $(x,\theta,\bar{\theta}) \to (y,\theta,\bar{\theta})$ they change to

$$Q_A^{(y)} = -i\partial_A \, , \tag{4.20a}$$

$$\bar{Q}_{\dot{A}}^{(y)} = i\bar{\partial}_{\dot{A}} - 2\theta^A \sigma^\mu_{A\dot{A}} \partial_\mu^{(y)} \tag{4.20b}$$

and

$$Q_A^{(\bar{y})} = -i\partial_A + 2\sigma^\mu_{A\dot{B}} \bar{\theta}^{\dot{B}} \partial_\mu \, , \tag{4.21a}$$

$$\bar{Q}_{\dot{A}}^{(\bar{y})} = i\bar{\partial}_{\dot{A}} \, . \tag{4.21b}$$

These may be compared with (4.6). It may also be verified directly from the representations (4.20) and (4.21) that $Q^{(y)}$, $\bar{Q}^{(y)}$ together on one hand and $Q^{(\bar{y})}$, $\bar{Q}^{(\bar{y})}$ on the other obey the supersymmetry algebra.

Returning to (4.19), we can substitute $y^\mu = x^\mu - i\theta\sigma^\mu\bar{\theta}$, and $\bar{y}^\mu = x^\mu + i\theta\sigma^\mu\bar{\theta}$ and Taylor expand the component fields. The expansion stops at a finite number of terms once cubes or higher powers in θ or $\bar{\theta}$ begin to appear. After the Taylor expansion one can make use of (3.19) to recast (4.19) as

$$\Phi(y,\theta) = \phi(x) - i\theta\sigma^\mu\bar{\theta}\partial_\mu\phi(x) - \frac{1}{4}\theta\theta\,\bar{\theta}\bar{\theta}\partial^\mu\partial_\mu\phi(x) + \sqrt{2}\theta\xi(x)$$

$$+ \frac{i}{\sqrt{2}}\theta\theta\,\partial_\mu\xi\sigma^\mu\bar{\theta} + \theta\theta F(x) \, , \tag{4.22a}$$

$$\Phi^\dagger(\bar{y},\bar\theta) \;=\; \phi^\star(x) + i\theta\sigma^\mu\bar\theta\partial_\mu\phi^\star(x) - \frac{1}{4}\theta\theta\,\bar\theta\bar\theta\partial^\mu\partial_\mu\phi^\star(x) + \sqrt{2}\,\bar\theta\bar\xi(x)$$

$$-\frac{i}{\sqrt{2}}\bar\theta\bar\theta\,\theta\sigma^\mu\partial_\mu\bar\xi(x) + \bar\theta\bar\theta F^\star(x)\ . \tag{4.22b}$$

Evidently, (4.22b) is the hermitian conjugate of (4.22a). Moreover, one can explicitly verify by use of (4.9) and (4.10) that $\bar{\mathcal{D}}_{\dot{A}}\Phi = 0 = \mathcal{D}_A\Phi^\dagger$. We end this discussion with a comment on the difference between (4.19) and the more complicated (4.22). The former corresponds to the chiral representation for supersymmetry generators and covariant derivatives. For the latter, one uses the "symmetric" representation where $\bar{Q} = Q^\dagger$. However, both are equally valid representations of the supersymmetry algebra. The shift from x to y or $\bar{y}$ and the subsequent Taylor expansion in $\theta, \bar\theta$ is just a way to describe how to switch from one representation to the other.

One can isolate the components of Φ from Φ itself. Thus

$$\phi(x) = \Phi(x,\theta,\bar\theta)| \ , \tag{4.23a}$$

$$\sqrt{2}\xi_A(x) = \mathcal{D}_A\Phi(x,\theta,\bar\theta)| \ , \tag{4.23b}$$

$$F(x) = \frac{1}{4}\mathcal{D}\mathcal{D}\Phi(x,\theta,\bar\theta)\Big| \ , \tag{4.23c}$$

where the symbol $|$ at extreme right indicates that the quantity is evaluated at $\theta = 0 = \bar\theta$. Similarly,

$$\phi^\star(x) = \Phi^\dagger(x,\theta,\bar\theta)| \ , \tag{4.24a}$$

$$\sqrt{2}\bar\xi_{\dot{A}}(x) = \bar{\mathcal{D}}_{\dot{A}}\Phi^\dagger(x,\theta,\bar\theta)| \ , \tag{4.24b}$$

$$F^\star(x) = \frac{1}{4}\bar{\mathcal{D}}\bar{\mathcal{D}}\Phi^\dagger(x,\theta,\bar\theta)\Big| \ . \tag{4.24c}$$

With the complex scalars $\phi(x)$ and $F(x)$ comprising four real scalar Bose fields and the two component complex spinor $\xi_A(x)$ describing four real Fermi fields, the system has an equal number of bosonic and fermionic components and obeys the criterion of a full supersymmetric representation. This is, however, an off-shell description. In fact, the F-field is an **auxiliary component** and can be removed by a constraint equation leaving ϕ and ξ as the only dynamical fields, as will be shown later in §5.1. (Not all components of ξ are independent on-shell[3] because of the Dirac equation). Thus, when on-shell, the two chiral superfields correspond to two chiral supermultiplets which are conjugates of each other.

Compare (4.22a) with (4.7) and map $f \to \phi,\ \xi \to \xi,\ \bar\chi \to 0,\ M \to F,\ N \to 0,\ A_\mu \to -i\partial_\mu\phi,\ \zeta \to 0,\ \lambda \to \frac{i}{\sqrt{2}}\partial_\mu\xi\sigma^\mu,\ \bar\lambda \to -\frac{i}{\sqrt{2}}\bar\sigma^\mu\partial_\mu\xi,\ D \to -\frac{1}{2}\partial^\mu\partial_\mu\phi.$ (4.8) then yields the infinitesimal supersymmetry transformations of the component fields in Φ:

$$\delta\phi = \sqrt{2}\epsilon\xi\ , \tag{4.25a}$$

---

[3]Going on-shell, half of the propagating degrees of freedom of ξ are eliminated. This happens since the fermionic Lagrangian is linear in the time derivative. Consequently, the canonical momenta can be rewritten in terms of configuration variables without any time derivative. The former are therefore not independent dynamical degrees of freedom.

$$\delta \xi_A = \sqrt{2}\epsilon_A F - \sqrt{2}i(\sigma^\mu \bar{\epsilon})_A \partial_\mu \phi \,, \tag{4.25b}$$

$$\delta F = i\partial_\mu(\sqrt{2}\xi \sigma^\mu \bar{\epsilon}) \,. \tag{4.25c}$$

Component fields in $\Phi^\dagger$ obey the corresponding hermitian conjugate transformations. The explicit representations (4.6) of Q and $\bar{Q}$ enable one to verify that (4.25) are equivalent to $\delta\Phi = i(\epsilon Q + \bar{\epsilon}\bar{Q})\Phi$. One can also see that the commutator of two infinitesimal supersymmetry transformations gives back the gradient of the original field, i.e.

$$(\delta_{\epsilon_2}\delta_{\epsilon_1} - \delta_{\epsilon_1}\delta_{\epsilon_2})X = 2i(\epsilon_1 \sigma^\mu \bar{\epsilon}_2 - \epsilon_2 \sigma^\mu \bar{\epsilon}_1)\partial_\mu X$$

for $X = \phi, \phi^\star, \xi, \bar{\xi}, F$ and $F^\star$. The verification of this result is left to the reader as an exercise. We make the following additional points.

- The F-component of Φ, denoted by $[\Phi]_F \equiv \frac{1}{4}\mathcal{D}\mathcal{D}\Phi\big|$, transforms into itself plus a spacetime derivative. Hence, such a term in the Lagrangian density, called an F-term, leads to a supersymmetry invariant action when surface terms can be discarded.

- Constants vanish both under $\bar{\mathcal{D}}$ and $\mathcal{D}$ and are chiral superfields themselves. The shifts $\Phi_i \to \Phi_i + a_i$, a_i being constants, are permissible.

Products of chiral superfields $\Phi_1\Phi_2\cdots\Phi_l$ or $\Phi_1^\dagger\Phi_2^\dagger\cdots\Phi_l^\dagger$ are also chiral superfields themselves. For instance, one can explicitly verify from (4.19a) that

$$\Phi_i\Phi_j = \phi_i\phi_j + \sqrt{2}\theta(\xi_i\phi_j + \phi_i\xi_j) + \theta\theta(\phi_i F_j + \phi_j F_i - \xi_i\xi_j) \,, \tag{4.26}$$

where the component fields in the RHS are *all functions of y*. Thus the product has the same type of decomposition as Φ itself in (4.19a) and obeys $\bar{\mathcal{D}}_{\dot{A}}(\Phi_i\Phi_j) = 0$. One can also make the corresponding statement in terms of $\bar{y}$ for $\Phi_i^\dagger\Phi_j^\dagger$, i.e. $\mathcal{D}_A(\Phi_i^\dagger\Phi_j^\dagger) = 0$. Again, in terms of component fields which are functions of y, we have the triple superfield product

$$\begin{aligned}
\Phi_i\Phi_j\Phi_k = {} & \phi_i\phi_j\phi_k + \sqrt{2}\theta(\xi_i\phi_j\phi_k + \xi_j\phi_k\phi_i + \xi_k\phi_i\phi_j) \\
& + \theta\theta(F_i\phi_j\phi_k + F_j\phi_k\phi_i + F_k\phi_i\phi_j - \xi_i\xi_j\phi_k - \xi_j\xi_k\phi_i - \xi_k\xi_i\phi_j)
\end{aligned} \tag{4.27}$$

as functions of y with $\bar{\mathcal{D}}_{\dot{A}}(\Phi_i\Phi_j\Phi_k) = 0$ and so on. The important point to note is that in the Lagrangian density the F-term in any polynomial of chiral superfields would yield a supersymmetric action. Such is not the case for the product of a chiral superfield with its hermitian conjugate:

$$\begin{aligned}
\Phi_i^\dagger\Phi_j = {} & \phi_i^\star\phi_j + \sqrt{2}\theta\xi_j\phi_i^\star + \sqrt{2}\bar{\theta}\bar{\xi}_i\phi_j + \theta\theta\phi_i^\star F_j + \bar{\theta}\bar{\theta}F_i^\star\phi_j + 2\theta\bar{\xi}_i\,\theta\xi_j \\
& + \sqrt{2}\theta\theta\,\bar{\theta}_{\dot{A}}\left(i\bar{\sigma}^{\mu\dot{A}B}\xi_{jB}[\partial_\mu]\phi_i^\star + \bar{\xi}_i^{\dot{A}}F_j\right) \\
& - 2i\theta\sigma^\mu\bar{\theta}\phi_i^\star[\partial_\mu]\phi_j \\
& + \sqrt{2}\bar{\theta}\bar{\theta}\,\theta^A\left(i\sigma^\mu_{A\dot{B}}\bar{\xi}_i^{\dot{B}}[\partial_\mu]\phi_j + \xi_{jA}F_i^\star\right) \\
& + \theta\theta\,\bar{\theta}\bar{\theta}\left(F_i^\star F_j + \frac{1}{2}\partial_\mu\phi_i^\star[\partial^\mu]\phi_j - \frac{1}{2}\phi_i^\star[\partial_\mu]\partial^\mu\phi_j + i\xi_j\sigma^\mu[\partial_\mu]\bar{\xi}_i\right).
\end{aligned} \tag{4.28}$$

In (4.28) the operator $[\partial_\mu]$, taken between two fields X and Y, is defined as $X[\partial_\mu]Y \equiv \frac{1}{2}X\partial_\mu Y - \frac{1}{2}(\partial_\mu X)Y$. The RHS of (4.28) is written in terms of *functions of x* and is evidently not the decomposition of a chiral superfield (cf. 4.19), though it is covered by the most general superfield (4.7). Interestingly, though $\int d^8 z \Phi_1(z) \cdots \Phi_n(z) = 0 = \int d^8 z \Phi_1^\dagger(z) \cdots \Phi_n^\dagger(z)$, $\int d^8 z \Phi_i^\dagger \Phi_j$ is in general nonzero. The latter is precisely the last component in (4.24), being the coefficient of $\theta\theta\,\bar\theta\bar\theta$. It is a D-like term (cf. 4.7) and, by virtue of (4.8i), transforms into itself plus a spacetime derivative under a supersymmetric transformation. We shall call it a D-term and designate it by the subscript D on the superfield product. It is an appropriate term to appear in a Lagrangian density yielding a supersymmetric action. We shall later make use of the fact that, for a single chiral superfield Φ, $[\Phi^\dagger\Phi]_D$ contains – apart from the extra $|F|^2$ term – the kinetic energy terms for a complex scalar field ϕ and a Weyl fermion field ξ.

4.4　Vector Superfields

The reality condition characterizes a vector superfield $V(x,\theta,\bar\theta)$:

$$V = V^\dagger \ . \tag{4.29}$$

Identifying V with $\mathcal{F}$ of (4.7), we can conclude that $f = f^\star \equiv C$, $\chi = \bar\xi$, $N = M^\star$, $A_\mu = A_\mu^\star$, $\zeta = \lambda$, $D = D^\star$. Thus, we can try to decompose

$$V(x,\theta,\bar\theta) \ \sim \ C(x) + \sqrt{2}\theta\xi(x) + \sqrt{2}\bar\theta\bar\xi(x) + \theta\theta M(x) + \bar\theta\bar\theta M^\star(x) + \theta\sigma^\mu\bar\theta A_\mu(x)$$

$$+\theta\theta\,\bar\theta\bar\lambda(x) + \bar\theta\bar\theta\,\theta\lambda(x) + \frac{1}{2}\theta\theta\,\bar\theta\bar\theta D(x) \ , \tag{4.30}$$

where $C(x)$, $A_\mu(x)$ and $D(x)$ must be **real fields** whereas $M(x)$ is a complex scalar field and $\xi(x)$, $\lambda(x)$ are complex two component spinor fields. (4.30), taken in conjunction with (4.7) and (4.8), implies that – under an infinitesimal supersymmetry transformation – $D(x)$ transforms into itself plus a spacetime four divergence. The D-term, which is a suitable candidate for a supersymmetric Lagangian density is denoted by $[V]_D \equiv \int d^4\theta V(x,\theta,\bar\theta) = \frac{1}{2}D(x)$.

Vector superfields, e.g. $\Phi + \Phi^\dagger$ and $\Phi^\dagger\Phi$, can be constructed from a chiral superfield Φ. Thus from (4.22), replacing Φ and ξ by $i\Lambda$ and χ respectively, we have

$$i\Lambda - i\Lambda^\dagger \ = \ 2\mathrm{Re}\ \phi(x) + \sqrt{2}\theta\chi(x) + \sqrt{2}\bar\theta\bar\chi(x) + \theta\theta F(x) + \bar\theta\bar\theta F^\star(x)$$

$$-2\theta\sigma^\mu\bar\theta\partial_\mu\Im m\ \phi(x) - \frac{i}{\sqrt{2}}\theta\theta\,\bar\theta\bar\sigma^\mu\partial_\mu\chi - \frac{i}{\sqrt{2}}\bar\theta\bar\theta\,\theta\sigma^\mu\partial_\mu\bar\chi(x)$$

$$-\frac{1}{2}\theta\theta\,\bar\theta\bar\theta\partial^\mu\partial_\mu\Re e\ \phi(x) \ . \tag{4.31}$$

A comparison between (4.30) and (4.31) is quite instructive. One has no problem identifying $C(x)$ with $2\Re e\ \phi(x)$, $\xi(x)$ with $\chi(x)$, $M(x)$ with $F(x)$, $A_\mu(x)$ with $-2\partial_\mu\Im m\ \phi(x)$. But in

(4.31), as compared with (4.30), the D and λ terms are constrained in terms of $\Re e\ \phi$ and χ respectively. These terms can be covered generally if we substitute $\lambda - i\sigma^\mu\partial_\mu\bar\xi/\sqrt2$ for λ, $\bar\lambda - i\bar\sigma^\mu\partial_\mu\xi/\sqrt2$ for $\bar\lambda$ and $D - \frac12\partial^\mu\partial_\mu C$ for D. Thus we properly define a more general vector superfield as

$$
\begin{aligned}
V(z) \;=\;& C(x) + \sqrt2\theta\xi(x) + \sqrt2\bar\theta\bar\xi(x) + \theta\theta M(x) + \bar\theta\bar\theta M^\star(x) + \theta\sigma^\mu\bar\theta A_\mu(x) \\
&+ \theta\theta\,\bar\theta\left\{\bar\lambda(x) - \frac{i}{\sqrt2}\bar\sigma^\mu\partial_\mu\xi(x)\right\} + \bar\theta\bar\theta\,\theta\left\{\lambda(x) - \frac{i}{\sqrt2}\sigma^\mu\partial_\mu\bar\xi(x)\right\} \\
&+ \frac12\theta\theta\,\bar\theta\bar\theta\left\{D(x) - \frac12\partial^\mu\partial_\mu C(x)\right\}.
\end{aligned}
\tag{4.32}
$$

(4.31) is a special case of (4.32) with λ and D put to zero and the specified identifications.

The above discussion brings out an important point. We are defining a vector superfield by postulating an invariance under a linear transformation in the space of vector superfields. Since $i\Lambda - i\Lambda^\dagger$ is a vector superfield, we can have an abelian "supergauge transformation"[4]

$$
V \longrightarrow V + i\Lambda - i\Lambda^\dagger ,
\tag{4.33}
$$

i.e., from (4.31) and (4.32), $C \to C + 2\Re e\ \phi$, $\xi \to \xi + \chi$, $M \to M + F$, $A_\mu \to A_\mu - 2\partial_\mu\Im m\ \phi$, $\lambda \to \lambda$, $D \to D$. The term "supergauge transformation" is apt since all these are local transformations. In particular, A_μ transforms like an Abelian gauge field, i.e. $F_{\mu\nu} = \partial_\mu A_\nu - \partial_\nu A_\mu$ is left invariant under the transformation. So are $\lambda(x)$ and $D(x)$. Moreover, the beauty of the D-term is that it is not only supersymmetry invariant, but also supergauge invariant.

One now has the freedom to choose a particular gauge, called the **Wess-Zumino gauge** [4.3], where C,ξ,M all vanish and the vector superfield as a supermultiplet of component fields reduces to the form $V = (0,0,0,A_\mu,\lambda,D)$. This is really a supergauge choice. Comparing with (4.32), one can see that this is trivially achieved by the choice $C = -2\Re e\ \phi$, $\xi = -\chi$ and $F = -M$. Note that we do not have to require anything for $\Im m\ \phi$. This freedom in $\Im m\ \phi$ is the ordinary $U(1)$ gauge freedom that is still retained in the W-Z gauge. Thus we can write the Wess-Zumino gauge fixed vector superfield as

$$
V_{WZ}(z) = \theta\sigma^\mu\bar\theta A_\mu(x) + \theta\theta\,\bar\theta\bar\lambda(x) + \bar\theta\bar\theta\,\theta\lambda(x) + \frac12\theta\theta\,\bar\theta\bar\theta D(x) .
\tag{4.34}
$$

A vector superfield has mass dimension zero and can be exponentiated.

We summarize the three main points learnt above.

- A vector superfield obeys the reality condition and has a vanishing mass dimension.

- An abelian "supergauge transformation", simply adding a real combination of chiral superfields, can transform a vector superfield to the Wess-Zumino gauge where it has only the gauge boson, gaugino and auxiliary D field components.

- $[V]_D \equiv \int d^4\theta V(x,\theta,\bar\theta)$, occuring in the Lagrangian density, would yield both a supersymmetric and supergauge invariant action.

---

[4]The interactions of V are taken to keep physics invariant under this transformation.

We can identify the LHS of (4.34) as a gauge superfield in the Wess-Zumino gauge. Among its components, A_μ can be designated as a gauge field with three independent real field components (there is a gauge restriction, say $\partial^\mu A_\mu = 0$) and $\lambda(x)$ the corresponding gaugino (cf §3.3), which is a two-component complex spinor with four real independent fermionic fields. $D(x)$ is an **auxiliary field** since it is not a dynamical field like $A_\mu(x)$ and $\lambda(x)$; it does not have a kinetic energy term in $\mathcal{L}$ and can be eliminated from the system through the equations of motion. In terms of particle helicity components, one has two helicities of the massless gauge boson and two of the gaugino and hence the gauge supermultiplet of Table 3.1. However, the above "gauge-fixing" breaks manifest supersymmetry since the relations $C(x) = \xi(x) = M(x) = 0$ cannot be maintained under a supersymmetry transformations. Nonetheless, after every supersymmetry transformation, the Wess-Zumino gauge can be restored by a field-dependent compensating gauge transformation. One interesting property of the Wess-Zumino gauge is the ease with which powers of V can be computed. Thus, from (4.34),

$$V_{WZ}^2 = \frac{1}{2}\theta\theta\,\bar\theta\bar\theta A^\mu A_\mu, \quad V_{WZ}^n = 0 \ \forall \ n \geq 3 \ . \tag{4.35}$$

Moreover,

$$\begin{aligned}
\exp V_{WZ} &= 1 + V_{WZ} + \frac{1}{2}V_{WZ}^2 \\[2mm]
&= 1 + \theta\sigma^\mu\bar\theta A_\mu + \theta\theta\,\bar\theta\bar\lambda + \bar\theta\bar\theta\,\theta\lambda + \frac{1}{2}\theta\theta\,\bar\theta\bar\theta\left(D + \frac{1}{2}A^\mu A_\mu\right).
\end{aligned} \tag{4.36}$$

For later utility, we provide a couple of additional superfield products involving V_{WZ}. First, consider the triple product between a chiral superfield, its hermitian conjugate and a vector superfield in the Wess-Zumino gauge. The result is

$$\begin{aligned}
\Phi_i^\dagger V_{WZ}\Phi_j &= \theta\sigma^\mu\bar\theta A_\mu\phi_i^\star\phi_j + \frac{1}{\sqrt{2}}\theta\theta(\bar\theta\bar\sigma^\mu\xi_j A_\mu\phi_i^\star + \sqrt{2}\bar\theta\bar\lambda\phi_i^\star\phi_j) \\[2mm]
&\quad + \frac{1}{\sqrt{2}}\bar\theta\bar\theta(-\theta\sigma^\mu\bar\xi_i A_\mu\phi_j + \sqrt{2}\theta\lambda\phi_i^\star\phi_j) \\[2mm]
&\quad + \frac{1}{2}\theta\theta\,\bar\theta\bar\theta(D\phi_i^\star\phi_j - 2iA^\mu\phi_i^\star[\partial_\mu]\phi_j - \bar\xi_i\bar\sigma^\mu\xi_j A_\mu \\[2mm]
&\quad - \sqrt{2}\phi_j\lambda\bar\xi_i - \sqrt{2}\phi_i^\star\lambda\xi_j) \ .
\end{aligned} \tag{4.37}$$

Next, let us write a quadruple product:

$$\Phi_i^\dagger V_{WZ}V_{WZ}'\Phi_j = \frac{1}{2}\theta\theta\,\bar\theta\bar\theta A^\mu A_\mu'\phi_i^\star\phi_j \ . \tag{4.38}$$

Given the properties of $V(z)$, one can view it as the supersymmetric generalization of an abelian gauge potential. The natural question arises: what is the field strength that remains invariant under an abelian supergauge transformation? For an arbitrary $V(z)$, *not necessarily in the Wess-Zumino gauge*, one can construct left and right chiral spinorial field-strength superfields:

$$W_A = -\frac{1}{4}\bar{\mathcal{D}}\bar{\mathcal{D}}\,\mathcal{D}_A V \ , \tag{4.39a}$$

$$\overline{W}_{\dot{A}} = -\frac{1}{4}\mathcal{D}\mathcal{D}\,\bar{\mathcal{D}}_{\dot{A}}V \tag{4.39b}$$

as superfield generalizations of the abelian field-strength tensor. While V (as well as Φ, $\Phi^\dagger$) is even under the Z_2-grading [cf.(3.4)], W and $\bar{W}$ are odd. Moreover, the following observations on (4.39) can be made readily.

- Since the product of any 3 $\mathcal{D}$'s or $\bar{\mathcal{D}}$'s vanishes, $\bar{\mathcal{D}}_{\dot{B}}W_A = 0$ and $\mathcal{D}_B\overline{W}_{\dot{A}} = 0$ in conformity with the claim that W_A and $\overline{W}_{\dot{A}}$ are indeed left and right chiral superfields respectively.

- Both W and $\overline{W}$ are seen to have the mass dimension of $3/2$.

- W_A and $\overline{W}_{\dot{A}}$ are invariant under the supergauge transformation

$$V \to V + i\Lambda - i\Lambda^\dagger.$$

This follows as a consequence of Λ and $\Lambda^\dagger$ being left chiral and right chiral superfields respectively with $\bar{\mathcal{D}}_{\dot{A}}\Lambda = 0 = \mathcal{D}_A\Lambda^\dagger$ and the facts that[5]

$$\bar{\mathcal{D}}_{\dot{B}}\bar{\mathcal{D}}^{\dot{B}}\,\mathcal{D}_A\Lambda = \bar{\mathcal{D}}_{\dot{B}}\big[\bar{\mathcal{D}}^{\dot{B}},\mathcal{D}_A\big]_+\Lambda = \big[\bar{\mathcal{D}}^{\dot{B}},\mathcal{D}_A\big]_+\bar{\mathcal{D}}_{\dot{B}}\Lambda = 0\,,$$
$$\mathcal{D}^B\mathcal{D}_B\,\bar{\mathcal{D}}_{\dot{A}}\Lambda^\dagger = \mathcal{D}^B\big[\mathcal{D}_B,\bar{\mathcal{D}}_{\dot{A}}\big]_+\Lambda^\dagger = \big[\mathcal{D}_B,\bar{\mathcal{D}}_{\dot{A}}\big]_+\mathcal{D}^B\Lambda^\dagger = 0\,.$$

- $\bar{\mathcal{D}}_{\dot{A}}\overline{W}^{\dot{A}} = \mathcal{D}^A W_A$, which can be derived immediately on applying the covariant chiral derivatives to (4.39) and using (4.11k).

Since $W_A(\overline{W}_{\dot{A}})$ is a left- (right-) chiral superfield, it is convenient to use the variable $y^\mu = x^\mu - i\theta\sigma^\mu\bar{\theta}$ ($\bar{y}^\mu = x^\mu + i\theta\sigma^\mu\bar{\theta}$) to calculate its component expansion. Moreover, since either of them is supergauge invariant, we need work only in the Wess-Zumino gauge. We take (4.34), substitute $x^\mu = y^\mu + i\theta\sigma^\mu\bar{\theta} = \bar{y}^\mu - i\theta\bar{\sigma}^\mu\theta$, Taylor-expand and make use of (3.19k) to get

$$V_{WZ}(y,\theta,\bar{\theta}) = \theta\sigma^\mu\bar{\theta}A_\mu(y) + \theta\theta\,\bar{\theta}\bar{\lambda}(y) + \bar{\theta}\bar{\theta}\,\theta\lambda(y) + \frac{1}{2}\theta\theta\,\bar{\theta}\bar{\theta}\left\{D(y) + i\partial^{(y)}_\mu A^\mu(y)\right\},\quad (4.40a)$$

$$V_{WZ}(\bar{y},\theta,\bar{\theta}) = \theta\sigma^\mu\bar{\theta}A_\mu(\bar{y}) + \theta\theta\,\bar{\theta}\bar{\lambda}(\bar{y}) + \bar{\theta}\bar{\theta}\,\theta\lambda(\bar{y}) + \frac{1}{2}\theta\theta\,\bar{\theta}\bar{\theta}\left\{D(\bar{y}) - i\partial^{(\bar{y})}_\mu A^\mu(\bar{y})\right\},\quad (4.40b)$$

Now we can utilize the forms of the covariant derivatives $\mathcal{D}$ and $\bar{\mathcal{D}}$ written in terms of y and $\bar{y}$ in (4.17) and (4.18) respectively. Thus, employing (4.39a), we can write

$$W_A = -\frac{1}{4}\bar{\mathcal{D}}^{(y)}\bar{\mathcal{D}}^{(y)}\mathcal{D}_A^{(y)}V_{WZ}(y,\theta,\bar{\theta})\,. \tag{4.41}$$

---

[5]From (4.11b), $[\bar{\mathcal{D}}^{\dot{B}},\mathcal{D}_A]_+$ is proportional to $\sigma^\mu\partial_\mu$ which commutes with $\bar{\mathcal{D}}_{\dot{B}}$. Similarly $[\mathcal{D}_B,\bar{\mathcal{D}}_{\dot{A}}]_+$ commutes with $\mathcal{D}^B$.

The evaluation of the component expansion of the RHS of (4.41) is straightforward, though tedious. First, one needs to compute $\mathcal{D}_A^{(y)} V_{WZ}(y,\theta,\bar{\theta})$. The result can be given in terms of the gauge field strength $F_{\mu\nu} \equiv \partial_\mu A_\nu - \partial_\nu A_\mu$.

$$
\begin{aligned}
\mathcal{D}_A^{(y)} V_{WZ}(y,\theta,\bar{\theta}) = {} & \sigma_{A\dot{B}}^\mu \bar{\theta}^{\dot{B}} A_\mu(y) + 2\theta_A \bar{\theta}\bar{\lambda}(y) + \bar{\theta}\bar{\theta}\lambda_A(y) \\
& + \bar{\theta}\bar{\theta}\left\{\delta_A{}^B D(y) - (\sigma^{\mu\nu})_A{}^B F_{\mu\nu}(y)\right\}\theta_B \\
& + i\theta\theta\,\bar{\theta}\bar{\theta}\left\{\sigma^\mu \partial_\mu^{(y)}\bar{\lambda}(y)\right\}_A .
\end{aligned}
$$

The application of $-\frac{1}{4}\bar{\mathcal{D}}^{(y)}\bar{\mathcal{D}}^{(y)}$ leads to

$$
W_A = \lambda_A(y) + D(y)\theta_A - (\sigma^{\mu\nu}\theta)_A F_{\mu\nu}(y) + i\theta\theta\sigma_{A\dot{B}}^\mu \partial_\mu^{(y)}\bar{\lambda}^{\dot{B}}(y). \tag{4.42}
$$

Similar manipulations in the $(\bar{y},\theta,\bar{\theta})$ superspace yield

$$
\overline{W}_{\dot{A}} = \bar{\lambda}_{\dot{A}}(\bar{y}) + D(\bar{y})\bar{\theta}_{\dot{A}} - \epsilon_{\dot{A}\dot{B}}(\bar{\sigma}^{\mu\nu}\bar{\theta})^{\dot{B}} F_{\mu\nu}(\bar{y}) - i\bar{\theta}\bar{\theta}\left\{\partial_\mu^{(\bar{y})}\lambda(\bar{y})\sigma^\mu\right\}_{\dot{A}} . \tag{4.43}
$$

Thus $W_A, \overline{W}_{\dot{A}}$ contain only the component fields λ, D and $F_{\mu\nu}$; each has a total of $4+1+3 = 8$ real components.

$W^A W_A$ also has to be a left chiral superfield in the light of the discussion in §4.2. In fact, one finds after some algebra that

$$
\begin{aligned}
W^A W_A = {} & \lambda(y)\lambda(y) + 2\theta\left\{D(y)\lambda(y) + \sigma^{\mu\nu}\lambda(y)F_{\mu\nu}(y)\right\} \\
& + \theta\theta\left\{D^2(y) + 2i\lambda(y)\sigma^\mu \partial_\mu^{(y)}\bar{\lambda}(y) - \frac{1}{2}F_{\mu\nu}(y)F^{\mu\nu}(y)\right. \\
& \qquad\quad \left. - \frac{i}{2}\tilde{F}_{\mu\nu}(y)F^{\mu\nu}(y)\right\},
\end{aligned}
\tag{4.44}
$$

where $\tilde{F}_{\mu\nu}$ is the dual field strength $\frac{1}{2}\epsilon_{\mu\nu\rho\sigma}F^{\rho\sigma}$. It follows similarly that

$$
\begin{aligned}
\overline{W}_{\dot{A}}\overline{W}^{\dot{A}} = {} & \bar{\lambda}(\bar{y})\bar{\lambda}(\bar{y}) + \left\{2D(\bar{y})\bar{\lambda}(\bar{y}) + 2\bar{\lambda}(\bar{y})\bar{\sigma}^{\mu\nu}F_{\mu\nu}(\bar{y})\right\}\bar{\theta} \\
& + \bar{\theta}\bar{\theta}\left\{D^2(\bar{y}) - 2i\partial_\mu^{(\bar{y})}\lambda(\bar{y})\sigma^\mu\bar{\lambda}(\bar{y}) - \frac{1}{2}F_{\mu\nu}(\bar{y})F^{\mu\nu}(\bar{y})\right. \\
& \qquad\quad \left. + \frac{i}{2}\tilde{F}_{\mu\nu}(\bar{y})F^{\mu\nu}(\bar{y})\right\}.
\end{aligned}
\tag{4.45}
$$

An important and interesting consequence of (4.44) and (4.45) is the following relation:

$$
\frac{1}{4}\left[W^A W_A + \overline{W}_{\dot{A}}\overline{W}^{\dot{A}}\right]_F = \frac{1}{2}D^2(x) - \frac{1}{4}F_{\mu\nu}(x)F^{\mu\nu}(x) + i\lambda(x)\sigma^\mu[\partial_\mu]\bar{\lambda}(x) . \tag{4.46}
$$

Here $[\partial_\mu]$ is as defined after (4.28) and in the last step we have utilized the evident fact that, in the F-terms of $W^A W_A$ $(\overline{W}_{\dot{A}}\overline{W}^{\dot{A}})$, the coordinates y^μ $(\bar{y}^\mu)$ can be replaced by x^μ. Since the last two RHS terms stand respectively for the kinetic energy (in a Lagrangian density) of an Abelian gauge boson and of a Majorana fermion (gaugino), (4.46) can serve as the kinetic energy term for a vector superfield. We conclude by noting an interesting identity, namely

$$\int d^4x (W^A \mathcal{D}_A V + \overline{W}_{\dot{A}} \bar{\mathcal{D}}^{\dot{A}} V)_D = \int d^4x (W^A W_A + \overline{W}_{\dot{A}} \overline{W}^{\dot{A}})_F \ . \tag{4.47}$$

The integrands in (4.47) are equal modulo surface terms.

Before ending this section, we want to comment on the quantization of various classical superfields that we have introduced above. One can construct a supersymmetric quantum field theory in component language by postulating canonical (anti-) commutation relations for the physical (fermionic) bosonic component fields. This is most easily accomplished after the auxiliary fields have been removed through their equations of constraint. Alternatively, one could try to construct quantum superfields 'from scratch'. The starting point of such a construction would be to define an infinitesimal supersymmetry transformation of a quantum superfield $\mathcal{F}$ **not** by $\delta\mathcal{F} \equiv i(\epsilon Q + \bar{\epsilon}\bar{Q})\mathcal{F}$ but rather by [4.4]

$$\delta\mathcal{F} \equiv i[\epsilon Q + \bar{\epsilon}\bar{Q}, \mathcal{F}] \ .$$

In view of the well-known problems, with which the canonical quantization of gauge theories is beset, we shall not pursue this approach any further in this book. Instead, in Ch.6, we shall discuss the extension of the functional integral quantization formalism to classical superfields and show how it leads to the notion of supergraphs.

4.5 Matter Parity and R-parity

In §3.1 we already introduced R-symmetry as a global $U(1)$-invariance of the supersymmetry algebra. Under phase rotations of the Grassmann coordinates $\theta \to e^{i\varphi}\theta$ and $\bar{\theta} \to e^{-i\varphi}\bar{\theta}$, from (3.26f, 3.26g) one has that

$$Q_A \to e^{i\varphi R}Q_A e^{-i\varphi R} = e^{-i\varphi}Q_A \ ,$$

$$\bar{Q}_{\dot{A}} \to e^{i\varphi R}\bar{Q}^{\dot{A}} e^{-i\varphi R} = e^{i\varphi}\bar{Q}^{\dot{A}}.$$

This means that θ, $\bar{\theta}$, Q and $\bar{Q}$ have the R-charges $1, -1, -1$ and 1 respectively. We can define an R-transformation on a general left chiral superfield as $\Phi \to \Phi'$, where

$$\Phi'(x, e^{i\varphi}\theta, e^{-i\varphi}\bar{\theta}) = e^{i\varphi R_\Phi}\Phi(x, \theta, \bar{\theta}) \ , \tag{4.48}$$

i.e. the R-charge of Φ is R_Φ. Correspondingly, $\Phi^\dagger \to \Phi'^\dagger$ where

$$\Phi'^\dagger(x, e^{i\varphi}\theta, e^{-i\varphi}\bar{\theta}) = e^{-i\varphi R_\Phi}\Phi^\dagger(x, \theta, \bar{\theta}) \ , \tag{4.49}$$

so that the R-charge of $\Phi^\dagger$ is $- R_\Phi$. The consistency of the R-transformations of both sides of (4.19) would moreover dictate the following R-charges for the component fields:

$$R(\phi) = R_\Phi \ , \tag{4.50a}$$

$$R(\xi) = -R(\bar{\xi}) = R_\Phi - 1 \, , \tag{4.50b}$$

$$R(F) = R_\Phi - 2 \, . \tag{4.50c}$$

For products of chiral superfields, the R-charges will add algebraically. Turning to a vector superfield, we can see that its reality requires a vanishing R-charge, i.e. $R(V) = 0$. A comparison with (4.34) makes it evident that

$$R(A_\mu) = 0 \, , \tag{4.51a}$$

$$R(\lambda) = -R(\bar{\lambda}) = 1 \, , \tag{4.51b}$$

$$R(D) = 0 \, . \tag{4.51c}$$

Eqs. (4.51) are true in any gauge, as can be checked by comparing with (4.30). In a general gauge [cf. (4.32)] we have the additional R-charge values

$$\begin{aligned} R(C) &= 0 \, , \\ R(\xi) &= -R(\bar{\xi}) = -1 \, , \\ R(M) &= -R(M^\star) = -2 \, . \end{aligned}$$

There has been much speculation in the literature on whether R-invariance can be a symmetry of nature. This possibility arises since the kinetic energy term $\int d^8 z \Phi_i^\dagger \Phi_i$ (cf. Ch.5) in the action for a system of interacting chiral superfields Φ_i is invariant under $U(1)_R$. So are gauge interaction terms $\int d^8 z \mathrm{Tr}(\Phi^\dagger e^V \Phi)$, discussed in Ch.5. One could therefore speculate that the Yukawa and other interaction terms in the action might also be R-invariant. Indeed, the requirement of R-invariance restricts the form of such interaction terms. It needs to be pointed out that, even with an R-invariant classical action, $U(1)_R$ could **not** be an exact global symmetry of the space of supersymmetric states. This is because the conservation of the corresponding Noether current, though classically valid, would be destroyed [4.5] in a gauge theory by quantum anomalies. Moreover, gaugino Majorana mass terms violate such a continuous symmetry and it is known from negative searches at LEP/TEVATRON that electroweak/strong gauginos are massive. Thus $U(1)_R$ has to be abandoned as a symmetry, but its discrete Z_2 subgroup (for $\varphi = \pi$) could be retained[6]. The latter has been conjectured [4.6] to be an exact symmetry in the multiplicative sense in the minimal supersymmetric extension of the Standard Model. Such a symmetry allows a nonzero gaugino mass term.

The value of $e^{i\pi R} = (-1)^R$ for any superfield is called its **matter parity** (M_p), while that of each component field is called the **R-parity** (R_p) of that field. Thus R_p is **always positive** for a vector field, whereas the R-parity of its fermionic partner is **necessarily negative**. For a chiral superfield Φ, R_Φ is conventionally chosen to be either ± 1 or 0. This respectively specifies whether $\Phi \to -\Phi$ or $\Phi \to \Phi$ under $\theta \to -\theta$. The corresponding scalar component then has either negative or positive R-parity while R_p for its fermionic partner is correspondingly positive or negative. These two options (i.e. $R_\Phi = \pm 1$ or 0) pertain to the

---

[6]A discrete symmetry, which is the invariant remnant of some gauge symmetry that gets destroyed by anomalies, is called a *discrete gauge symmetry*. It has been argued [4.7] that quantum gravity effects destroy not only all global symmetries but also all discrete symmetries except those that are discrete gauge symmetries.

cases when Φ is matterlike or quantalike. Here we are calling a chiral superfield **matterlike** if its fermionic component can be identified as a known elementary particle such as a quark or lepton; the corresponding scalar will then be a sparticle such as a squark or slepton. In contrast, we call a chiral superfield **quantalike** if its scalar component is identifiable as an elementary particle like a Higgs boson; now the corresponding fermion is a sparticle like a higgsino. In this way, *particles always turn out to have positive R-parity whereas the same for a sparticle is always negative* (see Table 4.1). This property of R_Φ can be ensured in the Standard Model by identifying it with $3(B - L)$ where B = baryon number and L = lepton number associated with Φ. Thus we can say that the matter parity M_p of a superfield is $(-1)^{3(B-L)}$. One is then led, with S being the spin, to the following formulae [4.8]

Supermultiplet	Name	Spin	R_p	Superfield nature
Particle	Quark q	1/2	+	Chiral, matterlike,
Sparticle	Squark $\tilde{q}$	0	−	odd matter parity
Particle	Lepton l	1/2	+	Chiral, matterlike,
Sparticle	Slepton $\tilde{l}$	0	−	odd matter parity
Particle	Higgs h	0	+	Chiral, quantalike,
Sparticle	Higgsino $\tilde{h}$	1/2	−	even matter parity
Particle	Gauge boson g	1	+	Vector
Sparticle	Gaugino $\tilde{g}$	1/2	−	

Table 4.1. Information on four classes of supermultiplets

$$M_p = (-1)^{3(B-L)} \, , \tag{4.52a}$$

$$R_p = (-1)^{3(B-L)+2S} \, , \tag{4.52b}$$

the latter being for the R-parity of any Standard Model particle and its sparticle partner(s) in a supermultiplet. Since L is always zero or an integer, $3(B-L)$ in the RHS of (4.52) can also be replaced by $3B - L$ or $3B + L$. Note that the R-parity of a state containing several particles and sparticles is the product of the individual R-parities. Evidently, for the particles and sparticles under discussion, matter parity conservation and R_p conservation in any interaction vertex (that conserves angular momentum) are the same thing since the product of $(-1)^{2S}$ factors for all the particles involved in the vertex must equal $+1$. In case R_p is exactly conserved, a starting assumption of the Minimal Supersymmetric Standard Model (Ch.8), the *Lightest Supersymmetric Particle* (**LSP**), being the lightest R-odd state in nature, has to be stable. As will be elaborated in Ch.16, this has rather fundamental implications for cosmology in that the LSP becomes a very attractive candidate for cold dark matter, supposedly pervading the Universe. An additional consequence of R_p invariance is that sparticle production processes must produce them in even numbers (usually a pair) while every sparticle other than the LSP will eventually decay into particles plus an odd number of LSPs (usually one). Each LSP, being weakly interacting like neutrinos, will escape detection, leading to a mismatch in the total momentum. Such a situation in particle collision at high energies will lead to spectacular effects in the laboratory like events with a large missing transverse energy or $\not{E}_T$, as will be explored in Ch.15.

R-parity need not be an exact symmetry of nature, though such may indeed turn out to be the case. This question has been investigated [4.9] in connection with high scale physics. The issue is whether or not an exact discrete symmetry can survive at low energies from of the discrete symmetries of the manifold on which a higher dimensional fundamental theory, such as superstring theory, needs to be compactified or as the protected subgroup of a gauge symmetry destroyed [4.10] by gravitational anomalies. If so, can this discrete symmetry be R_p or is it something like $(-1)^{3B}$? These investigations have been inconclusive [4.9, 4.10] so far – the answer depending on the specific manifold chosen. There is also a rich phenomenology of models with R_p violating interactions about which we shall have more to say in Ch.14.

We summarize this discussion of R-parity as follows.

- R_p is a discrete Z_2 symmetry which equals $(-1)^{3(B-L)+2S} = (-1)^{3B-L+2S}$ for Standard Model particles and their superpartners and is positive for particles but negative for sparticles; it is related to the matter parity $M_p = (-1)^R$ of the corresponding superfield.

- The matter parity of a vector superfield V is always positive. For a chiral superfield Φ it is positive or negative depending on whether Φ is quantalike or matterlike.

- R_p has been conjectured to be an exact symmetry of nature in which case the LSP is absolutely stable. This fact has deep cosmological and spectacular laboratory consequences. However, there are models with R_p violation.

References

[4.1] D.V. Volkov and V.P. Akulov, Phys. Lett. **B46** (1973) 109. S. Samuel and J. Wess, Nucl. Phys. **B226** (1983) 289; **B233** (1984) 488. S. Deser and B. Zumino, Phys. Rev. Lett. **38** (1977) 1433.

[4.2] A. Salam and J. Strathdee, Nucl. Phys. **B76** (1974) 477; Phys. Rev. **D11** (1975) 1521.

[4.3] J. Wess and B. Zumino, Nucl. Phys. **B78** (1974) 1. J. Wess and J. Bagger, *op. cit.*, *Bibl.* P.P. Srivistava, *op. cit.*, *Bibl.*

[4.4] S. Weinberg #3, *op. cit.*, *Bibl.*

[4.5] G.R. Farrar and S. Weinberg, Phys. Rev. **D27** (1983) 2732.

[4.6] A. Salam and J. Strathdee, Nucl. Phys. **B87** (1975) 85. P. Fayet, Nucl. Phys. **B90** (1975) 104. G.R. Farrar and P. Fayet, Phys. Lett. **B76** (1978) 575.

[4.7] L.M. Krauss and F. Wilczek, Phys. Rev. Lett. **62** (1989) 1221. T. Banks, Nucl. Phys. **B323** (1989) 90.

[4.8] S. Weinberg, Phys. Rev. **D26** (1982) 287. N. Sakai and T. Yanagida, Nucl. Phys. **B197** (1982) 533.

[4.9] M.C. Bento, L.J. Hall and G.G. Ross Nucl. Phys. **B292** (1987) 400. L.E. Ibáñez and G.G. Ross, Phys. Lett. **B260** (1991) 291. L.E. Ibáñez, Nucl. Phys. **B368** (1992) 3. A.E. Faraggi, Phys. Lett. **B398** (1995) 97.

[4.10] T. Banks and M. Dine, Phys. Rev. **D45** (1992) 424.

Chapter 5

INTERACTING SUPERFIELDS

5.1 System of Interacting Chiral Superfields

For a supersymmetric action, the Lagrangian density – describing a set of interacting super-fields – can at most change under a supersymmetry transformation into itself plus a total spacetime derivative. As explained in §4.1, that is precisely how the highest (D-) component of a general superfield transforms. So does the F-component of a chiral superfield. The construction of a supersymmetric Lagrangian density is thus a straightforward exercise. The D-term needs to be isolated from a product of superfields that contains such a component. If some products are chiral superfields, their F-terms are also eligible.

In general, a renormalizable Lagrangian density admits only terms with mass dimension not more than four [5.1]. A general renormalizable supersymmetric Lorentz invariant La-grangian density, involving only polynomials of left chiral superfields Φ_i (i referring to the type with repeated indices being summed), is therefore

$$\mathcal{L} = \left[\Phi_i^\dagger \Phi_i\right]_D + [\mathcal{W}(\Phi_i) + h.c.]_F , \qquad (5.1a)$$

$$\mathcal{W}(\Phi_i) = h_i \Phi_i + \frac{1}{2} m_{ij} \Phi_i \Phi_j + \frac{1}{3!} f_{ijk} \Phi_i \Phi_j \Phi_k . \qquad (5.1b)$$

The first term in the RHS of (5.1a) is the D-term of a vector superfield, namely $\Phi_i^\dagger \Phi_i$. The second is the F-term of a chiral superfield (plus its hermitian conjugate) defined as a polyno-mial of Φ_i (upto cubic terms so as to retain renormalizability) and called the *superpotential* $\mathcal{W}$. The couplings m_{ij} and f_{ijk} are symmetric in their indices. Though the explicit form of $\mathcal{W}$ appears in (5.1b), we shall try to keep a general $\mathcal{W}(\Phi_i)$ in what follows. Invariance under supersymmetry dictates that $\mathcal{W}$ be an analytic function of Φ_i; in particular, it cannot involve $\Phi_i^\dagger$. Similarly, $\mathcal{W}^\dagger$ involves the right chiral $\Phi_i^\dagger$ and not Φ_i. This property of the superpotential in any interacting chiral superfield system is known as **holomorphy**. Note that (5.1a) can be understood with either y or x (cf. §4.3) in the argument of each Φ since it is invariant[1] under the map $x \to y$ or $y \to x$. In other words, $\int d^6z\, \Phi \equiv \int d^4y\, d^2\theta\, \Phi$ and this integral has no $\bar\theta$-dependence. Similarly, $\int d^6\bar z\, \Phi^\dagger \equiv \int d^4\bar y\, d^2\bar\theta\, \Phi^\dagger$ has no θ-dependence.

---

[1]This is true for any D- or F- term since $\theta\theta G(y) = \theta\theta G(x)$ and $\theta\theta\,\bar\theta\bar\theta H(y) = \theta\theta\,\bar\theta\bar\theta H(x)$ $\forall$ functions G and H.

We only need to remember that $\Phi(x,\theta,\bar\theta) = \Phi(y,\theta)$ and $\Phi^\dagger(x,\theta,\bar\theta) = \Phi^\dagger(\bar y,\bar\theta)$. Finally, the supersymmetric action is

$$S = \int d^8 z \left[\Phi_i^\dagger \Phi_i + \mathcal{W}(\Phi_i)\delta^{(2)}(\bar\theta) + \mathcal{W}^\dagger(\Phi_i)\delta^{(2)}(\theta) \right].$$

We shall find it notationally convenient to use [5.2] two subspaces ϕ_i and $\bar\phi^i = \phi_i^\star$ for the complex scalar field and define:

$$\mathcal{W}_i(\phi) \equiv \left. \frac{\partial \mathcal{W}}{\partial \Phi_i} \right|, \tag{5.2a}$$

$$\overline{\mathcal{W}}^i(\bar\phi) \equiv \left. \frac{\partial \mathcal{W}^\dagger}{\partial \Phi_i^\dagger} \right|, \tag{5.2b}$$

$$\mathcal{W}_{ij}(\phi) \equiv \left. \frac{\partial^2 \mathcal{W}}{\partial \Phi_i \partial \Phi_j} \right|, \tag{5.2c}$$

$$\overline{\mathcal{W}}^{ij}(\bar\phi) \equiv \left. \frac{\partial^2 \mathcal{W}^\dagger}{\partial \Phi_i^\dagger \partial \Phi_j^\dagger} \right|, \tag{5.2d}$$

$$\mathcal{W}_{ijk}(\phi) \equiv \left. \frac{\partial^3 \mathcal{W}}{\partial \Phi_i \partial \Phi_j \partial \Phi_k} \right|, \tag{5.2e}$$

$$\overline{\mathcal{W}}^{ijk}(\bar\phi) \equiv \left. \frac{\partial^3 \mathcal{W}^\dagger}{\partial \Phi_i^\dagger \partial \Phi_j^\dagger \partial \Phi_k^\dagger} \right|. \tag{5.2f}$$

In each RHS of (5.2) the symbol $|$ on the right is as defined after (4.23).

The superfield products in the RHS of (5.1) can be decomposed into their components in the manner shown in §4.4. In particular, one can make use of (4.26), (4.27) and (4.28). Then, ignoring total spacetime derivatives, one can rewrite the Lagrangian density in terms of component fields as

$$\begin{aligned}
\mathcal{L} =\ & i\xi_i \sigma^\mu [\partial_\mu] \bar\xi_i + \partial^\mu \phi_i^\star \partial_\mu \phi_i + F_i^\star F_i \\
& + \left[\left(h_k + m_{ik}\phi_i + \frac{1}{2} f_{ijk}\phi_i\phi_j \right) F_k - \frac{1}{2}\xi_i\xi_j \left(m_{ij} + f_{ijk}\phi_k \right) + h.c. \right].
\end{aligned} \tag{5.3}$$

(5.3) makes an "off-shell" description of the system since it includes the auxiliary fields F_i which do not possess kinetic energy terms. It is an instructive but straightforward exercise to see that $\int d^4 x \mathcal{L}(x)$, constructed from the above, is invariant under infinitesimal supersymmetry transformations (4.25) applied to each field of type i. Indeed, the usual Noether procedure goes through. This means that one can define a conserved spinorial supercurrent density $K_{\mu A}$ and its conjugate $\bar K_\mu^A$ such that[2]

$$\epsilon K^\mu + \bar\epsilon \bar K^\mu = \sum_X \delta X \frac{\partial \mathcal{L}}{\partial(\partial_\mu X)} - L^\mu ,$$

---

[2] In precise terms, $K_A^\mu = -\sqrt{2}(\sigma^\nu \bar\sigma^\mu \xi_i)_A \partial_\nu \bar\phi^i - i\sqrt{2}(\sigma^\mu \bar\xi_i)_A \overline{\mathcal{W}}^i(\bar\phi)$ for the system of (5.3) .

with the summation covering all fields X present and $\delta\mathcal{L} = \partial^\mu L_\mu$ being a total derivative. Finally, the supercharge $Q_A = \int d^3x \, K_{0A}$ and its conjugate $\bar{Q}^{\dot{A}} = \int d^3x \, \bar{K}_0^{\dot{A}}$, constructed from these currents, can be checked to consistently obey

$$\delta X = i(\epsilon Q + \bar{\epsilon}\bar{Q})X \ .$$

The verification of these statements is left as an exercise for the reader. Let us just remark that the concept of a four spinor supercurrent density

$$K_a^\mu = \begin{pmatrix} K_A^\mu \\ \bar{K}^{\mu\dot{A}} \end{pmatrix}$$

goes beyond this particular set of interacting chiral superfields and is generally valid for any supersymmetric system.

Returning to (5.3), the auxiliary fields can be eliminated by using the equations of constraint $0 = \partial\mathcal{L}/\partial F_i^\star$ and $0 = \partial\mathcal{L}/\partial F_i$ which respectively yield

$$F_i = -\overline{\mathcal{W}}^i(\bar{\phi}) = -h_i^\star - m_{ij}^\star \bar{\phi}^j - \frac{1}{2}f_{ijk}^\star \bar{\phi}^j \bar{\phi}^k \ , \tag{5.4a}$$

$$F_i^\star = -\mathcal{W}_i(\phi) = -h_i - m_{ij}\phi_j - \frac{1}{2}f_{ijk}\phi_j\phi_k \ . \tag{5.4b}$$

Utilizing (5.4), (5.3) can be recast in the "on-shell" form, i.e. purely in terms of propagating dynamical fields, as

$$\mathcal{L} = i\xi_i\sigma^\mu[\partial_\mu]\bar{\xi}_i + \partial^\mu\phi_i^\star\partial_\mu\phi_i - \left(\frac{1}{2}\xi_i\xi_j\mathcal{W}_{ij}(\phi) + h.c.\right) - V(\phi_i,\phi_j^\star) \ , \tag{5.5a}$$

$$V(\phi_i,\phi_j^\star) = \mathcal{W}_i(\phi)\overline{\mathcal{W}}^i(\bar{\phi}) = F_i^\star F_i \ . \tag{5.5b}$$

Thus $\mathcal{W}_{ij}(\phi)$, the double superfield derivative of the superpotential evaluated with nonzero scalar fields only, contains **both the fermion mass terms and the Yukawa interaction terms**. Indeed, one can write

$$\mathcal{L}_{\text{Yukawa}} = -\frac{1}{2}\left(\xi_i\xi_j\frac{\partial^2\mathcal{W}}{\partial\Phi_i\partial\Phi_j}\bigg| + h.c.\right) \ .$$

Turning to the scalars, we find that the potential $V(\phi_i,\phi_j^\star)$ is the sum of the modulus squares of all the auxiliary fields. If they exist, configurations with $F_i = 0 \ \forall i$ correspond to absolute minima of the potential. Such configurations ($\phi_i = \langle\phi_i\rangle$) define the supersymmetric ground state, cf. §2.2. If no such solution exists, supersymmetry is broken spontaneously – as will be discussed in §7.4.

We will now discuss the mass terms in the Lagrangian density (5.5a) with $\phi_i = \langle\phi_i\rangle$ and develop a formalism in which squared masses of the bosons and fermions get related to various scalar derivatives of the superpotential. Of course, for exact supersymmetry (as discussed so far), all supermultiplet members are degenerate in mass. However, this degeneracy gets split in any realistic broken supersymmetric case. Later, when we come to that, this formalism will prove useful. The coefficient of $-\frac{1}{2}\xi_i\xi_j$ in $\mathcal{L}$, namely $\mathcal{W}_{ij}(\langle\phi\rangle)$, measured in these configurations, yields the fermion mass matrix:

$$m_{Fij} = \mathcal{W}_{ij}(\langle\phi\rangle)\ . \tag{5.6}$$

(5.6) follows from the fermionic equations of motion which are

$$i\sigma^\mu\partial_\mu\bar{\xi}_i - \mathcal{W}_{ij}(\phi)\xi_j = 0\ ,$$

$$i\bar{\sigma}^\mu\partial_\mu\xi_i - \overline{\mathcal{W}}^{ij}(\bar{\phi})\bar{\xi}_j = 0\ .$$

In discussing scalar masses, we can employ the notation

$$\frac{\partial^2 V}{\partial\phi_i\partial\phi_j} \equiv V_{ij}\ , \tag{5.7a}$$

$$\frac{\partial^2 V}{\partial\bar{\phi}^i\partial\phi_j} \equiv V^i{}_j\ , \tag{5.7b}$$

$$\frac{\partial^2 V}{\partial\phi_i\partial\bar{\phi}^j} \equiv V_i{}^j\ , \tag{5.7c}$$

$$\frac{\partial^2 V}{\partial\bar{\phi}^i\partial\bar{\phi}^j} \equiv V^{ij}\ . \tag{5.7d}$$

The scalar mass terms (SMT) in the Lagrangian density can then be written as

$$\mathcal{L}_{SMT} = -\frac{1}{2}(\bar{\phi}^i\ \phi_i)\,\underset{\sim}{m}{}^2_{Sij}\begin{pmatrix}\phi_j\\\bar{\phi}^j\end{pmatrix}\ , \tag{5.8a}$$

$$\underset{\sim}{m}{}^2_{Sij} = \begin{pmatrix}V^i{}_j & V^{ij}\\V_{ij} & V_i{}^j\end{pmatrix}_{\phi=\langle\phi\rangle}\ . \tag{5.8b}$$

Employing (5.5b) and (5.6) in (5.8b), we have the sum of squared masses of all scalars $\sum m_0^2$ and that of all fermions $\sum m_{1/2}^2$ in the chiral supermultiplets related by

$$\sum m_0^2 = \text{Tr}\ \underset{\sim}{m}{}^2_{Sii} = V^i_i(\langle\phi\rangle) + V_i^i(\langle\phi\rangle) = \overline{\mathcal{W}}^{ik}(\langle\bar{\phi}\rangle)\mathcal{W}_{ki}(\langle\phi\rangle) + \mathcal{W}_{ik}(\langle\phi\rangle)\overline{\mathcal{W}}^{ki}(\langle\bar{\phi}\rangle)$$

$$= \text{Tr}[(m_F^\dagger m_F) + (m_F m_F^\dagger)] = 2\sum m_{1/2}^2\ . \tag{5.9}$$

Thus follows the **supertrace mass sum rule** [5.3]

$$S\text{Tr}\ \underset{\sim}{m}{}^2 \equiv \sum_{J=0}^{1/2}(-1)^{2J}(2J+1)m_J^2 = 0\ . \tag{5.10}$$

In other words, the spin weighted graded trace of the squared mass matrix, taken over all chiral supermultiplets vanishes. In fact, we will derive below a stronger result, namely that (5.10) is true over each chiral supermultiplet.

In detail, the diagonal blocks of $m\,^2_{\sim Sij}$ are given by

$$(V^i_{\ j})_{\phi=\langle\phi\rangle} = (m^\dagger_F m_F)_{ij}, \ (V_i^{\ j})_{\phi=\langle\phi\rangle} = (m_F m^\dagger_F)_{ij} \ ,$$

while the off-diagonal elements can be written as

$$V_{ij}(\langle\phi\rangle) = \mathcal{W}_{ijk}(\langle\phi\rangle)\overline{\mathcal{W}}^k(\langle\bar\phi\rangle), \ V^{ij}(\langle\phi\rangle^\star) = \overline{\mathcal{W}}^{ijk}(\langle\bar\phi\rangle)\mathcal{W}_k(\langle\phi\rangle) \ .$$

For a supersymmetric ground state, V is at its absolute minimum with a vanishing $\langle F_k\rangle$, i.e. $\mathcal{W}_k(\langle\phi\rangle) = 0 = \overline{\mathcal{W}}^k(\langle\bar\phi\rangle)$. In this configuration the off-diagonal blocks in $m\,^2_{\sim Sij}$ vanish. In fact, we have

$$m\,^2_{\sim Sij} = \begin{pmatrix} (m^\dagger_F m_F)_{ij} & 0 \\ 0 & (m_F m^\dagger_F)_{ij} \end{pmatrix},$$

so that (5.10) is indeed true for mass eigenstates, (chiral) supermultiplet by supermultiplet. Indeed, at the moment, this is just the evident supersymmetric fact that fermions and bosons in each chiral supermultiplet are degenerate in mass. Strictly, in this situation, the content of the sum rule (5.10) is empty[3]. However, even if $\langle F_k\rangle \neq 0$ (as happens with supersymmetry broken spontaneously, à la O'Raifeartaigh, Ch.7) and those off-diagonal elements in $m\,^2_{\sim Sij}$ are nonvanishing, the supertrace sum rule (5.10) will be shown to be still valid. We have given a somewhat elaborate discussion of scalar and fermion masses here to lay the foundation for the derivation of that result, which will come in Ch.7.

Wess-Zumino model

A celebrated special case is the **Wess-Zumino model** [5.4] which consists of only one chiral superfield Φ. We impose time reversal invariance so that all couplings are real, but do not insist on the invariance of matter parity (cf. §4.5). For this specific case, the action can be written as

$$S = \int d^8 z \left[\Phi^\dagger\Phi \ + \ \left(h\Phi + \frac{1}{2}m\Phi^2 + \frac{f}{3!}\Phi^3 \right) \delta^{(2)}(\bar\theta) \right.$$

$$\left. + \ \left(h\Phi^\dagger + \frac{1}{2}m\Phi^{\dagger 2} + \frac{f}{3!}\Phi^{\dagger 3} \right) \delta^{(2)}(\theta) \right].$$

The corresponding "on-shell" Lagrangian density in component form is

$$\mathcal{L} = i\xi\sigma^\mu[\partial_\mu]\bar\xi - \phi^\star\partial^2\phi \ - \ \frac{1}{2}m(\xi\xi + \bar\xi\bar\xi)$$

$$- \ \frac{f}{2}(\bar\xi\bar\xi\phi^\star + \xi\xi\phi) - |h + m\phi + \frac{f}{2}\phi^2|^2. \tag{5.11}$$

---

[3]The relative factor of 2 in front of the fermionic contribution to (5.10) comes from the following fact: whereas ϕ_i and ϕ^i are counted as two separate scalar states, ξ_i is counted as a single two component fermion state. It is important to realize that this way of counting does not correspond to the number of degrees of freedom described by those fields.

This Lagrangian density makes it evident that the Wess-Zumino model is the supersymmetric generalization of the theory of a massive scalar field with upto quartic self interactions.

The equations of motion following from (5.11) are

$$i\sigma^\mu \partial_\mu \bar{\xi} - m\xi = f\xi\phi \ , \tag{5.12a}$$

$$i\bar{\sigma}^\mu \partial_\mu \xi - m\bar{\xi} = f\bar{\xi}\phi^\star \ , \tag{5.12b}$$

$$(\partial^2 + m^2)\phi = -mh - \frac{f}{2}(\bar{\xi}\bar{\xi} + 2m|\phi|^2 + m\phi^2 + f|\phi|^2\phi^\star + 2h\phi^\star) \ . \tag{5.12c}$$

In the limit of vanishing h and f, (5.12a) and (5.12b) combine to yield, cf §3.1 and (3.6) $f\!f$, the free Majorana equation in the Weyl representation while (5.12c) yields the free Klein-Gordon equation respectively:

$$(i\gamma^\mu \partial_\mu - m)\lambda_M = 0 \ , \ \lambda_M = \begin{pmatrix} \xi \\ \bar{\xi}^T \end{pmatrix} \ , \tag{5.13a}$$

$$(\partial^2 + m^2)\phi = 0 \ . \tag{5.13b}$$

Eqs. (5.13) describe free fields with identical masses corresponding to a Majorana spin half fermion and two spinless bosons which make up (cf §3.3) the on-shell **Wess-Zumino supermultiplet**. Of the spinless bosons, the linear combination $(\phi + \phi^\star)/\sqrt{2}$ corresponds to the scalar while the orthogonal $(\phi - \phi^\star)/i\sqrt{2}$ stands for the pseudoscalar. With two Wess-Zumino superfields, Φ_+ and Φ_-, say, one could make a Dirac fermion field ψ with left chiral component $\psi_L = \xi_+$ and right chiral component $\psi_R = \bar{\xi}_-$ as well as complex left and right chiral sfermion scalar fields ϕ_+ and $\phi_-^\star$ respectively. This will become clearer in §5.3 when we take up SQED. Even in the interacting case, we can rewrite (5.11) as

$$\mathcal{L} = \partial_\mu \phi^\star \partial^\mu \phi - |h + m\phi + \frac{f}{2}\phi^2|^2 + \frac{i}{2}\bar{\lambda}_M \gamma^\mu \partial_\mu \lambda_M - \frac{1}{2}m\bar{\lambda}_M \lambda_M$$

$$- \frac{f}{4}\bar{\lambda}_M (1 - \gamma_5)\lambda_M \phi - \frac{f}{4}\bar{\lambda}_M (1 + \gamma_5)\lambda_M \phi^\star. \tag{5.14}$$

5.2   Abelian Gauge Interactions

A $U(1)$ gauge transformation of the chiral superfields Φ_i, $\Phi_i^\dagger$ can be defined by

$$\Phi_i' = e^{-2igt_i \Lambda(z)}\Phi_i, \quad \bar{\mathcal{D}}^{\dot{A}}\Lambda = 0 \ , \tag{5.15a}$$

$$\Phi_i'^\dagger = \Phi_i^\dagger e^{2igt_i \Lambda^\dagger(z)}, \quad \mathcal{D}_A\Lambda^\dagger = 0 \ . \tag{5.15b}$$

Here g is the gauge coupling, t_i is a real number which is the $U(1)$ charge of Φ_i, Λ is a complex function of $z = (x, \theta, \bar{\theta})$ specifying the local gauge transformation and there is no sum on i. The requirements of vanishing chiral covariant derivatives have been imposed on Λ and $\Lambda^\dagger$ to maintain the chiral nature of the superfields Φ_i and $\Phi_i^\dagger$ under gauge transformations. A sufficient condition for these is for Λ to be function of y only (cf. 4.16a). Λ and $\Lambda^\dagger$ can be thought of as chiral superfunctions in superspace. *The factor 2 appears in the exponents* *of (5.15) because the imaginary part of the scalar component of the superfield $i\Lambda$ is minus half the usual gauge transformation function introduced in an abelian gauge field theory: $A_\mu \to A_\mu - 2\partial_\mu \, \Im m\phi$, as explained immediately after (4.33).*

Since $\Lambda \neq \Lambda^\dagger$, the kinetic energy D-term in (5.1) is not gauge invariant by itself. This is rectified by introducing a gauge vector superfield V (cf. §4.4) with the gauge transformation

$$V' = V + i(\Lambda - \Lambda^\dagger) \tag{5.16}$$

and by generalizing the said K.E. term to $[\Phi_i^\dagger e^{2gt_i V}\Phi_i]_D$. The introduction of the exponent does not destroy renormalizability, as can be simply seen by going to the Wess-Zumino gauge; from (4.35) $(V_{WZ})^n$ vanishes for $n \geq 3$. In addition, the D-component of V being gauge invariant, we can add that multiplied by a real constant η. We can also introduce, following (4.47), the kinetic energy term for the vector superfield in terms of the gauge invariant supersymmetric field strengths W and $\bar{W}$ of (4.39). The full Lagrangian density for a system of self-interacting chiral superfields Φ_i coupling to an abelian gauge vector superfield V reads

$$\mathcal{L} = \frac{1}{4}\left[W^A W_A + \bar{W}_{\dot{A}}\bar{W}^{\dot{A}}\right]_F + \left[\Phi_i^\dagger e^{2gt_i V}\Phi_i + 2\eta V\right]_D + [\mathcal{W}(\Phi_i) + \text{h.c.}]_F \,, \tag{5.17}$$

where the superpotential $\mathcal{W}(\Phi_i)$ has to respect gauge invariance. The corresponding action is

$$S = \int d^8 z \left(\Phi_i^\dagger e^{2gt_i V}\Phi_i + 2\eta V\right) \;+\; \int \delta^6 z \left\{\frac{1}{4}W^A W_A + \mathcal{W}(\Phi_i)\right\}$$

$$+ \int \delta^6 \bar{z} \left\{\frac{1}{4}\overline{W}_{\dot{A}}\overline{W}^{\dot{A}} + \mathcal{W}^\dagger(\Phi_i^\dagger)\right\}.$$

In component form and with the superpotential of §5.1, the Lagrangian density of (5.17) is given in the WZ gauge by use of (4.36), (4.37), (4.38) and (4.46) to be

$$\mathcal{L} \;=\; \frac{1}{2}D^2 + \eta D - \frac{1}{4}F_{\mu\nu}F^{\mu\nu} + i\lambda\sigma^\mu\partial_\mu\bar{\lambda} + i\xi_i\sigma^\mu\Delta_{i\mu}^\dagger\bar{\xi}_i + |\Delta_{i\mu}\phi_i|^2 - F_i^\star F_i$$

$$-\left(\frac{1}{2}\xi_i\xi_j\mathcal{W}_{ij}(\phi) + \text{h.c.}\right) - \sqrt{2}gt_i(\bar{\lambda}\bar{\xi}_i\phi_i + \text{h.c.}) + gt_i|\phi_i|^2 D \,. \tag{5.18}$$

In (5.18) $\Delta_{i\mu}$ is the gauge covariant derivative $\partial_\mu + igt_i A_\mu$ and we have discarded a total derivative. *We see that the coefficient 2 in the exponents of (5.15) has been taken care of by factors of $1/2$ generated in (4.37) and (4.38).* Moreover, the real field D acts as another auxiliary field in addition to the complex F_i given by (5.4). The corresponding equation of constraint $\partial\mathcal{L}/\partial D = 0$ yields

$$D = -g\phi_i^\star t_i\phi_i - \eta \,. \tag{5.19}$$

(5.11) is now generalized to

$$\mathcal{L} \;=\; i\xi_i\sigma^\mu\Delta_{i\mu}^\dagger\bar{\xi}_i + |\Delta_{i\mu}\phi_i|^2 - \frac{1}{4}F_{\mu\nu}F^{\mu\nu} + i\lambda\sigma^\mu[\partial_\mu]\bar{\lambda}$$

$$-\sqrt{2}gt_i(\bar{\lambda}\bar{\xi}_i\phi_i + \text{h.c.}) - \left(\frac{1}{2}\xi_i\xi_j\mathcal{W}_{ij}(\phi) + \text{h.c.}\right) - V(\phi_i, \phi_j^\star) \,, \tag{5.20a}$$

$$V(\phi_i, \phi_j^\star) = F_i F_i^\star + \frac{1}{2} D^2 \,, \tag{5.20b}$$

where F_i remains $-\dfrac{\partial \mathcal{W}^\dagger}{\partial \Phi_i^\dagger}\Big|$.

If the abelian gauge symmetry group is a product of several $U(1)$ factors, one uses the index a to refer to a particular factor. Now $D^a = -\eta^a - g^a \phi_i^\star (t^a \phi)_i$, a not summed in the second term, whereas the linear D-term in the Lagrangian density is $\eta^a D^a$ with repeated a summed as usual. Moreover, the gauge covariant derivative $\Delta_{i\mu}$ is $\partial_\mu + ig^a(t_i A_\mu)^a$ while the gauge kinetic energy terms and the supersymmetric gaugino-scalar-chiral fermion $(g\text{–}s\text{–}cf)$ interaction respectively are

$$\mathcal{L}_{\text{gauge}} = -\frac{1}{4} F_{\mu\nu}^a F^{\mu\nu a} + i\lambda^a \sigma^\mu [\partial_\mu] \bar{\lambda}^a \,, \tag{5.21a}$$

$$\mathcal{L}_{g\text{–}s\text{–}cf} = -\sqrt{2} g^a \left[(t_i \bar{\lambda})^a \bar{\xi}_i \phi_i + h.c. \right] \,. \tag{5.21b}$$

The scalar potential generalizes to

$$V(\phi_i, \phi_j^\star) = F_i F_i^\star + \frac{1}{2} D^a D^a \,, \tag{5.22a}$$

$$F_i = -\frac{\partial \mathcal{W}^\dagger}{\partial \Phi_i^\dagger}\Big|, \quad D^a = -\eta^a - g^a \phi_i^\star (t^a \phi)_i \,. \tag{5.22b}$$

Our remaining discussion in this section addresses this more general situation. First, though, let us remark on the ground state configuration with $\phi = \langle \phi \rangle$. The absolute minimum of V may yield a supersymmetric vacuum (cf. §2.2) with $\langle F_i \rangle = 0 = \langle D^a \rangle$. $(t^a \langle \phi \rangle)_i$ may not vanish, but any nonzero value of it would imply a spontaneous breakdown of the Abelian $U(1)$ symmetry corresponding to a and a mass (via the Higgs mechanism) for the concerned gauge boson. Let us consider this more general situation and direct our attention to the mass terms of the vector bosons and their superpartners.

We employ the ϕ, $\bar{\phi}$ subspace notation, introduced in §5.1, to investigate the gauge invariance of the superpotential $\mathcal{W}$. Now $D^{ai} \equiv \partial D^a / \partial \bar{\phi}^i = -g^a(t^a \phi)_i$ and $D_i^a \equiv \partial D^a / \partial \phi_i = -g^a(t^a \bar{\phi})^i$, a not summed. The vector-boson mass term (VMT) in the Lagrangian density, referred above, may now be written as

$$\mathcal{L}_{VMT} = \frac{1}{2} m_V^2{}^{ab} A_\mu^a A^{\mu b} \,, \tag{5.23a}$$

$$m_V^2{}^{ab} = (D_i^a D^{bi} + D^{ai} D_i^b)_{\phi = \langle \phi \rangle} \,. \tag{5.23b}$$

Furthermore, considering an infinitesimal gauge transformation of the superpotential, we have

$$\mathcal{W}_j(\phi) D^{aj}(\phi) \equiv -F_j^\star D^{aj} = 0 \,. \tag{5.24}$$

Differentiation with respect to ϕ_i yields

$$\mathcal{W}_{ij}(\phi) D^{aj}(\phi) - F_j^\star D_i^{aj} = 0 \,. \tag{5.25}$$

For a supersymmetric ground state, $\langle F_j \rangle$ vanishes and (5.25) becomes

$$\mathcal{W}_{ij}(\langle \phi \rangle) D^{aj}(\langle \phi \rangle) = 0 \ . \tag{5.26}$$

It follows that, for $\langle \phi \rangle \neq 0$ (otherwise $\langle D^{aj} \rangle$ also vanishes), the matrix

$$(\mathcal{M}^2)^i{}_j \equiv \overline{\mathcal{W}}^{ik}(\langle \bar{\phi} \rangle) \mathcal{W}_{kj}(\langle \phi \rangle) \tag{5.27}$$

has a null eigenvalue on the eigenvector $D^{aj}(\langle \phi \rangle)$; thus det $\mathcal{M}^2 = 0$. A similar statement holds for $\bar{\mathcal{M}}_i^{2\ j} \equiv \mathcal{W}_{ik}(\langle \phi \rangle) \overline{\mathcal{W}}^{kj}(\langle \bar{\phi} \rangle)$.

We now discuss the masses of the members of the matter supermultiplets. Once again, though they are mass degenerate in exact supersymmetry, we wish to develop a formalism like that in §5.1 which can be easily generalized to the case of supersymmetry breaking. Concerning scalar masses, one needs to be somewhat more general. For scalars with abelian gauge interactions, (5.8b) is still true. However, on using (5.22a), one now obtains rather different expressions for elements of the scalar mass squared matrix:

$$\underset{\sim}{m}{}^2{}_{Sij} = \begin{pmatrix} [\overline{\mathcal{W}}\mathcal{W}]^i{}_j + D^{ai} D_j^a & D^{ai} D^{aj} \\ D_i^a D_j^a & [\mathcal{W}\overline{\mathcal{W}}]_i{}^j + D_i^a D^{aj} \end{pmatrix}_{\phi = \langle \phi \rangle} , \tag{5.28}$$

where we have used the supersymmetric vacuum condition that $\mathcal{W}_k(\langle \phi \rangle) = \overline{\mathcal{W}}^k(\langle \bar{\phi} \rangle) = D^a(\langle \phi \rangle) = 0$. Now it is easy to see that (5.28) has a vanishing eigenvalue with eigenvector

$$\begin{pmatrix} D^{aj} \\ -D_j^a \end{pmatrix} ,$$

i.e. for each broken symmetry component (labelled by a), there is a massless scalar which is the Goldstone mode that gets eliminated by the Higgs mechanism.

Turning to fermion masses, (5.6) needs to be generalized to include the mixing between gauginos and chiral fermions. The latter arises from the gaugino-scalar-chiral fermion interaction (5.21b) as soon as the scalar develops a VEV. Identifying the fermion mass terms (FMT) from (5.20a) and (5.21b), we can now write

$$\mathcal{L}_{FMT} = -\sqrt{2} g^a (t^i \lambda)^a \xi_i \langle \bar{\phi}_i \rangle - \frac{1}{2} \xi_i \xi_j \mathcal{W}_{ij}(\langle \phi \rangle) + h.c.$$

$$= -\frac{1}{2} (\xi_i\ \lambda^a)\, \underset{\sim}{m}{}^a_{Fij} \begin{pmatrix} \xi_j \\ \lambda^a \end{pmatrix} + h.c. \ , \tag{5.29}$$

where a is summed and

$$\underset{\sim}{m}{}^a_{Fij} = \begin{pmatrix} \mathcal{W}_{ij} & -\sqrt{2}\ D_i^a \\ -\sqrt{2}\ D_j^a & 0 \end{pmatrix}_{\phi = \langle \phi \rangle} . \tag{5.30}$$

Thus

$$\mathrm{Tr} \left(m_F^{a\dagger} m_F^a \right)^i{}_j = \left[(\overline{\mathcal{W}}\mathcal{W})^i{}_j + 2\, D^{ai} D_{aj} + 2\, D_i^a D^{aj} \right]_{\phi = \langle \phi \rangle} , \tag{5.31a}$$

$$\mathrm{Tr} \left(\underset{\sim}{m}{}^a_F\, \underset{\sim}{m}{}^{a\dagger}_F \right)^j{}_i = \left[(\mathcal{W}\overline{\mathcal{W}})_i^j + 2\, D_i^a D^{aj} + 2\, D^{ai} D_j^a \right]_{\phi = \langle \phi \rangle} . \tag{5.31b}$$

Finally, from (5.23), (5.28) and (5.31), we obtain

$$\text{Tr } \underset{\sim}{m}{}^2_{Sii} - \text{Tr}(\underset{\sim}{m}{}^{a\dagger}_F \, \underset{\sim}{m}{}^a_F + \underset{\sim}{m}{}^a_F \, \underset{\sim}{m}{}^{a\dagger}_F)^i_i + 3 \, m^2_{V}{}^{aa} = 0 \; . \tag{5.32}$$

(5.32) is now the more general statement of the supertrace mass sum rule (5.10), as applied to the supermultiplet containing a massive vector boson. Note that here a chiral supermultiplet and a massless vector supermultiplet combine to form a massive one with members that are mass degenerate in the limit of exact supersymmetry. Once again, the vanishing of the supertrace is not only valid in exact supersymmetry but survives F-type spontaneous supersymmetry breaking, as will be shown in Ch.7. The latter makes (5.32) even more nontrivial.

5.3   Supersymmetric Quantum Electrodynamics (SQED)

The spectrum of the supersymmetric generalization of QED includes the electron which is a four component Dirac fermion. It can be thought of as a complex linear combination of two degenerate Weyl fermion fields. This obliges us to start with two left chiral superfields Φ_+, Φ_- with respective charges (multiplied by the QED coupling strength) $q, -q$ and the gauge transformations

$$\Phi'_+ = e^{-2iq\Lambda(z)}\Phi_+ \; , \quad \Phi'_- = e^{+2iq\Lambda(z)}\Phi_- \; . \tag{5.33}$$

Here Φ_+ contains the left chiral component of the fermion field and Φ_- that of the antifermion field, i.e. $\Phi^\dagger_-$ contains the right chiral fermion field. The vector superfield, of course, remains the same and obeys the transformation (5.16). Now (5.17) gets specialized (with a real mass M and setting $\eta = 0$ for definiteness) to

$$\mathcal{L} = \frac{1}{4}\left[W^A W_A + \overline{W}_{\dot{A}}\overline{W}^{\dot{A}}\right]_F + \left[\Phi^\dagger_+ e^{2qV}\Phi_+ + \Phi^\dagger_- e^{-2qV}\Phi_-\right]_D + M\left[\Phi_+\Phi_- + h.c.\right]_F \; . \tag{5.34}$$

The corresponding action is

$$S = \int d^8 z \left[\Phi^\dagger_+ e^{2qV}\Phi_+ \; + \; \Phi^\dagger_- e^{-2qV}\Phi_- + \delta^{(2)}(\bar{\theta})\left(\frac{1}{4}W^A W_A + M\Phi_+\Phi_-\right)\right.$$

$$\left. + \; \delta^{(2)}(\theta)\left(\frac{1}{4}\overline{W}_{\dot{A}}\overline{W}^{\dot{A}} + M\Phi^\dagger_+\Phi^\dagger_-\right)\right] \; .$$

(5.34) is manifestly gauge invariant. Furthermore, if each of the above chiral superfields is assigned a matter parity (§4.5) of value -1 and the vector superfield a value $+1$, the Lagrangian density contains no term that is odd under matter parity. In the light of the discussion in §4.5, this means that the Lagrangian conserves R-parity.

On using the equations of constraint, in place of (5.4) and (5.19), we have

$$F_{+,-} = -M\phi^\star_{-,+} \; , \tag{5.35a}$$

$$D = -q(|\phi_+|^2 - |\phi_-|^2) \; . \tag{5.35b}$$

Finally, the "on-shell" Lagrangian density for supersymmetric quantum electrodynamics is

$$
\begin{aligned}
\mathcal{L}_{SQED} \;=\; & -\frac{1}{4}F_{\mu\nu}F^{\mu\nu} + i\lambda\sigma^\mu\partial_\mu\bar\lambda + i\xi_+\sigma^\mu\Delta_\mu^\dagger\bar\xi_+ \\[4pt]
& +i\xi_-\sigma^\mu\Delta_\mu\bar\xi_- + |\Delta_\mu\phi_+|^2 + |\Delta_\mu^\dagger\phi_-|^2 - M^2\left(|\phi_+|^2 + |\phi_-|^2\right) \\[4pt]
& -M\left(\xi_+\xi_- + \bar\xi_+\bar\xi_-\right) - \frac{q^2}{2}\left(|\phi_+|^2 - |\phi_-|^2\right)^2 \\[4pt]
& -\sqrt{2}q\left[\bar\lambda\left(\bar\xi_+\phi_+ - \bar\xi_-\phi_-\right) + \lambda\left(\xi_+\phi_+^\star - \xi_-\phi_-^\star\right)\right],
\end{aligned}
\tag{5.36}
$$

with $\Delta_\mu = \partial_\mu + iqA_\mu$.

Next, we can define the four component Dirac fermion and gaugino fields $\psi(x)$, $\lambda_M(x)$ by [cf. (3.20) and (3.24)]

$$
\psi_a = \begin{pmatrix} \xi_{+A} \\ \bar\xi_-^{\dot A} \end{pmatrix}, \quad
\bar\psi_b = \begin{pmatrix} \xi_-^B & \bar\xi_{+\dot B} \end{pmatrix}, \quad
\lambda_{Ma} = \begin{pmatrix} \lambda_A \\ \bar\lambda^{\dot A} \end{pmatrix}, \quad
\bar\lambda_{Mb} = \begin{pmatrix} \lambda^B & \bar\lambda_{\dot B} \end{pmatrix}.
$$

The definitions are consistent with $\bar\psi = \psi^\dagger\gamma_W^0$, $\bar\lambda_M = \lambda_M^\dagger\gamma_W^0$ with $\gamma_W^0 = \begin{pmatrix} 0 & I \\ I & 0 \end{pmatrix}$ in the Weyl representation and $\psi_a^\dagger = \begin{pmatrix} \xi_{+A}^\dagger & \bar\xi_-^{\dagger\dot A} \end{pmatrix} = \begin{pmatrix} \bar\xi_{+\dot A} & \xi_-^A \end{pmatrix}$ and $\lambda_{Ma}^\dagger = \begin{pmatrix} \bar\lambda_{\dot A} & \lambda^A \end{pmatrix}$. By using (3.19a), (3.28c) and (3.29c), the gauge invariant fermion K.E. terms in (5.36) can be rewritten modulo total spacetime derivatives as

$$
\frac{i}{2}\bar\lambda\bar\sigma^\mu\partial_\mu\lambda + \frac{i}{2}\lambda\sigma^\mu\partial_\mu\bar\lambda + i\bar\xi_+\bar\sigma^\mu\Delta_\mu\xi_+ + i\xi_-\sigma^\mu\Delta_\mu\bar\xi_- = \frac{i}{2}\bar\lambda_M\gamma^\mu\partial_\mu\lambda_M + i\bar\psi\gamma^\mu\Delta_\mu\psi ,
$$

the invariance of which under the transformation $\psi(x) \to \psi'(x) = e^{-iq\Lambda(x)}\psi(x)$ is transparent. Furthermore, by use of (3.30), (5.36) can be recast as a whole:

$$
\begin{aligned}
\mathcal{L}_{SQED} \;=\; & -\frac{1}{4}F_{\mu\nu}F^{\mu\nu} + \frac{i}{2}\bar\lambda_M\gamma^\mu\partial_\mu\lambda_M + i\bar\psi\gamma^\mu\partial_\mu\psi - q\bar\psi\gamma^\mu A_\mu\psi - M\bar\psi\psi \\[4pt]
& +|\partial_\mu\phi_+|^2 + |\partial_\mu\phi_-|^2 - M^2(|\phi_+|^2 + |\phi_-|^2) \\[4pt]
& +q^2A^2(|\phi_+|^2 + |\phi_-|^2) - 2iqA^\mu\phi_+^\star[\partial_\mu]\phi_+ \\[4pt]
& +2iqA^\mu\phi_-^\star[\partial_\mu]\phi_- - \frac{q^2}{2}(|\phi_+|^2 - |\phi_-|^2)^2 \\[4pt]
& -\sqrt{2}q(\phi_+^\star\bar\lambda_M\psi_L + \overline{\psi_L}\lambda_M\phi_+ - \phi_-\bar\lambda_M\psi_R - \overline{\psi_R}\lambda_M\phi_-^\star) ,
\end{aligned}
\tag{5.37}
$$

with $\psi_{L,R} \equiv \frac{1}{2}(1 \mp \gamma_5)\psi$ and $[\partial_\mu]$ as defined after (4.28).

(5.37) describes our supersymmetric quantum electrodynamic system. We shall refer to the supersymmetric partner of a fermionic particle as a **sfermion** and that of a bosonic particle as a **bosino**. The spectrum of the matter sector of this theory thus consists of a Dirac fermion (particle f and its antiparticle $\bar f$) as described by the Dirac field ψ, the left chiral sfermion $\tilde f_L$ (plus its antiparticle $\overline{\tilde f_L}$) as described by the complex scalar field ϕ_+ (plus

$\phi_+^\star$) and the right chiral sfermion $\tilde{f}_R$ (plus its antiparticle $\overline{\tilde{f}_R}$) as described by the complex scalar field $\phi_-^\star$ (plus ϕ_-); $f, \tilde{f}_L$ and $\tilde{f}_R$ all carry charge q and their conjugates have charge $-q$. In the gauge sector the spectrum consists of the photon γ as described by the vector field A_μ and the photino $\tilde{\gamma}$ as described by the four component Majorana field λ_M. The supermultiplet structure is as shown below:

$$\begin{pmatrix} f_L \\ \tilde{f}_L \end{pmatrix}, \quad \begin{pmatrix} f_L^{C} \\ \tilde{f}_L \end{pmatrix}, \quad \begin{pmatrix} f_R \\ \tilde{f}_R \end{pmatrix}, \quad \begin{pmatrix} f_R^{C} \\ \tilde{f}_R \end{pmatrix}, \quad \begin{pmatrix} \gamma \\ \tilde{\gamma} \end{pmatrix}.$$

The fermion and the photon are to be regarded as ordinary particles with positive R-parity (cf. §4.5). But their superpartner sfermions (left as well as right chiral) and the photino are to be regarded as sparticles with negative R_p. The conservation of R_p is automatic in SQED and dictates that sparticles appear in a pair at every vertex. The vertex Feynman rules, following from the interaction part of $i\mathcal{L}_{SQED}$, are shown in Fig. 5.1 with the vertex factors, to be inserted between spinors, given explicitly. N.B. Here $P_L = \frac{1}{2}(1 - \gamma_5)$, $P_R = \frac{1}{2}(1 + \gamma_5)$ and the vertex factors are to be written between final state Dirac spinors $\bar{u}$, v and initial state ones $u, \bar{v}$. In this chapter we do not put arrows on any Majorana fermion line. Therefore, one is free to choose either u or $\bar{v}$ for a Majorana fermion linking to an initial state Dirac fermion and similarly $\bar{u}$ or v for one linking to a Dirac fermion in the final state. A more definitive formulation of rules for Majorana fermions will be given in Ch.9.

Fig.5.1

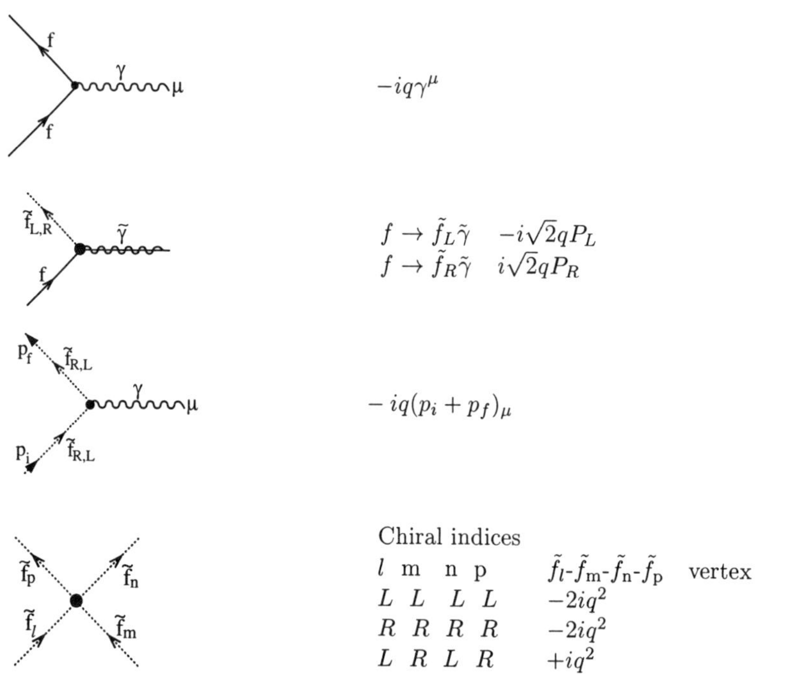

Chiral indices

l	m	n	p	$\tilde{f}_l$-$\tilde{f}_m$-$\tilde{f}_n$-$\tilde{f}_p$	vertex
L	L	L	L	$-2iq^2$	
R	R	R	R	$-2iq^2$	
L	R	L	R	$+iq^2$	

Fig.5.1 contd.

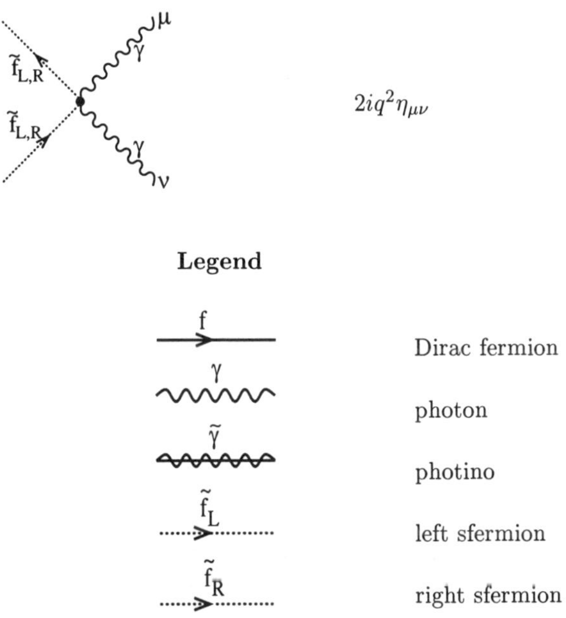

$$2iq^2\eta_{\mu\nu}$$

Legend

f	Dirac fermion
γ	photon
$\tilde{\gamma}$	photino
$\tilde{f}_L$	left sfermion
$\tilde{f}_R$	right sfermion

Fig.5.1. Feynman rules for vertices in SQED.

The four main features of SQED to remember are:

- Two left chiral superfields with opposite electric charges are needed to describe the supersymmetric quantum electrodynamics of a Dirac fermion.

- A sfermion, appearing at a vertex, has a specific chirality. Chirality is conserved in all vertices. Multisfermion vertices involve two or four sfermions, grouped in chiral pairs.

- R_p conservation is automatic in SQED and sparticles appear in a pair (or quartet) at any vertex.

- The coupling strength of the fermion-chiral sfermion-photon vertex is $\sqrt{2}$ times that of the gauge coupling strength and the strengths of the four sfermion vertices are also proportional to the square of the latter.

Indeed, these features are also true of the two gauge theories described later in this chapter.

5.4　Nonabelian Gauge Interactions

To start with, let us confine ourselves to a simple gauge group G with a universal gauge coupling constant g since the extension to a semisimple group or a product group with $U(1)$ factors is straightforward. The nonabelian gauge transformations on a left chiral superfield Φ, belonging to the unitary irreducible representation $\mathcal{R}$ of G, are nonabelian generalizations of (5.15). They can be described as

$$\Phi'_I = \left[e^{-2ig\Lambda^a(z)T^a} \right]_{IJ} \Phi_J \, , \quad \bar{D}^{\dot{A}}\Lambda^a T^a = 0 \, , \tag{5.38a}$$

$$\Phi^{\dagger\prime}_I = \Phi^{\dagger}_J \left[e^{+2ig\Lambda^{a\dagger}(z)T^a} \right]_{JI} , \quad D_A \Lambda^{a\dagger} T^a = 0 \, . \tag{5.38b}$$

Here $\Lambda^a(x,\theta,\bar{\theta})$ are the complex chiral superfunctions specifying the gauge transformation and T^a are the hermitian generators of G in the space of the representation $\mathcal{R}$ whose basis indices are denoted by subscripts I, J. Unless otherwise specified, we work in a Cartesian basis. Thus

$$[T^a, T^b] = it^{abc}T^c; \quad \mathrm{Tr}\, T^a T^b = T(\mathcal{R})\delta^{ab} \, , \tag{5.39}$$

where t^{abc} are the completely antisymmetric real structure constants and $T(\mathcal{R})$ is the representation constant of $\mathcal{R}$. For the adjoint representation, T is a positive constant k which is the quadratic Casimir $C_2(G)$ defined by $t^{abc}t^{dbc} = C_2(G)\delta^{ad}$. More generally, $(T^aT^a)_{IJ} = C_2(\mathcal{R})\delta_{IJ}$.

It is convenient to define

$$\Lambda_{IJ} \equiv 2g\Lambda^a T^a_{IJ}, \quad V_{IJ} \equiv 2gV^a T^a_{IJ} \, . \tag{5.40}$$

In (5.40) $V^a(z)$ are the vector superfields transforming in the adjoint representation of G. Now the extension of (5.18) to a nonabelian gauge invariant and supersymmetric Lagrangian density is simple provided the gauge transformation of V is generalized from (5.16) to

$$e^{V'} = e^{-i\Lambda^\dagger} e^V e^{i\Lambda} \, , \tag{5.41a}$$

$$e^{-V'} = e^{-i\Lambda} e^{-V} e^{i\Lambda^\dagger} \, . \tag{5.41b}$$

The transformations (5.41a) and (5.41b) are inverses of each other and imply that $\mathrm{Tr}[\Phi^\dagger e^V \Phi]$ is gauge invariant.

The transformations (5.41) are, in fact, valid for any representation of the generators provided they have a consistent solution $V' = 2gV'^a T^a$. It is sufficient for us to consider infinitesimal gauge transformations neglecting $O(\Lambda^2, \Lambda^\dagger\Lambda, \Lambda\Lambda^\dagger, \Lambda^{\dagger 2})$ terms. Then, writing $V' = V + \delta V$ and using the Baker-Campbell-Hausdorff formula [5.5], we have

$$e^{V'} = e^{V + i(\Lambda - \Lambda^\dagger) + \frac{i}{2}[V, \Lambda + \Lambda^\dagger] + \frac{i}{12}[V, [V, \Lambda - \Lambda^\dagger]] + \cdots} \, , \tag{5.42a}$$

$$\delta V \equiv V' - V = i(\Lambda - \Lambda^\dagger) + \frac{i}{2}[V, \Lambda + \Lambda^\dagger] + \frac{i}{12}[V, [V, \Lambda - \Lambda^\dagger]] + \cdots \, . \tag{5.42b}$$

In terms of the superfields V^a, (5.42) works out to

$$V'^a - V^a = +i(\Lambda^a - \Lambda^{a\dagger}) - gt^{abc}V^b(\Lambda^c + \Lambda^{c\dagger}) - \frac{i}{3}g^2 t^{abc}t^{cde}V^bV^d(\Lambda^e - \Lambda^{e\dagger}) + \cdots . \tag{5.43}$$

As before, we choose to work in the Wess-Zumino gauge of the superfields V^a where triple products of V's or those with a higher number of V-factors vanish. Since, in this gauge, a vector superfield has no θ-independent term, cf. (4.34), the θ-independent component of Λ^a must be real. A general gauge superfunction Λ^a will lead to a V'^a which is not in the WZ gauge, though an appropriate supergauge transformation can transform it to that gauge. Suppose one insists on both V^a and V'^a being in the WZ gauge. Then, if $V^aV^bV^c$ vanishes, one must consistently choose Λ^a in a way such that $V'^aV'^bV'^c$ is zero. This means not only that $(\Lambda^a - \Lambda^{a\dagger})(\Lambda^b - \Lambda^{b\dagger})(\Lambda^c - \Lambda^{c\dagger})$ and $V^a(\Lambda^b - \Lambda^{b\dagger})(\Lambda^c - \Lambda^{c\dagger})$ vanish, but also that the third and successive terms in the RHS of (5.43) consequently vanish, leading to the infinitesimal transformation of a vector superfield obeying the WZ condition:

$$\delta V^{WZ} = i(\Lambda - \Lambda^\dagger) + \frac{i}{2}\left[V^{WZ}, \Lambda + \Lambda^\dagger\right] . \tag{5.44}$$

The definitions (4.39) of the abelian supersymmetric field strengths W_A and $\overline{W}^{\dot{A}}$ need to be extended to the nonabelian case. The generalized definitions are

$$W_A = -\frac{1}{4}\bar{\mathcal{D}}\bar{\mathcal{D}}e^{-V}\mathcal{D}_A e^V , \tag{5.45a}$$

$$\overline{W}^{\dot{A}} = -\frac{1}{4}\mathcal{D}\mathcal{D}e^V\bar{\mathcal{D}}^{\dot{A}}e^{-V} . \tag{5.45b}$$

Unlike the gauge invariant abelian field strengths, these transform nontrivially (covariantly) under gauge transformations, i.e. $W_A \to W'_A$, with

$$W'_A = -\frac{1}{4}\bar{\mathcal{D}}\bar{\mathcal{D}}e^{-i\Lambda}e^{-V}e^{i\Lambda^\dagger}\mathcal{D}_A e^{-i\Lambda^\dagger}e^V e^{i\Lambda} = -\frac{1}{4}e^{-i\Lambda}\bar{\mathcal{D}}\bar{\mathcal{D}}e^{-V}\mathcal{D}_A e^V e^{i\Lambda} . \tag{5.46}$$

In the second step of (5.46), use has been made of the chiral properties of Λ and $\Lambda^\dagger$, i.e. $\bar{\mathcal{D}}\Lambda = 0 = \mathcal{D}\Lambda^\dagger$. These two relations enable one to commute $e^{-i\Lambda}$ through $\bar{\mathcal{D}}\bar{\mathcal{D}}$ as well as $e^{i\Lambda^\dagger}$ through $\mathcal{D}_A$ and write the gauge transformed field strengths as

$$W'_A = e^{-i\Lambda}W_A e^{i\Lambda} , \tag{5.47a}$$

$$\overline{W}'^{\dot{A}} = e^{i\Lambda}\overline{W}^{\dot{A}}e^{-i\Lambda} . \tag{5.47b}$$

Expanding in powers of V, one has

$$e^{-V}\mathcal{D}_A e^V = \mathcal{D}_A V + \frac{1}{2}[\mathcal{D}_A V, V] + \frac{1}{6}[[\mathcal{D}_A V, V], V] + O(V^4) + \cdots . \tag{5.48}$$

In the WZ gauge only the first two RHS terms in (5.48) survive and (5.45a) becomes

$$W_A = -\frac{1}{4}\bar{\mathcal{D}}\bar{\mathcal{D}}\,\mathcal{D}_A V_{WZ} - \frac{1}{8}\bar{\mathcal{D}}\bar{\mathcal{D}}[\mathcal{D}_A V_{WZ}, V_{WZ}] . \tag{5.49}$$

In analogy with V in (5.40), we can define W_A^a by

$$W_A = 2g W_A^a T^a \; .\tag{5.50}$$

(5.49) and (5.50) lead to the expression

$$W_A^a = -\frac{1}{4}\bar{\mathcal{D}}\bar{\mathcal{D}}[\mathcal{D}_A V_{WZ}^a + igt^{abc}(\mathcal{D}_A V_{WZ}^b)V_{WZ}^c] \; .\tag{5.51}$$

On expanding the superfield W_A^a in terms of its components, one obtains the nonabelian generalization of (4.42), namely

$$W_A^a = \lambda_A^a(y) + D^a(y)\theta_A - (\sigma^{\mu\nu}\theta)_A F_{\mu\nu}^a(y) + i\theta\theta\sigma^\mu_{A\dot{B}}\Delta_\mu\bar{\lambda}^{a\dot{B}}(y) \; ,\tag{5.52}$$

with

$$F_{\mu\nu}^a = \partial_\mu A_\nu^a - \partial_\nu A_\mu^a - gt^{abc}A_\mu^b A_\nu^c \; ,\tag{5.53a}$$

$$\Delta_\mu\bar{\lambda}^{a\dot{B}} = \partial_\mu\bar{\lambda}^{a\dot{B}} - gt^{abc}A_\mu^b\bar{\lambda}^{c\dot{B}} \; .\tag{5.53b}$$

The nonabelian component expansion of $\bar{W}_A^a$ can be made in a similar fashion. The final Lagrangian density is a nonabelian generalization of (5.17). It can be written as

$$\mathcal{L} = \frac{1}{16g^2 k}\mathrm{Tr}\left[W^A W_A + \overline{W}_A \overline{W}^A\right]_F + \left[\Phi_i^\dagger(e^V)_{ij}\Phi_j\right]_D + [\mathcal{W}(\Phi_i) + \text{h.c.}]_F \; ,\tag{5.54}$$

where $k = T(\mathcal{R})$ and $\mathcal{W}(\Phi_i)$ is a gauge invariant superpotential. **We have reverted to the use of i as a general type index which covers the gauge representation space index I and can include other types such as generations.** Note that the first RHS term above is simply $\frac{1}{4}(W^{Aa}W_A^a + \bar{W}_A^a\bar{W}^{Aa})_F$. Component expansion plus the use of the equation of constraint $\partial\mathcal{L}/\partial D^a = 0$ allow (5.54) to be rewritten as

$$\begin{aligned}\mathcal{L} = &\; i\xi_j\sigma_\mu\Delta_{ij}^{\dagger\mu}\bar{\xi}_i + (\Delta_{ij}^\mu\phi_j)^\dagger(\Delta_{\mu ik}\phi_k) - \frac{1}{4}F_{\mu\nu}^a F^{\mu\nu a} + i\lambda^a\sigma^\mu\Delta_\mu\bar{\lambda}^a \\[2mm] &-\sqrt{2}g(\bar{\lambda}^a\bar{\xi}_i T_{ij}^a\phi_j + h.c.) - V(\phi_i,\phi_j^\star) - \left[\frac{1}{2}\xi_i\xi_j\mathcal{W}_{ij}(\phi) + h.c.\right]\end{aligned}\tag{5.55}$$

with[4]

$$\Delta_{ij}^\mu = \delta_{ij}\partial^\mu + igA^{\mu a}T_{ij}^a \; ,\tag{5.56a}$$

$$V(\phi_i,\phi_j^\star) = F_i F_i^\star + \frac{1}{2}D^a D^a \; ,\tag{5.56b}$$

$$F_i = -\left.\frac{\partial\mathcal{W}^\dagger}{\partial\Phi_i^\dagger}\right|, \quad D^a = -g\phi_i^\dagger T_{ij}^a\phi_j \; .\tag{5.56c}$$

If we generalize the above discussion to include gauge groups that are products of factor groups (including $U(1)$ factors), $G = \prod_\alpha G_\alpha$, we need to put a superscript α to the gauge

---

[4]The reader can verify that the conserved spinorial supercurrent density for the system of (5.55) is
$K_A^\mu = -\sqrt{2}(\sigma_\nu\bar{\sigma}^\mu\xi_i)_A(\Delta_{ij}^\nu\phi_j)^\dagger - i\sqrt{2}(\sigma^\mu\bar{\xi}_i)\mathcal{W}_i^\dagger(\phi^\dagger) - \frac{1}{2}(\sigma^\nu\bar{\sigma}^\rho\sigma^\mu\bar{\lambda}^a)_A F_{\nu\rho}^a - ig(\phi_i^\dagger T_{ij}^a\phi_j)(\sigma^\mu\bar{\lambda}^a)_A.$

coupling and make it g^α and characterize T^a by $T^{a(\alpha)}$, α referring to the factor. Moreover, if we use the index h to refer generically to the generators of abelian factor groups, we can admit the term $\eta^h D^h$ in the Lagrangian density of (5.55). Now we have $D^{a(\alpha)} = -\eta^a - g^\alpha \phi_i^\dagger T_{ij}^{a(\alpha)} \phi_j$, α not summed, where $\eta^a = \eta^h \delta^{ah}$. It should be noted that, unlike in the Abelian case, a term proportional to D^a is not gauge invariant and hence cannot appear in the Lagrangian.

The scalar, fermion and vector masses can be discussed by generalizing the corresponding considerations §5.2 to a nonabelian gauge symmetry. Employing the ϕ - $\bar\phi$ subspace notation of §5.1 once again, we have $D^{a(\alpha)i} = -g^\alpha T_{ij}^{a,\alpha} \phi_j$ and $D_i^{a(\alpha)} = -g^\alpha \bar\phi^j T_{ji}^{a,\alpha}$, α not summed. For a supersymmetric vacuum, the rest of the discussion on masses proceeds identically to that of §5.2. Thus (5.28) and (5.31) hold here also and the derivation of the supertrace mass sum rule (5.32) follows exactly as in the abelian case.

5.5 Supersymmetric Quantum Chromodynamics (SQCD)

Let us take one quark flavor transforming as an $SU(3)$ color triplet and eight gluons transforming as the adjoint color octet. In order to supersymmetrize this uniflavor QCD theory, we proceed in analogy with electrodynamics and introduce two chiral superfields Φ_+ and Φ_- transforming as an $SU(3)$ color triplet and antitriplet respectively:

$$\Phi_+' = e^{-2ig_s T^a \Lambda^a} \Phi_+ , \tag{5.57a}$$

$$\Phi_-' = e^{2ig_s \bar{T}^a \Lambda^a} \Phi_- , \tag{5.57b}$$

where g_S is the QCD coupling, the generators T^a equal one-half times the Gell-Mann $SU(3)$ λ-matrices in their fundamental representation and $\bar{T}^a$ are the corresponding complex conjugate matrices. Since T^a are hermitian, $\Phi_-'^T = \Phi_-^T e^{2ig_s T^a \Lambda^a(x)}$. Thus the mass term

$$\mathcal{W}(\Phi_+, \Phi_-) = M \Phi_-^T \Phi_+ \tag{5.58}$$

qualifies to be a color invariant superpotential term and the full scalar potential at the tree level then is

$$V(\phi_+, \phi_-) = M^2(\phi_+^\dagger \phi_+ + \phi_-^\dagger \phi_-) + \frac{1}{2} g_s^2 \sum_a (\phi_+^\dagger T^a \phi_+ - \phi_-^\dagger \bar{T}^a \phi_-)^2 . \tag{5.59}$$

Proceeding as in the SQED case, one can write the onshell Lagrangian density in terms of the four component Dirac field ψ as

$$\begin{aligned}
\mathcal{L}_{SQCD} =\ & -\frac{1}{4} F_{\mu\nu}^a F^{\mu\nu a} + \frac{i}{2} \bar\lambda_M^a \gamma^\mu (\partial_\mu \delta^{ac} + g_s t^{abc} A_\mu^b) \lambda_M^c + |\Delta^\mu \phi_+|^2 \\
& + (\Delta^\mu \phi_-^\star)^\dagger (\Delta_\mu \phi_-^\star) + i\bar\psi \gamma^\mu \Delta_\mu \psi - M\bar\psi\psi - M^2(\phi_+^\dagger \phi_+ + \phi_-^\dagger \phi_-) \\
& - \sqrt{2} g_s (\overline{\psi_L} T^a \lambda_M^a \phi_+ + \phi_+^\dagger \bar\lambda_M^a T^a \psi_L - \phi_-^T \bar\lambda_M^a T^a \psi_R - \overline{\psi_R} \lambda_M^a T^a \phi_-^\star) \\
& - \frac{1}{2} g_s^2 \sum_a (\phi_+^\dagger T^a \phi_+ - \phi_-^\dagger \bar{T}^a \phi_-)^2 ,
\end{aligned} \tag{5.60}$$

where $\Delta_\mu = \partial_\mu + ig_s T^a A_\mu^a$, with $(T^a)^{bc} = -it^{abc}$ in the adjoint representation, where t^{abc} are the structure constants for $SU(3)$ in the standard form. The hermitian conjugation and the transposition in (5.60) are in color space.

We can now use (5.60) to describe our uniflavor supersymmetric quantum chromodynamics. The quark q (as well as the antiquark $\bar{q}$) is described by the color triplet Dirac field ψ, the left squark $\tilde{q}_L$ (together with its antiparticle $\bar{\tilde{q}}_L$) is denoted by the color triplet complex field ϕ_+ (plus $\phi_+^\star$) while the similar field $\phi_-^\star$ (plus ϕ_-) refers to the right squark $\tilde{q}_R$ (along with its antiparticle $\bar{\tilde{q}}_R$). The color octet gluons, described by the fields A_μ^a ($a = 1, \cdots, 8$), are accompanied by the corresponding gluinos which are denoted by the four component Majorana fields λ_M^a. Though the formalism developing a supersymmetric nonabelian gauge theory is more cumbrous than in the Abelian case, the supermultiplet structure and interactions are simple nonabelian generalizations of those in SQED. Here, quarks and gluons have positive R_p while R_p is negative for squarks and gluinos. Then R-parity conservation is automatic and ensures that all sparticles connect to any vertex in pairs. The vertex Feynman rules for SQCD interactions from $i\mathcal{L}_{SQCD}$ are given in Fig. 5.2, with the convention that any colour matrix T^a or product of such matrices, as appear, has to be taken between outgoing color states on the left and incoming ones on the right. Once again, $P_L = \frac{1}{2}(1 - \gamma_5)$ and $P_R = \frac{1}{2}(1 + \gamma_5)$ and we do not put arrows on Majorana fermion lines. Note that the coefficient $1/2$ multiplying the gluino-gluon-gluino interaction term in (5.60) gets removed from the Feynman rule for the corresponding vertex because of the extra factor of 2 from contracting $\bar{\lambda}_M^a \gamma^\mu \lambda_M^c$ with a pair of external Majorana fermions.

Fig.5.2

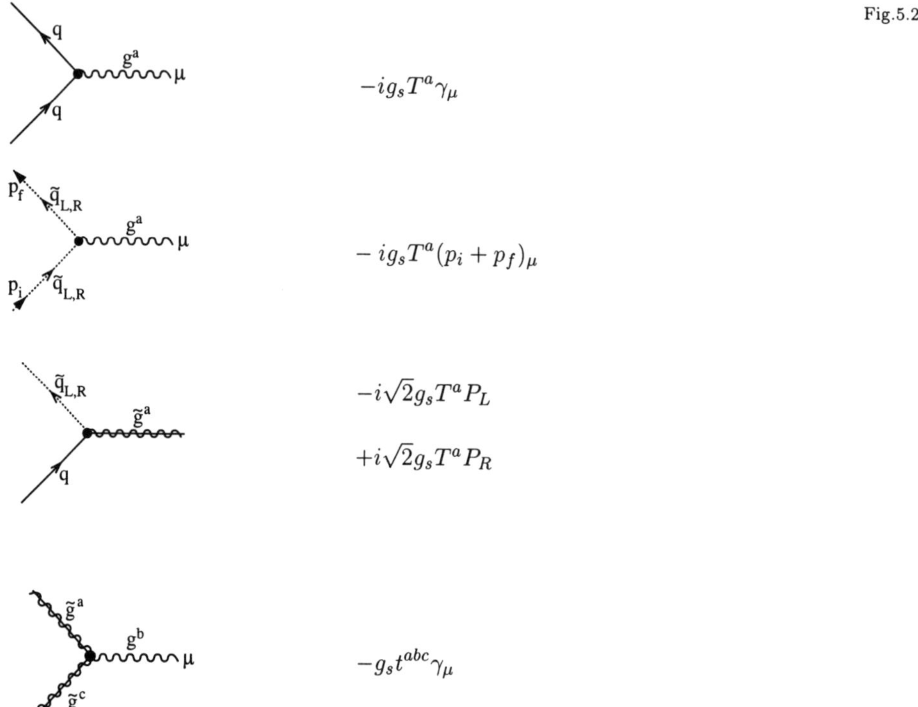

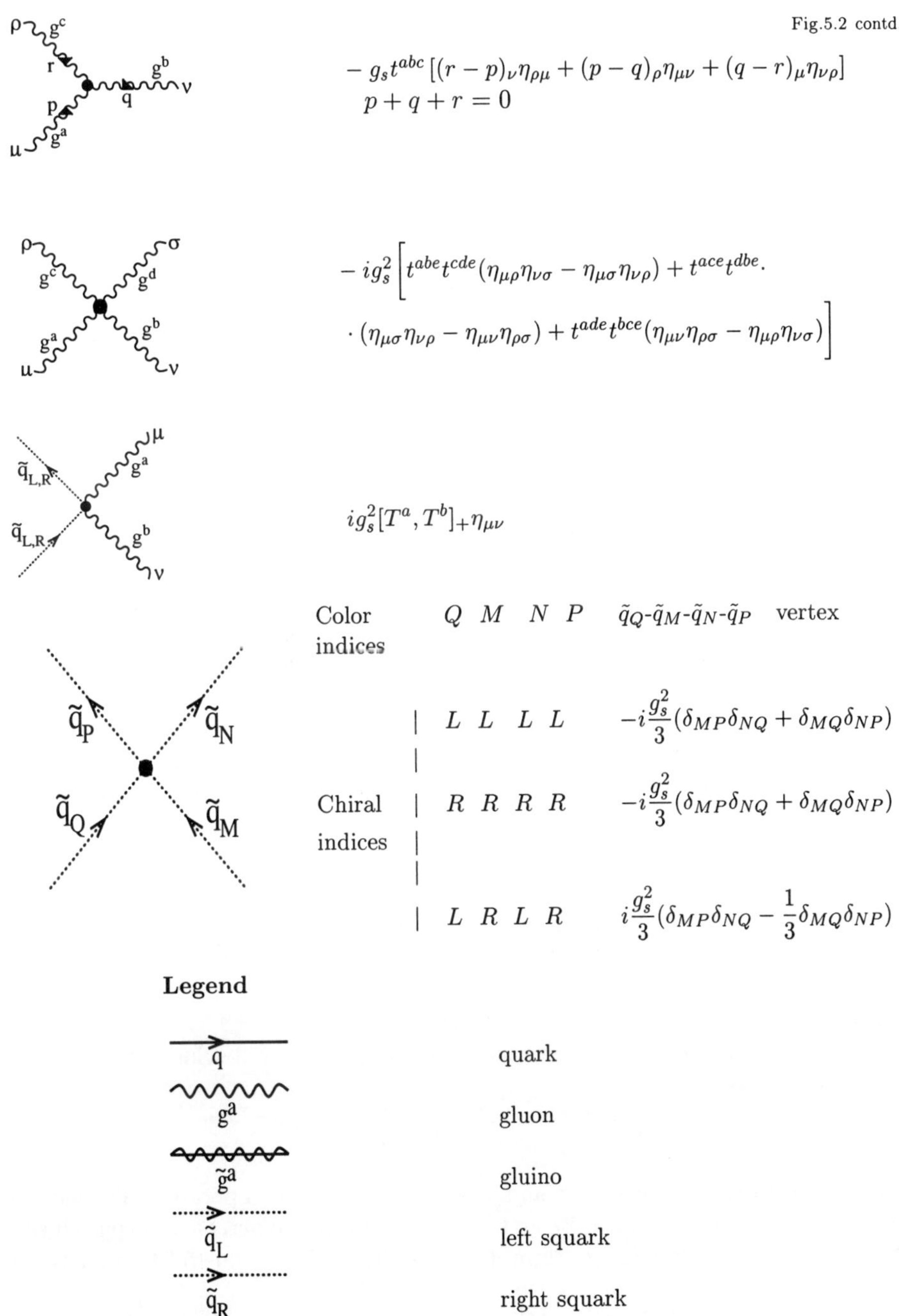

$$- g_s t^{abc}\left[(r-p)_\nu \eta_{\rho\mu} + (p-q)_\rho \eta_{\mu\nu} + (q-r)_\mu \eta_{\nu\rho}\right]$$
$$p+q+r=0$$

$$- i g_s^2 \left[t^{abe}t^{cde}(\eta_{\mu\rho}\eta_{\nu\sigma} - \eta_{\mu\sigma}\eta_{\nu\rho}) + t^{ace}t^{dbe}\cdot\right.$$
$$\left. \cdot (\eta_{\mu\sigma}\eta_{\nu\rho} - \eta_{\mu\nu}\eta_{\rho\sigma}) + t^{ade}t^{bce}(\eta_{\mu\nu}\eta_{\rho\sigma} - \eta_{\mu\rho}\eta_{\nu\sigma}) \right]$$

$$i g_s^2 [T^a, T^b]_+ \eta_{\mu\nu}$$

Color indices	Q	M	N	P	$\tilde{q}_Q$-$\tilde{q}_M$-$\tilde{q}_N$-$\tilde{q}_P$ vertex
	L	L	L	L	$-i\dfrac{g_s^2}{3}(\delta_{MP}\delta_{NQ} + \delta_{MQ}\delta_{NP})$
Chiral indices	R	R	R	R	$-i\dfrac{g_s^2}{3}(\delta_{MP}\delta_{NQ} + \delta_{MQ}\delta_{NP})$
	L	R	L	R	$i\dfrac{g_s^2}{3}(\delta_{MP}\delta_{NQ} - \dfrac{1}{3}\delta_{MQ}\delta_{NP})$

Legend

q	quark
g^a	gluon
$\tilde{g}^a$	gluino
$\tilde{q}_L$	left squark
$\tilde{q}_R$	right squark

Fig.5.2. Feynman rules for SQCD vertices with[5] $T^a = \frac{1}{2}\lambda^a$

[5]Here λ^a are the Gell-Mann $SU(3)$ matrices. Thus now $T^a_{ML}T^a_{NP} = \frac{1}{2}(\delta_{MP}\delta_{NL} - \frac{1}{3}\delta_{ML}\delta_{NP})$.

We can end this discussion of SQCD with the following comments.

- SQCD vertices include those that are straightforward generalizations of SQED ones.

- The intrinsically nonabelian triple gluon coupling of QCD has a supersymmetric analog in the gluino-gluon-gluino coupling carrying the same strength.

- The quadruple gluon coupling does not have such a supersymmetric analog; this is easy to see in terms of mass dimensions since any such analog would mean a term in the Lagrangian density with a dimension exceeding four.

- The quark-squark-gluino vertices, as given above, are in pure SQCD in which quark/squark mass and interaction eigenstates are identical. When supersymmetry is broken, differences in general arise between quark and squark flavor rotations effecting mass diagonalization. In consequence, there are some additional mixing factors in these vertices which will be detailed in Ch.9.

5.6   Supersymmetric chiral gauge theory (SχGT)

We now discuss the supersymmetric generalization of a chiral gauge theory based on the $SU(2)_L \times U(1)_Y$ gauge group of the electroweak theory of Glashow, Weinberg and Salam. For simplicity, we restrict ourselves to one fermionic generation without worrying about ABJ anomalies [5.6] and consider the purely massless gauge theory without the Higgs mechanism. A more realistic description with spontaneous symmetry breakdown will follow in Ch.8 when we discuss the Minimal Supersymmetric Standard Model. Here we start with one left chiral $SU(2)_L$ doublet (up type and down type) and two $SU(2)_L$ singlet left chiral superfields:

$$\Phi_+ = \begin{pmatrix} \Phi_{u+} \\ \Phi_{d+} \end{pmatrix}; \quad \Phi_{u-}, \Phi_{d-} \; . \tag{5.61}$$

Let the $SU(2)_L$ and $U(1)_Y$ gauge (vector) superfields be V^a ($a = 1, 2, 3$) and V^Y respectively. The gauge transformations of the superfields in (5.61) are straightforward applications of (5.15) and (5.38). We do not give them explicitly but simply specify the $SU(2)_L$ and $U(1)_Y$ gauge couplings to be g_2 and g_Y respectively.

The matter kinetic energy term can now be written, following the discussions given in §5.2 and §5.4, as

$$\left[\Phi_+^\dagger e^{(g_2 V^a \tau^a + g_Y V^Y Y)} \Phi_+ + \sum_{r=u,d} \Phi_{r-}^\dagger e^{g_Y V^Y Y} \Phi_{r-} \right]_D , \tag{5.62}$$

where τ^a are the Pauli matrices and Y is the hypercharge operator. The electric charge is $Q = (\tau_3 + Y)/2$. We shall use Y_L and Y_{rR} to denote the hypercharge eigenvalues for Φ_+ and $\Phi_{r-}^\dagger$. The gauge kinetic energy term, following from (5.17) and (5.54) similarly is

$$\frac{1}{4} \left[W^{YA} W_A^Y + \overline{W}_A^Y \overline{W}^{Y\dot{A}} \right]_F + \frac{1}{4} \left[W^{aA} W_A^a + \overline{W}_A^a \overline{W}^{a\dot{A}} \right]_F . \tag{5.63}$$

The sum of (5.62) and (5.63) leads us to the full Lagrangian density. This can be written in component notation by using (5.22) and (5.55). Let us use a more specialized notation in the gauge sector, namely $\vec{W}_\mu$, $\vec{W}_{\mu\nu}$, $\vec{\tilde{\lambda}}$ (the vector sign covering Cartesian subscripts 1,2,3) to denote A_μ^a, $F_{\mu\nu}^a$, λ_M^a respectively in the $SU(2)_L$ factor and B_μ, $B_{\mu\nu}$, $\tilde{\lambda}_0$ to denote A_μ, $F_{\mu\nu}$, λ_M respectively in the $U(1)_Y$ part. Furthermore, we use Δ_μ^{21} and Δ_μ^1 for the respective $SU(2)_L \times U(1)_Y$ and $U(1)_Y$ gauge covariant derivatives:

$$\Delta_\mu^{21} \equiv \partial_\mu + ig_2 \vec{W}_\mu \cdot \frac{\vec{\tau}}{2} + ig_Y B_\mu \frac{Y}{2} \ ,$$

$$\Delta_\mu^1 \equiv \partial_\mu + ig_Y B_\mu \frac{Y}{2} \ .$$

Now we can write the Lagrangian density of this theory as

$$
\begin{aligned}
\mathcal{L}_{S\chi GT} \ = \ & -\frac{1}{4}\vec{W}_{\mu\nu} \cdot \vec{W}^{\mu\nu} - \frac{1}{4}B_{\mu\nu}B^{\mu\nu} + \frac{i}{2}\Big(\vec{\bar{\tilde{\lambda}}}\gamma^\mu \partial_\mu \vec{\tilde{\lambda}} - g_2 \vec{\bar{\tilde{\lambda}}}\gamma^\mu \times \vec{W}_\mu \cdot \vec{\tilde{\lambda}}\Big) \\
& + \frac{i}{2}\bar{\tilde{\lambda}}_0 \gamma^\mu \partial_\mu \tilde{\lambda}_0 + i\overline{\psi_L}\gamma^\mu \Delta_\mu^{21}\psi_L + i\sum_{r=u,d}\overline{\psi_{rR}}\Delta_\mu^1 \psi_{rR} + |\Delta^{21\mu}\phi_+|^2 \\
& + \sum_{r=u,d} |\Delta^{1\mu}\phi_{r-}|^2 - \frac{1}{2}g_2^2\Big(\phi_+^\dagger \frac{\vec{\tau}}{2}\phi_+\Big)^2 - \frac{1}{2}g_Y^2\Big(\frac{1}{2}Y_L \phi_+^\dagger \phi_+ \\
& - \frac{1}{2}\sum_{r=u,d} Y_{rR}^r |\phi_{r-}|^?\Big)^2 - \sqrt{2}y_2\Big(\psi_+^\dagger \vec{\bar{\tilde{\lambda}}} \cdot \frac{\vec{\tau}}{2}\psi_L + \overline{\psi_L}\frac{\vec{\tau}}{2} \cdot \vec{\tilde{\lambda}}\phi_+\Big) \\
& - \sqrt{2}g_Y\Big[\frac{Y_L}{2}\Big(\phi_+^\dagger \bar{\tilde{\lambda}}_0 \psi_L + \overline{\psi_L}\tilde{\lambda}_0 \phi_+\Big) \\
& - \frac{1}{2}\sum_{r=u,d} Y_{rR}\Big(\phi_{r-}^\star \overline{\psi_R}\tilde{\lambda}_0 + \phi_{r-}\bar{\tilde{\lambda}}_0 \psi_R\Big)\Big].
\end{aligned}
\tag{5.64}
$$

The relative minus sign between the terms in (5.64) containing Y_L and those with Y_R occurs since, while Y_L is the hypercharge of ϕ_+, Y_R is the hypercharge of $\phi_{r-}^\star$.

As done with SQED and SQCD earlier, the vertex Feynman rules for this SχGT can also be given from (5.64). This time we use the fields

$$f_L = \begin{pmatrix} f_{uL} \\ f_{dL} \end{pmatrix}, \quad \tilde{f}_L = \begin{pmatrix} \tilde{f}_{uL} \\ \tilde{f}_{dL} \end{pmatrix}, \quad f_{uR}, \ f_{dR}, \ \tilde{f}_{uR}, \ \tilde{f}_{dR} \tag{5.65}$$

in place of ψ_L, ϕ_+, ψ_{uR}, ψ_{dR}, $\phi_{u-}^\star$, $\phi_{d-}^\star$ respectively. Charged gauge boson and gaugino fields respectively are $W^{\mu,\pm} = (W_1^\mu \mp iW_2^\mu)/\sqrt{2}$ and $\tilde{\lambda}^\pm = (\tilde{\lambda}_1 \mp i\tilde{\lambda}_2)/\sqrt{2}$. The weak isospin up, down fermionic particles will be designated f_q with q = u, d respectively. The electroweak vector boson particles will be described by the symbols $W^\pm$, W^3 and B, while the sparticles, corresponding to the fields $\tilde{\lambda}^\pm$, $\tilde{\lambda}_3$ and $\tilde{\lambda}_0$, will be referred to as $\tilde{w}^\pm$, $\tilde{w}^3$ and $\tilde{B}$ respectively. As before, states without (with) tilde will have positive (negative) R-parity. The vertices are given below in Fig. 5.3 in terms of the chiral projection matrices $P_{L,R} = (1 \mp \gamma_5)/2$. Note that no arrows have been put on Majorana fermion lines here. Once again, we draw attention to an extra factor of 2 for the $\tilde{w}^3$-$\tilde{w}^\pm$-$w^\mp$ and $\tilde{w}^\pm$-$\tilde{w}^\mp$-w^3 vertices.

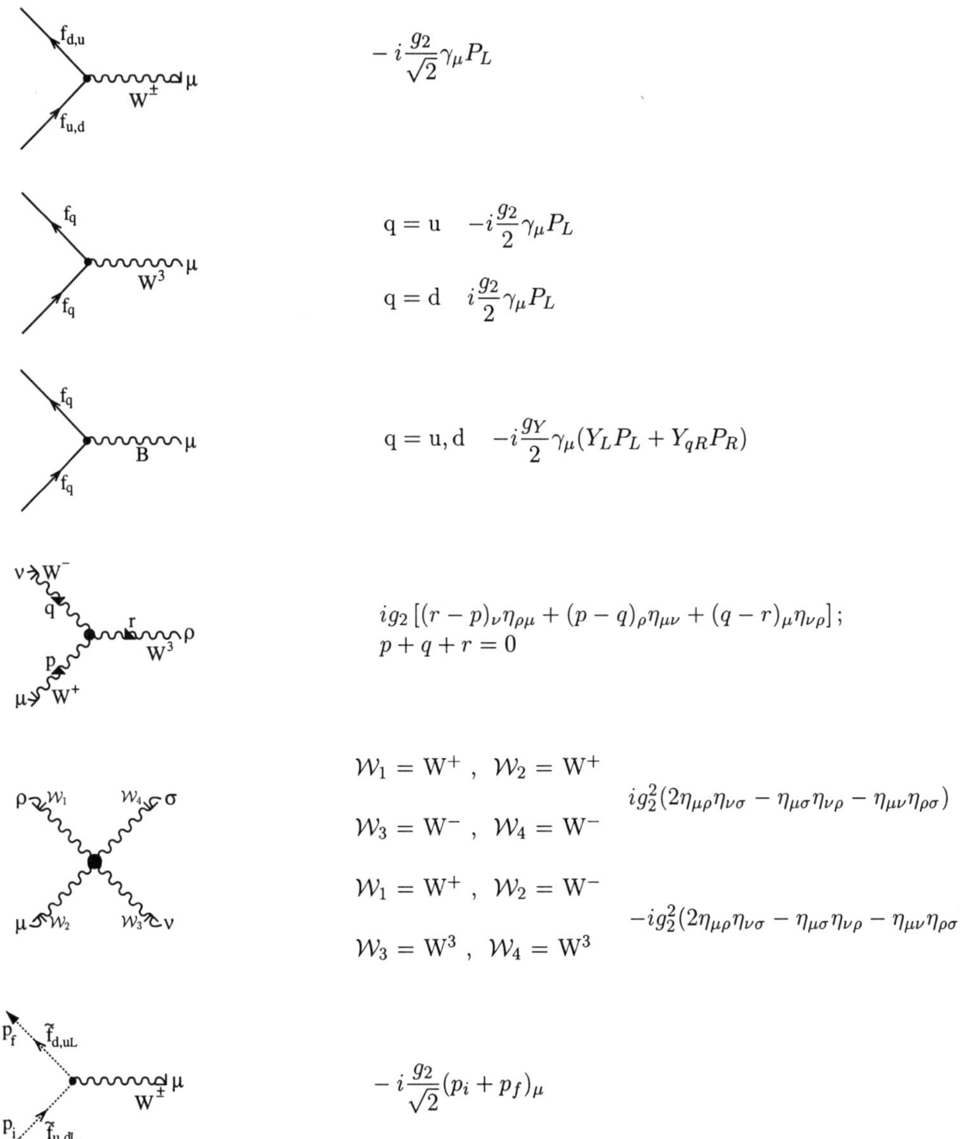

$$-i\frac{g_2}{\sqrt{2}}\gamma_\mu P_L$$

$$q = u \quad -i\frac{g_2}{2}\gamma_\mu P_L$$

$$q = d \quad i\frac{g_2}{2}\gamma_\mu P_L$$

$$q = u, d \quad -i\frac{g_Y}{2}\gamma_\mu(Y_L P_L + Y_{qR} P_R)$$

$$ig_2\left[(r-p)_\nu\eta_{\rho\mu} + (p-q)_\rho\eta_{\mu\nu} + (q-r)_\mu\eta_{\nu\rho}\right];$$
$$p + q + r = 0$$

$$\mathcal{W}_1 = W^+ , \quad \mathcal{W}_2 = W^+$$
$$\mathcal{W}_3 = W^- , \quad \mathcal{W}_4 = W^-$$
$$ig_2^2(2\eta_{\mu\rho}\eta_{\nu\sigma} - \eta_{\mu\sigma}\eta_{\nu\rho} - \eta_{\mu\nu}\eta_{\rho\sigma})$$

$$\mathcal{W}_1 = W^+ , \quad \mathcal{W}_2 = W^-$$
$$\mathcal{W}_3 = W^3 , \quad \mathcal{W}_4 = W^3$$
$$-ig_2^2(2\eta_{\mu\rho}\eta_{\nu\sigma} - \eta_{\mu\sigma}\eta_{\nu\rho} - \eta_{\mu\nu}\eta_{\rho\sigma})$$

$$-i\frac{g_2}{\sqrt{2}}(p_i + p_f)_\mu$$

Fig.5.3 contd.

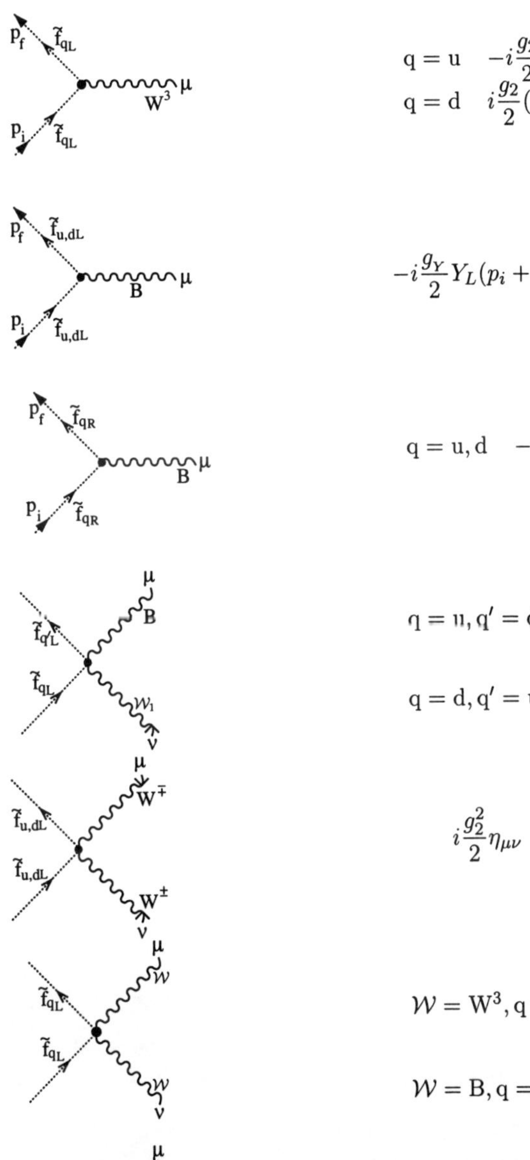

$$q = u \quad -i\frac{g_2}{2}(p_i + p_f)_\mu$$
$$q = d \quad i\frac{g_2}{2}(p_i + p_f)_\mu$$

$$-i\frac{g_Y}{2}Y_L(p_i + p_f)_\mu$$

$$q = u, d \quad -i\frac{g_Y}{2}Y_{qR}(p_i + p_f)_\mu$$

$$q = u, q' = d, \mathcal{W}_1 = W^- \quad i\frac{g_Y g_2}{\sqrt{2}}Y_L\eta_{\mu\nu}$$

$$q = d, q' = u, \mathcal{W}_1 = W^+ \quad i\frac{g_Y g_2}{\sqrt{2}}Y_L\eta_{\mu\nu}$$

$$i\frac{g_2^2}{2}\eta_{\mu\nu}$$

$$\mathcal{W} = W^3, q = u, d \quad i\frac{g_2^2}{2}\eta_{\mu\nu}$$

$$\mathcal{W} = B, q = u, d \quad i\frac{g_Y^2}{2}Y_L^2\eta_{\mu\nu}$$

$$q = u \quad i\frac{g_Y g_2}{2}Y_L\eta_{\mu\nu}$$

$$q = d \quad -i\frac{g_Y g_2}{2}Y_L\eta_{\mu\nu}$$

Fig.5.3 contd.

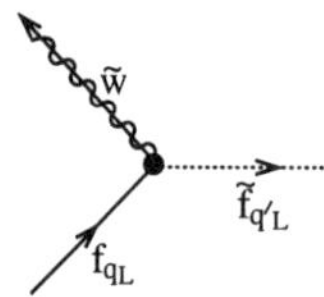

$q = u, d \qquad i\dfrac{g_2^2}{2} Y_{qR}^2 \eta_{\mu\nu}$

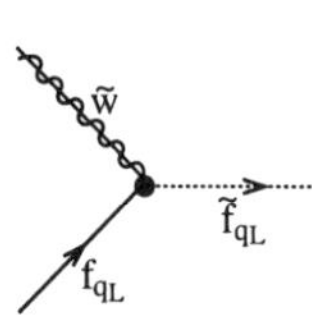

$q = u, q' = d, \tilde{w} = \tilde{w}^+ \qquad -ig_2 P_L$

$q = d, q' = u, \tilde{w} = \tilde{w}^- \qquad -ig_2 P_L$

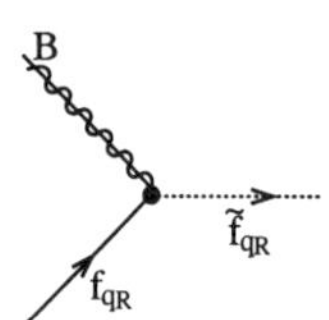

$\tilde{w} = \tilde{w}^3, q = u \qquad -i\dfrac{g_2}{\sqrt{2}} P_L$

$\tilde{w} = \tilde{w}^3, q = d \qquad i\dfrac{g_2}{\sqrt{2}} P_L$

$\tilde{w} = \tilde{B}, q = u, d \qquad -i\dfrac{g_Y}{\sqrt{2}} Y_L P_L$

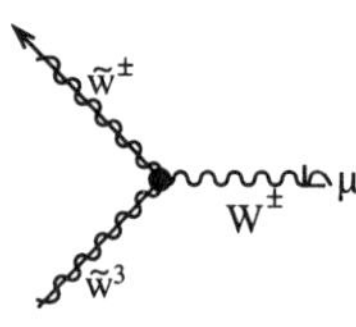

$q = u, d \qquad i\dfrac{g_Y}{\sqrt{2}} Y_{qR} P_R$

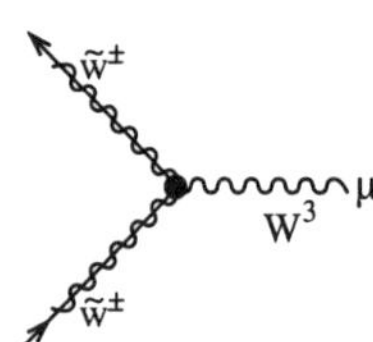

$\pm ig_2 \gamma_\mu$

$\mp ig_2 \gamma_\mu$

$q = q' = u, d \qquad -\dfrac{i}{2}\left(g_2^2 + g_Y^2 Y_L^2\right)$

$q = u, q' = d \qquad -\dfrac{i}{4}\left(g_2^2 + g_Y^2 Y_L^2\right)$

Fig.5.3 contd.

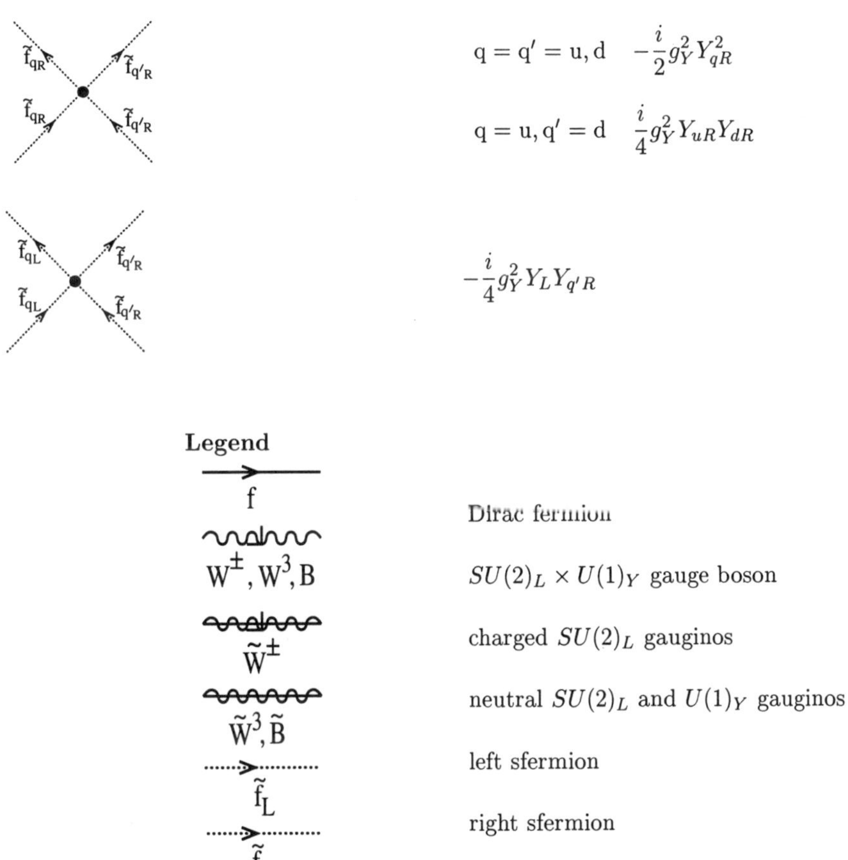

$$q = q' = u, d \qquad -\frac{i}{2} g_Y^2 Y_{qR}^2$$

$$q = u, q' = d \qquad \frac{i}{4} g_Y^2 Y_{uR} Y_{dR}$$

$$-\frac{i}{4} g_Y^2 Y_L Y_{q'R}$$

Legend

f — Dirac fermion

$W^\pm, W^3, B$ — $SU(2)_L \times U(1)_Y$ gauge boson

$\tilde{W}^\pm$ — charged $SU(2)_L$ gauginos

$\tilde{W}^3, \tilde{B}$ — neutral $SU(2)_L$ and $U(1)_Y$ gauginos

$\tilde{f}_L$ — left sfermion

$\tilde{f}_R$ — right sfermion

Fig.5.3. Vertex Feynman rules in $SχGT$

We end with the following comments.

- One needs to be careful about a factor of 2 in going from (5.64) to the vertices in so far as Majorana fermions are concerned.

- Mixed $SU(2)_L$ and $U(1)_Y$ vertices involve the coupling product $g_Y g_2$ as well as the squared sum $g_2^2 + g_Y^2 Y_L^2$. This is a characteristic of quartic vertices with complex scalars which transform nontrivially under the two gauge groups.

- Since chiral symmetry is unbroken here, there exist no couplings which can take left chiral (s)fermions to right chiral ones or vice versa. Indeed, in the nonsupersymmetric case, there are no couplings involving both f_L and f_R. But, in the scalar sector of SχGT, there is a mixed chiral scalar vertex, namely the last quartic one shown above, arising from the $-\frac{1}{2}(D^Y)^2$ term. However, its structure ensures 'chirality' conservation even for sfermions.

References

[5.1] S. Coleman, *op. cit.*, *Bibl.*

[5.2] S. Dimopoulos and H. Georgi, Nucl. Phys. **B193** (1981) 150.

[5.3] S. Ferrara, L. Girardello and F. Palumbo, Phys. Rev. **D20** (1979) 403.

[5.4] J. Wess and B. Zumino, Phys. Lett. **B49** (1974) 52. J. Iliopoulos and B. Zumino, Nucl. Phys. **B76** (1974) 310. J. Wess and J. Bagger, *op. cit.*, *Bibl.*

[5.5] W. Miller, *op. cit.*, *Bibl.*

[5.6] S.B. Treiman, R. Jackiw and D. Gross, *op. cit.*, *Bibl.*

Chapter 6

SUPERSPACE PERTURBATION THEORY AND SUPERGRAPHS

6.1   Nonrenormalization of Superpotential Terms

One of the striking properties of any supersymmetric theory pertains to the superpotential which is controlled by the **nonrenormalization theorem** [6.1]. That theorem has the conseqence that the superpotential does not receive any renormalization to any order[1] in perturbation theory. This means that the wave function renormalizations of the superfields appearing in superpotential terms must match oppositely with the renormalizations of the coefficient masses and couplings in those terms. If the former are regarded as independent, the latter will be dependent on them. Thus, for renormalized perturbation calculations, no counterterms need to be added to those terms in the Lagrangian density which are derived from the superpotential. To one loop order, the nonrenormalization theorem has been illustrated in a toy supersymmetric model in §1.3. A technical point, that was made (cf. ftnt.8, Ch.1) in connection with this illustration, was the need to avoid using the customary dimensional regularization [6.3] procedure but to employ modified dimensional reduction or $\overline{\text{DR}}$ [6.4]. Here the subtraction procedure is as in the $\overline{\text{MS}}$ scheme (cf. §1.3) and the momentum integrals, with propagator factors in the denominator, are evaluated in $d = 2\omega$ dimensions before eventually taking the limit $\omega \to 2$. But the γ-algebra and the $\mathcal{D}$-algebra in the numerator are done strictly in four dimensions since it is only there that the numbers of fermions and bosons match in the case of a supersymmetric system. The components of a gauge field A_μ^a, with $2\omega < \mu \le 4$, act as scalar fields ("ϵ-scalars") in the adjoint representation of the gauge group, but a variant of the $\overline{\text{DR}}$ scheme exists[2] where they completely decouple from physical degrees of freedom. Even the $\overline{\text{DR}}$ scheme in superspace is not completely free of trouble; there are problems [6.4] when the number of loops gets large (> 4), though they may not cause any practical difficulties. This, however, is the most convenient supersymmetric regularization scheme available so far and we shall adopt it.

A more important point, earlier alluded to in Ch.1, is the following. The nonrenormaliza-

---

[1] Certain deviations can arise in higher loops [6.2] when massless fields are present, but we ignore such subtleties here.

[2] I. Jack, D.R.T. Jones, S.P. Martin, M.T. Vaughn and Y. Yamada, *loc. cit.*, Ref. [6.4].

tion, mentioned above, is the key argument behind the perturbative stability of the hierarchy between the electroweak scale ($\sim 10^2$ GeV) and a high unification scale ($\sim 10^{16}$ GeV) or the scale where gravity becomes strong ($\sim 10^{19}$ GeV) in a supersymmetric theory. Once this type of a hierarchy is fixed at the tree level, it does not get destabilized by perturbative radiative corrections. In order to understand the origin of this result, one needs to employ supergraphs. The latter are also needed to efficiently perform renormalization group calculations in supersymmetric field theories, which will be required later. Therefore, in this chapter, we aim to introduce supergraph techniques with super-Feynman rules of perturbation theory in superspace. We have to develop, for this purpose, a formalism for the quantization of superfields. To that end, we first need to generalize functional methods to superspace. This is done below by largely following the two works cited in Ref. [6.5].

6.2 Functional Methods in Superspace

Points 1 and 2 in superspace can be represented by supercoordinates $z_1 = (x_1, \theta_1, \bar{\theta}_1)$ and $z_2 = (x_2, \theta_2, \bar{\theta}_2)$ respectively. We shall always use the notations $d^8z \equiv d^4x d^4\theta$, $d^6z \equiv d^4x d^2\theta$ and $d^6\bar{z} \equiv d^4x d^2\bar{\theta}$, introduced in Ch.4. Referring back to (4.4i,j), we can define a fermionic subspace δ-function and describe its properties[3]:

$$\delta_{12} \equiv \delta^{(4)}(\theta_1 - \theta_2) \equiv \delta^{(2)}(\theta_1 - \theta_2)\delta^{(2)}(\bar{\theta}_1 - \bar{\theta}_2) = (\theta_1 - \theta_2)(\theta_1 - \theta_2)\,(\bar{\theta}_1 - \bar{\theta}_2)(\bar{\theta}_1 - \bar{\theta}_2)\,, \quad (6.1a)$$

$$\int d^4\theta_l \delta_{12} = 1, \quad (l = 1, 2)\,, \tag{6.1b}$$

$$\frac{1}{16}\partial_l \partial_l\,\bar{\partial}_m \bar{\partial}_m \delta_{12} = 1 \quad (l, m = 1 \text{ or } 2, \text{ not summed})\,. \tag{6.1c}$$

It follows from (4.4) as well as from (4.9 - 4.11), by ignoring surface terms, that – under a spacetime integral $\int d^4x$ – the operator $\mathcal{D}_l \mathcal{D}_l\,\bar{\mathcal{D}}_m \bar{\mathcal{D}}_m$ behaves like $\partial_l \partial_l\,\bar{\partial}_m \bar{\partial}_m$, l, m not summed. (6.1a) and (6.1c) then imply that

$$\frac{1}{16}\mathcal{D}_l \mathcal{D}_l\,\bar{\mathcal{D}}_m \bar{\mathcal{D}}_m \delta_{12} \doteq 1\,, \tag{6.2a}$$

$$\frac{1}{16}\delta_{12}\mathcal{D}_l \mathcal{D}_l\,\bar{\mathcal{D}}_m \bar{\mathcal{D}}_m \delta_{12} = \delta_{12}\,, \tag{6.2b}$$

l, m not being summed. The symbol $\doteq$ in (6.2a) means that the equality is valid only under a spacetime integral $\int d^4x$; multiplication by a general superfunction from the right, but not from the left, is permitted. No such restriction is necessary for (6.2b).

Take a general superfield $\mathcal{F}(z) \equiv \mathcal{F}(x, \theta, \bar{\theta})$ which spans *all* of superspace, e.g. a vector superfield V. (Chiral superfields, which span only parts of superspace are therefore not covered here; they will be treated in the next paragraph). A functional derivative of $\mathcal{F}(z)$ can be defined with the following properties:

$$\frac{\delta \mathcal{F}(2)}{\delta \mathcal{F}(1)} = \delta^{(4)}(x_1 - x_2)\delta_{12} \equiv \delta^{(8)}(z_1 - z_2)\,. \tag{6.3a}$$

---

[3]The subscript l or m here standing for 1,2, refer to two points in superspace and should not be confused with the spinorial components 1,2 of θ which are denoted by A, B or with those of $\bar{\theta}$, as denoted by $\dot{A}, \dot{B}$.

$$\int d^8 z_2 \frac{\delta \mathcal{F}(2)}{\delta \mathcal{F}(1)} = 1 \ . \tag{6.3b}$$

The functional products between two superfields $\mathcal{F}_1(z)$, $\mathcal{F}_2(z)$ and more specifically between a superfield $\mathcal{F}$ and its source $\mathcal{J}$ are given by

$$\mathcal{F}_1 \cdot \mathcal{F}_2 \equiv \int d^8 z \mathcal{F}_1(z) \mathcal{F}_2(z), \tag{6.4a}$$

$$\mathcal{F} \cdot \mathcal{J} \equiv \int d^8 z \mathcal{F}(z) \mathcal{J}(z) \tag{6.4b}$$

respectively.

We now turn to chiral superfields $\Phi(x, \theta, \bar{\theta}) \equiv \Phi(y, \theta)$ and $\Phi^\dagger(x, \theta, \bar{\theta}) \equiv \Phi^\dagger(\bar{y}, \bar{\theta})$ with $y, \bar{y}$ as defined just before (4.16). It is clear from the action (cf. §5.1) of any theory, incorporating chiral superfields, that the latter do not span all of superspace. Instead of connecting to the full superspace integral $\int d^8 z$, they span the parts picked by a factor $\delta^{(2)}(\theta)$ or $\delta^{(2)}(\bar{\theta})$. The equivalents of (6.3b) are now

$$\int d^6 z_2 \frac{\delta \Phi(2)}{\delta \Phi(1)} = 1 \ , \tag{6.5a}$$

$$\int d^6 \bar{z}_2 \frac{\delta \Phi^\dagger(2)}{\delta \Phi^\dagger(1)} = 1 \ , \tag{6.5b}$$

while those of (6.4a) are[4]

$$\Phi_1 \cdot \Phi_2 \equiv \int d^6 z \Phi_1 \Phi_2 \ , \tag{6.6a}$$

$$\Phi_1^\dagger \cdot \Phi_2^\dagger \equiv \int d^6 \bar{z} \Phi_1^\dagger \Phi_2^\dagger \ . \tag{6.6b}$$

For a general supersymmetric action, the superpotential term, which is a chiral superfield, will have this limitation. In contrast, the kinetic energy and other terms, that allow being clubbed under a vector superfield, connect to the full superspace and are therefore not covered here; they can be treated functionally by using (6.3) or (6.4).

In order to overcome this difficulty and develop a common formalism, we introduce a left chiral superspace projector Π_+ and a right chiral superspace projector Π_-, defined by

$$\Pi_+ \equiv -\frac{1}{16\partial^2} \bar{\mathcal{D}}\bar{\mathcal{D}} \, \mathcal{D}\mathcal{D} \ , \tag{6.7a}$$

$$\Pi_- \equiv -\frac{1}{16\partial^2} \mathcal{D}\mathcal{D} \, \bar{\mathcal{D}}\bar{\mathcal{D}} \ , \tag{6.7b}$$

where $\partial^2 \equiv \partial^\mu \partial_\mu$ is the d'Alambertian in spacetime and the nonlocal operator $1/\partial^2$ is defined as in (4.14e,f). The choice of the forms of $\Pi_\pm$ is motivated by (4.14a,b) which also imply that

$$\Pi_+ \Phi = \Phi, \ \Pi_- \Phi^\dagger = \Phi^\dagger \ .$$

---

[4] $\Phi_1 \cdot \Phi_2 = \Phi_2 \cdot \Phi_1$ and $\Phi_1^\dagger \cdot \Phi_2^\dagger = \Phi_2^\dagger \cdot \Phi_1^\dagger$, assuming that the Φ's are even under the Z_2-grading mentioned in §3.1.

Evidently, acting on a general superfield $\mathcal{F}$, Π_+ yields a left chiral superfield since $\bar{\mathcal{D}}^{\dot{A}}\Pi_+\mathcal{F} = 0$ while Π_- yields a right chiral one since $\mathcal{D}_A\Pi_-\mathcal{F} = 0$. Finally,

$$\Pi_+^2 = \left(\frac{1}{16\partial^2}\right)^2 \bar{\mathcal{D}}\bar{\mathcal{D}}\,\mathcal{D}\mathcal{D}\,\bar{\mathcal{D}}\bar{\mathcal{D}}\,\mathcal{D}\mathcal{D} = -\frac{1}{16\partial^2}\bar{\mathcal{D}}\bar{\mathcal{D}}\,\mathcal{D}\mathcal{D} = \Pi_+ \,, \tag{6.8a}$$

$$\Pi_-^2 = \left(\frac{1}{16\partial^2}\right)^2 \mathcal{D}\mathcal{D}\,\bar{\mathcal{D}}\bar{\mathcal{D}}\,\mathcal{D}\mathcal{D}\,\bar{\mathcal{D}}\bar{\mathcal{D}} = \Pi_- \,, \tag{6.8b}$$

where we have used (4.11l,m). Moreover, $\Pi_+\Pi_- = \Pi_-\Pi_+ = 0$. Thus Π_+ satisfies all the requisite properties of a left chiral projector and Π_- satisfies those of a right chiral one. One can also introduce a third projector

$$\Pi_0 \equiv \frac{\mathcal{D}^A\bar{\mathcal{D}}\bar{\mathcal{D}}\mathcal{D}_A}{8\partial^2} = \frac{\bar{\mathcal{D}}_{\dot{A}}\mathcal{D}\mathcal{D}\bar{\mathcal{D}}^{\dot{A}}}{8\partial^2} \,, \tag{6.9}$$

cf. (4.11k). The set of projectors $\Pi_i = (\Pi_+, \Pi_0, \Pi_-)$, $i = 1, 2, 3$, may be seen to satisfy the relations

$$\sum_{i=1}^{3} \Pi_i = 1, \quad \Pi_i\Pi_j = \delta_{ij}\Pi_j \,. \tag{6.10}$$

We can now rewrite (6.6a,b) as

$$\Phi_1\cdot\Phi_2 = \int d^6z\,\Phi_1\Pi_+\Phi_2 = -\int d^6z\,\Phi_1\frac{\bar{\mathcal{D}}\bar{\mathcal{D}}}{4}\frac{\mathcal{D}\mathcal{D}}{4\partial^2}\Phi_2 = -\int d^6z\,\frac{\bar{\mathcal{D}}\bar{\mathcal{D}}}{4}\left(\Phi_1\frac{\mathcal{D}\mathcal{D}}{4\partial^2}\Phi_2\right),$$

$$\Phi_1^\dagger\cdot\Phi_2^\dagger = \int d^6\bar{z}\,\Phi_1^\dagger\Pi_-\Phi_2^\dagger = -\int d^6\bar{z}\,\Phi_1^\dagger\frac{\mathcal{D}\mathcal{D}}{4}\frac{\bar{\mathcal{D}}\bar{\mathcal{D}}}{4\partial^2}\Phi_2^\dagger = -\int d^6\bar{z}\,\frac{\mathcal{D}\mathcal{D}}{4}\left(\Phi_1^\dagger\frac{\bar{\mathcal{D}}\bar{\mathcal{D}}}{4\partial^2}\Phi_2\right),$$

where we have used the facts that $\bar{\mathcal{D}}^{\dot{A}}\Phi_1 = 0 = \mathcal{D}_A\Phi_1^\dagger$. Now one can employ (4.12d) and footnote 4 to rewrite the above equations as

$$\Phi_1\cdot\Phi_2 = \int d^8z\,\Phi_1\frac{\mathcal{D}\mathcal{D}}{4\partial^2}\Phi_2 = \int d^8z\left(\frac{\mathcal{D}\mathcal{D}}{4\partial^2}\Phi_1\right)\Phi_2 \,, \tag{6.11a}$$

$$\Phi_1^\dagger\cdot\Phi_2^\dagger = \int d^8z\,\Phi_1^\dagger\frac{\bar{\mathcal{D}}\bar{\mathcal{D}}}{4\partial^2}\Phi_2^\dagger = \int d^8z\left(\frac{\bar{\mathcal{D}}\bar{\mathcal{D}}}{4\partial^2}\Phi_1^\dagger\right)\Phi_2^\dagger \,, \tag{6.11b}$$

though one still has $\Phi_1^\dagger\cdot\Phi_2 = \int d^8z\,\Phi_1^\dagger\Phi_2$. The shifting of the nonlocal operators $\mathcal{D}\mathcal{D}/\partial^2$, $\bar{\mathcal{D}}\bar{\mathcal{D}}/\partial^2$ in (6.11) from one chiral superfield to the other (by applying $\Pi_\pm$ on the latter instead) is a useful trick. Now, returning to (6.5a) and comparing with (4.12c), we see that the former, as well as (6.5b), is solved by the following equations for the integrands:

$$\frac{\delta\Phi(2)}{\delta\Phi(1)} = -\frac{1}{4}\bar{\mathcal{D}}_1\bar{\mathcal{D}}_1\delta^{(8)}(z_1 - z_2) \,, \tag{6.12a}$$

$$\frac{\delta\Phi^\dagger(2)}{\delta\Phi^\dagger(1)} = -\frac{1}{4}\mathcal{D}_1\mathcal{D}_1\delta^{(8)}(z_1 - z_2) \,. \tag{6.12b}$$

Eqs.(6.12) are the left and right chiral superfield equivalents of (6.3a). We may note here some consequences of the definitions (4.9) and (4.10) of $\mathcal{D}$ and $\bar{\mathcal{D}}$ and of $\delta^{(8)}(z_1 - z_2)$ as $(2\pi)^{-4} \int d^4 p e^{-ip\cdot(x_1-x_2)} \delta_{12}$, namely

$$\mathcal{D}_{1A} \delta^{(8)}(z_1 - z_2) = -\mathcal{D}_{2A} \delta^{(8)}(z_1 - z_2) \,, \tag{6.13a}$$

$$\bar{\mathcal{D}}_1^{\dot{A}} \delta^{(8)}(z_1 - z_2) = -\bar{\mathcal{D}}_2^{\dot{A}} \delta^{(8)}(z_1 - z_2) \,, \tag{6.13b}$$

$$\mathcal{D}_1 \mathcal{D}_1 \delta^{(8)}(z_1 - z_2) = \mathcal{D}_2 \mathcal{D}_2 \delta^{(8)}(z_1 - z_2) \,, \tag{6.13c}$$

$$\bar{\mathcal{D}}_1 \bar{\mathcal{D}}_1 \delta^{(8)}(z_1 - z_2) = \bar{\mathcal{D}}_2 \bar{\mathcal{D}}_2 \delta^{(8)}(z_1 - z_2) \,. \tag{6.13d}$$

6.3 Functional Formulation of Superfield Theory

We start with a general superfield (cf. §4.1) $\mathcal{F}$, chiral superfields Φ, $\Phi^\dagger$ and a generic action

$$S = \int d^8 z K_0(\mathcal{F}, \Phi, \Phi^\dagger) + \left(\int d^6 z \mathcal{W}(\Phi) + h.c. \right). \tag{6.14}$$

In (6.14) K_0 is a vector superfield containing all the kinetic energy as well as other allowed interaction terms, while $\mathcal{W}(\Phi)$ is the superpotential. The generating superfunctional – with source superfields j, J matching onto $\mathcal{F}$, Φ respectively – is given by the following functional integral over superfields:

$$Z[j, J, J^\dagger] = N \int [d\mathcal{F}][d\Phi][d\Phi^\dagger] \exp i(S + j\cdot\mathcal{F} + J\cdot\Phi + J^\dagger\cdot\Phi^\dagger) \,, \tag{6.15}$$

where N is the normalization constant.

Let us use a square bracket] to the right to mean an evaluation at $0 = J = j$. We can then write the m-point function corresponding to (6.15) as

$$\frac{\delta^m Z[j, J, J^\dagger]}{\delta j(1) \cdots \delta j(n) \delta J(n+1) \cdots \delta J(k) \delta J^\dagger(k+1) \cdots \delta J^\dagger(m)} \bigg]$$

$$= i^m \langle \Omega | T \mathcal{F}(1) \cdots \mathcal{F}(n) \Phi(n+1) \cdots \Phi(k) \Phi^\dagger(k+1) \cdots \Phi^\dagger(m) | \Omega \rangle.$$

The connected Green's function $\langle \mathcal{F}(1) \cdots \mathcal{F}(n) \Phi(n+1) \cdots \Phi(k) \Phi^\dagger(k+1) \cdots \Phi^\dagger(m) \rangle_C$ is given in terms of

$$W[j, J, J^\dagger] \equiv -i \ln Z[j, J, J^\dagger]$$

by

$$i \frac{\delta^m W[j, J, J^\dagger]}{\delta j(1) \cdots \delta j(n) \delta J(n+1) \cdots \delta J(k) \delta J^\dagger(k+1) \cdots \delta J^\dagger(m)} \bigg]$$

$$= i^m \langle \mathcal{F}(1) \cdots \mathcal{F}(n) \Phi(n+1) \cdots \Phi(k) \Phi^\dagger(k+1) \cdots \Phi^\dagger(m) \rangle_C \,. \tag{6.16}$$

The "classical" superfields Φ, $\Phi^\dagger$ and $\mathcal{F}$, which are solutions to the equations of motion, relate to W via

$$\Phi = \frac{\delta W}{\delta J} \,, \quad \Phi^\dagger = \frac{\delta W}{\delta J^\dagger} \,, \quad \mathcal{F} = \frac{\delta W}{\delta j} \,.$$

These relations can be inverted to obtain expressions for J, $J^\dagger$ and j in terms of Φ, $\Phi^\dagger$ and $\mathcal{F}$. The substitution of those in W, plus the explicit subtraction of the $j\cdot\mathcal{F} + J\cdot\Phi + J^\dagger\cdot\Phi^\dagger$ piece, lead to [6.5] the effective action $\Gamma(\Phi, \Phi^\dagger, \mathcal{F})$ which is the quantum analog of the classical action $S[\Phi]$:

$$\Gamma(\Phi, \Phi^\dagger, \mathcal{F}) = W\left[j(\Phi, \Phi^\dagger, \mathcal{F}), J(\Phi, \mathcal{F}), J^\dagger(\Phi^\dagger, \mathcal{F})\right] - j\cdot\mathcal{F} - J\cdot\Phi - J^\dagger\cdot\Phi^\dagger \ . \tag{6.17}$$

Just as W generates the m-point connected Green's function, cf. (6.16), Γ generates one particle irreducible supergraphs $\Gamma^m(1, \cdots, m)$ $\forall$ positive integral m. If the classical action (6.14) splits additively into a free part S^0 and an interacting part $S^{\text{int.}}$ as

$$S = S^0 + S^{\text{int.}} \ ,$$

then (6.15) can be rewritten as

$$Z[j, J, J^\dagger] = \exp\left(iS^{\text{int.}}\left[\frac{1}{i}\frac{\delta}{\delta j}, \frac{1}{i}\frac{\delta}{\delta J}, \frac{1}{i}\frac{\delta}{\delta J^\dagger}\right]\right) Z_0[j, J, J^\dagger] \tag{6.18}$$

with

$$Z_0[j, J, J^\dagger] = N_0 \int [d\mathcal{F}][d\Phi][d\Phi^\dagger] \exp i(S^0 + j\cdot\mathcal{F} + J\cdot\Phi + J^\dagger\cdot\Phi^\dagger) \ , \tag{6.19a}$$

$$W_0[j, J, J^\dagger] = -i\ln Z_0[j, J, J^\dagger] \ , \tag{6.19b}$$

N_0 being the normalization and W_0 being the generating functional for free, connected Green's functions. Eq. (6.18) can be inverted [6.5] and rewritten as

$$Z_0[j, J, J^\dagger] = \exp\left(-iS^{\text{int.}}\left[\frac{1}{i}\frac{\delta}{\delta j}, \frac{1}{i}\frac{\delta}{\delta J}, \frac{1}{i}\frac{\delta}{\delta J^\dagger}\right]\right) Z[j, J, J^\dagger] \ .$$

The perturbative procedure goes as follows. Free, connected propagators of the theory are first obtained by explicitly evaluating the integrals in $W_0[j, J, J^\dagger]$ and then substituting that expression in place of W in the LHS of (6.16) for $m = 2$. The exponent in the RHS of (6.18) is then expanded in a perturbation series and the Feynman factors obtained order by order. This can be seen more concretely in the explicit case of the chiral superfield, as illustrated below.

Chiral Superfield: connected free Green's functions

The main technically nontrivial point in developing the superperturbation expansion for a chiral superfield is the need to keep using equations such as (6.11a) or (6.11b) in converting the subspace integral $\int d^6 z$, as in (6.6a), or $\int d^6 \bar{z}$, as in (6.6b), to the full superspace one, namely $\int d^8 z$. Thus the free part of a chiral superfield generating functional, corresponding to that a Wess-Zumino model (cf. §5.1), with vanishing couplings $h = f = 0$, is

$$Z_0[J, J^\dagger] = N_0 \int [d\Phi][d\Phi^\dagger] \exp i\left\{\Phi^\dagger\cdot\Phi + \left(J\cdot\Phi + \frac{1}{2}m\Phi\cdot\Phi + h.c.\right)\right\} \ .$$

It can be recast as

$$
\begin{aligned}
Z_0[J, J^\dagger] &= N_0 \int [d\Phi][d\Phi^\dagger]\exp i\left[\int d^8z\left\{\Phi^\dagger(z)\Phi(z) + \frac{m}{8}\Phi(z)\frac{\mathcal{D}\mathcal{D}}{\partial^2}\Phi(z)\right.\right.\\
&\quad \left.\left. + \frac{m}{8}\Phi^\dagger(z)\frac{\bar{\mathcal{D}}\bar{\mathcal{D}}}{\partial^2}\Phi^\dagger(z) + \frac{1}{4}\Phi(z)\frac{\mathcal{D}\mathcal{D}}{\partial^2}J(z) + \frac{1}{4}\Phi^\dagger(z)\frac{\bar{\mathcal{D}}\bar{\mathcal{D}}}{\partial^2}J^\dagger(z)\right\}\right]\\
&= N_0 \int [d\Phi][d\Phi^\dagger]\exp i\left[\int d^8z\left\{\frac{1}{2}\left(\Phi(z)\ \ \Phi^\dagger(z)\right)\begin{pmatrix} \dfrac{m}{4}\dfrac{\mathcal{D}\mathcal{D}}{\partial^2} & 1 \\ 1 & \dfrac{m}{4}\dfrac{\bar{\mathcal{D}}\bar{\mathcal{D}}}{\partial^2}\end{pmatrix}\begin{pmatrix}\Phi(z)\\ \Phi^\dagger(z)\end{pmatrix}\right.\right.\\
&\quad \left.\left. + \left(\Phi(z)\ \ \Phi^\dagger(z)\right)\begin{pmatrix}\dfrac{\mathcal{D}\mathcal{D}}{4\partial^2}J(z)\\ \dfrac{\bar{\mathcal{D}}\bar{\mathcal{D}}}{4\partial^2}J^\dagger(z)\end{pmatrix}\right\}\right].
\end{aligned}
$$

(6.20)

In the last step of (6.20), we have employed a complex two dimensional linear space of chiral superfields spanned by the row vector $\left(\Phi(z)\ \ \Phi^\dagger(z)\right)$ and the corresponding column vector. We can now use the identity

$$
\int dwdw^\dagger \exp i\left[\frac{1}{2}(w\ \ w^\dagger)\underset{\sim}{Q}\begin{pmatrix}w\\ w^\dagger\end{pmatrix} + (w\ \ w^\dagger)\begin{pmatrix}y\\ y^\dagger\end{pmatrix}\right]
$$

$$
= \text{const.}\exp\left[-\frac{i}{2}\left(y\ \ y^\dagger\right)\underset{\sim}{Q}{}^{-1}\begin{pmatrix}y\\ y^\dagger\end{pmatrix}\right],
$$

(6.21)

where $\underset{\sim}{Q}$ is any symmetric 2×2 matrix with nonzero determinant acting in the complex two dimensional linear space spanned by the variable vectors w, $w^\dagger$ and the fixed vectors y, $y^\dagger$.

For

$$
\underset{\sim}{Q} = \begin{pmatrix}\dfrac{m}{4}\dfrac{\mathcal{D}\mathcal{D}}{\partial^2} & 1 \\ 1 & \dfrac{m}{4}\dfrac{\bar{\mathcal{D}}\bar{\mathcal{D}}}{\partial^2}\end{pmatrix},
$$

acting between the row and column vectors of chiral superfields, as in (6.20), we have

$$
\underset{\sim}{Q}{}^{-1} = \begin{bmatrix} -\dfrac{m\bar{\mathcal{D}}\bar{\mathcal{D}}}{4(\partial^2 + m^2)} & 1 + \dfrac{m^2\bar{\mathcal{D}}\bar{\mathcal{D}}\,\mathcal{D}\mathcal{D}}{16\partial^2(\partial^2 + m^2)} \\ 1 + \dfrac{m^2\mathcal{D}\mathcal{D}\,\bar{\mathcal{D}}\bar{\mathcal{D}}}{16\partial^2(\partial^2 + m^2)} & -\dfrac{m\mathcal{D}\mathcal{D}}{4(\partial^2 + m^2)} \end{bmatrix}.
$$

That $\underset{\sim}{Q}{}^{-1}\underset{\sim}{Q} = \underset{\sim}{Q}\,\underset{\sim}{Q}{}^{-1} = \underset{\sim}{I}$ can be verified by means of (4.11 l,m). Now the use of (6.21) in (6.20) enables us to rewrite the latter as

$$
Z_0[J, J^\dagger] = Z_0[0, 0]\exp\left\{-\frac{i}{2}\int d^8z\left(\frac{\mathcal{D}\mathcal{D}}{4\partial^2}J(z)\ \ \frac{\bar{\mathcal{D}}\bar{\mathcal{D}}}{4\partial^2}J^\dagger(z)\right)\underset{\sim}{Q}{}^{-1}\begin{pmatrix}\dfrac{\mathcal{D}\mathcal{D}}{4\partial^2}J(z)\\ \dfrac{\bar{\mathcal{D}}\bar{\mathcal{D}}}{4\partial^2}J^\dagger(z)\end{pmatrix}\right\}.
$$

After straightforward algebra this equation can be simplified to

$$Z_0[J, J^\dagger] = Z_0[0,0] \exp\left\{ -\frac{1}{2} \int d^8 z \left[\frac{m}{4} \left(\frac{\mathcal{DD}}{\partial^2} J \right) \frac{1}{\partial^2 + m^2} J + \frac{m}{4} \left(\frac{\bar{\mathcal{D}}\bar{\mathcal{D}}}{\partial^2} J^\dagger \right) \frac{1}{\partial^2 + m^2} J^\dagger \right. \right.$$

$$\left. \left. + \frac{1}{16} \left(\frac{\bar{\mathcal{D}}\bar{\mathcal{D}}}{\partial^2} J^\dagger \right) \frac{\mathcal{DD}}{\partial^2 + m^2} J + \frac{1}{16} \left(\frac{\mathcal{DD}}{\partial^2} J \right) \frac{\bar{\mathcal{D}}\bar{\mathcal{D}}}{\partial^2 + m^2} J^\dagger \right] \right\}.$$

The number of covariant derivatives in each of the terms of the RHS exponent above can be reduced. Observe that J, $(\partial^2 + m^2)^{-1}J$ and $\bar{\mathcal{D}}\bar{\mathcal{D}}(\partial^2 + m^2)^{-1}J^\dagger$ are all left chiral superfields while each of $J^\dagger$, $(\partial^2 + m^2)^{-1}J^\dagger$ and $\mathcal{DD}(\partial^2 + m^2)^{-1}J$ is right chiral. The shifting of the nonlocal operators $\mathcal{DD}/\partial^2$, $\bar{\mathcal{D}}\bar{\mathcal{D}}/\partial^2$ from the first superfield factor to the second, as per (6.11), in each RHS term above and the use of (4.14) lead to the following expression for Z_0:

$$Z_0[J, J^\dagger] = Z_0[0,0] \exp\left\{ -\frac{1}{2} \int d^8 z \left(J(z) \ J^\dagger(z) \right) \frac{1}{\partial^2 + m^2} \begin{pmatrix} \dfrac{m\mathcal{DD}}{4\partial^2} & -1 \\ -1 & \dfrac{m\bar{\mathcal{D}}\bar{\mathcal{D}}}{4\partial^2} \end{pmatrix} \begin{pmatrix} J(z) \\ J^\dagger(z) \end{pmatrix} \right\}.$$

We can then write

$$Z_0[J, J^\dagger] = Z_0[0,0] \exp\left\{ -\frac{1}{2} \int d^8 z d^8 z' \left(J(z) \ J^\dagger(z) \right) \Delta_F(z - z') \begin{pmatrix} J(z') \\ J^\dagger(z') \end{pmatrix} \right\}$$

$$= Z_0[0,0] \exp\left\{ -\frac{1}{2} (J \ J^\dagger) \cdot \Delta_F \cdot \begin{pmatrix} J \\ J^\dagger \end{pmatrix} \right\}, \tag{6.22a}$$

$$\Delta_F(z - z') = \frac{1}{\partial^2 + m^2} \begin{pmatrix} \dfrac{m\mathcal{DD}}{4\partial^2} & -1 \\ -1 & \dfrac{m\bar{\mathcal{D}}\bar{\mathcal{D}}}{4\partial^2} \end{pmatrix} \delta^{(8)}(z - z') . \tag{6.22b}$$

One can try to draw a direct analogy between (6.22a) and the corresponding expression [6.6] for the generating functional of free connected Green's functions in a scalar quantum field theory (QFT). That would suggest that Δ_F should be the chiral superfield propagator matrix, i.e. $i\Delta_F$ should be the matrix for the connected two point Green's functions in the $(\Phi(z) \ \Phi^\dagger(z))$ linear space. Unfortunately, such is not the case. Unlike in nonsupersymmetric scalar QFT, the source superfields $J, J^\dagger$ are chiral. Therefore, $\delta J(2)/\delta J(1)$ and $\delta J^\dagger(2)/\delta J^\dagger(1)$, instead of being just $\delta^{(8)}(z_1 - z_2)$, are given by

$$\frac{\delta J(2)}{\delta J(1)} = -\frac{1}{4}\bar{\mathcal{D}}_1\bar{\mathcal{D}}_1 \delta^{(8)}(z_1 - z_2) , \tag{6.23a}$$

$$\frac{\delta J^\dagger(2)}{\delta J^\dagger(1)} = -\frac{1}{4}\mathcal{D}_1\mathcal{D}_1 \delta^{(8)}(z_1 - z_2) . \tag{6.23b}$$

Now, on using (6.22) in (6.16), we see that the only nonzero connected free particle Green's functions are

$$\langle \Phi(1)\Phi^\dagger(2) \rangle_C \equiv -i \frac{\delta^2 W_0[J]}{\delta J(1)\delta J^\dagger(2)} \right] = -\frac{i}{16} \frac{\bar{\mathcal{D}}_1\bar{\mathcal{D}}_1 \ \mathcal{D}_1\mathcal{D}_1}{\partial_1^2 + m^2} \delta^{(8)}(z_1 - z_2) , \tag{6.24a}$$

$$\langle \Phi(1)\Phi(2)\rangle_C \equiv -i\frac{\delta^2 W_0[J]}{\delta J(1)\delta J(2)}\bigg] = -\frac{i}{4}\frac{m\bar{\mathcal{D}}_1\bar{\mathcal{D}}_1}{\partial_1{}^2 + m^2}\delta^{(8)}(z_1 - z_2)\,, \tag{6.24b}$$

$$\langle \Phi^\dagger(1)\Phi^\dagger(2)\rangle_C = -\frac{i}{4}\frac{m\mathcal{D}_1\mathcal{D}_1}{\partial_1{}^2 + m^2}\delta^{(8)}(z_1 - z_2)\,. \tag{6.24c}$$

Note that, in deriving (6.24b,c), use has been made of (4.11ℓ,m) and (6.13c,d). These chiral superfield propagators, as described by (6.24), are originally due to Salam and Strathdee [6.7]. In comparison with (6.22b), the corresponding matrix form of the above chiral super-propagators in the $(\Phi(z)\ \Phi^\dagger(z'))$ linear space is $\Delta_F^{SS}(z - z')$ where

$$\Delta_F^{SS} = -\frac{1}{4(\partial^2 + m^2)}\begin{pmatrix} m\bar{\mathcal{D}}\bar{\mathcal{D}} & \dfrac{\bar{\mathcal{D}}\bar{\mathcal{D}}\,\mathcal{D}\mathcal{D}}{4} \\ \dfrac{\mathcal{D}\mathcal{D}\,\bar{\mathcal{D}}\bar{\mathcal{D}}}{4} & m\mathcal{D}\mathcal{D} \end{pmatrix}\delta^{(8)}(z - z') \tag{6.25}$$

and not the Δ_F of (6.22b).

By setting $\theta_{1,2}$ and $\bar{\theta}_{1,2}$ in (6.24a) to zero, one can replace $\mathcal{D}_{1A}$, $\bar{\mathcal{D}}_1^{\dot{A}}$ in its RHS by ∂_{1A}, $\bar{\partial}_1^{\dot{A}}$ respectively. Making use of (6.1c), one is then able to obtain the free connected scalar two point function:

$$\langle \phi(1)\phi^\star(2)\rangle_C = -\frac{i}{\partial_1{}^2 + m^2}\delta^{(4)}(x_1 - x_2)\,. \tag{6.26}$$

The corresponding Green's functions with higher components in the superfields $\Phi_{1,2}$ can be extracted by the application of appropriate covariant derivatives on (6.24a). For instance, suppose we apply $\mathcal{D}_{1A}\bar{\mathcal{D}}_{2\dot{B}}$, use (6.1a) and then put $\theta_{1,2} = 0 = \bar{\theta}_{1,2}$, utilizing (4.23b) and (4.24b). We are then left with

$$\langle \xi_A(1)\bar{\xi}_{\dot{B}}(2)\rangle_C = -\frac{i}{16}\mathcal{D}_{1A}\frac{\bar{\mathcal{D}}_1\bar{\mathcal{D}}_1\bar{\mathcal{D}}_{2\dot{B}}\mathcal{D}_1\mathcal{D}_1}{\partial_1{}^2 + m^2}\delta^{(8)}(z_1 - z_2)\bigg| = \frac{2\sigma^\mu_{A\dot{B}}\partial_{1\mu}\delta^{(4)}(x_1 - x_2)}{\partial_1{}^2 + m^2}\,, \tag{6.27}$$

where we have used (4.11e) in the last step of (6.27).

As an immediate demonstration of the power of the Green's functions (6.24), we can state the following result [6.8]:

> *Every L-loop ($L \geq 1$) in a supergraph, containing either only*
> *left chiral or only right chiral vertices, must vanish[5].*

It is sufficient to prove the above in the left chiral case. Note that such a loop will involve only left chiral internal lines and will therefore be proportional to a product of superpropagators of the type of (6.24b), namely:

$$\langle \Phi(1)\Phi(2)\rangle_C\langle \Phi(2)\Phi(3)\rangle_C \cdots \langle \Phi(k)\Phi(1)\rangle_C\,. \tag{6.28}$$

We can now employ the superspace variables $y, \theta, \bar{y}, \bar{\theta}$ of §4.3 and the fact that $\Phi \equiv \Phi(y, \theta)$, $\Phi^\dagger \equiv \bar{\Phi}^\dagger(\bar{y}, \bar{\theta})$. Thus, on using (4.17b), (6.24b) becomes

$$\langle \Phi(1)\Phi(2)\rangle_C = \frac{i}{4}\frac{m\bar{\partial}_1^{\dot{A}}\bar{\partial}_{1\dot{A}}}{\partial_{y_1}^2 + m^2}\delta^{(4)}(y_1 - y_2)\delta_{12} = \delta^{(2)}(\theta_1 - \theta_2)\frac{im}{\partial_{y_1}^2 + m^2}\delta^{(4)}(y_1 - y_2)\,,$$

---

[5]In particular, the one loop tadpole for Φ or $\Phi^\dagger$ is zero.

where (4.4j) and (4.3) have been used in putting $\bar{\partial}_1\bar{\partial}_1\delta^{(2)}(\bar{\theta}_1 - \bar{\theta}_2) = 4$. On substituting the last displayed equation into the expression (6.28), one obtains a string of δ-function products in θ, namely

$$\delta^{(2)}(\theta_1 - \theta_2)\delta^{(2)}(\theta_2 - \theta_3)\cdots\delta^{(2)}(\theta_{k-1} - \theta_k)\delta^{(2)}(\theta_k - \theta_1)$$

$$= \delta^{(2)}(\theta_1 - \theta_2)\delta^{(2)}(\theta_1 - \theta_3)\cdots\delta^{(2)}(\theta_1 - \theta_k)\delta^{(2)}(\theta_k - \theta_1)$$

$$= 0 \, ,$$

since, for any function f, $\delta^{(2)}(\theta_1-\theta_2)f(\theta_1) = \delta^{(2)}(\theta_1-\theta_2)f(\theta_2)$ and moreover $\delta^{(2)}(\theta)\delta^{(2)}(-\theta) = 0$ as a consequence of (4.4j). It follows that, for a system of only chiral superfields, nontrivial supergraphs must involve both left and right chiral vertices.

We can summarize the important physics points brought out in this section as follows.

- The effective action $\Gamma(\Phi, \Phi^\dagger, \mathcal{F})$ of (6.17) in superspace generates one particle irreducible supergraphs.

- The connected two point functions of chiral superfields are given in the $(\Phi(z)\Phi^\dagger(z'))$ linear space by $i\Delta_F^{SS}(z - z')$ of (6.25).

- Every L-loop ($L \geq 1$) in a supergraph, containing either only left chiral or only right chiral vertices, vanishes.

6.4   GRS Feynman Rules for the Wess-Zumino Model

Let us consider the Wess-Zumino model which was introduced in §5.1. We wish to develop for it a formalism of supergraphs with superfield lines and vertices in superspace in analogy with Feynman graphs. We first return to (6.18) and note that factors of

$$iS_{\text{int.}}\left(\frac{\delta}{\delta J_k}, \frac{\delta}{\delta J_k^\dagger}\right) = i\int d^4x\,\mathcal{L}_{\text{int.}}\left(\frac{\delta}{\delta J_k}, \frac{\delta}{\delta J_k^\dagger}\right)$$

generate vertices at superspace points z_k. Specifically, $\mathcal{L}_{\text{int.}}$ contains functional derivatives which act on more such derivatives to the right and eventually on Z_0 itself. Each such derivative connects an existing propagator to z_k in a way that every new vertex is properly linked with the requisite propagators. These functional derivatives constitute the key operations. So let us refocus our attention on the properties of chiral superfield functional derivatives (6.12), as compared with those of something like a vector superfield V, namely (6.3a). The latter are evidently more convenient since they span the entire superspace, whereas the partial spanning of superspace by the former leads to complications. Specifically, the matrix form of the chiral superpropagator is Δ_F^{SS} rather than the suggestive Δ_F of (6.22b). Difficulties increase when one turns to interaction vertices. An interaction, such as $\int d^8z\lambda\Phi^\dagger V\Phi$, yields the vertex contribution $i\lambda\int d^8z\cdots$ to a transition amplitude. In contrast, the Wess-Zumino cubic self interaction (cf. §5.1) $\frac{f}{3!}\int d^6z\Phi^3$ contributes $if\int d^6z\cdots$, whereas its hermitian conjugate contributes $if^\star \int d^6\bar{z}\cdots$. As we saw at the end of the last section, both left and right chiral vertices will be present in a nontrivial supergraph with chiral lines. The former vertex can be recast as $if \int d^8z \dfrac{\mathcal{DD}}{4\partial^2}\cdots$ and the latter as $if^\star \int d^8z \dfrac{\bar{\mathcal{D}}\bar{\mathcal{D}}}{4\partial^2}\cdots$ by using (4.14e,f). But the nonlocal operators and the asymmetric way of treating different vertices differently make life hard. All of this can be avoided: in effect, (1) Δ_F can be made the superpropagator matrix and (2) every vertex contribution can be given as i times the coupling strength times the full superspace integral $\int d^8z\cdots$. What we need to do is adopt a modified set of conventions. These are due to **Grisaru, Roček and Siegel (GRS)** [6.9] and lead to an improved set of super-Feynman rules.

The "problem" with the original convention of Salam and Strathdee can be traced back to the occurrence of the operator factor $-\frac{1}{4}\bar{\mathcal{D}}\bar{\mathcal{D}}$ and $-\frac{1}{4}\mathcal{DD}$ in the RHS of (6.23a) and (6.23b) respectively. *Suppose we agree to drop each such factor from those functional derivatives, modify the corresponding superpropagator accordingly, but agree to insert back a $-\frac{1}{4}\bar{\mathcal{D}}\bar{\mathcal{D}}$ $(-\frac{1}{4}\mathcal{DD})$ factor on each internal left chiral (right chiral) superfield line attached to a vertex.* Then we ought to get the same amplitude as before. Thus, in this new improved convention, the chiral superpropagators – now marked with primes – can be rewritten "as though derived functionally from (6.22) with $\delta J(2)/\delta J(1) = \delta J^\dagger(2)/\delta J^\dagger(1) = \delta^{(8)}(z_1 - z_2)$," so long as additional chiral derivative factors get inserted on internal lines at the time of writing the interaction vertex. The new GRS-improved chiral superpropagators are

$$\langle \Phi(1)\Phi^\dagger(2)\rangle'_C = -\frac{i}{\partial_1{}^2 + m^2}\delta^{(8)}(z_1 - z_2)\,, \tag{6.29a}$$

$$\langle \Phi(1)\Phi(2)\rangle'_C = \frac{i}{4}\frac{m\mathcal{D}_1\mathcal{D}_1}{\partial_1{}^2(\partial_1{}^2 + m^2)}\delta^{(8)}(z_1 - z_2)\,, \tag{6.29b}$$

$$\langle \Phi^\dagger(1)\Phi^\dagger(2)\rangle'_C = \frac{i}{4}\frac{m\bar{\mathcal{D}}_1\bar{\mathcal{D}}_1}{\partial_1{}^2(\partial_1{}^2 + m^2)}\delta^{(8)}(z_1 - z_2)\,, \tag{6.29c}$$

corresponding to the matrix operator form $i\Delta_F$ of (6.22b). This Δ_F is sometimes written [6.5] as Δ_F^{GRS} to distinguish it from Δ_F^{SS}.

An extra modification is, however, called for in the insertion of such $-\frac{1}{4}\bar{\mathcal{D}}\bar{\mathcal{D}}$ (or $-\frac{1}{4}\mathcal{DD}$) factors on internal left (right) chiral superfield lines when three of them meet at a vertex describing a cubic term of the superpotential, say in an interacting Wess-Zumino model (cf. §5.1) with $h = 0$, $f \neq 0$:

$$S_{\text{int.}}[\Phi] = \frac{f}{3!}\int d^6z\,\Phi^3 + h.c. \tag{6.30}$$

In this situation (a) none, (b) one, (c) two or (d) all the three of the superfield lines could be internal. The corresponding number of insertions would be (a) zero, (b) zero, (c) one or (d) two (i.e. one less than the number of internal lines) respectively. We demonstrate this below explicitly for case (d), i.e. when three new chiral superpropagators are created at the superspace point z. At the lowest order, this part of the supergraph is generated from

$$iS_{\text{int.}}\left[\frac{1}{i}\frac{\delta}{\delta J}\right]Z_0[J, J^\dagger] = \frac{if}{3!}\int d^6z\frac{1}{i^3}\frac{\delta^3}{\delta^3 J(z)}\exp\left\{-\frac{1}{2}(J\ J^\dagger)\cdot\Delta_F\cdot\begin{pmatrix}J\\J^\dagger\end{pmatrix}\right\}Z_0[0,0]\,.$$

Let us use the "old" functional derivative formula (6.23a) first, generate the corresponding $-\frac{1}{4}\bar{\mathcal{D}}\bar{\mathcal{D}}$ factors and then associate them with appropriate internal superfield lines. We can keep using (6.13c,d) and $\delta J(z_k)/\delta J(z) = -(1/4)\bar{\mathcal{D}}\bar{\mathcal{D}}\delta^{(8)}(z_k - z)$. The above expression then becomes

$$\frac{if}{3!}\int d^6z \prod_{k=1}^{3} d^8z_k \left(-\frac{1}{4}\bar{\mathcal{D}}\bar{\mathcal{D}}\right)\frac{1}{\partial^2 + m^2}\left(i\frac{m}{4}\frac{\mathcal{D}\mathcal{D}}{\partial^2}\delta^{(8)}(z - z_k)J(z_k)\right.$$

$$\left.\left.-i\delta^{(8)}(z - z_k)J^\dagger(z_k)\right)Z_0[0,0]\right].$$

We need to convert the $\int d^6z \cdots$ integration into a $\int d^8z \cdots$ one and can do so by means of (4.12c), provided any one of the three $\left(-\frac{1}{4}\bar{\mathcal{D}}\bar{\mathcal{D}}\right)$ factors is absorbed. It is simplest when we select the first one, for instance, in which case we have

$$\frac{if}{3!}\int d^8z \, d^8z_1 \frac{1}{\partial^2 + m^2}\left(i\frac{m}{4}\frac{\mathcal{D}\mathcal{D}}{\partial^2}\delta^{(8)}(z - z_1)J(z_1) - i\delta^{(8)}(z - z_1)J^\dagger(z_k)\right) \cdot$$

$$\prod_{k=2}^{3} d^8z_k \left(-\frac{1}{4}\bar{\mathcal{D}}\bar{\mathcal{D}}\right)\frac{1}{\partial^2 + m^2}\left(i\frac{m}{4}\frac{\mathcal{D}\mathcal{D}}{\partial^2}\delta^{(8)}(z - z_k)J(z_k)\right.$$

$$\left.\left.-i\delta^{(8)}(z - z_k)J^\dagger(z_k)\right)Z_0[0,0]\right],$$

It is now evident that we need not include a factor $-\frac{1}{4}\bar{\mathcal{D}}\bar{\mathcal{D}}$ for one of the three propagators at the vertex so long as we do so for each of the other two and include a contribution $if\int d^8z \cdots$ from the vertex. Thus the procedure would be to insert a factor of $-\frac{1}{4}\bar{\mathcal{D}}\bar{\mathcal{D}}$ each for two of the three left chiral internal lines in a cubic Wess-Zumino vertex: the rule for case (d). If any of the lines is an external leg, no such factor is to be inserted and in this way we also understand the rules for cases (a), (b) and (c) enumerated earlier. For the hermitian conjugate vertex, the same rules are valid, except that the insertion factor is $-\frac{1}{4}\mathcal{D}\mathcal{D}$. In case of the $\lambda\Phi^\dagger V\Phi$ vertex, with both chiral superfields corresponding to internal lines, the factor to be inserted would be $i\lambda\int d^8z\frac{1}{16}\mathcal{D}\mathcal{D}\,\bar{\mathcal{D}}\bar{\mathcal{D}}\cdots$; once again, if any of the chiral lines is external, the corresponding insertion factor, i.e. $-\frac{1}{4}\bar{\mathcal{D}}\bar{\mathcal{D}}$ or $-\frac{1}{4}\mathcal{D}\mathcal{D}$, is to be dropped. Of course, we need to mark each particular internal line on which the factor is inserted.

GRS super-Feynman rules in momentum space

The standard practice is to use the modified super-Feynman rules in momentum space and it would be useful to first discuss some preliminaries in that space. We define $\mathcal{F}(p,\theta,\bar{\theta})$

as the Fourier transform of the superfield[6] $\mathcal{F}(x, \theta, \bar{\theta})$:

$$\mathcal{F}(x, \theta, \bar{\theta}) = \int \frac{d^4 p}{(2\pi)^4} e^{-ip \cdot x} \mathcal{F}(p, \theta, \bar{\theta}) \; .$$

We further adopt the convention that $\mathcal{F}(p)$ corresponds to a superfield line carrying a four momentum p into a vertex so that ∂_μ on $\mathcal{F}(p)$ becomes $-ip_\mu \mathcal{F}(p)$. Turning to the action of covariant derivatives in momentum space, we can define

$$\mathcal{D}_A^p \mathcal{F}(p) \equiv \int d^4 x\, e^{ip \cdot x} \mathcal{D}_A \mathcal{F}(x) = (\partial_A - \sigma_{A\dot{B}}^\mu \bar{\theta}^{\dot{B}} p_\mu) \mathcal{F}(p) \; , \qquad (6.31a)$$

$$\mathcal{D}^{p,A} \mathcal{F}(p) \equiv \int d^4 x\, e^{ip \cdot x} \mathcal{D}^A \mathcal{F}(x) = \left(-\partial^A + \bar{\theta}_{\dot{B}} \bar{\sigma}^{\mu \dot{B} A} p_\mu \right) \mathcal{F}(p) \; , \qquad (6.31b)$$

$$\bar{\mathcal{D}}_{\dot{A}}^p \mathcal{F}(p) \equiv \int d^4 x\, e^{ip \cdot x} \bar{\mathcal{D}}_{\dot{A}} \mathcal{F}(x) = \left(-\bar{\partial}_{\dot{A}} + \theta^B \sigma_{B\dot{A}}^\mu p_\mu \right) \mathcal{F}(p) \; , \qquad (6.31c)$$

$$\bar{\mathcal{D}}^{p,\dot{A}} \mathcal{F}(p) \equiv \int d^4 x\, e^{ip \cdot x} \bar{\mathcal{D}}^{\dot{A}} \mathcal{F}(x) = \left(\bar{\partial}^{\dot{A}} - \bar{\sigma}^{\mu \dot{A} B} \theta_B p_\mu \right) \mathcal{F}(p) \; . \qquad (6.31d)$$

Furthermore,

$$\left[\mathcal{D}_A^p, \bar{\mathcal{D}}_{\dot{B}}^p \right]_+ \mathcal{F}(p) = 2 \sigma_{A\dot{B}}^\mu p_\mu \mathcal{F}(p) \; ,$$

which agrees with (4.11b) and our convention of $\mathcal{F}(p)$ as an annihilating superfield (denoted in Fig. 6.1 by an inward ordinary arrow) with $\partial_\mu \equiv -ip_\mu$. According to (6.31), in momentum space, for a superfield line flowing with four momentum p^μ (denoted by a bold arrow) into a superspace point with coordinates $(x_1, \theta_1, \bar{\theta}_1)$ and labelled 1 (Fig. 6.1), $\mathcal{D}_{1A}^p$ can be replaced

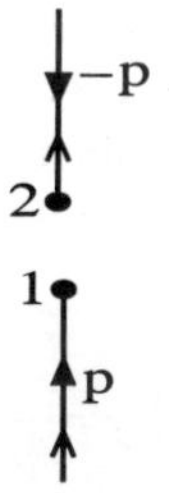

$$\partial_2^\mu \to -ip^\mu$$

$$\mathcal{D}_{2A}^{-p} = \partial_{2A} - \sigma_{A\dot{B}}^\mu \bar{\theta}_2^{\dot{B}} (-p_\mu)$$

$$\partial_1^\mu \to -ip^\mu$$

$$\mathcal{D}_{1A}^p = \partial_{1A} - \sigma_{A\dot{B}}^\mu \bar{\theta}_1^{\dot{B}} p_\mu$$

Fig.6.1. Flow into point 1 and out of point 2 in superspace.

by $\partial_{1A} - \sigma_{A\dot{B}}^\mu \bar{\theta}_1^{\dot{B}} p_\mu$. On the other hand, for a superfield created at superspace point 2 with coordinates $(x_2, \theta_2, \bar{\theta}_2)$ (denoted by an outward ordinary arrow) but with p^μ flowing out, consistency demands that the covariant derivative be taken as $\partial_{2A} - \sigma_{A\dot{B}}^\mu \bar{\theta}_2^{\dot{B}} (-p^\mu)$ and labeled $\mathcal{D}_{2A}^{-p}$. This is why we have shown the bold arrow there as inward with label $-p$. When there is a propagator between 1 and 2, the multiplying factor δ_{12} ensures that $\bar{\theta}_1 = \bar{\theta}_2$, whereas

---

[6]N.B. Since $\mathcal{F}$ can be either a left chiral superfield Φ or its right chiral conjugate $\Phi^\dagger$, $\Phi^\dagger(-p)$ is the conjugate of $\Phi(p)$ in momentum space.

(6.1a) implies that $\partial_{1A}\delta_{12} = -\partial_{2A}\delta_{12}$. Thus the validity of (6.13a) is retained in momentum space via

$$\mathcal{D}^p_{1A}\delta_{12} = -\mathcal{D}^{-p}_{2A}\delta_{12} , \qquad (6.32a)$$

$$\bar{\mathcal{D}}^p_{1\dot{A}}\delta_{12} = -\bar{\mathcal{D}}^{-p}_{2\dot{A}}\delta_{12} . \qquad (6.32b)$$

There is also the **super-Leibniz rule**

$$\mathcal{D}^p_A(\mathcal{F}\mathcal{G}) = (\mathcal{D}^q_A\mathcal{F})\mathcal{G} \pm \mathcal{F}(\mathcal{D}^{p-q}_A\mathcal{G}) ,$$

$$\bar{\mathcal{D}}^p_{\dot{A}}(\mathcal{F}\mathcal{G}) = (\bar{\mathcal{D}}^q_{\dot{A}}\mathcal{F})\mathcal{G} \pm \mathcal{F}(\bar{\mathcal{D}}^{p-q}_{\dot{A}}\mathcal{G}) ,$$

where the $+, -$ signs correspond to $\mathcal{F}, \mathcal{G}$ being superfields/superfunctions that are even, odd under the Z_2 grading of §3.1. Moreover,

$$\int d^4\theta (\mathcal{D}^q_A\mathcal{F})\mathcal{G} = \mp \int d^4\theta \mathcal{F}(\mathcal{D}^{-q}_A\mathcal{G}) ,$$

$$\int d^4\theta (\bar{\mathcal{D}}^q_{\dot{A}}\mathcal{F})\mathcal{G} = \mp \int d^4\theta \mathcal{F}(\bar{\mathcal{D}}^{-q}_{\dot{A}}\mathcal{G}) .$$

These results generalize further to products of three or more superfields/superfunctions. For instance, if $\mathcal{H}$ is a third member of the latter category, we can write

$$\int d^4\theta (\mathcal{D}^{q_1}_A\mathcal{F})\mathcal{G}\mathcal{H} + \int d^4\theta \mathcal{F}(\mathcal{D}^{q_2}_A\mathcal{G})\mathcal{H} + \int d^4\theta \mathcal{F}\mathcal{G}(\mathcal{D}^{q_3}_A\mathcal{H}) = 0 , \qquad (6.33a)$$

$$\int d^4\theta (\bar{\mathcal{D}}^{q_1}_{\dot{A}}\mathcal{F})\mathcal{G}\mathcal{H} + \int d^4\theta \mathcal{F}(\bar{\mathcal{D}}^{q_2}_{\dot{A}}\mathcal{G})\mathcal{H} + \int d^4\theta \mathcal{F}\mathcal{G}(\bar{\mathcal{D}}^{q_3}_{\dot{A}}\mathcal{H}) = 0 , \qquad (6.33b)$$

provided $q_1 + q_2 + q_3 = 0$ and $\mathcal{F}, \mathcal{G}, \mathcal{H}$ are all even under the Z_2 grading.

Next, we turn to superfield propagators in momentum space. We need to take Fourier transforms with respect to the spatial coordinates of both points 1 and 2 multiplying by one set of delta functions of $\delta^{(4)}(x_1 - x_2)$. In the modified convention [6.9], the momentum space superpropagators are

$$\int d^4x_1 d^4x_2 e^{ip\cdot(x_1-x_2)}\delta^{(4)}(x_1 - x_2)\langle\Phi(1)\Phi^\dagger(2)\rangle'_C = \frac{i\delta_{12}}{p^2 - m^2} , \qquad (6.34a)$$

$$\int d^4x_1 d^4x_2 e^{ip\cdot(x_1-x_2)}\delta^{(4)}(x_1 - x_2)\langle\Phi(1)\Phi(2)\rangle'_C = \frac{i}{4}\frac{m}{p^2(p^2 - m^2)}(\mathcal{D}^p_1\mathcal{D}^p_1\delta_{12}) , \qquad (6.34b)$$

$$\int d^4x_1 d^4x_2 e^{ip\cdot(x_1-x_2)}\delta^{(4)}(x_1 - x_2)\langle\Phi^\dagger(1)\Phi^\dagger(2)\rangle'_C = \frac{i}{4}\frac{m}{p^2(p^2 - m^2)}(\bar{\mathcal{D}}^p_1\bar{\mathcal{D}}^p_1\delta_{12}) . \qquad (6.34c)$$

We can now list below the super-Feynman rules for computing the contribution to the effective action from a particular supergraph in the Wess-Zumino model. Recall that for any line, internal or external, an ordinary arrow keeps track of annihilation or creation (i.e. into a vertex for Φ and out of a vertex for $\Phi^\dagger$) while, wherever needed, a bold arrow is put to denote momentum flow. The direction of the internal four momentum p_{int} is to be fixed as per the convention of Fig. 6.1. One further comment can be made regarding the momentum labels of covariant derivatives appearing in supergraphs. Apart from the sign (which is controlled by the direction of momentum flow), the momentum in the covariant derivative can always be taken as the momentum of the superfunction/superpropagator on which it acts. This property can be maintained even after partial integration through a judicious use of the super-Leibniz rule. Finally, the different factors associated with various parts of a supergraph, with the action of $\mathcal{D}^p$ or $\bar{\mathcal{D}}^p$ as given in (6.31), can be enumerated as follows:

- $\dfrac{i\delta_{12}}{p_{\text{int}}^2 - m^2}$, for the superpropagator

- $\dfrac{im}{4p_{\text{int}}^2(p_{\text{int}}^2 - m^2)}(\mathcal{D}^{p_{\text{int}}}\mathcal{D}^{p_{\text{int}}}\delta_{12})$, for the superpropagator

- $\dfrac{im}{4p_{\text{int}}^2(p_{\text{int}}^2 - m^2)}(\bar{\mathcal{D}}^{p_{\text{int}}}\bar{\mathcal{D}}^{p_{\text{int}}}\delta_{12})$, for the superpropagator

- $\int d^4\theta_{\text{vert}}$ with each vertex;

- an overall vertex factor of if times a factor of $-\dfrac{1}{4}\bar{\mathcal{D}}^{p_{\text{int}}}\bar{\mathcal{D}}^{p_{\text{int}}}$ (or $-\dfrac{1}{4}\mathcal{D}^{p_{\text{int}}}\mathcal{D}^{p_{\text{int}}}$) each[7] for $n-1$ of the n left (or right) chiral *internal* lines, each marked by a double bar perpendicular to the line, which meet at a Φ^3 vertex (denoted by a closed circle) or a $\Phi^{\dagger 3}$ vertex (denoted by an open circle for which the vertex factor is $if^\star$), but no such factor should be put for any external line;

- $\displaystyle\int \dfrac{d^4p_{\text{int}}}{(2\pi)^4}$ with each independent loop;

- $\displaystyle\int \prod_{p_{ext}} \dfrac{d^4p_{\text{ext}}}{(2\pi)^4}\left[(2\pi)^4\delta^{(4)}(\sum_{\text{ext}} p_{\text{ext}})\right]$ for the external momenta;

- $\Phi(p_{\text{ext}}, \theta_{\text{vert}})$ or $\Phi^\dagger(-p_{\text{ext}}, \bar{\theta}_{\text{vert}})$ for an external left chiral superfield line respectively entering or leaving a vertex with four momentum p_{ext} and the reverse for a right chiral superfield;

- usual symmetry and combinatorial factors;

- since the above rules yield a supergraph corresponding to the quantum version of $iS^{int}[\Phi]$, an overall multiplicative factor of $-i$ is needed to obtain the effective action.

An immediate convenient consequence of these ("improved") GRS super-Feynman rules is in the following fact. In a graph with v vertices one can readily integrate out $\theta, \bar{\theta}$ for $v-1$ of them. We shall have a greater appreciation of the power of the above rules when we do explicit loop supergraph calculations later.

---

[7]Under the Grassmann integrations, generated by vertices, the order of these $\mathcal{D}\mathcal{D}$ (or $\bar{\mathcal{D}}\bar{\mathcal{D}}$ terms) turns out not to matter. However, $p_{\text{int.}}$ is always defined as the momentum flowing *into* the vertex.

6.5 Feynman Rules for Nonabelian Supergauge Theories

Let us first start with the pure super Yang-Mills theory. We shall take its action from (5.54) but will **rescale** the gauge coupling strength $g \to \sqrt{2}g$ and the vector superfields $V^a \to \frac{1}{\sqrt{2}}V^a$, so that V (cf. 5.40), W etc, are unchanged. The original normalization was useful in making contact with the component content of the theory, but the present one is more convenient for perturbative superfield computations [6.8, 6.9]. Thus we write the super-YM action as

$$S_{SYM} = \frac{1}{32g^2k}\mathrm{Tr}\left(\int d^6z\, WW + \int d^6\bar{z}\,\bar{W}\bar{W}\right). \tag{6.35}$$

Each chiral superspace integral in the RHS of (6.35) is real[8] and hence equal to the other on account of the constraints on W_A, $\bar{W}^{\dot{A}}$. The definitions (5.45a,b), as well as the chiral properties of W_A, $\bar{W}^{\dot{A}}$, can now be utilized to convert the partial superspace integrals $\int d^6z$, $\int d^6\bar{z}$ in (6.35) into the full $\int d^8z$ by means of (4.12c,d). This leads to

$$S_{SYM} = \frac{1}{16g^2k}\int d^8z\,\mathrm{Tr}\left(W^A e^{-V}\mathcal{D}_A e^{V}\right). \tag{6.36}$$

On using (5.45) and (5.48), we can expand the RHS of (6.36) in powers of V:

$$S_{SYM} = \frac{1}{64g^2k}\int d^8z\,\mathrm{Tr}\Big\{V\mathcal{D}^A\bar{\mathcal{D}}\bar{\mathcal{D}}\mathcal{D}_A V - (\bar{\mathcal{D}}\bar{\mathcal{D}}\mathcal{D}^A V)\,[\mathcal{D}_A V, V] - \frac{1}{4}\,[\mathcal{D}^A V, V]\,\bar{\mathcal{D}}\bar{\mathcal{D}}\,[\mathcal{D}_A V, V]$$

$$-\frac{1}{3}\,(\bar{\mathcal{D}}\bar{\mathcal{D}}\mathcal{D}^A V)\,[[\mathcal{D}_A V, V], V] + O(V^5)\Big\}. \tag{6.37}$$

In writing (6.37), we have used the formulae (4.12a,b). Since $V = 2gV^a T^a$ (cf. 5.40), each power of V brings in one power of the gauge coupling strength g. The super Yang-Mills action has now acquired the form

$$S_{SYM} = S^{(2)}_{SYM} + O(g), \tag{6.38}$$

$S^{(2)}_{SYM}$ – the part that is quadratic in V – being given by

$$S^{(2)}_{SYM} = \frac{1}{64g^2k}\int d^8z\,\mathrm{Tr}\left(V\mathcal{D}^A\bar{\mathcal{D}}\bar{\mathcal{D}}\mathcal{D}_A V\right) = \frac{1}{8g^2k}\int d^8z\,\mathrm{Tr}(V\Pi_0\partial^2 V), \tag{6.39}$$

where Π_0 is as in (6.9). In terms of vector superfields V^a,

$$S^{(2)}_{SYM} = \frac{1}{2}\int d^8z\,V^a\Pi_0\partial^2 V^a. \tag{6.40}$$

---

[8]We are here ignoring $\tilde{F}F$ terms (cf. §4.4) which, in fact, cancel out in (6.35).

(6.40) makes it explicitly clear that the $S^{(2)}_{SYM}$ is actually of zeroth order in g, so that the remainder in the RHS of (6.38) can be identified as being $O(g)$.

Gauge fixing and superpropagator

In defining propagators of gauge fields, one requires [6.6] the addition of a gauge fixing term to the Yang-Mills action. A common choice is

$$-(8\alpha g^2 k)^{-1} \int d^4x \mathrm{Tr}(\partial^\mu A_\mu \partial^\nu A_\nu) \, , \tag{6.41}$$

α being the gauge parameter and A_μ being $2gA^a_\mu T^a$. The supersymmetric generalization [6.8] of this term, covering gaugino fields as well, is (cf. 6.55 below for justification)

$$S_{GF} = -\frac{1}{64\alpha g^2 k} \int d^8z \mathrm{Tr} \left(\bar{\mathcal{D}}\bar{\mathcal{D}} V \mathcal{D}\mathcal{D} V \right) = -\frac{1}{128\alpha g^2 k} \int d^8z \mathrm{Tr} \left(V \left[\bar{\mathcal{D}}\bar{\mathcal{D}}, \mathcal{D}\mathcal{D} \right]_+ V \right) . \tag{6.42}$$

The second step in (6.42) follows from the trace property and partial integration. We shall justify this form of S_{GF} when we discuss the Faddeev-Popov superdeterminant later in this section.

The sum of (6.39) and (6.42) can be recast, by means of (4.11f,g,k), as

$$S^{(2)}_{SYM} + S_{GF} = \frac{1}{8g^2 k} \int d^8z \mathrm{Tr} \left\{ V\partial^2 V + \frac{1}{16} \left(1 - \frac{1}{\alpha} \right) V \left[\bar{\mathcal{D}}\bar{\mathcal{D}}, \mathcal{D}\mathcal{D} \right]_+ V \right\} . \tag{6.43}$$

The complications of fourth order derivatives in the RHS of (6.43) are avoided in the super-symmetrized Feynman gauge (sFg) with $\alpha = 1$. Henceforth this is our chosen gauge for the rest of this discussion. Thus, with repeated superscripts understood as summed, we have

$$S^{(2)}_{SYM} + S^{sFg}_{GF} = \frac{1}{8g^2 k} \int d^8z \mathrm{Tr}(V\partial^2 V) = \frac{1}{2} \int d^8z V^a \partial^2 V^a. \tag{6.44}$$

In order to derive the super Yang-Mills propagator, it is sufficient to retain only that part of the action which is quadratic in V, together with the gauge fixing term. The free field generating superfunctional, which pertains to a vanishing value of g, can be written as

$$Z_0[j] = Z_0[0] \int [dV] \exp i(S^{(2)}_{SYM} + S_{GF} + j^a \cdot V^a) . \tag{6.45}$$

On substituting (6.44) into (6.45) and doing the functional integration in V^a, one derives the final result for $W_0[j] = -i \ln Z_0[j]$:

$$W_0[j]^{sFg} = -\frac{1}{2} \int d^8z j^a \frac{1}{\partial^2} j^a \, . \tag{6.46}$$

The two point connected Green's function obtains easily from (6.46):

$$\langle V^a(1)V^b(2)\rangle^{sFg}_C = -i\frac{\delta^2 W_0[j]}{\delta j^a(1)\delta j^b(2)} = \frac{i}{\partial^2}\delta^{ab}\delta^{(8)}(z_1 - z_2) \, . \tag{6.47}$$

In momentum space (6.47) becomes

$$G(p)^{sFg} = -\frac{i}{p^2}\delta^{ab}\delta_{12} \ . \tag{6.48}$$

The overall sign in the RHS of (6.48) is opposite to that of the chiral superpropagator, as expected. We can diagrammatically represent it as

$$-\frac{i}{p^2}\delta^{ab}\delta_{12} \ .$$

Unitarity and ghosts

The gauge fixing conditions need to be introduced into the functional integral defining $Z[j]$ via δ-functional constraints. In this respect, however, there is a nontrivial generalization from ordinary Yang-Mills theory to the super-YM one. We need to first quickly review the former case [6.6]. A gauge transformation there involves one gauge parametric function of spacetime, namely $\omega(x)$, constructed out of real gauge functions $\omega^a(x)$: $\omega(x) = 2g\omega^a(x)T^a$, defined in analogy with $A_\mu = 2gA_\mu^a(x)T^a$. An infinitesimal gauge transformation is now given by

$$\delta A_\mu^\omega = A_\mu^\omega - A_\mu = \partial_\mu\omega + \frac{i}{2}[A_\mu,\omega] \ .$$

There is, in consequence, one gauge fixing function $F(A_\mu^\omega)$ whose vanishing is ensured by the insertion of the δ-functional $\delta[F(A_\mu^\omega)]$ under the functional integral $[dA_\mu]$.

Returning to super-YM field theory, however, we see that here the gauge transformation (5.41) of the vector superfield involves two chiral superfunctions: the left chiral $\Lambda(z)$ and the right chiral $\Lambda^\dagger(z)$. Hence two gauge fixing superfunctions, one left chiral and the other right chiral, need to be introduced. These are chosen to be [6.8]

$$\mathcal{K}(V) = -\frac{1}{4}\bar{\mathcal{D}}\bar{\mathcal{D}}V(z) + f(z) \ , \tag{6.49a}$$

$$\mathcal{K}^\dagger(V) = -\frac{1}{4}\mathcal{D}\mathcal{D}V + f^\dagger(z) \ , \tag{6.49b}$$

where $f(z)$ and $f^\dagger(z)$ are external left chiral and right chiral source superfields respectively, with

$$f(z) = 2gf^a(z)T^a \ . \tag{6.50}$$

The two δ-functionals, that are needed to ensure the vanishing of the lefthand sides of (6.49), are thus $\delta(\mathcal{K}[V])$ and $\delta(\mathcal{K}^\dagger[V])$. Note furthermore that $\mathcal{K}$ has unit mass dimension though that of V is zero.

One can now introduce a Faddeev-Popov superdeterminant Δ_{FP} in analogy with that in ordinary Yang-Mills field theory [6.9]. The free field generating superfunctional can actually be written as a functional integral over the vector superfield V as follows:

$$Z_0[j] = Z[0]\int [dV]\exp\left[i\left(S_{SYM}^{(2)} + j^a\cdot V^a\right)\right]\Delta_{FP}\delta(\mathcal{K}[V])\delta(\mathcal{K}^\dagger[V]) \ . \tag{6.51}$$

In (6.51), $S_{SYM}^{(2)}$ is that of (6.39) and (6.40). Moreover, we have introduced the measure $[dV] \propto \Pi_a[dV^a]$, while the δ-functionals are given by $\delta[f - \frac{1}{4}\bar{\mathcal{D}}\bar{\mathcal{D}}V] \propto \Pi_a\delta[f^a - \frac{1}{4}\bar{\mathcal{D}}\bar{\mathcal{D}}V^a]$ etc. Here each product Π_a extends over the dimension of the gauge group and proportionality constants are immaterial. The Faddeev-Popov superdeterminant is defined by

$$\Delta_{FP}^{-1} \equiv \int [d\Lambda^+][d\Lambda]\delta[\mathcal{K}(V^\lambda)]\delta[\mathcal{K}^\dagger(V^\lambda)] , \qquad (6.52)$$

with V^λ as the vector superfield that is gauge transformed (cf. 5.43) away from V via the chiral superfunctions $\Lambda(z)$ and $\Lambda^\dagger(z)$.

Since, in (6.51), Δ_{FP} multiplies the two δ-functionals, we need only its specific value Δ'_{FP} when $\mathcal{K}[V]$ and $\mathcal{K}^\dagger[V]$ vanish. Suppose V represents a solution to the equations $\frac{1}{4}\bar{\mathcal{D}}\mathcal{D}V = f$ and $\frac{1}{4}\mathcal{D}\mathcal{D}V = f^\dagger$. For the purpose of (6.51), the only contributions to Δ'_{FP} from the functional integrations over $\Lambda, \Lambda^\dagger$ in (6.52) evidently arise when $\Lambda, \Lambda^\dagger$ are quite close to zero. This means that we can use (5.42b) and keep only linear terms in $\Lambda, \Lambda^\dagger$. In other words, we can take

$$\delta V^\lambda \equiv V^\lambda - V \simeq \hat{H}(V)\Lambda + \hat{H}^\dagger(V)\Lambda^\dagger . \qquad (6.53)$$

The linear operations $\hat{H}(V), \hat{H}^\dagger(V)$ on $\Lambda, \Lambda^\dagger$ in (6.52) are defined (cf. 5.41b) by

$$\hat{H}(V)\Lambda \equiv i\Lambda + \frac{i}{2}[V, \Lambda] + \frac{i}{12}[V, [V, \Lambda]] + \cdots , \qquad (6.54a)$$

$$\hat{H}^\dagger(V)\Lambda^\dagger \equiv -i\Lambda^\dagger + \frac{i}{2}[V, \Lambda^\dagger] - \frac{i}{12}[V, [V, \Lambda^\dagger]] + \cdots . \qquad (6.54b)$$

Thus, we can rewrite (6.52) as

$$\Delta'^{-1}_{FP} = \int [d\Lambda][d\Lambda^\dagger]\delta\left[-\frac{1}{4}\bar{\mathcal{D}}\bar{\mathcal{D}}\left\{\hat{H}(V)\Lambda + \hat{H}^\dagger(V)\Lambda^\dagger\right\}\right]$$
$$\delta\left[-\frac{1}{4}\mathcal{D}\mathcal{D}\left\{\hat{H}(V)\Lambda + \hat{H}^\dagger(V)\Lambda^\dagger\right\}\right] .$$

Finally, since the RHS of this equation is independent of the external chiral superfields $f(z), f^\dagger(z)$, one can – as in ordinary YM theory [6.6] – average over the latter by integrating the RHS of (6.51) functionally over $\int [df][df^\dagger]$ with $\exp[-i(4\alpha g^2 k)^{-1}\mathrm{Tr}\{f^\dagger(z)f(z)\}]$ as the weight factor[9]. Such a procedure gets rid of the δ-functionals from the RHS of (6.51) and results in the introduction of the additional factor $\exp iS_{GF}$ where

$$S_{GF} = -\frac{1}{64\alpha g^2 k} \int d^8z \; \mathrm{Tr}(\bar{\mathcal{D}}\bar{\mathcal{D}}V\mathcal{D}\mathcal{D}V) . \qquad (6.55)$$

This justifies the first step of (6.42) since we have shown (6.55) to be the supersymmetric generalization of (6.41). Going to the super-Feynman gauge with $\alpha = 1$, we now have the

---

[9]Note that $\int [df^\dagger][df] \exp\left[i(4\alpha g^2 k)^{-1}\mathrm{Tr}\left\{f^\dagger(z)f(z)\right\}\right]$ is just a field independent constant which does not contribute to any of the n-point functions.

full Yang-Mills generating functional as

$$Z[j] = N \int [dV] \exp\left[i\left(S_{SYM} + S_{GF}^{sFg} + j^a \cdot V^a \right) \right] \triangle'_{FP} \,, \tag{6.56}$$

N being the normalization.

For the restoration of unitarity, ghost superfields need to be introduced as supersymmetric generalizations of Faddeev-Popov ghost fields in quantized Yang-Mills theory [6.6]. Let us briefly review the latter. One needs there to introduce one complex pair of anticommuting spin zero Grassmann ghost fields $c^a(x)$, $c^{\dagger a}(x)$, for each gauge field component $A^a_\mu(x)$, to tackle a single gauge fixing δ-functional $\delta[F(A^\omega_\mu)]$ involving the gauge parametric function $\omega(x)$. Indeed, one defines

$$c(x) \equiv 2g c^a(x) T^a$$

and replaces the Faddeev-Popov determinant by (cf. 2.9)

$$det\left(\frac{\delta F}{\delta \omega} \right) = \int [dc^\dagger][dc] \exp\left[\frac{i}{4g^2 k} \mathrm{Tr} \left\{ c^\dagger \left(\frac{\delta F}{\delta A^\omega_\mu} \delta A^\omega_\mu \right)_{\omega=c} \right\} \right] .$$

Since there are two δ-functionals in the supersymmetric generalization, we need to introduce two pairs of hermitian conjugate chiral superfields, each of unit mass dimension, which are odd under the Z_2-grading (cf. §3.1). Let us denote these by $C(z)$, $C^\dagger(z)$ and $C'(z)$, $C'^\dagger(z)$ with $C(z) = 2gC^a(z)T^a$ etc. and $\bar{\mathcal{D}}^{\dot{A}}C = \bar{\mathcal{D}}^{\dot{A}}C' = \mathcal{D}_A C^\dagger = \mathcal{D}_A C'^\dagger = 0$. Now $\triangle'_{FP}$ in (6.56) can be written as

$$\triangle'_{FP} = \int [dC'^\dagger][dC'][dC^\dagger][dC] e^{iS_{GH}} \,, \tag{6.57}$$

with the ghost contribution to the action S_{GH} being given by[10] [6.7]

$$S_{GH} = \frac{1}{4g^2 k} \left[\int d^6 z \mathrm{Tr}\left\{ C'\left(\frac{\delta \mathcal{K}}{\delta V^\lambda} \delta V^\lambda \right)_{\substack{i\Lambda=C, \\ -i\Lambda^\dagger=C^\dagger}} \right\} + \int d^6 \bar{z} \mathrm{Tr}\left\{ \left(\frac{\delta \mathcal{K}^\dagger}{\delta V^\lambda} \delta V^\lambda \right)_{\substack{i\Lambda=C \\ -i\Lambda^\dagger=C^\dagger}} C'^\dagger \right\} \right]$$

$$= \frac{1}{4g^2 k} \left[\int d^6 z \left(-\frac{1}{4}\bar{\mathcal{D}}\bar{\mathcal{D}} \right) \mathrm{Tr}\left\{ C'(\delta V^\lambda)_{\substack{i\Lambda=C \\ -i\Lambda^\dagger=C^\dagger}} \right\} \right.$$

$$\left. + \int d^6 \bar{z} \left(-\frac{1}{4}\mathcal{D}\mathcal{D} \right) \mathrm{Tr}\left\{ (\delta V^\lambda)_{\substack{i\Lambda=C \\ -i\Lambda^\dagger=C^\dagger}} C'^\dagger \right\} \right] .$$

On substituting δV^λ from (6.53) into the RHS above, we obtain

$$S_{GH} = \frac{1}{4g^2 k} \int d^8 z \mathrm{Tr}\left[C'\{\hat{H}(V)(-iC) + \hat{H}^\dagger(V)(iC^\dagger)\} + \text{h.c.} \right] .$$

---

[10]The evaluation of $C, C^\dagger$ or of $C', C'^\dagger$ at the values $i\Lambda$, $-i\Lambda^\dagger$ is due to the fact that, to linear order, $\delta V^\lambda \simeq i(\Lambda - \Lambda^\dagger)$, cf. (6.53,4). N.B. Twice the imaginary part of the scalar component of Λ equals $-\omega$.

The substitution of $\hat{H}(V)$, $\hat{H}^\dagger(V)$ from (6.53) enables us to write the ghost contribution to the action (upto quadratic terms in V) as

$$S_{GH}^{(2)} = \frac{1}{4g^2 k} \int d^8 z \, \mathrm{Tr}\Big\{ C^{\dagger\prime} C + C^\dagger C' + \frac{1}{2}(C' + C^{\dagger\prime})[V, C - C^\dagger]$$

$$+ \frac{1}{12}(C' + C^{\dagger\prime})[V, [V, C + C^\dagger]]\Big\} \qquad (6.58)$$

N.B. The purely left chiral terms $C'C$ and the purely right chiral terms $C^\dagger C^{\prime\dagger}$ have vanished under $\int d^8 z \cdots$. By utilizing (6.57) and (6.58), the generating superfunctional of (6.56) can now be rewritten, with $j = 2gj^a T^a$, as

$$Z[j] = N \int [dV][dC^{\dagger\prime}][dC'][dC^\dagger][dC] e^{i[S_{TOT} + \frac{1}{4g^2 k}\mathrm{Tr}(j \cdot V)]} , \qquad (6.59\mathrm{a})$$

$$S_{TOT} = S_{SYM} + S_{GF}^{sFg} + S_{GH} . \qquad (6.59\mathrm{b})$$

Let S_0 be the interaction free quadratic part of the total action S_{TOT} and let us define $S_{\mathrm{int.}} \equiv S_{TOT} - S_0$ as the interaction part. As in (6.18), we can then write

$$Z[j] = \exp\left(iS_{\mathrm{int.}} \left[\frac{1}{i}\frac{\delta}{\delta j}, \frac{1}{i}\frac{\delta}{\delta \eta'}, \frac{1}{i}\frac{\delta}{\delta \eta}, \frac{1}{i\delta\eta^{\prime\dagger}}, \frac{1}{i\delta\eta^\dagger} \right] \right) Z_0[j, \eta', \eta, \eta^{\prime\dagger}, \eta^\dagger]_{\eta'=\eta=\eta^{\prime\dagger}=\eta^\dagger=0} , \qquad (6.60)$$

where

$$Z_0[j, \eta', \eta, \eta^{\prime\dagger}, \eta^\dagger] = \int [dV][dC^{\dagger\prime}][dC'][dC^\dagger][dC] \exp i \Big(S_0[V, C', C, C^{\prime\dagger}, C^\dagger]$$

$$+ \frac{1}{4g^2 k} \mathrm{Tr}\{ j \cdot V + (\eta' \cdot C' + \eta \cdot C + \mathrm{h.c.}) \} \Big) \qquad (6.61)$$

with

$$j \cdot V \equiv \int d^8 z j(z) V(z), \quad \eta \cdot C \equiv \int d^6 z \eta(z) C(z), \quad \eta' \cdot C \equiv \int d^6 z \eta'(z) C(z)$$

and similarly for the hermitian conjugates of the last two dot products. In (6.60) and (6.61), we have introduced chiral source superfields η', $\eta^{\prime\dagger}$, η and $\eta^\dagger$, with $\eta = 2g\eta^a T^a$ etc., which are odd under the Z_2-grading (cf. §3.1). They obey variational relations analogous to (6.12), namely

$$\frac{\delta \eta^{\prime b}(z_2)}{\delta \eta^{\prime a}(z_1)} = \frac{\delta \eta^b(z_2)}{\delta \eta^a(z_1)} = \delta^{ab}\left(-\frac{1}{4}\bar{\mathcal{D}}_1\bar{\mathcal{D}}_1 \right) \delta^{(8)}(z_1 - z_2) , \qquad (6.62\mathrm{a})$$

$$\frac{\delta \eta^{\dagger\prime b}(z_2)}{\delta \eta^{\dagger\prime a}(z_1)} = \frac{\delta \eta^{\dagger b}(z_2)}{\delta \eta^{\dagger a}(z_1)} = \delta^{ab}\left(-\frac{1}{4}\mathcal{D}_1\mathcal{D}_1 \right) \delta^{(8)}(z_1 - z_2) , \qquad (6.62\mathrm{b})$$

as contrasted with that obeyed by the source superfields $j^a(x)$:

$$\frac{\delta j^b(z_2)}{\delta j^a(z_1)} = \delta^{ab}\delta^{(8)}(z_1 - z_2) . \qquad (6.63)$$

Utilizing (6.37), we can now write[11]

$$(S_0)^{sFg} = \frac{1}{4g^2k} \int d^8z \mathrm{Tr}\left(\frac{1}{2}V\partial^2 V + C'^\dagger C + C^\dagger C'\right), \tag{6.64a}$$

$$(S_{\text{int.}})^{sFg} = \frac{1}{8g^2k} \int d^8z \mathrm{Tr}\left\{\frac{1}{8}(\bar{D}\bar{D}D^A V)[V,\mathcal{D}_A V] + (C' + C'^\dagger)[V, C - C^\dagger]\right.$$

$$+ \frac{1}{6}(C' + C'^\dagger)[V,[V, C + C^\dagger]] - \frac{1}{32}[V,\mathcal{D}^A V]\bar{D}\bar{D}[V,\mathcal{D}_A V]$$

$$\left. - \frac{1}{24}(\bar{D}\bar{D}D^A V)[V,[V,\mathcal{D}_A V]] + O(g^5)\right\}. \tag{6.64b}$$

The evaluation of the RHS of (6.61), by substituting (6.64a), involves the use of standard functional integration methods and yields the following generalization of (6.46):

$$W_0[j,\eta',\eta,\eta'^\dagger,\eta^\dagger]^{sFg} = \int d^8z \left(-\frac{1}{2}j^a\frac{1}{\partial^2}j^a + \eta^{\dagger a}\frac{1}{\partial^2}\eta^a + \eta^{\dagger a}\frac{1}{\partial^2}\eta'^a\right). \tag{6.65}$$

Thus, alongside the Yang-Mills superpropagator, we have two sets of chiral ghost superpropagators which can be treated with the methodology of §6.4. The only point of difference is that the ghost superfields have negative metric, no mass terms and propagate not freely but only within loops and have just the two superpropagators $\langle C(1)C'^\dagger(2)\rangle_C$ and $\langle C'(1)C^\dagger(2)\rangle_C$ which can be written in the GRS-improved form. Thus, in momentum space, the ghost superpropagators appear as shown below.

$$\underset{1}{C} \xrightarrow{\; p \;} \underset{2}{C'^\dagger} \qquad -\frac{i}{p^2}\delta_{12}\;,$$

$$\underset{1}{C'} \xrightarrow{\; p \;} \underset{2}{C^\dagger} \qquad -\frac{i}{p^2}\delta_{12}\;.$$

A comparison between the different terms of (6.65) shows that the ghost superpropagator has a sign opposite to that of a massless vector superpropagator. However, since ghosts always propagate within loops and S_{GH} is bilinear in ghost superfields, their superpropagators must occur in pairs so that this sign is inconsequential. Coming to interactions, the self interacting V^3 and V^4 vertices, as well as the V-ghost-ghost and V^2-ghost-ghost vertices, are included in $(S_{int.})^{sFg}$. The explicit forms can be read off from (6.64b).

Matter fields

The above considerations can be readily extended to a more general supersymmetric nonabelian gauge theory comprising both gauge and matter fields. The latter are contained in vector and chiral superfields respectively, as described by the Lagrangian density of (5.54).

---

[11]Whereas in Ch.5 we had fixed the supergauge to the WZ gauge and kept the ordinary gauge degree of freedom, here we have fixed the latter to the supersymmetrized Feynman gauge and kept the former free.

Indeed, the total action S_{TOT} now consists of super-YM, gauge fixing, ghost and matter parts, i.e.

$$S_{TOT} = S_{SYM} + S_{GF}^{sFg} + S_{GH} + S_{MAT} . \tag{6.66}$$

In (6.66)

$$S_{MAT} = \int d^8z\ \Phi_i^\dagger (e^V)_{ij} \Phi_j + \left[\int d^6z\ \mathcal{W}(\Phi_i) + \text{h.c.} \right], \tag{6.67}$$

cf. (5.54), while the rest of the RHS remains as in the matterless case. We are once again using i, j as type indices which include I, J in the space of the gauge group representation $\mathcal{R}$ to which Φ belongs. The generating superfunctional now is simply that of (6.15) with $\mathcal{F}, j$ specified to be V^a, j^a. Furthermore, S in (6.15) needs to be replaced by the total action S_{TOT} of (6.66). For the sake of being definite, let us suppose $\mathcal{W}(\Phi_i)$ is given by (5.1b) with $m_{ij} = m\delta_{ij}$. Then m can be identified as a common mass parameter of the chiral superfields. The total action S_{TOT} now splits easily between the free part S_0 and the interaction part S_{int}. In the super-Feynman gauge

$$S_0 = \int d^8z \left[\left(\frac{1}{2} V^a \partial^2 V^a + C'^{\dagger a} C^a + C^{\dagger a} C'^a \right) + \Phi_i^\dagger \Phi_i \right] + \frac{1}{2} \left[\left(m \int d^6z\ \Phi_i \Phi_i + \text{h.c.} \right) \right], \tag{6.68a}$$

$$S_{\text{int.}} = (S_{\text{int.}})^{sFg} + \int d^8z\ \Phi_i^\dagger \left[(e^V)_{ij} - \delta_{ij} \right] \Phi_j + \frac{1}{3!} \left(\int d^6z\ f_{ijk} \Phi_i \Phi_j \Phi_k + \text{h.c.} \right), \tag{6.68b}$$

with $(S_{\text{int.}}^{SYM})^{sFg}$ as given by (6.64b). Of course, the last RHS terms in (6.68a,b) must be gauge singlets.

It is now straightforward to derive the superpropagators and the vertex rules for super-graphs in this theory. They are just a combination of the rules for chiral superfield theory and super-YM theory. Consider the lowest order gauge-matter-matter vertex, with one type of matter and the interaction

$$S_{\text{int.}}^{(\Phi, V)} = 2g \int d^8z\ \Phi_I^\dagger (T^a)_{IJ} \Phi_J V^a + \cdots . \tag{6.69}$$

In (6.69) we have restricted the type index i to only the $\mathcal{R}$-space basis index I and the ellipsis stands for terms of order g^2 and higher order corrections, which will not be needed for the explicit calculations done later in this chapter. One puts in, for this vertex, a factor of $2ig \int d^4\theta_{vert}$ times an insertion of $-\frac{1}{4}\bar{\mathcal{D}}^{Pint}\bar{\mathcal{D}}^{Pint}$ or $-\frac{1}{4}\mathcal{D}^{Pint}\mathcal{D}^{Pint}$ for each left (or right) chiral internal line, shown by an ordinary arrow flowing into (or out of) the gauge vertex, but not for such an external line. Similarly, to order g^2, there is a $\Phi^\dagger V V \Phi$ vertex and so on[12]. Ghost superpropagators, which only enter into loops, involving gauge-ghost-ghost or gauge-gauge-ghost-ghost vertices, are to be treated as massless left or right chiral superpropagators in the improved GRS framework described in §6.4. One point of difference is that, since ghost chiral superfields are odd under the Z_2-grading of §3.1, there needs to be an additional -1 factor for each ghost superloop. Finally, one should not forget to put in a $V(p_{ext}, \theta_{vert})$ or $V(-p_{ext}, \theta_{vert})$ for an external vector superfield line entering or leaving a vertex.

---

[12]In the Wess-Zumino gauge one need consider only these two vertices.

6.6　Sample One Loop Supergraph Calculations

One loop chiral superpropagator

We shall illustrate the utility of the Feynman rules, derived in the two previous sections, with several one loop calculations. First, let us calculate the one loop correction to the superpropagator (6.34a) in the Wess-Zumino model. Recall (cf. Ch.5) that the action of this model is given by

$$S_{WZ} = \int d^8 z \, \Phi^\dagger \Phi + \left[\int d^6 z \left(\frac{f}{3!}\Phi^3 + \frac{m}{2}\Phi^2 \right) + \text{h.c.} \right] .$$

The concerned supergraph appears in Fig. 6.2, 1 and 2 being the Φ and $\Phi^\dagger$ ends respectively. The closed circle represents the Φ^3 vertex and the open circle stands for the $\Phi^{\dagger 3}$ vertex. The contribution to the effective action, after setting $i^4 = 1$ for the product of the i factors from the two vertices and the two propagators, reads

$$-i\frac{1}{2}|f|^2 \int \frac{d^4 p}{(2\pi)^4} \frac{d^4 k}{(2\pi)^4} d^4\theta_2 d^4\theta_1 \Phi^\dagger(-p, \bar\theta_2) \frac{\delta_{12}}{(p-k)^2 - m^2} \cdot$$

$$\cdot \left((-)\frac{\mathcal{D}_2^k \mathcal{D}_2^k}{4}(-)\frac{\bar{\mathcal{D}}_1^{-k}\bar{\mathcal{D}}_1^{-k}}{4}\frac{\delta_{12}}{k^2 - m^2} \right) \Phi(p, \theta_1) ,$$

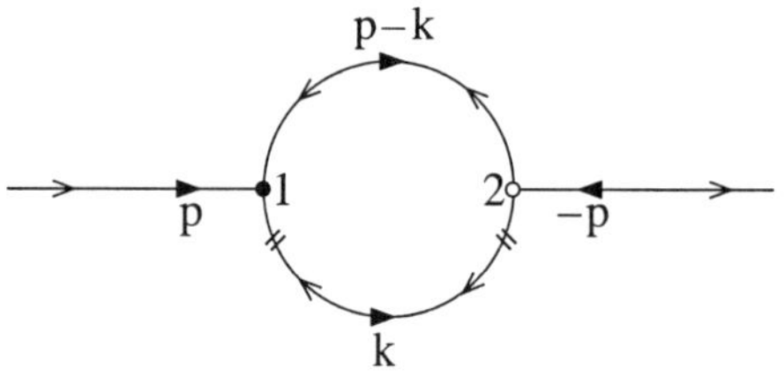

Fig.6.2. One loop correction to the superpropagator of (6.34a) in momentum space.

the overall 1/2 being the combinatoric factor arising from $(3 \times 3 \times 2)/(3!3!)$. The internal four momentum k flows into vertex 2 and out of vertex 1; so, as per our convention (cf. ftnt. 7), the momentum label on the covariant derivative at 2 is k while the same on the anticovariant derivative at 1 is $-k$. On replacing $\bar{\mathcal{D}}_1^{-k}$ by $\bar{\mathcal{D}}_2^k$ (cf. 6.32b) and then[13] using (6.2b), the above expression simplifies and can be written in terms of the Passarino-Veltman (cf. §1.3) function $B_0\,(p^2, m^2, m^2)$ [6.10] as

$$-\frac{|f|^2}{2} \int \frac{d^4 p}{(2\pi)^4} B_0(p^2, m^2, m^2) \int d^4\theta \Phi^\dagger(-p, \bar\theta)\Phi(p, \theta) , \qquad (6.70a)$$

$$B_0(p^2, m_1^2, m_2^2) \equiv i \int \frac{d^4 k}{(2\pi)^4} \frac{1}{k^2 - m_1^2} \frac{1}{(p-k)^2 - m_2^2} = B_0(p^2, m_2^2, m_1^2) . \qquad (6.70b)$$

It is evident from (6.70b) that $B_0(p^2, m^2, m^2)$ of (6.70a) has a logarithmic ultraviolet divergence which needs to be regulated by the dimensional reduction procedure, as explained in

---

[13]Recall that all the γ-matrix algebra, and thus the $\mathcal{D}$-algebra, must be done in four dimensional space.

§1.3 and §6.1. An evaluation, along the lines discussed there, yields [6.8, 6.10] the following expression

$$B_0(p^2, m^2, m^2) = -\frac{1}{(4\pi)^2}\frac{1}{2-\omega} + \text{finite terms} ,$$

as $\omega \to 2$. (Recall that the number of spacetime dimensions is $d = 2\omega$). Therefore, the divergent contribution to the effective action from the one loop superpropagator, corresponding to (6.34a), is given by

$$S_E^\Phi(\text{one loop}) = \frac{|f|^2\mu^{2\omega-4}}{(4\pi)^2(4-2\omega)} \int d^8z \Phi^\dagger\Phi , \tag{6.71}$$

where we have introduced the renormalization scale μ in order to ensure that the RHS of (6.71) has a vanishing mass dimension for $2\omega \neq 4$. We end this calculation with two remarks. (1) Were we to evaluate the one loop correction to (6.34b) or (6.34c), we would have needed the corresponding superpropagator in the internal lines of the loop. The loop then would be a purely chiral loop, thereby vanishing on account of the result proved at the end of §6.3. (2) Suppose the final $\Phi^\dagger$ line were missing from Fig. 6.2, i.e. one was computing the one loop Φ-tadpole supergraph. The result of the computation would be similar to the RHS of (6.70a), but with one factor of if and one propagator less and with the $\Phi^\dagger(-p,\bar\theta)$ factor missing. Hence the answer would be proportional to $\int d^4\theta\Phi(p,\theta)$ which vanishes explicitly demonstrating (cf. ftnt. 5) the absence of a Φ-tadpole.

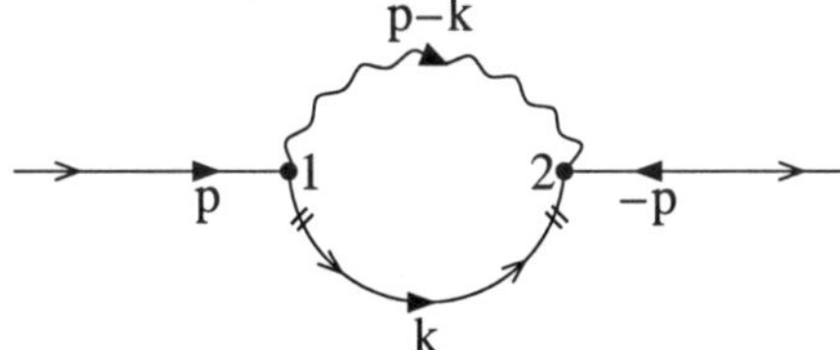

Fig.6.3. Gauge superfield contribution at one loop to the superpropagator of (6.34a).

As another exercise, let us calculate the one loop gauge superfield correction to the superpropagator of (6.34a) in the Wess-Zumino model incorporated within a super-YM theory. The action of this theory is given by (6.66) and (6.67) where the superpotential is as discussed in §5.1 and S_{SYM} is as given in (6.35). The particular contribution to the effective action shown in Fig. 6.3 can be written as

$$-4ig^2 \int \frac{d^4p}{(2\pi)^4}\frac{d^4k}{(2\pi)^4}d^4\theta_2 d^4\theta_1 \Phi_I^\dagger(-p,\bar\theta_2)T_{IL}^a \frac{-\delta_{12}}{(p-k)^2} \cdot$$

$$\left((-)\frac{\bar{\mathcal{D}}_2^k\bar{\mathcal{D}}_2^k}{4}(-)\frac{\mathcal{D}_1^{-k}\mathcal{D}_1^{-k}}{4}\frac{\delta_{12}}{k^2-m^2}\right)T_{LJ}^a\Phi_J(p,\theta_1) .$$

By manipulations, similar to those detailed for the one loop correction to the chiral superpropagator, the above expression can be simplified to

$$4g^2C_2(\mathcal{R}) \int \frac{d^4p}{(2\pi)^4}B_0(p^2,0,m^2) \int d^4\theta\Phi_I^\dagger(-p,\bar\theta)\Phi_I(p,\theta) .$$

The relative minus sign here, in comparison with the contribution of Fig. 6.2, is due to the sign difference between the $\langle VV \rangle$ and $\langle \Phi\Phi^\dagger \rangle$ superpropagators. Moreover, $C_2(\mathcal{R})$ is the quadratic Casimir constant of the representation $\mathcal{R}$ of the gauge group to which the Φ's belong:

$$(T^a T^a)_{IJ} = C_2(\mathcal{R})\delta_{IJ} . \tag{6.72}$$

The expression for $B_0(p^2, 0, m^2)$, in the limit $\omega \to 0$, is (cf. 6.70b):

$$B_0(p^2, 0, m^2) \equiv i \int \frac{d^4k}{(2\pi)^4} \frac{1}{(p-k)^2} \frac{1}{k^2 - m^2} = -\frac{1}{(4\pi)^2} \frac{1}{2-\omega} + \text{finite terms}$$

in the $\overline{\text{DR}}$ scheme. Comparing $B_0(p^2, 0, m^2)$ with $B_0(p^2, m^2, m^2)$ of (6.70b), we see that any difference is only in the finite piece. Thus the one loop contribution to the effective action from the present divergence is given by

$$S_E^{(\Phi,V)}(\text{one loop}) = -\frac{g^2 C_2(\mathcal{R})\mu^{2\omega-4}}{2\pi^2(4-2\omega)} \int d^8z \; \Phi^\dagger \Phi . \tag{6.73}$$

One loop vector superpropagator

As the next example, consider the one loop correction to the vector superpropagator $\langle V^a V^b \rangle$, due to internal Φ-lines, as illustrated in the supergraph of Fig.6.4. The corresponding contribution to the effective action reads

$$-4ig^2 \, T(\mathcal{R})\frac{1}{2}\delta^{ab} \int \frac{d^4p}{(2\pi)^4} \frac{d^4k}{(2\pi)^4} d^4\theta_2 d^4\theta_1 V^a(-p, \theta_2, \bar\theta_2)$$

$$\left((-)\frac{\bar{\mathcal{D}}_2^{p+k}\bar{\mathcal{D}}_2^{p+k}}{4} (-)\frac{\mathcal{D}_1^{-p-k}\mathcal{D}_1^{-p-k}}{4} \frac{\delta_{12}}{(p+k)^2 - m^2} \right)$$

$$\left((-)\frac{\mathcal{D}_2^{-k}\mathcal{D}_2^{-k}}{4} (-)\frac{\bar{\mathcal{D}}_1^{k}\bar{\mathcal{D}}_1^{k}}{4} \frac{\delta_{12}}{k^2 - m^2} \right) V^b(p, \theta_1, \bar\theta_1) , \tag{6.74}$$

where $T(\mathcal{R})$ is the representation constant (cf. 5.39) of $\mathcal{R}$ and an extra factor of $1/2$ has been inserted because the term in the effective action contains V^2. We can first replace $\bar{\mathcal{D}}_1^k$ by $-\bar{\mathcal{D}}_2^{-k}$ and $\mathcal{D}_1^{-p-k}$ by $-\mathcal{D}_2^{p+k}$ in the expression (6.74), cf. (6.32). This makes all the covariant derivatives act at point 2 and in partial integrations $V^b(p, \theta_1, \bar\theta_1)$ can always be factored out.

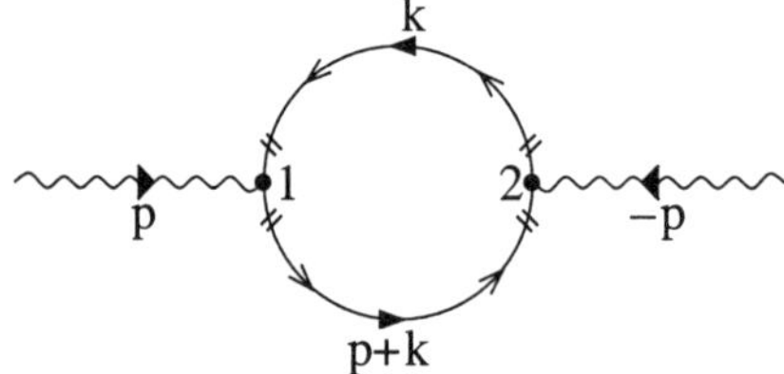

Fig.6.4. One loop internal chiral superfield contribution to the vector superpropagator.

Now one can partially integrate the $\bar{\mathcal{D}}_2^{p+k}\bar{\mathcal{D}}_2^{p+k}$ factor off the left most δ_{12} by utilizing the super-Leibniz rule of §6.4 while ensuring that *the momentum label on any covariant derivative remains (upto a sign) that of the superfunction it is acting on*. Thus (6.74) changes to

$$-i\frac{g^2}{128}T(\mathcal{R})\delta^{ab}\int\frac{d^4p}{(2\pi)^4}\frac{d^4k}{(2\pi)^4}d^4\theta_2 d^4\theta_1$$

$$\left[\bar{\mathcal{D}}_2^{-p}\bar{\mathcal{D}}_2^{-p}V^a(-p,\theta_2,\bar{\theta}_2)\left(\mathcal{D}_2^{p+k}\mathcal{D}_2^{p+k}\frac{\delta_{12}}{(p+k)^2-m^2}\mathcal{D}_2^{-k}\mathcal{D}_2^{-k}\ \bar{\mathcal{D}}_2^{-k}\bar{\mathcal{D}}_2^{-k}\frac{\delta_{12}}{k^2-m^2}\right)\right.$$

$$+2\bar{\mathcal{D}}_{2\dot{A}}^{-p}V^a(-p,\theta_2,\bar{\theta}_2)\left(\mathcal{D}_2^{p+k}\mathcal{D}_2^{p+k}\frac{\delta_{12}}{(p+k)^2-m^2}\right)\cdot$$

$$\left(\bar{\mathcal{D}}_2^{-k,\dot{A}}\ \mathcal{D}_2^{-k}\mathcal{D}_2^{-k}\ \bar{\mathcal{D}}_2^{-k}\bar{\mathcal{D}}_2^{-k}\frac{\delta_{12}}{k^2-m^2}\right)$$

$$+V^a(-p,\theta_2,\bar{\theta}_2)\left(\mathcal{D}_2^{p+k}\mathcal{D}_2^{p+k}\frac{\delta_{12}}{(p+k)^2-m^2}\right)\cdot$$

$$\left.\left(\bar{\mathcal{D}}_2^{-k}\bar{\mathcal{D}}_2^{-k}\ \mathcal{D}_2^{-k}\mathcal{D}_2^{-k}\ \bar{\mathcal{D}}_2^{-k}\bar{\mathcal{D}}_2^{-k}\frac{\delta_{12}}{k^2-m^2}\right)\right]V^b(p,\theta_1,\bar{\theta}_1)\ . \tag{6.75}$$

In going from (6.74) to (6.75), identities like (6.33b) have been repeatedly used with $q_1=-p$, $q_2=p+k$ and $q_3=-k$. The use of (4.11m) simplifies the last factor within parentheses in the third term within the square brackets of (6.75) to $16k^2\bar{\mathcal{D}}_2^{-k}\bar{\mathcal{D}}_2^{-k}\delta_{12}/(k^2-m^2)$. Similarly, the second term can be simplified, by utilizing the commutator (cf. 4.11h) $[\bar{\mathcal{D}}_2^{\dot{A},-k},\mathcal{D}_2^{-k}\mathcal{D}_2^{-k}]=-4k^\mu\bar{\sigma}_\mu^{\dot{A}B}\mathcal{D}_{2B}^{-k}$ and also (3.18a), to

$$8\bar{\mathcal{D}}_2^{-p,\dot{C}}V^a(-p,\theta_2,\bar{\theta}_2)\left(\mathcal{D}_2^{p+k}\mathcal{D}_2^{p+k}\frac{\delta_{12}}{(p+k)^2-m^2}\right)k_\mu\sigma_{\dot{C}}^{\mu B}\left(\mathcal{D}_{2B}^{-k}\ \bar{\mathcal{D}}_2^{-k}\bar{\mathcal{D}}_2^{-k}\frac{\delta_{12}}{k^2-m^2}\right)\ .$$

We are now in a position to partially integrate the $\mathcal{D}_2^{p+k}\mathcal{D}_2^{p+k}$ factor in each of the terms of (6.75) by sequential use of the super-Leibniz rule (6.33). Furthermore, we can employ the trick of manipulating with partial integrations of covariant derivatives by the use of relations like[14] $0=\delta_{12}\delta_{12}=\delta_{12}\mathcal{D}_A\delta_{12}=\delta_{12}\bar{\mathcal{D}}^{\dot{B}}\delta_{12}=\delta_{12}\mathcal{D}\mathcal{D}\delta_{12}=\delta_{12}\bar{\mathcal{D}}\bar{\mathcal{D}}\delta_{12}=\delta_{12}\mathcal{D}_A\ \bar{\mathcal{D}}\bar{\mathcal{D}}\delta_{12}=\delta_{12}\bar{\mathcal{D}}^{\dot{B}}\ \mathcal{D}\mathcal{D}\delta_{12}$ etc. and the most effective $\delta_{12}\mathcal{D}\mathcal{D}\ \bar{\mathcal{D}}\bar{\mathcal{D}}\delta_{12}=16\delta_{12}$ till the product of two δ_{12}'s reduces to a single δ_{12} in every term. Consequently, the θ_1-integration can be performed and we are left with

$$-i\frac{g^2}{8}T(\mathcal{R})\delta^{ab}\int\frac{d^4p}{(2\pi)^4}\frac{d^4k}{(2\pi)^4}\frac{d^4\theta}{[(p+k)^2-m^2](k^2-m^2)}V^a(-p,\theta,\bar{\theta})$$

$$\left[\bar{\mathcal{D}}^p\bar{\mathcal{D}}^p\ \mathcal{D}^p\mathcal{D}^p+8k^\mu\bar{\mathcal{D}}^{p\dot{c}}\sigma_{\dot{c}}^{\mu B}\mathcal{D}_B^p+16k^2\right]V^b(p,\theta,\bar{\theta})\ . \tag{6.76}$$

In going from (6.75) to (6.76), identities such as (6.33a) with $q_1=-p$, $q_2=p+k$ and $q_3=-k$ have been repeatedly used. The first term within the square bracket of (6.76) can

---

[14]The basic point here is that *anything less* than two $\mathcal{D}$'s and two $\bar{\mathcal{D}}$'s between two δ_{12}'s yields zero.

be rewritten by means of the relation $\bar{\mathcal{D}}\bar{\mathcal{D}}\,\mathcal{D}\mathcal{D} = \bar{\mathcal{D}}_{\dot{A}}\mathcal{D}\mathcal{D}\bar{\mathcal{D}}^{\dot{A}} + 4i\bar{\mathcal{D}}\bar{\sigma}^{\mu}\mathcal{D}\partial_{\mu}$, which follows by contracting $\bar{\mathcal{D}}_{\dot{A}}$ with both sides of (4.11h). Thus (6.76) now appears in the notation of (6.9) as

$$-i\frac{g^2}{8}T(\mathcal{R})\int \frac{d^4p}{(2\pi)^4}\frac{d^4k}{(2\pi)^4}\frac{d^4\theta}{[(p+k)^2-m^2]}\frac{1}{k^2-m^2}V^a(-p,\theta,\bar{\theta})\cdot$$

$$\left[-8\Pi_0 p^2 + 4(p+2k)_\mu \bar{\mathcal{D}}^{p\dot{C}}\sigma^{\mu B}_{\dot{C}}\mathcal{D}^p_B + 16k^2\right]V^a(p,\theta,\bar{\theta})\ . \qquad (6.77)$$

Suppose now that we are interested only in the infinite part of the contribution of Fig. 6.4 to the effective action. Then we can set the mass m, associated with the chiral superfield, to zero in (6.77). We are now able to use the following formulae in the $\overline{\text{DR}}$ scheme [6.4,6.5]:

$$0 = \int \frac{d^{2\omega}k}{(p+k)^2} = \int \frac{d^{2\omega}k}{k^2} = \int \frac{d^{2\omega}k}{(p+k)^2 k^2}(p+2k)_\mu$$

in isolating the infinite part of (6.77) as

$$g^2 T(\mathcal{R})\int \frac{d^4p}{(2\pi)^4}d^4\theta B_0(p^2)V^a(-p,\theta)\Pi_0 p^2 V^a(p,\theta)\ , \qquad (6.78)$$

where $B_0(p^2)$ is the $m_{1,2}\to 0$ limit of $B_0(p^2,m_1^2,m_2^2)$ introduced in (6.70b).

There are two more contributions to the vector superpropagator at one loop. The first comes from internal V-lines effected by a V^3 vertex at either end (Fig. 6.5). (The V-loop from a quartic V-vertex does not have a divergent part in dimensional reduction). The calculation for this involves lengthy algebra. There are six terms at each vertex from six

Fig.6.5. One loop contribution from V internal lines to the vector superpropagator.

possible contractions in $(64g^2 k)^{-1}\text{Tr}\int d^8z\,\bar{\mathcal{D}}\bar{\mathcal{D}}\mathcal{D}^A V[V,\mathcal{D}_A V]$, cf.(6.64b), producing thirty six terms in total. We know that, in the string of covariant derivatives, there must be at least four involving the product of $\mathcal{D}\mathcal{D}$ and $\bar{\mathcal{D}}\bar{\mathcal{D}}$ between the two δ_{12}'s to produce a nonzero result. Owing to the symmetry between the two lines and the two vertices, there are, in fact, only seven separate contributions with at least four covariant derivatives in the form mentioned above. The long algebra leads to the following result [6.9] for the infinite part:

$$g^2 C_2(G)\int \frac{d^4p}{(2\pi)^4}d^4\theta B_0(p^2)V^a(-p,\theta,\bar{\theta})\left(-\frac{5}{2}\Pi_0 + \frac{1}{2}\Pi_+ + \frac{1}{2}\Pi_-\right)p^2 V^a(p,\theta,\bar{\theta})\ , \qquad (6.79)$$

where $\Pi_\pm$ are the same as in (6.7). Recall from the discussion in §5.4 that $C_2(G)$ is the quadratic Casimir of the gauge group G. Finally, there is the ghost loop contribution of Fig. 6.6 which is the only divergent ghost contribution at one loop. The infinite part of this can be evaluated, by methods similar to those outlined earlier, as

$$-\frac{g^2}{2}C_2(G)\int \frac{d^4p}{(2\pi)^4}d^4\theta B_0(p^2)V^a(-p,\theta,\bar{\theta})p^2 V^a(p,\theta,\bar{\theta})\ . \qquad (6.80)$$

Adding (6.78), (6.79) and (6.80) and utilizing the result $1 = \Pi_0 + \Pi_+ + \Pi_-$, the infinite part of the net one loop contribution to the vector superpropagator can be rewritten as

$$-g^2\left[3C_2(G) - T(\mathcal{R})\right] \int \frac{d^4p}{(2\pi)^4} d^4\theta B_0(p^2) V^a(-p,\theta,\bar{\theta}) \Pi_0 p^2 V^a(p,\theta,\bar{\theta}) \ .$$

Once again, cf. (6.70b),

$$B_0(p^2) = -\frac{1}{8\pi^2}\frac{1}{4-2\omega} + \text{finite part} \ .$$

Taking into account the change of sign in going from p^2 in momentum space to ∂^2 in configuration space, the divergent contribution to the effective action from the one loop vector superpropagator is

$$S_E^V(\text{one loop}) = -\frac{g^2\mu^{2\omega-4}}{4\pi^2(4-2\omega)}\left\{3C_2(G) - T(\mathcal{R})\right\}\frac{1}{2}\int d^8z\, V^a\Pi_0\partial^2 V^a. \tag{6.81}$$

We end this section with a remark. If we rescale the definitions of the gauge coupling and gauge superfield back (cf.§6.5), i.e. $g \to \frac{1}{\sqrt{2}}g$ and $V^a \to \sqrt{2}V^a$, so as to recover the standard normalizations introduced in Ch.4, (6.81) remains unchanged. Indeed, we shall now **revert back to those standard normalizations** for g and V^a and use them henceforth.

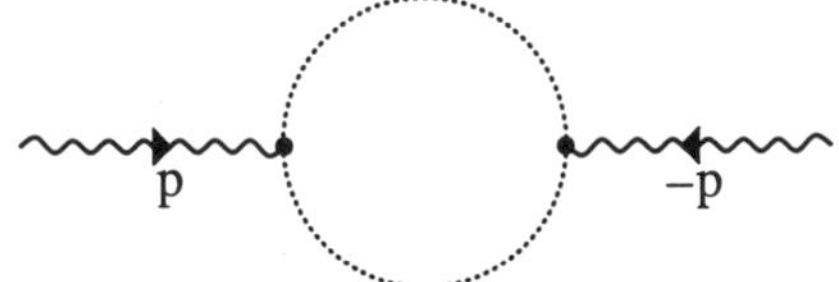

Fig.6.6. Ghost contribution to the one loop vector superpropagator.

6.7 The Nonrenormalization Theorem

The one loop contributions to the two point vertex functions, calculated in the previous section, have one thing in common. They finally involve only one integral $\int d^4\theta \cdots$ of an integrand that is a *local* function of the external superfields[15] and their covariant derivatives in the $\theta,\bar{\theta}$ subspace of superspace. This integrand also has a factor $\int d^4k \cdots$ of a nonpolynomial function of the internal loop four momentum k. When higher loops are taken into account, the latter form may change to multiple integrals of this sort, involving several internal loop four momenta k_i or equivalently several spacetime coordinates y_i or several Feynman parameters α_i. But they can all be expressed as a single integral $\int d^4\theta \cdots$ of an integrand with the property mentioned above. This is the content of the **nonrenormalization theorem** [6.1], the technical version of which is enunciated below.

Any perturbative quantum contribution to the effective action is expressible as a single integral $\int d^4\theta \cdots$ over the entire $\theta,\bar{\theta}$ subspace of superspace.

---

[15]These, in turn, are polynomials in $\theta,\bar{\theta}$.

In other words, the above contribution must have the form

$$\Gamma = \sum_n \prod_{i=1}^{n} \int d^4 x_i d^4\theta \, G_n(x_1, \cdots, x_n) F_i(\Phi, \Phi^\dagger, V, \mathcal{D}^A\Phi, \bar{\mathcal{D}}^{\dot{A}}\Phi^\dagger, \mathcal{D}^A V, \cdots) \, . \qquad (6.82)$$

In (6.82) the G_n's are translationally invariant functions on Minkowski spacetime and the F_i are local functions of possible external superfields $\Phi, \Phi^\dagger, V$ and their covariant derivatives.

Proof. This results in a simple and direct way from the super-Feynman rules given in the earlier sections. Consider the $\theta, \bar{\theta}$-structure of an L-loop supergraph. We recall that each propagator – linking the ith and jth vertices, say – yields δ_{ij} while each vertex (ith, say) yields $\int d^4\theta_i \cdots$ as well as $\mathcal{D}$'s and/or $\bar{\mathcal{D}}$'s acting on the δ_{ij} factors of the propagators. Suppose we take up a fixed irreducible loop containing n vertices and n propagators. Let us focus on one particular propagator connecting the i_1th and i_2th vertices. We can partially integrate any $\mathcal{D}$'s or $\bar{\mathcal{D}}$'s, acting on the $\delta_{i_1 i_2}$ factor, onto other propagators or external fields/sources. Once these partial integrations are done, a residual naked $\delta_{i_1 i_2}$ remains which can be "killed" by doing the $\int d^4\theta_i$ integration. After the latter, all i_1 indices have to be replaced by i_2 at all other points in the supergraph, i.e. the propagator gets contracted in the $\theta, \bar{\theta}$-subspace to one point at the i_2 vertex. We can continue this procedure $n-1$ times, i.e. for all but one of the propagators. The loop has now shrunk to contain only two vertices (1 and 2, say), with its $\theta, \bar{\theta}$-structure given by the expression

$$\prod_E \int d^4\theta_E d^4\theta_2 d^4\theta_1 \delta_{12} \bar{\mathcal{D}}_1 \cdots \bar{\mathcal{D}}_1 \cdots \mathcal{D}_1 \cdots \mathcal{D}_1 \cdots \delta_{12} \, .$$

In this expression all vertices external to the loop are covered by the label E. Note that the appearance of fewer than two $\mathcal{D}$'s and two $\bar{\mathcal{D}}$'s – between two δ_{12}'s in this, or any other earlier occurring expression, will make the whole supergraph vanish on account of relations described earlier. On the other hand, more than two factors of $\mathcal{D}$'s or $\bar{\mathcal{D}}$'s can always be reduced to two such factors by the $\mathcal{D}$-, $\bar{\mathcal{D}}$-algebra. Finally, for the remaining two $\mathcal{D}$'s and two $\bar{\mathcal{D}}$'s, we can use the result

$$\delta_{12} \bar{\mathcal{D}}_1^{\dot{A}} \bar{\mathcal{D}}_1^{\dot{B}} \mathcal{D}_{1C} \mathcal{D}_{1D} \delta_{12} = 4\epsilon^{\dot{A}\dot{B}} \epsilon_{CD} \delta_{12} \qquad (6.83)$$

to eliminate one of the δ_{12}'s and perform the $d^4\theta_2$ with the remaining one. The entire loop, in consequence, gets contracted to a single point in the $\theta, \bar{\theta}$-subspace. We can now continue this procedure, loop by loop, till we are left with an expression involving loop as well as external four momenta, external superfields, covariant derivatives acting on external superfields and one final $\int d^4\theta \cdots$ vertex integration, i.e. an expression like that in the RHS of (6.82). QED.

Corollary. All vacuum supergraphs vanish.

Proof. If a supergraph has no external legs, at the end of the procedure outlined above, we are left with an expression $C \int d^4\theta$ where C represents one or more integrals with loop four momenta and there are no $\theta, \bar{\theta}$-dependent terms inside the integral. This is evidently zero. In consequence, the generating superfunctional of (6.15) is naturally normalized by $Z[0,0,0] = 1$.

A somewhat more pointed statement can be made as follows.

The quantum contribution to the absolute minimum of the effective potential for classically supersymmetric configurations vanishes in perturbation theory.

Proof. The said configuration must have zero vacuum energy at the classical level. This was already demonstrated earlier and will be discussed in more detail in Ch.7. Indeed, all the auxiliary F- and D-fields, whose absolute squares add to make the classical potential (cf. 5.56b), must individually vanish in this configuration. The perturbative quantum contribution to the effective potential, which is the momentum independent part of the effective action, can only have the following form

$$\int d^8z F\left(\langle\Phi\rangle, \langle\mathcal{D}\Phi\rangle, \langle\Phi^\dagger\rangle, \langle\bar{\mathcal{D}}\Phi^\dagger\rangle, \langle V\rangle, \langle\mathcal{D}V\rangle, \cdots\right),$$

where F is some function of the superfields and their covariant derivatives in that configuration. However, in classically supersymmetric configurations, the auxiliary fields vanish while spinorial fields can never have nonzero VEVs. This means that all the VEVs in the argument of F are independent of $\theta, \bar{\theta}$ and so is F. As a result, the integral vanishes, and the statement follows. Of course, nonvanishing quantum corrections to the effective potential do exist away from the minimum.

This statement has an important implication. The perturbatively quantum corrected effective potential will be at the same absolute minimum of field configurations as it is for a classically supersymmetric theory. Specifically, any degeneracies at that minimum will not be removed by these corrections. This means that vacuum configurations, which were classically different, will remain different after quantum fluctuations are perturbatively switched on. Moreover, if supersymmetry is unbroken at the tree level, it will not be broken perturbatively by quantum corrections. The latter cannot shift the minimum of the potential away from zero where it already is located classically. Only nonperturbative quantum effects may be able to achieve such a shift.

Remark on the nonrenormalization theorem

Each term in the RHS of (6.82) involves the full integral $\int d^4\theta \cdots$ in $\theta, \bar{\theta}$-subspace but not the subintegrals $\int d^2\theta \cdots$ or $\int d^2\bar{\theta} \cdots$ separately. This means that counterterms of the form $\int d^6z \mathcal{W}'(\Phi_i)$ or $\int d^6\bar{z}\overline{\mathcal{W}}'(\Phi_i^\dagger)$, where $\mathcal{W}'(\Phi_i)$ is a local analytic function of chiral superfields Φ_i and $\overline{\mathcal{W}}'$ its hermitian conjugate, cannot arise in perturbation theory. The holomorphy property of the superpotential protects it from any perturbative renormalization and is the basis of the following alternative and generally more practical statement of the nonrenormalization theorem that can be made for any supersymmetric field theory:

No superpotential term gets renormalized in perturbation theory[16].

---

[16] Nonperturbative renormalization of a superpotential term is, however, possible; cf. §7.6.

More specifically, counterterms – added to the Lagrangian density after renormalization – cannot have the structure of superpotential terms. There will, of course, be wave function renormalizations of the chiral superfields. Thus renormalizations of couplings and masses in each term of the superpotential must balance the corresponding chiral wave function renormalizations. The latter, being dimensionless, can at most have a logarithmic dependence on the mass scale of the theory. Therefore, the same must be true of the former.

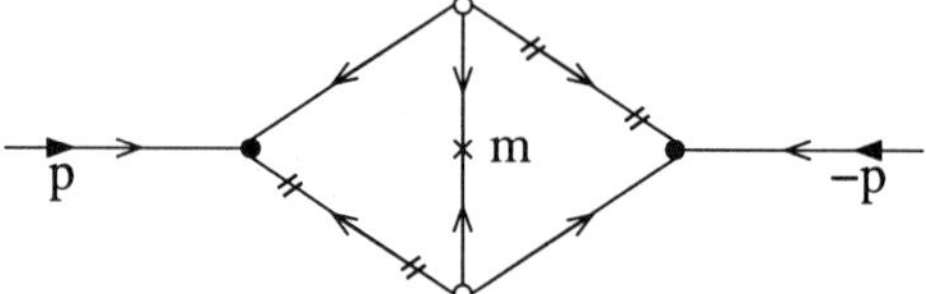

Fig.6.7. A two loop contribution to the $\langle\Phi\Phi\rangle$ superpropagator.

In this context, a clarification is perhaps necessary. A purely left or right chiral contribution to the effective action involving $\int d^6z \cdots$ or $\int d^6\bar{z} \cdots$ can arise with *nonlocal* combinations of chiral superfields, the nonlocality being in coordinate space[17]. For instance, the two loop supergraph of Fig. 6.7 contributes the following expression as a higher order correction to the superpropagator of (6.34b):

$$m \int \frac{d^4p}{(2\pi)^4} B(p^2) d^2\theta \Phi(-p,\theta)\Phi(p,\theta) = -m \int \frac{d^4p}{(2\pi)^4} \frac{B(p^2)}{p^2} d^4\theta \Phi(-p,\theta) \frac{\mathcal{D}\mathcal{D}}{4} \Phi(p,\theta) \ .$$

Here $B(p^2)$ is a Lorentz invariant function of p that satisfies $dB/dp^2 = 0$ for $p^2 \to 0$ and (4.14e) has been used to derive the equality. Superpotential terms are, however, local analytic functions and remain unaffected.

We focus on two important aspects and consequences of the nonrenormalization theorem.

- No term in the superpotential can be perturbatively renormalized; wave function renormalizations of chiral superfields appearing in any term must be cancelled by the renormalization of the coefficient coupling/mass.

- Given a classically supersymmetric configuration, there are no perturbative quantum corrections to the absolute minimum of the corresponding effective potential.

6.8   One Loop Infinities and β-,γ-functions

We return to the classical Lagrangian density of a supersymmetric nonabelian gauge theory, as described by (5.54). Let us, for the moment, suppress the general type index i. We are then left with a generic chiral superfield Φ along with the gauge superfield V. Let us also

---

[17]This is simply because $\int d^6z f(\Phi) = -\frac{1}{4}\int d^8z \mathcal{D}\mathcal{D}/\partial^2 f(\Phi)$ for any analytic function $f(\Phi)$ and similarly $\int d^6\bar{z} f(\Phi^\dagger) = -\frac{1}{4}\int d^8z \bar{\mathcal{D}}\bar{\mathcal{D}}/\partial^2 f(\Phi^\dagger)$, cf. (4.14e,f). Nonlocality in coordinate space shows up as a nonpolynomial factor in momentum space.

specify the superpotential $\mathcal{W}(\Phi)$ to be that of the Wess-Zumino model. The classical action, cf. (5.54), is

$$S = \int d^8z \Phi^\dagger e^V \Phi + \frac{1}{16g^2 k} \cdot$$

$$\mathrm{Tr}\left(\int d^6z\, WW + \int d^6\bar{z}\, \bar{W}\bar{W}\right) + \left\{\int d^6z \left(\frac{1}{2}m\Phi\Phi + \frac{f}{3!}\Phi\Phi\Phi\right) + \text{h.c.}\right\}. \quad (6.84)$$

After renormalization, we can put subscripts R on all superfields and parameters in (6.84), referring to renormalized versions of the same. But then there will be additional counterterms for the gauge covariant kinetic energy terms in the RHS of (6.84), though not for the V-independent curly bracketted set of terms which come from the superpotential. We shall therefore have a counterterm[18] for each chiral superfield Φ which is $\Delta Z \int d^8z \Phi_R^\dagger e^{V_R}\Phi_R$ and one for each gauge superfield V that is $\Delta Z_V \frac{1}{16g_R^2 k}\mathrm{Tr}\left(\int d^6z\, W_R W_R + \int d^6\bar{z}\, \bar{W}_R\bar{W}_R\right)$. Defining $Z \equiv 1 + \Delta Z$ and $Z_V \equiv 1 + \Delta Z_V$, the action can be cast in terms of renormalized superfields Φ_R, V_R and the renormalized parameters m_R, f_R, g_R as

$$S = Z\int d^8z \Phi_R^\dagger e^{V_R}\Phi_R + Z_V\frac{1}{16g_R^2 k}\mathrm{Tr}\left(\int d^6z\, W_R W_R + \int d^6\bar{z}\, \bar{W}_R\bar{W}_R\right)$$

$$+ \left\{\int d^6z \left(\frac{1}{2}m_R\Phi_R\Phi_R + \frac{f_R}{3!}\Phi_R\Phi_R\Phi_R\right) + \text{h.c.}\right\}. \quad (6.85)$$

In (6.85) W_{RA} and $\bar{W}_R^{\dot{A}}$ are renormalized spinorial supersymmetric field strengths with

$$W_{RA} = -\frac{1}{4}\bar{\mathcal{D}}\bar{\mathcal{D}}e^{-V_R}\mathcal{D}_A e^{V_R}\,,$$

$$\bar{W}_R^{\dot{A}} = -\frac{1}{4}\mathcal{D}\mathcal{D}e^{V_R}\bar{\mathcal{D}}^{\dot{A}}e^{V_R}\,.$$

A comparison between (6.84) and (6.85) leads to relations between renormalized and unrenormalized superfields, namely

$$\Phi = Z^{1/2}\Phi_R\,, \quad (6.86a)$$

$$V^a = Z_V^{1/2}V_R^a\,, \quad (6.86b)$$

as well as between renormalized and unrenormalized parameters, i.e.

$$V = V_R \Rightarrow gV^a = g_R V_R^a\,, \quad (6.87a)$$

$$m\Phi\Phi = m_R\Phi_R\Phi_R\,, \quad (6.87b)$$

$$\lambda\Phi\Phi\Phi = \lambda_R\Phi_R\Phi_R\Phi_R\,. \quad (6.87c)$$

---

[18]The fact that renormalizations constants ΔZ and ΔZ_V are only logarithmically divergent can be shown by an analysis of the superficial degree of divergence of the concerned one loop supergraphs [6.8].

Relations (6.87) indeed follow directly from the nonrenormalization theorem and supersymmetric gauge invariance[19]. We can formally introduce renormalization constants Z_m, Z_f and Z_g for the mass and the two coupling strengths f and g respectively:

$$m = Z_m m_R \,, \tag{6.88a}$$

$$f = Z_f f_R \,, \tag{6.88b}$$

$$g = Z_g g_R \,. \tag{6.88c}$$

It follows from (6.86-6.88) that

$$Z_f Z^{3/2} = 1 \,, \tag{6.89a}$$

$$Z_m Z = 1 \,, \tag{6.89b}$$

$$Z_g Z_V^{1/2} = 1 \,. \tag{6.89c}$$

Finally then, there are only two independent renormalization constants: Z and Z_V.

Let us now consider the formally divergent expressions for the infinite parts of these renormalization constants. That for ΔZ can be obtained easily. Expand the factor e^{V_R} in the covariant chiral superfield kinetic energy term in (6.84) as $1 + V_R + \frac{1}{2} V_R^2 + \cdots$ and treat the contribution from the linear V_R term as the lowest order perturbative gauge interaction on par with the trilinear self coupling term. Now (6.71) and (6.73) already tell us what the divergent one loop contributions from these two couplings are. To one loop, this is all there is for the divergent part. The one loop counterterm $\Delta Z^{(1)} \int d^8 z \Phi_R^\dagger \Phi_R$ must be chosen to cancel this divergence. Hence, taking due note of the rescaling of the gauge coupling strength $g \to \frac{1}{\sqrt{2}} g$, we can write

$$\Delta Z^{(1)} = -\frac{1}{16\pi^2} \left\{ |f|^2 - 4g^2 C_2(\mathcal{R}) \right\} \frac{\mu^{2\omega - 4}}{4 - 2\omega} \,. \tag{6.90}$$

One can now display the one loop expression for the wavefunction renormalization constant $Z^{1/2}(\mu)$ of (6.86a), including the dependence [6.6] on the renormalization mass scale (cf. §1.3) μ, as

$$Z^{1/2}(\mu)^{(1)} = 1 - \frac{1}{32\pi^2} \left\{ |f|^2 - 4g^2 C_2(\mathcal{R}) \right\} \frac{\mu^{2\omega - 4}}{4 - 2\omega} \,, \tag{6.91}$$

i.e. the one loop wavefunction renormalization constant vanishes if $|f|^2 = 4g^2 C_2(\mathcal{R})$. The anomalous dimension of the chiral superfield $\gamma = \lim_{\omega \to 2} \mu \partial \ln Z^{1/2}(\mu)/\partial \mu$ can be immediately

---

[19] A caveat is called for here. Counterterms in quantized gauge theories generally become noninvariant under classical gauge transformations of the original action because of gauge fixing. However, they have to obey the Slavnov-Taylor identities [6.6] which ensure gauge invariance of the effective action. This is manifestly attainable in the background field method [6.5]. For supersymmetric nonabelian gauge theories, the viability of the background field method was demonstrated by Grisaru, Roček and Siegel [6.9], i.e. they formulated it as a framework within which the effective action is invariant under the initial gauge transformations. It is in this framework, i.e. within the background field method, that relations like (6.87a) and (6.89c) can be shown to be valid.

computed to be

$$\gamma^{(1)} = \frac{1}{32\pi^2}\left\{|f|^2 - 4g^2 C_2(\mathcal{R})\right\} = -\frac{1}{2}\lim_{\omega\to 2}(4 - 2\omega)\Delta Z^{(1)}\,. \tag{6.92}$$

The superscript (1) in (6.90), (6.91) and (6.92) refers to the fact that the RHS has been calculated only upto one loop.

Turning to Z_V, let us refer back to (6.40) and rescale $V^a \to \sqrt{2}V^a$. The counterterm $\Delta Z_V^{(1)} \int d^8 z V_R^a \Pi_0 \partial^2 V_R^a$ must be chosen to cancel the divergence of (6.81). This allows us to deduce that

$$\Delta Z_V^{(1)} = 1 + \frac{g^2 \mu^{2\omega-4}}{8\pi^2(4 - 2\omega)}\left\{3C_2(G) - T(\mathcal{R})\right\}, \tag{6.93a}$$

$$Z_V^{1/2}(\mu)^{(1)} = 1 + \frac{g^2}{16\pi^2}[3C_2(G) - T(\mathcal{R})]\frac{\mu^{2\omega-4}}{4 - 2\omega}\,. \tag{6.93b}$$

It follows from (6.89c) and (6.93b) that

$$Z_g(\mu)^{(1)} = 1 - \frac{g^2}{16\pi^2}[3C_2(G) - T(\mathcal{R})]\frac{\mu^{2\omega-4}}{4 - 2\omega}\,. \tag{6.94}$$

If we make use of (6.88c) now, we can derive for this supersymmetric nonabelian gauge theory the β-function[20] which is the derivative of the gauge coupling strength with respect to the logarithm of the renormalization mass scale μ. Suppose the lowest order nontrivial (one loop) part of the β-function is written as $\beta^{(1)}(g)$. Then, from (6.88c), we can write

$$\beta^{(1)}(g) = \lim_{\omega\to 2}\mu\frac{\partial}{\partial\mu}g_R = g\lim_{\omega\to 2}\mu\frac{\partial}{\partial\mu}Z_g^{-1}(\mu)^{(1)}\,. \tag{6.95}$$

Now (6.94) and (6.95) imply that

$$\beta^{(1)}(g) = -\frac{g^3}{16\pi^2}[3C_2(G) - T(\mathcal{R})]\,. \tag{6.96}$$

Let us consider the more general case with a type index i. Eq. (6.86b) remains unchanged, but (6.86a) generalizes to

$$\Phi_i = (Z^{1/2})_{ii'}\Phi_{Ri}\,. \tag{6.97}$$

The corresponding superpotential, containing symmetric tensor couplings m_{ij} and f_{ijk}, is

$$\mathcal{W} = \frac{1}{2}m_{ij}\Phi_i\Phi_j + \frac{1}{3!}f_{ijk}\Phi_i\Phi_j\Phi_k\,.$$

Again, (6.88c) remains unaffected while (6.88a) and (6.88b) get changed to

$$m_{ij} = (Z_m)_{ij,i'j'}(m_R)_{i'j'}\,, \tag{6.98a}$$

$$f_{ijk} = (Z_f)_{ijk,i'j'k'}(f_R)_{i'j'k'} \tag{6.98b}$$

---

[20]See §6.9 for a more elaborate discussion of the β-function.

respectively. Furthermore, (6.89c) remains the same but (6.89a) and (6.89b) do not. The extended versions of the latter respectively are

$$(Z_f)_{ijk,i'j'k'}(Z^{1/2})_{i'i''}(Z^{1/2})_{j'j''}(Z^{1/2})_{k'k''}$$

$$= \frac{1}{6}\left(\delta_{ii''}\delta_{jj''}\delta_{kk''} + \delta_{ii''}\delta_{jk''}\delta_{kj''} + \delta_{ij''}\delta_{ji''}\delta_{kk''} + \delta_{ij''}\delta_{jk''}\delta_{kj''} \right.$$

$$\left. + \delta_{ik''}\delta_{jj''}\delta_{ki''} + \delta_{ik''}\delta_{ji''}\delta_{kj''} \right), \tag{6.99a}$$

$$(Z_M)_{ij,i'j'}(Z^{1/2})_{i'i''}(Z^{1/2})_{j'j''} = \frac{1}{2}\left(\delta_{ii''}\delta_{jj''} + \delta_{ij''}\delta_{ji''} \right). \tag{6.99b}$$

Finally, the generalizations of (6.92) and (6.95) respectively are

$$\gamma_{ij}^{(1)} = \mu\frac{\partial}{\partial\mu}\ln(Z^{1/2})_{ij}^{(1)} = \frac{1}{32\pi^2}\left\{ f_{ik\ell}^{\star}f_{jk\ell} - 4g^2\sum_i C_2(\mathcal{R}_i)\delta_{ij} \right\}, \tag{6.100a}$$

$$\beta^{(1)}(g) = -\frac{g^3}{16\pi^2}\left[3C_2(G) - \sum_i T(\mathcal{R}_i) \right]. \tag{6.100b}$$

We can also write the β-functions for quantities in the superpotential, namely the mass m_{ij} and coupling strength f_{ijk} as follows:

$$\beta(m)_{ij} \equiv \mu\frac{\partial}{\partial\mu}(m_R)_{ij}(\mu) = \gamma_{ii'}m_{i'j} + \gamma_{jj'}m_{ij'}, \tag{6.101a}$$

$$\beta(f)_{ijk} \equiv \mu\frac{\partial}{\partial\mu}(f_R)_{ijk}(\mu) = \gamma_{ii'}f_{i'jk} + \gamma_{jj'}f_{ij'k} + \gamma_{kk'}f_{ijk'}. \tag{6.101b}$$

Before ending this discussion, we wish to comment on the gauge (in-)dependence that can arise when calculating β-functions in a supersymmetric theory with gauge interactions. Consider the coupling coefficient of a superpotential term that is a product of three different left chiral superfields. Let the fermionic component of one of these (F_1, say) be combined with the charge conjugate of that of another (F_2, say) to form a Dirac fermion f. If ϕ is the scalar component of the third, called Φ, the Yukawa interaction $h\bar{f}_L f_R \phi +$ h.c. will then be derived from the said superpotential term. The β-function of the coupling strength h of this Yukawa interaction should be obtainable from the wavefunction renormalizations of the superfields $F_{1,2}$ and Φ on account of the nonrenormalization theorem. Despite contributions from gauge interactions in the loop, $\beta(h)$ is gauge invariant to one loop. This, however, does not mean the gauge invariance of the wavefunction renormalizations of the *component fields* of the left chiral superfields. Only in a manifestly supersymmetric gauge, such as the sFg (cf. §6.5), will the wavefunction renormalizations of the scalar and fermionic components be the same; in such a gauge the sum of all proper vertex corrections will vanish. However, in such a gauge there will generally be loops involving the auxiliary components of the vector superfield, in addition to those involving gauge bosons and gauginos. The story is different with a choice such as the Wess-Zumino supergauge, followed by the R_ξ-gauge for the gauge boson propagator. Here the auxiliary components of the vector superfield have been gauged away, but such a gauge fixing procedure violates manifest supersymmetry. In this case the gauge dependence of the divergent part of the proper vertex part of the one (gauge boson) loop correction to the Yukawa interaction will need to cancel that of the fermionic wavefunction renormalization. In other words, it is no longer possible here to compute $\beta(h)$ from the wavefunction renormalizations of the component fields alone. Moreover, those wavefunction renormalizations will in general be different for the different component fields of a single superfield. However, the one loop $\beta(h)$, even if calculated with this gauge choice, will come out the same as calculated in the sFg from the supersymmetric wavefunction renormalization constants.

To conclude this section, let us summarize its contents:

- The calculation of the β-function of the gauge coupling in a super-YM theory bears some resemblances to the corresponding calculation in ordinary YM theory; owing to Ward identities, the β-function can be computed solely from vacuum polarization (super-) graphs, though the numerical coefficients will differ in the two cases.

- For β-functions of Yukawa and scalar four point couplings, there *is* a qualitative difference between theories with and without supersymmetry. While, for the former, all β-functions of superpotential couplings orginate from the wave function renormalizations of the chiral superfields and are computable solely from two point functions, such is generally not the case for the latter.

- All superpotential parameters renormalize multiplicatively, the multiplication being in a matrix sense (cf. 6.101), in the presence of nondiagonal wave function renormalization. Since scalar self couplings in a supersymmetric theory are related to either superpotential (Yukawa) or gauge couplings, this statement applies to them, too, unlike[21] in nonsupersymmetric theories.

6.9 Renormalization Group Evolution

The β- and γ-functions, derived above, control the *renormalization group evolution* (RGE) of any supersymmetric system. A treatment of RGE can be found in most modern books [6.6] on quantum field theory. We need therefore give only a brief introduction here into its basic methodology – always with a supersymmetric field theory at the back of our mind. Our approach would be somewhat intuitive rather than rigorous. The main assumption is that we are dealing with a problem containing only two basic scales that matter: an "external" energy or momentum scale Q and the renormalization scale μ. The specific implication is that all masses, pertinent to the problem, are much smaller than Q and μ.

---

[21]For instance, the β-function of the coefficient λ of the scalar quartic coupling of the Standard Model receives [6.11] a contribution proportional to the fourth power of the top Yukawa coupling, independent of λ.

Such a situation can always be ensured by integrating out fields with masses which are larger than those two scales. Of course, matrix elements or connected Green's functions, describing physical processes, cannot really depend on μ; the latter is a pure artifice which gets introduced in the process of removing loop divergences from quantum field theory. The parameters of the Lagrangian such as couplings and masses, that appear in expressions for concerned matrix elements or Green's functions, do nonetheless acquire μ-dependence through radiative corrections. In physical amplitudes this dependence gets cancelled by explicit μ-dependent terms emerging from loops. This cancellation is quantitatively described by the renormalization group equation.

To one loop order, a connected Green's function Γ can be schematically written as

$$\Gamma = \sum_i [X_i(\mu)]^n \left[c_i + \sum_j d_{ij} X_j(\mu) \left(\ln \frac{Q}{\mu} + \text{finite terms} \right) \right] . \tag{6.102}$$

Here $\{X_i(\mu)\}$ are a set of renormalized Lagrangian parameters while c_i and d_{ij} are coefficients describing tree level and one loop contributions respectively. On renormalization, the logarithmic divergences in the latter generate logarithmic Q-dependent terms. These must involve $\ln Q/\mu$ since, by assumption, Q and μ are the only dimensional quantities available. Two important consequences of (6.102) can be highlighted right away. First, it allows the deduction of the μ-dependence of X_i from the requirement that $d\Gamma/d\mu$ vanish; in a renormalizable theory the answer is guaranteed to be the same for all processes, i.e. for all connected Green's functions Γ. Second, the solution to the resulting differential equation automatically resums all leading logarithmic corrections of the form $(X_i \ln Q/\mu)^m$, $\forall m$. Moreover, it is easy to see from (6.102) that (most) one loop corrections to Γ can be absorbed into the "running parameters" $X_i(Q)$ by the choice[22] $\mu = Q$. The resummation of leading loop corrections is now included, by simply plugging the running parameters $X_i(Q)$, taken at the external momentum scale Q, into the *tree level* expression for Γ.

Suppose we introduce the evolution variable t as

$$t \equiv \ln \frac{\mu}{\mu_0} , \tag{6.103}$$

μ_0 being a fixed but arbitrary reference mass. The latter does not contribute to $dt = d\mu/\mu$. Now the **beta-function** for the specific parameter X_i, defined by

$$\mu \frac{\partial}{\partial \mu} X_i \equiv \beta(X_i) , \tag{6.104}$$

controls [6.6] the t-evolution of X_i:

$$\frac{dX_i}{dt} = \beta(X_i(t)) . \tag{6.105}$$

This dependence can be computed from any matrix element or connected Green's function to any loop order. In practice, the complexity of the calculation is reduced by finding the

---

[22]Setting $\mu = Q$ exactly, one removes only the logarithmic corrections. The "finite terms" f can be absorbed by setting $\mu = Qe^f$, but this evidently requires a calculation of these terms which vary from process to process.

simplest Green's function which contains the required information. The resulting expressions for the running parameters can then be used to describe the Q-dependence of far more complicated matrix elements. The RGE equations constitute a very powerful tool to study the dependence of physical quantities on the external scale Q. They have the general form of a loop expanded series:

$$
\begin{aligned}
\beta(X_i) &= \beta^{(1)}(X_i) + \beta^{(2)}(X_i) + \cdots \\
&= \frac{1}{8\pi^2}\sum_{j,k}(b^{(1)}_{X_i})_{jk}X_j(\mu)X_k(\mu) + \frac{1}{(8\pi^2)^2}\sum_{j,k,\ell}(b^{(2)}_{X_i})_{jk\ell}X_j(\mu)X_k(\mu)X_\ell(\mu) + \cdots .
\end{aligned}
$$

$$(6.106)$$

In (6.106) $(8\pi^2)^{-n}$ is the numerical denominator associated with the nth loop term which is proportional to the coefficient $(b^{(n)}_{X_i})_{jk\ell\cdots}$ carrying $n+1$ indices. The first RHS term in (6.106) with the coefficient $(b^{(1)}_{X_i})_{jk}$, which is a bilinear in X, is the lowest order nontrivial term coming from one loop. This is the term that we shall be concerned with since we work only upto one loop and neglect higher ones.

If we compare (6.106) with the β-functions, calculated to one loop order in §6.8, we notice something interesting. Since the first RHS term in that equation is quadratic in X, the square of the gauge coupling strength g^2, rather than g itself, is the right choice for our X. Thus, upto one loop, we can use (6.100b), (6.105) and (6.106) to write the evolution equation for g^2 **in a supersymmetric nonabelian gauge theory** as

$$
8\pi^2\frac{dg^2}{dt} = b^{(1)}_{g^2}\Big|_{\mathrm{SUSY}}g^4 = \left[-3C_2(G) + \sum_i T_i(\mathcal{R})\right]g^4. \tag{6.107}
$$

Let us recall that $C_2(G)$ is the quadratic Casimir of the gauge group G, being N for $SU(N)$, and $T_i(\mathcal{R})$ is the representation constant of the representation $\mathcal{R}$ of the gauge group according to which the left chiral superfield of type i in the loop transforms. For comparison, the equation corresponding to (6.107) in the same nonabelian gauge theory, but **without supersymmetry and sparticles**, is [6.6]:

$$
8\pi^2\frac{dg^2}{dt} = \left[-\frac{11}{3}C_2(G) + \frac{2}{3}\sum_i T_i(\mathcal{R}) + \frac{1}{3}\sum_\alpha T_\alpha(\mathcal{R})\right]g^4, \tag{6.108}
$$

where i now refers to the ith chiral fermion and α to the αth complex scalar in the loop.

It is easy to see how (6.108) is related to (6.107). Suppose we start with the former, impose supersymmetry and introduce the requisite sparticles into the system. First, each cartesian gauge boson component in the loop will need the addition of one gaugino. The latter counts as a single chiral fermion field in the adjoint representation with $T(\mathcal{R}) = C_2(G)$, since its two chirality states are related through the Majorana condition. One then obtains $(-11/3 + 2/3)C_2(G) = -3C_2(G)$ as the coefficient of g^4, yielding the first RHS term of (6.107). Next, each chiral fermion will now have an accompanying complex sfermion in the loop. For each i (and with $\alpha = i$), this would yield $\sum_i(2/3 + 1/3)T_i(\mathcal{R}) = \sum_i T_i(\mathcal{R})$ as the coefficient of g^4, viz. the second RHS term of (6.107). Of course, no third RHS term is needed now.

Finally, turning to the masses and Yukawa couplings of a supersymmetric theory, (6.101a) and (6.101b) translate to the evolution equations

$$\frac{d}{dt}m_{ij} = \gamma_{ii'}m_{i'j} + \gamma_{jj'}m_{ij'} , \tag{6.109a}$$

$$\frac{d}{dt}f_{ijk} = \gamma_{ii'}f_{i'jk} + \gamma_{jj'}f_{ij'k} + \gamma_{kk'}f_{ijk'} , \tag{6.109b}$$

the γ's being given to the one loop order by (6.100a). Eqs. (6.107) and (6.109) will be of considerable utility to us in Chs. 11–13.

References

[6.1] J. Wess and B. Zumino, Phys. Lett. **49B** (1974) 52. J. Iliopoulos and B. Zumino, Nucl. Phys. **B76** (1974) 310. S. Ferrara, J. Iliopoulos and B. Zumino, Nucl. Phys. **B77** (1974) 41. B. Zumino, Nucl. Phys. **B89** (1975) 535. M. Grisaru, M. Roček and W. Siegel, Nucl. Phys. **B159** (1979) 429.

[6.2] P.S. Howe and P. West, Phys. Lett. **227B** (1989) 379. I. Jack and D.R.T. Jones, *ibid* **258B** (1991) 382.

[6.3] G. 't Hooft and M. Veltman, Nucl. Phys. **B44** (1972) 189. C. Bollini and J. Giambiagi, Nuov. Cim. **12B** (1972) 20.

[6.4] W. Siegel, Phys. Lett. **84B** (1979) 193; **94B** (1980) 37. D.M. Capper, D.R.T. Jones and P. van Nieuwenhuizen, Nucl. Phys. **167** (1980) 479. L.V. Avdeev, G.V. Chochia and A.A. Vladimirov, Phys. Lett. **105B** (1981) 272. L.V. Avdeev and A.A. Vladimirov, Nucl. Phys. **B219** (1983) 262. I. Jack, D.R.T. Jones, S.P. Martin, M.T. Vaughn and Y. Yamada, Phys. Rev. **D50**, 5481 (1994). I. Jack and D.R.T. Jones in (G.L. Kane, ed.) *Perspectives on Supersymmetry, op. cit., Bibl.*, p149.

[6.5] P. West, *op. cit., Bibl.* J. Wess and J. Bagger, *op. cit., Bibl.*

[6.6] M.E. Peskin and D.V. Schroeder, *op. cit., Bibl.* A. Zee, *op. cit., Bibl.*

[6.7] A. Salam and J. Strathdee, Nucl. Phys. **B86** (1975) 142.

[6.8] I.L. Buchbinder and S.M. Kuzenko, *op. cit., Bibl.*

[6.9] M. Grisaru, M. Roček and W. Siegel, *loc. cit.*, Ref. [6.1].

[6.10] M. Drees, K. Hagiwara and A. Yamada, Phys. Rev. **D45** (1992) 1725, cf. Appendix.

[6.11] M. Sher, Phys. Lett. **B317** (1993) 159; *ibid.* **331** (1994) 448.

Chapter 7

GENERAL ASPECTS OF SUPERSYMMETRY BREAKING

7.1 Initial Remarks

Sparticles would be degenerate in mass with particles in an exactly supersymmetric world and would have the same abundance. Such is not the case in reality; no sparticle has yet been observed. In fact, experiment already tells us that (1) the superpartner of the 0.511 MeV mass electron has to have a mass at least of the order of 100 GeV and (2) the superpartner of the notionally massless gluon is obliged (in the minimal theory) to weigh more than about 200 GeV. Any supersymmetry, present in nature, must be broken severely indeed! All global continuous symmetries can be broken [7.1] in broadly two different ways – (1) explicitly in the Heisenberg-Wigner mode or (2) spontaneously in the Nambu-Goldstone mode. (A combination is also possible). In the former a "small" part of the Lagrangian breaks a symmetry of the remaining larger part, but the vacuum is left invariant; multiplets are split in mass but are otherwise kept intact. In the latter the Lagrangian is unchanged, but the vacuum does not remain symmetric and one or more massless Goldstone particles will occur; the multiplet structure is destroyed. If the symmetry in question is bosonic, the Goldstone particles will have spin zero. When one has a combination of both modes, the Goldstone particles acquire masses and are called pseudo-Goldstone particles. In QCD the breaking of flavor $SU(n)$ by different quark masses is an example of (1), while the violation of the chiral symmetry $SU(n) \times SU(n) \to SU(n)$, for massless quarks, illustrates (2) with pseudoscalar mesons as massless Goldstone bosons. When quark masses are switched on, one has a combination of (1) and (2); consequently, the pseudoscalar mesons become massive pseudo-Goldstone bosons. We shall explore both possibilities for global $N=1$ supersymmetry, taking up spontaneous violation first. The supersymmetry algebra is assumed to be unaffected in either case.

7.2 Spontaneous Supersymmetry Breaking : Some Generalities

Supersymmetry can be broken spontaneously, but there are specific constraints. We came across some such restrictions in §2.2 during our discussions of supersymmetric quantum mechanics. These generally carry over *mutatis mutandis* to field theory. In particular, the conclusion regarding the ground state – namely its energy being zero for exact supersymmetry but turning positive under spontaneous breaking – is valid for the vacuum in a supersymmetric field theory. So is the positive semidefiniteness property of energy for any state. We already know from (3.26a) that

$$\left[Q_A, \bar{Q}_{\dot{B}}\right]_+ = 2\sigma^\mu_{A\dot{B}} P_\mu \ .$$

This can be inverted, via the trace relation $\mathrm{Tr}(\sigma^\mu \bar{\sigma}^\nu) = 2\eta^{\mu\nu}$, to yield

$$P_\mu = \frac{1}{4}\bar{\sigma}_\mu^{\dot{B}A} \left[Q_A, \bar{Q}_{\dot{B}}\right]_+ \ .$$

Thus the Hamiltonian $H = P_0$ can be written as

$$H = \frac{1}{4}\left[Q_1, \bar{Q}_{\dot{1}}\right]_+ + \frac{1}{4}\left[Q_2, \bar{Q}_{\dot{2}}\right]_+ \ . \tag{7.1}$$

Since $\bar{Q}_{\dot{A}}$ is the hermitian conjugate of Q_A, it follows from (7.1) that

$$\langle\alpha|H|\alpha\rangle = \frac{1}{4}\sum_{A=1}^{2}\sum_n \left(|\langle\alpha|Q_A|n\rangle|^2 + |\langle\alpha|\bar{Q}_{\dot{A}}|n\rangle|^2\right) \tag{7.2}$$

for any state $|\alpha\rangle$. The positivity of the energy of any eigenstate of the Hamiltonian with a nonzero energy is evident from (7.2) and is inevitable so long as the supersymmetry algebra is intact. A supersymmetric vacuum state $|\Omega\rangle$ is defined by the condition that it remains invariant under any supersymmetry transformation. For an infinitesimal transformation, the condition requires

$$\delta|\Omega\rangle = i\left(\epsilon^A Q_A + \bar{\epsilon}_{\dot{A}}\bar{Q}^{\dot{A}}\right)|\Omega\rangle = 0 \ . \tag{7.3}$$

Since ϵ_A, $\bar{\epsilon}_{\dot{A}}$ are arbitrary Grassmann parameters, (7.3) is equivalent to the statement that a supersymmetric vacuum is annihilated by the supersymmetry generators Q_A, $\bar{Q}^{\dot{A}}$, i.e. $Q_a|\Omega\rangle = 0$ or, in two component form,

$$Q_A|\Omega\rangle = 0 \ , \tag{7.4a}$$

$$\bar{Q}_{\dot{A}}|\Omega\rangle = 0 \ . \tag{7.4b}$$

It follows then from (7.2) that such a vacuum must have zero energy.

Consider the scalar sector of the theory described generically by a complex field ϕ. The physical vacuum configuration corresponds to a minimum of the effective potential density $V(\phi)$, including quantum corrections. The supersymmetric vacuum must correspond to a global minimum of $V(\phi)$ with a zero value as shown in Fig. 7.1a. On the other hand, supersymmetry is spontaneously broken if and only if any one of the generators Q_A, $\bar{Q}^A$ does not annihilate the vacuum state, thereby forcing that state – via (7.2) – to have a positive definite energy. In other words, such a violation occurs if and only if the said global minimum has a positive value. This situation is described by Fig. 7.1b. These situations may be contrasted with the corresponding ones in the spontaneous breakdown of an ordinary global internal symmetry as discussed below.

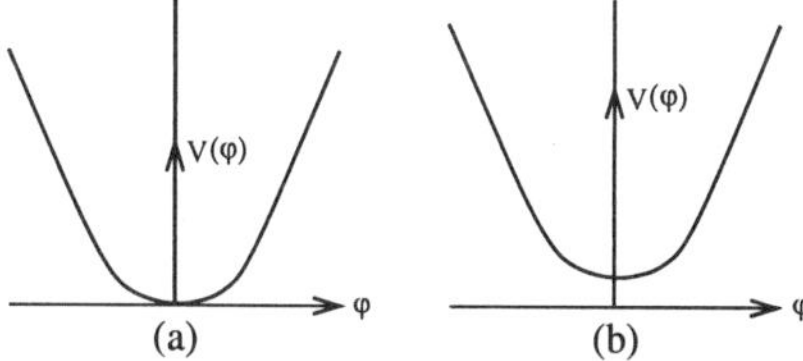

Fig.7.1. Vacua in supersymmetric theories: (a) exactly supersymmetric case, (b) spontaneously broken case.

In general, in a field theory with a global internal symmetry, the zero point of the energy is not well defined (the null point energy is usually a **divergent integral**). In case one exercises the choice to make the ground state energy zero, one subtracts any nonzero constants from the Hamiltonian density θ_{00} and appropriately normal orders the field dependent terms so that the VEV of θ_{00} vanishes. However, this does not have to be the case and the ground state energy could very well have been nonzero. Not so for global supersymmetry, where the ground state in the unbroken case must necessarily have a vanishing energy. For instance, the famous Mexican hat potential in the complex ϕ-plane, as shown in Fig. 7.2a, spontaneously violates the global $U(1)$ internal symmetry $\phi \to e^{i\varphi}\phi$ but preserves supersymmetry. In contrast, the configuration shown in Fig. 7.2b leads to the spontaneous breakdown of both the said internal symmetry and supersymmetry.

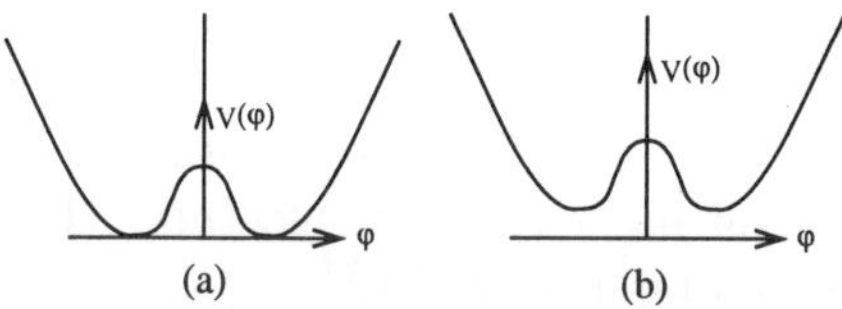

Fig.7.2. Mexican hat supersymmetric potential (a) without and (b) with spontaneous supersymmetry breaking.

A bosonic continuous internal symmetry, which is unbroken at the tree level, is usually stable under small[1] loop corrections; only the existence of some kind of degeneracy in the potential can lead to a (dynamical) symmetry breakdown – arbitrarily induced by such tiny perturbations. A particularly well known example is the Coleman-Weinberg potential [7.2]. Here the vanishing of the curvature near the origin at the tree level leads to an $O(\alpha)$ one loop violation of the global $U(1)$ symmetry, α being $(4\pi)^{-1}$ times the square of a

---

[1] Of course, 'large' quantum corrections can always break a classical bosonic symmetry.

small coupling strength. Similarly, there are examples with special degeneracies [7.3] where a global symmetry, that is spontaneously broken at the tree level, gets restored by loop corrections, i.e. the curvature of the potential becomes positive everywhere instead of being negative over a finite region. Without such degeneracies, though, this type of restoration is not possible. But in a supersymmetric system, once supersymmetry is spontaneously broken at the tree level (i.e. $V_{\min}^{\mathrm{tree}} > 0$), it can never be restored (cf. [7.4]) for any range of couplings. Also, given unbroken supersymmetry at the tree level and even with any amount of classical degeneracy, a potential is immune to supersymmetry breaking by perturbative quantum corrections on account of the corollary to the nonrenormalization theorem (§6.7). It has been shown, however, that strong nonperturbative effects can lead to a dynamical breakdown of supersymmetry in a classically supersymmetric field theory; an explicit example will be given in §7.6.

Vacuum expectation values

We know that the spontaneous breakdown of a bosonic internal symmetry arises from [7.1] a nonzero VEV acquired by a scalar or pseudoscalar operator which transforms nontrivially under the symmetry. Similarly, the spontaneous violation of global supersymmetry can be expected to arise from a nonzero VEV accruing to some operator that changes under a supersymmetry transformation. Consider first a chiral superfield Φ with component fields ϕ, ξ and F. Their infinitesimal supersymmetry transformations are given by (4.25). Examining the right hand sides of (4.25a-c), we can see that ξ cannot acquire a nonzero VEV without compromising Lorentz invariance. Furthermore, the VEVs of $\partial_\mu \phi$, $\partial_\mu \xi$ must be zero since the vacuum carries a vanishing four momentum. Only the auxiliary component F and hence $\delta\xi$ can have this property. Thus

$$|\langle\Omega|F|\Omega\rangle| \equiv \Lambda_s^2 \neq 0$$

will lead to a spontaneous violation of supersymmetry. Since F has mass dimension 2, Λ_s can be identified as a mass scale, namely the scale of supersymmetry breaking. This type of spontaneous supersymmetry breaking is called F-type breaking or the O'Raifeartaigh mechanism [7.5].

Turning to a vector superfield in the Wess-Zumino gauge, with component fields A_μ, λ and D, we can again apply the criterion of the preservation of Lorentz invariance and make use of the fact that the vacuum has no four momentum. Then it follows that only the auxiliary component D could have a nonzero VEV:

$$|\langle\Omega|D|\Omega\rangle| \equiv \tilde{\Lambda}_s^2 \neq 0 \ .$$

In a theory with exact gauge symmetry, this is permitted **only for abelian vector superfields** which are gauge invariant. (Since nonabelian vector superfields V^a are only gauge covariant instead, a nonzero VEV for D^a would mean a vacuum that is no longer gauge invariant.) The kind of spontaneous breakdown of supersymmetry, effected through the equation above, is called D-type supersymmetry breaking or the Fayet-Iliopoulos mechanism [7.6]. It may be noted from (5.22) that, for either F-type or D-type breaking, the potential energy in the ground state becomes positive definite, as may be expected from our general arguments.

We summarize below the main points of the above discussion on spontaneous supersymmetry breaking (SSB).

- The necessary and sufficient condition for SSB is the failure of the supercharge to annihilate the vacuum.

- The energy of the ground state is always positive after SSB.

- Two kinds of SSB have been outlined: F-type or D-type with a nonzero VEV of an auxiliary F or (abelian) D field respectively.

7.3 The goldstino

Just as the spontaneous violation of any ordinary bosonic symmetry leads to [7.1] a massless spinless Goldstone boson, that of global supersymmetry generates [7.6] a massless spin half Majorana fermion called the goldstino[2]. We had introduced in §5.1 the two component spinorial supercurrent density $K_{\mu A}$ and its conjugate $\bar{K}_\mu^{\dot{A}}$ for a system of interacting chiral superfields. Let us now define a general four component Majorana fermionic supercurrent density K_a^μ as the four spinor

$$K_a^\mu(x) \equiv \begin{pmatrix} K_A^\mu(x) \\ \bar{K}^{\mu\dot{A}}(x) \end{pmatrix},$$

with the space integral of its timelike component being the supercharge Q_a. Thus

$$Q_a = \int d^3x \; K_a^0(x) = Q_a^{\mathcal{C}} \;, \tag{7.5a}$$

$$\partial_\mu K_a^\mu(x) = 0 \;. \tag{7.5b}$$

Of course, supercurrent conservation (7.5b) holds only if supersymmetry is not broken explicitly. Since Q_a has mass dimension $1/2$, that of K_a^μ is $7/2$. As with bosonic global symmetries, the supersymmetry algebra (3.12) involving Q_a and the Poincaré generators can be promoted to an algebra of the corresponding densities with the addition of [7.7] Schwinger terms ($S.T.$). The latter involve derivatives of delta functions and hence drop out of VEVs. Thus, corresponding to (3.26a), we have

$$\left[K_A^\mu(x), \bar{Q}_{\dot{B}} \right]_+ = 2\sigma_{\nu A\dot{B}}\theta^{\mu\nu}(x) + S.T. \;, \tag{7.6}$$

with $\theta^{\mu\nu}$ as the stress energy tensor, i.e. $P_\mu = \int d^3x\theta_{o\mu}(x)$. (7.6) can be recast as

$$\theta^{\mu\nu}(x) = \frac{1}{4}\bar{\sigma}^{\nu\dot{A}B}\left[\bar{Q}_{\dot{A}}, K_B^\mu(x) \right]_+ + S.T. \tag{7.7}$$

Relations that are similar to (7.6) and (7.7) can easily be written with the anticommutator between Q_A and $\bar{K}_\mu^{\dot{B}}(x)$.

---

[2]This terminology deviates from our usual custom of employing the suffix -ino to the name of a known boson in order to denote its superpartner. Here the goldstino is not the superpartner of a Goldstone boson, but is itself a Goldstone fermion. However, we choose to use this term because of its prevalent occurrence in the literature. Any superpartner of a Goldstone boson will instead be explicitly described as such. See the discussion at the end of this section.

(7.7) makes two things quite clear. The VEV of $\theta^{\mu\nu}(x)$ vanishes when $\bar{Q}_{\dot{A}}$ (and correspondingly Q_A) annihilates $|\Omega\rangle$, i.e. when the vacuum is supersymmetric. Also, the same VEV acquires a nonzero value when, on account of spontaneous supersymmetry breakdown, $Q_a|\Omega\rangle \neq 0$. In this case we may write

$$\langle\Omega|\theta^{\mu\nu}(x)|\Omega\rangle = E_\Omega^4 \eta_{\mu\nu} , \tag{7.8}$$

where E_Ω is an energy scale associated with the vacuum state. Taking VEVs of either side in (7.7) and using (7.8), we have

$$E_\Omega^4 \eta^{\mu\nu} = \frac{1}{4}\bar{\sigma}^{\nu\dot{A}B}\langle\Omega|[\bar{Q}_{\dot{A}}, K^\mu{}_B]_+|\Omega\rangle. \tag{7.9}$$

(7.9) demonstrates the reverse of the result derived earlier, namely that whenever the vacuum has a nonzero (necessarily positive) energy, it fails to be annihilated by the supercharge, i.e. $Q_A|\Omega\rangle \neq 0$, $\bar{Q}^{\dot{A}}|\Omega\rangle \neq 0$ and supersymmetry has been spontaneously broken.

Since, according to (3.26c), the supercharge commutes with the Hamiltonian, it is conserved as a consequence of Heisenberg's equation, i.e.

$$\frac{dQ_a}{dt} = 0 . \tag{7.10}$$

Therefore, for any local fermionic operator $A(x')$ taken at the spacetime point $x'^\mu = (t', \vec{x}')$, we can write

$$\frac{d}{dt}\langle\Omega| [Q_a(t), A(x')]_+ |\Omega\rangle = 0 . \tag{7.11}$$

Let us insert a complete set of states $\sum_n |n\rangle\langle n|$, which are eigenstates of the operator P^μ, between the two operators in (7.11). Then the replacement of Q_a by $\int d^3x K_a^0(x)$ and the use of the four translational properties (with $x^0 = t$) of the operators

$$K_a^0(x) = e^{ix\cdot P} K_a^0(0)e^{-ix\cdot P},$$

$$A(x') = e^{ix'\cdot P} A(0)e^{-ix'\cdot P},$$

followed by a differentiation with respect to time, yield the result

$$\sum_n E_n \delta^{(3)}(\vec{p}_n) \big[\langle\Omega|K_a^0(0)|n\rangle\langle n|A(0)|\Omega\rangle e^{-iE_n(t-t')} \\ -\langle\Omega|A(0)|n\rangle\langle n|K_a^0(0)|\Omega\rangle e^{iE_n(t-t')}\big] = 0 , \tag{7.12}$$

with E_n defined as the energy of the state $|n\rangle$.

Eq. (7.12) has a very important consequence. Because of the different time dependences of the two coefficients in the square bracketted expression, the latter can vanish $\forall\, t, t'$ if and only if

$$E_n \delta^{(3)}(\vec{p}_n)\langle\Omega|K_a^0(0)|n\rangle\langle n|A(0)|\Omega\rangle = 0 . \tag{7.13}$$

Hence, if $\exists$ some $|n\rangle$ for which $\langle\Omega|K_a^0|n\rangle \neq 0$, the validity of (7.13) would require the condition

$$E_n \delta^{(3)}(\vec{p}_n) \equiv \delta^{(3)}(\vec{p}_n)\sqrt{\vec{p}_n^2 + m_n^2} = 0 ,$$

where m_n is the "mass" $\sqrt{E_n^2 - \vec{p}_n^2}$ of the state $|n\rangle$. However, this implies that

$$m_n = 0 \ . \tag{7.14}$$

In other words, the state $|n\rangle$, which must be a fermion since it admits a nonzero matrix element $\langle\Omega|K_0^a|n\rangle$, has to be massless on account of (7.14). Note that such a state *must* exist once supersymmetry is spontaneously broken, since $\langle\Omega|Q_a \neq 0 \Rightarrow \langle\Omega|K_a^0 \neq 0$ and $\exists$ some state $|n\rangle$ for which $\langle\Omega|K_a^0|n\rangle \neq 0$; the operator A can always be chosen in a way such that $\langle\Omega|A(0)|n\rangle$ is nonzero. This state then is the goldstino. Because of (3.12b) and the commutation relation

$$[P^2, Q_a] = 0$$

following from it, the state $Q_a|\Omega\rangle$ must have the same four momentum as $|\Omega\rangle$ and can be regarded as another degenerate and fermionic ground state comprising $|\Omega\rangle$ and a goldstino of vanishing energy and momentum.

There is another (perhaps more physical) way to see the existence of the goldstino. Let us return to four component notation and start from (3.12a). Eq. (7.9) can be recast, with $\bar{K}_b^\mu$ taken at the origin, as follows.

$$\begin{aligned}
E_\Omega^4 \eta^{\mu\nu} &= \frac{1}{8}\langle\Omega|[Q_a, \bar{K}_b^\mu(0)]_+|\Omega\rangle(\gamma^\nu)_{ba} \\
&= \frac{1}{8}\int d^4x \ \partial_\rho\langle\Omega|TK_a^\rho(x)\bar{K}_b^\mu(0)|\Omega\rangle(\gamma^\nu)_{ba} \\
&= \frac{1}{8}\int_S dS_\rho\langle\Omega|TK_a^\rho(x)\bar{K}_b^\mu(0)|\Omega\rangle(\gamma^\nu)_{ba} \ . \tag{7.15}
\end{aligned}$$

We have used the result $\partial_\rho\theta(x^0) = \delta_{\rho 0}\delta(x^0)$ in the second step of (7.15), whereas in the last step Gauss' theorem has been used, employing a spacelike three dimensional surface S at spatial infinity. For a finite surface integral in (7.15), the two point supercurrent correlator needs to fall off as $|\vec{x}|^{-3}$ for large spatial distances $|\vec{x}|$. Since this falloff behavior can come only from the three dimensional Fourier transform of a massless spin half propagator, the existence of an intermediate massless spin half state, viz. the goldstino $|\lambda_g\rangle$, is established. From the fact that it contributes through the nonzero matrix element $\langle\Omega|K_{\mu a}|\lambda_g\rangle$, it is evidently a Majorana fermion. In fact, one can define a dimensional goldstino constant f_λ in the limit of vanishing three momentum via

$$(2\pi)^{3/2}\langle\Omega|K_{\mu a}(0)|\lambda_g(\vec{o}, b)\rangle = f_\lambda(\gamma_\mu)_{ab} \ , \tag{7.16}$$

where b is the free spinor index of the goldstino state in (7.16), defined at zero four momentum. If the goldstino had a nonzero four momentum p, there would be a term proportional to $p_\mu\delta_{ab}$ in the RHS of (7.16); but that part would lead to a spatial fall off of the correlator of (7.15) faster than $|\vec{x}|^{-3}$ causing a vanishing contribution to E_Ω. The only nonvanishing contribution to E_Ω comes from an intermediate $|\lambda_g\rangle$ at zero four momentum and then (7.15) and (7.16) imply that

$$E_\Omega^2 = f_\lambda \ . \tag{7.17}$$

The matrix element of the spinorial supercurrent density $K_{\mu a}$, between bosonic and fermionic states in the same supermultiplet is of considerable importance. Let the bosonic (fermionic) state $|b\rangle$ ($|f\rangle$) carry the four momentum p_b (p_f) with $p_f = p_b - q$. One can then write

$$(2\pi)^3 \langle b(p_b)|K_{\mu a}(0)|f(p_f)\rangle = F_1(q^2)(\gamma_\mu \psi)_a + F_2(q^2)(p_b + p_f)_\mu \psi_a + F_3(q^2) q_\mu \psi_a \,,$$

where ψ_a is the four spinor from $|f\rangle$ and $F_{1,2,3}(q^2)$ are form factors. If there is a spontaneous breakdown of supersymmetry, the above matrix element will have a goldstino pole occuring in the form factor F_1 on account of (7.16). Suppose $h_{fb\lambda}$ is defined as the strength of the λ_g-f-b Yukawa coupling. One can then write

$$F_1(q^2) = \frac{h_{fb\lambda} f_\lambda}{\not{q}} + \cdots \,,$$

the ellipsis standing for terms which behave smoothly when $q \to 0$. On contracting the matrix element above with q^μ and taking the limit $q \to 0$, we can use the conservation law (7.5b) to write

$$m_f^2 - m_b^2 \simeq \frac{h_{fb\lambda} f_\lambda}{F_2(0)} \,,$$

where $m_{f,b}$ refers to the mass of f, b. Evidently, the spontaneous breakdown of supersymmetry lifts the fermion boson mass degeneracy which, however, gets restored when $f_\lambda \to 0$, i.e. when supersymmetry breaking disappears.

A related interesting issue emerges when $|f\rangle$ and $|b\rangle$, as defined in the above paragraph, do not describe single particles but represent multiparticle states. In such a situation one can replace $h_{fb\lambda}$ by $\mathcal{M}_{fb\lambda}$, the amplitude describing a goldstino scattering process $f + \lambda_g \to b$ or $f \to b + \lambda_g$. Now the matrix element of the vanishing four divergence $\partial^\mu K_{\mu\alpha} = 0$ between $\langle b|$ and $|f\rangle$ implies that

$$\mathcal{M}_{fb\lambda} f_\lambda F_1(q^2) + q \cdot (p_b + p_f) F_2(q^2) + q^2 F_3(q^2) = 0 \,.$$

On letting $q \to 0$, we derive the result that the amplitude $\mathcal{M}_{fb\lambda}$ vanishes in the limit of a 'soft' goldstino. This is sometimes called the **soft goldstino theorem** and has important physical consequences. For instance, no leptonic or semileptonic weak decay amplitude vanishes in the limit when the neutrino four momentum goes to zero. That proves, therefore, that the neutrino is not a goldstino.

Let us conclude this section with the following comments:

- Not every massless fermion in a supersymmetric theory need be a goldstino, a gaugino being a counterexample. In particular, one should distinguish between the goldstino and the superpartner of a Goldstone boson (cf. ftnt. 2).

- An exactly supersymmetric theory may be obliged to have a massless fermionic superpartner of a Goldstone boson. For instance, take a theory with a spontaneously broken global $U(1)$ symmetry and containing one massless Goldstone boson. If the system is supersymmetrized, the Goldstone particle must necessarily acquire a massless fermionic partner. The latter is not a goldstino since the system remains exactly invariant under supersymmetry.

- The feature characterizing a goldstino is not so much its massless fermionic nature but rather its coupling (7.16) to the spinorial supercurrent density K_a^μ.

7.4 Model of F-type Supersymmetry Breaking

As explained in §7.2, an F-type spontaneous supersymmetry breaking is characterized by the nonzero VEV of some auxiliary F-component of a chiral superfield. A field theory exhibiting supersymmetry, broken by an F-type mechanism, needs to have the following feature. It must admit a solution of the equations of motion with some $F_i \neq 0$. If one insists (and this may not be strictly necessary) that supersymmetry be broken in the global minimum of the scalar potential $V(\phi)$, one needs to require in addition that the choice $F_j = 0$, $\forall j$, does *not* represent a solution of these equations. In the following discussion we shall assume that such is indeed the case. An immediate corollary of this is the fact that, in a system of interacting chiral superfields[3] Φ_i, any superpotential of the form $\mathcal{W} = \frac{1}{2}m_{ij}\Phi_i\Phi_j + \frac{1}{6}f_{ijk}\Phi_i\Phi_j\Phi_k$ without a linear Φ term will fail to produce F-type supersymmetry breaking. In such a case $F_i^\star \equiv -\partial W/\partial\Phi_i|$ can always be made zero with the choice of all expectation values $\langle\phi_i\rangle = 0$. The latter correspond to a perfectly realizable configuration which is a global minimum of $V(\phi)$. In order to effect an F-type supersymmetry breaking, we need to have a linear term in the superpotential. Take the Wess-Zumino model of one self interacting chiral superfield (cf §5.1). Replacing f by λ, we get

$$F^\star = -h - m\phi - \frac{\lambda}{2}\phi^2. \tag{7.18}$$

Now the condition $F = 0$ can always be achieved by the choice

$$\phi_\pm = -\frac{m}{\lambda} \pm \frac{1}{\lambda}(m^2 - 2h\lambda)^{1/2} \tag{7.19}$$

and evidently it does correspond to the vanishing minimum of $V = |F|^2$. Hence this model is also incapable of attaining the spontaneous breakdown of supersymmetry. One needs at least three chiral superfields $\Phi_{1,2,3}$ ($\phi_{1,2,3}$ with $\xi_{1,2,3}$ and $F_{1,2,3}$), as was noted by O'Raifeartaigh [7.5].

In O'Raifeartaigh's model the superpotential is chosen to be

$$\mathcal{W}(\Phi_1, \Phi_2, \Phi_3) = m\Phi_2\Phi_3 + \lambda\Phi_1(\Phi_3^2 - \mu^2) , \tag{7.20}$$

where μ is a fixed mass parameter. For simplicity, we assume that m, λ and μ are all real. The equations of constraint yield the relations

$$F_1^\star = -\lambda(\phi_3^2 - \mu^2) , \tag{7.21a}$$

$$F_2^\star = -m\phi_3 , \tag{7.21b}$$

$$F_3^\star = -m\phi_2 - 2\lambda\phi_1\phi_3 . \tag{7.21c}$$

---

[3] If there are gauge interactions present, we assume here the Φ_i are gauge singlets.

There is no consistent set of solutions to (7.21a-c) which can make all the F's vanish simultaneously, i.e. some F is always nonzero everywhere in the ϕ-space including the minimum of $V(\phi)$. Consequently, supersymmetry gets spontaneously broken.

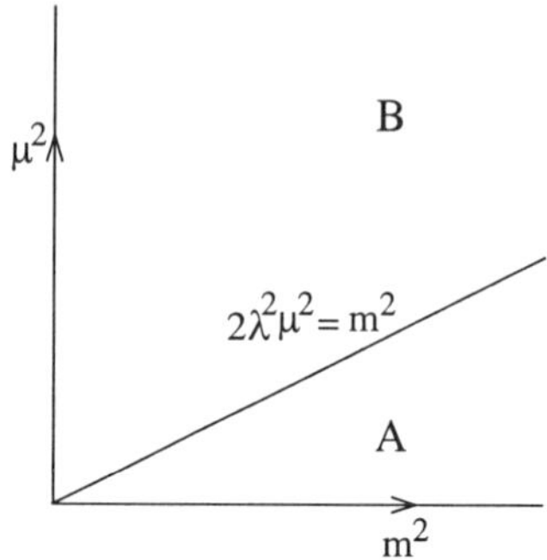

Fig.7.3. Two regions in the parameter space of O'Raifearteigh's model.

The scalar potential in O'Raifeartaigh's model is given by

$$V(\phi_1, \phi_2, \phi_3) = |\lambda(\phi_3^2 - \mu^2)|^2 + |m\phi_3|^2 + |m\phi_2 + 2\lambda\phi_1\phi_3|^2 \,, \tag{7.22}$$

with λ, μ and m as real parameters. The minima of this potential are different in two distinct regions (Fig. 7.3) of the μ^2, m^2 parametric plane: (A) $\mu^2 < m^2/2\lambda^2$ and (B) $\mu^2 > m^2/2\lambda^2$. The difference in behavior can be easily seen under the simplifying assumption that $\langle\phi_1\rangle = 0$.

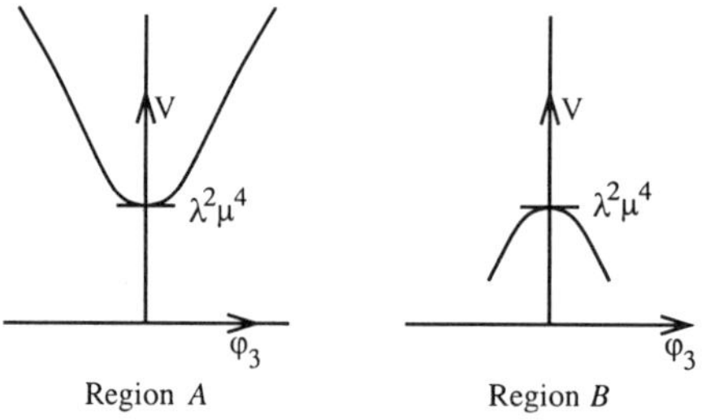

Fig.7.4. Scalar potential of O'Raifeartaigh's model near $\phi_3 = 0$.

For $0 = \partial V/\partial\phi_1 = \partial V/\partial\phi_2 = \partial V/\partial\phi_3$, $\langle\phi_3\rangle$ has to be real and, at $\langle\phi_1\rangle = \langle\phi_3\rangle = 0$, $\frac{1}{2}\langle\partial^2 V/\partial(\text{Re}\phi_3)^2\rangle = m^2 - 2\lambda^2\mu^2$, i.e. that is a minimum of V in region A but a maximum in region B (Fig. 7.4). It turns out to be a global minimum for region A but a local maximum for region B which also admits [7.8] two symmetrically placed global minima at nonzero values of $\langle\phi_3\rangle$. For region A, we have one global minimum of the potential with $V_{min} = \lambda^2\mu^4 > 0$, $\langle\phi_2\rangle = \langle\phi_3\rangle = 0$ and $\langle\phi_1\rangle$ actually undetermined. Restricting ourselves to region A, we see a spontaneous supersymmetry breaking and can define $\sqrt{|\langle F_1\rangle|} = \sqrt{|\lambda\mu^2|} \equiv \Lambda_s$ as the supersymmetry breaking scale. That $\langle\phi_1\rangle$ is undertermined means that the minimum has a "flat direction"[4] along $\langle \phi_1 \rangle$. Since the nonzero value of V_{min} comes only from $|\langle F_1 \rangle|^2$, ξ_1 – the fermionic component of Φ_1 – has to be the goldstino[5]. Its masslessness is shown below.

Generalizing from (5.3) and using (5.5a), the fermion mass terms (FMT) in a Lagrangian density of interacting chiral superfields, with some scalar fields acquiring VEVs, can be written as

$$\mathcal{L}_{FMT} = -\frac{1}{2} m_{ij} \xi_i \xi_j - \frac{1}{2} f_{ijk} \langle \phi_k \rangle \xi_i \xi_j + h.c. \tag{7.23}$$

For the present model (7.23) becomes

$$\mathcal{L}_{FMT} = -m \xi_2 \xi_3 - 2\lambda \langle \phi_1 \rangle \xi_3 \xi_3 + h.c. \tag{7.24}$$

(7.24) is consistent with ξ_1 being the massless goldstino while ξ_2 and ξ_3 combine to yield a massive state. Suppose we choose $\langle \phi_1 \rangle = 0$. The mass m is associated with both ξ_2 and ξ_3 which can be combined into a single massive Dirac fermion

$$\psi = \begin{pmatrix} \xi_2 \\ \bar{\xi}_3^T \end{pmatrix}$$

with

$$\mathcal{L}_{FMT} = -m \bar{\psi} \psi \; . \tag{7.25}$$

Turning our attention to the bosonic spectrum for the case of the parametric region A, we can isolate the quadratic terms in boson fields from (7.22) and identify the boson mass terms in the Lagrangian density as

$$\begin{aligned} \mathcal{L}_{BMT} &= \lambda^2 \mu^2 (\phi_3^2 + \phi_3^{\star 2}) - m^2 \left(|\phi_2|^2 + |\phi_3|^2 \right) \\ &= -m^2 |\phi_2|^2 - \frac{1}{2} \left(m^2 - 2\lambda^2 \mu^2 \right) A^2 \\ &\quad - \frac{1}{2} \left(m^2 + 2\lambda^2 \mu^2 \right) B^2 \; . \end{aligned} \tag{7.26}$$

In (7.26) we have introduced real scalar fields A, B defined as $A \equiv (\phi_3 + \phi_3^\star)/\sqrt{2}$, $B \equiv (\phi_3 - \phi_3^\star)/\sqrt{2i}$. This equation implies a bosonic spectrum consisting of a massless scalar field ϕ_1, a complex scalar field ϕ_2 of mass $|m|$ and two real scalar fields A, B of masses $m_A = |m|\sqrt{1 - 2\lambda^2 \mu^2/m^2} = \sqrt{m^2 - 2\lambda \Lambda_s^2}$, $m_B = |m|\sqrt{1 + 2\lambda^2 \mu^2/m^2} = \sqrt{m^2 + 2\lambda \Lambda_s^2}$.

The physical significance of the above mass spectrum is clear. The complex scalar ϕ_3 has a mass $|m|$ in the supersymmetric limit $\Lambda_s \to 0$. However, under supersymmetry breaking, it splits into two real scalars lying symmetrically in mass on either side of $|m|$ with a difference in squared mass $\delta^2 \equiv 4\lambda \Lambda_s^2$. In detail, this is caused by the operator $\lambda [\langle \Phi_1 \rangle \Phi_3^2]_F = \lambda \langle F_1 \rangle (A^2 - B^2)$. Moreover, we have the sum rule (5.10), namely

$$S\mathrm{Tr}\; \underset{\sim}{m}^2 \equiv 2m^2 + m_A^2 + m_B^2 - 4m_\psi^2 = 0 \tag{7.27}$$

---

[4] "Flat directions" are noncompact lines and surfaces in the space of scalar fields along which the scalar potential vanishes. The present flat direction is an accidental feature of the classical potential and gets removed by quantum corrections.

[5] This fact, namely that the goldstino is the fermionic component of the superfield whose auxiliary field develops a VEV, is also true of D-type supersymmetry breaking, cf. §7.5.

in this case with m as the mass of the complex scalar field ϕ_2 and m_ψ $(= m)$ as the mass of the Dirac field ψ. Thus the real scalar components of the chiral superfield Φ_3 get split in mass from their fermionic superpartner, one becoming heavier and the other lighter. Their fermionic superpartners, together with those of the complex scalar field ϕ_2, make up the Dirac fermion field ψ. It is interesting to see how this simple model leads us to quite important conclusions which will be pertinent when we discuss supersymmetry breaking for the MSSM in the hidden sector in Ch.12. First, all the physical states in the superfields $\Phi_{2,3}$ can be made very massive by choosing m to be very large without affecting either the scale of supersymmetry breaking Λ_s or the mass splitting characterized by δ. Second, if $\lambda \ll 1$, the 'mass splitting' δ will be much smaller than the supersymmetry breaking scale Λ_s. Thus all the three scales m, Λ_s and δ in this model can be very different from one another. Finally, the masses of the physical states of the model obey (7.27). The graded trace of the mass squared operator over a supermultiplet vanishes at the tree level in the O'Raifeartaigh model. This sum rule is, however, subject to modifications involving any explicit supersymmetry breaking terms. It also changes under D-type supersymmetry breaking which is treated in §7.5.

We would also like to clarify, in the context of F-type supersymmetry breaking, the interrelation between the auxiliary field VEV $\langle F \rangle$, the spinorial supercurrent K_A^μ and the (two component) goldstino field λ_g which is its superpartner. Referring to ftnt. 3 of Ch.5, we can characterize the second RHS term in K_A^μ as the contribution from $\langle F \rangle$ and λ_g. Identifying, in particular, ξ with λ_g and utilizing (5.4a), we can then rewrite the supercurrent conservation condition (7.5b) as

$$0 = \partial_\mu K_A^\mu = i\sqrt{2}\langle F \rangle (\sigma^\mu \partial_\mu \bar\lambda_g)_A + \partial^\mu k_A^\mu + \cdots . \tag{7.28}$$

In (7.28) k_A^μ generically represents the contribution to the supercurrent from all other supermultiplets (cf. ftnt. 3, Ch.5), i.e.

$$k_A^\mu = -\sqrt{2}(\sigma_\nu \bar\sigma^\mu \xi_i)_A (\Delta_{ij}^\nu \phi_j)^\dagger - i\sqrt{2}(\sigma^\mu \bar\xi_i)\mathcal{W}_i^\dagger - \frac{1}{2}(\sigma^\nu \bar\sigma^\rho \sigma^\mu \bar\lambda^a)_A F_{\nu\rho}^a - ig(\phi_i^\dagger T_{ij}^a \phi_j)(\sigma^\mu \bar\lambda^a)_A , \tag{7.29}$$

where ξ_i, ϕ_i now represent components of other chiral superfields. Finally, the ellipsis in (7.28) stands for the contribution to the divergence of the supercurrent from the remaining members of the goldstino supermultiplet. As a result of (7.28), we can write an effective Lagrangian density involving the goldstino and its interactions with all other fermionic and bosonic fields. This reads

$$\mathcal{L}_{\tilde g} = i\lambda_g \sigma^\mu [\partial_\mu]\bar\lambda_g + \frac{1}{\sqrt{2}\langle F \rangle}(\lambda_g \partial_\mu k^\mu + \text{h.c.}) . \tag{7.30}$$

The appearance of the inverse of a VEV $\langle F \rangle^{-1}$ as a coupling strength in (7.30) is a characteristic feature of the interactions of Goldstone particles (bosons or fermions) with others. Eqs. (7.29) and (7.30) do show that the goldstino field has $\lambda_g \xi_i \phi_i$ and $\lambda_g \lambda A$ vertices with (matter) fermion-sfermion and with gaugino-gauge boson pairs. The strength of the former was denoted by $h_{fb\lambda}$ in §7.3.

7.5   Model of D-type Supersymmetry Breaking

A D-type spontaneous supersymmetry breaking is effected by the nonzero value of the VEV of the auxiliary D-component of a vector superfield (cf. §7.2). The simplest model exhibiting this mechanism, due to Fayet and Iliopoulos [7.6], is a supersymmetric $U(1)$ gauge theory with a single chiral superfield of charge q in which the auxiliary component $D(x)$ of the vector superfield V develops a nonzero VEV. Tracing back to (4.8h) and using $\zeta = \lambda$, we see the consequent implication that

$$|(\epsilon\epsilon)^{-1}\langle\Omega|\epsilon\delta\lambda(x)|\Omega\rangle| = |\langle D(x)\rangle| \equiv \tilde{\Lambda}_s^2 \neq 0 \ . \tag{7.31}$$

(7.31) means that the gaugino field has a nonzero variation in the vacuum configuration. Moreover taking $\langle F\rangle$ to be zero, it is evident from (5.20b) that

$$V_{min} = \frac{1}{2}\tilde{\Lambda}_s^4 \tag{7.32}$$

is indeed positive, as required for spontaneous supersymmetry breakdown. In this model (5.19) becomes

$$D = -\eta - q|\phi|^2 \ . \tag{7.33}$$

On account of a nonvanishing η, a nonzero VEV of D does not require a nonvanishing $\langle\phi\rangle$.

If the sign of ηq is negative, it becomes possible through a nonzero VEV of ϕ to make $V_{min} = \frac{1}{2}D^2$ vanish and keep supersymmetry unbroken, though the $U(1)$ symmetry suffers a spontaneous breakdown. Hence the sign of ηq is chosen to be positive so that the scalar potential density

$$V(\phi) = \frac{1}{2}\left(\eta + q|\phi|^2\right)^2 \tag{7.34}$$

requires $\langle\phi\rangle = 0$ for minimization and

$$V_{min} = \frac{1}{2}\eta^2 \ ,$$

i.e. supersymmetry does get violated spontaneously while the $U(1)$ symmetry remains intact. However, the scalar component ϕ of the chiral superfield Φ becomes massive, in consequence, with the mass

$$m_\phi = \sqrt{q\eta} \ , \tag{7.35}$$

while its fermionic superpartner ξ remains massless and is the goldstino. In other words, the members of the supermultiplet are split in mass. Moreover, this means that $\sum m_0^2$ no longer equals $2\sum m_{1/2}^2$, unlike in (5.9). In fact, the sum rule (5.10) has changed to

$$S\text{Tr}\, \underset{\sim}{m}^2 = 2q\eta = -2q\langle D\rangle \ . \tag{7.36}$$

Gauge invariance has, however, been maintained so that both the gauge boson, described by A_μ, and the gaugino, described by λ, remain massless. One problem with this $U(1)$ gauge model, however, is that it suffers from the occurrence of an ABJ anomaly [7.7]. Suppose we then consider SQED (cf. §5.3) which, with two chiral superfields, is anomaly free. Here the introduction of the Fayet-Iliopoulos term simply changes $V(\phi)$ to

$$V(\phi_+, \phi_-) = \frac{1}{2}\left[\eta + q(|\phi_+|^2 - |\phi_-|^2)\right]^2 . \tag{7.37}$$

But now the potential of (7.37) vanishes at its minimum independently of the sign of $q\eta$. The latter merely determines whether it is ϕ_+ or ϕ_- which develops a VEV. This example illustrates the difficulty [7.9] of implementing the Fayet-Iliopoulos mechanism in an anomaly free gauge theory. In order to produce a nonzero $\langle D \rangle$ in a field theory of the latter kind, one needs to introduce *both* extra fields and extra mass terms in the superpotential. We shall not go into this discussion any further but only refer the reader to Ref. [7.9] for details.

7.6　Dynamical Model of Supersymmetry Breaking

There is a model [7.10] constructed to achieve supersymmetry breaking via an interplay between perturbative and nonperturbative effects. It consists of a supersymmetric nonabelian gauge theory (cf. §5.4) with the semisimple gauge group $SU(3) \times SU(2)$, i.e. eleven gauge superfield components, characterized by gauge coupling strengths (g_3, g_2) and containing left chiral superfields $Q(\mathbf{3}, \mathbf{2})$, $\bar{U}(\bar{\mathbf{3}}, \mathbf{1})$, $\bar{D}(\bar{\mathbf{3}}, \mathbf{1})$ and $\Phi(\mathbf{1}, \mathbf{2})$ in the matter sector[6]. The bold pair of integers in the parenthesis of each chiral superfield refers to the dimensions of the representations of $SU(3)$, $SU(2)$ that the superfield transforms as. Let us also assume an additional global $U(1)$ symmetry, called H, and take the corresponding H-charges of $Q, \bar{U}, \bar{D}$ and Φ respectively to be $1, -4, 2$ and -3. Now there is only one possible polynomial form for the superpotential of the model which can be written as

$$\mathcal{W}_p = \lambda(Q{\cdot}\Phi\bar{D}) . \tag{7.38}$$

In (7.38) λ is a coupling constant, the subscript p in the LHS stands for the designation 'perturbative', $Q{\cdot}\Phi = \epsilon_{AB}Q^A\Phi^B$ with A, B being $SU(2)$ indices and the brackets imply an $SU(3)$ index contraction.

　　Had (7.38) been the only contribution to the superpotential, all VEVs would vanish and supersymmetry would remain unbroken. However, $SU(3)$ instantons [7.11] generate [7.10] a contribution (that is of a nonpolynomial form in the chiral superfields) whose nonperturbative origin we undescore with the subscript np:

$$\mathcal{W}_{np} = \frac{\Lambda_3^7}{(Q\bar{U}) \cdot (Q\bar{D})} . \tag{7.39}$$

Here Λ_3 is the mass scale at which the $SU(3)$ coupling strength g_3 becomes strong, though the $SU(2)$ gauge coupling strength g_2 is assumed to remain in the weak coupling regime. Thus the nonperturbative effects are determined by the dynamics of the $SU(3)$ gauge group

---

[6]We are using notation similar to that of quark and antiquark superfields of the Minimal Supersymmetric Standard Model, to be described in Ch.8, because we later use an anology with multiflavored SQCD.

alone. The actual derivation [7.12] of (7.39) makes use of instanton calculus and is beyond the scope of this book, but we provide below a qualitative symmetry argument justifying the form of the RHS.

By assumption, $SU(3)$ instantons dominate and determine $\mathcal{W}_{np}$. One can then draw an analogy with multiflavored supersymmetric quantum chromodynamics with N_c $(= 3)$ colors and N_f $(= 2)$ flavors. The effective action of this kind of a theory is greatly constrained by its full flavor related global symmetry which happens [7.10] to be $SU(N_f)_\mathrm{L} \times SU(N_f)_\mathrm{R} \times U(1)_V \times U(1)_H \times U(1)_R$. The subscripts L, R and V for the three factor groups refer respectively to left chiral, right chiral and vectorial representations while the $U(1)_R$ factor is a consequence of supersymmetry (cf. Ch.3). Anomaly cancellation requires [7.10] each chiral superfield to carry an R-charge $(N_c - N_f)/N_f$ $(= 1/2)$. The only product combination of chiral superfields, that is invariant under transformations of the complete gauge and flavor group is $(Q\bar{U}){\cdot}(Q\bar{D})$, i.e. that appearing in the right hand denominator of (7.39). However, the R-charge of this combination is 2 while that of the superpotential, on the grounds that $\int d^2\theta \mathcal{W}$ must be R-invariant, is -2. Hence the only allowed contribution to $\mathcal{W}$ is proportional to $[(Q\bar{U}){\cdot}(Q\bar{D})]^{-1}$, while the power of Λ_3 in the numerator follows from the requirement that $\mathcal{W}$ must have a mass dimension of 3. Moreover, the power of Λ_3 has to be positive in order to facilitate a smooth transition to the weak coupling regime which corresponds to the limit $\Lambda_3 \to 0$. The unit coefficient in the RHS of (7.39) cannot be determined by this argument and has to be computed by instanton calculus [7.10]. There is also a caveat. The perturbative superpotential in (7.38) breaks $SU(2)_\mathrm{R}$ explicitly. For the above argument to go through, we need to assume that the magnitude of λ is small enough not to change the nonperturbative behavior of the theory.

Eq. (7.39) clearly violates the nonrenormalization theorem of §6.7. However, the latter only applies to perturbative contributions to $\mathcal{W}$. Perturbation theory in the trivial vacuum cannot [7.11] describe instanton effects which are entirely nonperturbative in origin. Hence (7.39) is outside the purview of the said theorem. A more important feature of the RHS of (7.39) is that it is singular in the vacuum configuration for vanishing VEVs. This signals the breakdown of $SU(3)$ perturbation theory in the infrared region where the perturbative part of $\mathcal{W}$ vanishes. Conversely, the contribution (7.39) by itself would drive some (or all) VEVs to infinite values, but there the perturbative contribution (7.38) diverges. A simple scaling argument then shows [7.10] that the complete scalar potential, computed from $\mathcal{W} = \mathcal{W}_p + \mathcal{W}_{np}$, has a minimum where all fields have VEVs of order $v \sim \Lambda_3 \lambda^{-1/7}$. All of these VEVs point in a direction in field space along which the entire set of D-term contributions to the scalar potential vanishes.

The system is characterized by the spontaneous breakdown, not only of supersymmetry, but also of the $SU(3) \times SU(2)$ symmetry. The latter is completely destroyed, i.e. every generator fails to annihilate the vacuum. The Higgs mechanism becomes operative via the VEVs. In consequence, all eleven gauge supermultiplets become massive, eight with masses $\sim g_3 v$ and three with $\sim g_2 v$. The VEVs also generate masses for eleven of the fourteen chiral superfields; only the $\bar{U}(\mathbf{3}, \mathbf{1})$ are left massless on account of its nonappearance in (7.38). Let us understand the presence of the massless fermions contained in the latter in terms of symmetries. There is a massless goldstino (as always when supersymmetry is spontaneously broken) plus the superpartner of a massless "axion" associated with the spontaneous breakdown of the global axial $U(1)$ symmetry. There is an additional massless fermion associated with the composite superfield $X = (\bar{U}Q)\cdot\Phi$ which is charged under $U(1)_H$. The latter is required so that the 't Hooft anomaly matching condition [7.13] is satisfied. The supersymmetry violating VEVs of F-components are of order λv^2, i.e. $\langle F \rangle \sim \lambda^{5/7}\Lambda_3^2$ while the mass splitting between members of a given supermultiplet is characterized by $\delta \sim \lambda v$. If the gauge coupling strengths are of order unity and yet $|\lambda| \ll 1$, the supersymmetry breaking scale $\sqrt{|\langle F \rangle|}$ and the intra-supermultiplet mass splitting scale δ can again (cf. §7.4) be quite different from each other. As in the earlier case, this will be shown to have a great deal of relevance to discussion in Ch.12 of supersymmetry breaking in the hidden sector.

Dynamical supersymmetry breaking[7], i.e. of the kind that is induced by nonperturbative dynamics, as illustrated in the present model, is sometimes considered to be more attractive than the purely perturbative mechanisms considered in §7.4 and §7.5. Whereas the mass scales in the O'Raifeartaigh and Fayet-Iliopoulos models have been introduced "by hand", the scale Λ_3 of (7.39) has been created dynamically. One thing should be kept in mind, though. For a given particle content, there is a one-to-one correspondence between Λ_3 and the (unknown) value of the coupling strength g_3 at some high energy scale where the $SU(3)$ interactions become perturbative. In that sense Λ_3 is as much a free model parameter as the scales appearing in O'Raifeartaigh's superpotential or in the scalar potential of Fayet and Iliopoulos.

7.7　Soft Explicit Supersymmetry Breaking

We now discuss the violation of supersymmetry via the Heisenberg-Wigner mode mentioned in §7.1, namely through explicit supersymmetry breaking (SSB) terms in the Lagrangian density:

$$\mathcal{L} = \mathcal{L}_{SUSY} + \mathcal{L}_{SSB} \ .$$

As remarked earlier, the supersymmetry breaking terms, lumped under $\mathcal{L}_{SSB}$, need to be "small" compared to the supersymmetric part $\mathcal{L}_{SUSY}$. More important, however, is the mass dimensional constraint that these terms have to obey. The desired convergent behavior of the supersymmetric theory at high energies and the nonrenormalization of its superpotential couplings need to be retained in the presence of explicit supersymmetry breaking. By a supersymmetric generalization of Symanzik's rule [7.1], this turns out to be possible [7.15] (to all orders in perturbation theory), provided that the explicit supersymmetry breaking terms are **soft**. The latter means that every field operator, occurring in $\mathcal{L}_{SSB}$, needs to have a mass dimension [7.1] less than four. (This is a necessary but not a sufficient condition and one needs to check explicitly in perturbation theory whether a particular operator does have the desired properties). The above equation for $\mathcal{L}$ can then be rewritten as

$$\mathcal{L} = \mathcal{L}_{SUSY} + \mathcal{L}_{SOFT} \ . \tag{7.40}$$

As an illustration, let us consider a supersymmetric gauge model with a simple gauge group. (Our considerations can easily be extended to a product of any number of such simple

---

[7]A review of ideas connected with dynamical supersymmetry breaking may be found in Ref.[7.14].

groups). We have the gauge group G with generators T^a, a set of chiral superfields Φ_i in representations R_i of G and a gauge vector superfield $V \equiv 2gV^a T^a$ with g as the gauge coupling. A generally renormalizable supersymmetric Lagrangian density can be written in the notation of §5.4 as

$$\mathcal{L}_{\text{SUSY}} = \frac{1}{4}\left(W^{aA}W^a_A + \bar{W}^a_{\dot{A}}\bar{W}^{\dot{A}a}\right)_F + [\Phi^\dagger_i(e^{2gV^a T^a})_{ij}\Phi_j]_D$$

$$+[h_i\Phi_i + \frac{1}{2}\mu_{ij}\Phi_i\Phi_j + \frac{1}{3!}f_{ijk}\Phi_i\Phi_j\Phi_k + \text{h.c.}]_F \ . \tag{7.41}$$

Written in component form, (7.41) would look identical to (5.55) with $W(\Phi)$ given by (5.1b) except that we have a made a change of notation from m_{ij} to μ_{ij} since we shall shortly use m to mean something different.

The most general [7.16] soft supersymmetry breaking gauge invariant terms, that can be added to (7.41), may be written as

$$\mathcal{L}_{SOFT} = -\phi^\star_i(m^2)_{ij}\phi_j \ + \ \left(\frac{1}{3!}\mathcal{A}_{ijk}\phi_i\phi_j\phi_k - \frac{1}{2}\mathcal{B}_{ij}\phi_i\phi_j + \mathcal{C}_i\phi_i + \text{h.c.}\right)$$

$$- \frac{1}{2}(M\lambda^a\lambda^a + \text{h.c.}) \ . \tag{7.42}$$

In (7.42) ϕ_i is the scalar component of the superfield Φ_i of type i. Furthermore, λ^a, $\bar{\lambda}^a$ are two component gaugino fields. (Type as well as gauge indices are summed when repeated). This most general[8] soft supersymmetry breaking piece contains a gaugino Majorana mass term, M being the mass, plus a set of trilinear scalar interaction ($\mathcal{A}$-) terms, each with mass dimension three. There are also terms with mass dimension two: a scalar squared mass matrix term and a bilinear scalar interaction ($\mathcal{B}$-) term[9]. Needless to add, the constant tensors $\mathcal{A}_{ijk}$ and $\mathcal{B}_{ij}$, carrying mass dimensions one and two respectively are symmetric in their indices and $(m^2)_{ij}$ is a hermitian matrix. These are assumed to be able to ensure that only gauge invariant combinations are kept. Finally, the terms linear in the scalar fields ($\mathcal{C}$-terms) have unit mass dimension and the constants $\mathcal{C}_i$, carrying mass dimension three, are nonzero only for those scalar components ϕ_i that are themselves gauge invariant. The specific signs and coefficients chosen for the $\mathcal{A}-$, $\mathcal{B}-$ and $\mathcal{C}-$terms are a matter of convention, but those of the m^2 and M terms are determined by the latter being interpreted as scalar and gaugino mass terms respectively.

We can rewrite the explicit supersymmetry breaking terms (7.42) in the Lagrangian density in a superfield form analogous to that of (7.41) by employing a trick. Let us introduce

---

[8]A term such as $\mathcal{D}_{ijk}\phi^\star_i\phi_j\phi_k$ in $\mathcal{L}_{SOFT}$, though of mass dimension three, generally tends to generate quadratic divergences from loops and is hence excluded as not being a genuine soft term. (This illustrates the fact that the mass dimension being less than four is a *necessary but not a sufficient* condition for the softness of any operator.) If there are no chiral gauge singlet superfields, such a term does become soft [7.17]. Nevertheless, this kind of a term is not generated in the most popular methods implementing supersymmetry breaking in the MSSM, cf. Chs.12 and 13.

[9]In the absence of chiral gauge singlet superfields, a bilinear like $\xi_i\xi_j + \bar{\xi}_j\bar{\xi}_i$, where ξ_i is the two component fermionic partner of ϕ_i, can be rewritten as a linear combination of $[\Phi_i\Phi_j]_F$, $\phi_i\phi_j$ and their hermitian conjugates. On the other hand, if such singlet superfields are present, explicit mass terms for chiral fermions are no longer soft.

an external "spurion" [7.16] chiral superfield defined by $\eta = \theta\theta$ and $\bar{\eta} \equiv \bar{\theta}\bar{\theta}$. Now (7.42) can be rewritten as

$$-\mathcal{L}_{SOFT} = \left[\eta \left(-\frac{1}{3!} \mathcal{A}_{ijk} \Phi_i \Phi_j \Phi_k + \frac{1}{2} \mathcal{B}_{ij} \Phi_i \Phi_j - \mathcal{C}_i \Phi_i + \text{h.c.} \right. \right.$$

$$\left. \left. + \frac{1}{2} (M W^{aA} W_A^a + \text{h.c.}) \right) \right]_F$$

$$+ \left[\bar{\eta}\eta \; \Phi_i^\dagger (m^2)_{ij} (e^V)_{jk} \Phi_k \right]_D . \tag{7.43}$$

It is customary in the literature to define the parameters[10] A_{ijk}, B_{ij} and C_i, dividing out $\mathcal{A}_{ijk}$, $\mathcal{B}_{ij}$ and $\mathcal{C}_i$ by the superpotential couplings f_{ijk}, μ_{ij} and h_i respectively, i.e.

$$A_{ijk} \equiv \mathcal{A}_{ijk} / f_{ijk} ,$$

$$B_{ij} \equiv \mathcal{B}_{ij} / \mu_{ij} ,$$

$$C_i = \mathcal{C}_i / h_i .$$

Because of this, $\mathcal{A}_{ijk}$, $\mathcal{B}_{ij}$ and $\mathcal{C}_i$ are sometimes written as $(fA)_{ijk}$, $(\mu B)_{ij}$ and $(hC)_i$. Thus the action corresponding to (7.43) reads

$$S_{SOFT} = \int d^6 z \; \eta \left[\frac{1}{6} (fA)_{ijk} \Phi_i \Phi_j \Phi_k - \frac{1}{2} (\mu B)_{ij} \Phi_i \Phi_j + (hC)_i \Phi_i \right.$$

$$\left. - \frac{1}{2} (M W^{aA} W_A^a + \text{h.c.}) \right]$$

$$- \int d^8 z \; \bar{\eta}\eta \; \Phi_i^\dagger (m^2)_{ij} (e^V)_{jk} \Phi_k . \tag{7.44}$$

We would like to make the following two comments on (7.43) and (7.44).

- Suppose G is not a simple group but a product of invariant subgroup factors G_α, i.e. $G = \prod_\alpha G_\alpha$, where G_α is either a simple group or a $U(1)$. Now these two equations have to have only minor modifications. First, $V = 2g V^a T^a$ changes to $2\sum_\alpha g_\alpha V^\alpha T^\alpha$, where g_α is the gauge coupling for the factor group G_α and *a summation of the gauge index within each factor group, with the coupling strength held fixed, is also implied.*

---

[10]The A-terms contribute to off-diagonal elements of sfermion squared mass matrices, cf. §9.4. Even if they are taken to be zero at the tree level, they will be induced at one loop in case the gaugino mass M is nonzero. The B-terms play an essential role via the Higgs potential in spontaneous electroweak symmetry breakdown, cf. §10.1, §10.2. The MSSM does not have C-terms, cf. §9.1.

Next, $W^{aA}W_A^a$ ($\bar{W}_{\dot{A}}^a\bar{W}^{a\dot{A}}$) is changed to $W^{\alpha A}W_A^{\alpha}$ ($\bar{W}_{\dot{A}}^{\alpha}\bar{W}^{\alpha\dot{A}}$). Finally, the gaugino mass term in (7.42) gets changed to

$$\frac{1}{2}\sum_{\alpha}(M_{\alpha}\lambda^{\alpha}\lambda^{\alpha} + \text{h.c.}) \ ,$$

where M_{α} is the gaugino mass pertaining to G_{α}; again, a summation within each factor group with M_{α} held fixed is included.

- Eq. (7.44) describes the soft part of the tree level action only. Soft supersymmetry breaking terms are unprotected from loop corrections (that are of course, at worst, logarithmically divergent) by any nonrenormalization theorem (cf. Ch.6). Such corrections lead to [7.18] additional terms of the form

$$\int d^8z \left[\eta \ \Phi^{\dagger}\Phi, \ \bar{\eta} \ \Phi, \ \bar{\eta}\eta \ \Phi, \ \bar{\eta}\eta \ \mathcal{D}^A W_A^{\alpha}\right]$$

in (7.44) after renormalization. Gauge invariance disallows the last term within the square bracket above, except for a $U(1)$ factor in the gauge group. Hermitian conjugates of the operators displayed within the square brackets above also appear. However, these terms can be reabsorbed [7.18] into the conventional terms of S_{SOFT} by appropriate θ-dependent superfield redefinitions and linearly redefined superpotential parameters. There will be a more detailed discussion of those issues in Ch.11.

7.8 The General Mass Sum Rule

We shall now give a more general discussion, cf. Dimopoulos and Georgi [7.15], of the mass sum rule (5.10) in the presence of spontaneous supersymmetry breakdown with a general superpotential $\mathcal{W}(\Phi_i)$ and arbitrary gauge interactions. We are interested in the vacuum configurations of the scalar fields $\phi_i = \langle\phi_i\rangle$. However, unlike in Ch.5, $V(\langle\phi_i\rangle, \langle\bar{\phi}^j\rangle)$ no longer vanishes. Rather, the extremization $(\partial V/\partial\phi_i)_{\phi=\langle\phi\rangle} = 0$ is realized by the condition

$$\left(\mathcal{W}_{ij}\overline{\mathcal{W}}^j + D_i^a D^a\right)_{\phi=\langle\phi\rangle} = 0 \ , \tag{7.45}$$

where we have employed the notation introduced in §5.1 and §5.2.

The mass squared matrix of the vector bosons (generated via the Higgs mechanism) is given by (5.23). As before:

$$\underset{\sim}{m}{}_V^{2\ ab} = (D_i^a D^{bi} + D^{ai}D_i^b)_{\phi=\langle\phi\rangle} \ . \tag{7.46}$$

The mass matrix of the fermions, enriched by the mixing of gauginos and chiral fermions, is still given by (5.30). Thus

$$\underset{\sim}{m}{}_{Fij}^a = \begin{pmatrix} \mathcal{W}_{ij} & -\sqrt{2}D_i^a \\ -\sqrt{2}D_j^a & 0 \end{pmatrix}_{\phi=\langle\phi\rangle} , \tag{7.47a}$$

$$(\underset{\sim}{m} {}^{a\dagger}_{F} \underset{\sim}{m} {}^{a}_{F})_{ij} = \begin{pmatrix} [\overline{\mathcal{W}}\mathcal{W}]^{i}{}_{j} + 2D^{ai}D^{a}_{j} & -\sqrt{2}\,\overline{\mathcal{W}}^{ik}D^{a}_{k} \\ -\sqrt{2}\,D^{ak}\mathcal{W}_{kj} & 2D^{ai}D^{a}_{j} \end{pmatrix}_{\phi=\langle\phi\rangle} , \tag{7.47b}$$

$$(\underset{\sim}{m} {}^{a}_{F} \underset{\sim}{m} {}^{a\dagger}_{F})_{ij} = \begin{pmatrix} [\mathcal{W}\overline{\mathcal{W}}]_{i}{}^{j} + 2D^{a}_{i}D^{aj} & -\sqrt{2}\,\mathcal{W}_{ik}D^{ak} \\ -\sqrt{2}\,D^{a}_{k}\overline{\mathcal{W}}^{kj} & 2D^{a}_{i}D^{aj} \end{pmatrix}_{\phi=\langle\phi\rangle} . \tag{7.47c}$$

However, the scalar mass squared matrix is now changed from that of (5.28) because of the nonzero values acquired by some of the auxiliary fields in terms of $\langle F_i \rangle$, $\langle F_i^\star \rangle$ and $\langle D^a \rangle$. (Here the index a spans only those $U(1)$ subgroups of the full gauge group for which D^a acquire a nonzero VEV, cf.§5.2). On substituting (5.22a) in (5.8b), we obtain

$$\underset{\sim}{m} {}^{2}_{Sij} = \begin{pmatrix} [\overline{\mathcal{W}}\mathcal{W}]^{i}{}_{j} + D^{ai}D^{a}_{j} + D^{ai}_{j}D^{a} & \overline{\mathcal{W}}^{ijk}\mathcal{W}_{k} + D^{ai}D^{aj} \\ \mathcal{W}_{ijk}\overline{\mathcal{W}}^{k} + D^{a}_{i}D^{a}_{j} & [\mathcal{W}\overline{\mathcal{W}}]^{i}{}_{j} + D^{a}_{i}D^{aj} + D^{a\,j}_{i}D^{a} \end{pmatrix}_{\phi=\langle\phi\rangle} . \tag{7.48}$$

(7.46), (7.47b) and (7.48) yield

$$\mathrm{Tr}\ \underset{\sim}{m} {}^{2}_{Sii} - \mathrm{Tr}(\underset{\sim}{m} {}^{a\dagger}_{F} \underset{\sim}{m} {}^{a}_{F} + \underset{\sim}{m} {}^{a}_{F} \underset{\sim}{m} {}^{a\dagger}_{F})_{ii} + 3\,m^{2}_{Vaa} = -2\mathrm{Tr}\,(T^{a}\langle D^{a}\rangle) , \tag{7.49}$$

where we have used the result that $D^{ai}_i = -g^a t^a \equiv -T^a$, cf. (5.22b). Eq. (7.49) is now the generalized supertrace mass sum rule with the RHS covering any $U(1)$ subgroup of the gauge group that admits a Fayet-Iliopoulos type of supersymmetry breaking. In fact, for any chiral supermultiplet, the sum rule now reads

$$S\mathrm{Tr}\ \underset{\sim}{m} {}^{2} = -2(T^{a}\langle D^{a}\rangle) , \tag{7.50}$$

the index a spanning the appropriate abelian gauge subgroups. Thus, even with spontaneous supersymmetry breaking, the supertrace mass sum rule holds modulo a nonzero RHS, as given in (7.50), when $\langle T^a D^a \rangle \neq 0$. The following remark is pertinent here. In theories with a Fayet-Iliopoulos type of supersymmetry breaking, it is desirable to have [7.9] a vanishing trace of any $U(1)$ operator, i.e. the sum over all Abelian charges should be zero. This ensures the absence of a quadratically divergent contribution to the Fayet-Iliopoulos term as well as the cancellation of mixed gauge-gravitational anomalies. In this case the total supertrace over all fields in (7.49) vanishes, but the D-terms can still contribute to the mass splitting within a single supermultiplet, as in (7.50). Ref. [7.9] may be consulted for further details on this point.

We end this discussion with the following points.

- The supertrace mass sum rule (7.50) suffers significant modifications when there are explicit soft supersymmetry breaking terms. This is apparent already at the tree level. For instance, with an explicit gaugino mass M, (7.47a) changes to

$$\underset{\sim}{m} {}^{a}_{Fij} = \begin{pmatrix} \mathcal{W}_{ij} & -\sqrt{2}D^{a}_{i} \\ -\sqrt{2}D^{a}_{j} & M\delta_{ij} \end{pmatrix} .$$

Similarly, there are extra contributions to (7.48) and the argument, leading to (7.49), does not go through. Furthermore, these modifications are unprotected from loop corrections. Indeed, this is how explicit soft supersymmetry breaking is utilized to describe the real world at or near weak scale energies.

- A purely spontaneous breaking of global supersymmetry, invoked to cause mass splittings between particles and sparticles, will imply (7.50). As will be explained later in §9.1, in the minimal supersymmetric extension of the Standard Model (SM) of particle physics, such a sum rule is incompatible with the observed requirement of most (as yet undiscovered) sparticles having to be heavier than the corresponding particles.

- Soft explicit supersymmetry breaking terms can modify the sum rule (7.50) so as to make it compatible with observation. The latter could, however, originate from the spontaneous breakdown of supersymmetry induced at very high energies by (as yet) undiscovered heavy superfields lying in a hidden sector (which is invariant under the known gauge interactions) and behaving as singlets under the gauge transformations of the SM. There will be more discussion of this issue in Ch.12.

References

[7.1] S. Coleman, *op. cit.*, *Bibl.*

[7.2] S. Coleman and E.J. Weinberg, Phys. Rev. **D7** (1973) 1888.

[7.3] R. Rajaraman and M. Raj Lakshmi, Phys. Rev. **D23** (1981) 2399.

[7.4] E. Witten, *loc. cit.*, Ref. [2.3].

[7.5] L. O'Raifeartaigh, Nucl. Phys. **B96** (1975) 331.

[7.6] P. Fayet and J. Iliopoulos, Phys. Lett. **51B** (1974) 461. P. Fayet, Nuov. Cim. **A31** (1976) 31.

[7.7] S.B. Treiman, R. Jackiw and D. Gross, *op. cit.*, *Bibl.*

[7.8] H.J. Müller-Kirsten and A. Wiedemann, *op. cit.*, *Bibl.*

[7.9] H.P. Nilles, *loc. cit.*, *Bibl.*

[7.10] G.F. Giudice and R. Rattazzi, *loc. cit.*, *Bibl.*

[7.11] R. Rajaraman, *op. cit.*, *Bibl.*

[7.12] I. Affleck, M. Dine and N. Seiberg, Nucl. Phys. **B241** (1984) 493.

[7.13] G. 't Hooft in *Recent Developments in Gauge Theories* (eds. G. 't Hooft et al., Plenum, New York, 1980) reprinted in *Dynamical Gauge Symmetry Breaking* (eds. E. Farhi and R. Jackiw, World Scientific, Singapore, 1982) and in *Under the Spell of the Gauge Principle* (G. 't Hooft, World Scientific, Singapore, 1994). I. Affleck, M. Dine and N. Seiberg, Phys. Lett. **B137** (1984) 187.

[7.14] E. Poppitz and S.P. Trivedi, *loc. cit.*, *Bibl.*

[7.15] S. Dimopoulos and H. Georgi, *loc. cit.*, Ref. [5.2]. N. Sakai, *loc. cit.* Ref. [1.20].

[7.16] L. Girardello and M. Grisaru, Nucl. Phys. **B194** (1980) 65. P. Fayet and S. Ferrara, *loc. cit.*, *Bibl.*

[7.17] L.J. Hall and L. Randall, Phys. Rev. Lett. **65** (1990) 239.

[7.18] Y. Yamada, Phys. Rev. **D50** (1994) 3537. A. Pomerol and S. Dimopoulos, Nucl. Phys. **B453** (1995) 83.

PART TWO

SUPERSYMMETRY PHENOMENOLOGY

Chapter 8

BASIC STRUCTURE OF THE MSSM

8.1 Brief Review of the Standard Model

We discuss in this chapter the minimal extension of the Standard Model (SM) [8.1] that is needed to incorporate softly broken $N=1$ global supersymmetry in the latter. This is called the *Minimal Supersymmetric Standard Model* (MSSM) [8.2]. The prefix "minimal" is used to distinguish from nonminimal extensions which we shall come to in Ch.14. In order to supersymmetrize the SM, we need (cf. Chs. 1,3) to introduce for every particle a superpartner. The latter differs from the former in spin by half and in mass generally by some positive amount $O(M_s)$, but with all other internal quantum numbers kept identical. In the SM all matter fields (pertaining to quarks and leptons) are spin half fermionic fields while gauge bosons have spin one. The superpartners of the former cannot have spin one. Since they are supposed to be matter fields, they are not gauge bosons while the only known consistent relativistic field theories of spin one particles are those of gauge bosons. Thus superpartners of matter fermions are taken to be spin zero scalars and are described, along with the latter, by chiral superfields. These scalars are called **sfermions** and they can be classfied into scalar leptons or **sleptons** and scalar quarks or **squarks**. Similarly, since even at the classical level, the only consistent interacting field theory of spin 3/2 particles has to include [8.3] gravity, the superpartner fields of the SM gauge bosons are chosen to have spin 1/2; they are called **gauginos**. Gauge bosons and gauginos are described by vector superfields. Gauginos can be further classified into the strongly interacting **gluinos** as well as the electroweak **zino** (corresponding to the Z boson) and **winos** (corresponding to the W bosons). Spin zero Higgs bosons are described, along with their spin half superpartners (called **higgsinos**), by chiral superfields. We shall later see that electroweak symmetry breaking mixes the EW gauginos with the higgsinos making physical **charginos** and **neutralinos**.

To begin with, let us set up the notation by briefly summarizing some basic ingredients of the SM itself. The gauge symmetry group is $SU(3)_C \times SU(2)_L \times U(1)_Y$, with subscripts C, L, Y referring respectively to color, left chirality and weak hypercharge. All matter (quark and lepton) fields are fermion fields with left chiral ones transforming as doublets and right chiral ones as singlets of $SU(2)_L$. The hypercharge Y_f of each fermion field is related to its electromagnetic charge Q_f and the third component of its left chiral weak isospin T_{3L}^f by

$$Q_f = T_{3L}^f + \frac{Y_f}{2} \, . \tag{8.1}$$

The electroweak gauge transformation properties of the left chiral, right chiral fermion fields $f_L = \frac{1}{2}(1 - \gamma_5)f$, $f_R = \frac{1}{2}(1 + \gamma_5)f$ are:

$$f_L(x) \to e^{-ig_Y \alpha_Y(x)Y/2} \, e^{-ig_2 \vec{\alpha}_2(x)\cdot\vec{\tau}/2} f_L(x) \, , \tag{8.2a}$$

$$f_R(x) \to e^{-ig_Y \alpha_Y(x)Y/2} f_R(x) \, , \tag{8.2b}$$

where g_Y, $\alpha_Y(x)$ and g_2, $\vec{\alpha}_2(x)$ are the $U(1)_Y$ and $SU(2)_L$ gauge couplings, functions respectively. Moreover, Y is the hypercharge operator and the Pauli matrices $\vec{\tau}$ act in the weak isospin doublet representation space.

Fields for the three generations (generation index $i = 1, 2, 3$) of leptons and quarks, along with the dimension of the corresponding $SU(2)_L$ representation and the Y quantum number are listed below.

$$\ell_{iL} = \begin{pmatrix} \nu_i \\ e_i \end{pmatrix}_L, \text{ i.e. } \ell_{1L} = \begin{pmatrix} \nu_e \\ e^- \end{pmatrix}_L, \ \ell_{2L} = \begin{pmatrix} \nu_\mu \\ \mu^- \end{pmatrix}_L, \ \ell_{3L} = \begin{pmatrix} \nu_\tau \\ \tau^- \end{pmatrix}_L : (\mathbf{2}, -1) \, ,$$

$$e_{1R} = e_R^-, \ e_{2R} = \mu_R^-, \ e_{3R} = \tau_R^- : (\mathbf{1}, -2),$$

$$q_{iL} = \begin{pmatrix} u_i \\ d_i \end{pmatrix}_L, \text{ i.e. } q_{1L} = \begin{pmatrix} u \\ d \end{pmatrix}_L, \ q_{2L} = \begin{pmatrix} c \\ s \end{pmatrix}_L, \ q_{3L} = \begin{pmatrix} t \\ b \end{pmatrix}_L : \left(\mathbf{2}, \frac{1}{3}\right), \tag{8.3}$$

$$u_{1R} = u_R, \ u_{2R} = c_R, \ u_{3R} = t_R : \left(\mathbf{1}, \frac{4}{3}\right),$$

$$d_{1R} = d_R, \ d_{2R} = s_R, \ d_{3R} = b_R : \left(\mathbf{1}, -\frac{2}{3}\right).$$

The color gauge transformations of quark (q) and lepton (ℓ) fields are:

$$q_{L,R}(x) \to e^{-ig_s \alpha_s^a(x)\lambda^a/2} q_{L,R}(x), \ \ell_{L,R}(x) \to \ell_{L,R}(x) \, , \tag{8.4}$$

with g_s, α_s^a being the $SU(3)_C$ gauge coupling, functions and λ^a being the Gell-Mann $SU(3)$ lambda matrices acting in the triplet ($\mathbf{3}$) representation space. The quark fields of (8.4) transform as color triplets ($\mathbf{3}$) of $SU(3)_C$ whereas the lepton fields of (8.3) are color singlets. The $SU(2)_L$ singlet right chiral fermion fields can be converted into left chiral ones by charge conjugation. For instance, $u_R^C = (u^C)_L$ is such a field with $T_f^L = 0$, $Y = -\frac{4}{3}$ and transforming as a color antitriplet ($\bar{\mathbf{3}}$). Again, $e_R^- = (e_L^+)^C$ and so on.

The gauge fields g_μ^a ($a = 1, \cdots, 8$), $\vec{W}_\mu$ and B_μ transform according to the adjoint representations of $SU(3)_C$, $SU(2)_L$ and $U(1)_Y$ respectively. The eight gluons g^a are always massless while the three $SU(2)_L$ gauge bosons $W_{1,2,3}$ and the one $U(1)_Y$ gauge boson B are massless only in the limit of exact electroweak symmetry. At the weak scale, the $SU(2)_L \times U(1)_Y$ electroweak (EW) symmetry gets spontaneously broken to $U(1)_{em}$. The unbroken symmetry group at energies lower than the weak scale is thus $SU(3)_C \times U(1)_{em}$. This spontaneous symmetry breakdown is driven by an $SU(2)_L$ doublet of scalar Higgs fields $\phi = \begin{pmatrix} \phi^+ \\ \phi^0 \end{pmatrix}$ with $Y = 1$ and is signaled by a real nonzero vacuum expectation value (VEV) for this field, arising from the minimization of the Higgs potential term $V(\phi)$ and given by

$$\langle \phi \rangle = \frac{1}{\sqrt{2}} \begin{pmatrix} 0 \\ v \end{pmatrix}. \tag{8.5}$$

While the photon γ remains massless, the weak bosons $W^\pm$ and Z acquire masses though the VEV v in (8.5). The latter is related to the masses $M_{W,Z}$ and the couplings $g_{2,Y}$ as well as to the Fermi constant G_F by

$$M_W = \frac{1}{2} g_2 v, \ M_Z = \frac{1}{2} \sqrt{g_Y^2 + g_2^2}\, v, \ v = \left(\frac{1}{\sqrt{2} G_F} \right)^{1/2} \simeq 246 \text{ GeV}. \tag{8.6}$$

The fields $W_\mu^\pm$, Z_μ and A_μ, which are mass eigenstates, are given respectively in terms of the fields $\vec{W}_\mu$ and B_μ, introduced earlier, as

$$W^{\mu,\pm} = \frac{1}{\sqrt{2}} (W_1^\mu \mp i W_2^\mu)\,, \tag{8.7a}$$

$$Z^\mu = \frac{g_2}{\sqrt{g_Y^2 + g_2^2}} W_3^\mu - \frac{g_Y}{\sqrt{g_Y^2 + g_2^2}} B^\mu$$
$$= -\sin\theta_W B^\mu + \cos\theta_W W_3^\mu\,, \tag{8.7b}$$
$$A^\mu = \cos\theta_W B^\mu + \sin\theta_W W_3^\mu\,, \tag{8.7c}$$

with
$$e = g_2 \sin\theta_W = g_Y \cos\theta_W\,. \tag{8.8}$$

The nonzero VEV, introduced in (8.5), is also responsible in the SM for generating fermion masses through Yukawa interaction terms characterized by coupling strengths f and generation indices[1] i, j. For the latter, we can write:

$$\mathcal{L}_Y^1 = -f_{ij}^{e\star} \overline{\ell_{iL}} \phi e_{jR} - f_{ij}^{d\star} \overline{q_{iL}} \phi d_{jR} + \text{h.c.} \tag{8.9}$$

in case of "down type" right chiral fermions (e_{jR}, d_{jR}) and

$$\mathcal{L}_Y^2 = -f_{ij}^{u\star} \overline{q_{iL}} \phi^C u_{jR} + \text{h.c.} \tag{8.10}$$

for "up type" right chiral fermions u_{jR}. The complex conjugate $f^\star$ has been chosen here for convenience in later supersymmetric generalization (cf. 8.33) and i, j are summed on repetition. Furthermore,

$$\phi^C = i\tau_2 \phi^\star = \begin{pmatrix} \phi^{0\star} \\ -\phi^- \end{pmatrix}$$

---

[1]We shall not use here the type subspace formalism, introduced in Ch.5, with both superscripts and subscripts. Thus all generation indices will henceforth be subscripts.

is the "charge conjugated" Higgs doublet field. Note that leptonic couplings are absent from (8.10) since there is no ν_R. The substitution of (8.5) into (8.9) and (8.10) leads to the fermion mass terms. Suppose, for a set of Dirac fermions ψ_i, we define the mass matrix m_{ij} by writing the fermion mass term in the Lagrangian density as

$$\mathcal{L}_{FMT} = -\left(\overline{\psi_{iL}}m_{ij}\psi_{Rj} + \text{h.c.}\right).$$

Then one can write the charged lepton, down type quark, up type quark mass matrices [8.1] in generation space as

$$(\mathbf{m}_e)_{ij} = \frac{1}{\sqrt{2}}f_{ij}^{e\star}v = m_{e_i}\delta_{ij}, \ (\mathbf{m}_d)_{ij} = \frac{1}{\sqrt{2}}f_{ij}^{d\star}v, \ (\mathbf{m}_u)_{ij} = \frac{1}{\sqrt{2}}f_{ij}^{u\star}v \, , \tag{8.11}$$

the first being brought into a real diagonal form without loss of generality on account of the assumed masslessness[2] of the neutrinos. However, the up type and down type quark mass matrices do not have this advantage and can be put into real diagonal forms only by biunitary transformations. Thus if the mass eigenstate left, right u- and d-quark fields are unitarily transformed to the corresponding flavor eigenstate ones by $\mathbf{U}^{u_L}$, $\mathbf{U}^{u_R}$ and $\mathbf{U}^{d_L}$, $\mathbf{U}^{d_R}$, the quark mass matrices transform as

$$(\mathbf{U}^{u\dagger}_L \mathbf{m}_u \mathbf{U}^{u_R})_{ij} = [\mathbf{m}_u^{(D)}]_{ij} \equiv m_{u_i}\delta_{ij} \, , \tag{8.12a}$$

$$(\mathbf{U}^{d\dagger}_L \mathbf{m}_d \mathbf{U}^{d_R})_{ij} = [\mathbf{m}_d^{(D)}]_{ij} \equiv m_{d_i}\delta_{ij} \, . \tag{8.12b}$$

In (8.12) $\mathbf{m}_u^{(D)}$ and $\mathbf{m}_d^{(D)}$ are the physical real diagonal mass matrices for up and down type quarks respectively.

Baryon number B and lepton type numbers $L_{e,\mu,\tau}$ (and hence lepton number $L \equiv L_e + L_\mu + L_\tau$) are conserved in the SM. These 'accidental' global symmetries are a consequence of the particle content and the gauge group. As will be discussed in more detail later, the situation is quite different for the MSSM. The latter can accommodate several types of renormalizable interactions which violate some or all of these symmetries. For the time being, let us nonetheless restrict ourselves to a version of the MSSM where these symmetries are conserved by the assumption of R-parity invariance (cf. §4.5).

8.2 Superfields of the MSSM

We now proceed to introduce a chiral superfield for every chiral fermion of the SM. Apart from these chiral fermions and auxiliary fields, such superfields will contain new scalar fields. For the first generation, these scalar fields can be enumerated as

$$\tilde{\ell}_{1L} = \begin{pmatrix} \tilde{\nu} \\ \tilde{e}^- \end{pmatrix}_L , \ \tilde{e}_{1R} = \tilde{e}_R, \ \tilde{q}_{1L} = \begin{pmatrix} \tilde{u} \\ \tilde{d} \end{pmatrix}_L , \ \tilde{u}_{1R} = \tilde{u}_R, \ \tilde{d}_{1R} = \tilde{d}_R \, . \tag{8.13}$$

---

[2]An important example of a term that violates the lepton number symmetry is a Majorana neutrino mass. Since neutrino masses can be introduced without significantly altering the specific supersymmetric aspects of particle phenomenology, we postpone a detailed discussion of this point to Ch.14.

Here $\tilde{\ell}_{1L}$ are called left sleptons (more specifically, left selectron and sneutrino) while $\tilde{e}_R$ is called the right selectron. Let us denote by L_1 (Q_1) and $\bar{E}_1$ ($\bar{U}_1$, $\bar{D}_1$) the left chiral lepton (quark) doublet and antilepton (antiquark) singlet chiral superfields respectively. Thus, for the first generation of leptons and sleptons, we can take the superfields

$$L_1 = \begin{pmatrix} L_{\nu_e} \\ L_e \end{pmatrix}, \ \bar{E}_1 . \tag{8.14}$$

Contained in these are the fields ℓ_{1L}, $\tilde{\ell}_{1L}$, $e_{1R}{}^C = e_R^C$ and $\tilde{e}_{1R}{}^\star = \tilde{e}_R^\star$ corresponding[3] respectively to ψ_{e1L}, $\phi_{\ell 1+}$, $\psi_{eR}{}^C$ and ϕ_{e-} in the notation of Ch.5. There is no singlet neutrino superfield since the SM does not contain any left chiral antineutrino. Similarly, the first quark (and squark) generation is represented by the superfields

$$Q_1 = \begin{pmatrix} Q_u \\ Q_d \end{pmatrix}; \ \bar{U}_1, \ \bar{D}_1 . \tag{8.15}$$

These contain the fields q_{1L}, $\tilde{q}_{1L}$, $u_{1R}{}^C = u_R^C$, $d_{1R}{}^C = d_R^C$, $\tilde{u}_{1R}{}^\star = \tilde{u}_R^\star$ and $\tilde{d}_{1R}{}^\star = \tilde{d}_R^\star$ corresponding to ψ_{q1L}, ϕ_{q1+}, $\psi_{uR}{}^C$, $\psi_{dR}{}^C$, ϕ_{u-} and ϕ_{d-} respectively.

The above procedure can be repeated for the second and third generations. Thus we denote matter superfields corresponding to these generations by L_i, $\bar{E}_i$, Q_i, $\bar{U}_i$ and $\bar{D}_i$ with $i = 2, 3$. So we have

$$L_2 = \begin{pmatrix} L_{\nu_\mu} \\ L_\mu \end{pmatrix}, \ \bar{E}_2, \ Q_2 = \begin{pmatrix} Q_c \\ Q_s \end{pmatrix}, \ \bar{U}_2, \ \bar{D}_2 , \tag{8.16}$$

respectively containing the fields ℓ_{2L}, $\tilde{\ell}_{2L}$, $e_{2R}{}^C = \mu_R^C$, $\tilde{e}_{2R}{}^\star = \tilde{\mu}_R^\star$, q_{2L}, $\tilde{q}_{2L}$, $u_{2R}{}^C = c_R^C$, $\tilde{u}_{2R}{}^\star = \tilde{c}_R^\star$, $d_{2R}{}^C = s_R^C$, $\tilde{d}_{2R}{}^\star = \tilde{s}_R^\star$. Furthermore, there are

$$L_3 = \begin{pmatrix} L_{\nu_\tau} \\ L_\tau \end{pmatrix}, \ \bar{E}_3, \ Q_3 = \begin{pmatrix} Q_t \\ Q_b \end{pmatrix}, \bar{U}_3, \bar{D}_3 , \tag{8.17}$$

respectively containing the fields ℓ_{3L}, $\tilde{\ell}_{3L}$, $e_{3R}{}^C = \tau_R^C$, $\tilde{e}_{3R}{}^\star = \tilde{\tau}_R^\star$, q_{3L}, $\tilde{q}_{3L}$, $u_{3R}{}^C = t_R^C$, $\tilde{u}_{3R}{}^\star = \tilde{t}_R^\star$, $d_{3R}{}^C = b_R^C$, $\tilde{d}_{3R}{}^\star = \tilde{b}_R^\star$.

Supersymmetry, by itself, does not provide any clear answer to the generation or family problem and, in the MSSM, one simply replicates the superfields thrice for the three generations. Within each family, however, the counting of fermionic and bosonic degrees of freedom must match for every supermultiplet, as described by a chiral superfield. Corresponding to a massive Dirac fermion field, f_u say, with four on-shell degrees of freedom (two spin states for the particle and two for the antiparticle, as embodied in the complex chiral fields f_{uL} and f_{uR}), there are two corresponding complex scalar fields $\tilde{f}_{uL}$ and $\tilde{f}_{uR}$. Each of the latter, together with its complex conjugate, stands for particle and antiparticle fields; thus the components match. Note further that $\tilde{f}_{uL}$ and $\tilde{f}_{uR}$ have different $SU(2)_L \times U(1)_Y$ quantum numbers just as f_{uL} and f_{uR} do. Another point needs to be emphasized here. Since the superpotential $\mathcal{W}$ can contain only left chiral superfields, one is obliged to use the left chiral

---

[3] Cf. §5.6, except that we have dropped the $+, -$ subscripts and used overbars for singlets.

charge conjugates of the $SU(2)_L$ singlet right chiral fermion fields, i.e. $f_{uR}{}^C = (f_u^C)_L$ etc., and the complex conjugates of their superpartner right sfermion fields, i.e. $\tilde{f}_{uR}^*$ etc. These are contained in left chiral superfields with quantum numbers of the conjugate representations. Finally, all matter superfields are taken to have *odd* matter parity (cf. §4.5).

In the gauge sector we introduce one vector superfield corresponding to each gauge field in the gauge group $SU(3)_C \times SU(2)_L \times U(1)_Y$. Thus we have the $U(1)_Y$, $SU(2)_L$, $SU(3)_C$ gauge fields B_μ, $\vec{W}_\mu$, g_μ^a and the corresponding spin half (four component) Majorana gaugino fields $\tilde{\lambda}_0$, $\vec{\tilde{\lambda}}$, $\tilde{g}^a$ contained in the superfields

$$\{V^Y, \vec{V}^W, V_g^a\} \tag{8.18}$$

respectively. Every gaugino field, like its gauge boson partner, transforms as the adjoint representation of the corresponding gauge group. Moreover, each such field has left chiral and right chiral components which are charge conjugates of each other:

$$(\tilde{\lambda}_{0L})^C = \tilde{\lambda}_{0R} . \tag{8.19}$$

Next, we turn to the supersymmetrization of the Higgs sector of the SM. The latter has only one $SU(2)_L$ doublet field ϕ with a hypercharge $Y_\phi = 1$. As discussed earlier, the same Higgs VEV v can be used to give masses to the $T_{3L} = 1/2$ and $T_{3L} = -1/2$ fermions via the Yukawa interaction terms of (8.9) and (8.10). In particular, (8.10) has been made possible only by use of the conjugate Higgs field ϕ^C which has $Y_{\phi^C} = -1$. Such a term, however, will not be allowed in a supersymmetric theory. There the Yukawa interactions are derived from the superpotential $\mathcal{W}$ which has to be an *analytic* function of left chiral superfields (see §5.1). Hence interaction terms, derived from the same superpotential, cannot contain both ϕ and ϕ^C. Therefore, in order to make the $T_{3L} = -1/2$ fermions massive, a second Higgs doublet is needed. We must then have – in a supersymmetric theory – two Higgs doublets with hypercharges $Y = -1$ and 1 which we shall denote by h_1 (down type) and h_2 (up type) respectively. If the superscript D is an $SU(2)$ doublet index taking values $1, 2$, we can write for $D = 1$, $h_1^1 = h_1^0$ and $h_2^1 = h_2^+$ while, for $D = 2$, we can write $h_1^2 = h_1^-$, $h_2^2 = h_2^0$:

$$h_1 \equiv \begin{pmatrix} h_1^1 \\ h_1^2 \end{pmatrix} = \begin{pmatrix} h_1^0 \\ h_1^- \end{pmatrix} ; \quad h_2 \equiv \begin{pmatrix} h_2^1 \\ h_2^2 \end{pmatrix} = \begin{pmatrix} h_2^+ \\ h_2^0 \end{pmatrix} . \tag{8.20}$$

Their Yukawa interactions can be written down simply by replacing ϕ and ϕ^C by $-i\tau_2 h_1^*$ and $i\tau_2 h_2^*$ respectively in (8.9) and (8.10). The Higgs VEVs, after the spontaneous breakdown of electroweak symmetry, are now given by real, positive quantities (cf. §10.2) $v_{1,2}$ which arise from the minimization of the Higgs potential term $V(h_1, h_2)$ and are shown below:

$$\langle h_1 \rangle = \frac{1}{\sqrt{2}} \begin{pmatrix} v_1 \\ 0 \end{pmatrix} , \quad \langle h_2 \rangle = \frac{1}{\sqrt{2}} \begin{pmatrix} 0 \\ v_2 \end{pmatrix} . \tag{8.21}$$

It is well known [8.4] that this two Higgs doublet extension of the SM, with the up and down type fermions coupling to separate Higgs doublets, is perfectly compatible with all FCNC constraints[4] since it obeys the Glashow-Weinberg/Paschos condition [8.1]. The only change

---

[4]This is true even including one loop corrections.

is that (8.6) and (8.11) are now respectively modified to

$$M_W = \frac{1}{2}g_2\sqrt{v_1^2 + v_2^2}, \; M_Z = \frac{1}{2}\sqrt{g_Y^2 + g_2^2}\sqrt{v_1^2 + v_2^2}, \; \sqrt{v_1^2 + v_2^2} = \left(\frac{1}{\sqrt{2}G_F}\right)^{1/2} \simeq 246 \text{ GeV} \tag{8.22}$$

and

$$(\mathbf{m}_e)_{ij} = m_{e_i}\delta_{ij} = \frac{1}{\sqrt{2}}f_{ij}^{e\star}v_1, \; (\mathbf{m}_d)_{ij} = \frac{1}{\sqrt{2}}f_{ij}^{d\star}v_1, \; (\mathbf{m}_u)_{ij} = \frac{1}{\sqrt{2}}f_{ij}^{u\star}v_2 \,, \tag{8.23a}$$

$$f_{ij}^{e\star} = \frac{g_2}{\sqrt{2}M_W \cos\beta}(\mathbf{m}_e)_{ij}, \; f_{ij}^{d\star} = \frac{g_2}{\sqrt{2}M_W \cos\beta}(\mathbf{m}_d)_{ij}, \; f_{ij}^{u\star} = \frac{g_2}{\sqrt{2}M_W \sin\beta}(\mathbf{m}_u)_{ij} \,. \tag{8.23b}$$

The relations in (8.23b) have been obtained by inverting those in (8.23a). The ratio

$$\frac{v_2}{v_1} = \tan\beta \tag{8.24}$$

becomes a free parameter of the theory in so far as fermion masses are concerned.

The left chiral fermionic partners of the Higgs bosons of (8.20) are given by

$$\tilde{h}_{1L} \equiv \begin{pmatrix} \tilde{h}_1^1 \\ \tilde{h}_1^2 \end{pmatrix} = \begin{pmatrix} \tilde{h}_1^0 \\ \tilde{h}_1^- \end{pmatrix}_L ; \; \tilde{h}_{2L} \equiv \begin{pmatrix} \tilde{h}_2^1 \\ \tilde{h}_2^2 \end{pmatrix} = \begin{pmatrix} \tilde{h}_2^+ \\ \tilde{h}_2^0 \end{pmatrix}_L . \tag{8.25}$$

In (8.25) we have defined higgsino fields $\tilde{h}_{1L}^0, \tilde{h}_{1L}^-, \tilde{h}_{2L}^+$ and $\tilde{h}_{2L}^0$, which are two component spinorial fields in the $(\frac{1}{2}, 0)$ representation (cf. §3.2) and identified with $\tilde{h}_1^1, \tilde{h}_1^2, \tilde{h}_2^1$ and $\tilde{h}_2^2$ respectively. Generalizing, we can denote the left chiral superfields containing $h_1, \tilde{h}_{1L}$ and $h_2, \tilde{h}_{2L}$ by H_1, H_2 respectively. So we have

$$H_1 = \begin{pmatrix} H_1^1 \\ H_1^2 \end{pmatrix}, \; H_2 = \begin{pmatrix} H_2^1 \\ H_2^2 \end{pmatrix} \tag{8.26}$$

as the down type, up type Higgs superfields with $Y = -1, 1$ respectively. They are assigned *even* matter parity since they are perceived to be quantalike (cf. Table 4.1). Note that, for quarks and leptons, the need to have a massive Dirac fermion makes it necessary for us to introduce $SU(2)_L$ doublet and singlet chiral superfields. This is unnecessary in the case of the Higgs superfields since $\tilde{h}_{1L}^0$ and $(\tilde{h}_{2L}^0)^C$ can combine to form a four component spinorial field and ditto $\tilde{h}_{1L}^-$ and $(h_{2L}^+)^C$. There is therefore only one four component neutral higgsino field and similarly only one four component charged higgsino field[5]. The two Higgs superfields of (8.26) are thus sufficient. These, together with those in (8.14) – (8.18), comprise all the superfields of MSSM. They are all listed in Tables 8.1a and 8.1b.

---

[5]This is true with unbroken electroweak symmetry. The broken symmetric case is more complicated and will be discussed later.

LEFT CHIRAL MATTER SUPERFIELDS			
Lepton doublets	(color multiplet, T_{3L}, Y)	Quark doublets	(color multiplet, T_{3L}, Y)
$L_1 = \begin{pmatrix} L_{\nu_e} \\ L_e \end{pmatrix}$	$\left(1, \frac{1}{2}, -1\right)$ $\left(1, -\frac{1}{2}, -1\right)$	$Q_1 = \begin{pmatrix} Q_u \\ Q_d \end{pmatrix}$	$\left(3, \frac{1}{2}, \frac{1}{3}\right)$ $\left(3, -\frac{1}{2}, \frac{1}{3}\right)$
$L_2 = \begin{pmatrix} L_{\nu_\mu} \\ L_\mu \end{pmatrix}$	$\left(1, \frac{1}{2}, -1\right)$ $\left(1, -\frac{1}{2}, -1\right)$	$Q_2 = \begin{pmatrix} Q_c \\ Q_s \end{pmatrix}$	$\left(3, \frac{1}{2}, \frac{1}{3}\right)$ $\left(3, -\frac{1}{2}, \frac{1}{3}\right)$
$L_3 = \begin{pmatrix} L_{\nu_\tau} \\ L_\tau \end{pmatrix}$	$\left(1, \frac{1}{2}, -1\right)$ $\left(1, -\frac{1}{2}, -1\right)$	$Q_3 = \begin{pmatrix} Q_t \\ Q_b \end{pmatrix}$	$\left(3, \frac{1}{2}, \frac{1}{3}\right)$ $\left(3, -\frac{1}{2}, \frac{1}{3}\right)$
Antilepton singlets	(color multiplet, T_{3L}, Y)	Antiquark singlets	(color multiplet, T_{3L}, Y)
$\bar{E}_e$	$(1, 0, 2)$	$\bar{U}_1, \bar{D}_1$	$\left(\bar{3}, 0, -\frac{4}{3}\right), \left(\bar{3}, 0, \frac{2}{3}\right)$
$\bar{E}_\mu$	$(1, 0, 2)$	$\bar{U}_2, \bar{D}_2$	$\left(\bar{3}, 0, -\frac{4}{3}\right), \left(\bar{3}, 0, \frac{2}{3}\right)$
$\bar{E}_\tau$	$(1, 0, 2)$	$\bar{U}_3, \bar{D}_3$	$\left(\bar{3}, 0, -\frac{4}{3}\right), \left(\bar{3}, 0, \frac{2}{3}\right)$

Table 8.1a. Matter superfield content of the MSSM.

GAUGE SUPERFIELDS		LEFT CHIRAL HIGGS SUPERFIELDS		
Notation	Name	Doublets	Name	Y
V^Y	Hypercharge	$H_1 = \begin{pmatrix} H_1^0 \\ H_1^- \end{pmatrix}$	Down type	-1
$\vec{V}^W$	Weak isospin			
V_g^a	Color	$H_2 = \begin{pmatrix} H_2^+ \\ H_2^0 \end{pmatrix}$	Up type	1

Table 8.1b. Gauge and Higgs Superfield content of the MSSM.

A question can be raised at this point as to whether one could have been more economical with the contents of superfields in the MSSM. The requirement that all the component fields in each superfield must carry the same internal quantum numbers would quickly convince anyone that the above is necessarily the minimum set. The components of H_1 and L_i, for instance, have the same electromagnetic charges, but they differ in lepton number (including lepton type) and matter parity. We have already given the *raison d'être* for the existence of two Higgs superfield doublets with $Y = -1$ and $Y = 1$, namely the generation of masses for both $T_{3L} = -1/2$ and $T_{3L} = 1/2$ fermions respectively. In fact, even in the supersymmetric extension of a matterless SM (with only gauge and Higgs fields), the two Higgs doublet superfields H_1 and H_2 are necessary for self-consistency. The condition of anomaly cancellation [8.5] in the higgsino sector, a requirement of renormalizability, demands in particular that $\Sigma_{\tilde{h}} Y_{\tilde{h}}^3 = 0$ where $Y_{\tilde{h}}$ is the hypercharge of each higgsino field $\tilde{h}$. Thus one doublet $\tilde{h}_2$ with $Y_{\tilde{h}_2} = 1$ has to be compensated by another $\tilde{h}_1$ doublet with $Y_{\tilde{h}_1} = -1$. (Gauginos, which are another set of new fermions in the supersymmetric theory, are in the safe adjoint representations and do not cause anomaly problems.) We see finally that all the superfields, introduced above and tabulated in Tables 8.1a,b are indeed necessary for the minimal extension of the SM keeping intact its local symmetries, such as electromagnetic charge and color, as well as its global symmetries through the conservation of baryon (B) and lepton (L) number (including lepton type). As stated earlier, the exact conservation of R-parity is an assumed additional requirement. Within the MSSM the assumption of B and L (including lepton type L_i) conservation[6] is equivalent to that of R-parity conservation[7]. But, for superpotential terms and supersymmetry breaking operators in the Lagrangian density, this is a highly constraining requirement.

Of course, states corresponding to all component fields of the superfields, described above, are only 'interaction' eigenstates. In the real world, the absence of mass degenerate particle-sparticle pairs requires supersymmetry to be broken. We shall discuss in the next chapter why such a breaking cannot be spontaneous within the framework of the MSSM itself. Suffice it to say here that it has to be explicit and soft (cf. §7.7). This breaking of supersymmetry in the MSSM can be parametrized in terms of a few explicit soft terms added to the Lagrangian density. We choose the most general terms of this kind. But they are first introduced in an ad hoc manner, though some rationale for them will be given in Chs.12 and 13 on the basis of high scale physics. The contents of these terms will be discussed in detail in Ch.9. Let us remark, for the moment, that they can induce mixing between different sparticles with the same charge and color. Indeed, even without supersymmetry breaking, electroweak symmetry breaking alone causes mixings between gauginos and higgsinos (cf. 5.30). Thus, for instance, charged gauginos mix with charged higgsinos through a 2×2 mixing matrix. The two physical mass eigenstates from that are called **charginos** $\tilde{\chi}_{1,2}^{\pm}$, the subscript 1 (2) conventionally referring to the lighter (heavier) sparticle. A more elaborate discussion will appear in §9.2.

We can immediately see yet another need for two Higgs doublets in this theory. The two doublet superfields H_1, H_2 are left chiral ones and they contain the left chiral higgsinos of (8.25); the conjugate superfields $H_1^\dagger, H_2^\dagger$ contain the corresponding right chiral ones. The left chiral charginos comprise four orthogonal states: the positively charged $\tilde{\chi}_{1L}^+, \tilde{\chi}_{2L}^+$ and the negatively charged $\tilde{\chi}_{1L}^-, \tilde{\chi}_{2L}^-$. Let us define charged gaugino (wino) fields

$$\tilde{\lambda}^{\pm} = \frac{1}{\sqrt{2}} \left(\tilde{\lambda}_1 \mp i\tilde{\lambda}_2 \right) = \tilde{\lambda}_L^{\pm} + \tilde{\lambda}_R^{\pm} , \tag{8.27}$$

where the superscripts $1, 2$ are Cartesian $SU(2)_L$ indices. The massive $\tilde{\chi}_{1L}^+$ and $\tilde{\chi}_{2L}^+$ are orthogonal linear combinations of $\tilde{\lambda}_L^+$ and $\tilde{h}_{2L}^+$ while $\tilde{\chi}_{1L}^-$ and $\tilde{\chi}_{2L}^-$ are formed by similarly

---

[6]Strictly speaking, even B and L (also L_i) are violated at the loop level through anomalies both in the SM and the MSSM, only $\frac{1}{3}B - L_i$ is exactly conserved. But these violations are very tiny in a zero temperature field theory.

[7]This equivalence is not necessarily valid in extensions of the MSSM. Of course, supersymmetric Grand Unified Theories usually violate B and L but may respect R-parity.

combining $\tilde{\lambda}_L^-$ and $\tilde{h}_{1L}^-$. (N.B. there is *no* $\tilde{h}_{2L}^-$ or $\tilde{h}_{1L}^+$!) Correspondingly, the right chiral charginos $\tilde{\chi}_{1R}^-$, $\tilde{\chi}_{2R}^-$ and $\tilde{\chi}_{1R}^+$, $\tilde{\chi}_{2R}^+$ are orthogonal linear combinations of the charge conjugates of the above pairs of gauginos and higgsinos, viz. $\tilde{\lambda}_R^-$, $\tilde{h}_{2R}^-$ and $\tilde{\lambda}_R^+$, $\tilde{h}_{1R}^+$ respectively. Evidently, we require both $\tilde{h}_{2L}^+$ and $\tilde{h}_{1L}^-$, as appear in the two Higgs doublets, otherwise some chargino field, lacking a partner to make a Dirac mass term in the Lagrangian density, would remain massless. Thus we see how the two higgsino doublet fields in the MSSM are used, in combination with the charged winos, to generate two massive Dirac charginos.

Similarly, there is mixing among the neutral gauginos, which can be described by four component Majorana fields. There are two, namely $\tilde{\lambda}_0$ and $\tilde{\lambda}_3$, which mix with the neutral higgsinos $\tilde{h}_2^0$ and $\tilde{h}_1^0$ through a 4×4 mixing matrix. In this case the four physical mass eigenstate Majorana fermions are called **neutralinos** $\tilde{\chi}_i^0$ ($i = 1, \cdots, 4$), the subscripts being monotonically ordered in the direction of increasing mass, by convention. Once again, a detailed description of the mixing among charge neutral gauginos and higgsinos, forming mass eigenstate neutralinos, will be given in §9.2. In fact, similar mixings can occur among different squark generations or among different slepton generations (if lepton type number gets violated) as well. Also, one can (and does) have left right sfermion mixing. Not much more can be said a priori about mixing between different interaction eigenstates in the sparticle sector. These depend on the detailed structure of the supersymmetry breaking terms and their relationship with *EW* symmetry breaking. Such details about sparticle mass eigenstates will be taken up in the next chapter after we have discussed the soft supersymmetry breaking terms at length.

An enumeration has been given below (Table 8.2) of sparticle fields in the minimal globally supersymmetric extension of the SM which follows from the construction described earlier.

Sfermions		Gauginos and higgsinos	
Name	Symbol	Name	Symbol
(left, right) selectron	$\tilde{e}_{L,R}$	gluinos	$\tilde{g}^a$
(left, right) smuon	$\tilde{\mu}_{L,R}$		
(left, right) stau	$\tilde{\tau}_{L,R}$	lighter charginos	$\tilde{\chi}_1^\pm$
e-sneutrino	$\tilde{\nu}_e$		
μ-sneutrino	$\tilde{\nu}_\mu$	heavier charginos	$\tilde{\chi}_2^\pm$
τ-sneutrino	$\tilde{\nu}_\tau$		
(left, right) u-squark	$\tilde{u}_{L,R}$	lightest neutralino	$\tilde{\chi}_1^0$
(left, right) d-squark	$\tilde{d}_{L,R}$		
(left, right) c-squark	$\tilde{c}_{L,R}$	next-to-lightest neutralino	$\tilde{\chi}_2^0$
(left, right) s-squark	$\tilde{s}_{L,R}$		
(left, right) stop	$\tilde{t}_{L,R}$	next-to-heaviest neutralino	$\tilde{\chi}_3^0$
(left, right) sbottom	$\tilde{b}_{L,R}$	heaviest neutralino	$\tilde{\chi}_4^0$

Table 8.2. List of sparticle fields in the MSSM. Antisfermion fields have not been listed.

Sfermions of the third generation are likely to have strong *L-R* mixing; the mass eigenstate sfermion fields are denoted as $\tilde{\tau}_{1,2}$, $\tilde{t}_{1,2}$ and $\tilde{b}_{1,2}$. Antisfermionic fields are denoted by conjugations of sfermionic fields, e.g. $\tilde{e}_{L,R}^\star$ from $\tilde{e}_{L,R}$ and $\tilde{q}_{L,R}^\dagger$ from $\tilde{q}_{L,R}$. However, this is a notation that we shall use for fields only, while an antisfermionic particle – the superpartner of an antifermion – will be labeled $\bar{\tilde{f}}$, i.e. $\bar{\tilde{e}}_L$ for the right spositron and $\bar{\tilde{u}}_L$ for the right u-antisquark. Additional particles and sparticles may be needed by theoretical schemes which go beyond this minimal extension. For instance, the gravitino $\tilde{G}$, which is needed in a spontaneously broken $N{=}1$ supergravity (SUGRA) theory, has not been included here.

8.3 Supersymmetric Part of the MSSM

In this section we will introduce and discuss those interaction and mass terms in the Lagrangian density $\mathcal{L}_{\mathrm{MSSM}}$ which come from the exact supersymmetrization of the SM. Soft interaction terms with mass dimensions less than four as well as mass terms, which describe the heavier masses of sparticles as different from those of their particle partners, arise from supersymmetry breaking and will be addressed in a later section. The general form of the Lagrangian density is

$$\mathcal{L}_{\mathrm{MSSM}} = \mathcal{L}_{\mathrm{SUSY}} + \mathcal{L}_{\mathrm{SOFT}} \tag{8.28}$$

and in this section we will give explicit expressions for $\mathcal{L}_{\mathrm{SUSY}}$ only. In order to write down the supersymmetric interactions among the dynamical fields enumerated in §8.1, we will essentially use the forms of the Lagrangian densities of SQED, SQCD and SχGT of Chapter 5, but covering three families of quarks and leptons. The only really new addition is the contribution from the Higgs sector. The gauge couplings are the same as in the SM. There is no need to give the explicit gauge transformations of the matter superfields enumerated in §8.1. These can be obtained by a straightforward extension of (5.15) and (5.38). But we can decompose the supersymmetric part of the MSSM Lagrangian density as follows:

$$\mathcal{L}_{\mathrm{SUSY}} = \mathcal{L}_g + \mathcal{L}_M + \mathcal{L}_H \ , \tag{8.29}$$

where $\mathcal{L}_g, \mathcal{L}_M$ and $\mathcal{L}_H$ are the pure gauge, matter and Higgs-Yukawa parts respectively.

The **pure gauge part** of $\mathcal{L}_{\mathrm{SUSY}}$ can be written, in terms of field strength spinorial superfields W_g^a, $\vec{W}_W$ and W_Y, constructed respectively via (4.39) and (5.45) from V_g^a, $\vec{V}_W$ and V^Y, according to (5.17) and (5.54):

$$\mathcal{L}_g = \frac{1}{4} \int d^2\theta \left(W_g^{aA} W_{gA}^a + \vec{W}_W^A \cdot \vec{W}_{WA} + W_Y^A W_{YA} \right) + \mathrm{h.c.} \ , \tag{8.30}$$

where the color index a has been summed on repetition. Similarly, the matter contribution can be given by the generalization of (5.62) as

$$\mathcal{L}_M = \int d^4\theta \left[L_i^\dagger \ e^{(g_2 \vec{V}^W \cdot \vec{\tau} + g_Y V^Y Y)} L_i + \bar{E}_i^\dagger \ e^{g_Y V^Y Y} \bar{E}_i + \bar{U}_i^\dagger \ e^{(g_s V_g^a \bar{\lambda}^a + g_Y V^Y Y)} \bar{U}_i \right.$$

$$\left. + \bar{D}_i^\dagger \ e^{(g_s V_g^a \bar{\lambda}^a + g_Y V^Y Y)} \bar{D}_i + Q_i^\dagger \ e^{(g_s V_g^a \lambda^a + g_2 \vec{V}^W \cdot \vec{\tau} + g_Y V^Y Y)} Q_i \right] . \tag{8.31}$$

In (8.31) the Pauli matrices $\vec{\tau}$ act in the weak isospin doublet representation space while the Gell-Mann matrices λ^a (and their complex conjugates $\bar{\lambda}^a$) act in the color triplet **3** (and antitriplet $\bar{\mathbf{3}}$) representation spaces. The subscript i is a family index, summed over $1, 2, 3$ on repetition. Finally, the Higgs contribution can be written as

$$\mathcal{L}_H = \sum_{p=1}^{2} \int d^4\theta \left[H_p^\dagger \, e^{(g_2 \vec{V}_W \cdot \vec{\tau} + g_Y V^Y Y)} H_p + \mathcal{W}_{\text{MSSM}} \delta^{(2)}(\bar{\theta}) + \mathcal{W}_{\text{MSSM}}^\dagger \delta^{(2)}(\theta) \right], \quad (8.32)$$

where the superpotential $\mathcal{W}_{\text{MSSM}}$ is given by

$$\mathcal{W}_{\text{MSSM}} = \mu H_1 \cdot H_2 - f_{ij}^e H_1 \cdot L_i \bar{E}_j - f_{ij}^d H_1 \cdot Q_i \bar{D}_j - f_{ij}^u Q_i \cdot H_2 \bar{U}_j . \quad (8.33)$$

(We use the notation $A \cdot B \equiv \epsilon_{DE} A^D B^E$ for two $SU(2)$-doublet superfields or fields A, B with D, E being indices in the doublet representation space with the same superscript/subscript conventions as for two component spinors in Ch.3). The signs in (8.33) have been chosen so that the f_{ij}'s here are the same as of those in (8.9) and (8.10), as can be checked by use of (5.3) and (3.28a,b). The second, third and fourth terms in the RHS of (8.33) are just the supersymmetric generalizations of the Yukawa couplings in (8.9) and (8.10). Only the first RHS term of (8.33) is new. This term, containing the parameter μ, which has the dimension of mass, can be thought of as a supersymmetric generalization of a higgsino mass term. We shall later see that a consistent incorporation of spontaneous electroweak symmetry breakdown requires μ to be of the order of the weak scale. The choice of terms in $\mathcal{W}_{\text{MSSM}}$ has been constrained by the **requirement** of R-parity (R_p) conservation (cf. §4.5) which is one of the assumptions of the MSSM. Let us remark here that, since baryon number B and lepton number L are conserved in the SM Lagrangian, the conservation of R_p may be posited as a natural assumption in a minimal supersymmetric extension of the SM which may be expected to preserve the conservation laws of the latter. Additional terms, that are gauge invariant with respect to SM gauge transformations, could be admitted to the RHS of (8.33) if R-parity were violated explicitly. We postpone a discussion of this possibility to Ch.14. For the moment, we take the **conservation of R-parity to be a central assumption** of the MSSM. The terms in $\mathcal{L}_{\text{MSSM}}$, that are generated from $\mathcal{W}_{\text{MSSM}}$, are obtained from a generalization of (5.5) with the Higgs VEVs from (8.21) taken into account to properly incorporate spontaneous electroweak symmetry breaking.

Let us concentrate first on the auxiliary F and D fields following from (8.31)-(8.33). By use of (5.56c), we can identify seventeen (including $i = 1, 2, 3$) F fields from (8.33). For the $SU(2)$ doublet representation space, we can employ the two spinor subscript/superscript notation of Ch.3, i.e. $H_{1D} = \epsilon_{DE} H_1^E$ and $F_{H_1}^{\star D} = -\partial \mathcal{W}/\partial H_{1D}|$ etc. This enables us to write

$$F_{H_1}^{\star D} = -\mu h_2^D + f_{ij}^e \tilde{e}_{jR}^\star \tilde{\ell}_{iL}^D + f_{ij}^d \tilde{d}_{jR}^\dagger \tilde{q}_{iL}^D , \quad (8.34a)$$

$$F_{H_2}^{\star D} = \mu h_1^D - f_{ij}^u \tilde{u}_{jR}^\dagger \tilde{q}_{iL}^D , \quad (8.34b)$$

$$F_{L_i}^{\star D} = -f_{ij}^e h_1^D \tilde{e}_{jR}^\star , \quad (8.34c)$$

$$F_{\bar{E}_i}^\star = f_{ji}^e h_1 \cdot \tilde{\ell}_{jL} , \quad (8.34d)$$

$$F_{Q_{i\alpha}}^{\star D} = -f_{ij}^d h_1^D \tilde{d}_{jR\alpha}^\dagger + f_{ij}^u h_2^D \tilde{u}_{jR\alpha}^\dagger , \quad (8.34e)$$

$$F_{\bar{D}_{i\alpha}}^\star = f_{ji}^d h_1 \cdot \tilde{q}_{jL\alpha} , \quad (8.34f)$$

$$F^{\star}_{\tilde{U}_{i\alpha}} \;=\; f^{u}_{ji}\tilde{q}_{jL\alpha}\!\cdot\!h_2 \,. \tag{8.34g}$$

In (8.34e-g) the subscript α is the floating color index, whereas in (8.34a,b) appropriate color contractions are implied. Now the three D fields, corresponding to the three factors $U(1)_Y, SU(2)_L$ and $SU(3)_C$ of the gauge group and ignoring a possible field independent term in D_Y, cf. (5.22b), are given respectively from (5.56c) by

$$D^Y \;=\; -\frac{1}{2}g_Y\left(h_2^\dagger h_2 - h_1^\dagger h_1 + \frac{1}{3}\tilde{q}_{iL}^\dagger\tilde{q}_{iL} - \frac{4}{3}\tilde{u}_{iR}\tilde{u}_{iR}^\dagger + \frac{2}{3}\tilde{d}_{iR}\tilde{d}_{iR}^\dagger\right.$$
$$\left. -\tilde{\ell}_{iL}^\dagger\tilde{\ell}_{iL} + 2\tilde{e}_{iR}\tilde{e}_{iR}^\star\right), \tag{8.35a}$$

$$\vec{D} \;=\; -\frac{1}{2}g_2\left(h_1^\dagger\vec{\tau}h_1 + h_2^\dagger\vec{\tau}h_2 + \tilde{q}_{iL}^\dagger\vec{\tau}\tilde{q}_{iL} + \tilde{\ell}_{iL}^\dagger\vec{\tau}\tilde{\ell}_{iL}\right), \tag{8.35b}$$

$$D^a \;=\; -\frac{1}{2}g_s\left(\tilde{q}_{iL}^\dagger\lambda^a\tilde{q}_{iL} + \tilde{u}_{iR}^T\bar{\lambda}^a\tilde{u}_{iR}^\star + \tilde{d}_{iR}^T\bar{\lambda}^a\tilde{d}_{iR}^\star\right)$$
$$=\; -\frac{1}{2}g_s\left(\tilde{q}_{iL}^\dagger\lambda^a\tilde{q}_{iL} + \tilde{u}_{iR}^\dagger\lambda^a\tilde{u}_{iR} + \tilde{d}_{iR}^\dagger\lambda^a\tilde{d}_{iR}\right). \tag{8.35c}$$

Here a is a color index and we have utilized the hermiticity of λ^a in the last step. It may be noted that, in both (8.35a) and (8.35c), $\tilde{u}_{iR}^\star$, $\tilde{d}_{iR}^\star$ and $\tilde{e}_{iR}^\star$ are the equivalents of ϕ in (5.56c). Finally, the supersymmetric scalar potential is given (cf. 5.56b) by

$$V_{\text{SUSY}} = F_k^\star F_k + \frac{1}{2}\left[\vec{D}^2 + (D^Y)^2 + D^a D^a\right]. \tag{8.36}$$

k referring to the type of superfield (including any internal symmetry index) and repeated k and a being summed.

The interaction part can be written down in terms of component fields in four component notation in much the same way as shown in Ch.5. The major difference now is that we want to incorporate the spontaneous electroweak symmetry breakdown $SU(2)_L \times U(1)_Y \to U(1)_{em}$ and obtain the consequent mass terms and mass eigenstates, i.e. equivalents of (8.5) to (8.10). Let us consider non-Higgs vertices for the moment. We postpone all discussions of interactions involving Higgs bosons to Ch.10; in particular, these include Yukawa, Higgs-gauge and Higgs-Higgs interactions and some of their supersymmetric generalizations. Furthermore, those vertices with physical sparticles, which involve supersymmetry breaking, will be treated in Ch.9. In the next section of this chapter we consider (A) fermion-fermion-gauge boson, (B) triple gauge boson, and (C) quadruple gauge boson vertices in the Standard Model. We also discuss from the MSSM those (D) sfermion-sfermion-gauge boson, (E) gauge boson-gaugino-gaugino, (F) fermion-sfermion-gaugino, (G) gauge boson-gauge boson-sfermion-sfermion and (H) sfermion quartic vertices which have to do with only the purely supersymmetric part of $\mathcal{L}_{\text{MSSM}}$ and without left right mixing. Some subsets of these as well as other non-Higgs vertices crucially involve supersymmetry breaking and both generation as well as left right mixing in a physical situation. Those will be covered in Ch.9, which will contain the corresponding final physical vertices with the said mixings.

8.4   Some Non-Higgs Vertices of the MSSM

First, we recount the non-Higgs SM vertices in (A), (B) and (C). Subsections (D), (E), (F) and (G) contain the new supersymmetric extensions.

(A) *Fermion-fermion-gauge boson vertices*

   We can discuss the strong and electroweak vertices separately.

(i)   Quark-quark-gluon vertices

These are the same as in QCD, vide §5.5. The only additional remark is that all six flavors of quarks ($p = u, d, c, s, t, b$) have to be included with all interactions being diagonal in flavour space. Thus, with p henceforth summed on repetition, we have

$$\mathcal{L}_{q\bar{q}g} = -g_s g_\mu^a \bar{q}_p T^a \gamma^\mu q_p \ .$$

This form is valid in any basis for the quarks that can be reached from the current basis by a unitary rotation in generation space. Thus flavor mixings of quark mass eigenstates are inconsequential here.

(ii)   Fermion-fermion-electroweak vector boson vertices

These follow exactly those given in §5.6. The only differences arise on account of $\gamma - Z$ mixing, cf. (8.7) and (8.8). Furthermore, one has to replicate for three generations. We can now employ the notation of (5.65), understanding f_{u_i,d_i} to be either a quark or a lepton of generation i with $f_{u_i L} = P_L f_{u_i}$, $f_{u_i R} = P_R f_{u_i}$, $f_{d_i L} = P_L f_{d_i}$, $f_{d_i R} = P_R f_{d_i}$ and $P_{L,R} = \frac{1}{2}(1 \mp \gamma_5)$. Then, with f, f' and v respectively chosen as two fermions and one EW vector boson generically, we can write

$$
\begin{aligned}
\mathcal{L}_{f\bar{f}'v} = {}& -\frac{g_2}{\sqrt{2}} \left(W_\mu^+ \bar{f}_{u_i} \gamma^\mu P_L f_{d_i} + W_\mu^- \bar{f}_{d_i} \gamma^\mu P_L f_{u_i} \right) \\[2mm]
& -eA_\mu \left(Q_{f_u} \bar{f}_{u_i} \gamma^\mu f_{u_i} + Q_{f_d} \bar{f}_{d_i} \gamma^\mu f_{d_i} \right) - \frac{g_2}{2\cos\theta_W} Z_\mu \cdot \\[2mm]
& \left[\bar{f}_{u_i} \gamma^\mu \left\{ (1 - 2Q_{f_u} \sin^2\theta_W)P_L - 2Q_{f_u}\sin^2\theta_W P_R \right\} f_{u_i} \right. \\[2mm]
& \left. - \bar{f}_{d_i} \gamma^\mu \left\{ (1 + 2Q_{f_d}\sin^2\theta_W)P_L + 2Q_{f_d}\sin^2\theta_W P_R \right\} f_{d_i} \right] .
\end{aligned}
\tag{8.37}
$$

In (8.37) Q_{f_u} and Q_{f_d} are the electromagnetic charges of the up type and down type fermions f_{u_i} and f_{d_i} respectively in units of the charge of the positron. Thus $Q_{u_i} = \frac{2}{3}$, $Q_{d_i} = -\frac{1}{3}$, $Q_{e_i} = -1$, $Q_{\nu_i} = 0$. Referring back to (8.3) and comparing with (5.65), we note that for quarks, we can write

$$q_{iL} = \begin{pmatrix} f_{u_i L} \\ f_{d_i L} \end{pmatrix}, \quad u_{iR} = f_{u_i R}, \quad d_{iR} = f_{d_i R} \ . \tag{8.38}$$

Similarly, for leptons, the notation is

$$\ell_{iL} = \begin{pmatrix} f_{u_{iL}} \\ f_{d_{iL}} \end{pmatrix}, \quad e_{iR} = f_{d_{iR}}. \tag{8.39}$$

However, the above quarks are gauge interaction or "current" basis eigenstates. When we go to physical mass eigenstates, we will need to incorporate the Cabibbo-Kobayashi-Maskawa (CKM) matrix for charged current couplings in the quark sector. This part can be written as

$$\mathcal{L}_{q\bar{q}'W^{\pm}} = -\frac{g_2}{\sqrt{2}} \left(W_{\mu}^{+} \bar{u}_i \gamma^{\mu} P_L V_{ij}^{q_L} d_j + \text{h.c.} \right), \tag{8.40}$$

where the u_i, d_j etc. are now *understood to be mass eigenstate* quark fields. In (8.40), $V_{ij}^{q_L}$ are the elements of the CKM matrix [8.1] $\mathbf{V}^{q_L} = \mathbf{U}^{u_L^{\dagger}} \mathbf{U}^{d_L}$ in the notation of (8.12). Electromagnetic and neutral current vertices, of course, do not involve these on account of the GIM mechanism [8.1]. Finally, the vertices and Feynman rules for $i\mathcal{L}_{ff'v}$ can be written as in Fig. 8.1 below.

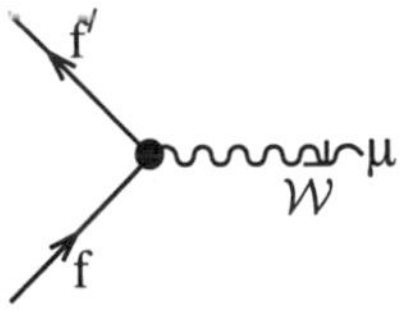

$$\mathcal{W} = W^-, f = d_j, f' = u_i \quad -\frac{ig_2}{\sqrt{2}} \gamma_{\mu} P_L V_{ij}^{q_L}$$

$$\mathcal{W} = W^+, f = u_j, f' = d_i \quad -\frac{ig_2}{\sqrt{2}} \gamma_{\mu} P_L V_{ij}^{q_L^{\dagger}}$$

$$\mathcal{W} = W^+, f = \nu_j, f' = e_i \quad -\frac{ig_2}{\sqrt{2}} \gamma_{\mu} P_L \delta_{ij}$$

$$\mathcal{W} = W^-, f = e_j, f' = \nu_i \quad -\frac{ig_2}{\sqrt{2}} \gamma_{\mu} P_L \delta_{ij}$$

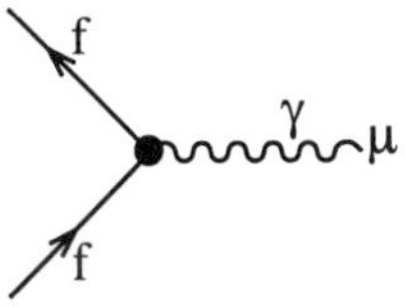

$$-ieQ_f \gamma_{\mu}$$

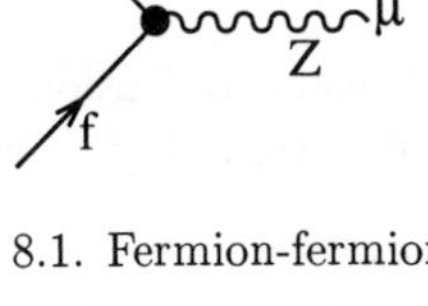

$$-\frac{ig_2}{\cos\theta_W} T_{3L}^f \left[(1 - 4T_{3L}^f Q_f \sin^2\theta_W)\gamma_{\mu} P_L - 4T_{3L}^f Q_f \sin^2\theta_W \gamma_{\mu} P_R \right]$$

$$\equiv -\frac{ig_2}{\cos\theta_W} (g_L^f \gamma_{\mu} P_L - g_R^f \gamma_{\mu} P_R)$$

Fig. 8.1. Fermion-fermion-electroweak vector boson vertices with T_{3L}^f, Q_f as in (8.1).

Note that in the lowermost vertex g_L^f stands for $T_{3L}^f(1 - 4T_{3L}^f Q_f \sin^2\theta_W)$ and g_R^f for $4(T_{3L}^f)^2 Q_f \sin^2\theta_W$.

(B) *Triple gauge boson vertices*

Again, the strong and electroweak cases can be distinguished.

(i) Triple gluon vertex

This is exactly the same as in QCD, vide (5.60) and Fig. 5.2.

(ii) Triple electroweak vector boson vertices

These are generalized from the $W^+W^-W^3$ vertex of SχGT in §5.6, as shown in Fig. 8.2.

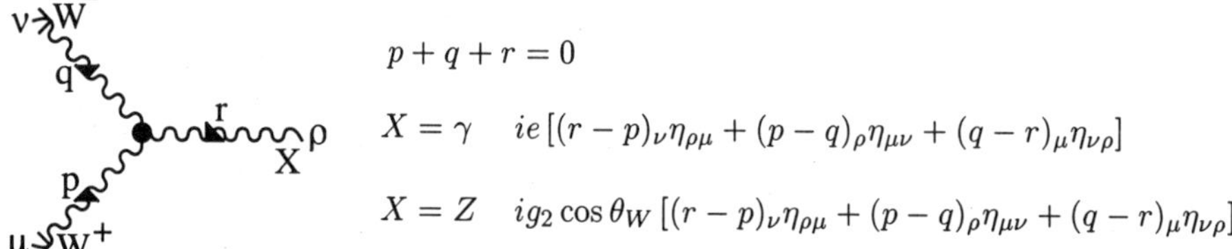

$$p + q + r = 0$$

$$X = \gamma \quad ie\,[(r-p)_\nu \eta_{\rho\mu} + (p-q)_\rho \eta_{\mu\nu} + (q-r)_\mu \eta_{\nu\rho}]$$

$$X = Z \quad ig_2 \cos\theta_W\,[(r-p)_\nu \eta_{\rho\mu} + (p-q)_\rho \eta_{\mu\nu} + (q-r)_\mu \eta_{\nu\rho}]$$

Fig. 8.2. Triple electroweak vector boson vertices

(C) *Quadruple gauge boson vertices*

Once more, we can consider the strong and electroweak vertices in different categories.

(i) Quadruple gluon vertex

This is identical to that in QCD, vide §5.5 and Fig. 5.2.

(ii) Quadruple electroweak vector boson vertices

The $W^+W^-W^+W^-$ vertex is identical to that given in Fig. 5.3. The $W^+W^-W^3W^3$ vertex, shown there, generalizes to three cases here, as given in Fig. 8.3.

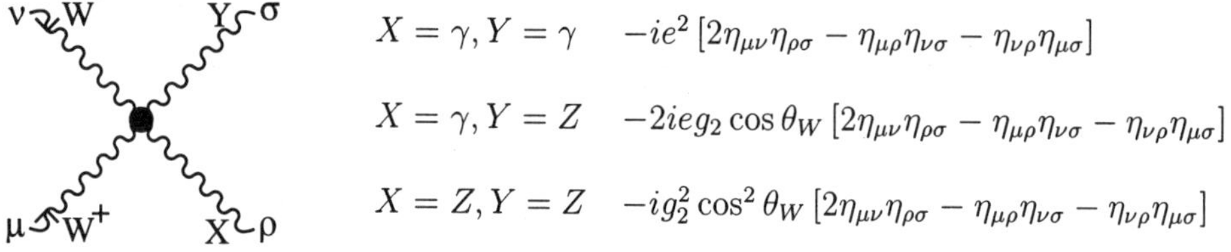

$$X = \gamma, Y = \gamma \quad -ie^2\,[2\eta_{\mu\nu}\eta_{\rho\sigma} - \eta_{\mu\rho}\eta_{\nu\sigma} - \eta_{\nu\rho}\eta_{\mu\sigma}]$$

$$X = \gamma, Y = Z \quad -2ieg_2 \cos\theta_W\,[2\eta_{\mu\nu}\eta_{\rho\sigma} - \eta_{\mu\rho}\eta_{\nu\sigma} - \eta_{\nu\rho}\eta_{\mu\sigma}]$$

$$X = Z, Y = Z \quad -ig_2^2 \cos^2\theta_W\,[2\eta_{\mu\nu}\eta_{\rho\sigma} - \eta_{\mu\rho}\eta_{\nu\sigma} - \eta_{\nu\rho}\eta_{\mu\sigma}]$$

Fig. 8.3. Quadruple electroweak vector boson vertices

(D) *Sfermion-sfermion-gauge boson vertices*

We shall work in the $\tilde{f}_L$-$\tilde{f}_R$ basis, deferring a discussion of left right sfermion mixing to Ch.9. Again, we can consider two cases, pertaining to strong and electroweak interactions.

(i) Squark-squark-gluon vertices

These are the same as the SQCD (vide §5.5) except for the generalization to six diagonal flavors (index p summed on repetition). Then, in the notation of §5.5, we can write

$$\mathcal{L}_{\tilde{q}\tilde{q}g} = -2ig_S A_\mu^a \tilde{q}_p^\star T^a [\partial^\mu]\tilde{q}_p \,,$$

where the operator $[\partial^\mu]$ is as defined in (4.28). Each vertex is precisely the same as that in Fig. 5.2 with $\tilde{q}$ generalized to $\tilde{q}_p$. As with quarks, neither flavor nor left right mixing in squark mass eigenstates will matter here.

(ii) Sfermion-sfermion-electroweak vector boson vertices

Once more, we generalize from the corresponding interactions of $S\chi GT$ in §5.6 and write (with $v = W, Z, \gamma$ and $Q_{f_{u,d}}$ as electric charge of $f_{u,d}$ in units of the positron charge) the terms containing an electroweak vector boson as follows:

$$
\mathcal{L}_{\tilde{f}\tilde{f}'v} = -i\sqrt{2}g_2 \left\{ W_\mu^+ \tilde{f}_{u_iL}^\star [\partial^\mu] \tilde{f}_{d_iL} + W_\mu^- \tilde{f}_{d_iL}^\star [\partial^\mu] \tilde{f}_{u_iL} \right\}
$$

$$
-2ieA_\mu \left\{ q_{f_u} \left(\tilde{f}_{u_iL}^\star [\partial^\mu] \tilde{f}_{u_iL} + \tilde{f}_{u_iR}^\star [\partial^\mu] \tilde{f}_{u_iR} \right) + q_{f_d} \left(\tilde{f}_{d_iL}^\star [\partial^\mu] \tilde{f}_{d_iL} + \tilde{f}_{d_iR}^\star [\partial^\mu] \tilde{f}_{d_iR} \right) \right\}
$$

$$
-\frac{ig_2}{\cos\theta_W} Z_\mu \Big\{ \tilde{f}_{u_iL}^\star \left(1 - 2Q_{f_u} \sin^2\theta_W \right) [\partial^\mu] \tilde{f}_{u_iL} - 2Q_{f_u} \sin^2\theta_W \tilde{f}_{u_iR}^\star [\partial^\mu] \tilde{f}_{u_iR}
$$

$$
-\tilde{f}_{d_iL}^\star \left(1 + 2Q_{f_d} \sin^2\theta_W \right) [\partial^\mu] \tilde{f}_{d_iL} - 2Q_{f_d} \sin^2\theta_W \tilde{f}_{d_iR}^\star [\partial^\mu] \tilde{f}_{d_iR} \Big\}, \quad (8.41)
$$

with the repeated generation index i summed. As done for quarks and leptons, the expressions

$$
\tilde{q}_{iL} = \begin{pmatrix} \tilde{f}_{u_iL} \\ \tilde{f}_{d_iL} \end{pmatrix}, \quad \tilde{u}_{iR} = \tilde{f}_{u_iR}, \quad \tilde{d}_{iR} = \tilde{f}_{d_iR} \tag{8.42}
$$

can be written for squark fields and

$$
\tilde{\ell}_{iL} = \begin{pmatrix} \tilde{f}_{u_iL} \\ \tilde{f}_{d_iL} \end{pmatrix}, \quad \tilde{e}_{iR} = \tilde{f}_{d_iR} \tag{8.43}
$$

for slepton ones. For charged current couplings of mass eigenstate squarks, we can, in analogy with (8.40), use left chiral flavor rotation matrix elements $V_{ij}^{\tilde{q}_L}$ in generation space. Here we have used a symbol different[8] from that of the CKM matrix $\mathbf{V}^{q_L}$ to take account of the general situation with supersymmetry breaking which may make $\mathbf{V}^{\tilde{q}_L} \neq \mathbf{V}^{q_L}$. Again, in the coupling of the charged W to two sleptons of different flavor too, to account for different generation dependent masses for charged sleptons and sneutrinos, we put in the left chiral flavor rotation matrix element $V_{ij}^{\tilde{\ell}_L}$ in generation space, though such a matrix element is absent in the leptonic sector. Thus we have

$$
\mathcal{L}_{\tilde{q}\tilde{q}'W} = -i\sqrt{2}g_2 \left\{ W_\mu^+ \tilde{u}_{iL}^\star V_{ij}^{\tilde{q}_L} [\partial^\mu] \tilde{d}_{jL} + \text{h.c.} \right\}, \tag{8.44a}
$$

$$
\mathcal{L}_{\tilde{\ell}\tilde{\ell}'W} = -i\sqrt{2}g_2 \left\{ W_\mu^+ \tilde{\nu}_{iL}^\star V_{ij}^{\tilde{\ell}_L} [\partial^\mu] \tilde{e}_{jL} + \text{h.c.} \right\}, \tag{8.44b}
$$

$$
\mathcal{L}_{\tilde{f}\tilde{f}\gamma} = -2ieA_\mu Q_{\tilde{f}} \left(\tilde{f}_{iL}^\star [\partial^\mu] \tilde{f}_{iL} + \tilde{f}_{iR}^\star [\partial^\mu] \tilde{f}_{iR} \right), \tag{8.44c}
$$

$$
\mathcal{L}_{\tilde{f}\tilde{f}Z} = -\frac{ig_2}{\cos\theta_W} Z_\mu \left\{ 2T_{3L}^{\tilde{f}} \left(1 - 4T_{3L}^{\tilde{f}} Q_{\tilde{f}} \sin^2\theta_W \right) \tilde{f}_{iL}^\star [\partial^\mu] \tilde{f}_{iL} - 2\sin^2\theta_W Q_{\tilde{f}} \tilde{f}_{iR}^\star [\partial^\mu] \tilde{f}_{iR} \right\}.
$$

$$
\tag{8.44d}
$$

---

[8]Of course, in the limit of exact supersymmetry, $\mathbf{V}^{\tilde{q}_L}$ equals $\mathbf{V}^{q_L}$ of (8.40) and $\mathbf{V}^{\tilde{\ell}_L}$ becomes the unit matrix.

Here $\tilde{f}_{i(L,R)}$ covers $\tilde{u}_{i(L,R)}$, $\tilde{d}_{i(L,R)}$, $\tilde{e}_{i(L,R)}$ and $\tilde{\nu}_{iL}$. Eq. (8.44) describes the sfermion-sfermion-EW vector boson couplings with the left chiral or right chiral squark and slepton fields understood as mass eigenstates in the limit of no left right sfermionic mixing. The latter, to be treated in Ch.9, will generate additional complications in these equations except for the photon vertex. We defer an enumeration of the final physical vertices and Feynman rules in this case till that discussion.

(E) *Gauge boson-gaugino-gaugino vertices*

Here also strong and electroweak vertices are distinctly separate.

(1) **Gluon-gluino-gluino vertices**

These are identical to those in SQCD, as discussed in §5.5 (vide Fig. 5.2).

(2) **EW gauge boson-neutralino/chargino-neutralino/chargino vertices**

Even in the supersymmetric limit, these will not be similar to those of SχGT, Fig. 5.3. This is because mass eigenstate charginos and neutralinos will involve combinations of gauginos and higgsinos on account of the breakdown of EW symmetry. Moreover, in reality, supersymmetry breaking has a significant influence. We shall discuss those aspects in detail in Ch.9, and give the final physical vertices there.

(F) *Fermion-sfermion-gaugino vertices*

(1) **Quark-squark-gluino vertex**

In the supersymmetric approximation of neglecting the differences between squark flavor rotations and quark flavor rotations from eigenstates of mass to those of gauge interactions, these vertices will be the same as in SQCD (§5.5, Fig. 5.2). In reality, however, these differences need to be recognized. In a broken supersymmetric world the two flavor rotation matrices will be different. We have already introduced the matrices $\mathbf{U}^{u_{L,R}}$ and $\mathbf{U}^{d_{L,R}}$ in (8.12) for quarks. Let us define analogous flavor rotation matrices $\mathbf{U}^{\tilde{u}_{L,R}}$ and $\mathbf{U}^{\tilde{d}_{L,R}}$ for u- and d-squarks respectively. These take mass diagonal squarks to flavor eigenstate ones. If we define u, d as three component column vectors in color space, then the relevant terms in (5.60) can be generalized to

$$\mathcal{L}_{\bar{q}\tilde{q}\tilde{g}} = -\sqrt{2}g_s \left[\bar{u}_i P_R T^a \tilde{g}^a \left(\mathbf{U}^{u_L^\dagger} \mathbf{U}^{\tilde{u}_L} \right)_{ij} \tilde{u}_{jL} - \tilde{u}_{iR}^\dagger \left(\mathbf{U}^{\tilde{u}_R^\dagger} \mathbf{U}^{u_R} \right)_{ij} \bar{\tilde{g}}^a T^a P_R u_j \right]$$

$$-\sqrt{2}g_s \left[\bar{d}_i P_R T^a \tilde{g}^a \left(\mathbf{U}^{d_L^\dagger} \mathbf{U}^{\tilde{d}_L} \right)_{ij} \tilde{d}_{jL} - \tilde{d}_{iR}^\dagger \left(\mathbf{U}^{\tilde{d}_R^\dagger} \mathbf{U}^{d_R} \right)_{ij} \bar{\tilde{g}}^a T^a P_R d_j \right] + \text{h.c.}, (8.45)$$

i, j being generation indices. The interaction $\mathcal{L}_{\bar{\tilde{q}}\tilde{q}\tilde{g}}$ is obviously the hermitian conjugate of (8.45). These forms are valid in the *absence of left-right mixing* for squarks. The corresponding physical vertices will be given Ch.9 after accounting for the latter.

(2) Fermion-sfermion-neutralino/chargino vertices

Once again, we postpone a treatment of these to Ch.9 because of their essential dependence on supersymmetry breaking via both flavor rotations and gaugino higgsino mixings.

(G) *Gauge boson-gauge boson-sfermion-sfermion vertices*

These will be given in three categories since there are mixed strong and electroweak vertices apart from purely strong and purely electroweak ones.

(i) Gluon-gluon-squark-squark vertex

Since the two squarks at this four point vertex have the same flavor, the corresponding flavor rotations cancel out. Thus this vertex is exactly the same as in SQCD (§5.5, Fig. 5.2) with a trivial flavor generalization $\tilde{q} \to \tilde{q}_i$.

(ii) Electroweak vector boson-electroweak vector boson-sfermion-sfermion vertices

These are present in SχGT and can, therefore, be read off from (5.64) with $\tilde{f}$ generalized to cover three generations of sleptons and squarks. The neutral gauge bosons W^3 and B get transformed to Z and A via (8.7b,c). As before, we work in the $\tilde{f}_L$-$\tilde{f}_R$ basis, neglecting left right sfermion mixing for the moment. The corresponding interaction terms in the Lagrangian density can be written as

$$
\mathcal{L}_{\tilde{f}\tilde{f}'\mathrm{vv'}} =
$$

$$
\frac{g_2^2}{2} W_\mu^+ W^{\mu-} \left(\tilde{f}_{u_{iL}}^\star \tilde{f}_{u_{iL}} + \tilde{f}_{d_{iL}}^\star \tilde{f}_{d_{iL}} \right)
$$

$$
+ \frac{g_2}{\sqrt{2}} \left(eA_\mu - \frac{g_2 \sin^2\theta_W}{\cos\theta_W} Z_\mu \right) Y_{f_L} \left(\tilde{f}_{u_{iL}}^\star \tilde{f}_{d_{iL}} W^{\mu+} + \tilde{f}_{d_{iL}}^\star \tilde{f}_{u_{iL}} W^{\mu-} \right)
$$

$$
+ e^2 A_\mu A^\mu \left\{ Q_{f_u}^2 \left(\tilde{f}_{u_{iL}}^\star \tilde{f}_{u_{iL}} + \tilde{f}_{u_{iR}}^\star \tilde{f}_{u_{iR}} \right) + Q_{f_d}^2 \left(\tilde{f}_{d_{iL}}^\star \tilde{f}_{d_{iL}} + \tilde{f}_{d_{iR}}^\star \tilde{f}_{d_{iR}} \right) \right\}
$$

$$
+ \frac{g_2^2}{4\cos^2\theta_W} Z_\mu Z^\mu \left\{ \tilde{f}_{u_{iL}}^\star \left(1 - 2Q_{f_u} \sin^2\theta_W \right)^2 \tilde{f}_{u_{iL}} + 4Q_{f_u}^2 \sin^4\theta_W \tilde{f}_{u_{iR}}^\star \tilde{f}_{u_{iR}} \right.
$$

$$
\left. + \tilde{f}_{d_{iL}}^\star \left(1 + 2Q_{f_d} \sin^2\theta_W \right)^2 \tilde{f}_{d_{iL}} + 4Q_{f_d}^2 \sin^4\theta_W \tilde{f}_{d_{iR}}^\star \tilde{f}_{d_{iR}} \right\}
$$

$$
+ \frac{g_2 e}{\cos\theta_W} A_\mu Z^\mu \left[Q_{f_u} \left\{ \tilde{f}_{u_{iL}}^\star \left(1 - 2Q_{f_u} \sin^2\theta_W \right) \tilde{f}_{u_{iL}} - 2Q_{f_u} \sin^2\theta_W \tilde{f}_{u_{iR}}^\star \tilde{f}_{u_{iR}} \right\} \right.
$$

$$
\left. - Q_{f_d} \left\{ \tilde{f}_{d_{iL}}^\star \left(1 + 2Q_{f_d} \sin^2\theta_W \right) \tilde{f}_{d_{iL}} + 2Q_{f_d} \sin^2\theta_W \tilde{f}_{d_{iR}}^\star \tilde{f}_{d_{iR}} \right\} \right].
$$

$$
(8.46)
$$

A summation over the generation index i is understood. The relations between $\tilde{f}_{u_{iL,R}}$, $\tilde{f}_{d_{iL,R}}$ and the corresponding squark/slepton flavor eigenstate fields are given in (8.42-3). Elements of the CKM type matrix $\mathbf{V}^{\tilde{q}_L}$ arising out of flavor rotations between sfermionic mass and flavor eigenstate fields (cf. 8.44), enter the $\bar{\tilde{f}}\tilde{f}'\gamma W$ and $\bar{\tilde{f}}\tilde{f}'ZW$ interactions but not the[9] $\bar{\tilde{f}}\tilde{f}\gamma\gamma$, $\bar{\tilde{f}}\tilde{f}ZZ$ and $\bar{\tilde{f}}\tilde{f}\gamma Z$ ones. All these can be rewritten in terms of mass eigenstate sfermion fields. They then read

$$\mathcal{L}_{\bar{\tilde{q}}\tilde{q}'\gamma W} = \frac{g_2 e}{3\sqrt{2}} A_\mu \left(\tilde{u}_{iL}^\dagger V_{ij}^{\tilde{q}_L} \tilde{d}_{jL} W^{\mu+} + \tilde{d}_{iL}^\dagger V_{ij}^{\tilde{q}_L^\dagger} \tilde{u}_{jL} W^{\mu-} \right), \tag{8.47a}$$

$$\mathcal{L}_{\bar{\tilde{q}}\tilde{q}'ZW} = -\frac{g_2^2 \sin^2\theta_W}{3\sqrt{2}\cos\theta_W} Z_\mu \left(\tilde{u}_{iL}^\dagger V_{ij}^{\tilde{q}_L} \tilde{d}_{jL} W^{\mu+} + \tilde{d}_{iL}^\dagger V_{ij}^{\tilde{q}_L^\dagger} \tilde{u}_{jL} W^{\mu-} \right), \tag{8.47b}$$

$$\mathcal{L}_{\bar{\tilde{\ell}}\tilde{\ell}'\gamma W} = -\frac{g_2 e}{\sqrt{2}} A_\mu \left(\tilde{\nu}_{iL}^\star V_{ij}^{\tilde{\ell}_L} \tilde{e}_{jL} W^{\mu+} + \tilde{e}_{iL}^\star V_{ij}^{\tilde{\ell}_L^\dagger} \tilde{\nu}_{jL} W^{\mu-} \right), \tag{8.47c}$$

$$\mathcal{L}_{\bar{\tilde{\ell}}\tilde{\ell}'ZW} = \frac{g_2^2 \sin^2\theta_W}{\sqrt{2}\cos\theta_W} Z_\mu \left(\tilde{\nu}_{iL}^\star V_{ij}^{\tilde{\ell}_L} \tilde{e}_{jL} W^{\mu+} + \tilde{e}_{iL}^\star V_{ij}^{\tilde{\ell}_L^\dagger} \tilde{\nu}_{jL} W^{\mu-} \right), \tag{8.47d}$$

$$\mathcal{L}_{\bar{\tilde{f}}\tilde{f}\gamma\gamma} = e^2 A_\mu Q_{\tilde{f}}^2 \left(\tilde{f}_{iL}^\dagger \tilde{f}_{iR} + \tilde{f}_{iL}^\dagger \tilde{f}_{iR} \right) A^\mu, \tag{8.47e}$$

$$\mathcal{L}_{\bar{\tilde{f}}\tilde{f}ZZ} = \frac{g_2^2}{4\cos^2\theta_W} Z_\mu \Big\{ \left(1 - 4T_{3L}^{\tilde{f}} Q_{\tilde{f}} \sin^2\theta_W \right)^2 \tilde{f}_{iL}^\dagger \tilde{f}_{iL}$$
$$+ 4Q_{\tilde{f}}^2 \sin^4\theta_W \tilde{f}_{iR}^\dagger \tilde{f}_{iR} \Big\} Z^\mu, \tag{8.47f}$$

$$\mathcal{L}_{\bar{\tilde{f}}\tilde{f}\gamma Z} = \frac{2g_2 e}{\cos\theta_W} A_\mu Q_{\tilde{f}} \Big\{ T_{3L}^{\tilde{f}} \left(1 - 4T_{3L}^{\tilde{f}} Q_{\tilde{f}} \sin^2\theta_W \right) \tilde{f}_{iL}^\dagger \tilde{f}_{iL}$$
$$- Q_{\tilde{f}} \sin^2\theta_W \tilde{f}_{iR}^\dagger \tilde{f}_{iR} \Big\} Z^\mu, \tag{8.47g}$$

$$\mathcal{L}_{\bar{\tilde{f}}\tilde{f}WW} = \frac{g_2^2}{2} W_\mu^+ W^{\mu-} \tilde{f}_{iL}^\dagger \tilde{f}_{iL}. \tag{8.47h}$$

In (8.47e–h) $\tilde{f}_{i(L,R)}$ covers[10] $\tilde{u}_{i(L,R)}$, $\tilde{d}_{i(L,R)}$, $\tilde{e}_{i(L,R)}$ and $\tilde{\nu}_{iL}$ which, as in (8.44), are now understood as mass diagonal fields *in the limit of no left right mixing*. Since we have yet to include the left right mixing of sfermions and this will be done in Ch.9, we postpone a listing of the vertices and Feynman rules till then.

(iii) **Electroweak vector boson-gluon-squark-squark vertices**
These mixed terms can be written, with the CKM-type matrix elements $V_{ij}^{\tilde{q}_L}$ put in and with $Q_q = 2/3$ or $-1/3$ for $q = u$ or d respectively, as

$$\mathcal{L}_{\bar{\tilde{q}}\tilde{q}'gv} = \sqrt{2} g_2 g_s A_\mu^a (W^{\mu+} \tilde{u}_{iL}^\dagger T^a V_{ij}^{\tilde{q}_L} \tilde{d}_{jL} + W^{\mu-} \tilde{d}_{iL}^\dagger T^a V_{ij}^{\tilde{q}_L^\dagger} \tilde{d}_{jL})$$
$$+ 2g_s e Q_q A^\mu A_\mu^a \tilde{q}_i^\dagger T^a \tilde{q}_i$$
$$+ 2g_s g_2 (\cos\theta_W)^{-1} Z^\mu A_\mu^a \tilde{q}_i^\star (T_{3L}^{\tilde{q}} - Q_q \sin^2\theta_W) T^a \tilde{q}_i. \tag{8.48}$$

---

[9]Sfermion flavor mixings do not matter here because of the GIM-mechanism [8.1].
[10]Read $\tilde{f}^\star$ for $\tilde{f}^\dagger$ in case of sleptons.

In (8.48) g_2, g_s are the $SU(2)_L$, $SU(3)_C$ coupling strengths, as before, while v can be W or Z or γ, and i in $\tilde{q}_i$ is summed over all flavors as well as left and right chiral fields. Here all squark fields are supposed to be mass eigenstates in the limit of no L-R mixing. The corresponding vertices and Feynman rules will be listed in Ch.9 along with the proper L-R mixing factors put in.

(H) *Scalar quartic vertices without Higgs*

These can be picked out from the supersymmetric potential (8.36) and the detailed expressions for the F- and D-terms given in (8.34) and (8.35) respectively. One can then write the corresponding interaction term in the Lagrangian density in the limit of zero left right mixing as follows.

$$\mathcal{L}_{\tilde{f}_1\tilde{f}_2\tilde{f}_3\tilde{f}_4} =$$

$$-\frac{g_2^2}{2M_W^2\sin^2\beta}\left(|\tilde{u}_L^\dagger \mathbf{U}^{\tilde{u}_L^\dagger}\mathbf{U}^{u_L}\mathbf{m}_u^{(D)}\mathbf{U}^{u_R^\dagger}\mathbf{U}^{\tilde{u}_R}\tilde{u}_R|^2\right.$$

$$\left.+|\tilde{d}_L^\dagger \mathbf{U}^{\tilde{d}_L^\dagger}\mathbf{U}^{u_L}\mathbf{m}_u^{(D)}\mathbf{U}^{u_R^\dagger}\mathbf{U}^{\tilde{u}_R}\tilde{u}_R|^2\right)$$

$$-\frac{g_2^2}{2M_W^2\cos^2\beta}\left(|\tilde{e}_L^\dagger \mathbf{U}^{\tilde{e}_L^\dagger}\mathbf{m}_e^{(D)}\mathbf{U}^{\tilde{e}_R}\tilde{e}_R + \tilde{d}_L^\dagger \mathbf{U}^{\tilde{d}_L^\dagger}\mathbf{U}^{d_L}\mathbf{m}_d^{(D)}\mathbf{U}^{d_R^\dagger}\mathbf{U}^{\tilde{d}_R}\tilde{d}_R|^2\right.$$

$$\left.+|\tilde{\nu}^\dagger \mathbf{U}^{\tilde{\nu}^\dagger}\mathbf{m}_e^{(D)}\mathbf{U}^{\tilde{e}_R}\tilde{e}_R + \tilde{u}_L^\dagger \mathbf{U}^{\tilde{u}_L^\dagger}\mathbf{U}^{d_L}\mathbf{m}_d^{(D)}\mathbf{U}^{d_R^\dagger}\mathbf{U}^{\tilde{d}_R}\tilde{d}_R|^2\right)$$

$$-\frac{g_s^2}{4}\left[\sum_{i,j}\left(|\tilde{u}_{iL}^\dagger\tilde{u}_{jL}|^2 + |\tilde{d}_{iL}^\dagger\tilde{d}_{jL}|^2 + |\tilde{u}_{iR}^\dagger\tilde{u}_{jR}|^2 + |\tilde{d}_{iR}^\dagger\tilde{d}_{jR}|^2 + 2|\tilde{u}_{iL}^\dagger\tilde{d}_{jL}|^2 - 2|\tilde{u}_{iL}^\dagger\tilde{u}_{jR}|^2\right.\right.$$

$$\left.\left.-2|\tilde{u}_{iL}^\dagger\tilde{d}_{jR}|^2 - 2|\tilde{d}_{iL}^\dagger\tilde{u}_{jR}|^2 - 2|\tilde{d}_{iL}^\dagger\tilde{d}_{jR}|^2 + 2|\tilde{u}_{iR}^\dagger\tilde{d}_{jR}|^2\right)\right.$$

$$\left.-\frac{1}{3}\left\{\sum_i\left(|\tilde{u}_{iL}|^2 + |\tilde{d}_{iL}|^2 - |\tilde{u}_{iR}|^2 - |\tilde{d}_{iR}|^2\right)\right\}^2\right]$$

$$-\frac{g_2^2}{8}\left[\left\{\sum_i\left(|\tilde{u}_{iL}|^2 - |\tilde{d}_{iL}|^2 + |\tilde{\nu}_i|^2 - |\tilde{e}_{iL}|^2\right)\right\}^2 + 4|\tilde{u}_L^\dagger\mathbf{V}^{\tilde{q}_L}\tilde{d}_L + \tilde{\nu}^\dagger\mathbf{V}^{\tilde{\ell}_L}\tilde{e}_L|^2\right]$$

$$-\frac{g_2^2\tan^2\theta_W}{8}\left\{\sum_i\left(\frac{1}{3}|\tilde{u}_{iL}|^2 + \frac{1}{3}|\tilde{d}_{iL}|^2 - \frac{4}{3}|\tilde{u}_{iR}|^2 + \right.\right.$$

$$\left.\left.\frac{2}{3}|\tilde{d}_{iR}|^2 - |\tilde{\nu}_i|^2 - |\tilde{e}_{iL}|^2 + 2|\tilde{e}_{iR}|^2\right)\right\}^2. \tag{8.49}$$

In (8.49) i,j are generation indices. Moreover, $\mathbf{m}_e^{(D)}$ is the physical real diagonal charged lepton mass matrix of (8.11) in the generation space; the unitary flavor rotation matrix $\mathbf{U}^{\tilde{e}_R}$ transforms the mass eigenstate charged right slepton fields to the corresponding flavor eigenstate ones, while $\mathbf{U}^{\tilde{e}_L}$ and $\mathbf{U}^{\tilde{\nu}}$ do the same for charged and neutral left slepton fields

$\tilde{e}_L$ and $\tilde{\nu}$ respectively. Similarly, $\mathbf{m}_u^{(D)}$ and $\mathbf{m}_d^{(D)}$ are the real diagonal quark mass matrices of (8.12). Thus we needed to take into account not only the unitary left, right squark flavor rotation matrices $\mathbf{U}^{\tilde{u}_L}$, $\mathbf{U}^{\tilde{u}_R}$, $\mathbf{U}^{\tilde{d}_L}$, $\mathbf{U}^{\tilde{d}_R}$, defined in analogy with $\mathbf{U}^{\tilde{\ell}_{L,R}}$ and $\mathbf{U}^{\tilde{\nu}}$, but also the corresponding ones for quarks, namely $\mathbf{U}^{u_L}$, $\mathbf{U}^{u_R}$, $\mathbf{U}^{d_L}$, $\mathbf{U}^{d_R}$ of (8.12). Finally, $\mathbf{V}^{\tilde{q}_L}$ and $\mathbf{V}^{\tilde{\ell}_L}$ are respectively the left squark and left slepton versions of the Kobayashi-Maskwa matrix $\mathbf{V}^{q_L}$, cf (8.40). Thus

$$\mathbf{V}^{\tilde{q}_L} \;=\; \mathbf{U}^{\tilde{u}_L \dagger}\mathbf{U}^{\tilde{d}_L}, \tag{8.50a}$$

$$\mathbf{V}^{\tilde{\ell}_L} \;=\; \mathbf{U}^{\tilde{\nu}\dagger}\mathbf{U}^{\tilde{e}_L}. \tag{8.50b}$$

These may be called 'super-CKM' matrices. We shall enumerate the physical quartic sfermion vertices in Ch.9 with left right mixing taken into account.

This brings us to the end of our discussion of supersymmetric vertices in the MSSM except for the ones which are significantly affected by supersymmetry breaking parameters as well as *L-R* mixing and possible gaugino-higgsino mixing. Those will be discussed extensively in the next chapter.

References

[8.1] T.P. Cheng and L-F. Li, *op. cit.*, *Bibl.* C. Quigg, *op. cit.*, *Bibl.* S. Pokorski, *op. cit.*, *Bibl.* E. Leader and E. Predazzi, *op. cit.*, *Bibl.*

[8.2] H.E. Haber and G.L. Kane, *loc. cit.*, *Bibl.* C. Csaki, *op. cit.*, *Bibl.* I. Simonsen, hep-ph/9506369. K. Hikasa, Lecture Notes: *Supersymmetric Standard Model for Collider Physicists*, unpublished. A. Djouadi et al., (MSSM working group), hep-ph/9901246. N. Polonsky, *op. cit.*, *Bibl.*

[8.3] P. van Nieuwenhuizen, *loc. cit.*, *Bibl.*

[8.4] L.J. Hall and M.B. Wise, Nucl. Phys. **B157** (1981) 31. R. Flores and M. Sher, Ann. Phys. (N.Y.) **148** (1983) 95.

[8.5] S.B. Treiman, D. Gross and R. Jackiw, *op. cit.*, *Bibl.*

Chapter 9

SOFT SUPERSYMMETRY BREAKING IN THE MSSM

9.1 The Content of $\mathcal{L}_{\text{SOFT}}$

We now turn to that part of $\mathcal{L}_{\text{MSSM}}$ through which supersymmetry breaking is explicitly introduced. But first we need to demonstrate [9.1] the impossibility of effecting a spontaneous breakdown of global supersymmetry purely within the framework of the MSSM. We follow the *reductio ad absurdum* procedure in assuming such a spontaneous breaking and applying the supertrace mass sum rule (7.50). Let us separately consider the mass squared matrices M_e^2 for the charge -1, M_u^2 for the charge $+\frac{2}{3}$, M_d^2 for the charge $-\frac{1}{3}$ and M_ν^2 for the neutral matter fermion sfermion supermultiplets of any generation. Assuming charge and color conservation, the RHS of (7.50) now can receive possible contributions from the generators T_3 and $Y/2$ only. We can sum over all possible left and right chiral supermultiplets in the supertrace, except that the latter have to be conjugated since (7.50) has been written for a left chiral supermultiplet. We can then use the results $(T_3)_{e_L} = -\frac{1}{2}$, $(Y/2)_{e_L} = -\frac{1}{2}$, $(T_3)_{e_L^C} = 0$, $(Y/2)_{e_L^C} = 1$, $(T_3)_{u_L} = \frac{1}{2}$, $(Y/2)_{u_L} = \frac{1}{6}$, $(T_3)_{u_L^C} = 0$, $(Y/2)_{u_L^C} = -\frac{2}{3}$, $(T_3)_{d_L} = -\frac{1}{2}$, $(Y/2)_{d_L} = \frac{1}{6}$, $(T_3)_{d_L^C} = 0$, $(Y/2)_{d_L^C} = \frac{1}{3}$, $(T_3)_{\nu_L} = \frac{1}{2}$, $(Y/2)_{\nu_L} = -\frac{1}{2}$. Thus we have[1]

$$S\text{Tr}M_e^2 \;=\; g_2\langle D^3\rangle - g_Y\langle D^Y\rangle, \tag{9.1a}$$

$$S\text{Tr}M_u^2 \;=\; -g_2\langle D^3\rangle + g_Y\langle D^Y\rangle, \tag{9.1b}$$

$$S\text{Tr}M_d^2 \;=\; g_2\langle D^3\rangle - g_Y\langle D^Y\rangle, \tag{9.1c}$$

$$S\text{Tr}M_\nu^2 \;=\; -g_2\langle D^3\rangle + g_Y\langle D^Y\rangle. \tag{9.1d}$$

Two positive combinations of the above four supertraces are seen to have vanishing RHS, namely

---

[1]These M's are the mass matrices of (5.10) and (7.50) taken for each generation and summed over left and right chiral supermultiplets. Thus $S\text{Tr}M_e^2 \equiv S\text{Tr}M_{e_L}^2 + S\text{Tr}M_{e_R}^2 = m_{\tilde{e}_L}^2 + m_{\tilde{e}_R}^2 - 2m_e^2$ etc. Though we have not included generation mixing in this argument, (9.2) can be generalized to cover generation space.

$$S\mathrm{Tr}M_e^2 + S\mathrm{Tr}M_\nu^2 = S\mathrm{Tr}M_u^2 + S\mathrm{Tr}M_d^2 = 0 \ . \tag{9.2}$$

Eq. (9.2) can be satisfied only if in each family *some* slepton/squark is lighter than the corresponding fermion. This is manifestly contrary to observation, except possibly for the third squark family. Hence the starting assumption is wrong and, if one sticks to MSSM fields alone, supersymmetry has to be explicitly broken.

In principle, one could introduce an extra $U(1)_{Y'}$ factor in the gauge group in a way such that all left chiral fermionic fields carried the quantum number $Y' = 1$. The $D_{Y'}$-term (cf. 7.50), corresponding to this $U(1)_{Y'}$, could be given a nonzero VEV, spontaneously breaking supersymmetry. (9.2) would now have a nonzero RHS and the compulsion of having some sfermions lighter than the corresponding fermions could be evaded. But then there will be an additional weak neutral gauge boson Z', mixing with the Z, whereas such mixing is now severely constrained by experiment. Moreover, an extra $U(1)_{Y'}$ gauge factor would introduce [9.2] uncancelled ABJ anomalies [9.3] and make the theory nonrenormalizable. A great many extra superfields would be needed to cancel all anomalies and it would be difficult in general to keep all sfermions heavier than extant lower mass bounds. Furthermore, gauginos would not acquire masses at the tree level.

We can then conclude that, though the spontaneous breakdown of supersymmetry is a theoretically desirable feature, such a mechanism will have to involve fields beyond those of the MSSM. Phenomenological constraints point to such fields being significantly heavier than the electroweak scale and hence carrying masses much larger than those of the MSSM sparticles. Much theoretical speculation has taken place so far regarding the specifics of such a mechanism and the current wisdom on it will be elaborated in Chs.12 and 13. Two broad characteristics can, however, be mentioned at this juncture. Spontaneous Supersymmetry Breakdown (SSB) needs to be effected in a sector of fields which are singlets with respect to the SM gauge group and known as the **hidden** or **secluded sector**. SSB can take place there at a distinct scale denoted by Λ_s, say. Supersymmetry breaking is then transmitted to the gauge nonsinglet **observable** or **visible sector** by a messenger sector (associated with a typical mass scale M_M that could, but need not, be as high as the Planck mass M_{Pl}); this may or may not require the introduction of additional gauge nonsinglet messenger superfields. Fig. 9.1 is a cartoon depicting this.

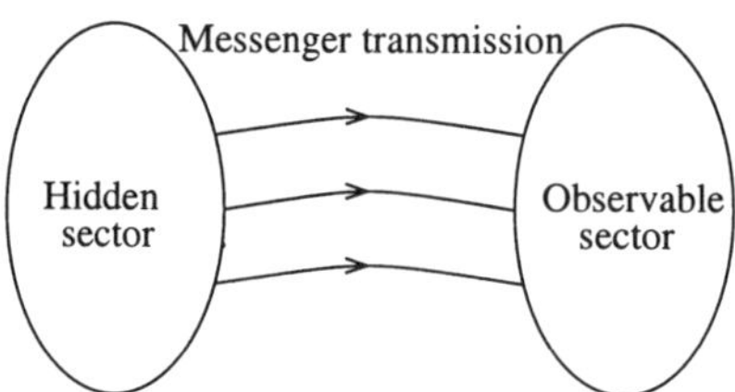

Fig. 9.1. Cartoon showing the transmission of supersymmetry breaking from the hidden to the observable sector.

It is nonetheless true that this messenger scale must be at least two (and perhaps many more) orders of magnitude above the mass of the MSSM fields. Hence, when the former are integrated out at lower (electroweak) energies, the residual theory is described (cf. 7.40) by the supersymmetric Lagrangian density of the MSSM, namely $\mathcal{L}_{SUSY}$ plus some soft explicit supersymmetry breaking terms, collected in $\mathcal{L}_{SOFT}$ and characterized by the supermultiplet splitting mass parameter M_s (cf. Ch.1). In Chs. 12 and 13 we shall discuss in detail two alternative broad scenarios in which the messenger sector consists of

(1) higher dimensional operators [9.4] suppressed by inverse powers of the Planck mass,

or,

(2) fields with gauge interactions [9.5] at lower energy scales.

For (1), the mechanism of Fig. 9.1 can generally proceed at the tree level leading to $M_s \sim \Lambda_s^2/M_{Pl}$. For (2), the origin of M_s may be seen in terms of a one loop supergraph such as that of Fig. 9.2, in which the letters V, M and H refer to superfields in the visible, messenger and hidden sectors respectively, yielding $M_s \sim$ (gauge coupling)^2 Λ_s^2/M_M. The occurrence of the square of Λ_s in the numerator in either case is easy to understand if supersymmetry breaking in the hidden sector arises through the VEV of an auxilary F- or D-field (cf. §7.4–§7.6). Finally, then, a total Lagrangian density of the form of (7.40) can provide a phenomenologically realistic description at least for a range of energies above the EW scale. That will be our starting point here.

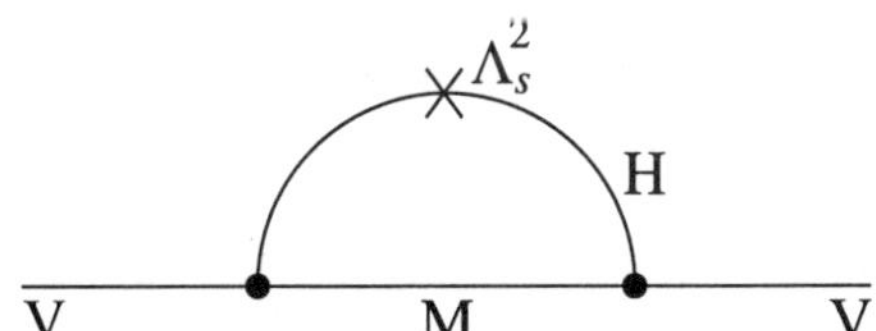

Fig. 9.2. Possible one loop supergraph implementing the scheme of Fig. 9.1.

We wrote the most general form of $\mathcal{L}_{SOFT}$ in (7.42) for a supersymmetric gauge theory. An appraisal of the different terms in it shows that, for the MSSM, $\mathcal{L}_{SOFT}$ can have no C_i-type terms. This is due to the fact that the model does not contain any scalar field that is invariant under $SU(3)_C \times SU(2)_L \times U(1)_Y$ gauge transformations. All other types of terms, shown in (7.42), are possible. Thus we can write

$$
\begin{aligned}
-\mathcal{L}_{\text{SOFT}} =\ & \tilde{q}_{iL}^\star(\mathcal{M}_{\tilde{q}}^2)_{ij}\tilde{q}_{jL} + \tilde{u}_{iR}^\star(\mathcal{M}_{\tilde{u}}^2)_{ij}\tilde{u}_{jR} + \tilde{d}_{iR}^\star(\mathcal{M}_{\tilde{d}}^2)_{ij}\tilde{d}_{jR} + \tilde{\ell}_{iL}^\star(\mathcal{M}_{\tilde{\ell}}^2)_{ij}\tilde{\ell}_{jL} \\[4pt]
& + \tilde{e}_{iR}^\star(\mathcal{M}_{\tilde{e}}^2)_{ij}\tilde{e}_{jR} + \left[h_1\cdot\tilde{\ell}_{iL}(f^e A^e)_{ij}\tilde{e}_{jR}^\star + h_1\cdot\tilde{q}_{iL}(f^d A^d)_{ij}\tilde{d}_{jR}^\star \right. \\[4pt]
& \left. + \tilde{q}_{iL}\cdot h_2(f^u A^u)_{ij}\tilde{u}_{jR}^\star + \text{h.c.} \right] + m_1^2|h_1|^2 + m_2^2|h_2|^2 + (B\mu h_1\cdot h_2 + \text{h.c.}) \\[4pt]
& + \frac{1}{2}(M_1\bar{\tilde{\lambda}}_0 P_L\tilde{\lambda}_0 + M_1^\star\bar{\tilde{\lambda}}_0 P_R\tilde{\lambda}_0) + \frac{1}{2}(M_2\bar{\vec{\tilde{\lambda}}} P_L\vec{\tilde{\lambda}} + M_2^\star\bar{\vec{\tilde{\lambda}}} P_R\vec{\tilde{\lambda}}) \\[4pt]
& + \frac{1}{2}(M_3\bar{\tilde{g}}^a P_L\tilde{g}^a + M_3^\star\bar{\tilde{g}}^a P_R\tilde{g}^a) \\[4pt]
\equiv\ & V_{\text{SOFT}} + \text{gaugino mass terms.}
\end{aligned}
\tag{9.3}
$$

In (9.3) $M_{1,2,3}$ are the (generally complex) gaugino (Majorana) mass parameters in the Lagrangian density pertaining to $\tilde{\lambda}_0$, $\vec{\tilde{\lambda}}$ and $\tilde{g}^a$ which are (cf. Ch.8) the $U(1)_Y$, $SU(2)_L$ and $SU(3)_C$ gaugino fields respectively while $m_{1,2}$ are the real Higgs scalar mass parameters. Furthermore, i, j are generation indices with summation implied by repetition. Thus the squared left squark mass $\mathcal{M}_{\tilde{q}}^2$ and the squared right squark masses $\mathcal{M}_{\tilde{u}}^2$, $\mathcal{M}_{\tilde{d}}^2$ as well as those for left sleptons $\mathcal{M}_{\tilde{\ell}}^2$ and those for charged right sleptons $\mathcal{M}_{\tilde{e}}^2$ are all 3×3 hermitian matrices in generation space. The products $f^e A^e, f^d A^d$ and $f^u A^u$, which form coefficients of the trilinear scalar terms in (9.3), are general 3×3 complex matrices in the same space. These are the soft supersymmetry breaking $\mathcal{A}$ terms of (7.43), each written as a product of a superpotential coupling f of (7.41) times an A parameter with the dimension of mass, cf. (7.44). Similarly, we have scaled the coefficient of the $SU(2)_L \times U(1)_Y$-invariant Higgs bilinear term by the supersymmetry invariant Higgsino mass μ. This ensures that the soft supersymmetry breaking parameter B (cf. 7.44) also has the dimension of mass. Note further the absence of any linear term in the Higgs fields, which would have been a $\mathcal{C}$-type term, cf. (7.43). If we allow all the new parameters, introduced in (9.3), to be complex, we would be dealing with some one hundred and twenty four [9.5] unknown real constants of which nineteen were already in the SM and one hundred and five are new. Fortunately, many processes are sensitive only to a small subset of these parameters, at least at the tree level[2]. In fact, in practical calculations in the MSSM (e.g. those for supersymmetry searches at colliders) several simplifying assumptions are usually made in order to drastically reduce the number of these additional parameters to only a handful. The final set of parameters is determined by the specific assumptions made. Different assumptions (usually motivated by different scenarios of supersymmetry breaking) result in different versions of the *Constrained Minimal Supersymmetric Standard Model* (**CMSSM**). Let us remark that, though well motivated, these assumptions do need to be tested in experiments and such tests form an important part of supersymmetry phenomenology at colliders. Of course, once again R_p conservation has been assumed in (9.3). The introduction of R_p violation in the soft supersymmetry breaking part of $\mathcal{L}$, without R_p nonconserving supersymmetric terms present in the superpotential $\mathcal{W}$ of (8.33), generally makes the scalar potential unbounded from below. We shall consider the latter kind of terms in Ch.14, when dealing with extensions of the MSSM.

Yet another issue confronting us is that of phases. As mentioned earlier, many of the new parameters in the part $\mathcal{L}_{SOFT}$ of (9.3) can, in general, be complex in a CP noninvariant theory. Two of these can be chosen to be real by appropriate phase rotations of the fields appearing in $\mathcal{L}_{SOFT}$ without compromising the form of $\mathcal{L}_{SUSY}$ in (7.41). However, many different nontrivial (i.e. in principle measurable) phases remain in the MSSM in addition to the single CP violating phase of the CKM matrix of the SM. On the other hand, some of these phases are subject to strong phenomenological constraints [9.6, 9.7] which come from the lack of observation of any additional, beyond-Standard-Model CP violation in low energy experiments so far. For example, if the phases in the gaugino/higgsino sector are large, effective cancellation mechanisms need to be devised [9.6] to meet those constraints. The

---

[2]For instance, negative search results from LEP, cf. Ch.15, already imply that both $|M_2|$ and $|\mu|$ must exceed M_W. Herein lies the origin of the μ problem about which we shall have more to say in §13.4 and §14.2.

simplest way to satisfy the experimental bounds on new sources of CP violation is to assume [9.8] that the phases of all soft supersymmetry breaking parameters are small. In fact, most analyses of CP conserving processes in softly broken supersymmetric scenarios have been performed under this assumption. Our phenomenological discussions will be mostly based on such a framework, but the mass matrices and couplings, given in this chapter, allow for the possibility of CP-violating phases.

Once supersymmetry and (at a lower energy scale) EW symmetry get broken, different sparticles of the same electric charge can mix. The sparticles, listed in Table 8.2, then no longer remain eigenstates of mass. Left squarks (sleptons) mix with right squarks (sleptons); there can be generation mixing as well. The EW gauginos and higgsinos mix too, as mentioned in Ch.8. The mixing patterns and mass values of sparticle mass eigenstates depend crucially on the manner of supersymmetry breaking. These masses and mixing angles, in turn, determine the experimental signals of supersymmetry. This is true both for sparticle production as well as decay analyses and for low energy signatures caused by the exchange of virtual sparticles in loops. We therefore need to study all nontrivial restrictions on sparticle mass matrices implied by low energy physics constraints, mainly from the absence [9.8] of FCNC processes in nature. These constraints also play a crucial role in relating softly broken supersymmetry to some higher scale physics which causes the transmission of supersymmetry breaking to the MSSM fields in the observable sector. The mass values of matter sfermions as well as of nonmatter fermions (i.e. gauginos and higgsinos) are controlled by the explicitly supersymmetry breaking soft operators, introduced at this higher scale. One then needs to consider the subsequent modification of these via renormalization group evolution down to electroweak energies. This scale dependence of the mass spectrum of sparticles will be discussed in Ch.11 whereas here we concentrate on the extra masses and mixing angles of the MSSM at laboratory energies. Let us note meanwhile that there is no really satisfactory theory of soft supersymmetry breaking terms at this point; only speculative models exist. Thus low energy constraints are the only phenomenological pointers to them that we have at present and these merit careful attention.

The next section contains a discussion of the masses of higgsinos and electroweak gauginos as well as of the two cases of mixing among them: one for charged ones and another for neutral ones. In subsequent sections we shall consider the general mass matrices for sleptons and squarks incorporating various supersymmetric and nonsupersymmetric mass terms. We shall also address the different cases of mixing among them and what effects these have on their interaction vertices.

9.2 Electroweak Gauginos and Higgsinos

We concentrate here on the spin half supersymmetric partners of the electroweak gauge and Higgs bosons: the electroweak gauginos and higgsinos. While gaugino mass terms are part of the soft supersymmetry breaking $\mathcal{L}_{SOFT}$ of (9.3), the spontaneous symmetry breaking $SU(2)_L \times U(1)_Y \to U(1)_{em}$ forces the gaugino fields $\tilde{\lambda}^{\pm}$ of (8.27) to mix with the higgsino fields $\tilde{h}_i^{\pm}$ of (8.25), leading to physical mass eigenstate charginos $\tilde{\chi}_{1,2}^{\pm}$. This fact, already mentioned in §8.2, will receive our attention first. Similar mixings exist in the sector of neutral EW gauginos and higgsinos and will be discussed later. The soft supersymmetry breaking gaugino mass parameters $M_a(a = 1, 2)$ and the supersymmetry preserving higgsino mass parameter μ of (8.33), plus the ratio $\tan\beta$ of Higgs VEVs, cf. (8.24), are the only parameters of the model that are relevant to our present discussion. In case M_a $(a = 1, 2)$ and μ are complex, then M_2 can be chosen to be real and positive without loss of generality. In this situation two additional parameters enter the game, viz. Φ_μ and Φ_M – the relative phases between M_2 and μ and between M_2 and M_1 respectively. However, in some (though not all) of our discussions below we shall assume these phases to be zero.

The chargino mass matrix

Starting from (5.55), we can isolate in the Lagrangian density the matter-gaugino-Higgs coupling terms that generate chargino masses. They can be written generically in **two component notation** as

$$-\sqrt{2}g_2(T^a)_{ij}\lambda^a\xi_j\phi_i^* + \text{h.c.}$$

Here λ^a stands for a gaugino field, while ξ and ϕ stand for the fermionic and bosonic components of a Higgs chiral superfield respectively; T^a is a gauge group generator acting in the representation space of ξ and ϕ typified by indices i, j. Here ξ_i are two component spinorial fields in the $(\frac{1}{2}, 0)$ representation (cf. Ch.3) while their barred versions are the corresponding conjugate fields in the $(0, \frac{1}{2})$ representation. Once the fields $h^0_{1,2}$ of (8.20) acquire VEVs $v_{1,2}$ on the spontaneous breakdown of the EW symmetry, the above expression generates a sum of mixed gaugino and higgsino mass terms. We further need to add to the above the supersymmetry breaking gaugino mass terms from (9.3) and the supersymmetric bilinear higgsino mixing terms contained in the $\mu H_1 \cdot H_2$ part of the superpotential (8.33). Thus the mass terms of the nonmatter charged fermions can finally be written as

$$\mathcal{L}^c_{MASS} = -\frac{g_2}{\sqrt{2}}(v_1\lambda^+\tilde{h}^2_1 + v_2\lambda^-\tilde{h}^1_2 + \text{h.c.}) - (M_2\lambda^+\lambda^- + \mu\tilde{h}^2_1\tilde{h}^1_2 + \text{h.c.}) \ . \tag{9.4}$$

In (9.4) $\tilde{h}^2_1$ and $\tilde{h}^1_2$ are two component spinorial higgsino fields in the $(\frac{1}{2}, 0)$ representation carrying $Y = -1, Q = -1$ and $Y = 1, Q = 1$ respectively, cf. (8.25). Moreover, the two component charged gaugino fields $\lambda^\pm$ are defined as $(\sqrt{2})^{-1}(\lambda_1 \mp i\lambda_2)$. The mass term of (9.4) can now be rewritten in terms of a 2×2 matrix $\mathbf{X}$ as follows. Define two column vectors $\psi^\pm$, each consisting of one gaugino field component and one higgsino field component, as

$$\psi^+ \equiv \begin{pmatrix} \lambda^+ \\ \tilde{h}^1_2 \end{pmatrix}, \ (\psi^+)^T \equiv (\lambda^+ \ \tilde{h}^1_2) \ , \tag{9.5a}$$

$$\psi^- \equiv \begin{pmatrix} \lambda^- \\ \tilde{h}^2_1 \end{pmatrix}, \ (\psi^-)^T \equiv (\lambda^- \ \tilde{h}^2_1) \ . \tag{9.5b}$$

Let us denote the components of $\psi^\pm$ by $\psi^\pm_m$ with $m = 1, 2$, i.e. $\psi^+_1 = \lambda^+$ etc. Now we can make use of (8.22) and (8.24) to rewrite (9.4) as

$$-\mathcal{L}^c_{MASS} = (\psi^-)^T\mathbf{X}\psi^+ + \text{h.c.} \ , \tag{9.6}$$

with

$$\mathbf{X} = \begin{pmatrix} M_2 & \sqrt{2}M_W\sin\beta \\ \sqrt{2}M_W\cos\beta & \mu \end{pmatrix} \ . \tag{9.7}$$

One can find unitary matrices $\mathcal{U}$ and $\mathcal{V}$ such that

$$\mathcal{U}^{\star}\mathbf{X}\mathcal{V}^{-1} = \mathbf{M}_{c}^{D}, \tag{9.8}$$

where $\mathbf{M}_{c}^{D}$ is a diagonal matrix with real nonnegative entries $\tilde{M}_1$ and $\tilde{M}_2$. The two component **chargino** mass eigenstate fields can then be identified as

$$\chi_{k}^{+} = \mathcal{V}_{km}\psi_{m}^{+}, \tag{9.9a}$$

$$\chi_{k}^{-} = \mathcal{U}_{km}\psi_{m}^{-}, \tag{9.9b}$$

with $k = 1, 2$. These two component $\chi^{\pm}$ fields enable one to recast (9.6) as

$$-\mathcal{L}_{MASS}^{c} = \chi_{k}^{-}(\mathbf{M}_{D}^{c})_{km}\chi_{m}^{+} + \text{h.c.} \tag{9.10}$$

We are now in a position to define four component Dirac chargino fields

$$\tilde{\chi}_{1}^{+} \equiv \begin{pmatrix} \tilde{\chi}_{1}^{+} \\ \overline{\chi_{1}^{-}}^{T} \end{pmatrix}, \tag{9.11a}$$

$$\tilde{\chi}_{2}^{+} \equiv \begin{pmatrix} \tilde{\chi}_{2}^{+} \\ \overline{\chi_{2}^{-}}^{T} \end{pmatrix}. \tag{9.11b}$$

By using (3.28a), the mass term (9.10) can be rewritten in terms of these Dirac chargino fields as

$$-\mathcal{L}_{MASS}^{c} = \widetilde{M_1}\overline{\tilde{\chi}_{1}^{+}}\tilde{\chi}_{1}^{+} + \widetilde{M_2}\overline{\tilde{\chi}_{2}^{+}}\tilde{\chi}_{2}^{+}. \tag{9.12}$$

By convention, $\tilde{\chi}_{1}^{+}$ is chosen to be lighter than $\tilde{\chi}_{2}^{+}$, i.e. $\widetilde{M_1} < \widetilde{M_2}$. $\widetilde{M}_{1,2}$ are actually the positive square roots of the eigenvalues of the matrix $\mathbf{X}^{\dagger}\mathbf{X}$. From (9.8) we see that

$$(\mathbf{M}_{c}^{D})^{2} = \mathcal{V}\mathbf{X}^{\dagger}\mathbf{X}\mathcal{V}^{-1} = \mathcal{U}^{\star}\mathbf{X}\mathbf{X}^{\dagger}(\mathcal{U}^{\star})^{-1}, \tag{9.13}$$

i.e. $\mathcal{U}, \mathcal{V}$ are the unitary matrices which diagonalize the hermitian matrices $\mathbf{X}\mathbf{X}^{\dagger}$ and $\mathbf{X}^{\dagger}\mathbf{X}$ respectively. For such 2×2 matrices, the eigenvalues and mixing matrices are easy to write down analytically. The squared masses are given by

$$\widetilde{M}_{2,1}^{2} = \frac{1}{2}\Bigg[|M_{2}^{2}| + |\mu^{2}| + 2M_{W}^{2} \pm \Big\{ (|M_{2}^{2}| - |\mu^{2}|)^{2}$$

$$+ 4M_{W}^{4}\cos^{2}2\beta + 4M_{W}^{2}(|M_{2}^{2}| + |\mu^{2}| + 2\Re e(M_{2}\mu)\sin 2\beta)\Big\}^{1/2}\Bigg]. \tag{9.14}$$

If the phases of M_2 and μ are ignored, all the entries of $\mathbf{X}$ become real. We work in the convention where M_2 is positive, but μ can have either sign (N.B. $\tan\beta$ is always positive, cf. §10.2). Then the mixing matrices can be written as

$$\mathcal{U} = \mathbf{O}_{u}, \tag{9.15a}$$

$$\mathcal{V} = \begin{cases} \mathbf{O}_{v} \text{ for } \det\mathbf{X} > 0, \\ \sigma_{3}\mathbf{O}_{v} \text{ for } \det\mathbf{X} < 0, \end{cases} \tag{9.15b}$$

$$\mathbf{O}_{v,u} = \begin{pmatrix} \cos\phi_{v,u} & \sin\phi_{v,u} \\ -\sin\phi_{v,u} & \cos\phi_{v,u} \end{pmatrix}, \tag{9.15c}$$

where

$$\tan 2\phi_u = \frac{2\sqrt{2}M_W(\mu\sin\beta + M_2\cos\beta)}{M_2^2 - \mu^2 - 2M_W^2\cos 2\beta} , \qquad (9.16a)$$

$$\tan 2\phi_v = \frac{2\sqrt{2}M_W(\mu\cos\beta + M_2\sin\beta)}{M_2^2 - \mu^2 + 2M_W^2\cos 2\beta} . \qquad (9.16b)$$

The corresponding expressions for complex $\mathbf{X}$ can be found in Ref. [9.9]. Eqs. (9.16) are invariant under the change $\phi \to \phi + \pi/2$. However, these solutions are not equivalent. One has to check whether (9.8) holds in order to decide which of the four solutions of (9.16) is the correct one.

It is convenient at this stage to relate the starting two component charged gaugino[3] and higgsino fields to the four component weak interaction eigenstate ones of (8.25) and (8.27):

$$\tilde{\lambda}^+ = \begin{pmatrix} \lambda^+ \\ \overline{\lambda^-}^T \end{pmatrix} , \qquad (9.17a)$$

$$\tilde{h}^+ = \begin{pmatrix} \tilde{h}_2^1 \\ \overline{\tilde{h}_1^2}^T \end{pmatrix} . \qquad (9.17b)$$

The relations between these and the four component mass eigenstate chargino fields $\tilde{\chi}^{\pm}$ are:

$$P_L\tilde{\lambda}^+ = V_{k1}^\star P_L \tilde{\chi}_k^+ , \qquad (9.18a)$$

$$P_R\tilde{\lambda}^+ = U_{k1} P_R \tilde{\chi}_k^+ , \qquad (9.18b)$$

$$P_L\tilde{h}^+ = V_{k2}^\star P_L \tilde{\chi}_k^+ , \qquad (9.18c)$$

$$P_R\tilde{h}^+ = U_{k2} P_R \tilde{\chi}_k^+ . \qquad (9.18d)$$

Using these equations, we can also derive similar relations for the charge conjugate and adjoint spinors,

$$P_R(\tilde{\lambda}^+)^C = V_{k1} P_R (\tilde{\chi}_k^+)^C, \qquad (9.19a)$$

$$P_L(\tilde{h}^+)^C = U_{k2}^* P_L (\tilde{\chi}_k^+)^C, \qquad (9.19b)$$

$$\overline{\tilde{\lambda}^+} P_L = U_{k1}^\star \overline{\tilde{\chi}_k^+} P_L , \qquad (9.19c)$$

$$\overline{\tilde{h}^+} P_R = V_{k2} \overline{\tilde{\chi}_k^+} P_R . \qquad (9.19d)$$

Eqs. (9.18,19) and similar relations will prove useful later in deriving the interaction vertices involving various particles/sparticles and charginos.

---

[3]We should emphasize that our convention on gaugino field components is different from that of Haber and Kane [9.10]. However, our Feynman rules are the same as theirs except that β is the complement of their θ_v. Our $\mathcal{V}$ and $\mathcal{U}$ matrices are the same as the V and U respectively of Gunion and Haber [9.10].

The neutralino mass matrix

Let us now take up the issue of mass eigenstates for neutral non matter fermions. Again, the corresponding mass terms receive contributions from V_{SOFT}, from the superpotential as well as from the matter-gauge-Higgs couplings with the neutral Higgs fields replaced by their VEVs. Retaining only terms relevant to the neutral sector, the mass term in two component notation reads

$$\mathcal{L}^n_{MASS} = -\frac{g_2}{2}\lambda_3\left(v_1\tilde{h}^1_1 - v_2\tilde{h}^2_2\right) + \frac{g_Y}{2}\lambda_0\left(v_1\tilde{h}^1_1 - v_2\tilde{h}^2_2\right) + \mu\tilde{h}^1_1\tilde{h}^2_2$$
$$-\frac{1}{2}M_2\lambda_3\lambda_3 - \frac{1}{2}M_1\lambda_0\lambda_0 + \text{h.c.} \tag{9.20}$$

In (9.20) we have extended the notation of (9.4) for two component EW charged gaugino and higgsino fields to the corresponding neutral ones. In general, the three mass parameters M_1, M_2 and μ, which determine the neutral nonmatter fermionic mass matrix and the mixing contained therein, are completely arbitrary. However, in simple grand unified theories M_1 and M_2 are related to each other. Such theories predict that $M_1 = M_2$ at the high scale where the gauge couplings are presumed to unify. The gaugino mass M_α will be shown in Ch.11 to evolve (at one loop) with the momentum scale in a way identical to that of the square of the corresponding gauge coupling strength g_α, the subscript α referring to one of the factors of the SM gauge group. The unification condition then implies

$$M_1(M_Z) = \frac{5}{3}\tan^2\theta_W M_2(M_Z) \simeq \frac{1}{2}M_2(M_Z) , \tag{9.21}$$

θ_W being the Weinberg angle. As explained more clearly in Ch.11, the factor $5/3$ appears in (9.21) from the difference between the normalization of generators in a simple unifying gauge group and that of the electroweak hypercharge generator in the SM.

Define a row vector $(\psi^0)^T$ with two gaugino field components and two higgsino field components:

$$(\psi^0)^T \equiv (\lambda_0 \ \lambda_3 \ \tilde{h}^1_1 \ \tilde{h}^2_2) . \tag{9.22}$$

Eq. (9.20) can then be recast as

$$\mathcal{L}^n_{MASS} = -\frac{1}{2}\left(\psi^0\right)^T \mathcal{M}^n\psi^0 + \text{h.c.} \tag{9.23}$$

In (9.23) the 4×4 mass matrix $\mathcal{M}^n$ is given by

$$\mathcal{M}^n = \begin{pmatrix} M_1 & 0 & -M_Z c_\beta s_W & M_Z s_\beta s_W \\ 0 & M_2 & M_Z c_\beta c_W & -M_Z s_\beta c_W \\ -M_Z c_\beta s_W & M_Z c_\beta c_W & 0 & -\mu \\ M_Z s_\beta s_W & -M_Z s_\beta c_W & -\mu & 0 \end{pmatrix}, \tag{9.24}$$

where $s_W \equiv \sin\theta_W, c_W \equiv \cos\theta_W, s_\beta \equiv \sin\beta, c_\beta \equiv \cos\beta$ in the notation of Ch.8. Let us denote the components of ψ^0 in (9.22) as ψ^0_n, with $n = 1, 2, 3, 4$, i.e. $\psi^0_1 = \lambda_0$ etc. Now we can define two component neutralino mass eigenstate fields χ^0_l by

$$\chi^0_l = Z_{ln}\psi^0_n , \tag{9.25}$$

where $l = 1, 2, 3, 4$ and $\mathbf{Z}$ is a 4×4 unitary matrix, as defined by Gunion and Haber [9.10], satisfying

$$\mathbf{Z}^* \mathcal{M}^n \mathbf{Z}^{-1} = \mathbf{M}_n^D \ , \tag{9.26}$$

$\mathbf{M}_n^D$ being a diagonal matrix with only nonnegative entries. The latter can be computed from

$$(\mathbf{M}_n^D)^2 = \mathbf{Z} \mathcal{M}^{n\dagger} \mathcal{M}^n \mathbf{Z}^{-1} \ . \tag{9.27}$$

Sometimes, for simplicity of calculation, the possible phases in the entries of $\mathcal{M}^n$ are ignored. Now the rows of $\mathbf{Z}$ can be either purely real or purely imaginary. A common practice in the literature is to choose a real, orthogonal $\mathbf{Z}$. In this case, however, the eigenvalues of M_n^D can sometimes be negative. *Then those neutralino mass eigenstates, which correspond to such negative mass eigenvalues, need to be redefined with chiral rotations so as to make the latter positive.* It is difficult to keep track of this during calculations, since one has to introduce an explicit $i\gamma_5$ factor whenever a neutralino corresponding to a negative eigenvalue of $\mathcal{M}^n$ appears at a vertex. So we shall not make such an assumption. There is one point to be noted, though. In many applications, it is sufficient to keep the sign of the neutralino mass in the neutralino propagator and in neutralino spin sums without any modification of Feynman rules.

As with charginos, the masses and mixing angles of the neutralinos are completely determined in terms of a few parameters; here these are $M_{1,2}$, μ and $\tan\beta$. We can choose to introduce four component Majorana spinorial fields $\tilde{\chi}_l^0$:

$$\tilde{\chi}_l^0 = \begin{pmatrix} \chi_l^0 \\ \overline{\chi_l^0}^T \end{pmatrix} . \tag{9.28}$$

Now the mass term of (9.23) takes a simple four component Majorana form, namely

$$\mathcal{L}_{MASS}^n = -\frac{1}{2} \sum_l \widetilde{M}_l^n \overline{\tilde{\chi}_l^0} \tilde{\chi}_l^0 \ , \tag{9.29}$$

where $\widetilde{M}_l^n \equiv M_{\tilde{\chi}_l^0}$ are the nonnegative diagonal elements of $\mathbf{M}_n^D$. The eigenvalues $\widetilde{M}_l^n$ and the matrix $\mathbf{Z}$ *can most easily be obtained numerically.* If all entries of $\mathcal{M}^n$ are real, an analytical calculation of the former is possible [9.11]. However, the expressions are quite cumbrous and will not be given here. The neutralino eigenstates are labeled[4] in the mass order $M_{\tilde{\chi}_1^0} < M_{\tilde{\chi}_2^0} < M_{\tilde{\chi}_3^0} < M_{\tilde{\chi}_4^0}$ by convention. In most phenomenological discussions of the MSSM (unless there is a lighter gravitino or a violation of R-parity), the lightest neutralino $\tilde{\chi}_1^0$ is assumed to be the Lightest Supersymmetric Particle (LSP).

It is instructive to relate the mass eigenstate neutralino fields $\tilde{\chi}_l^0$ to four component gaugino and higgsino fields which are weak interaction eigenstates. Let us consider the latter first. They are the Majorana spinors

$$\tilde{\lambda}_3 = \begin{pmatrix} \lambda_{3,} \\ \overline{\lambda}_3^T \end{pmatrix} , \tag{9.30a}$$

---

[4]**Caution:** the subscripts 1,2,3 in $\tilde{\chi}_{1,2,3}^0$ do not have any specific association with the subscripts of the gaugino mass parameters $M_{1,2,3}$.

$$\tilde{\lambda}_0 = \begin{pmatrix} \lambda_0 \\ \bar{\lambda}_0^T \end{pmatrix}, \tag{9.30b}$$

$$\tilde{h}_1^0 = \begin{pmatrix} \tilde{h}_1^1 \\ \bar{\tilde{h}}_1^{1T} \end{pmatrix}, \tag{9.30c}$$

$$\tilde{h}_2^0 = \begin{pmatrix} \tilde{h}_2^2 \\ \bar{\tilde{h}}_2^{2T} \end{pmatrix}. \tag{9.30d}$$

Then the desired relations can be given as follows:

$$P_L \tilde{\lambda}_0 = P_L Z_{l1}^* \tilde{\chi}_l^0 , \tag{9.31a}$$

$$P_R \tilde{\lambda}_0 = P_R Z_{l1} \tilde{\chi}_l^0 , \tag{9.31b}$$

$$P_L \tilde{\lambda}_3 = P_L Z_{l2}^* \tilde{\chi}_l^0 , \tag{9.31c}$$

$$P_R \tilde{\lambda}_3 = P_R Z_{l2} \tilde{\chi}_l^0 , \tag{9.31d}$$

$$P_L \tilde{h}_s^0 = P_L Z_{l,s+2}^* \tilde{\chi}_l^0 , \tag{9.31e}$$

$$P_R \tilde{h}_s^0 = P_R Z_{l,s+2} \tilde{\chi}_l^0 . \tag{9.31f}$$

Note that the index l in (9.31) spans the values $1, 2, 3, 4$, while the index s covers $1, 2$ only. Similar relations can be written for $\overline{\tilde{\lambda}^0} P_L$ etc. using (9.31).

We can study and comment on the nature of the chargino and neutralino sectors in some limiting cases. If $|\mu| \gg |M_{1,2}| \gg M_Z$, the two lightest neutralinos $\tilde{\chi}_{1,2}^0$ are gaugino dominated. **If (9.21) is assumed**, it follows that $\tilde{\chi}_1^0$ is mostly the $U(1)_Y$ gaugino ("bino") field $\tilde{\lambda}_0$ and $\tilde{\chi}_2^0$ is largely the neutral $SU(2)_L$ gaugino ("wino") $\tilde{\lambda}^3$. The two higher mass neutralinos $\tilde{\chi}_{3,4}^0$ are then predominantly higgsinos. Similarly, the lighter chargino $\tilde{\chi}_1^\pm$ is more or less the charged "wino" and the heavier chargino is largely the charged higgsino. Furthermore, the magnitude of the μ parameter and the masses of the chargino and neutralino masses are roughly related by $M_{\tilde{\chi}_1^\pm} \simeq M_{\tilde{\chi}_2^0} \simeq 2M_{\tilde{\chi}_1^0}$ and $|\mu| \simeq M_{\tilde{\chi}_3^0} \simeq M_{\tilde{\chi}_4^0} \simeq M_{\tilde{\chi}_2^\pm} \gg M_{\tilde{\chi}_1^\pm}$. In the opposite limit $|\mu| \ll |M_{1,2}|$, the lighter neutralinos and the lighter chargino are mostly higgsinos with masses close to $|\mu|$, whereas the heavier chargino is predominantly the charged "wino". Finally, when $|\mu| \simeq |M_2|$ or $|M_1|$, strong kinds of mixing occur between gauginos and higgsinos in the formation of physical nonmatter fermions; in general, the masses are no longer related in any simple way. If $|\mu, M_2| \gg M_Z$ and (9.21) is assumed, then the approximate relation $M_{\tilde{\chi}_2^0} \simeq M_{\tilde{\chi}_1^\pm}$ holds [9.12] irrespective of the ordering and the relative magnitudes of $|\mu|$ and $|M_{1,2}|$. All these statements are insensitive to variations in $\tan\beta$ within the usually covered range (cf. Chs. 10 and 11).

9.3   Chargino and Neutralino Interactions with Gauge Bosons

Chargino-Neutralino-$W^\pm$ interactions

These receive contributions from two sources: (1) the analog of the fourth term in the RHS of (5.55) corresponding to the $SU(2)_L$ gauge group and (2) the analog of the first term, for the two Higgs superfields, for the $SU(2)_L \times U(1)$ gauge group. It is clear that only the gauge field part of the covariant derivative will contribute to the interaction. With I, J as gauge group representation indices and subscript s $(= 1, 2)$ distinguishing the two higgsino two component spinors, the latter reads,

$$-\overline{\tilde{h}_{sI}}\, \bar{\sigma}^\mu \left(\frac{g_Y}{2} Y_{h_s} \delta_{IJ} B_\mu + \frac{g_2}{2} (\tau^a)_{IJ} W_\mu^a \right) \tilde{h}_{sJ} \ .$$

The resulting charged weak boson terms in the Lagrangian density, expressed in the four component notation and in the weak basis after using (3.28c,d), read:

$$\mathcal{L}_{\tilde{\chi}_k^\pm \tilde{\chi}_l^0 W^\pm} = g_2 W_\mu^- \left[\overline{\tilde{\lambda}_3} \gamma^\mu \tilde{\lambda}^+ - \frac{1}{\sqrt{2}} \left(\overline{\tilde{h}_2^0} \gamma^\mu P_L \tilde{h}^+ - \overline{\tilde{h}_1^0} \gamma^\mu P_R \tilde{h}^+ \right) \right] + \text{h.c.} \qquad (9.32)$$

One can rewrite the interaction (9.32) in terms of chargino and neutralino fields by using the chargino and neutralino mixing matrices using (9.17,18) and (9.30,31). The final expression is

$$\mathcal{L}_{\tilde{\chi}_k^\pm \tilde{\chi}_l^0 W^\pm} = g_2 W_\mu^- \overline{\tilde{\chi}_l^0} \gamma^\mu \left(C_{lk}^L P_L + C_{lk}^R P_R \right) \tilde{\chi}_k^+ + \text{h.c.} \ , \qquad (9.33)$$

where the couplings C_{lk}^L and C_{lk}^R are given by

$$C_{lk}^L = -\frac{1}{\sqrt{2}} Z_{l4} V_{k2}^* + Z_{l2} V_{k1}^* \ , \qquad (9.34a)$$

$$C_{lk}^R = \frac{1}{\sqrt{2}} Z_{l3}^* U_{k2} + Z_{l2}^* U_{k1} \ . \qquad (9.34b)$$

In (9.33) and (9.34) the subscript k takes values $1, 2$ while l goes from 1 to 4. The generic vertex corresponding to (9.33) is shown in Fig. 9.3. Note that an arrow has been put on the Majorana fermion line also in accordance with the convention in Appendix D of the first paper of Ref. [9.10].

Fig. 9.3 is included in Appendix A

Neutralino-Neutralino-Z and Chargino-Chargino-(Z, γ) interactions

In the four component basis of (9.17) and (9.30), after using the Majorana identities (3.29c,d) and the definitions (8.7), we can write in analogy with the previous case

$$\mathcal{L}_{Z(\gamma)\tilde{\chi}\tilde{\chi}} = \frac{g_2}{c_W} Z_\mu \left(-c_W^2 \overline{\tilde{\lambda}^+} \gamma^\mu \tilde{\lambda}^+ - \frac{1}{2}\cos 2\theta_W \overline{\tilde{h}^+} \gamma^\mu \tilde{h}^+ \right)$$

$$+ \frac{g_2}{4c_W} Z_\mu \left(\overline{\tilde{h}_1^0} \gamma^\mu \gamma_5 \tilde{h}_1^0 - \overline{\tilde{h}_2^0} \gamma^\mu \gamma_5 \tilde{h}_2^0 \right)$$

$$- e A_\mu \left(\overline{\tilde{\lambda}^+} \gamma^\mu \tilde{\lambda}^+ + \overline{\tilde{h}^+} \gamma^\mu \tilde{h}^+ \right) . \tag{9.35}$$

The second line of (9.35), when rewritten in terms of the mass eigenstates $\tilde{\chi}_\ell^0$, yields the $Z\tilde{\chi}_l^0 \tilde{\chi}_n^0$ interaction. The use of (9.31) leads to

$$\mathcal{L}_{Z\tilde{\chi}_l^0 \tilde{\chi}_n^0} = \frac{g_2}{2c_W} Z_\mu \overline{\tilde{\chi}_l^0} \gamma^\mu \left(N_{ln}^L P_L + N_{ln}^R P_R \right) \tilde{\chi}_n^0 . \tag{9.36}$$

Here the couplings $N_{ln}^{L,R}$ are given by

$$N_{ln}^L = -\frac{1}{2} Z_{l3} Z_{n3}^* + \frac{1}{2} Z_{l4} Z_{n4}^* , \tag{9.37a}$$

$$N_{ln}^R = - \left(N_{ln}^L \right)^* . \tag{9.37b}$$

Referring back to (9.27), note that, under the assumption of a real $\mathcal{M}^n$, the $Z\tilde{\chi}_l^0 \tilde{\chi}_{l'}^0$ interaction will always involve a pure vector (axial vector) coupling, for a negative (positive) value of $\cos[2\mathrm{Arg}(Z_{ln}Z_{l'n})]$. In this situation the cosine is just a signature factor.

Turning to charginos, the last term in the RHS of (9.35) gives the $\gamma\tilde{\chi}_k^+ \tilde{\chi}_k^-$ interaction as

$$\mathcal{L}_{\gamma\tilde{\chi}_k^- \tilde{\chi}_k^+} = -e A_\mu \left[\overline{\tilde{\chi}_m^+} \gamma^\mu \left(\mathcal{V}_{m1}\mathcal{V}_{k1}^* + \mathcal{V}_{m2}\mathcal{V}_{k2}^* \right) P_L \right.$$

$$\left. + \overline{\tilde{\chi}_m^+} \gamma^\mu \left(\mathcal{U}_{m1}^*\mathcal{U}_{k1} + \mathcal{U}_{l2}^*\mathcal{U}_{k2} \right) P_R \right] \tilde{\chi}_k^+$$

$$= -e A_\mu \overline{\tilde{\chi}_k^+} \gamma^\mu \tilde{\chi}_k^+ , \tag{9.38}$$

where we have used $\mathcal{U}^\dagger\mathcal{U} = \mathcal{V}^\dagger\mathcal{V} = \mathbb{1}$. Finally, the $Z\tilde{\chi}_m^- \tilde{\chi}_k^+$ interaction follows from the first RHS term of (9.35). Rewritten in terms of the mixing angles, it reads

$$\mathcal{L}_{Z\tilde{\chi}_m^- \tilde{\chi}_k^+} = \frac{g_2}{c_W} Z_\mu \overline{\tilde{\chi}_m^+} \gamma^\mu \left(O_{mk}^L P_L + O_{mk}^R P_R \right) \tilde{\chi}_k^+ , \tag{9.39}$$

with the couplings $O_{mk}^{L,R}$ given by

$$O_{mk}^L = -\mathcal{V}_{m1}\mathcal{V}_{k1}^* - \frac{1}{2}\mathcal{V}_{m2}\mathcal{V}_{k2}^* + \delta_{mk} s_W^2 , \tag{9.40a}$$

$$O_{mk}^R = -\mathcal{U}_{m1}^*\mathcal{U}_{k1} - \frac{1}{2}\mathcal{U}_{m2}^*\mathcal{U}_{k2} + \delta_{mk} s_W^2 . \tag{9.40b}$$

The unitarity properties of the $\mathcal{V}, \mathcal{U}$ matrices have again been used in deriving (9.39). The vertices corresponding to (9.36), (9.38) and (9.39) are given in Fig. 9.4. Those corresponding to (9.36) have an additonal factor of 2 in the Feynman rules [9.10] which appears due to $\tilde{\chi}_l^0$ being Majorana fermions. Once again, we have put arrows [9.10] on lines corresponding to the latter.

$$\boxed{\text{Fig. 9.4 is included in Appendix A}}$$

9.4　Masses and Mixing Patterns of Sfermions

Slepton mass terms

There are three sources of slepton mass terms in the Lagrangian density: 1) explicit mass terms as well as trilinear A-terms from the soft part of the scalar potential V_{SOFT}, cf. (9.3), 2) the contribution to the scalar potential by the F-terms of (8.34), which arise out of the superpotential W of (8.33) and 3) the contribution to the scalar potential from the D-terms given by (8.35). The F- and D-contributions, as well as those from the trilinear terms in V_{SOFT}, materialize after the neutral Higgs fields acquire nonvanishing VEVs on the spontaneous breakdown of the $SU(2)_L \times U(1)_Y$ symmetry. On the other hand, each sfermion mass term in V_{SOFT} is invariant under $SU(2)_L \times U(1)_Y$ transformations. If all sfermions are heavier than the electroweak gauge bosons, as indicated by present null search experiments, their large masses could be due to these terms. The pieces in the sfermion mass terms due to trilinear scalar couplings, as well as the terms which orginate from the higgsino mass term in the superpotential W, mix the left and right sleptons $\tilde{e}_{iR}$ and $\tilde{e}_{jL}$. Depending upon the nature of V_{SOFT}, there can also be generation mixing for charged sleptons. However, we will show later that, under some simple assumptions about V_{SOFT}, one can often neglect some of the generation mixing in the slepton sector, once one has imposed the restrictions implied by strong experimental limits that exist on the nonconservation of lepton flavor.

The relevant terms in V (and hence in $-\mathcal{L}$), which contribute to slepton masses, can be written, using (9.3),(8.33)-(8.35) and (8.36), as

$$V^{\tilde{\ell}} = V^{\tilde{\ell}}_{SOFT} + V^{\tilde{\ell}}_F + V^{\tilde{\ell}}_D \ . \tag{9.41}$$

The different terms in the RHS of (9.41) can be shown, with repeated indices summed, as follows:

$$V^{\tilde{\ell}}_{SOFT} = \tilde{\ell}^*_{iL}(\mathcal{M}^2_{\tilde{\ell}})_{ij}\tilde{\ell}_{jL} + \tilde{e}^*_{iR}(\mathcal{M}^2_{\tilde{e}})_{ij}\tilde{e}_{jR} + \left[h_1\tilde{\ell}_{iL}(f^e A^e)_{ij}\tilde{e}^*_{jR} + \text{h.c.}\right], \tag{9.42a}$$

$$V^{\tilde{\ell}}_F = \left|\mu^\star h_2^- - \tilde{\nu}^*_i f^{e*}_{ij}\tilde{e}_{jR}\right|^2 + \left|\mu^\star h_2^{0\star} - \tilde{e}^*_{iL} f^{e*}_{ij}\tilde{e}_{jR}\right|^2$$

$$+ \sum_i \left|f^e_{ji}h_1\tilde{\ell}_{jL}\right|^2 + f^e_{ij}f^{e*}_{ij'}\tilde{e}^*_{jR}\tilde{e}_{j'R}\left(|h_1^0|^2 + h_1^+ h_1^-\right), \tag{9.42b}$$

$$V^{\tilde{\ell}}_D = \frac{1}{4}g^2_Y(|h_1|^2 - |h_2|^2)\sum_i\left(|\tilde{\ell}_{iL}|^2 - 2|\tilde{e}_{iR}|^2\right)$$

$$+ \frac{1}{4}g^2_2\left(h_1^\dagger\vec{\tau}h_1 + h_2^\dagger\vec{\tau}h_2\right)\tilde{\ell}^\dagger_{iL}\vec{\tau}\tilde{\ell}_{iL} \ . \tag{9.42c}$$

When the neutral Higgs fields acquire vacuum expectation values, as per (8.21), (9.42) lead to the following mass terms in the Lagrangian density.

$$-\mathcal{L}^{\tilde{\ell}}_m = \tilde{\nu}^*_i\left(\mathcal{M}^2_{\tilde{\ell}} + M^2_Z\cos 2\beta\,(1/2)\mathbb{1}\right)_{ij}\tilde{\nu}_j$$

$$+ \tilde{e}^*_{iL}\left[\mathcal{M}^2_{\tilde{\ell}} - M^2_Z\cos 2\beta\,(1/2 - \sin^2\theta_W)\mathbb{1} + m^2_{e_i}\mathbb{1}\right]_{ij}\tilde{e}_{jL}$$

$$+ \tilde{e}^*_{iR}\left(\mathcal{M}^2_{\tilde{e}} - M^2_Z\cos 2\beta\sin^2\theta_W\mathbb{1} + m^2_{e_i}\mathbb{1}\right)_{ij}\tilde{e}_{jR}$$

$$- \left[\tilde{e}^*_{iL}(m_{e_i}A^{e*}_{ij} + m_{e_i}\delta_{ij}\mu\tan\beta)\tilde{e}_{jR} + \text{h.c.}\right]. \tag{9.43}$$

In writing the above mass term, we have absorbed the electroweak couplings and VEVs $v_{1,2}$ in M_Z, β and θ_W; m_{e_i} stands for the mass of the charged lepton e_i (cf. 8.23) and $\mathbb{1}_{ij} = \delta_{ij}$. Moreover, we have used (8.23a) for f_{ij}^e. The choice of the signs in front of the $f^e A^e$ etc. terms in (9.3) was made in accordance with the convention established in §7.7 and determines the sign of the A^e-term in (9.43). The opposite signs of the terms proportional to $M_Z^2 \cos 2\beta \sin^2 \theta_W$ in the second and third lines of (9.43) are noteworthy. The coefficient of this term is essentially decided by the electric charge of the slepton field. The chiral superfield $\bar{E}_i$ contains $\tilde{e}_{iR}^*$ and hence carries the electric charge of the positron, unlike L_i containing $\tilde{e}_{iL}$ with the opposite charge. Furthermore, the term containing $M_Z^2 \cos 2\beta$ is proportional to T_{3f}^L and hence changes sign between the left selectron and the sneutrino. Clearly, the states $\tilde{e}_{iL}$ $\tilde{e}_{iR}^*$ and $\tilde{\nu}_i$, which appear in (9.43), are the interaction eigenstates; the corresponding mass eigenstates will be linear combinations of these. In principle, both lepton flavor mixing[5] as well as L-R mixing are now possible.

Squark mass terms

The supersymmetric and nonsupersymmetric mass terms for squark fields can be written in a manner analogous to that for slepton ones with the correspondence $\tilde{\ell}_L \to \tilde{q}_L$, $\tilde{\nu}_i \to \tilde{u}_{iL}$, $\tilde{e}_{iL,R} \to \tilde{d}_{iL,R}$. Just the additional singlet fields $\tilde{u}_{iR}$, that are present, need to be included. Moreover, the nontrivial CKM mixing, present in the quark sector, needs to be taken into account. Expressions similar to those appearing in (9.42) can be written for the squark scalar potential. In the following we first write the relevant part of the squark scalar potential which will contribute to squark masses as

$$V^{\tilde{q}} = V_{SOFT}^{\tilde{q}} + V_F^{\tilde{q}} + V_D^{\tilde{q}} , \tag{9.44}$$

without any specific assumptions about the supersymmetry breaking parametric matrices A^d, A^u. We then have

$$
\begin{aligned}
V_{SOFT}^{\tilde{q}} =\ & \tilde{q}_{iL}^\dagger (\mathcal{M}_{\tilde{q}}^2)_{ij} \tilde{q}_{jL} + \tilde{d}_{iR}^\dagger (\mathcal{M}_{\tilde{d}}^2)_{ij} \tilde{d}_{jR} + \tilde{u}_{iR}^\dagger (\mathcal{M}_{\tilde{u}}^2)_{ij} \tilde{u}_{jR} \\
& + \left[h_1 \cdot \tilde{q}_{iL} (f^d A^d)_{ij} \tilde{d}_{jR}^\star + \tilde{q}_{iL} \cdot h_2 (f^u A^u)_{ij} \tilde{u}_{jR}^\star + \text{h.c.} \right] ,
\end{aligned}
\tag{9.45a}
$$

$$
\begin{aligned}
V_F^{\tilde{q}} =\ & \left| \mu^\star h_2^- - \tilde{u}_{iL}^\dagger f_{ij}^{d\star} \tilde{d}_{jR} \right|^2 + \left| \mu^\star h_2^{0\star} - \tilde{d}_{iL}^\dagger f_{ij}^{d\star} \tilde{d}_{jR} \right|^2 + \sum_i \left| f_{ji}^d h_1 \cdot \tilde{q}_{jL} \right|^2 \\
& + \left| -\mu^\star h_1^+ + \tilde{d}_{iL}^\dagger f_{ij}^u \tilde{u}_{jR} \right|^2 + \left| -\mu^\star h_1^{0\star} + \tilde{u}_{iL}^\dagger f_{ij}^{u\star} \tilde{u}_{jR} \right|^2 + \sum_i \left| f_{ji}^{u\star} h_2 \cdot \tilde{q}_{jL} \right|^2 \\
& + \sum_i \left| f_{ij}^{d\star} h_1^{0\star} \tilde{d}_{jR} - f_{ij}^{u\star} h_2^- \tilde{u}_{jR} \right|^2 + \sum_i \left| f_{ij}^{d\star} h_1^+ \tilde{d}_{jR} - f_{ij}^{u\star} h_2^{0\star} \tilde{u}_{jR} \right|^2 ,
\end{aligned}
\tag{9.45b}
$$

$$
\begin{aligned}
V_D^{\tilde{q}} =\ & \tfrac{1}{4} g_Y^2 \left(|h_1|^2 - |h_2|^2 \right) \left[-\tfrac{1}{3} \tilde{q}_{iL}^\dagger \tilde{q}_{iL} + \sum_i 2 \left(Q_u |\tilde{u}_{iR}|^2 + Q_d |\tilde{d}_{iR}|^2 \right) \right] \\
& + \tfrac{1}{4} g_2^2 \left(h_1^\dagger \vec{\tau} h_1 + h_2^\dagger \vec{\tau} h_2 \right) \cdot \tilde{q}_{iL}^\dagger \vec{\tau} \tilde{q}_{iL} ,
\end{aligned}
\tag{9.45c}
$$

---

[5]There are currently some scenarios, going beyond the MSSM, which anticipate a large $\tilde{\nu}_\mu$-$\tilde{\nu}_\tau$ mixing in analogy with what is observed in the ν_μ-ν_τ sector by the super-Kamiokande experiment.

where $Q_{u,d}$ are the electric charges of the u, d-type squarks in units of the positron charge. From (9.49) the squark mass terms in the Lagrangian density can be written as

$$
\begin{aligned}
-\mathcal{L}_m^{\tilde{q}} \;=\; & \tilde{u}_{iL}^* \left[\mathcal{M}_{\tilde{q}}^2 + M_Z^2 \cos 2\beta \, (1/2 - Q_u \sin^2 \theta_W)\, \mathbb{1} + (\mathbf{m_u m_u}^\dagger) \right]_{ij} \tilde{u}_{jL} \\
& + \tilde{d}_{iL}^* \left[\mathcal{M}_{\tilde{q}}^2 - M_Z^2 \cos 2\beta \, (1/2 + Q_d \sin^2 \theta_W)\, \mathbb{1} + (\mathbf{m_d m_d}^\dagger) \right]_{ij} \tilde{d}_{jL} \\
& + \tilde{u}_{iR}^* \left[\mathcal{M}_{\tilde{u}}^2 + Q_u M_Z^2 \cos 2\beta \sin^2 \theta_W\, \mathbb{1} + (\mathbf{m_u}^\dagger \mathbf{m_u}) \right]_{ij} \tilde{u}_{jR} \\
& + \tilde{d}_{iR}^* \left[\mathcal{M}_{\tilde{d}}^2 + Q_d M_Z^2 \cos 2\beta \sin^2 \theta_W\, \mathbb{1} + (\mathbf{m_d}^\dagger \mathbf{m_d}) \right]_{ij} \tilde{d}_{jR} \\
& - \tilde{u}_{iL}^* \left[(\mathbf{m_u} A^{u^*})_{ij} + \mu(\mathbf{m_u})_{ij} \cot \beta \right] \tilde{u}_{jR} + \text{h.c.} \\
& - \tilde{d}_{iL}^* \left[(\mathbf{m_d} A^{d^*})_{ij} + \mu(\mathbf{m_d})_{ij} \tan \beta \right] \tilde{d}_{jR} + \text{h.c.}
\end{aligned}
\tag{9.46}
$$

In (9.50) $\mathbf{m}_u$ and $\mathbf{m}_d$ are the up and down type quark mass matrices respectively in generation space (cf. 8.11). One may note that , just as with sleptons, the squarks are massive even in the limit of unbroken $SU(2)_L \times U(1)_Y$ symmetry. Once more, there is a relative negative sign between the mass terms for left squarks and right squarks for the pieces proportional to the charge Q_u and Q_d. The mixing between the left and right squark fields, given in the last two RHS terms, is caused by the trilinear A-terms as well as by the higgsino mass contribution to the F-terms. Because of extant mixing in the quark sector, both L-R mixing and generation mixing are nontrivial and complicated for squarks.

Sfermion mixing: some generalities

Let us define a six component vector

$$
\tilde{\mathbf{f}} = \begin{pmatrix} \tilde{f}_L \\ \tilde{f}_R \end{pmatrix},
\tag{9.47}
$$

where $\tilde{f}_L, \tilde{f}_R$ are each a three component column vector in generation space with components $\tilde{f}_{iL}, \tilde{f}_{iR}, \tilde{f}$ being the superpartner of any matter fermion field f, quark or lepton. Thus $\tilde{f}$ can be $\tilde{\nu}, \tilde{e}, \tilde{u}, \tilde{d}$ except that we put $\tilde{\nu}_R = 0$. The general squared mass matrix for such sfermions can then be written as a 2×2 Hermitian matrix of 3×3 blocks in the space spanned by the vector of (9.47):

$$
\mathcal{M}_{\tilde{\mathbf{f}}}^2 = \begin{pmatrix} \mathcal{M}_{\tilde{f}_{LL}}^2 & \mathcal{M}_{\tilde{f}_{LR}}^2 \\ \mathcal{M}_{\tilde{f}_{LR}}^{2\dagger} & \mathcal{M}_{\tilde{f}_{RR}} \end{pmatrix}.
\tag{9.48}
$$

In (9.48) $\mathcal{M}_{\tilde{f}_{LL}}^2$ and $\mathcal{M}_{\tilde{f}_{RR}}^2$ are hermitian in generation space. Now all the sfermion mass terms of (9.43) and (9.46) can be collected under

$$
-\mathcal{L}_{SFERMION\ MASS} = \sum_{\tilde{f}} \tilde{\mathbf{f}}^\dagger \mathcal{M}_{\tilde{\mathbf{f}}}^2 \tilde{\mathbf{f}} \, .
\tag{9.49}
$$

Specifically, for sneutrinos, charged sleptons, u-squarks and d-squarks, we can respectively write from (9.43) and (9.46) the 6×6 squared mass matrices in terms of 3×3 submatrix blocks as

$$\mathcal{M}_{\tilde{\nu}}^2 = \begin{pmatrix} \mathcal{M}_{\tilde{\ell}}^2 + M_Z^2 T_{3L}^{\tilde{\nu}} \cos 2\beta \, \mathbb{1} & 0 \\ 0 & 0 \end{pmatrix}, \tag{9.50a}$$

$$\mathcal{M}_{\tilde{e}}^2 =$$
$$\begin{pmatrix} \mathcal{M}_{\tilde{\ell}}^2 + M_Z^2 (T_{3L}^{\tilde{e}} - Q_e \sin^2 \theta_W) \cos 2\beta \, \mathbb{1} + \mathbf{m}_e \mathbf{m}_e^\dagger & -\mathbf{m}_e (A^{e^*} + \mu \tan \beta) \\ -(A^{e^T} + \mu^\star \tan \beta) \mathbf{m}_e^\dagger & \mathcal{M}_{\tilde{e}}^2 + Q_e M_Z^2 \cos 2\beta \sin^2 \theta_W \, \mathbb{1} + \mathbf{m}_e^\dagger \mathbf{m}_e \end{pmatrix}, \tag{9.50b}$$

$$\mathcal{M}_{\tilde{u}}^2 =$$
$$\begin{pmatrix} \mathcal{M}_{\tilde{q}}^2 + M_Z^2 (T_{3L}^{\tilde{u}} - Q_u \sin^2 \theta_W) \cos 2\beta \, \mathbb{1} + \mathbf{m}_u \mathbf{m}_u^\dagger & -\mathbf{m}_u (A^{u^*} + \mu \cot \beta) \\ -(A^{u^T} + \mu^\star \cot \beta) \mathbf{m}_u^\dagger & \mathcal{M}_{\tilde{u}}^2 + Q_u M_Z^2 \cos 2\beta \sin^2 \theta_W \, \mathbb{1} + \mathbf{m}_u^\dagger \mathbf{m}_u \end{pmatrix}, \tag{9.50c}$$

$$\mathcal{M}_{\tilde{d}}^2 =$$
$$\begin{pmatrix} \mathcal{M}_{\tilde{q}}^2 + M_Z^2 (T_{3L}^{\tilde{d}} - Q_d \sin^2 \theta_W) \cos 2\beta \, \mathbb{1} + \mathbf{m}_d \mathbf{m}_d^\dagger & -\mathbf{m}_d (A^{d^*} + \mu \tan \beta) \\ -(A^{d^T} + \mu^\star \tan \beta) \mathbf{m}_d^\dagger & \mathcal{M}_{\tilde{d}}^2 + Q_d M_Z^2 \cos 2\beta \sin^2 \theta_W \, \mathbb{1} + \mathbf{m}_d^\dagger \mathbf{m}_d \end{pmatrix}. \tag{9.50d}$$

In (9.50) $T_{3L}^{\tilde{f}}$ is the third component of the weak isospin of $\tilde{f}_L$, Q_f the electromagnetic charge of f and $\mathbf{m}_f$ the mass matrix (cf. 8.11 and 8.12) for f in generation space, with $(m_e)_{ij}$ being of course $m_{e_i} \delta_{ij}$. However, $\mathcal{M}_{\tilde{f}}^2$ involves not only $\mathbf{m}_f$ but also the soft supersymmetry breaking squared mass matrices $\mathcal{M}^2$ both for the $SU(2)_L$ doublet left sfermions and for the $SU(2)_L$ singlet right sfermions plus the matrix A^f in generation space and finally the supersymmetric higgsino mass parameter μ. Note that A^f is in general a complex 3×3 matrix and μ can be complex too. Observe furthermore that the D-term contributions are diagonal in generation space. The offdiagonal LR mixing terms are proportional to fermion masses and hence appreciable only for the third generation. Otherwise, generation mixing is really controlled by the soft supersymmetry breaking terms.

Referring back to (9.47), we can define mass eigenstate sfermions through the six component column vector $\tilde{\mathbf{f}}^m$ which is unitarily transformed from $\tilde{\mathbf{f}}$:

$$\tilde{\mathbf{f}}^m = \mathbf{W}^{\tilde{f}\dagger} \tilde{\mathbf{f}} . \tag{9.51}$$

The 6×6 unitary matrices $\mathbf{W}^{\tilde{f}}$ then diagonalize the squared mass matrices $\mathcal{M}_{\tilde{f}}^2 \; \forall \tilde{f}$:

$$\mathbf{M}_{\tilde{f}}^{2(D)} = \mathbf{W}^{\tilde{f}\dagger} \mathcal{M}_{\tilde{f}}^2 \mathbf{W}^{\tilde{f}}. \tag{9.52}$$

Let us introduce the indices s, t running from 1 to 6 while we keep the generation indices as i, j running from 1 to 3. We make a convention to order the sfermions by mass, $\tilde{f}_1^m$ being the lightest and $\tilde{f}_6^m$ the heaviest among sfermions of a given charge. Eq. (9.51) can then be rewritten as

$$\tilde{f}_s^m = W_{ts}^{\tilde{f}^\star} \tilde{f}_t = W_{is}^{\tilde{f}^\star} \tilde{f}_{iL} + W_{i+3\,s}^{\tilde{f}^\star} \tilde{f}_{iR} , \tag{9.53}$$

the generation index i being summed on repetition. The second step of (9.53) shows the decomposition of a mass eigenstate sfermion field into left and right chiral interaction eigenstate sfermions. The latter can be written, by inverting (9.53), as

$$\tilde{f}_{iL} = W^{\tilde{f}}_{is} \tilde{f}^m_s, \tag{9.54a}$$

$$\tilde{f}_{iR} = W^{\tilde{f}}_{i+3\,s} \tilde{f}^m_s. \tag{9.54b}$$

Two limiting cases of the above most general sfermion mass mixing are also quite transparent.

(a) No _L-R_ mixing

In this case $\mathcal{M}_{\tilde{f}LR}$ vanishes and (9.48) reduces to

$$\mathcal{M}^2_{\tilde{f}} = \begin{pmatrix} \mathcal{M}^2_{\tilde{f}LL} & 0 \\ 0 & \mathcal{M}^2_{\tilde{f}RR} \end{pmatrix}. \tag{9.55}$$

Now the 6×6 unitary matrix $\mathbf{W}^{\tilde{f}}$ has the chiral block diagonal form

$$\mathbf{W}^{\tilde{f}} = \begin{pmatrix} \mathbf{U}^{\tilde{f}_L} & 0 \\ 0 & \mathbf{U}^{\tilde{f}_R} \end{pmatrix}, \tag{9.56}$$

where $\mathbf{U}^{\tilde{f}_L}$ and $\mathbf{U}^{\tilde{f}_R}$ are unitary submatrices for the distinct left and right sfermion sectors. In terms of explicit generation indices i, j $(= 1, 2, 3)$ we can write

$$W^{\tilde{f}}_{i\,j+3} = W^{\tilde{f}}_{i+3\,j} = 0 , \tag{9.57a}$$

$$W^{\tilde{f}}_{ij} = U^{\tilde{f}_L}_{ij}, \tag{9.57b}$$

$$W^{\tilde{f}}_{i+3\,j+3} = U^{\tilde{f}_R}_{ij}. \tag{9.57c}$$

The 3×3 unitary matrices $\mathbf{U}^{\tilde{f}_L}$ and $\mathbf{U}^{\tilde{f}_R}$ in generation space, appearing in (9.56) and (9.57), are sfermionic generalizations of the flavor rotation matrices $\mathbf{U}^{f_L}, \mathbf{U}^{f_R}$ for a chiral fermion f that we introduced for $f = u, d$ in Ch.8 to put the quark mass matrices $\mathbf{m}_u, \mathbf{m}_d$ into diagonal form via biunitary transformations. The chiral block submatrices of (9.50), for $f = \tilde{\nu}, \tilde{e}, \tilde{u}, \tilde{d}$, now have the following respective expressions after diagonalization.

$$\mathbf{M}^{2(D)}_{\tilde{\nu}} = \mathbf{U}^{\tilde{\nu}\dagger}(\mathcal{M}^2_{\tilde{\ell}} - M^2_Z \cos 2\beta\, T^{\tilde{\nu}}_{3L}\mathbb{1})\mathbf{U}^{\tilde{\nu}}, \tag{9.58a}$$

$$\mathbf{M}^{2(D)}_{\tilde{e}LL} = \mathbf{U}^{\tilde{e}_L\dagger} \left[\mathcal{M}^2_{\tilde{\ell}} + M^2_Z \cos 2\beta(T^{\tilde{e}}_{3L} - \sin^2\theta_W)\mathbb{1} + \mathbf{m}^{2(D)}_e\right] \mathbf{U}^{\tilde{e}_L}, \tag{9.58b}$$

$$\mathbf{M}^{2(D)}_{\tilde{e}RR} = \mathbf{U}^{\tilde{e}_R\dagger} \left[\mathcal{M}^2_{\tilde{e}} + Q_e M^2_Z \cos 2\beta \sin^2\theta_W \mathbb{1} + \mathbf{m}^{2(D)}_e\right] \mathbf{U}^{\tilde{e}_R}, \tag{9.58c}$$

$$\mathbf{M}^2_{\tilde{u}LL} = \mathbf{U}^{\tilde{u}_L\dagger} \left[\mathcal{M}^2_{\tilde{u}} + M^2_Z \cos 2\beta(T^{\tilde{u}}_{3L} - Q_u \sin^2\theta_W)\mathbb{1} + \mathbf{m}^\dagger_u\mathbf{m}_u\right] \mathbf{U}^{\tilde{u}_L}, \tag{9.58d}$$

$$\mathbf{M}^{2(D)}_{\tilde{u}RR} = \mathbf{U}^{\tilde{u}_R\dagger} \left[\mathcal{M}^2_{\tilde{u}} + Q_u M^2_Z \cos 2\beta \sin^2\theta_W \mathbb{1} + \mathbf{m}^\dagger_u\mathbf{m}_u\right] \mathbf{U}^{\tilde{u}_R}, \tag{9.58e}$$

$$\mathbf{M}^{2(D)}_{\tilde{d}LL} = \mathbf{U}^{\tilde{d}_L\dagger} \left[\mathcal{M}^2_{\tilde{d}} + M^2_Z \cos 2\beta(T^{\tilde{d}}_{3L} - Q_d \sin^2\theta_W)\mathbb{1} + \mathbf{m}^\dagger_d\mathbf{m}_d\right] \mathbf{U}^{\tilde{d}_L}, \tag{9.58f}$$

$$\mathbf{M}^{2(D)}_{\tilde{d}RR} = \mathbf{U}^{\tilde{d}_R\dagger} \left[\mathcal{M}^2_{\tilde{d}} + Q_{\tilde{d}} M^2_Z \cos 2\beta \sin^2\theta_W \mathbb{1} + \mathbf{m}^\dagger_d\mathbf{m}_d\right] \mathbf{U}^{\tilde{d}_R}. \tag{9.58g}$$

Note that mass eigenstate sfermions will now be ordered by mass within $s = 1, 2, 3$ for left sfermions and within $s = 4, 5, 6$ for right sfermions, i.e. now we have

$$\text{mass}(f_1^m) < \text{mass}(f_2^m) < \text{mass}(f_3^m), \quad \text{(left sfermions)} , \tag{9.59a}$$

$$\text{mass}(f_4^m) < \text{mass}(f_5^m) < \text{mass}(f_6^m), \quad \text{(right sfermions)} , \tag{9.59b}$$

without any definite ordering between the two groups. Thus a program, made to diagonalize the original 6×6 matrix, will not automatically return a block diagonal mixing matrix as in (9.56) since the program will insist on all mass eigenstate sfermions being ordered according to their masses. The latter can be obtained just by interchanging certain rows and columns of $\mathbf{W}^{\tilde{f}}$ without affecting physics.

(b) <u>No flavor mixing</u>

In this limit the 6×6 mixing matrix only couples the two sfermionic states labelled by the indices i and $i + 3$, i.e. the left and the right states of a given flavor. For a real mass matrix, one has

$$W_{ii}^{\tilde{f}} = W_{i+3\ i+3}^{\tilde{f}} = \cos \theta_{\tilde{f}_i} , \tag{9.60a}$$

$$W_{i\ i+3}^{\tilde{f}} = -W_{i+3\ i}^{\tilde{f}} = -\sin \theta_{\tilde{f}_i} . \tag{9.60b}$$

Thus, for instance, mass eigenstate charged sleptons will now be described by

$$\mathbf{\tilde{f}}^m = \begin{pmatrix} \tilde{e}_1 \\ \tilde{\mu}_1 \\ \tilde{\tau}_1 \\ \tilde{e}_2 \\ \tilde{\mu}_2 \\ \tilde{\tau}_2 \end{pmatrix}, \tag{9.61}$$

i.e. the mass ordering is enforced between f_i^m and f_{i+3}^m and not between different flavor states.

Before closing this section, we want to comment specifically on the squared mass matrices of staus, sbottoms and stops. These third generation sleptons and squarks are somewhat special. It is reasonable to take them to be decoupled from other sleptons and squarks i.e. assume no flavor mixing for them. On the other hand, they do involve substantial L-R mixing on account of the nonnegligible masses of their fermion partners. Indeed, they physically manifest themselves as the mass eigenstates $\tilde{\tau}_{1,2}, \tilde{b}_{1,2}$ and $\tilde{t}_{1,2}$. In this picture the corresponding squared mass matrices can be written approximately in 2×2 form

$$\mathcal{M}_{\tilde{\tau}}^2 = \begin{pmatrix} m_{\tilde{\ell}_3}^2 - (1/2 - \sin^2 \theta_W) M_Z^2 \cos 2\beta + m_\tau^2 & -m_\tau(A^{\tau^*} + \mu \tan \beta) \\ -m_\tau(A^\tau + \mu^* \tan \beta) & m_{\tilde{\tau}}^2 - M_Z^2 \cos 2\beta \sin^2 \theta_W + m_\tau^2 \end{pmatrix}, \tag{9.62a}$$

$$\mathcal{M}_{\tilde{b}}^2 = \begin{pmatrix} m_{\tilde{q}_3}^2 - (1/2 - 1/3 \sin^2 \theta_W) M_Z^2 \cos 2\beta + m_b^2 & -m_b(A^{b^*} + \mu \tan \beta) \\ -m_b(A^b + \mu^* \tan \beta) & m_{\tilde{b}}^2 - 1/3 M_Z^2 \cos 2\beta \sin^2 \theta_W + m_b^2 \end{pmatrix},$$

$$\tag{9.62b}$$

$$\mathcal{M}_{\tilde{t}}^2 = \begin{pmatrix} m_{\tilde{q}3}^2 + (1/2 - 2/3\sin^2\theta_W)M_Z^2\cos 2\beta + m_t^2 & -m_t(A^{t\star} + \mu\cot\beta) \\ -m_t(A^t + \mu^\star\cot\beta) & m_{\tilde{t}}^2 + 2/3M_Z^2\cos 2\beta\sin^2\theta_W + m_t^2 \end{pmatrix}.$$

(9.62c)

The off-diagonal L-R mixing term is particularly large in the stop case, being proportional to the mass of the top quark. This can in principle make $\tilde{t}_1$ the lightest sfermion.

9.5   The Flavor Problem in Supersymmetry

Many discussions in previous sections have hinted that there is a generic flavor problem [9.8] in supersymmetric theories. The origin of the problem is in the occurrence of sizable flavor dependence in sfermion mass matrices. The latter naturally leads to large induced FCNC amplitudes which are, however, unobserved by experiment. The lack of observation of the decay $\mu \to e\gamma$ puts some constraints on the lepton-slepton sector. Though processes like $D^0 \leftrightarrow \bar{D}^0$ and $B^0 \leftrightarrow \bar{B}^0$ transitions as well as $b \to s\gamma$ decay yield constraints on the quark-squark sector, the most stringent restrictions here come from what is already known about K^0-$\bar{K}^0$ mixing. Let us elaborate on this last statement by following the treatment of Hagelin et al [9.8]. At the one loop level the box diagram of Fig. 9.5 can induce an operator such as $\bar{d}_L\gamma_\mu s_L\,\bar{s}_L\gamma^\mu d_L$

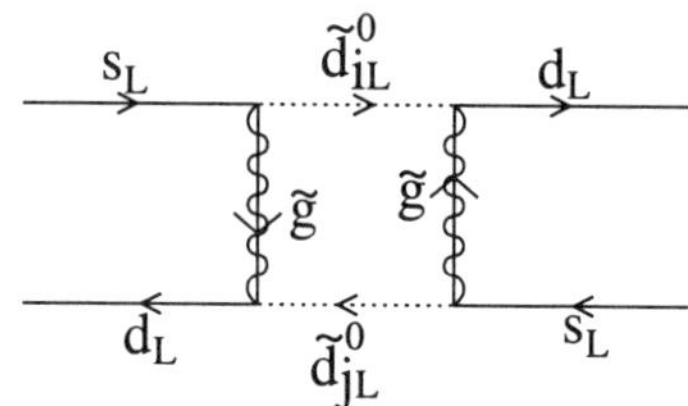

Fig. 9.5 One loop squark induced K^0-$\bar{K}^0$ mixing

into the effective Lagrangian density contributing to the said mixing. From the product of two squark propagators and four elements of the matrix $\mathbf{U}^{\tilde{d}_L}$ of (9.56) in this diagram, the transition amplitude for $\bar{s}_L d_L \to \bar{d}_L s_L$ picks up a factor[6]

$$\sum_i \frac{U_{di}^{\tilde{d}_L} U_{is}^{\tilde{d}_L\dagger}}{k^2 - m_{\tilde{d}_i}^2 + i\epsilon} \sum_j \frac{U_{dj}^{\tilde{d}_L} U_{js}^{\tilde{d}_L\dagger}}{k^2 - m_{\tilde{d}_j}^2 + i\epsilon},$$

k being the loop momentum. We have set all external momenta to zero because $m_K^2 \ll m_{\tilde{q}}^2$. Since the unitarity of $\mathbf{U}^{\tilde{d}_L}$ makes this factor vanish in case $m_{\tilde{d}_i}^2$ is the same for $\tilde{d}_i = \tilde{d}, \tilde{s}, \tilde{b}$, it can be rewritten as

$$\frac{1}{(k^2 - m_{\tilde{d}}^2 + i\epsilon)^4} \left| \sum_i U_{si}^{\tilde{d}_L} U_{id}^{\tilde{d}_L\dagger} \Delta m_{\tilde{d}_i}^2 \right|^2 + \mathcal{O}\left([k^2 - m_{\tilde{d}}^2]^{-5}\right),$$

---

[6]We work in an interaction basis where the down quark mass matrix is diagonal.

where $m_{\tilde{d}}^2$ is an average mass squared for charge $-1/3$ squarks and $m_{\tilde{d}_i}^2 = m_{\tilde{d}}^2 + \Delta m_{\tilde{d}_i}^2$. The aforementioned transition amplitude has the dimensionality of an inverse mass squared. So after inserting the product of the two gluino propagators and four powers of the QCD coupling strength g_s and performing the loop integration, one is left with an amplitude proportional to

$$\frac{g_s^4}{\tilde{m}^6} \left| \sum_i U_{di}^{\tilde{d}_L} U_{is}^{\tilde{d}_L^\dagger} \Delta m_{\tilde{d}_i}^2 \right|^2 ,$$

where $\tilde{m} = \max(m_{\tilde{q}}, M_{\tilde{g}})$, i.e. the larger of the squark and gluino masses. With $|\Delta m_{\tilde{d}_i}| \sim m_{\tilde{d}_i} = \mathcal{O}(10^2)$ GeV, this yields a contribution which is three orders of magnitude larger than that from the SM. The latter obtains through the replacement of the gluino lines by $W^\pm$ lines and of the squark lines by u_i-quark ones in Fig. 9.5 and reproduces the observed value of the K_L-K_S mass difference [9.13] rather well.

The above discussion raises an important question : how can such undesirable amplitudes be suppressed in supersymmetric theories? The structure of the expression in the above paragraph implies that there are basically three ways in which a suppression of the desired nature can be achieved. One may also consider various combinations of these options. We shall describe these three possibilities one by one. Note that we keep our focus on the quark-squark sector here. Analogous arguments do apply to the lepton-slepton sector, though with a certain simplification; flavor mixing among leptons can be neglected – at least in the limit of vanishing neutrino masses. Thus constraints, from the yet unobserved $\mu \to e\gamma$ decay and muon conversion to electron in atoms, can also be taken care of.

The first choice is to make the prefactor in the said expression small, i.e. to take [9.14] the masses of sfermions of the first two generations to be very large, in the multi-TeV range. Of course, the naturalness argument, discussed in Ch.1, requires one to keep third generation sfermion and Higgs boson masses at or below the TeV scale. However, the smallness of first and second generation Yukawa couplings allows the choice of quite large masses for the corresponding sfermions without destabilizing the hierarchy[7]. This is a "brute force" solution of the flavor problem, since all loop corrections involving internal first or second generation sfermions and external fermion or gauge boson legs are then suppressed, including in particular those corrections that give rise to FCNC transitions. The prevention of unacceptably large loop corrections from the hypercharge $U(1)_Y$ D-terms to Higgs masses requires the condition $\sum_i Y_i m_{\tilde{f}_i}^2 \lesssim \mathcal{O}(1)$ TeV^2. Another problem arises in any attempt to implement such a spectrum at a high energy scale: two loop contributions to the renormalization group equations due to $SU(3)_C$ interactions tend to drive the squared stop masses to negative values [9.15], leading to color and/or charge symmetry breaking. On the positive side, this kind of model also easily satisfies constraints on flavor conserving CP violating amplitudes. In particular, those from the yet unobserved electric dipole moments of the neutron and electron are respected even though all soft supersymmetry breaking parameters have CP violating phases of $\mathcal{O}(1)$. This kind of "inverted hierarchy" model sometimes goes under the name *more minimal* or *Effective Supersymmetry*, since first and second generation sfermions essentially decouple from physics at energies that will be accessible in the foreseeable future at collider experiments.

---

[7]If $\tan\beta$ is small, the $\tilde{b}_R, \tilde{\tau}$ and h masses can also be made large.

The second strategy [9.16] is to assume a (presumably dynamically generated) **alignment** between the fermion and sfermion mass matrices so that both can be made diagonal in the same basis. In fact, in that case the mixing matrix, appearing in the expression for the box graph of Fig. 9.5, is diagonal. The expression then vanishes and the problem is solved. This is actually only a partial solution, since, owing to nontrivial CKM mixing, $\mathcal{M}_{\tilde{q}}^2$ cannot commute simultaneously with the u and d quark mass matrices (except when $\mathcal{M}_{\tilde{q}}^2$ is proportional to the unit matrix; this case will be treated below). As already stated (see also Table 9.1 below), by far the most stringent constraints come from the kaon sector. Models of alignment hence usually assume that $\mathcal{M}_{\tilde{q}}^2$ is aligned with the d quark mass matrix. Since CKM mixing angles are in fact quite small, an approximate alignment with the u quark mass matrix then also obtains. However, generically one would expect nonnegligible D^0-$\bar{D}^0$ mixing in this class of models.

The third option is to assume a high degree of mass degeneracy, or **universality of masses**, among sfermions with given $SU(2)_L \times U(1)_Y$ quantum numbers (including electro-magnetic charge) but occurring in different generations. In this scenario the K^0-$\bar{K}^0$ mixing expression is suppressed because the $\Delta m_{\tilde{d}_i}^2$ are very small. Large flavor mixing is possible in this option if on-shell sparticles can be produced,[8] but FCNC amplitudes, involving only SM particles as external legs, are suppressed by a super-GIM mechanism. In practice, it suffices to assume a near mass degeneracy between sfermions of the first and second generations; experimental flavor mixing constraints on the third generation are weak, mostly because the SM contribution to B^0-$\bar{B}^0$ mixing is quite large, and has a sizable theoretical uncertainty. Indeed, with substantial L-R mixing, one may expect $\tilde{\tau}_1$, $\tilde{b}_1$ and $\tilde{t}_1$ to be significantly lighter than the corresponding mass degenerate charge -1, charge $-1/3$ and charge $2/3$ sfermions of the first two generations, respectively. Note that FCNC constraints do not lead to any relations between, say, $\mathcal{M}_{\tilde{u}}^2$, $\mathcal{M}_{\tilde{d}}^2$ and $\mathcal{M}_{\tilde{q}}^2$. As will be shown in more detail in Chs.12 and 13, specific models with high scale supersymmetry breaking nevertheless do usually imply a high degree of degeneracy between first and second generation squarks with different $SU(2)_L \times U(1)_Y$ quantum numbers. On the other hand, in such models exact universality only holds at a high scale. Quantum corrections will typically lead to deviations from universality at the weak scale. We shall see later that many such models, while still compatible with the present constraints, therefore predict significant new contributions to certain FCNC processes. In the remaining sections of this chapter we shall hence present Feynman rules for sfermion interactions allowing for a completely general mixing between all six sfermions of a given electric charge.

Before coming to the Feynman rules, mentioned above, however, we would like to give a more quantitative discussion of the bounds on flavor violation in the sfermion sector. This can most easily be done using the *mass insertion* method [9.18]. In this approach one works in a basis where the mass matrix of quarks of a given charge as well as the corresponding quark-squark-neutral gaugino couplings are diagonal in flavor space. As a result, different bases need to be used for problems involving external d-type or external u-type quarks. Flavor violation is then described by flavor nondiagonal entries $(\Delta_{ij}^{\tilde{f}})_{AB}$ of the sfermion squared mass matrices in that basis, where i and j are generation indices and $A, B \in \{L, R\}$

---

[8]The effects of such large mixing may be observable as slepton oscillations [9.17] in pp and $\ell^+\ell^-$ colliders.

labels the four 3×3 blocks in (9.48). These off-diagonal entries are treated as two point interactions in the perturbation expansion, leading to nondiagonal propagators with explicit flavor offdiagonal mass insertions. The experimental constraints can most conveniently be

quantity	$x = 0.3$	$x = 1.0$	measurable
$\sqrt{\left\|\Re e(\delta^{\tilde{d}}_{12})^2_{LL}\right\|}$	1.9×10^{-2}	4.0×10^{-2}	
$\sqrt{\left\|\Re e(\delta^{\tilde{d}}_{12})^2_{LR}\right\|}$	7.9×10^{-3}	4.4×10^{-3}	Δm_K
$\sqrt{\left\|\Re e(\delta^{\tilde{d}}_{12})_{LL}(\delta^{\tilde{d}}_{12})_{RR}\right\|}$	2.5×10^{-3}	2.5×10^{-3}	
$\sqrt{\left\|\Re e(\delta^{\tilde{d}}_{13})^2_{LL}\right\|}$	4.6×10^{-2}	9.8×10^{-2}	
$\sqrt{\left\|\Re e(\delta^{\tilde{d}}_{13})^2_{LR}\right\|}$	5.6×10^{-2}	3.3×10^{-2}	Δm_B
$\sqrt{\left\|\Re e(\delta^{\tilde{d}}_{13})_{LL}(\delta^{\tilde{d}}_{13})_{RR}\right\|}$	1.6×10^{-2}	1.8×10^{-2}	
$\sqrt{\left\|\Re e(\delta^{\tilde{u}}_{12})^2_{LL}\right\|}$	4.7×10^{-2}	1.0×10^{-1}	
$\sqrt{\left\|\Re e(\delta^{\tilde{u}}_{12})^2_{LR}\right\|}$	6.3×10^{-2}	3.1×10^{-2}	Δm_D
$\sqrt{\left\|\Re e(\delta^{\tilde{u}}_{12})_{LL}(\delta^{\tilde{u}}_{12})_{rr}\right\|}$	1.6×10^{-2}	1.7×10^{-2}	
$\Im m(\delta^{\tilde{d}}_{12})_{LL}$	1.0×10^{-1}	4.8×10^{-1}	ϵ'_K/ϵ_K
$\Im m(\delta^{\tilde{d}}_{12})_{LR}$	1.1×10^{-5}	2.0×10^{-5}	ϵ'_K/ϵ_K
$(\delta^{\tilde{d}}_{23})_{LL}$	4.4	8.2	$\mathrm{BR}(b \to s\gamma)$
$(\delta^{\tilde{d}}_{23})_{LR}$	1.3×10^{-2}	1.6×10^{-2}	$\mathrm{BR}(b \to s\gamma)$
$(\delta^{\tilde{\ell}}_{12})_{LL}$	4.1×10^{-3}	7.7×10^{-3}	$\mathrm{BR}(\mu \to e\gamma)$
$(\delta^{\tilde{\ell}}_{12})_{LR}$	1.4×10^{-6}	1.7×10^{-6}	$\mathrm{BR}(\mu \to e\gamma)$
$(\delta^{\tilde{\ell}}_{13})_{LL}$	15	29	$\mathrm{BR}(\tau \to e\gamma)$
$(\delta^{\tilde{\ell}}_{13})_{LR}$	8.9×10^{-2}	1.1×10^{-1}	$\mathrm{BR}(\tau \to e\gamma)$
$(\delta^{\tilde{\ell}}_{23})_{LL}$	2.8	5.3	$\mathrm{BR}(\tau \to \mu\gamma)$
$(\delta^{\tilde{\ell}}_{23})_{LR}$	1.7×10^{-2}	2.0×10^{-2}	$\mathrm{BR}(\tau \to \mu\gamma)$

Table 9.1. Experimental upper bounds [9.18] on flavor violation in the soft supersymmetry breaking terms of sfermions.

expressed as bounds on the dimensionless quantities $(\delta^{\tilde{f}}_{ij})_{AB}$. In the simplest case, $(\delta^{\tilde{f}}_{ij})_{AB} = (\Delta^{\tilde{f}}_{ij})_{AB}/(\overline{m^2}^{\tilde{f}}_{ij})_{AB}$, where the "average" sfermion squared mass is given by $(\overline{m^2}^{\tilde{f}}_{ij})_{AB} = \sqrt{(\mathcal{M}^2_{ii})_{AA}(\mathcal{M}^2_{jj})_{BB}}$. Moreover, this formalism also allows the inclusion of higher order expressed as bounds on the dimensionless quantities $(\delta^{\tilde{f}}_{ij})_{AB}$. In the simplest case, $(\delta^{\tilde{f}}_{ij})_{AB} = (\Delta^{\tilde{f}}_{ij})_{AB}/(\overline{m^2}^{\tilde{f}}_{ij})_{AB}$, where the "average" sfermion squared mass is given by $(\overline{m^2}^{\tilde{f}}_{ij})_{AB} = \sqrt{(\mathcal{M}^2_{ii})_{AA}(\mathcal{M}^2_{jj})_{BB}}$. Moreover, this formalism also allows the inclusion of higher order contributions. Thus, for instance, the second order contribution to $(\delta^{\tilde{f}}_{ij})_{RR}$ is given by

$$(\Delta^{\tilde{f}}_{ik})_{RL}(\Delta^{\tilde{f}}_{kj})_{LR}/\left((\overline{m^2}^{\tilde{f}}_{ik})_{RL}(\overline{m^2}^{\tilde{f}}_{kj})_{LR}\right).$$

The experimental bounds on the various off-diagonal entries $(\delta^{\tilde{f}}_{ij})_{AB}$ are summarized in Table 9.1, which has been extracted from Ref.[9.18]. It has been assumed here that each supersymmetric contribution separately satisfies the overall constraint on the quantity indicated, i.e. "accidental" cancellations between different kinds of contributions have not been considered. For simplicity, moreover, $(\delta^{\tilde{f}}_{ij})_{LR}$ and $(\delta^{\tilde{f}}_{ij})_{RL}$ are taken to be equal, though the assumption could be avoided. The bounds on the slepton sector have been computed from loop diagrams involving a photino, rather than treating the two neutral electroweak gauginos separately. It is in this limit that the bounds on $(\delta^{\tilde{\ell}}_{ij})_{RR}$ are identical to those on $(\delta^{\tilde{\ell}}_{ij})_{LL}$ that have been listed in the table. An analogous statement holds for the bounds in the squark sector, which come from diagrams involving gluinos. All these bounds scale inversely with the relevant sfermion mass. The numerical values, given in Table 9.1, assume a common squark mass of 500 GeV and a common slepton mass of 100 GeV. Thus the bounds on $\delta^{\tilde{\ell}}$ scale like $m_{\tilde{\ell}}/(100~\text{GeV})$ while those on $\delta^{\tilde{q}}$ ($q = u, d$) scale like $m_{\tilde{q}}/(500~\text{GeV})$. Note that we only quote bounds from contributions involving flavor changing couplings to neutral gauginos (gluinos or neutralinos) $\tilde{\lambda}^0$; values are given for two values of the ratio $x \equiv (m_{\tilde{\lambda}^0}/\overline{m}_{\tilde{f}})^2$. Each entry in the last column in this table indicates the physical measurable from which the corresponding bound has been derived. Note moreover that the bounds on $(\delta^{\tilde{f}}_{ij})_{RR}$ and $(\delta^{\tilde{f}}_{ij})_{RL}$ are equal to those on the corresponding $(\delta^{\tilde{f}}_{ij})_{LL}$ and $(\delta^{\tilde{f}}_{ij})_{RL}$ respectively. As mentioned earlier, the most severe constraints exist on the mixing between first and second generation charge $-1/3$ squarks. The constraint on the mixing between charge $2/3$ squarks of the first two generations is considerably milder. Furthermore, $\mathcal{O}(1)$ mixing between second and third generation squarks is allowed in the LL or RR sector. The constraint on mixing between left and right sfermions is often much more stringent than that on LL and RR mixing. The reason is that the relevant effective fermionic operators leading to radiative decays treated in the last eight rows of Table 9.5, break chiral symmetry, i.e. cause couplings between left chiral and right chiral fermions, facilitated by the transitions between the corresponding sfermions.

9.6 Interactions of Sfermions with Gauge Bosons

A sfermion participates in an MSSM gauge interaction in two ways: (1) as a member of a sfermion pair and (2) along with another fermion. We shall take up (2) in the next Section. Here we consider (1) and enumerate the different possibilities below.

Slepton-slepton-electroweak gauge boson interactions

In this category come cubic/quartic vertices involving a *pair* of sleptons and one/two EW gauge boson(s). The two sleptons could be various combinations of charged and neutral ones while the gauge boson(s) would be correspondingly neutral and/or charged. These interactions were covered earlier in (8.44b–d) and (8.47c–h), but we now describe physical vertices with mass eigenstate sleptons and general mixing as described at the end of §9.4. We can collect all such vertices in three groups.

(1) The first group (Fig. 9.6) consists of vertices which involve either only one (two) Z boson(s) interacting with a sneutrino pair or only one (two) photon(s). These vertices have the feature that the mixing matrices $\mathbf{W}^{\tilde{f}}$ cancel out. The crucial observation here is that $\mathbf{W}^{\tilde{f}\dagger}\mathbf{W}^{\tilde{f}} = \mathbb{1}$.

(2) The second group (Fig. 9.7), comprising either a W^+W^- pair interacting with two sleptons or a Z interacting with a charged slepton pair, shows a nontrivial dependence on the mixing matrices $\mathbf{W}^{\tilde{f}}$ only in the presence of left right mixing. Without such mixing, i.e. if the 6×6 slepton mixing matrix has the form (9.56), flavor mixing would again drop out, owing to the unitarity of the $\mathbf{U}$ matrices. On the other hand, if L-R mixing is present, a nontrivial dependence on the mixing angle emerges even in the absence of flavor mixing. The relevant Feynman rules for this case can be derived easily from the general rules listed in Fig. 9.7, using (9.60). Notice that we have replaced $\mathbf{W}^{\tilde{\nu}}$ by $\mathbf{U}^{\tilde{\nu}}$ since in the MSSM no righthanded (s)neutrinos exist at the weak scale.

(3) All the remaining vertices, which are in general affected by generation mixing even in the absence of L-R mixing, make up the third group (Fig. 9.8). In this case a nontrivial dependence on the mixing angle will survive in both simplified scenarios discussed in §9.4, i.e. (9.57) and (9.60). For the convenience of the reader we give both the $W^+\tilde{l}\tilde{\nu}^*$ and $W^-\tilde{\nu}\tilde{l}^*$ vertices, which are related to each other by complex conjugation.

$$\boxed{\text{Figs. 9.6, 9.7 and 9.8 are included in Appendix A}}$$

Squark-squark-gauge boson interactions

The simplest set of vertices in this category are those that involve only squarks and gluons in SQCD. The cubic $\tilde{q}\bar{\tilde{q}}g$ and the quartic $\tilde{q}\bar{\tilde{q}}gg$ vertices have already been fully discussed in §5.5 and §8.4. Nothing needs to be added to those discussions, since the mixing matrices will cancel out in these vertices.

Turning to cubic and quartic vertices of physical mass eigenstate squarks with electroweak gauge bosons, we can again collect them in three groups as in the case of sleptons. However, the first group – which is free from *any* mixing – now has only pure photon vertices, cf. Fig. 9.9. The second group (Fig. 9.10), involving a W^+W^- pair or one (two) neutral gauge boson(s), at least one being the Z, shows a nontrivial mixing dependence only in the presence of L-R mixing. Only the third group, containing a single W either by itself or in association with a neutral gauge boson interacting with a squark pair, has the complication of both types of mixing, i.e. generation as well as left right. These vertices are given in Fig. 9.11.

As with sleptons, it is straightforward to derive the corresponding Feynman rules if flavor or L-R mixing can be ignored, using (9.57) and (9.60), respectively.

$$\boxed{\text{Figs. 9.9, 9.10 and 9.11 are included in Appendix A}}$$

This brings us to the end of the discussion of sfermion-gauge boson vertices.

9.7　Fermion-sfermion-gaugino/higgsino interactions

Fermion-sfermion-chargino interactions

Let us first discuss the fermion-sfermion-chargino vertices in the "current" basis in which generically $\tilde{f}_{u_iL}$, $\tilde{f}_{d_iL}$ are the left sfermions of the up, down type, $\tilde{f}_{u_iR}$, $\tilde{f}_{d_iR}$ are the corresponding right sfermions and f_{u_i}, f_{d_i} the corresponding fermions following the notation introduced in §8.4. Our starting points are (1) the gaugino-sfermion-fermion interactions, as given by expressions analogous to (5.55) and (2) the higgsino-sfermion-fermion couplings arising from the superpotential (8.33). In the two component spinor notation used in previous sections, the relevant part of the Lagrangian density reads

$$
\begin{aligned}
\mathcal{L}_{f\tilde{f}^*\tilde{\chi}^{\pm}} = {} & -g_2\left(\lambda^-\xi^1_{Q_i}\tilde{d}^*_{iL} + \lambda^+\xi^2_{Q_i}\tilde{u}^*_{iL} + \lambda^-\xi^1_{L_i}\tilde{e}^*_{iL} + \lambda^+\xi^2_{L_i}\tilde{\nu}^*_{iL}\right) \\[2mm]
& + \frac{g_2(\mathbf{m}^*_u)_{ij}}{\sqrt{2}M_W\sin\beta}\left(\tilde{h}^1_2\xi^2_{Q_i}\tilde{u}^*_{jR} + \tilde{h}^1_2\xi_{\bar{U}_j}\tilde{d}_{iL}\right) \\[2mm]
& + \frac{g_2(\mathbf{m}^*_d)_{ij}}{\sqrt{2}M_W\cos\beta}\left(\tilde{h}^2_1\xi^1_{Q_i}\tilde{d}^*_{jR} + \tilde{h}^2_1\xi_{\bar{D}_j}\tilde{u}_{iL}\right) \\[2mm]
& + \frac{g_2(\mathbf{m}^*_e)_{ij}}{\sqrt{2}M_W\cos\beta}\left(\tilde{h}^2_1\xi^1_{L_i}\tilde{e}^*_{jR} + \tilde{h}^2_1\xi_{\bar{E}_j}\tilde{\nu}_{iL}\right) + \text{h.c.}
\end{aligned}
\tag{9.63}
$$

We have written out the squark and slepton terms separately. The first term in the RHS of (9.62) describe the gaugino-fermion-sfermion couplings, while the last three terms correspond to higgsino-fermion-sfermion interactions. The latter are proportional to fermion mass matrices and vanish in the limit of massless fermions. In this expression $\xi^{1(2)}_{Q_i}$ and $\xi^{1(2)}_{L_i}$ are the two component spinors representing the $T_{3L} = 1/2\ (-1/2)$ fermionic components of a chiral $SU(2)_L$ doublet superfield such as Q_i or L_i of (8.15)-(8.17). Furthermore, $\xi_{\bar{D}_j}$ and $\xi_{\bar{E}_j}$ are the fermionic components of $SU(2)_L$ singlet superfields. The four component Dirac spinor fields corresponding to the various matter fermions are constructed out of $\xi_{U_i}, \xi_{D_i}, \xi_{L_i}, \xi_{\bar{U}_i}, \xi_{\bar{D}_i}, \xi_{\bar{E}_i}$ (where *e.g.* $\xi_{U_i} \equiv \xi^1_{Q_i}$ and so on) as described in (3.20) of §3.2. For example, for the up type quarks

$$
u_i = \begin{pmatrix} \xi_{U_i} \\ \bar{\xi}^T_{\bar{U}_i} \end{pmatrix}.
\tag{9.64}
$$

Recall that each of the singlet superfields $\bar{E}_i$, $\bar{D}_i$ and $\bar{U}_i$ contains the left chiral component of the antifermion field. Let us define Dirac fields f_{u_i,d_i} for general up, down type matter fermions (covering both quarks and leptons) in analogy with the u_i of (9.64). In terms of these generic up, down fermions and sfermions and the four component wino and higgsino eigenstates defined in (9.17), we can rewrite (9.63) as

$$
\begin{aligned}
\mathcal{L}_{f\tilde{f}'^*\tilde{\chi}^\pm} &= -g_2\left[\bar{f}_{u_i}P_R\tilde{\lambda}^+\tilde{f}_{d_iL} + \bar{f}_{d_i}P_R(\tilde{\lambda}^+)^C\tilde{f}_{u_iL}\right] \\
&\quad + \frac{g_2(\mathbf{m}_{f_d})_{ij}}{\sqrt{2}M_W\cos\beta}\left[\bar{f}_{u_i}P_R\tilde{h}^+\tilde{f}_{d_jR} + \overline{(\tilde{h}^+)^C}P_Rf_{d_j}\tilde{f}^*_{u_iL}\right] \\
&\quad + \frac{g_2(\mathbf{m}_{f_u})_{ij}}{\sqrt{2}M_W\sin\beta}\left[\bar{f}_{d_i}P_R(\tilde{h}^+)^C\tilde{f}_{u_jR} + \overline{\tilde{h}^+}P_Rf_{u_j}\tilde{f}^*_{d_iL}\right] + \text{h.c.}
\end{aligned}
\tag{9.65}
$$

Of course, a sum over all fermions f_{u_i}, f_{d_i} covering quarks and leptons (and corresponding sfermions) is implied. On utilizing (9.18) and (9.19), this Lagrangian can be recast in terms of the chargino mass eigenstates $\tilde{\chi}^\pm_k$, $k = 1, 2$, as

$$
\begin{aligned}
\mathcal{L}_{f\tilde{f}'^*\tilde{\chi}^\pm} &= -g_2\left[\mathcal{U}_{k1}\bar{f}_{u_i}P_R\tilde{\chi}^+_k\tilde{f}_{d_iL} + \mathcal{V}_{k1}\bar{f}_{d_i}P_R(\tilde{\chi}^+_k)^C\tilde{f}_{u_iL}\right] \\
&\quad + \frac{g_2(\mathbf{m}_{f_d})_{ij}}{\sqrt{2}M_W\cos\beta}\mathcal{U}_{k2}\left[\bar{f}_{u_i}P_R\tilde{\chi}^+_k\tilde{f}_{d_jR} + \tilde{f}^*_{u_iL}\overline{(\tilde{\chi}^+_k)^C}P_Rf_{d_j}\right] \\
&\quad + \frac{g_2(\mathbf{m}_{f_u})_{ij}}{\sqrt{2}M_W\sin\beta}\mathcal{V}_{k2}\left[\bar{f}_{d_i}P_R(\tilde{\chi}^+_k)^C\tilde{f}_{u_jR} + \tilde{f}^*_{d_iL}\overline{\tilde{\chi}^+_k}P_Rf_{u_j}\right] + \text{h.c.}
\end{aligned}
\tag{9.66}
$$

In the supersymmetric limit the lepton-slepton-chargino vertices can be read off from this expression, using $(\mathbf{m}_e)_{ij} = m_{e_i}\delta_{ij}$, modulo $\tilde{e}_L$-$\tilde{e}_R$ mixing in the slepton sector. However, the existence of the soft supersymmetric breaking terms can change that. In case of the quark-squark-chargino interaction, there is also the additional complication of generation mixing which is present even in the supersymmetric limit. A further point to note in (9.66) is the occurrence of $(\tilde{\chi}^+_k)^C$. The appearance of charge conjugated fermion fields is generic in supersymmetric theories and gives rise to the explicit presence of the charge conjugation matrix C in Feynman rules. The basic reason for the necessity of introducing these ugly C-factors in Feynman rules is the following. In contrast with charged fermions in the SM, charginos do not carry a "fermion number" like lepton or baryon number. The same field can thus couple to $\bar{u}\tilde{d}$ and to $d\bar{\tilde{u}}$. If the first vertex is written in terms of an incoming (positive) chargino field, the second vertex has to be written in terms of the outgoing charge conjugate of that chargino field (or vice versa).

We are now in a position to write down the interaction terms of (9.66) explicitly for the quark/squark and lepton/slepton sectors in terms of mass diagonal matter fermion and sfermion fields. We use quark flavor rotation matrices $\mathbf{U}^{u_{L,R}}$ and $\mathbf{U}^{d_{L,R}}$, introduced in Ch. 8, as well as the sfermion rotation matrices $\mathbf{U}^{\tilde{\nu}}$, $\mathbf{W}^{\tilde{e}}$, $\mathbf{W}^{\tilde{u}}$ and $\mathbf{W}^{\tilde{d}}$ of §9.4. We employ $i, j, k = 1, 2, 3$ as indices in generation space, while $s = 1, \ldots, 6$ labels charged slepton or squark mass eigenstates. The physical quark masses are denoted by m_{d_i} and m_{u_i}. Finally, quark/squark fields are taken to be row or column vectors in color space. The quark/squark part of (9.66) then reads (for simplicity we omit the superscript m denoting mass eigenstates)

$$
\begin{aligned}
\mathcal{L}_{q\tilde{q}'\chi^\pm} &= \bar{u}_i C^L_{isk}P_R\tilde{d}_s\tilde{\chi}^+_k + \bar{d}_i D^L_{isk}P_R\tilde{u}_s(\tilde{\chi}^+_k)^C \\
&\quad + \tilde{u}^\dagger_s\overline{(\tilde{\chi}^+_k)^C}E^R_{isk}P_Rd_i + \tilde{d}^\dagger_s\overline{\tilde{\chi}^+_k}F^R_{isk}P_Ru_i + \text{h.c.} ,
\end{aligned}
\tag{9.67}
$$

with

$$C_{isk}^L = -g_2\mathcal{U}_{k1}\sum_{j=1}^{3}U_{ji}^{u_L*}W_{js}^{\tilde{d}} + \frac{g_2\mathcal{U}_{k2}}{\sqrt{2}M_W\cos\beta}\sum_{j,n=1}^{3}V_{in}^{q_L}m_{d_n}U_{jn}^{d_R*}W_{j+3\ s}^{\tilde{d}}\ , \qquad (9.68a)$$

$$D_{isk}^L = -g_2\mathcal{V}_{k1}\sum_{j=1}^{3}U_{ji}^{d_L*}W_{js}^{\tilde{u}} + \frac{g_2\mathcal{V}_{k2}}{\sqrt{2}M_W\sin\beta}\sum_{j,n=1}^{3}V_{ni}^{q_L*}m_{u_n}U_{jn}^{u_R*}W_{j+3\ s}^{\tilde{u}}\ , \qquad (9.68b)$$

$$E_{isk}^R = \frac{g_2\mathcal{U}_{k2}m_{d_i}}{\sqrt{2}M_W\cos\beta}\sum_{j=1}^{3}U_{ji}^{d_L}W_{js}^{\tilde{u}*}\ , \qquad (9.68c)$$

$$F_{isk}^R = \frac{g_2\mathcal{V}_{k2}m_{u_i}}{\sqrt{2}M_W\sin\beta}\sum_{j=1}^{3}U_{ji}^{u_L}W_{js}^{\tilde{d}*}\ . \qquad (9.68d)$$

The corresponding vertices are given in Fig. 9.12. It may be noted that left (right) fermions connect to the left (right) components of the sfermions through the gaugino components of the charginos, which are described by $\mathcal{U}_{\ell 1}$ and $\mathcal{V}_{\ell 1}$. In contrast, the terms coming from Yukawa couplings, which are proportional to a quark mass, couple a left (right) fermion to the right (left) component of the corresponding sfermion. If squarks and quarks could be aligned exactly (see §9.5), all combinations of quark and squark mixing matrices appearing in (9.68) would reduce either to the unit matrix (in the right handed sector) or to the standard KM matrix $\mathbf{V}^{q_L}$ (in the left handed sector); however, as discussed earlier, alignment cannot be exact in the u and d sectors simultaneously. Note finally that, as per the convention of Appendix D of Haber and Kane [9.10], a charge conjugation matrix $\underset{\sim}{C}$ to the right operates on the transposed $\bar{u}$-spinor $\bar{u}^T$ or $\bar{v}$-spinor $\bar{v}^T$ while a $\underset{\sim}{C}^{-1}$ to the left requires a transposed v-spinor v^T or u-spinor u^T to left multiply it.

$$\boxed{\text{Fig. 9.12 is included in Appendix A}}$$

We turn next to the lepton/slepton part of (9.63). It reads

$$\mathcal{L}_{\ell\tilde{\ell}\tilde{\chi}^\pm} = \bar{\nu}_i c_{isk}^L \tilde{e}_s P_R \tilde{\chi}_k^+ + d_{ijk}^L \bar{e}_i P_R(\tilde{\chi}_k^+)^C \tilde{\nu}_j + e_{ijk}^R \overline{(\tilde{\chi}_k^+)^C} P_R e_i \tilde{\nu}_j^* + \text{h.c.}\ , \qquad (9.69)$$

with

$$c_{isk}^L = -g_2\mathcal{U}_{k1}W_{is}^{\tilde{e}} + \frac{g_2 m_{e_i}}{\sqrt{2}M_W\cos\beta}\mathcal{U}_{k2}W_{i+3\ s}^{\tilde{e}}\ , \qquad (9.70a)$$

$$d_{ijk}^L = -g_2 U_{ij}^{\tilde{\nu}}\mathcal{V}_{k1}\ , \qquad (9.70b)$$

$$e_{ijk}^R = \frac{g_2 m_{e_i}}{\sqrt{2}M_W\cos\beta}\mathcal{U}_{k2}U_{ij}^{\tilde{\nu}*}\ . \qquad (9.70c)$$

The corresponding vertices are drawn in Fig. 9.13; they can be obtained from those of Fig. 9.12 with the replacements $u \to \nu$, $d \to e$, $\mathbf{V}^{q_L} \to \mathbb{1}$, $\mathbf{U}^{u_L}, \mathbf{U}^{d_L}, \mathbf{U}^{u_R}, \mathbf{U}^{d_R} \to \mathbb{1}$, $\mathbf{W}^{\tilde{u}} \to \mathbf{U}^{\tilde{\nu}}$, $\mathbf{W}^{\tilde{d}} \to \mathbf{W}^{\tilde{e}}$, and $m_{u_k} \to 0$.

$$\boxed{\text{Fig. 9.13 is included in Appendix A}}$$

Fermion-sfermion-neutralino interactions

The neutralino-fermion-sfermion interaction can be written down in a similar fashion. This time we need to isolate the $a = 3$ term from (5.55) for the $SU(2)_L$ gauge group and the $U(1)_Y$ analog of the terms in (5.36) and express them in terms of the four component matter fermions as well as the four component gauginos and higgsinos in the weak interaction basis, defined in (9.30):

$$
\begin{aligned}
\mathcal{L}_{f\bar{f}\tilde{\chi}^0} &= -\sqrt{2}g_2\tilde{f}_{iL}\sum_{f=e,\nu,u,d}\bar{f}_i P_R\left[T_3^f\tilde{\lambda}_3 + \tan\theta_W(Q_f - T_3^f)\tilde{\lambda}_0\right] \\
&\quad +\sqrt{2}g_2\tan\theta_W\, Q_f\tilde{f}_{iR}^*\overline{\tilde{\lambda}_0}P_R f_i - \frac{g_2}{\sqrt{2}M_W\cos\beta}(\mathbf{m}_d^*)_{ij}\left[\overline{\tilde{h}_1^0}P_L\tilde{d}_{jR}^\dagger d_i + \bar{d}_j P_L\tilde{h}_1^0\tilde{d}_{iL}\right] \\
&\quad -\frac{g_2}{\sqrt{2}M_W\sin\beta}(\mathbf{m}_u^*)_{ij}\left[\overline{\tilde{h}_2^0}P_L\tilde{u}_{jR}^\dagger u_i + \bar{u}_j P_L\tilde{h}_2^0\tilde{u}_{iL}\right] \\
&\quad -\frac{g_2}{\sqrt{2}M_W\cos\beta}(\mathbf{m}_e^*)_{ij}\left[\overline{\tilde{h}_1^0}P_L e_i\tilde{e}_{jR}^* + \bar{e}_j P_L\tilde{h}_1^0\tilde{e}_{iL}\right] + \text{h.c.}
\end{aligned}
\tag{9.71}
$$

In (9.71) T_{3L}^f and Q_f are respectively the third component of weak isospin and the electromagnetic charge of fermion type f and i,j are generation indices as before. In terms of the neutralino mass eigenstates $\tilde{\chi}_l^0$, (9.71) becomes

$$
\begin{aligned}
\mathcal{L}_{f\bar{f}\tilde{\chi}^0} &= \sum_{f=u,d,e,\nu}\overline{\tilde{\chi}_l^0}\left(G_l^{f_L}\tilde{f}_{iL}^*P_L + G_l^{f_R}\tilde{f}_{iR}^*P_R\right)f_i \\
&\quad -\frac{g_2}{\sqrt{2}M_W\sin\beta}\left[(\mathbf{m}_u^*)_{ij}Z_{l4}^*\tilde{u}_{jR}^\dagger\overline{\tilde{\chi}_l^0}P_L u_i + (\mathbf{m}_u)_{ij}Z_{l4}\tilde{u}_{iL}^\dagger\overline{\tilde{\chi}_l^0}P_R u_j\right] \\
&\quad -\frac{g_2}{\sqrt{2}M_W\cos\beta}\left[(\mathbf{m}_d^*)_{ij}Z_{l3}^*\tilde{d}_{jR}^\dagger\overline{\tilde{\chi}_l^0}P_L d_i + (\mathbf{m}_d)_{ij}Z_{l3}\tilde{d}_{iL}^\dagger\overline{\tilde{\chi}_l^0}P_R d_j\right] \\
&\quad -\frac{g_2}{\sqrt{2}M_W\cos\beta}\left[(\mathbf{m}_e^*)_{ij}Z_{l3}^*\tilde{e}_{jR}^*\overline{\tilde{\chi}_l^0}P_L e_i + (\mathbf{m}_e)_{ij}Z_{l3}\tilde{e}_{iL}^*\overline{\tilde{\chi}_l^0}P_R e_j\right] + \text{h.c.} ,
\end{aligned}
\tag{9.72}
$$

where we have used (9.31). The coupling strengths $G_l^{f_L}$ and $G_l^{f_R}$ in (9.71) can be written as

$$
G_l^{f_L} = -\sqrt{2}g_2\left[T_{3L}^f Z_{l2}^* + \tan\theta_W(Q_f - T_{3L}^f)Z_{l1}^*\right],
\tag{9.73a}
$$

$$
G_l^{f_R} = \sqrt{2}g_2\tan\theta_W Q_f Z_{l1} .
\tag{9.73b}
$$

Once more, we can rewrite the interactions of (9.72) in terms of mass diagonal quark and lepton fields by performing flavor rotations in generation space with indices i,j. Similarly, the squark and slepton interaction eigenstates appearing in (9.72) can be related to the corresponding mass eigenstates through (9.54). The quark and squark fields are also three component row or column vectors in color space. Altogether the relevant interaction terms for the quark/squark sector can be written as (we again suppress the superscript m indicating mass eigenstates):

$$\mathcal{L}_{q\tilde{q}'\tilde{\chi}^0} = \overline{\tilde{\chi}^0_l}\left[(G^{u_L}_{isl}P_L + G^{u_R}_{isl}P_R)\tilde{u}^\dagger_s u_i + (G^{d_L}_{isl}P_L + G^{d_R}_{isl}P_R)\tilde{d}^\dagger_s d_i\right] + \text{h.c.} \qquad (9.74)$$

In (9.74) we have defined the couplings

$$G^{u_L}_{isl} = G^{u_L}_l \sum_{j=1}^{3} W^{\tilde{u}*}_{js} U^{u_L}_{ji} - \frac{g_2}{\sqrt{2}M_W \sin\beta}m_{u_i}Z^*_{l4}\sum_{j=1}^{3} W^{\tilde{u}*}_{j+3\,s} U^{u_R}_{ji}\,, \qquad (9.75a)$$

$$G^{u_R}_{isl} = G^{u_R}_l \sum_{j=1}^{3} W^{\tilde{u}*}_{j+3\,s} U^{u_R}_{ji} - \frac{g_2}{\sqrt{2}M_W \sin\beta}m_{u_i}Z_{l4}\sum_{j=1}^{3} W^{\tilde{u}*}_{js} U^{u_L}_{ji}\,, \qquad (9.75b)$$

$$G^{d_L}_{isl} = G^{d_L}_l \sum_{j=1}^{3} W^{\tilde{d}*}_{js} U^{d_L}_{ji} - \frac{g_2}{\sqrt{2}M_W \cos\beta}m_{d_i}Z^*_{l3}\sum_{j=1}^{3} W^{\tilde{d}*}_{j+3\,s} U^{d_R}_{ji}\,, \qquad (9.75c)$$

$$G^{d_R}_{isl} = G^{d_R}_l \sum_{j=1}^{3} W^{\tilde{d}*}_{j+3\,s} U^{d_R}_{ji} - \frac{g_2}{\sqrt{2}M_W \cos\beta}m_{d_i}Z_{l3}\sum_{j=1}^{3} W^{\tilde{d}*}_{js} U^{d_L}_{ji}\,, \qquad (9.75d)$$

where the coefficients $G^{q_L}_l$ and $G^{q_R}_l$ are as in eqs.(9.73a) and (9.73b), respectively. Feynman rules for the vertices of (9.75) are given in Fig. 9.14. An arrow has been put on the neutralino line in conformity with the convention in Appendix D of the first paper of Ref. [9.10].

$$\boxed{\text{Fig. 9.14 is included in Appendix A}}$$

Let us remark once again that, in the limit of massless fermions, the higgsinos will decouple from the matter fermion/sfermion sector. Note also that the couplings of neutral higgsinos to quark mass eigenstates are proportional to the mass of that eigenstate. This is in contrast to the couplings of the charged higgsinos, where heavy quark masses contribute to the coupling of light quarks. However, due to the smallness of the KM elements mixing the third generation with the first two, in practice one can still often neglect the Yukawa contributions to chargino and neutralino couplings to first and second generation fermions. In the alignment option of §9.5 the products of flavor rotation matrices can be put equal to unity in either the up or down quark sector (but not for both simultaneously, as we noted earlier). On the other hand, if squarks of all three generations are degenerate and LR mixing can be ignored, all products of rotation matrices appearing in (9.75) collapse to Kronecker-δs, where either $i = s$ or $i + 3 = s$.

Let us now turn our attention to the lepton/slepton sector. The interaction terms with neutralinos can be written as

$$\mathcal{L}_{\ell\tilde{\ell}'\tilde{\chi}^0} = \overline{\tilde{\chi}^0_l}\left[G^{\nu}_{ijl}\tilde{\nu}^*_j P_L \nu_i + (G^{e_L}_{isl}P_L + G^{e_R}_{isl}P_R)\tilde{e}^*_s e_i\right] + \text{h.c.} \qquad (9.76)$$

In (9.76) we have introduced the couplings

$$G^{\nu}_{ijl} = G^{\nu}_l U^{\tilde{\nu}*}_{ij}, \tag{9.77a}$$

$$G^{e_L}_{isl} = G^{e_L}_l W^{\tilde{e}*}_{is} - \frac{g_2}{\sqrt{2}M_W \cos\beta} m_{e_i} Z^*_{l3} W^{\tilde{e}*}_{i+3\ s}, \tag{9.77b}$$

$$G^{e_R}_{isl} = G^{e_R}_l W^{\tilde{e}*}_{i+3\ s} - \frac{g_2}{\sqrt{2}M_W \cos\beta} m_{e_i} Z_{l3} W^{\tilde{e}*}_{is}, \tag{9.77c}$$

The vertex Feynman rules appear in Fig. 9.15. In the alignment option, or if sleptons are mass degenerate, the slepton flavor rotation matrices $\mathbf{U}^{\tilde{\nu}}$ and $\mathbf{W}^{\tilde{e}}$ can be put equal to the identity matrix, if $\tilde{e}_L$-$\tilde{e}_R$ mixing is negligible. LR mixing can, as usual, be included in these options by using (9.60). Once again an arrow has been put [9.10] on the neutralino line.

$$\boxed{\text{Fig. 9.15 is included in Appendix A}}$$

Quark-squark-gluino interactions

These are now different from the pure SQCD case, cf. (5.60) and Fig. 5.2. However, with the armory of quark and squark flavor rotation matrices that have been developed already, we can write the relevant interaction terms in a straightforward way as follows[9].

$$\mathcal{L}_{q\tilde{q}'*\tilde{g}} = -\sqrt{2}g_s \sum_{q=u,d} \bar{q}_i \left[U^{q_L*}_{ji} W^{\tilde{q}}_{js} P_R - U^{q_R*}_{ji} W^{\tilde{q}}_{j+3\ s} P_L \right] T^a \tilde{g}^a \tilde{q}_s + \text{h.c.} \tag{9.78}$$

We have again suppressed the superscript m denoting mass eigenstates, and have written (s)quark fields as vectors in color space. The corresponding Feynman rules are given in Fig. 9.16; we have used them already in §9.5, in the basis where $\mathbf{U}^{d_L} = \mathbf{U}^{d_R} = \mathbb{1}$.

$$\boxed{\text{Fig. 9.16 is included in Appendix A}}$$

Eqs. (9.68), (9.75) and (9.78) are in a general basis of the quark and squark interaction eigenstates. Not all the rotation matrices appearing in these equations are separately physical quantities. Despite the occurrence of the matrices $\mathbf{U}^{u_R}$ and $\mathbf{U}^{d_R}$ in some of these equations, one can only measure the products of quark and squark mixing matrices which appear in these couplings. Note that exactly one factor in these products is always the hermitian conjugate of a rotation matrix. This shows that only any *misalignment* between righthanded quarks and "righthanded" ($SU(2)$ singlet) squarks is measurable. That can also be seen by defining $U^{q_R*}_{ij} q_{iR}$ and $U^{q_R*}_{ij} \tilde{q}_{iR}$ as new "interaction eigenstates". This redefinition does not modify any of the gauge interactions in the MSSM Lagrangian. The righthanded quark mixing matrices would then disappear from (9.68), (9.75) and (9.78); more exactly, they would be absorbed in the squark rotation matrices $\mathbf{W}^{\tilde{q}}$, which are not invariant under this redefinition of the $\tilde{q}_R$ "interaction eigenstates".[10] Indeed, practical calculations are usually performed in this basis,

---

[9]The s subscript of g, referring to the strong coupling, should not be confused with the squark mass eigenstate label s.

[10]Of course, *products* of rotation matrices that appear in couplings of mass eigenstates are invariant under redefinitions of current eigenstates.

because the relevant couplings are simpler than in a general basis. One can even go one step further and chose the $SU(2)_L$ doublet (s)quark interaction eigenstates in such a way that either the up or the down quark mass matrix (but not both!) becomes diagonal. The only quark rotation matrix appearing in the quark squark chargino/neutralino/gluino couplings is then the KM matrix. Of course, such a procedure will yet again modify the squark rotation matrices. In these bases our interactions are modified as follows: $\mathbf{U}^{u_R}, \mathbf{U}^{d_R} \rightarrow \mathbb{1}$, and either $\mathbf{U}^{d_L} \rightarrow \mathbb{1}, \mathbf{U}^{u_L} \rightarrow (\mathbf{V}^{q_L})^\dagger$ (in the basis where $\mathbf{m}_d$ is diagonal), or $\mathbf{U}^{u_L} \rightarrow \mathbb{1}, \mathbf{U}^{d_L} \rightarrow \mathbf{V}^{q_L}$ (in the basis where $\mathbf{m}_u$ is diagonal).

Flavor mixing in the fermion-sfermion-bosino couplings is of much greater phenomenological importance than the "super-CKM mixing" introduced in §8.4. The latter appears in the coupling of W bosons to squarks and sleptons; the only process of current interest where these couplings play a role is slepton production at hadron colliders, which is however difficult to detect anyway (see §15.3). In contrast, the couplings listed in this section not only determine the constraints on flavor mixing described in §9.5; they also largely determine how sparticles decay. For example, the "flavor" of a squark is usually defined through the quark to which this squark decays. However, in the presence of significant flavor mixing this definition may not be unique: several different quarks might couple to the same squark mass eigenstate. The relative branching ratios into different quark flavors may even depend on the -ino that is produced in that decay. For example, different combinations of mixing matrices appear in squark to neutralino plus quark decays, described by the Lagrangian (9.74), than in squark to gluino plus quark decays described by (9.78). Conversely, these couplings determine which (combinations of) flavors are produced in the decays of gluinos, charginos and neutralinos. For example, (9.76) and (9.77) show that the observation of decays of the type $\tilde{\chi}_l^0 \rightarrow \tilde{\chi}_1^0 \ell^+ \ell'^-$, with $l > 1$ and $\ell \neq \ell'$, would be an unambiguous sign for slepton flavor mixing.

This completes our discussion of vertices with gauginos/higgsinos interacting with a fermion-sfermion combination.

9.8 Quartic Sfermion Vertices

The final nongauge and nonHiggs interaction that needs to be discussed is the interaction of four sfermions. These vertices appear e.g. in one loop corrections to sfermion pair production processes, and in two loop corrections to reactions without external superparticles. In (8.49) we gave the relevant part of the Lagrangian in the absence of $\tilde{f}_L$-$\tilde{f}_R$ mixing. In that case sfermion mixing matrices only appeared in the F-term (Yukawa) contributions, and in the part of the $SU(2)_L$ D-term that couples $\tilde{u}_L$ to $\tilde{d}_L$ squarks, and $\tilde{\nu}$ to $\tilde{e}_L$ sleptons. However, since $\tilde{f}_L$ and $\tilde{f}_R$ have different gauge quantum numbers, nonvanishing $\tilde{f}_L$-$\tilde{f}_R$ mixing means that sfermion mixing in general affects almost all terms in the quartic interaction Lagrangian. This is true even for the $SU(3)_C$ D-terms, since the $\tilde{q}_L$ squarks reside in left chiral superfields that transform as triplets under $SU(3)_C$, while the $\tilde{\bar{q}}_R^*$ reside in antitriplet left chiral superfields: The $SU(3)_C$ D-term contributions from the two therefore differ by a relative sign, as shown in (5.60). The only exception is the term involving four sneutrinos, since the MSSM assumes the absence of $SU(2)_L$ singlet sneutrinos with weak scale masses.

The relevant part of the Lagrangian can now be written as

$$-\mathcal{L}_{\tilde{f}^4} = \sum_{\tilde{f}_1,\tilde{f}_2,\tilde{f}_3,\tilde{f}_4} Y[\tilde{f}_1,\tilde{f}_2,\tilde{f}_3,\tilde{f}_4]\tilde{f}_1^*\tilde{f}_2\tilde{f}_3^*\tilde{f}_4 \,, \tag{9.79}$$

where the Y are constant (field independent) coefficients. In (9.79) the indices $\tilde{f}_i$ of Y have been written in the form of arguments, rather than as superscripts or subscripts, in order to avoid an excessive proliferation of subscripts. The sum in (9.79) runs over sfermion type ($\tilde{u}, \tilde{d}, \tilde{e}$ and $\tilde{\nu}$), mass eigenstate labels, and color indices.

The Y coefficients of (9.79) are given explicitly by

$$
\begin{aligned}
Y[\tilde{u}_s^{a*}, \tilde{u}_t^{a}, \tilde{u}_u^{b*}, \tilde{u}_v^{b}] \;=\;& \frac{g_2^2}{2M_W^2\sin^2\beta} \sum_{i,j,k,l,m,n=1}^{3} W_{is}^{\tilde{u}*}U_{ik}^{u_L}m_{u_k}U_{jk}^{u_R*}W_{j+3\ t}^{\tilde{u}}\,W_{lv}^{\tilde{u}}U_{lm}^{u_L*}m_{u_m}U_{nm}^{u_R}W_{n+3\ u}^{\tilde{u}*} \\[4pt]
&+\ \frac{g_s^2}{4}\left[\delta_{sv}\delta_{tu} - \frac{1}{3}\delta_{st}\delta_{uv} - 4\sum_{i,j=1}^{3} W_{is}^{\tilde{u}*}W_{j+3\ u}^{\tilde{u}*}\left(W_{iv}^{\tilde{u}}W_{j+3\ t}^{\tilde{u}} - \frac{1}{3}W_{it}^{\tilde{u}}W_{j+3\ v}^{\tilde{u}}\right)\right] \\[4pt]
&+\ \frac{g_2^2}{8}\left(1+\frac{\tan^2\theta_W}{9}\right)\sum_{i,j=1}^{3} W_{is}^{\tilde{u}*}W_{it}^{\tilde{u}}W_{ju}^{\tilde{u}*}W_{jv}^{\tilde{u}} \\[4pt]
&+\ \frac{g_2^2\tan^2\theta_W}{9}\sum_{i,j=1}^{3}\left(2W_{i+3\ s}^{\tilde{u}*}W_{i+3\ t}^{\tilde{u}} - W_{is}^{\tilde{u}*}W_{it}^{\tilde{u}}\right)W_{j+3\ u}^{\tilde{u}*}W_{j+3\ v}^{\tilde{u}} \,, \tag{9.80a}
\end{aligned}
$$

$$
\begin{aligned}
Y[\tilde{d}_s^{a*}, \tilde{d}_t^{a}, \tilde{d}_u^{b*}, \tilde{d}_v^{b}] \;=\;& \frac{g_2^2}{2M_W^2\cos^2\beta} \sum_{i,j,k,l,m,n=1}^{3} W_{is}^{\tilde{d}*}U_{ik}^{d_L}m_{d_k}U_{jk}^{d_R*}W_{j+3\ t}^{\tilde{d}}\,W_{lv}^{\tilde{d}}U_{lm}^{d_L*}m_{d_m}U_{nm}^{d_R}W_{n+3\ u}^{\tilde{d}*} \\[4pt]
&+\ \frac{g_s^2}{4}\left[\delta_{sv}\delta_{tu} - \frac{1}{3}\delta_{st}\delta_{uv} - 4\sum_{i,j=1}^{3} W_{is}^{\tilde{d}*}W_{j+3\ u}^{\tilde{d}*}\left(W_{iv}^{\tilde{d}}W_{j+3\ t}^{\tilde{d}} - \frac{1}{3}W_{it}^{\tilde{d}}W_{j+3\ v}^{\tilde{d}}\right)\right] \\[4pt]
&+\ \frac{g_2^2}{8}\left(1+\frac{\tan^2\theta_W}{9}\right)\sum_{i,j=1}^{3} W_{is}^{\tilde{d}*}W_{it}^{\tilde{d}}W_{ju}^{\tilde{d}*}W_{jv}^{\tilde{d}} \\[4pt]
&+\ \frac{g_2^2\tan^2\theta_W}{36}\sum_{i,j=1}^{3}\left(W_{i+3\ s}^{\tilde{d}*}W_{i+3\ t}^{\tilde{d}} - W_{is}^{\tilde{d}*}W_{it}^{\tilde{d}}\right)W_{j+3\ u}^{\tilde{d}*}W_{j+3\ v}^{\tilde{d}} \,, \tag{9.80b}
\end{aligned}
$$

$$
\begin{aligned}
Y[\tilde{u}_s^{a*}, \tilde{u}_t^{a}, \tilde{d}_u^{b*}, \tilde{d}_v^{b}] \;=\;& -\frac{g_s^2}{6}\left[\delta_{st}\delta_{uv} - 2\sum_{i,j=1}^{3}\left(W_{is}^{\tilde{u}*}W_{it}^{\tilde{u}}W_{j+3\ u}^{\tilde{d}*}W_{j+3\ v}^{\tilde{d}} + W_{i+3\ s}^{\tilde{u}*}W_{i+3\ t}^{\tilde{u}}W_{ju}^{\tilde{d}*}W_{jv}^{\tilde{d}}\right)\right] \\[4pt]
&-\ \frac{g_2^2}{4}\left(1-\frac{\tan^2\theta_W}{9}\right)\sum_{i,j=1}^{3} W_{is}^{\tilde{u}*}W_{it}^{\tilde{u}}W_{ju}^{\tilde{d}*}W_{jv}^{\tilde{d}} \\[4pt]
&+\ \frac{g_2^2\tan^2\theta_W}{18}\sum_{i,j=1}^{3}\left[W_{is}^{\tilde{u}*}W_{it}^{\tilde{u}}W_{j+3\ u}^{\tilde{d}*}W_{j+3\ v}^{\tilde{d}}\right.\\[4pt]
&\qquad\qquad\left.+\,2W_{i+3\ s}^{\tilde{u}*}W_{i+3\ t}^{\tilde{u}}\left(W_{ju}^{\tilde{d}*}W_{jv}^{\tilde{d}} + 2W_{j+3\ u}^{\tilde{d}*}W_{j+3\ v}^{\tilde{d}}\right)\right] \,, \tag{9.80c}
\end{aligned}
$$

$$Y[\tilde{u}_s^{a*}, \tilde{d}_t^a, \tilde{d}_u^{b*}, \tilde{u}_v^b] = \frac{g_2^2}{2M_W^2 \sin^2\beta} \sum_{i,j,k,l,m,n=1}^{3} W_{iu}^{\tilde{d}*} U_{ik}^{u_L} m_{u_k} U_{jk}^{u_R*} W_{j+3\ v}^{\tilde{u}} W_{lt}^{\tilde{d}} U_{lm}^{u_L*} m_{u_m} U_{nm}^{u_R} W_{n+3\ s}^{\tilde{u}*}$$

$$+ \frac{g_2^2}{2M_W^2 \cos^2\beta} \sum_{i,j,k,l,m,n=1}^{3} W_{is}^{\tilde{u}*} U_{ik}^{d_L} m_{d_k} U_{jk}^{d_R*} W_{j+3\ t}^{\tilde{d}} W_{lv}^{\tilde{u}} U_{lm}^{d_L*} m_{d_m} U_{nm}^{d_R} W_{n+3\ u}^{\tilde{d}*}$$

$$+ \frac{g_s^2}{2}\left[\delta_{sv}\delta_{tu} - 2\sum_{i,j=1}^{3}\left(W_{is}^{\tilde{u}*} W_{iv}^{\tilde{u}} W_{j+3\ u}^{\tilde{d}*} W_{j+3\ t}^{\tilde{d}} + W_{i+3\ s}^{\tilde{u}*} W_{i+3\ v}^{\tilde{u}} W_{ju}^{\tilde{d}*} W_{jt}^{\tilde{d}}\right)\right]$$

$$+ \frac{g_2^2}{2}\sum_{i,j=1}^{3} W_{is}^{\tilde{u}*} W_{it}^{\tilde{d}} W_{ju}^{\tilde{d}*} W_{jv}^{\tilde{u}}\ , \tag{9.80d}$$

$$Y[\tilde{u}_s^{a*}, \tilde{u}_t^a, \tilde{e}_u^*, \tilde{e}_v] = -\frac{g_2^2}{4}\left(1 + \frac{\tan^2\theta_W}{3}\right)\sum_{i,j=1}^{3} W_{is}^{\tilde{u}*} W_{it}^{\tilde{u}} W_{ju}^{\tilde{e}*} W_{jv}^{\tilde{e}}$$

$$+ \frac{g_2^2 \tan^2\theta_W}{6}\sum_{i,j=1}^{3}\Big[W_{is}^{\tilde{u}*} W_{it}^{\tilde{u}} W_{j+3\ u}^{\tilde{e}*} W_{j+3\ v}^{\tilde{e}}$$

$$+ 2W_{i+3\ s}^{\tilde{u}*} W_{i+3\ t}^{\tilde{u}}\left(W_{ju}^{\tilde{e}*} W_{jv}^{\tilde{e}} - 2W_{j+3\ u}^{\tilde{e}*} W_{j+3\ v}^{\tilde{e}}\right)\Big]\ , \tag{9.80e}$$

$$Y[\tilde{d}_s^{a*}, \tilde{d}_t^a, \tilde{e}_u^*, \tilde{e}_v] = \frac{g_2^2}{2M_W^2 \cos^2\beta}\sum_{i,j,k,l=1}^{3}\Big(W_{is}^{\tilde{d}*} U_{ik}^{d_L} m_{d_k} U_{jk}^{d_R*} W_{j+3\ t}^{\tilde{d}} W_{l+3\ u}^{\tilde{e}*} m_{e_l} W_{lv}^{\tilde{e}}$$

$$+ W_{i+3\ s}^{\tilde{d}*} U_{ik}^{d_R} m_{d_k} U_{jk}^{d_L*} W_{jt}^{\tilde{d}} W_{lu}^{\tilde{e}*} m_{e_l} W_{l+3\ v}^{\tilde{e}}\Big)$$

$$+ \frac{g_2^2}{4}\left(1 - \frac{\tan^2\theta_W}{3}\right)\sum_{i,j=1}^{3} W_{is}^{\tilde{d}*} W_{it}^{\tilde{d}} W_{ju}^{\tilde{e}*} W_{jv}^{\tilde{e}}$$

$$+ \frac{g_2^2 \tan^2\theta_W}{6}\sum_{i,j=1}^{3}\Big[W_{is}^{\tilde{d}*} W_{it}^{\tilde{d}} W_{j+3\ u}^{\tilde{e}*} W_{j+3\ v}^{\tilde{e}}$$

$$- W_{i+3\ s}^{\tilde{d}*} W_{i+3\ t}^{\tilde{d}}\left(W_{ju}^{\tilde{e}*} W_{jv}^{\tilde{e}} - 2W_{j+3\ u}^{\tilde{e}*} W_{j+3\ v}^{\tilde{e}}\right)\Big]\ , \tag{9.80f}$$

$$Y[\tilde{u}_s^{a*}, \tilde{u}_t^a, \tilde{\nu}_i^*, \tilde{\nu}_i] = \frac{g_2^2}{4}\sum_{j=1}^{3}\left[W_{js}^{\tilde{u}*} W_{jt}^{\tilde{u}}\left(1 - \frac{\tan^2\theta_W}{3}\right) + \frac{4\tan^2\theta_W}{3} W_{j+3\ s}^{\tilde{u}*} W_{j+3\ t}^{\tilde{u}}\right]\ , \tag{9.80g}$$

$$Y[\tilde{d}_s^{a*}, \tilde{d}_t^a, \tilde{\nu}_i^*, \tilde{\nu}_i] = -\frac{g_2^2}{4}\sum_{j=1}^{3}\left[W_{js}^{\tilde{d}*} W_{jt}^{\tilde{d}}\left(1 + \frac{\tan^2\theta_W}{3}\right) - \frac{2\tan^2\theta_W}{3} W_{j+3\ s}^{\tilde{d}*} W_{j+3\ t}^{\tilde{d}}\right]\ , \tag{9.80h}$$

$$Y[\tilde{u}_s^{a*}, \tilde{d}_t^a, \tilde{e}_u^*, \tilde{\nu}_i] = \frac{g_2^2}{2M_W^2 \cos^2\beta}\sum_{j,k,l,m=1}^{3} W_{js}^{\tilde{u}*} U_{jk}^{d_L} m_{d_k} U_{lk}^{d_R*} W_{l+3\ t}^{\tilde{d}} U_{mi}^{\tilde{\nu}} m_{e_m} W_{m+3\ u}^{\tilde{e}*}$$

$$+ \frac{g_2^2}{2}\sum_{j,k=1}^{3} W_{js}^{\tilde{u}*} W_{jt}^{\tilde{d}} U_{ki}^{\tilde{\nu}} W_{ku}^{\tilde{e}*}\ , \tag{9.80i}$$

$$Y[\tilde{d}_t^{a*}, \tilde{u}_s^a, \tilde{\nu}_i^*, \tilde{e}_u] = \left(Y[\tilde{u}_s^{a*}, \tilde{d}_t^a, \tilde{e}_u^*, \tilde{\nu}_i]\right)^*\ , \tag{9.80j}$$

$$Y[\tilde{e}_s^*, \tilde{e}_t, \tilde{e}_u^*, \tilde{e}_v] = \frac{g_2^2}{2M_W^2 \cos^2\beta} \sum_{i,j=1}^{3} W_{is}^{\tilde{e}*} m_{e_i} W_{i+3\ t}^{\tilde{e}} W_{j+3\ u}^{\tilde{e}*} m_{e_j} W_{jv}^{\tilde{e}}$$

$$+ \frac{g_2^2}{8}\left(1+\tan^2\theta_W\right) \sum_{i,j=1}^{3} W_{is}^{\tilde{e}*} W_{it}^{\tilde{e}} W_{ju}^{\tilde{e}*} W_{jv}^{\tilde{e}}$$

$$+ \frac{g_2^2 \tan^2\theta_W}{2} \sum_{i,j=1}^{3} \left(W_{i+3\ s}^{\tilde{e}*} W_{i+3\ t}^{\tilde{e}} - W_{is}^{\tilde{e}*} W_{it}^{\tilde{e}}\right) W_{j+3\ u}^{\tilde{e}*} W_{j+3\ v}^{\tilde{e}} \,, \tag{9.80k}$$

$$Y[\tilde{e}_s^*, \tilde{e}_t, \tilde{\nu}_i^*, \tilde{\nu}_j] = \frac{g_2^2}{2M_W^2 \cos^2\beta} \sum_{k,l=1}^{3} U_{ki}^{\tilde{\nu}*} m_{e_k} W_{k+3\ t}^{\tilde{e}} U_{lj}^{\tilde{\nu}} m_{e_l} W_{l+3\ s}^{\tilde{e}*}$$

$$+ \frac{g_2^2}{2} \sum_{k,l=1}^{3} W_{ks}^{\tilde{e}*} U_{kj}^{\tilde{\nu}} W_{lt}^{\tilde{e}} U_{li}^{\tilde{\nu}*}$$

$$- \delta_{ij}\frac{g_2^2}{4} \sum_{k=1}^{3} \left[\left(1-\tan^2\theta_W\right) W_{ks}^{\tilde{e}*} W_{kt}^{\tilde{e}} + 2\tan^2\theta_W W_{k+3\ s}^{\tilde{e}*} W_{k+3\ t}^{\tilde{e}}\right] \,, \tag{9.80l}$$

$$Y[\tilde{\nu}_i^*, \tilde{\nu}_i, \tilde{\nu}_j^*, \tilde{\nu}_j] = \frac{g_2^2}{8}\left(1+\tan^2\theta_W\right) . \tag{9.80m}$$

In (9.80) we have used $s, t, u, v = 1, \ldots, 6$ to label sfermion mass eigenstates, $i, j, k, l, m, n = 1, 2, 3$ are generation (or sneutrino mass eigenstate) labels, and superscripts $a, b = 1, 2, 3$ are $SU(3)$-color labels. Note that there are two different color connections for $\tilde{u}^*\tilde{u}\tilde{d}^*\tilde{d}$ vertices, as shown in (9.80c,d). These two color connections are equivalent for interactions of four squarks of the same type, since they can be transformed into each other by simply exchanging mass eigenstate labels, which are summed in (9.79). When computing the Feynman rules from (9.79) and (9.80), care must be taken to symmetrize properly. The result is displayed in Fig. 9.17.

$$\boxed{\text{Fig. 9.17 is included in Appendix A}}$$

We conclude with a remark (cf. ftnt.10). The mixing matrices appearing in (9.80) are again not separately invariant under redefinitions of the "current" eigenstates. However, the *products* of mixing matrices appearing in these expressions are invariant under such redefinitions, since they describe couplings of physical particles (i.e. mass eigenstates).

References

[9.1] S. Dimopoulos and H. Georgi, *loc. cit.*, Ref. [5.2].

[9.2] S. Weinberg, Phys. Rev. **D26** (1987) 287.

[9.3] S.B. Treiman, D. Gross and R. Jackiw, *op. cit.*, *Bibl.*

[9.4] A. Brignole, L.E. Ibáñez and C. Muñoz in *Perspectives on supersymmetry*, ed. G.L. Kane, *op. cit.*, *Bibl.*

[9.5] G.F. Giudice and R. Rattazzi, *loc. cit.*, *Bibl.*

[9.6] T. Ibrahim and P. Nath, Phys. Rev. **D58** (1999) 111301. M. Brhlik, G.J. Good and G.L. Kane, Phys. Rev. **D59** (1999) 115004.

[9.7] R. Garisto, Nucl. Phys. **B149** (1994) 279. A. Masiero and A. Silvestrini in G.L. Kane, *Perspectives on Supersymmetry*, *op. cit.*, *Bibl.* A. Masiero, in *Turin 1999: Neutrino Mixing* (ed. W.M. Alberico, World Scientific, Singapore, 2000). A. Masiero and O. Vives, Nucl. Phys. Proc. Suppl. **99B** (2001) 228.

[9.8] L.J. Hall, V.A. Kostelecky and S. Raby, Nucl. Phys. **B267** (1986) 415. M. Dine, hep-ph/9306328, Proc. *Supersymmetry and Unification of Fundamental Interactions*, SUSY 93 (Boston), p136. J.S. Hagelin, S. Kelly and T. Tanaka, Nucl. Phys. **B415** (1994) 293. S. Dimopoulos and D.W. Sutter, Nucl. Phys. **B452** (1995) 496. D.W. Sutter, hep-ph/9704390. H.E. Haber, Nucl. Phys. B Proc. Suppl. **62A–C** (1998) 469.

[9.9] S.Y. Choi, A. Djouadi, M. Guchait, J. Kalinowski, H.S. Song and P.M. Zerwas, Eur. Phys. J. **C14** (2000) 535.

[9.10] H.E. Haber and G.L. Kane, *loc. cit.*, *Bibl.* J.F. Gunion and H.E. Haber, Nucl. Phys. **B272** (1986) 1.

[9.11] M. Guchait, Z. Phys. **C57** (1993) 157; errtm. *ibid.* **C67** (1994) 178.

[9.12] S.P. Martin in G.L. Kane, *Perspectives on Supersymmetry*, *op. cit.*, *Bibl.* F. Gabbiani and A. Masiero, Phys. Lett. **B209** (1988) 289.

[9.13] J.F. Donoghue, E. Golowitcz and B. Holstein, *op. cit.*, *Bibl.*

[9.14] M. Dine, A. Kagan and S. Samuel, Phys. Lett. **B243** (1990) 250. K. Agashe and M. Graesser, Phys. Rev. **D59** (1999) 015007.

[9.15] N. Arkani-Hamed and H. Murayama, Phys. Rev. **D56** (1997) 6733.

[9.16] Y. Nir and N. Seiberg, Phys. Lett. **B309** (1993) 337. S. Dimopoulos, G.F. Giudice and N. Tetradis, Nucl. Phys. **B454** (1995) 59.

[9.17] N. Arkani-Hamed, J.L. Feng, L.J. Hall and H.C. Chang, Nucl. Phys. **B505** (1997) 3.

[9.18] F. Gabbiani, E. Gabrielli, A. Masiero and L. Silvestrini, Nucl. Phys. **B477** (1996) 321.

Chapter 10

HIGGS BOSONS IN THE MSSM

10.1 Higgs Potential in the MSSM

As discussed in Ch.1, low energy supersymmetry has been theoretically motivated to stabilize the mass and the VEV of the Standard Model Higgs boson with respect to higher scales. This makes the Higgs sector of a supersymmetric extension of the Standard Model especially interesting. We have already shown in Ch.8 that the minimal supersymmetric model requires two Higgs doublets $h_{1,2}$ (with D as an SU(2) doublet index and $Y = -1, 1$ respectively):

$$h_1^D \equiv \begin{pmatrix} h_1^1 \\ h_1^2 \end{pmatrix} = \begin{pmatrix} h_1^0 \\ h_1^- \end{pmatrix}, \quad h_2^D \equiv \begin{pmatrix} h_2^1 \\ h_2^2 \end{pmatrix} = \begin{pmatrix} h_2^+ \\ h_2^0 \end{pmatrix}. \tag{10.1}$$

We shall see in this chapter how these doublets lead to five physical Higgs particles $h, H, A, H^\pm$ and what one can say about their masses and couplings [10.1], [10.2]. A noteworthy feature, specific to this supersymmetric extension, is that all quartic self couplings of the Higgs fields get related to the gauge couplings of the electroweak theory. This is quite unlike in nonsupersymmetric theories where the former are a priori arbitrary. This restriction is the key to various mass bounds and relations [10.3] which exist for physical Higgs particles in the supersymmetric extension of the Standard Model. A second important feature is that the couplings of the neutral Higgs particles to quark mass eigenstates turn out to be flavor diagonal. This happens because up type quarks obtain their masses purely from the VEV $v_2/\sqrt{2}$ of h_2^0 while down type ones do so from the VEV $v_1/\sqrt{2}$ of h_1^0. In the language of Glashow and Weinberg [10.4] the Higgs sector of the MSSM is a special case of the 'type 2' two Higgs doublet model.

We have already given the MSSM superpotential and the soft explicit supersymmetry breaking terms in Chs. 8 and 9 respectively. The tree level scalar potential is

$$V = V_{SUSY} + V_{SOFT}, \tag{10.2}$$

where V_{SUSY} was defined in (8.36) and V_{SOFT} in (9.3). Recall that

$$F_k = -\partial \mathcal{W}_{\text{MSSM}}/\partial \Phi_k^\dagger\bigg|, \quad \mathcal{W}_{\text{MSSM}} = \mu H_1 \cdot H_2 - f_{ij}^\ell H_1 \cdot L_i \bar{E}_j - f_{ij}^d H_1 \cdot Q \bar{D}_j - f_{ij}^u Q \cdot H_2 \bar{U}_j. \tag{10.3}$$

As before, i, j are generation indices and, for any two SU(2)-doublet superfields A^D and B^E, $A \cdot B \equiv \epsilon_{DE} A^D B^E$. Moreover,

$$\vec{D}_H = -g_2 h_k^\dagger \frac{\vec{\tau}}{2} h_k \ , \tag{10.4a}$$

$$D_H^Y = -g_Y h_k^\dagger \frac{Y}{2} h_k \ , \tag{10.4b}$$

where we are now using the subscript H to refer exclusively to the Higgs sector and k is summed. Needless to say, both $\vec{D}$ and D^Y will have additional bilinear terms involving squarks as well as those with sleptons.

The tree level Higgs potential follows from (10.2) – (10.4) by inputting $V_{\rm SOFT}$ from (9.3) and utilizing the relation $\vec{\tau}_{AB} \cdot \vec{\tau}_{CD} = 2\delta_{AD}\delta_{BC} - \delta_{AB}\delta_{CD}$. Using the notation $h^\dagger h \equiv |h|^2$, it can be written as

$$V_H = \frac{1}{8}(g_Y^2 + g_2^2)(|h_1|^2 - |h_2|^2)^2 + \frac{g_2^2}{2}\,|h_1^\dagger h_2|^2 + |\mu|^2(|h_1|^2 + |h_2|^2) + V_{H,SOFT} \ , \tag{10.5a}$$

$$V_{H,SOFT} = m_1^2|h_1|^2 + m_2^2|h_2|^2 + (m_{12}^2 h_1 \cdot h_2 + h.c.) \ , \tag{10.5b}$$

with coefficients m_1^2, m_2^2 and $m_{12}^2 \equiv B\mu$, cf.(9.3), having the dimension[1] of squared mass. In following the steps to (10.5), it may be noted that $h_1 \cdot h_2 = \tilde{h}_1^\dagger h_2$ where $\tilde{h}_1 = i\tau_2 h_1^*$ is an SU(2) doublet with $Y = 1$. (10.5a) and (10.5b) can be rewritten as

$$V_H = \frac{1}{8}(g_Y^2 + g_2^2)(|h_1|^2 - |h_2|^2)^2 + \frac{g_2^2}{2}|h_1^\dagger h_2|^2 + m_{1h}^2|h_1|^2 + m_{2h}^2|h_2|^2 + (m_{12}^2 h_1 \cdot h_2 + h.c.), \tag{10.6}$$

where

$$m_{1,2h}^2 = m_{1,2}^2 + |\mu|^2 \ . \tag{10.7}$$

The sign of the last RHS term in (10.6) has been chosen with care. It will be seen later that $m_{12}^2 = B\mu$ is expected to be positive.

10.2   Spontaneous Symmetry Breakdown and VEVs

A Higgs induced spontaneous symmetry breaking will take place if the minimum of V_H is attained at nonzero values of the Higgs fields:

$$\langle h_1 \rangle = \frac{1}{\sqrt{2}} \begin{pmatrix} v_1 \\ 0 \end{pmatrix}, \quad \langle h_2 \rangle = \frac{1}{\sqrt{2}} \begin{pmatrix} 0 \\ v_2 \end{pmatrix}. \tag{10.8}$$

In (10.6) one can[2] always absorb a relative phase between h_1 and h_2 by redefining one of them with an additional phase; this freedom enables us to define $v_{1,2}$ as **real and positive**

---

[1] We remind the reader that B is a soft supersymmetry breaking parameter with the dimension of mass, while μ is a supersymmetry invariant (higgsino mass) parameter.

[2] Any VEV for one charged Higgs field can be rotated to zero by an $SU(2)$ transformation and then the minimization condition means a vanishing VEV for the other charged Higgs. This is a consequence of inbuilt $U(1)_{\rm em}$ invariance which thus remains unbroken.

and also to treat m_1^2, m_2^2 and m_{12}^2 as real. Recall from §8.2 that these VEVs can be related to the W and Z masses by

$$M_W = \frac{g_2}{2}(v_1^2 + v_2^2)^{1/2}, \quad M_Z = \frac{(g_Y^2 + g_2^2)^{1/2}}{2}(v_1^2 + v_2^2)^{1/2}, \tag{10.9}$$

i.e.

$$(v_1^2 + v_2^2)^{1/2} = (\sqrt{2}G_F)^{-1} \simeq 246 \text{ GeV}. \tag{10.10}$$

Let us consider the parameter $\tan\beta$, as introduced in (8.24), namely

$$\tan\beta \equiv v_2/v_1. \tag{10.11}$$

Now, our phase freedom to define $v_{1,2}$ as positive restricts β to the range

$$0 \leq \beta \leq \pi/2 \,.$$

Though $\tan\beta$ will generally be left undetermined in this book, current theoretical widsom suggests [10.5] that the value of $\tan\beta$ is restricted to the range $1 \leq \tan\beta \lesssim 60$. The lower and upper bounds both stem from the desired requirement (cf. Ch.11) of radiatively induced electroweak symmetry breakdown by which one of the eigenvalues of the neutral Higgs mass squared matrix, evaluated at $v_1 = 0 = v_2$, is driven to be negative by the top Yukawa coupling via Renormalization Group Evolution. They come also from the requirement of all the couplings participating in the RGE equations remaining perturbative upto a high grand unifying scale like 2×10^{16} GeV. These issues, including additional experimental constraints on $\tan\beta$, will be discussed more thoroughly in Ch.11.

Near the minimum, characterized by the VEVs $\langle h_{1,2}^0 \rangle = v_{1,2}/\sqrt{2}$, $\langle h_1^- \rangle = 0 = \langle h_2^+ \rangle$, it is sufficient to explore the Higgs potential retaining only the neutral Higgs fields. This part of the Higgs potential can be written from (10.6) as

$$V_H^0 = \frac{1}{8}(g_Y^2 + g_2^2)^2(|h_1^0|^2 - |h_2^0|^2)^2 + m_{1h}^2|h_1^0|^2 + m_{2h}^2|h_2^0|^2 - m_{12}^2(h_1^0 h_2^0 + \text{h.c.}) \,, \tag{10.12}$$

where the negative sign before the last RHS term proportional to m_{12}^2 has arisen because $\epsilon_{12} = -1$. The quartic terms in (10.12) vanish along $|h_1^0| = |h_2^0|$. By further choosing $h_1^0 = \pm h_2^0$, we see that the fact that V_H^0 must be bounded from below requires that

$$m_{1h}^2 + m_{2h}^2 = m_1^2 + m_2^2 + 2|\mu|^2 > 2|m_{12}^2| \,. \tag{10.13}$$

Because of quantum corrections and renormalization group evolution (cf. Ch.11), m_{1h}^2, m_{2h}^2 and m_{12}^2 become running quantities — varying with the energy scale, cf. §6.9. However, (10.13) has to be valid at all scales. On the other hand, the quadratic part of V_H^0 can be written as

$$V_H^{0,quadr.} = (h_1^{0*} \quad h_2^0) \begin{pmatrix} m_{1h}^2 & -m_{12}^2 \\ -m_{12}^2 & m_{2h}^2 \end{pmatrix} \begin{pmatrix} h_1^0 \\ h_2^{0*} \end{pmatrix}. \tag{10.14}$$

For the nonzero VEVs $v_{1,2}$ to develop, at least one of the eigenvalues of the mass squared matrix in (10.14) has to be negative. Since (10.13) requires the matrix to have a positive trace, one is led to the necessary condition for spontaneous symmetry breakdown that its determinant be negative, i.e.

$$m_{12}^4 > m_{1h}^2 m_{2h}^2 = (m_1^2 + |\mu|^2)(m_2^2 + |\mu|^2) \ . \tag{10.15}$$

(10.15) is valid only at and below the energy scale where the spontaneous breaking of electroweak symmetry becomes operative. Furthermore, (10.13) and (10.15) become mutually incompatible in the supersymmetry invariant limit when $m_{1h}^2 = m_{2h}^2 = \mu^2$. Hence there is **an intimate connection between the breaking of supersymmetry and that of electroweak symmetry in the MSSM**.

Let us return to (10.6) and explore V_H at its supposed minimum, i.e. at $h_{1,2} = \langle h_{1,2} \rangle$, as given by (10.8). Thus

$$V_H^{min} = \frac{1}{32}(g_Y^2 + g_2^2)(v_1^2 - v_2^2)^2 + \frac{1}{2}m_{1h}^2 v_1^2 + \frac{1}{2}m_{2h}^2 v_2^2 - m_{12}^2 v_1 v_2 \ . \tag{10.16}$$

The consistency conditions for the above mentioned minimum is the vanishing of $\partial V_H^{min}/\partial v_1$ and $\partial V_H^{min}/\partial v_2$. These respectively imply the relations

$$m_{1h}^2 = m_{12}^2 \frac{v_2}{v_1} - \frac{1}{8}(g_Y^2 + g_2^2)(v_1^2 - v_2^2) \ , \tag{10.17a}$$

$$m_{2h}^2 = m_{12}^2 \frac{v_1}{v_2} + \frac{1}{8}(g_Y^2 + g_2^2)(v_1^2 - v_2^2) \ . \tag{10.17b}$$

By using (10.7), (10.9) and (10.11) in (10.17), the latter can be recast into the following equations:

$$-2B\mu = -2m_{12}^2 = (m_1^2 - m_2^2)\tan 2\beta + M_Z^2 \sin 2\beta \ , \tag{10.18a}$$

$$|\mu|^2 = (\cos 2\beta)^{-1}(m_2^2 \sin^2 \beta - m_1^2 \cos^2 \beta) - \frac{1}{2}M_Z^2 \ . \tag{10.18b}$$

10.3   Higgs Masses at the Tree Level

Though we shall see in §10.6 that there are significant radiative corrections to Higgs masses in the MSSM, we first discuss their tree level values here. The mass squared matrix of the Higgs scalars can be obtained from the quadratic part of V_H, i.e. $V_H^{(2)} = \frac{1}{2}m_{lm}^2 \phi_l \phi_m$ with

$$m_{lm}^2 = \left\langle \frac{\partial^2 V_H}{\partial \phi_l \partial \phi_m} \right\rangle , \tag{10.19}$$

where ϕ_l is the generic notation for the real or imaginary part of any Higgs component field and the double derivative is evaluated at the minimum. The 8×8 Higgs mass squared matrix then breaks up diagonally into a set of 2×2 matrices.

Charged Goldstones and Higgs

The total charged Higgs mass term, obtained by using (10.8) in (10.6), is given by

$$V_{h^\pm}^{quadr} =$$

$$(h_1^+ \; h_2^+) \begin{pmatrix} m_{1h}^2 + \frac{1}{8}(g_Y^2 + g_2^2)(v_1^2 - v_2^2) + \frac{1}{4}g_2^2 v_2^2 & m_{12}^2 + \frac{1}{4}g_2^2 v_1 v_2 \\ m_{12}^2 + \frac{1}{4}g_2^2 v_1 v_2 & m_{2h}^2 - \frac{1}{8}(g_Y^2 + g_2^2)(v_1^2 - v_2^2) + \frac{1}{4}g_2^2 v_1^2 \end{pmatrix} \begin{pmatrix} h_1^- \\ h_2^- \end{pmatrix}$$

$$= \left(\frac{m_{12}^2}{v_1 v_2} + \frac{1}{4}g_2^2 \right) (h_1^+ \; h_2^+) \begin{pmatrix} v_2^2 & v_1 v_2 \\ v_1 v_2 & v_1^2 \end{pmatrix} \begin{pmatrix} h_1^- \\ h_2^- \end{pmatrix} , \tag{10.20}$$

where – in the last step – eqs. (10.17) have been used. The vanishing determinant and the nonvanishing trace of the matrix in the RHS of (10.20) imply massless as well as massive charged modes. The former are the Goldstone boson pair $G^\pm$ which combine with the massless $W^\pm$ to give them mass. The latter pertain to the physical charged Higgs particles $H^\pm$. Thus one has

$$m_{G^\pm}^2 = 0 , \tag{10.21a}$$

$$m_{H^\pm}^2 = \left(\frac{m_{12}^2}{v_1 v_2} + \frac{1}{4}g_2^2 \right) (v_1^2 + v_2^2) . \tag{10.21b}$$

It follows from (10.20) and (10.11) that the corresponding mass diagonal fields are

$$H^\pm = \sin\beta \; h_1^\pm + \cos\beta \; h_2^\pm , \tag{10.22a}$$

$$G^\pm = -\cos\beta \; h_1^\pm + \sin\beta \; h_2^\pm , \tag{10.22b}$$

The couplings of $G^\pm$ in a general R-gauge are given in Ref. [10.1]. However, we formulate our discussions in the unitary gauge where $G^\pm$ are set equal to zero.

Neutral Goldstone and CP odd Higgs

Choosing $\phi_{\ell,m}$ in (10.19) to be $\Im m \; h_{1,2}^0$, we have the corresponding mass squared matrix:

$$m_{\Im m \; h^0}^2 =$$

$$\begin{pmatrix} m_{1h}^2 + \frac{1}{8}(g_Y^2 + g_2^2)(v_1^2 - v_2^2) & m_{12}^2 \\ m_{12}^2 & m_{2h}^2 - \frac{1}{8}(g_Y^2 + g_2^2)(v_1^2 - v_2^2) \end{pmatrix} = m_{12}^2 \begin{pmatrix} v_2/v_1 & 1 \\ 1 & v_1/v_2 \end{pmatrix} , \tag{10.23}$$

once again using (10.17). As before, the vanishing determinant and the nonvanishing trace imply a massless neutral Goldstone mode G (which combines with the massless Z) and a neutral scalar which is CP odd on being a linear combination of the imaginary components of the neutral Higgs fields. In fact, we have

$$m_{G^0}^2 = 0 , \tag{10.24a}$$

$$m_A^2 = \frac{m_{12}^2}{v_1 v_2} (v_1^2 + v_2^2) = \frac{2m_{12}^2}{\sin 2\beta} . \tag{10.24b}$$

N.B. since $\sin 2\beta$ is restricted to be positive, (10.24b) makes sense only if m_{12}^2 is positive – at least at electroweak energy scales. This is the explanation of the choice of the sign of the last RHS term in (10.6). The mass diagonal fields corresponding to (10.24) are

$$\frac{A}{\sqrt{2}} = \Im m \, h_1^0 \sin\beta + \Im m \, h_2^0 \cos\beta \,, \tag{10.25a}$$

$$\frac{G^0}{\sqrt{2}} = -\Im m \, h_1^0 \cos\beta + \Im m \, h_2^0 \sin\beta \,. \tag{10.25b}$$

In the physical basis, the CP odd neutral Higgs mass term in the Lagrangian density becomes

$$\frac{1}{2}(G^0 \ \ A)\begin{pmatrix} 0 & \\ & m_A^2 \end{pmatrix}\begin{pmatrix} G^0 \\ A \end{pmatrix},$$

and the correct normalization of m_A^2 in (10.24b) can be checked from this. The couplings of G^0 in a general R-gauge can be found in Ref. [10.1], but again, in the U-gauge of ours, $G^0 = 0$.

Neutral CP even Higgs

Turning to the $\Re e \, h_{1,2}^0$ components, we find the corresponding mass squared matrix in an analogous way to be

$$m_{\Re e \, h^0}^2 = \frac{1}{2}\begin{pmatrix} 2m_{1h}^2 + \frac{1}{4}(g_Y^2 + g_2^2)(3v_1^2 - v_2^2) & -2m_{12}^2 - \frac{1}{2}v_1 v_2(g_Y^2 + g_2^2) \\ -2m_{12}^2 - \frac{1}{2}v_1 v_2(g_Y^2 + g_2^2) & 2m_{2h}^2 + \frac{1}{4}(g_Y^2 + g_2^2)(3v_2^2 - v_1^2) \end{pmatrix}$$

$$= \begin{pmatrix} m_A^2 \sin^2\beta + M_Z^2 \cos^2\beta & -(m_A^2 + M_Z^2)\sin\beta\cos\beta \\ -(m_A^2 + M_Z^2)\sin\beta\cos\beta & m_A^2 \cos^2\beta + M_Z^2 \sin^2\beta \end{pmatrix}, \tag{10.26}$$

where (10.10), (10.17) and (10.24b) have been used. The eigenvalues of the matrix in the RHS of (10.26), standing for the tree level physical squared masses of the two CP even Higgs scalars (H, h) of the MSSM, are

$$m_{H,h}^2 = \frac{1}{2}\left[m_A^2 + M_Z^2 \pm \{(m_A^2 + M_Z^2)^2 - 4M_Z^2 m_A^2 \cos^2 2\beta\}^{1/2}\right]. \tag{10.27}$$

In (10.27) we have defined H to be the heavier of the two, i.e. $m_h \leq m_H$. The corresponding mass diagonal fields are

$$\frac{1}{\sqrt{2}}H = (\Re e \, h_1^0 - \frac{v_1}{\sqrt{2}})\cos\alpha + (\Re e \, h_2^0 - \frac{1}{\sqrt{2}}v_2)\sin\alpha \,, \tag{10.28a}$$

$$\frac{1}{\sqrt{2}}h = -(\Re e \, h_1^0 - \frac{v_1}{\sqrt{2}})\sin\alpha + (\Re e \, h_2^0 - \frac{1}{\sqrt{2}}v_2)\cos\alpha \,. \tag{10.28b}$$

Referring back to the matrix of (10.26) as $\begin{pmatrix} A & B \\ B & C \end{pmatrix}$, the angle of rotation α in (10.28) is seen to obey the relations [10.1], [10.2]

$$\sin 2\alpha = \frac{2B}{\sqrt{(A-C)^2 + 4B^2}} = -\frac{m_H^2 + m_h^2}{m_H^2 - m_h^2}\sin 2\beta \,, \tag{10.29a}$$

$$\cos 2\alpha = \frac{A - C}{\sqrt{(A-C)^2 + 4B^2}} = -\frac{m_A^2 - M_Z^2}{m_H^2 - m_h^2} \cos 2\beta , \qquad (10.29b)$$

$$\tan 2\alpha = \frac{m_h^2 + m_H^2}{m_A^2 - M_Z^2} \tan 2\beta = \frac{m_A^2 + M_Z^2}{m_A^2 - M_Z^2} \tan 2\beta . \qquad (10.29c)$$

Since β is in the range $0 \leq \beta \leq \pi/2$, (10.29a) restricts α to the interval

$$-\pi/2 \leq \alpha \leq 0 .$$

A geometrical depiction is given in Fig. 10.1. Note that we always have

$$\sin(\beta - \alpha), \cos(\beta + \alpha) \geq 0 .$$

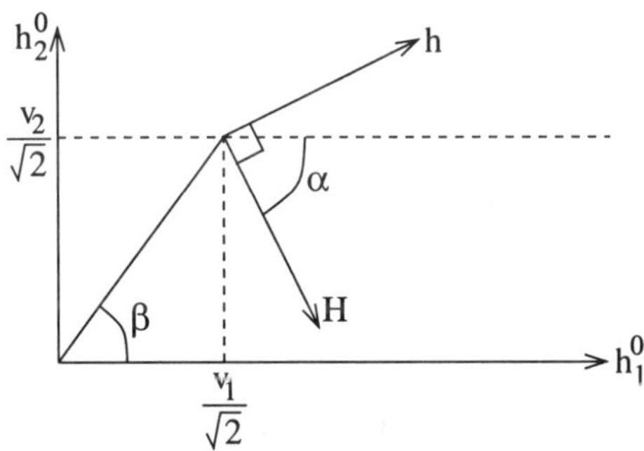

Fig.10.1. Geometrical depiction of physical CP even neutral Higgs states.

Relations and constraints

The Higgs mass spectrum is completely controlled by two new parameters which can be taken to be m_A and $\tan\beta$. These strongly influence the other masses, e.g. $m_h \to 0$ if $m_A \to 0$. The following tree level relations and constraints [10.1–10.3] emerge from (10.21b), (10.24b) and (10.27):

$$m_{H\pm}^2 = m_A^2 + M_W^2 > \max\left(M_W^2, m_A^2\right) , \qquad (10.30a)$$

$$m_h^2 + m_H^2 = m_A^2 + M_Z^2 , \qquad (10.30b)$$

$$m_h < \min\left(m_A, M_Z\right)|\cos 2\beta| < \min\left(m_A, M_Z\right) , \qquad (10.30c)$$

$$m_H > \max\left(m_A, M_Z\right) , \qquad (10.30d)$$

$$\cos^2(\beta - \alpha) = \frac{m_h^2(M_Z^2 - m_h^2)}{m_A^2(m_H^2 - m_h^2)} . \qquad (10.30e)$$

Thus the charged Higgs bosons $H^\pm$ are predicted to be heavier than the W. Of the CP even neutral ones, one light Higgs h is expected to be lighter than the Z and one heavier H is expected to exceed the Z in mass. The mass of the CP odd Higgs A is expected to be between those of the two CP even ones. The contents of (10.30c) and (10.30d) are illustrated in Fig. 10.2 below where $m_{h<}$ and $m_{H>}$ are the absolute (β-independent) upper and lower bounds on m_h and m_H respectively. For large $\tan\beta$ (i.e. $|\cos 2\beta| \to 1$) , m_h saturates $m_{h<}$ from below and m_H comes down to $m_{H>}$ from above. These are all **tree level predictions**; we discuss radiative effects on these mass bounds in §10.6.

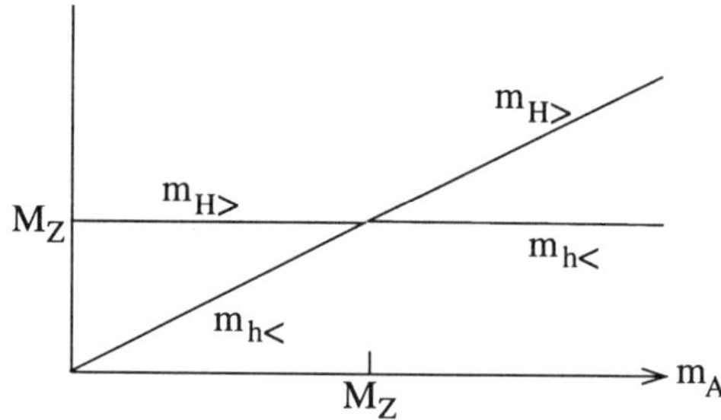

Fig.10.2. Tree level upper and lower mass bounds on h and H as a function of m_A.

10.4   Higgs-particle Vertices

The electroweak parameters of the Standard Model, together with $\tan\beta$ and α, completely determine the couplings of the physical Higgs particles to the Standard Model gauge bosons and fermions. We shall first discuss those and then come to Higgs self couplings. A discussion of Higgs couplings to sparticles is relegated to §10.5. In (8.32) we have already given the Higgs contribution to the supersymmetric part of the MSSM Lagrangian density. From this one can obtain all the Higgs couplings to fermions and gauge bosons in terms of the originally introduced but unphysical Higgs fields $h^0_{1,2}$ and $h^\pm_{1,2}$. The conversion to couplings with mass diagonal Higgs fields can be easily done through the transformations (10.22), (10.25) and (10.28). One should also put $G^\pm = 0 = G^0$ in the unitary gauge which we choose. For simplicity, we confine ourselves to one generation of up and down type fermions (masses m_u and m_d respectively): $f_L \equiv \begin{pmatrix} f_{uL} \\ f_{dL} \end{pmatrix}$, f_{uR}, f_{dR}, where f covers both quarks and leptons. Generation effects can be obtained by interpreting $m_{u,d}$ as 3×3 diagonal quark mass matrices and multiplying the charged Higgs coupling to fermions by the Cabibbo-Kobayashi-Maskawa matrix V.

The Higgs-fermion-antifermion Yukawa interactions can then be written as

$$
\begin{aligned}
\mathcal{L}_Y &= -\frac{g_2 m_d}{2M_W \cos\beta} \sum_f \bar{f}_d f_d (H \cos\alpha - h \sin\alpha) + \frac{i g_2 m_d \tan\beta}{2M_W} \sum_f \bar{f}_d \gamma_5 f_d A \\
&\quad -\frac{g_2 m_u}{2M_W \sin\beta} \sum_f \bar{f}_u f_u (H \sin\alpha + h \cos\alpha) + \frac{i g_2 m_u \cot\beta}{2M_W} \sum_f \bar{f}_u \gamma_5 f_u A \\
&\quad +\frac{g_2}{\sqrt{2}M_W} \sum_f \left[H^+ \bar{f}_u (m_u \cot\beta\, P_L + m_d \tan\beta\, P_R) f_d + h.c. \right],
\end{aligned}
\tag{10.31}
$$

with f being summed over quarks and leptons. The corresponding vertex couplings (i times the coefficients of the interaction terms in $\mathcal{L}$) are given in Fig. 10.3. We do the same for the trilinear gauge-gauge-Higgs and Higgs-gauge-Higgs as well as the quartic gauge-gauge-Higgs-Higgs vertices instead of writing out the algebraic expressions in $\mathcal{L}$.

$$\boxed{\text{Fig. 10.3 is included in Appendix B}}$$

We can make the following comments on the couplings of Fig. 10.3.

- Tree level Higgs couplings to fermions are parity conserving and that of A to matter fermions involves a γ_5. That is why, in contrast with the 'scalars' h and H, A is sometimes called the 'pseudoscalar' Higgs. But, in the presence of CP violation, loop effects can mix the 'scalar' and 'pseudoscalar' Higgs bosons, especially since the MSSM admits additional sources of CP violation beyond the CKM phase.

- The parameters m_u and m_d refer to masses of up and down type quarks respectively for each generation.

- Bose statistics forbids the ZHH and Zhh trilinear couplings, while any ZhH coupling is forbidden by CP invariance. Since the latter is violable, a ZhH coupling could exist.

- The absence of any tree level $ZW^{\pm}H^{\mp}$ or $\gamma W^{\pm}H^{+}$ coupling is not surprising since neither can occur [10.1] in any model containing just $SU(2)_L$ doublet and singlet Higgs fields.

- The couplings for the vertices $(W^+W^-h$ and $W^+W^-H)$, $(W^+HH^-$ and $W^+hH^-)$, $(ZHA$ and $ZhA)$, $(ZZH$ and $ZZh)$ and $(ZZh$ and $ZhA)$ are pairwise complementary, i.e. if one is suppressed by the combination of mixing angles, the other is nearly full strength.

- For large $\tan\beta$ and moderate α, the neutral CP even Higgs couplings with the down type fermions get enhanced relative to those with up type ones. For the CP odd Higgs, this statement is true independent of α.

Turning to the self couplings of the Higgs bosons, we notice that they follow from the Higgs potential V_H of (10.5) on using the formulae for the physical Higgs fields, namely (10.22a), (10.25a) and (10.28). Following Ref. [10.1], one can introduce the convenient differential operators

$$
\begin{aligned}
D_H &\equiv (\sqrt{2})^{-1}[\cos\alpha(\partial/\partial h_1^0 + \partial/\partial h_1^{0*}) + \sin\alpha(\partial/\partial h_2^0 + \partial/\partial h_2^{0*})]\,, \\
D_h &\equiv (\sqrt{2})^{-1}[-\sin\alpha(\partial/\partial h_1^0 + \partial/\partial h_1^{0*}) + \cos\alpha(\partial/\partial h_2^0 + \partial/\partial h_2^{0*})]\,, \\
D_A &\equiv (\sqrt{2})^{-1}i[\sin\beta(\partial/\partial h_1^0 - \partial/\partial h_1^{0*}) + \cos\beta(\partial/\partial h_2^0 - \partial/\partial h_2^{0*})]\,, \\
D_{H-} &\equiv \sin\beta\,\partial/\partial h_1^- + \cos\beta\,\partial/\partial h_2^-\,, \\
D_{H+} &\equiv \sin\beta\,\partial/\partial h_1^+ + \cos\beta\,\partial/\partial h_2^+\,.
\end{aligned}
$$

Now the cubic and quartic vertex factors listed beside each vertex below (with legs, a, b, c spanning h, H, A and a, b, c, d spanning $h, H, A, H^\pm$) can be obtained respectively from $D_a D_b D_c V_H$ and $D_a D_b D_c D_d V_H$ evaluated at $\langle h_1^0 \rangle = v_1/\sqrt{2}, \langle h_1^- \rangle = 0 = \langle h_2^+ \rangle, \langle h_2^0 \rangle = v_2/\sqrt{2}$. We list these cubic and quartic self coupling vertices of the physical Higgs bosons in Fig. 10.4.

$$\boxed{\text{Fig. 10.4 is included in Appendix B}}$$

The **decoupling limit** in the Higgs sector of the MSSM is attained [10.5] by taking m_A to be very large : $m_A \to \infty$. (In practice, this usually obtains once m_A exceeds 250 GeV). From (10.27) and (10.30) we now have the results

$$m_h \to M_Z |\cos 2\beta|, \quad \cos^2 2\beta \to m_h^2/M_Z^2 , \tag{10.32a}$$

$$m_H^2 \to m_A^2 + M_Z^2 \sin^2 2\beta , \tag{10.32b}$$

$$|\cos(\beta - \alpha)| \to M_Z^2 |\sin 4\beta|/(2m_A^2) . \tag{10.32c}$$

In this limit we have $m_A \sim m_H \sim m_{H^\pm}$ and $\cos(\beta - \alpha) \simeq 0$, i.e. $\beta - \alpha \to \pi/2$ and $\sin \alpha \simeq -\cos \beta$ up to corrections $\mathcal{O}(M_Z^2/m_A^2)$. Thus the lightest Higgs particle h saturates[3] its upper mass bound $M_Z |\cos 2\beta|$ while the other Higgses all become uniformly heavy. Moreover, a perusal of the gauge couplings of the Higgs particles, all described above, shows that the vertices HW^+W, HZZ, ZAh, $W^\pm H^\mp h$, $ZW^\pm H^\mp h$ and $\gamma W^\pm H^\mp h$ are all proportional to $\cos(\beta - \alpha)$ while the vertices hZZ, ZAH, $W^\mp H^\pm H$, $ZW^\pm H^\mp H$ and $\gamma W^\pm H^\mp H$ are all proportional to $\sin(\beta - \alpha)$. Hence any vertex involving at least one vector boson and *exactly* one heavy Higgs particle (H, A or $H^\pm$) vanishes as $\cos(\beta - \alpha)$ when $m_A \to \infty$. Turning to matter fermions, the coupling strengths of the CP even neutral Higgs scalars to down type and up type fermions – *relative to those of the Standard Model Higgs* – are given below

$$h f_d \bar{f}_d: \quad -\frac{\sin \alpha}{\cos \beta} = \sin(\beta - \alpha) - \tan \beta \cos(\beta - \alpha) , \tag{10.33a}$$

$$h f_u \bar{f}_u: \quad \frac{\cos \alpha}{\sin \beta} = \sin(\beta - \alpha) + \cot \beta \cos(\beta - \alpha) , \tag{10.33b}$$

$$H f_d \bar{f}_d: \quad \frac{\cos \alpha}{\cos \beta} = \cos(\beta - \alpha) + \tan \beta \sin(\beta - \alpha) , \tag{10.33c}$$

$$H f_u \bar{f}_u: \quad \frac{\sin \alpha}{\sin \beta} = \cos(\beta - \alpha) - \cot \beta \sin(\beta - \alpha) . \tag{10.33d}$$

Fig. 10.3 and (10.33) imply that, in the decoupling limit $|\beta - \alpha| \to \pi/2$, the couplings of the lightest Higgs scalar h to fermions and gauge boson pairs are identical to those of the Standard Model Higgs. Likewise, Fig. 10.4 shows that the self couplings hhh and $hhhh$ also reduce to their SM values in this limit. Thus, for a heavy A with $m_A \gg M_Z$, the effects of the extra scalars $H^\pm$, H and A in the MSSM decouple and the residual scalar h, while saturating its appear mass bound, looks just like the SM Higgs boson ϕ^0. The onset of decoupling is

---

[3] The reader is reminded that the present discussion is at the tree level.

controlled by (10.32c), and is depicted in Fig. 10.5 where the functions $\sin^2(\beta - \alpha)$ and $\cos^2(\beta - \alpha)$, i.e. the squared coupling strengths of h and H respectively to WW (cf. Fig. 10.3) relative to that of the SM Higgs, are plotted against m_A for two characteristic values of $\tan\beta$. Though the tree level results, mentioned above, change somewhat on account of radiative corrections (cf. §10.6), this last statement remains valid. The low energy effective scalar sector of the MSSM indeed becomes indistinguishable from that of the SM in the decoupling limit, except that, unlike in the latter, the mass of the lightest Higgs particle perforce remains bounded from above.

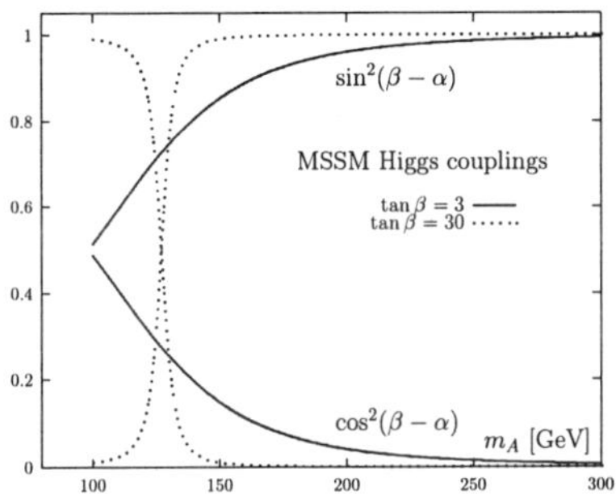

Fig.10.5. Squared coupling strengths of h and H to WW, relative to that of the SM Higgs, as functions of m_A, courtesy A. Djouadi.

10.5 Higgs-sparticle Vertices

Higgs couplings to neutralinos and charginos

The couplings of the Higgs bosons to the electroweak neutralinos and charginos originate from the gauge strength Yukawa couplings of gauginos to the scalar and fermionic component of a given chiral supermultiplet. In two component notation these are given by the last RHS term in (5.36) for abelian interactions, and by the fifth RHS term in (5.55) for nonabelian, presently $SU(2)_L$, interactions with the Higgs superfields H_1 and H_2 being the relevant chiral superfields. The corresponding terms in the interaction Lagrangian density can be rewritten in terms of the four component gaugino and higgsino fields of (9.17) and (9.30) by using the identities (3.28a,b) and (3.29a,b) to obtain the result

$$\mathcal{L}_{H\tilde{\chi}\tilde{\chi}} = -\frac{g_2}{\sqrt{2}} \left[h_1^0 \left(\overline{\tilde{h}_1^0} P_R \tilde{\lambda}_3 + \sqrt{2}\overline{\tilde{\lambda}^+} P_R \tilde{h}^+ \right) + h_1^- \left(\sqrt{2}\overline{\tilde{h}_1^0} P_R \tilde{\lambda}^+ - \overline{\tilde{\lambda}_3} P_R \tilde{h}^+ \right) \right.$$
$$\left. + h_2^0 \left(\sqrt{2}\overline{\tilde{h}^+} P_R \tilde{\lambda}^+ - \overline{\tilde{h}_2^0} P_R \tilde{\lambda}_3 \right) + h_2^+ \left(\overline{\tilde{h}^+} P_R \tilde{\lambda}_3 + \sqrt{2}\overline{\tilde{\lambda}^+} P_R \tilde{h}_2^0 \right) \right]$$
$$- \frac{g_Y}{\sqrt{2}} \left(h_2^+ \overline{\tilde{h}^+} P_R \tilde{\lambda}_0 + h_2^0 \overline{\tilde{h}_2^0} P_R \tilde{\lambda}_0 - h_1^0 \overline{\tilde{h}_1^0} P_R \tilde{\lambda}_0 - h_1^- \overline{\tilde{\lambda}_0} P_R \tilde{h}^+ \right) + h.c. \quad (10.34)$$

Finally, we use (9.18), (9.19), (9.27), (10.22a), (10.25a) and (10.28) to express (10.34) in terms of chargino, neutralino and Higgs mass eigenstates:

$$
\begin{aligned}
\mathcal{L}_{H\tilde{\chi}\tilde{\chi}} =\ & -g_2 \left(H\cos\alpha - h\sin\alpha\right)\left[\overline{\tilde{\chi}_k^+}\left(P_R Q_{km} + P_L Q^*_{mk}\right)\tilde{\chi}_m^+ + \frac{1}{2}\overline{\tilde{\chi}_n^0}\left(P_R Q''_{nl} + P_L Q''^*_{ln}\right)\tilde{\chi}_l^0\right] \\
& -g_2 \left(H\sin\alpha + h\cos\alpha\right)\left[\overline{\tilde{\chi}_k^+}\left(P_R S_{km} + P_L S^*_{mk}\right)\tilde{\chi}_m^+ - \frac{1}{2}\overline{\tilde{\chi}_n^0}\left(P_R S''_{nl} + P_L S''^*_{ln}\right)\tilde{\chi}_l^0\right] \\
& -ig_2 A\left\{\overline{\tilde{\chi}_k^+}\left[P_R\left(Q_{km}\sin\beta + S_{km}\cos\beta\right) - P_L\left(Q^*_{mk}\sin\beta + S^*_{mk}\cos\beta\right)\right]\tilde{\chi}_m^+\right. \\
& \left. + \frac{1}{2}\overline{\tilde{\chi}_n^0}\left[P_R\left(Q''_{nl}\sin\beta - S''_{nl}\cos\beta\right) + P_L\left(S''^*_{ln}\cos\beta - Q''^*_{ln}\sin\beta\right)\right]\tilde{\chi}_l^0\right\} \\
& -\left[g_2 H^-\overline{\tilde{\chi}_l^0}\left(P_R Q'^R_{lk} + P_L Q'^L_{lk}\right)\tilde{\chi}_k^+ + h.c.\right].
\end{aligned}
\tag{10.35}
$$

In (10.35) we have introduced the following quantities:

$$
Q_{km} \equiv \frac{1}{\sqrt{2}}\mathcal{V}_{k1}\mathcal{U}_{m2}\ ,
\tag{10.36a}
$$

$$
S_{km} \equiv \frac{1}{\sqrt{2}}\mathcal{V}_{k2}\mathcal{U}_{m1}\ ,
\tag{10.36b}
$$

$$
Q'^R_{lk} \equiv \sin\beta\left[Z_{l3}\mathcal{U}_{k1} - \frac{1}{\sqrt{2}}\mathcal{U}_{k2}\left(Z_{l2} + \tan\theta_W Z_{l1}\right)\right],
\tag{10.36c}
$$

$$
Q'^L_{lk} \equiv \cos\beta\left[Z^*_{l4}\mathcal{V}^*_{k1} + \frac{1}{\sqrt{2}}\mathcal{V}^*_{k2}\left(Z^*_{l2} + \tan\theta_W Z^*_{l1}\right)\right],
\tag{10.36d}
$$

$$
Q''_{nl} \equiv \frac{1}{2}\left[Z_{n3}\left(Z_{l2} - \tan\theta_W Z_{l1}\right) + Z_{l3}\left(Z_{n2} - \tan\theta_W Z_{n1}\right)\right],
\tag{10.36e}
$$

$$
S''_{nl} \equiv \frac{1}{2}\left[Z_{n4}\left(Z_{l2} - \tan\theta_W Z_{l1}\right) + Z_{l4}\left(Z_{n2} - \tan\theta_W Z_{n1}\right)\right].
\tag{10.36f}
$$

We have closely followed the notation of Ref.[10.2] in defining the above quantities. The only difference is an overall factor of g_2 in the definition of Q'^R and Q'^L which have been put in order to conform with the convention used for the other coefficients in the interaction Lagrangian density (10.35). The corresponding Feynman rules are given in Fig. 10.6 in the Appendix, the only nontrivial feature being an extra factor of two in vertices involving two Majorana (neutralino) fermions.

Fig. 10.6 is included in Appendix B

Recall from Ch.9 that $\mathcal{U}_{k1}$, $\mathcal{V}_{k1}$, Z_{k1} and Z_{k2} label gaugino components, while $\mathcal{U}_{k2}$, $\mathcal{V}_{k2}$, Z_{k3} and Z_{k4} label higgsino components. Thus eqs. (10.36) clearly reflect the origin of the quantities defined from Higgs–higgsino–gaugino interactions. These couplings are *not* proportional to the masses of the corresponding charginos and neutralinos. In fact, as discussed in §9.2, gaugino–higgsino mixing in the chargino and neutralino sectors is often suppressed. The neutral Higgs bosons will then predominantly couple to two *different* charginos and neutralinos. However, chargino and neutralino final states can nonetheless play a prominent role in the decays of the heavy neutral Higgs bosons A and H if $m_A, m_H \leq 2m_t$ and $\tan\beta$ is not large. Conversely, final states containing the light neutral Higgs boson h can play an important role in the decays of the heavier neutralinos and charginos into lighter ones. On the other hand, the lower bound $m_{\tilde{\chi}_1^\pm} > 100$ GeV, which comes from chargino searches at LEP, implies that the decays $H^- \to \tilde{\chi}_k^- \tilde{\chi}_l^0$ can dominate only over the small region of parameter space where $m_{\tilde{\chi}_1^+} + m_{\tilde{\chi}_1^0} < m_{H^+} < m_t + m_b$. Indeed, LEP searches imply that $m_{\tilde{\chi}_1^+} + m_{\tilde{\chi}_1^0} \geq 140$ GeV, if the "gaugino mass unification condition", cf. (9.21), holds. Finally, note that the couplings of h, H and A would be scalar and pseudoscalar respectively, were all rotation matrices in the chargino and neutralino sector strictly real.

Sfermion Higgs couplings

The couplings between Higgs bosons and sfermions receive contributions from the supersymmetric F- and D-terms in the scalar potential, as well as from trilinear soft supersymmetry breaking terms. The same terms also contribute to sfermion mass matrices and have been collected in (9.42) and (9.45) for sleptons and squarks respectively. We use (10.22a), (10.25a) and (10.28) to move to the Higgs mass eigenstate basis. The quartic F- and D-terms then also give rise to trilinear interactions of a single Higgs particle with two sfermions, due to the VEVs of the neutral components of the Higgs fields. We first present the relevant pieces of the interaction Lagrangian density in the current basis for sfermions. This allows easier comparison with results in the literature.. Moreover, as discussed in §9.5, in many realistic SUSY models intergeneration sfermion mixing can often be neglected, in which case the mass eigenstates are essentially equal to current eigenstates. For ease of presentation, we show the trilinear and quartic interactions of Higgs bosons with sleptons and squarks separately. For any angle ϕ, we use s_ϕ, c_ϕ, t_ϕ and $(ct)_\phi$ to mean $\sin\phi$, $\cos\phi$, $\tan\phi$ and $\cot\phi$ respectively, except that the corresponding symbols for θ_W are s_W, c_W, t_W and $(ct)_W$ respectively. The results for the relevant cubic and quartic interactions are

$$
\begin{aligned}
\mathcal{L}_{H\tilde{\ell}\tilde{\ell}} \;=\; & \frac{g_2}{\sqrt{2}M_W} H^+ \Big\{ \tilde{\nu}_i^* \tilde{e}_{jR} \left[\mu\,(\mathbf{m_e})_{ij} - (\mathbf{m_e}A^{e*})_{ij}\, t_\beta \right] \\
& \qquad\qquad + \tilde{\nu}_i^* \tilde{e}_{jL} \left[(\mathbf{m_e m_e^\dagger})_{ij}\, t_\beta - M_W^2 \delta_{ij} s_{2\beta} \right] \Big\} \\
& + \frac{g_2}{2M_W c_\beta} \tilde{e}_{iL}^* \tilde{e}_{jR} \Big[(\mathbf{m_e}A^{e*})_{ij} \left(Hc_\alpha - hs_\alpha - iAs_\beta \right) \\
& \qquad\qquad + \mu\,(\mathbf{m_e})_{ij} \left(Hs_\alpha + hc_\alpha + iAc_\beta \right) \Big] + h.c. \\
& + \frac{g_2}{M_W c_\beta} \left[\tilde{e}_{iL}^* \tilde{e}_{jL} \left(\mathbf{m_e m_e^\dagger} \right)_{ij} + \tilde{e}_{iR}^* \tilde{e}_{jR} \left(\mathbf{m_e^\dagger m_e} \right)_{ij} \right] \left(hs_\alpha - Hc_\alpha \right) \\
& + \frac{g_2 M_W}{2} \left[hs_{(\alpha+\beta)} - Hc_{(\alpha+\beta)} \right] \sum_i \left[|\tilde{\nu}_i|^2 \left(1 + t_W^2 \right) - |\tilde{e}_{iL}|^2 \left(1 - t_W^2 \right) \right. \\
& \qquad\qquad\qquad\qquad\qquad\qquad\qquad \left. - 2\,|\tilde{e}_{iR}|^2\, t_W^2 \right] .
\end{aligned}
\tag{10.37}
$$

$$
\begin{aligned}
\mathcal{L}_{H\tilde{q}\bar{\tilde{q}}} =\ & \frac{g_2}{\sqrt{2}M_W} H^+ \Big\{ \tilde{u}_{iL}^\dagger \tilde{d}_{jR} \left[\mu \left(\mathbf{m_d}\right)_{ij} - \left(\mathbf{m_d}A^{d*}\right)_{ij} t_\beta \right] \\
& + \tilde{u}_{iR}^\dagger \tilde{d}_{jL} \left[\mu^* \left(\mathbf{m_u^*}\right)_{ij} - \left(\mathbf{m_u^*}A^u\right)_{ij} (ct)_\beta \right] \\
& + \tilde{u}_{iL}^\dagger \tilde{d}_{jL} \left[\left(\mathbf{m_u m_u^\dagger}\right)_{ij} (ct)_\beta + \left(\mathbf{m_d m_d^\dagger}\right)_{ij} t_\beta - M_W^2 \delta_{ij} s_{2\beta} \right] \\
& + \tilde{u}_{iR}^\dagger \tilde{d}_{jR} \left(\mathbf{m_u^\dagger m_d}\right)_{ij} \left(t_\beta + (ct)_\beta \right) \Big\} \\
& + \frac{g_2}{2M_W c_\beta} \Big[\tilde{d}_{iL}^\dagger \tilde{d}_{jR} \Big\{ \left(\mathbf{m_d}A^{d*}\right)_{ij} \left(Hc_\alpha - hs_\alpha - iAs_\beta \right) \\
& \hspace{5em} + \mu \left(\mathbf{m_d}\right)_{ij} \left(Hs_\alpha + hc_\alpha + iAc_\beta \right) \Big\} + \text{h.c.} \Big] \\
& + \frac{g_2}{2M_W s_\beta} \Big[\tilde{u}_{iL}^\dagger \tilde{u}_{jR} \Big\{ \left(\mathbf{m_u}A^{u*}\right)_{ij} \left(Hs_\alpha + hc_\alpha - iAc_\beta \right) \\
& \hspace{5em} + \mu \left(\mathbf{m_u}\right)_{ij} \left(Hc_\alpha - hs_\alpha + iAs_\beta \right) \Big\} + \text{h.c.} \Big] \\
& + \frac{g_2}{M_W c_\beta} \left(hs_\alpha - Hc_\alpha \right) \left[\tilde{d}_{iL}^\dagger \tilde{d}_{jL} \left(\mathbf{m_d m_d^\dagger}\right)_{ij} + \tilde{d}_{iR}^\dagger \tilde{d}_{jR} \left(\mathbf{m_d^\dagger m_d}\right)_{ij} \right] \\
& - \frac{g_2}{M_W s_\beta} \left(Hs_\alpha + hc_\alpha \right) \left[\tilde{u}_{iL}^\dagger \tilde{u}_{jL} \left(\mathbf{m_u m_u^\dagger}\right)_{ij} + \tilde{u}_{iR}^\dagger \tilde{u}_{jR} \left(\mathbf{m_u^\dagger m_u}\right)_{ij} \right] \\
& + g_2(M_W/2) \sum_i \Big[\left|\tilde{u}_{iL}\right|^2 \left(1 - t_W^2/3 \right) - \left|\tilde{d}_{iL}\right|^2 \left(1 + t_W^2/3 \right) \\
& \hspace{5em} + (2t_W^2/3) \left(2\left|\tilde{u}_{iR}\right|^2 - \left|\tilde{d}_{iR}\right|^2 \right) \Big] \left[hs_{(\alpha+\beta)} - Hc_{(\alpha+\beta)} \right].
\end{aligned}
\tag{10.38}
$$

$$
\begin{aligned}
\mathcal{L}_{HH\tilde{\ell}\bar{\tilde{\ell}}} =\ & \frac{g_2^2}{2\sqrt{2}M_W^2} H^+ \tilde{\nu}_i^* \tilde{e}_{jL} \Big\{ \frac{s_\beta}{c_\beta^2} \left(\mathbf{m_e m_e^\dagger}\right)_{ij} \left(Hc_\alpha - hs_\alpha + iAs_\beta \right) \\
& \hspace{5em} - M_W^2 \delta_{ij} \left[Hs_{(\alpha+\beta)} + hc_{(\alpha+\beta)} - iAc_{2\beta} \right] \Big\} + \text{h.c.} \\
& - \frac{g_2^2}{4M_W^2 c_\beta^2} \left(H^2 c_\alpha^2 + h^2 s_\alpha^2 - Hh s_{2\alpha} + A^2 s_\beta^2 \right) \\
& \hspace{5em} \cdot \left[\tilde{e}_{iL}^* \tilde{e}_{jL} \left(\mathbf{m_e m_e^\dagger}\right)_{ij} + \tilde{e}_{iR}^* \tilde{e}_{jR} \left(\mathbf{m_e^\dagger m_e}\right)_{ij} \right] \\
& - \frac{g_2^2 t_\beta^2}{2M_W^2} H^+ H^- \left[\tilde{\nu}_i^* \tilde{\nu}_j \left(\mathbf{m_e m_e^\dagger}\right)_{ij} + \tilde{e}_{iR}^* \tilde{e}_{jR} \left(\mathbf{m_e^\dagger m_e}\right)_{ij} \right] \\
& + \frac{g_2^2}{8} \left[\left(h^2 - H^2 \right) c_{2\alpha} + 2Hh s_{2\alpha} + A^2 c_{2\beta} \right] \\
& \hspace{5em} \cdot \sum_i \left[\left|\tilde{\nu}_i\right|^2 \left(1 + t_W^2 \right) - \left|\tilde{e}_{iL}\right|^2 \left(1 - t_W^2 \right) - 2t_W^2 \left|\tilde{e}_{iR}\right|^2 \right] \\
& - \frac{g_2^2}{4} H^+ H^- c_{2\beta} \sum_i \left[\left|\tilde{\nu}_i\right|^2 \left(1 - t_W^2 \right) - \left|\tilde{e}_{iL}\right|^2 \left(1 + t_W^2 \right) + 2t_W^2 \left|\tilde{e}_{iR}\right|^2 \right].
\end{aligned}
\tag{10.39}
$$

$$
\begin{aligned}
\mathcal{L}_{HH\tilde{q}\tilde{q}} \;=\;& \frac{g_2^2}{2\sqrt{2}M_W^2} H^+ \Big\{ \tilde{u}_{iL}^\dagger \tilde{d}_{jL} \Big[\left(\mathbf{m_u m_u^\dagger}\right)_{ij} \frac{c_\beta}{s_\beta^2}\left(Hs_\alpha + hc_\alpha - iAc_\beta\right) \\
&\qquad\qquad + \left(\mathbf{m_d m_d^\dagger}\right)_{ij} \frac{s_\beta}{c_\beta^2}\left(Hc_\alpha - hs_\alpha + iAs_\beta\right) \\
&\qquad\qquad - M_W^2 \delta_{ij}\left(Hs_{(\alpha+\beta)} + hc_{(\alpha+\beta)} - iAc_{2\beta}\right) \Big] \\
&\quad + \tilde{u}_{iR}^\dagger \tilde{d}_{jR}\left[Hc_{(\alpha-\beta)} - hs_{(\alpha-\beta)}\right]\frac{2}{s_{2\beta}}\left(\mathbf{m_u^\dagger m_d}\right)_{ij} \Big\} + h.c. \\[4pt]
& -\frac{g_2^2}{4M_W^2 c_\beta^2}\left(H^2 c_\alpha^2 + h^2 s_\alpha^2 - Hh s_{2\alpha} + A^2 s_\beta^2\right) \\
&\qquad \cdot \left[\tilde{d}_{iL}^\dagger \tilde{d}_{jL}\left(\mathbf{m_d m_d^\dagger}\right)_{ij} + \tilde{d}_{iR}^\dagger \tilde{d}_{jR}\left(\mathbf{m_d^\dagger m_d}\right)_{ij}\right] \\[4pt]
& -\frac{g_2^2}{4M_W^2 s_\beta^2}\left(H^2 s_\alpha^2 + h^2 c_\alpha^2 + Hh s_{2\alpha} + A^2 c_\beta^2\right) \\
&\qquad \cdot \left[\tilde{u}_{iL}^\dagger \tilde{u}_{jL}\left(\mathbf{m_u m_u^\dagger}\right)_{ij} + \tilde{u}_{iR}^\dagger \tilde{u}_{jR}\left(\mathbf{m_u^\dagger m_u}\right)_{ij}\right] \\[4pt]
& -\frac{g_2^2}{2M_W^2} H^+ H^- \Big[\tilde{u}_{iL}^\dagger \tilde{u}_{jL}\left(\mathbf{m_d m_d^\dagger}\right)_{ij} t_\beta^2 + \tilde{d}_{iL}^\dagger \tilde{d}_{jL}\left(\mathbf{m_u m_u^\dagger}\right)_{ij}(ct)_\beta^2 \\
&\qquad\qquad + \tilde{u}_{iR}^\dagger \tilde{u}_{jR}\left(\mathbf{m_u^\dagger m_u}\right)_{ij}(ct)_\beta^2 + \tilde{d}_{iR}^\dagger \tilde{d}_{jR}\left(\mathbf{m_d^\dagger m_d}\right)_{ij} t_\beta^2\Big] \\[4pt]
& +\frac{g_2^2}{8}\left[\left(h^2 - H^2\right)c_{2\alpha} + 2Hh s_{2\alpha} + A^2 c_{2\beta}\right] \\
&\qquad \cdot \sum_i \Big[\left|\tilde{u}_{iL}\right|^2\left(1 - t_W^2/3\right) - \left|\tilde{d}_{iL}\right|^2\left(1 + t_W^2/3\right) \\
&\qquad\qquad + \left(2t_W^2/3\right)\left(2\left|\tilde{u}_{iR}\right|^2 - \left|\tilde{d}_{iR}\right|^2\right)\Big] \\[4pt]
& -\frac{g_2^2}{4} H^+ H^- c_{2\beta} \sum_i \Big[\left|\tilde{u}_{iL}\right|^2\left(1 + \tfrac{1}{3}t_W^2\right) - \left|\tilde{d}_{iL}\right|^2\left(1 - t_W^2/3\right) \\
&\qquad\qquad - \left(2t_W^2/3\right)\left(2\left|\tilde{u}_{iR}\right|^2 - \left|\tilde{d}_{iR}\right|^2\right)\Big].
\end{aligned}
\tag{10.40}
$$

The following points about (10.37)–(10.40) are noteworthy:

- The hermitian conjugation in these equations acts only on terms to the left of the h.c. as written, terms to the right being already hermitian after summation over the generation indices i and j.

- The coupling of one Higgs boson to two sfermions is again *not* proportional to the sfermion mass. In the case of third generation sfermions the usually most important contributions to such a trilinear coupling are those proportional to μ or one of the

$A-$parameters, since the absolute values of these quantities can be significantly larger than M_W. Indeed, in principle, such couplings offer the only *direct* experimental access to the $A-$parameter. In practice, however, these couplings are difficult to measure since they involve three as yet undiscovered particles.

- The only significant contributions to the Higgs couplings to first and second generation sfermions are the pure gauge terms. In contrast, the quartic interactions of third generation sfermions are often dominated by contributions proportional to m_f^2.

- The F-term contributions to the couplings of $SU(2)_L$-doublet, "left chiral" sfermions are proportional to $\mathbf{m_f m_f^\dagger}$, while those of $SU(2)_L$-singlet, "right chiral" sfermions are proportional to $\mathbf{m_f^\dagger m_f}$. This is analogous to the LL and RR entries of the squared squark mass matrix listed in (9.46).

- The relative sign between the $SU(2)$ and $U(1)_Y$ D-term contributions to the quartic interactions differs for neutral and charged Higgs boson pairs. For example, the $H^+ H^-$ pair couples more strongly to $\tilde{e}_L$ pairs than to $\tilde{\nu}$ pairs, while the opposite is true for pairs of neutral Higgs bosons.

In the final step, (9.51) and (9.54) are to be utilized to convert the current eigenstate sfermion fields in (10.37)–(10.40) into mass eigenstate ones. We can use a notation similar to that in §9.8. The final result for Higgs-sfermion interactions can then be written as

$$\mathcal{L}_{H\tilde{f}} = \sum_{\phi,\tilde{f},\tilde{f}'} C[\phi,\tilde{f},\tilde{f}']\phi\tilde{f}^*\tilde{f}' + \sum_{\phi,\phi',\tilde{f},\tilde{f}'} D[\phi,\phi',\tilde{f},\tilde{f}']\phi\phi'\tilde{f}^*\tilde{f}', \tag{10.41}$$

where ϕ and ϕ' stand for any of the five physical Higgs fields of the MSSM, while $\tilde{f}$ and $\tilde{f}'$ are sfermion fields. Invariance under $SU(3)_C$ implies that $\tilde{f}'$ must be a squark if and only if $\tilde{f}$ is a squark; both squarks must then have the same color index, which therefore need not be displayed. Moreover, we shall again assume that the superpotential is written in a basis where the leptonic Yukawa couplings are flavor diagonal. The coefficients describing slepton-Higgs interactions can then be written down. First, we display the coefficients of the various cubic **Higgs-slepton-slepton** terms. They are

$$C[H^+,\tilde{\nu}_i,\tilde{e}_s] = \frac{g_2}{\sqrt{2}M_W}\Bigg\{ \sum_{k=1}^{3} U_{ki}^{\tilde{\nu}*}\left[\mu m_{e_k} W_{k+3\ s}^{\tilde{e}} + \left(m_{e_k}^2 t_\beta - M_W^2 s_{2\beta}\right) W_{ks}^{\tilde{e}}\right]$$

$$- \sum_{j,k=1}^{3} t_\beta \left(\mathbf{m_e} A^{e*}\right)_{kj} U_{ki}^{\tilde{\nu}*} W_{j+3\ s}^{\tilde{e}} \Bigg\}, \tag{10.42a}$$

$$C[H^-,\tilde{e}_s,\tilde{\nu}_i] = \left(C[H^+,\tilde{\nu}_i,\tilde{e}_s]\right)^*, \tag{10.42b}$$

$$C[H,\tilde{\nu}_i,\tilde{\nu}_j] = -c_g[\tilde{\nu}]c_{\alpha+\beta}\delta_{ij}, \tag{10.42c}$$

$$C[h,\tilde{\nu}_i,\tilde{\nu}_j] = c_g[\tilde{\nu}]s_{\alpha+\beta}\delta_{ij}, \tag{10.42d}$$

$$C[H,\tilde{e}_s,\tilde{e}_t] = c_A[\tilde{e}_s,\tilde{e}_t]c_\alpha + c_\mu[\tilde{e}_s,\tilde{e}_t]s_\alpha - c_g[\tilde{e}_s,\tilde{e}_t]c_{\alpha+\beta}, \tag{10.42e}$$

$$C[h, \tilde{e}_s, \tilde{e}_t] = -c_A[\tilde{e}_s, \tilde{e}_t]s_\alpha + c_\mu[\tilde{e}_s, \tilde{e}_t]c_\alpha + c_g[\tilde{e}_s, \tilde{e}_t]s_{\alpha+\beta} \ , \tag{10.42f}$$

$$\begin{aligned}
C[A, \tilde{e}_s, \tilde{e}_t] = \ & \frac{ig_2}{2M_W}\Big\{ \sum_{i=1}^{3} \left(\mu m_{e_i} W^{\tilde{e}*}_{is} W^{\tilde{e}}_{i+3\ t} - \mu^* m_{e_i} W^{\tilde{e}*}_{i+3\ s} W^{\tilde{e}}_{it} \right) \\
& + t_\beta \sum_{i,j=1}^{3} \left[(\mathbf{m_e}A^e)_{ij}\, W^{\tilde{e}*}_{j+3\ s} W^{\tilde{e}}_{it} - (\mathbf{m_e}A^{e*})_{ij}\, W^{\tilde{e}*}_{is} W^{\tilde{e}}_{j+3\ t} \right] \Big\}.
\end{aligned} \tag{10.42g}$$

All coefficients $C[\phi, \tilde{\ell}, \tilde{\ell}']$ not listed in eqs.(10.42) vanish. One could rewrite the matrix $\mathbf{m_e}A_e^*$ in terms of its eigenvalues and corresponding 3×3 rotation matrices. We have not done so since the eigenvalues of this matrix have no special physical meaning, in contrast to those of the SM matter fermion mass matrices. Moreover, we have introduced the quantities

$$\begin{aligned}
c_A[\tilde{e}_s, \tilde{e}_t] \equiv \ & \frac{g_2}{M_W c_\beta}\Big\{ -\sum_{i=1}^{3} m_{e_i}^2 \left(W^{\tilde{e}*}_{is} W^{\tilde{e}}_{it} + W^{\tilde{e}*}_{i+3\ s} W^{\tilde{e}}_{i+3\ t} \right) \\
& + \frac{1}{2} \sum_{i,j=1}^{3} \left[(\mathbf{m_e}A^{e\dagger})_{ij}\, W^{\tilde{e}*}_{is} W^{\tilde{e}}_{j+3\ t} + (\mathbf{m_e}A^e)_{ij}\, W^{\tilde{e}*}_{j+3\ s} W^{\tilde{e}}_{it} \right] \Big\},
\end{aligned} \tag{10.43a}$$

$$c_\mu[\tilde{e}_s, \tilde{e}_t] \equiv \frac{g_2}{2M_W c_\beta} \sum_{i=1}^{3} m_{e_i} \left(\mu W^{\tilde{e}*}_{is} W^{\tilde{e}}_{i+3\ t} + \mu^* W^{\tilde{e}*}_{i+3\ s} W^{\tilde{e}}_{it} \right), \tag{10.43b}$$

$$c_g[\tilde{\nu}] \equiv \frac{g_2 M_W}{2} \left(1 + t_W^2 \right), \tag{10.43c}$$

$$c_g[\tilde{e}_s, \tilde{e}_t] \equiv \frac{g_2 M_W}{2} \sum_{i=1}^{3} \left[W^{\tilde{e}*}_{is} W^{\tilde{e}}_{it} \left(t_W^2 - 1 \right) - 2t_W^2 W^{\tilde{e}*}_{i+3\ s} W^{\tilde{e}}_{i+3\ t} \right]. \tag{10.43d}$$

The coefficients of the various quartic **Higgs-Higgs-slepton-slepton interactions** can also be displayed. They are

$$D[H^+, H^-, \tilde{\nu}_i, \tilde{\nu}_j] = \frac{g_2^2 c_{2\beta}}{4} \left(t_W^2 - 1 \right) \delta_{ij} - \frac{g_2^2 t_\beta^2}{2M_W^2} \sum_{k=1}^{3} U^{\tilde{\nu}*}_{ki} U^{\tilde{\nu}}_{kj} m_{e_k}^2 \ , \tag{10.44a}$$

$$\begin{aligned}
D[H^+, H^-, \tilde{e}_s, \tilde{e}_t] = \ & \sum_{i=1}^{3} \Big\{ \frac{g_2^2 c_{2\beta}}{4} \left[W^{\tilde{e}*}_{is} W^{\tilde{e}}_{it} \left(1 + t_W^2 \right) - 2t_W^2 W^{\tilde{e}*}_{i+3\ s} W^{\tilde{e}}_{i+3\ t} \right] \\
& - \frac{g_2^2 t_\beta^2}{2M_W^2} m_{e_i}^2 W^{\tilde{e}*}_{i+3\ s} W^{\tilde{e}}_{j+3\ t} \Big\},
\end{aligned} \tag{10.44b}$$

$$D[H^+, H, \tilde{\nu}_i, \tilde{e}_s] = -d_g[\tilde{\nu}_i, \tilde{e}_s]s_{\alpha+\beta} + d_Y[\tilde{\nu}_i, \tilde{e}_s]c_\alpha \ , \tag{10.44c}$$

$$D[H^-, H, \tilde{e}_s, \tilde{\nu}_i] = \left(D[H^+, H, \tilde{\nu}_i, \tilde{e}_s] \right)^* , \tag{10.44d}$$

$$D[H^+, h, \tilde{\nu}_i, \tilde{e}_s] = -d_g[\tilde{\nu}_i, \tilde{e}_s]c_{\alpha+\beta} - d_Y[\tilde{\nu}_i, \tilde{e}_s]s_\alpha \,, \tag{10.44e}$$

$$D[H^-, h, \tilde{e}_s, \tilde{\nu}_i] = \left(D[H^+, h, \tilde{\nu}_i, \tilde{e}_s]\right)^* \,, \tag{10.44f}$$

$$D[H^+, A, \tilde{\nu}_i, \tilde{e}_s] = id_g[\tilde{\nu}_i, \tilde{e}_s]c_{2\beta} + id_Y[\tilde{\nu}_i, \tilde{e}_s]s_\beta \,, \tag{10.44g}$$

$$D[H^-, A, \tilde{e}_s, \tilde{\nu}_i] = \left(D[H^+, A, \tilde{\nu}_i, \tilde{e}_s]\right)^* \,, \tag{10.44h}$$

$$D[H, H, \tilde{\nu}_i, \tilde{\nu}_j] = -d_g[\tilde{\nu}]c_{2\alpha}\,\delta_{ij} \,, \tag{10.44i}$$

$$D[H, h, \tilde{\nu}_i, \tilde{\nu}_j] = 2d_g[\tilde{\nu}]s_{2\alpha}\,\delta_{ij} \,, \tag{10.44j}$$

$$D[h, h, \tilde{\nu}_i, \tilde{\nu}_j] = d_g[\tilde{\nu}]c_{2\alpha}\,\delta_{ij} \,, \tag{10.44k}$$

$$D[A, A, \tilde{\nu}_i, \tilde{\nu}_j] = d_g[\tilde{\nu}]c_{2\beta}\,\delta_{ij} \,, \tag{10.44l}$$

$$D[H, H, \tilde{e}_s, \tilde{e}_t] = -d_Y[\tilde{e}_s, \tilde{e}_t]c_\alpha^2 - d_g[\tilde{e}_s, \tilde{e}_t]c_{2\alpha} \,, \tag{10.44m}$$

$$D[H, h, \tilde{e}_s, \tilde{e}_t] = d_Y[\tilde{e}_s, \tilde{e}_t]s_{2\alpha} + 2d_g[\tilde{e}_s, \tilde{e}_t]s_{2\alpha} \,, \tag{10.44n}$$

$$D[h, h, \tilde{e}_s, \tilde{e}_t] = -d_Y[\tilde{e}_s, \tilde{e}_t]s_\alpha^2 + d_g[\tilde{e}_s, \tilde{e}_t]c_{2\alpha} \,, \tag{10.44o}$$

$$D[A, A, \tilde{e}_s, \tilde{e}_t] = -d_Y[\tilde{e}_s, \tilde{e}_t]s_\beta^2 + d_g[\tilde{e}_s, \tilde{e}_t]c_{2\beta} \,. \tag{10.44p}$$

We again list only the nonvanishing coefficients; for example, $D[H^-, H^+, \tilde{f}, \tilde{f}'] \equiv 0$, for $\tilde{f} \neq \tilde{f}'$. The Lagrangian density in (10.41) is nonetheless hermitian. Moreover, we have introduced the quantities

$$d_g[\tilde{\nu}] \equiv \frac{g_2^2}{8}\left(1 + t_W^2\right), \tag{10.45a}$$

$$d_g[\tilde{\nu}_i, \tilde{e}_s] \equiv \frac{g_2^2}{2\sqrt{2}} \sum_{j=1}^{3} U_{ji}^{\tilde{\nu}*} W_{js}^{\tilde{e}} \,, \tag{10.45b}$$

$$d_g[\tilde{e}_s, \tilde{e}_t] \equiv -\frac{g_2^2}{8} \sum_{i=1}^{3} \left[2t_W^2 W_{i+3\ s}^{\tilde{e}*}\, W_{i+3\ t}^{\tilde{e}} + W_{is}^{\tilde{e}*} W_{it}^{\tilde{e}}\left(1 - t_W^2\right)\right], \tag{10.45c}$$

$$d_Y[\tilde{\nu}_i, \tilde{e}_s] \equiv \frac{g_2^2 s_\beta}{2\sqrt{2}M_W^2 c_\beta^2} \sum_{j=1}^{3} m_{e_j}^2 U_{ji}^{\tilde{\nu}*} W_{js}^{\tilde{e}} \,, \tag{10.45d}$$

$$d_Y[\tilde{e}_s, \tilde{e}_t] \equiv \frac{g_2^2}{4M_W^2 c_\beta^2} \sum_{i=1}^{3} m_{e_i}^2 \left(W_{is}^{\tilde{e}*} W_{it}^{\tilde{e}} + W_{i+3\ s}^{\tilde{e}*}\, W_{i+3\ t}^{\tilde{e}}\right). \tag{10.45e}$$

The analogous expressions for Higgs interactions with squarks are complicated by nontrivial quark flavor mixing. In addition to rotating the squarks into mass eigenstates using the matrices $\mathbf{W}^{\tilde{u}}$ and $\mathbf{W}^{\tilde{d}}$, we also need to diagonalize the quark mass matrices using (8.12). The coefficients of the various cubic **Higgs-squark-squark interaction** terms are given below. They are

$$C[H^+, \tilde{u}_s, \tilde{d}_t] = \frac{g_2}{\sqrt{2}M_W} \sum_{i,j=1}^{3} \Big\{ W_{is}^{\tilde{u}*} W_{j+3\ t}^{\tilde{d}} \Big[\sum_{k=1}^{3} \mu U_{ik}^{d_L} m_{d_k} U_{jk}^{d_R*} - \big(\mathbf{m_d} A^{d*}\big)_{ij} t_\beta \Big]$$

$$+ W_{j+3\ s}^{\tilde{u}*} W_{it}^{\tilde{d}} \Big[\sum_{k=1}^{3} \mu^* U_{ik}^{u_L*} m_{u_k} U_{jk}^{u_R} - \big(\mathbf{m_u^*} A^u\big)_{ij} (ct)_\beta \Big]$$

$$+ W_{is}^{\tilde{u}*} W_{jt}^{\tilde{d}} \Big[\sum_{k=1}^{3} \Big(U_{ik}^{u_L} m_{u_k}^2 U_{jk}^{u_L*} (ct)_\beta + U_{ik}^{d_L} m_{d_k}^2 U_{jk}^{d_L*} t_\beta \Big)$$

$$- M_W^2 \delta_{ij} s_{2\beta} \Big]$$

$$+ W_{i+3\ s}^{\tilde{u}*} W_{j+3\ t}^{\tilde{d}} (t_\beta + (ct)_\beta) \sum_{k,l=1}^{3} U_{ik}^{u_R} m_{u_k} V_{kl}^{q_L} m_{d_l} U_{jl}^{d_R*} \Big\},$$

$$\tag{10.46a}$$

$$C[H^-, \tilde{d}_t, \tilde{u}_s] = \Big(C[H^+, \tilde{u}_s, \tilde{d}_t] \Big)^*, \tag{10.46b}$$

$$C[H, \tilde{d}_s, \tilde{d}_t] = c_A[\tilde{d}_s, \tilde{d}_t] c_\alpha + c_\mu[\tilde{d}_s, \tilde{d}_t] s_\alpha - c_g[\tilde{d}_s, \tilde{d}_t] c_{\alpha+\beta}, \tag{10.46c}$$

$$C[h, \tilde{d}_s, \tilde{d}_t] = -c_A[\tilde{d}_s, \tilde{d}_t] s_\alpha + c_\mu[\tilde{d}_s, \tilde{d}_t] c_\alpha + c_g[\tilde{d}_s, \tilde{d}_t] s_{\alpha+\beta}, \tag{10.46d}$$

$$C[H, \tilde{u}_s, \tilde{u}_t] = c_A[\tilde{u}_s, \tilde{u}_t] s_\alpha + c_\mu[\tilde{u}_s, \tilde{u}_t] c_\alpha - c_g[\tilde{u}_s, \tilde{u}_t] c_{\alpha+\beta}, \tag{10.46e}$$

$$C[h, \tilde{u}_s, \tilde{u}_t] = c_A[\tilde{u}_s, \tilde{u}_t] c_\alpha - c_\mu[\tilde{u}_s, \tilde{u}_t] s_\alpha + c_g[\tilde{u}_s, \tilde{u}_t] s_{\alpha+\beta}, \tag{10.46f}$$

$$C[A, \tilde{d}_s, \tilde{d}_t] = \frac{ig_2}{2M_W} \sum_{i,j=1}^{3} \Big\{ t_\beta \Big[\big(\mathbf{m_d^*} A^d\big)_{ij} W_{j+3\ s}^{\tilde{d}*} W_{it}^{\tilde{d}} - \big(\mathbf{m_d} A^{d*}\big)_{ij} W_{is}^{\tilde{d}*} W_{j+3\ t}^{\tilde{d}} \Big]$$

$$+ \sum_{k=1}^{3} m_{d_k} \Big[\mu U_{ik}^{d_L} U_{jk}^{d_R*} W_{is}^{\tilde{d}*} W_{j+3\ t}^{\tilde{d}} - \mu^* U_{ik}^{d_L*} U_{jk}^{d_R} W_{j+3\ s}^{\tilde{d}*} W_{it}^{\tilde{d}} \Big] \Big\},$$

$$\tag{10.46g}$$

$$C[A, \tilde{u}_s, \tilde{u}_t] = \frac{ig_2}{2M_W} \sum_{i,j=1}^{3} \Big\{ (ct)_\beta \Big[\big(\mathbf{m_u^*} A^u\big)_{ij} W_{j+3\ s}^{\tilde{u}*} W_{it}^{\tilde{u}} - \big(\mathbf{m_u} A^{u*}\big)_{ij} W_{is}^{\tilde{u}*} W_{j+3\ t}^{\tilde{u}} \Big]$$

$$+ \sum_{k=1}^{3} m_{u_k} \Big[\mu U_{ik}^{u_L} U_{jk}^{u_R*} W_{is}^{\tilde{u}*} W_{j+3\ t}^{\tilde{u}} - \mu^* U_{ik}^{u_L*} U_{jk}^{u_R} W_{j+3\ s}^{\tilde{u}*} W_{it}^{\tilde{u}} \Big] \Big\}.$$

$$\tag{10.46h}$$

The quantity $V_{k\ell}^{q_L}$ appearing in (10.46a) is an element of the Cabibbo-Kobayashi–Maskawa matrix of (8.40). Moreover, we have introduced the quantities

$$c_A[\tilde{d}_s, \tilde{d}_t] \; \equiv \; \frac{g_2}{M_W c_\beta} \sum_{i,j=1}^{3} \left\{ \frac{1}{2} \left[W_{is}^{\tilde{d}*} W_{j+3\,t}^{\tilde{d}} \left(\mathbf{m_d} A^{d*} \right)_{ij} + W_{j+3\,s}^{\tilde{d}*} W_{it}^{\tilde{d}} \left(\mathbf{m_d^*} A^{d} \right)_{ij} \right] \right.$$

$$\left. - \sum_{k=1}^{3} m_{d_k}^2 \left[U_{ik}^{d_L} U_{jk}^{d_L*} W_{is}^{\tilde{d}*} W_{jt}^{\tilde{d}} + U_{ik}^{d_R} U_{jk}^{d_R*} W_{i+3\,s}^{\tilde{d}*} W_{j+3\,t}^{\tilde{d}} \right] \right\},$$

$$(10.47\mathrm{a})$$

$$c_A[\tilde{u}_s, \tilde{u}_t] \; \equiv \; \frac{g_2}{M_W s_\beta} \sum_{i,j=1}^{3} \left\{ \frac{1}{2} \left[W_{is}^{\tilde{u}*} W_{j+3\,t}^{\tilde{u}} \left(\mathbf{m_u} A^{u*} \right)_{ij} + W_{j+3\,s}^{\tilde{u}*} W_{it}^{\tilde{u}} \left(\mathbf{m_u^*} A^{u} \right)_{ij} \right] \right.$$

$$\left. - \sum_{k=1}^{3} m_{u_k}^2 \left[U_{ik}^{u_L} U_{jk}^{u_L*} W_{is}^{\tilde{u}*} W_{jt}^{\tilde{u}} + U_{ik}^{u_R} U_{jk}^{u_R*} W_{i+3\,s}^{\tilde{u}*} W_{j+3\,t}^{\tilde{u}} \right] \right\},$$

$$(10.47\mathrm{b})$$

$$c_\mu[\tilde{d}_s, \tilde{d}_t] \equiv \frac{g_2}{2 M_W c_\beta} \sum_{i,j,k=1}^{3} m_{d_k} \left[\mu U_{ik}^{d_L} U_{jk}^{d_R*} W_{is}^{\tilde{d}*} W_{j+3\,t}^{\tilde{d}} + \mu^* U_{ik}^{d_L*} U_{jk}^{d_R} W_{j+3\,s}^{\tilde{d}*} W_{it}^{\tilde{d}} \right], \quad (10.47\mathrm{c})$$

$$c_\mu[\tilde{u}_s, \tilde{u}_t] \equiv \frac{g_2}{2 M_W s_\beta} \sum_{i,j,k=1}^{3} m_{u_k} \left[\mu U_{ik}^{u_L} U_{jk}^{u_R*} W_{is}^{\tilde{u}*} W_{j+3\,t}^{\tilde{u}} + \mu^* U_{ik}^{u_L*} U_{jk}^{u_R} W_{j+3\,s}^{\tilde{u}*} W_{it}^{\tilde{u}} \right], \quad (10.47\mathrm{d})$$

$$c_g[\tilde{d}_s, \tilde{d}_t] \equiv -\frac{g_2 M_W}{2} \sum_{i=1}^{3} \left[W_{is}^{\tilde{d}*} W_{it}^{\tilde{d}} \left(1 + \frac{1}{3} t_W^2 \right) + \frac{2}{3} W_{i+3\,s}^{\tilde{d}*} W_{i+3\,t}^{\tilde{d}} t_W^2 \right], \qquad (10.47\mathrm{e})$$

$$c_g[\tilde{u}_s, \tilde{u}_t] \equiv \frac{g_2 M_W}{2} \sum_{i=1}^{3} \left[W_{is}^{\tilde{u}*} W_{it}^{\tilde{u}} \left(1 - \frac{1}{3} t_W^2 \right) + \frac{4}{3} W_{i+3\,s}^{\tilde{u}*} W_{i+3\,t}^{\tilde{u}} t_W^2 \right]. \qquad (10.47\mathrm{f})$$

Similarly the coefficients of the various quartic **Higgs-Higgs-squark-squark interaction** terms can be written down. They are

$$D[H^+, H^-, \tilde{d}_s, \tilde{d}_t] \; = \; -\frac{g_2^2}{2 M_W^2} \sum_{i,j,k=1}^{3} \left[U_{ik}^{d_R} m_{d_k}^2 U_{jk}^{d_R*} W_{i+3\,s}^{\tilde{d}*} W_{j+3\,t}^{\tilde{d}} t_\beta^2 \right.$$

$$\left. + U_{ik}^{u_L} m_{u_k}^2 U_{jk}^{u_L*} W_{is}^{\tilde{d}*} W_{jt}^{\tilde{d}} (ct)_\beta^2 \right]$$

$$+ \; \frac{g_2^2 c_{2\beta}}{4} \sum_{i=1}^{3} \left[W_{is}^{\tilde{d}*} W_{it}^{\tilde{d}} \left(1 - \frac{1}{3} t_W^2 \right) \right.$$

$$\left. - \frac{2}{3} W_{i+3\,s}^{\tilde{d}*} W_{i+3\,t}^{\tilde{d}} t_W^2 \right], \qquad (10.48\mathrm{a})$$

$$D[H^+, H^-, \tilde{u}_s, \tilde{u}_t] = -\frac{g_2^2}{2M_W^2} \sum_{i,j,k=1}^{3} \left[U_{ik}^{u_R} m_{u_k}^2 U_{jk}^{u_R*} W_{i+3\ s}^{\tilde{u}*}\, W_{j+3\ t}^{\tilde{u}} (ct)_\beta^2 \right.$$

$$\left. + U_{ik}^{d_L} m_{d_k}^2 U_{jk}^{d_L*} W_{is}^{\tilde{u}*} W_{jt}^{\tilde{u}} t_\beta^2 \right]$$

$$- \frac{g_2^2 c_{2\beta}}{4} \sum_{i=1}^{3} \left[W_{is}^{\tilde{u}*} W_{it}^{\tilde{u}} \left(1 + \frac{1}{3} t_W^2 \right) \right.$$

$$\left. - \frac{4}{3} W_{i+3\ s}^{\tilde{u}*}\, W_{i+3\ t}^{\tilde{u}}\, t_W^2 \right], \tag{10.48b}$$

$$D[H^+, H, \tilde{u}_s, \tilde{d}_t] = d_{Yu}[\tilde{u}_s, \tilde{d}_t] s_\alpha + d_{Yd}[\tilde{u}_s, \tilde{d}_t] c_\alpha$$

$$+ d_{Yud}[\tilde{u}_s, \tilde{d}_t] c_{\alpha-\beta} - d_g[\tilde{u}_s, \tilde{d}_t] s_{\alpha+\beta}, \tag{10.48c}$$

$$D[H^-, H, \tilde{d}_t, \tilde{u}_s] = \left(D[H^+, H, \tilde{u}_s, \tilde{d}_t] \right)^*, \tag{10.48d}$$

$$D[H^+, h, \tilde{u}_s, \tilde{d}_t] = d_{Yu}[\tilde{u}_s, \tilde{d}_t] c_\alpha - d_{Yd}[\tilde{u}_s, \tilde{d}_t] s_\alpha$$

$$- d_{Yud}[\tilde{u}_s, \tilde{d}_t] s_{\alpha-\beta} - d_g[\tilde{u}_s, \tilde{d}_t] c_{\alpha+\beta}, \tag{10.48e}$$

$$D[H^-, h, \tilde{d}_t, \tilde{u}_s] = \left(D[H^+, h, \tilde{u}_s, d_t] \right)^*, \tag{10.48f}$$

$$D[H^+, A, \tilde{u}_s, \tilde{d}_t] = -i d_{Yu}[\tilde{u}_s, \tilde{d}_t] c_\beta + i d_{Yd}[\tilde{u}_s, \tilde{d}_t] s_\beta + i d_g[\tilde{u}_s, \tilde{d}_t] c_{2\beta}, \tag{10.48g}$$

$$D[H^-, A, \tilde{d}_t, \tilde{u}_s] = \left(D[H^+, A, \tilde{u}_s, \tilde{d}_t] \right)^*, \tag{10.48h}$$

$$D[H, H, \tilde{d}_s, \tilde{d}_t] = -d_Y[\tilde{d}_s, \tilde{d}_t] c_\alpha^2 - d_g[\tilde{d}_s, \tilde{d}_t] c_{2\alpha}, \tag{10.48i}$$

$$D[H, h, \tilde{d}_s, \tilde{d}_t] = d_Y[\tilde{d}_s, \tilde{d}_t] s_{2\alpha} + 2 d_g[\tilde{d}_s, \tilde{d}_t] s_{2\alpha}, \tag{10.48j}$$

$$D[h, h, \tilde{d}_s, \tilde{d}_t] = -d_Y[\tilde{d}_s, \tilde{d}_t] s_\alpha^2 + d_g[\tilde{d}_s, \tilde{d}_t] c_{2\alpha}, \tag{10.48k}$$

$$D[A, A, \tilde{d}_s, \tilde{d}_t] = -d_Y[\tilde{d}_s, \tilde{d}_t] s_\beta^2 + d_g[\tilde{d}_s, \tilde{d}_t] c_{2\beta}, \tag{10.48l}$$

$$D[H, H, \tilde{u}_s, \tilde{u}_t] = -d_Y[\tilde{u}_s, \tilde{u}_t] s_\alpha^2 - d_g[\tilde{u}_s, \tilde{u}_t] c_{2\alpha}, \tag{10.48m}$$

$$D[H, h, \tilde{u}_s, \tilde{u}_t] = -d_Y[\tilde{u}_s, \tilde{u}_t] s_\alpha^2 + 2 d_g[\tilde{u}_s, \tilde{u}_t] s_{2\alpha}, \tag{10.48n}$$

$$D[h, h, \tilde{u}_s, \tilde{u}_t] = -d_Y[\tilde{u}_s, \tilde{u}_t] c_\alpha^2 + d_g[\tilde{u}_s, \tilde{u}_t] c_{2\alpha}, \tag{10.48o}$$

$$D[A, A, \tilde{u}_s, \tilde{u}_t] = -d_Y[\tilde{u}_s, \tilde{u}_t] c_\beta^2 + d_g[\tilde{u}_s, \tilde{u}_t] c_{2\beta}. \tag{10.48p}$$

In (10.48) we have introduced the quantities

$$d_{Yu}[\tilde{u}_s, \tilde{d}_t] \equiv \frac{g_2^2 c_\beta}{2\sqrt{2} M_W^2 s_\beta^2} \sum_{i,j,k=1}^{3} U_{ik}^{u_L} m_{u_k}^2 U_{jk}^{u_L*} W_{is}^{\tilde{u}*} W_{jt}^{\tilde{d}}, \tag{10.49a}$$

$$d_{Yd}[\tilde{u}_s, \tilde{d}_t] \equiv \frac{g_2^2 s_\beta}{2\sqrt{2} M_W^2 c_\beta^2} \sum_{i,j,k=1}^{3} U_{ik}^{d_L} m_{d_k}^2 U_{jk}^{d_L*} W_{is}^{\tilde{u}*} W_{jt}^{\tilde{d}}, \tag{10.49b}$$

$$d_{Yud}[\tilde{u}_s, \tilde{d}_t] \equiv \frac{g_2^2}{\sqrt{2}M_W^2 s_{2\beta}} \sum_{i,j,k,l=1}^{3} U_{ik}^{u_R} m_{u_k} V_{kl}^{q_L} m_{d_l} U_{jl}^{d_R*} W_{i+3\ s}^{\tilde{u}*} W_{j+3\ t}^{\tilde{d}} \,, \qquad (10.49c)$$

$$d_Y[\tilde{d}_s, \tilde{d}_t] \equiv \frac{g_2^2}{4M_W^2 c_\beta^2} \sum_{i,j,k=1}^{3} m_{d_k}^2 \left[U_{ik}^{d_L} U_{jk}^{d_L*} W_{is}^{\tilde{d}*} W_{jt}^{\tilde{d}} + U_{ik}^{d_R} U_{jk}^{d_R*} W_{i+3\ s}^{\tilde{d}*} W_{j+3\ t}^{\tilde{d}} \right], \qquad (10.49d)$$

$$d_Y[\tilde{u}_s, \tilde{u}_t] \equiv \frac{g_2^2}{4M_W^2 s_\beta^2} \sum_{i,j,k=1}^{3} m_{u_k}^2 \left[U_{ik}^{u_L} U_{jk}^{u_L*} W_{is}^{\tilde{u}*} W_{jt}^{\tilde{u}} + U_{ik}^{u_R} U_{jk}^{u_R*} W_{i+3\ s}^{\tilde{u}*} W_{j+3\ t}^{\tilde{u}} \right], \qquad (10.49e)$$

$$d_g[\tilde{u}_s, \tilde{d}_t] \equiv \frac{g_2^2}{2\sqrt{2}} \sum_{i=1}^{3} W_{is}^{\tilde{u}*} W_{jt}^{\tilde{d}} \,, \qquad (10.49f)$$

$$d_g[\tilde{d}_s, \tilde{d}_t] \equiv -\frac{g_2^2}{8} \sum_{i=1}^{3} \left[W_{is}^{\tilde{d}*} W_{it}^{\tilde{d}} \left(1 + \frac{1}{3}t_W^2 \right) + \frac{2}{3} W_{i+3\ s}^{\tilde{d}*} W_{j+3\ t}^{\tilde{d}} \tan^2 \theta_W \right], \qquad (10.49g)$$

$$d_g[\tilde{u}_s, \tilde{u}_t] \equiv \frac{g_2^2}{8} \sum_{i=1}^{3} \left[W_{is}^{\tilde{u}*} W_{it}^{\tilde{u}} \left(1 - \frac{1}{3}t_W^2 \right) + \frac{4}{3} W_{i+3\ s}^{\tilde{u}*} W_{j+3\ t}^{\tilde{u}} t_W^2 \right]. \qquad (10.49h)$$

The corresponding Feynman rules are shown in Fig. 10.7.

Fig. 10.7 is included in Appendix B

10.6 Radiative Effects on MSSM Higgs Particles

The properties of the Higgs particles in the MSSM and the relations among them, following naturally from supersymmetry, have been discussed in §10.3–§10.5 at the tree level. However, it is now known [10.6] that significant changes are induced radiatively in many of the expressions and relations, appearing in those sections, by quantum loop corrections. We shall discuss some of these effects at the one loop level, confining ourselves largely to the mass of the lightest Higgs h. That is where they are most spectacular and are of the greatest importance to experiment. Before going into the details, let us make three general points:

- One needs to be clear about the meaning of a physical Higgs mass when radiative effects are to be taken into account. The on-shell mass is defined as the square root of *that* value of q^2 for which the real part of the inverse scalar propagator $q^2 - m_{tree}^2 + \Sigma(q^2)$, $\Sigma(q^2)$ being the one loop self energy correction, vanishes. However, we will compute radiative corrections to static Higgs masses.

- Radiative corrections to the mass of h are dominated by loops involving the top quark t and its stop partners $\tilde{t}_{1,2}$, cf. Fig. 10.8. This domination occurs owing to the large Yukawa coupling that these states have with h. Contributions from loops mediated by other states are negligible by comparison and will be ignored.

- In the limit of exact supersymmetry, tree level Higgs masses are protected by the nonrenormalization theorem discussed in Ch.6. This explains why radiative corrections to those masses are controlled by M_s, the scale of soft supersymmetry breaking.

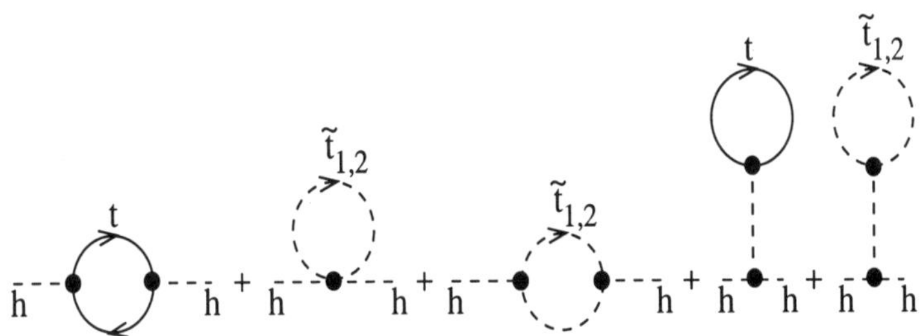

Fig.10.8. One loop self energy diagrams for h.

The radiatively corrected Higgs sector of the MSSM has been the subject of considerable study over several years. We do not go here into the initial and orginal works, but a detailed discussion with a historical perspective and a complete set of pertinent references may be found in the second paper cited in Ref. [10.1]. Three main tools have been employed in the literature: (1) direct diagrammatic calculations, (2) renormalization group methods and (3) effective potential techniques. Let us focus our attention on the correction to m_h as a sample case. The procedure in (1) is to adopt a straight computational approach by calculating the one loop self energy diagrams for h as given in Fig. 10.8.

In contrast, the methodology in (2) is that of Renormalization Group Evolution (RGE). For instance, when the sparticle mass spectrum (characterized by the scale M_s) is much heavier than the weak scale, i.e. $M_s \gg M_Z$, the quartic self coupling of h at the scale M_s is taken from (10.12) and (10.28b) to be

$$\frac{1}{32}\cos^2 2\alpha\, [g_Y^2(M_s) + g_2^2(M_s)]^2 \ .$$

It is then evolved to its value at the electroweak scale by means of the Standard Model RGE and used in the computation of the mass of h utilizing Standard Model expressions. However, we choose to present below an exposition of approach (3) – namely that of the effective potential – in calculating the correction to the tree level value of m_h. Though this method is numerically not as accurate as the diagrammatic one, it is pedagogically more interesting and gives a better theoretical insight into these loop induced radiative corrections. In addition, the inclusion of leading two loop corrections and the computation of corrections to static Higgs self couplings are more straightforward in this approach.

We start by considering the static approximation in which the effective action is approximated by the one loop effective potential. The actual calculation of the one loop effective potential can be found in standard text books [10.7–10.9]. The final expression reads

$$V_H^1(Q) = V_H^0(Q) + \Delta V_H^{(1)}(Q) \ , \tag{10.50a}$$

$$\Delta V_H^{(1)}(Q) = \frac{1}{64\pi^2} \mathrm{STr}\ \mathcal{M}^4(h) \left[\ln \frac{\mathcal{M}^2(h)}{Q^2} - \frac{3}{2} \right]. \qquad (10.50b)$$

In (10.50), $V_H^0(Q)$ is the tree level Higgs potential with its couplings renormalized at some scale Q, $\mathcal{M}(h)$ is the field dependent mass matrix and the supertrace, cf. 5.10, covers all supermultiplet fields whose masses depend on the VEVs of the Higgs fields.

Corrections in the absence of $\tilde{t}_L$-$\tilde{t}_R$ mixing

We have earlier noted that the most important loop corrections to the Higgs potential V_H come from the top-stop sector of the theory. To keep the discussion simple, we will consider only these. First, we neglect any mixing between the $SU(2)_L$ doublet ($\tilde{t}_L$) and singlet ($\tilde{t}_R$) squarks and assume equal soft supersymmetry breaking squared masses $\tilde{m}^2$ for those two fields. The relevant field dependent masses then are

$$m_t^2(h) = f_t^2 |h_2^0|^2 \,, \qquad (10.51a)$$

$$m_{\tilde{t}_1}^2(h) = m_{\tilde{t}_2}^2(h) = \tilde{m}^2 + f_t^2 |h_2^0|^2 \,, \qquad (10.51b)$$

where f_t is the top Yukawa coupling strength, being equal to $m_t(2\sqrt{2}G_F)^{1/2}/\sin\beta$. We have neglected D-term contributions to the stop masses since they are proportional to electroweak gauge couplings. They are thus suppressed by at least one power of $M_W^2 m_t^{-2}$ compared to the pure Yukawa contribution.

Each fermion or boson contributes to the supertrace of (10.50b) with a multiplicative weight factor equal to the number of independent degrees of freedom associated with it. Let us recall that each Dirac fermion contains four degrees of freedom, while each complex scalar has two. In addition, we have to include a color factor of three. Altogether, we thus have from (10.50b) that

$$\Delta V_{H,t-\tilde{t}}^{(1)}(Q) = \frac{3}{16\pi^2} \left[(\tilde{m}^2 + f_t^2 |h_2^0|^2)^2 \left(\ln \frac{\tilde{m}^2 + f_t^2 |h_2^0|^2}{Q^2} - \frac{3}{2} \right) - (f_t^2 |h_2^0|^2)^2 \left(\ln \frac{f_t^2 |h_2^0|^2}{Q^2} - \frac{3}{2} \right) \right],$$
$$(10.52)$$

where the overall factor of 3 in the numerator comes from color. As already advertised, the two terms in the RHS of (10.52) cancel exactly in the limit of unbroken supersymmetry.

In order to understand the physical significance of these corrections, we first have to redo the minimization of the Higgs potential. In §10.2 we had minimized the tree level expression which we now call $V_H^{(0)}$. Here we will do the same for $V_H^{(0)} + \Delta V_{H,t-\tilde{t}}^{(1)}$. As evident from (10.52), in the limit of vanishing $\tilde{t}_L$-$\tilde{t}_R$ mixing, corrections from the top-stop sector only involve the second Higgs doublet h_2. Therefore (10.17a) remains unchanged. On the other hand, (10.17b) now becomes

$$m_{2h}^2 = m_{12}^2 \cot\beta + \frac{M_Z^2}{2}\cos 2\beta - \frac{3f_t^2}{16\pi^2} \left[f(m_{\tilde{t}}^2) - f(m_t^2) \right], \qquad (10.53)$$

where we have introduced the function

$$f(m^2) \equiv 2m^2 \left(\ln \frac{m^2}{Q^2} - 1 \right) \qquad (10.54)$$

and m_t^2, $m_{\tilde{t}}^2$ are squared masses of the top and the stops respectively. We need not, for the moment, be bothered by the presence of the $\ln Q^2$ term since it can ultimately be absorbed in the renormalization of m_{2h}^2.

As in (10.19), we calculate the mass squared matrices now for the CP odd and CP even Higgs bosons by taking second derivatives of $V_H^{(0)} + \Delta V_{H,t-\tilde{t}}^{(1)}$ with respect to the imaginary and real parts of the neutral Higgs fields respectively. Once again, only the VEV v_2 (and not v_1) contributes to (10.53) as a consequence of our assumption of no $\tilde{t}_L$-$\tilde{t}_R$ mixing. Therefore, only the 2, 2 entries in the concerned matrices can possibly receive corrections from the top-stop sector. Moreover, since the VEVs v_1 and v_2 are real, the derivatives have to be taken at $\Im m\, h_1^0 = \Im m\, h_2^0 = 0$. It is then easy to see that the final result for the mass squared matrix of the CP odd states is the same as at the tree level, i.e. (10.23). The explicit correction to the 2,2 entry from the second derivative of (10.53) exactly cancels the correction to m_{2h}^2, given in (10.53). However, such is not the case for the 2,2 element of the mass squared matrix of the CP even Higgs scalars, (cf. 10.26). There we find the following finite and positive correction:

$$\Delta_{22}^{LL} = \frac{3 f_t^2 m_t^2}{4\pi^2} \ln \frac{m_{\tilde{t}}^2}{m_t^2} \equiv \frac{\epsilon_h}{\sin^2 \beta}\,, \tag{10.55}$$

where

$$\epsilon_h = \frac{3 G_F m_t^4}{\sqrt{2}\pi^2} \ln \frac{m_{\tilde{t}}^2}{m_t^2}\,. \tag{10.56}$$

We have put the superscript LL on Δ_{22} to denote the fact that (10.56) is a leading logarithm (in the ratio $m_{\tilde{t}}/m_t$) expression.

The one loop correction to (10.26), in this scenario, reads

$$\delta m_{\Re e\, h^0}^2 = \begin{pmatrix} 0 & 0 \\ 0 & \dfrac{\epsilon_h}{\sin^2 \beta} \end{pmatrix}, \tag{10.57}$$

so that (10.27), (10.29c) and (10.30b) extend respectively to

$$m_{h,H}^2 = \frac{1}{2}\left[m_A^2 + M_Z^2 + \frac{\epsilon_h}{\sin^2 \beta} \; \pm \; \left\{ (m_A^2 + M_Z^2)^2 \sin^2 2\beta \right.\right.$$

$$\left.\left. + \left[(M_Z^2 - m_A^2) \cos 2\beta + \frac{\epsilon_h}{\sin^2 \beta} \right]^2 \right\}^{1/2} \right], \tag{10.58a}$$

$$\tan 2\alpha = (m_A^2 + M_Z^2) \tan 2\beta \left(m_A^2 - M_Z^2 + \frac{\epsilon_h}{\sin^2 \beta \cos 2\beta} \right)^{-1}, \tag{10.58b}$$

$$m_h^2 + m_H^2 = m_A^2 + M_Z^2 + \frac{\epsilon_h}{\sin^2 \beta}\,. \tag{10.58c}$$

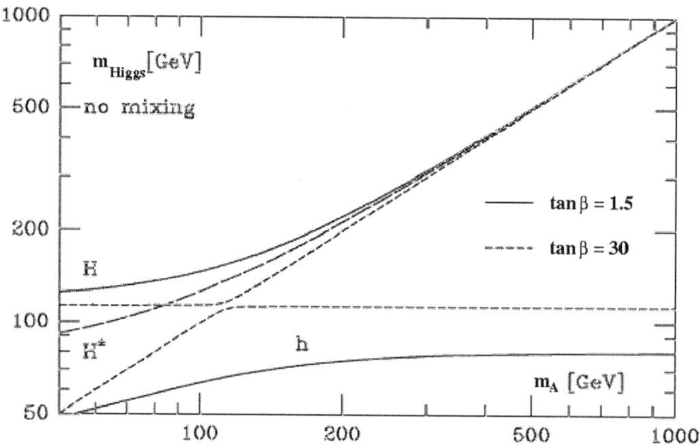

Fig.10.9. Other Higgs masses vs m_A for $\tan\beta = 1.5$ and 30 with $m_{\tilde{t}} \simeq 1$ TeV; adapted from Ref. [10.10].

In Fig. 10.9 the Higgs masses m_h, m_H and $m_{H^\pm}$, as given in (10.58a), are plotted [10.10] against m_A, the mass of the CP odd Higgs boson, for two rather extreme values of $\tan\beta$. For $\tan\beta > 1$, the mass eigenvalue of h increases monotonically with m_A, saturating to its maximum upper bound

$$m_h \;<\; (M_Z^2 \cos^2 2\beta + \epsilon_h)^{1/2} \tag{10.59}$$

for modest values of m_A, i.e. $m_A > 300$ GeV. For large $\tan\beta$ and $m_{\tilde{t}}$ taken to be $\mathcal{O}(\text{TeV})$, the RHS of (10.59) is ~ 110 GeV. We shall see later that the possibility of $\tilde{t}_L$-$\tilde{t}_R$ mixing can increase this upper bound[4]. At this level, the charged Higgs mass is still given by (10.30a) and is hence independent of $\tan\beta$, as shown in [10.10] Fig. 10.9. The tree level properties of the Higgs mass spectrum in the decoupling limit ($m_A \to \infty$) are still maintained. Now the $A, H, H^\pm$ Higgs particles remain nearly degenerate while the lightest h saturates its maximum mass value. The tree level mass orderings, $m_H > m_H^\pm > m_A$ remain valid for small $\tan\beta$. Otherwise, the larger $\tan\beta$ curves in Fig. 10.9 are fairly similar to the curves in Fig. 10.2, with M_Z replaced by $\{M_Z^2 + \epsilon_h^2\}^{1/2}$.

Eq. (10.56) represents the celebrated correction which has a quartic power dependence on the mass of the top quark. Note that it has only a logarithmic dependence on the stop mass squared $m_{\tilde{t}}^2$ which is characteristic of the square of the soft supersymmetry breaking scale M_s. This would seem to contradict our starting proposition that corrections to the masses of Higgs bosons should be proportional ,to supersymmetry breaking masses. This apparent contradiction is resolved by the fact that the shift in the tree level parameter m_{2h}^2 is indeed proportional to $m_{\tilde{t}}^2 - m_t^2$, c.f.(10.53). One would need to fine tune the parameters appearing in this equation if the tree level part were much smaller than the correction term. Furthermore, notice that the renormalization scale Q has disappeared from (10.56). This is to be expected since this equation describes the correction to a relation among *physical*

---

[4]Indeed, the final experimental lower bound on the mass of an SM-like Higgs boson of about 115 GeV from the completed runs at LEP indicates the need for some amount of $\tilde{t}_L$-$\tilde{t}_R$ mixing unless $\tilde{t}_{L,R}$-masses are much in excess of 1 TeV. This point will be discussed in more detail in §15.5.

quantities (masses of CP even Higgs bosons one hand and M_Z, m_A on the other). Indeed, it can be shown already at the level of the effective potential (10.50) that the explicit $\ln Q^2$ dependence of the one loop correction cancels against a similar dependence of the running quantities appearing in the tree level potential. In the simplified scenario, considered by us so far, it follows from (10.52) that the entire $\ln Q^2$ dependence collapses to

$$\frac{\partial \Delta V_H^{(1)}}{\partial \ln Q^2} = -\frac{3}{8\pi^2}\tilde{m}^2 \left(f_t^2 |h_2^0|^2 + \frac{1}{2}\tilde{m}^2 \right). \tag{10.60}$$

Thus the first term in the RHS of (10.60) exactly cancels the Q^2-dependence of $m_{2h}^2(Q)|h_2^0|^2$. The second term in the RHS of (10.60), a field independent constant, is of no immediate interest to particle physics, though it may contribute to the cosmological constant.

Corrections with $\tilde{t}_L$-$\tilde{t}_R$ mixing

Let us now introduce a nonzero $\tilde{t}_L$-$\tilde{t}_R$ mixing, described (c.f. 9.62c) by the off-diagonal matrix element[5] $-m_t(A^t + \mu \cot \beta)$ of the stop squared mass matrix. We will also allow the soft supersymmetry breaking $\tilde{t}_L$ and $\tilde{t}_R$ mass terms to differ. The eigenvalues of the field dependent $\tilde{t}$ squared mass matrix are then given by

$$m_{\tilde{t}_{1,2}}^2(h) = f_t^2|h_2^0|^2 + \frac{1}{2}\left[m_{\tilde{t}_L}^2 + m_{\tilde{t}_R}^2 \pm \sqrt{(m_{\tilde{t}_L}^2 - m_{\tilde{t}_R}^2) + 4f_t^2|A^t h_2^0 + \mu h_1^{0*}|^2} \right]. \tag{10.61}$$

Note that these eigenvalues depend on both neutral Higgs fields $h_{1,2}^0$. The corresponding one loop correction to the Higgs effective potential now becomes

$$\begin{aligned}
\Delta V_{H,t-\tilde{t}}^{(1)}(Q) &= \frac{3}{32\pi^2}\Bigg[m_{\tilde{t}_1}^4(h)\left\{ \ln \frac{m_{\tilde{t}_1}^2(h)}{Q^2} - \frac{3}{2} \right\} + m_{\tilde{t}_2}^4(h)\left\{ \ln \frac{m_{\tilde{t}_2}^2(h)}{Q^2} - \frac{3}{2} \right\} \\
&\quad - 2f_t^4|h_2^0|^4\left\{ \ln \frac{f_t^2|h_2^0|^2}{Q^2} - \frac{3}{2} \right\} \Bigg].
\end{aligned}$$

Both the minimization conditions $\partial V_H/\partial h_1^0 = 0$, $\partial V_H/\partial h_2^0 = 0$ are now affected by radiative corrections. Therefore, (10.17) change to

$$m_{1h}^2 = m_{12}^2 \tan \beta - \frac{1}{2}M_Z^2 \cos 2\beta - \frac{3f_t^2}{32\pi^2}\frac{\mu(\mu + A^t \tan \beta)}{m_{\tilde{t}_1}^2 - m_{\tilde{t}_2}^2}\left[f(m_{\tilde{t}_1}^2) - f(m_{\tilde{t}_2}^2) \right], \tag{10.62a}$$

$$\begin{aligned}
m_{2h}^2 &= m_{12}^2 \cot \beta + \frac{1}{2}M_Z^2 \cos 2\beta \\
&\quad - \frac{3f_t^2}{32\pi^2}\left\{ f(m_{\tilde{t}_1}^2) + f(m_{\tilde{t}_2}^2) - 2f(m_t^2) + \frac{A^t(A^t + \mu \cot \beta)}{m_{\tilde{t}_1}^2 - m_{\tilde{t}_2}^2}[f(m_{\tilde{t}_1}^2) - f(m_{\tilde{t}_2}^2)] \right\}.
\end{aligned} \tag{10.62b}$$

---

[5] We take A^t and μ to be real here.

Once again, the squared mass matrices for the neutral Higgs bosons can be computed from the second derivatives of the Higgs potential. The calculation for the CP odd case is greatly simplified by the observation that the first derivatives of any of the field dependent top (stop) masses with respect to the imaginary parts of h_1^0 and h_2^0 vanish in the (real) minimum of the Higgs potential. A straightforward calculation yields the result

$$m^2_{\Im m\ h^0} = (m_{12}^2 + \Delta) \begin{pmatrix} \tan\beta & 1 \\ 1 & \cot\beta \end{pmatrix}, \qquad (10.63)$$

with

$$\Delta = -\frac{3f_t^2}{32\pi^2} \frac{\mu A^t}{m_{\tilde{t}_1}^2 - m_{\tilde{t}_2}^2} [f(m_{\tilde{t}_1}^2) - f(m_{\tilde{t}_2}^2)] . \qquad (10.64)$$

As anticipated, this correction vanishes in the limit of no $\tilde{t}_L$-$\tilde{t}_R$ mixing ($\mu = A^t = 0$). The one loop corrected mass of the physical CP odd Higgs boson A thus becomes

$$m_A^2 = \frac{2(m_{12}^2 + \Delta)}{\sin 2\beta} . \qquad (10.65)$$

The explicit $\ln Q^2$ dependence can again be shown to cancel in this equation, if m_{12}^2 and $\tan\beta$ are understood to be running parameters. However, this cancellation works exactly only in one loop order; beyond that, terms of order $[f_t^2 m_{\tilde{t}}^2 \ln(m_{\tilde{t}}^2/Q^2)]^2$ remain in (10.65). In the interest of perturbative stability, one should therefore choose a renormalization scale Q close to the stop mass, e.g. $Q = \sqrt{m_{\tilde{t}_1} m_{\tilde{t}_2}}$. This is totally analogous to the choice made in perturbative QCD calculations (involving massless partons) of the renormalization scale to be close to the external momentum.

The generalization of (10.57), including $\tilde{t}_L$-$\tilde{t}_R$ mixing, now reads

$$\delta m^2_{\Re e\ h^0} = \begin{pmatrix} \Delta_{11} & \Delta_{12} \\ \Delta_{12} & \Delta_{22} \end{pmatrix}, \qquad (10.66)$$

with

$$\Delta_{11} = \frac{3G_F m_t^4}{2\sqrt{2}\pi^2 \sin^2\beta} \left[\frac{\mu(A^t + \mu\cot\beta)}{m_{\tilde{t}_1}^2 - m_{\tilde{t}_2}^2}\right]^2 \left(2 - \frac{m_{\tilde{t}_1}^2 + m_{\tilde{t}_2}^2}{m_{\tilde{t}_1}^2 - m_{\tilde{t}_2}^2} \ln\frac{m_{\tilde{t}_1}^2}{m_{\tilde{t}_2}^2}\right), \qquad (10.67a)$$

$$\Delta_{12} = \frac{3G_F m_t^4}{2\sqrt{2}\pi^2 \sin^2\beta} \frac{\mu(A^t + \mu\cot\beta)}{m_{\tilde{t}_1}^2 - m_{\tilde{t}_2}^2} \ln\frac{m_{\tilde{t}_1}^2}{m_{\tilde{t}_2}^2} + \frac{A^t}{\mu}\Delta_{11} , \qquad (10.67b)$$

$$\Delta_{22} = \frac{3G_F m_t^4}{\sqrt{2}\pi^2 \sin^2\beta} \left[\ln\frac{m_{\tilde{t}_1} m_{\tilde{t}_2}}{m_t^2} + \frac{A^t(A^t + \mu\cot\beta)}{m_{\tilde{t}_1}^2 - m_{\tilde{t}_2}^2} \ln\frac{m_{\tilde{t}_1}^2}{m_{\tilde{t}_2}^2}\right] + \left(\frac{A^t}{\mu}\right)^2 \Delta_{11}. \qquad (10.67c)$$

Again, each of eqs. (10.67) is independent of Q^2. Note also that the corrected value (10.65) of m_A^2 has to be used in the tree level squared mass matrix (10.26).

We have, so far, considered corrections only from the top-stop sector. If $\tan\beta$, the ratio of the Higgs VEVs, becomes very large, the bottom Yukawa coupling can be comparable to that of the top and make substantial additional corrections to (10.62), (10.65) and (10.66). These can be obtained from our expressions by the following three substitutions:- (1) interchange top (stop) masses and couplings with those of the bottom (sbottom); (2) interchange h_1^0 and h_2^0, i.e. $\tan\beta \leftrightarrow \cot\beta$; (3) interchange the shifts of m_{1h}^2 and m_{2h}^2, i.e. the leading logarithmic corrections from the bottom-sbottom sector only affect m_{1h}^2. Note that even if $\tan\beta$ is as high as m_t/m_b, the 'leading' logarithmic corrections from the b-$\tilde{b}$ sector to the squared mass matrix of the CP even Higgs bosons are suppressed by a factor $(m_b/m_t)^2$ as compared with those from the t-$\tilde{t}$ sector and thus can be safely neglected; however, the nonlogarithmic corrections from $\tilde{b}_L$-$\tilde{b}_R$ mixing can be significant in this case.

The question can be raised as to whether one can go beyond the one loop corrections from heavy quarks/squarks, presented above. Leading two loop corrections at $O(\alpha\alpha_s)$ to (10.67) can be incorporated with just a little more effort. This is done by treating the top mass in the overall m_t^4 factor as a scale dependent running quantity. In other words, m_t should be interpreted as the $\overline{\text{MS}}$ (or $\overline{\text{DR}}$) mass, *not* the pole mass. The two quantities are related by the boundary condition [10.11]

$$m_t(m_t) = m_t^{\text{pole}} \left(1 - \frac{4\alpha_S}{3\pi} \right) \tag{10.68}$$

plus higher order corrections. The scale dependence of m_t for scales $Q \leq m_{\tilde{t}}$ is the same as in the nonsupersymmetric SM:

$$m_t(Q) = m_t(m_t) \left\lfloor \frac{\alpha_S(Q)}{\alpha_S(m_t)} \right\rfloor^{12/23} . \tag{10.69}$$

The first (leading log) term in (10.67c) can be understood to have originated from the running of the SM Higgs self coupling from the scale[6]

$$M_s = \sqrt{m_{\tilde{t}_1} m_{\tilde{t}_2}} \tag{10.70}$$

to the scale m_t. Using this observation, the leading two loop corrections to this term can be easily incorporated by taking the factor m_t^4 at the intermediate scale $\sqrt{M_s m_t}$. All other terms in (10.67) can be absorbed in the boundary condition on the Higgs self coupling at the scale M_s; the m_t^4 factors in all such nonlogarithmic terms should therefore be taken at the high scale M_s.

By far, the most significant effect of the radiative corrections, discussed in this section, is that they relax the upper bound (10.30c) on the mass of the lighter CP even Higgs scalar h. We had already derived an upper bound ~ 110 GeV in the absence of $\tilde{t}_L$-$\tilde{t}_R$ mixing, but here we give the more general result when such a mixing is present. For a given value of $\tan\beta$, m_h is still maximal when m_A is large (the "decoupling limit", as discussed earlier), but the bound is now given by

$$m_h^2 < M_Z^2 \cos^2 2\beta + \Delta_{11} \cos^2\beta + \Delta_{12} \sin 2\beta + \Delta_{22} \sin^2\beta , \tag{10.71}$$

with the Δ's given by (10.67). Numerically, the correction Δ_{22} is usually the most important one. The absolute upper bound is still reached for $\tan\beta \gg 1$ (i.e. $|\cos 2\beta| \to 1$) just as at

---

[6]We assume here for simplicity that supersymmetry breaking is characterized by this one mass scale.

the tree level. For equal $\tilde{t}_L$ and $\tilde{t}_R$ soft supersymmetry breaking mass terms, a simple yet accurate formula for this upper bound obtains in the limit $|m_t A^t| \ll m_{\tilde{t}}^2$:

$$m_h^2 \leq M_Z^2 + \frac{3G_F}{\sqrt{2}\pi^2} \left[m_t^4(\sqrt{m_t M_s}) \ln(M_s^2/m_t^2) \right.$$

$$\left. + (A^t)^2 M_s^{-2} m_t^4(M_s)(1 - \frac{1}{12}(A^t)^2 M_s^{-2}) \right], \qquad (10.72)$$

with M_s as given by (10.70). We have explained why the two m_t^4 factors in the RHS of (10.72) have to be taken at different scales. Taking $M_s = 1$ TeV and $m_t^{\text{pole}} = 175$ GeV from direct TEVATRON experiments [10.12], one finds $m_t(\sqrt{m_t M_s}) \simeq 157$ GeV and $m_t(M_s) \simeq 150$ GeV. Since the last RHS term in (10.72) is maximal at $A^t = \sqrt{6}M_s$, one then obtains an absolute upper bound on m_h which is a critical test of MSSM, namely

$$m_h < 132 \text{ GeV}. \qquad (10.73)$$

Comparing with (10.59), we see that the effect of nonzero A^t, μ are quite significant and shifts the upper bound on the h-mass by more[7] than 20 GeV. Radiative corrections can therefore push m_h well beyond the reach of existing e^+e^- colliders. We finally mention that the treatment presented here has recently been extended by including corrections $\mathcal{O}(f_t^2 g^2)$ to the squared Higgs mass matrix and by allowing for large CP violating phases in the third generation squark sector [10.15]. These phases lead CP even and CP odd Higgs states to mix but do not alter the upper bound (10.73) on m_h. Later, in Ch.14, we shall discuss the generalization of (10.73) to cover extensions of the MSSM.

Concluding remarks

Before concluding this section, we want to make some brief general remarks on one loop radiative corrections to the charged Higgs mass and also to Higgs couplings in the MSSM. We have already shown that, in the absence of $\tilde{t}_L$-$\tilde{t}_R$ mixing (i.e. neglecting the effects of μ and A^t) – the charged Higgs mass is given by (10.30a) and is independent of $\tan\beta$. Even with $\tilde{t}_L$-$\tilde{t}_R$ mixing, the one loop corrections to $m_{H^\pm}^2$ remain small if the renormalization scale Q is chosen in a way such that perturbation theory is reliable. Explicit expressions for these corrections may be found in Ref. [10.6]. It is worth remarking here, though, that corrections from the top (stop)-bottom (sbottom) sector go to zero in the limit of a vanishing bottom Yukawa coupling. We further remind the reader that all D-term contributions to the squark mass matrices were neglected. The inclusion of such terms will introduce additional corrections of order $g_2^2 m_t^2/(8\pi^2)$ or $g_2^2 M_W^2/(8\pi^2)$. These corrections can be computed along

---

[7]We have presented here the results within the effective potential framework, implicitly working with $\overline{\text{MS}}$ renormalized parameters. A more recent analysis [10.13] shows that a diagrammatic calculation in the on-shell renormalization scheme, again including leading two loop corrections, almost exactly reproduces the result from the effective potential approach, once the difference between the two renormalization schemes has been taken into account. One should nonetheless assign a theoretical error of two to three GeV to the predicted value of m_h, due to higher order terms. An upward shift of such a magnitude has very recently been found [10.14] from two loop $\mathcal{O}(f_t^4)$ corrections.

the lines presented here. Though, strictly speaking, these modify [10.6] the relation (10.30a), they are numerically unimportant. Note also that a complete calculation of pure electroweak $O(g_2^2 M_W^2)$ corrections should include contributions from loops involving first and second generation sfermions as well as those from the gauge-Higgs-gaugino-higgsino sector. Turning to Higgs couplings, one loop radiative corrections, at the level discussed in this section, do not affect Higgs-gauge and Higgs-fermion couplings[8] directly. They only come in indirectly through a shift in the value of α, as indicated by (10.58b). Only in the case of Higgs self couplings are there some direct contributions [10.16]. For instance, ignoring $\tilde{t}_L$-$\tilde{t}_R$ mixing, the coefficients (denoted by $\lambda_{...}$) of $-ig_2 M_Z/(2\cos\theta_W)$ in the triple scalar Hhh and HAA vertices are changed from what appear in Fig. 10.4 to

$$\lambda_{Hhh} = 2\sin 2\alpha \sin(\alpha+\beta) - \cos 2\alpha \cos(\alpha+\beta) + 3\frac{\epsilon_h \sin\alpha}{M_Z^2 \sin^3\beta}\cos^2\beta \ , \tag{10.74a}$$

$$\lambda_{HAA} = -\cos 2\beta \cos(\alpha+\beta) + \frac{\epsilon_h \sin\alpha}{M_Z^2 \sin^3\beta}\cos^2\beta \ . \tag{10.74b}$$

Our final comment is on the static approximation. That may not work so well for Higgs bosons which are heavy, e.g. with masses comparable to those of the top/stop(s). Indeed, on-shell couplings of H and A then often develop imaginary dispersive parts from loops induced by the latter.

References

[10.1] J.F. Gunion, G. Kane, H.E. Haber and S. Dawson,, *op. cit.*, *Bibl.* M. Carena and H.E. Haber, *loc. cit.*, *Bibl.*

[10.2] J.F. Gunion and H.E. Haber, *loc. cit.*, *Bibl.*

[10.3] K. Inoue, A. Kakuto, H. Komatsu and S. Takeshita, Prog. Theor. Phys. **67** (1982) 1877. R. Flores and M. Sher, Ann. Phys. **148** (1983) 95.

[10.4] S.L. Glashow and S. Weinberg, Phys. Rev. **D15** (1977) 1958. E.A. Paschos, Phys. Rev. **D15** (1977) 3416.

---

[8] If $\tan\beta$ is $\gg 1$ and $|\mu M_{\tilde{g}}| = \mathcal{O}(m_{\tilde{q}}^2)$, $\tilde{g}$-$\tilde{q}$ loop diagrams can yield $\mathcal{O}(1)$ corrections to the couplings of down type quarks to neutral Higgs bosons. A reliable perturbative treatment is nonetheless possible after a resummation of these corrections [10.16]. Analogous corrections can also lead to significant flavor nondiagonal couplings of the neutral Higgs bosons to down type quarks [10.17]. One loop vertex corrections also exist, but are typically small and do not significantly alter the pattern of Higgs couplings. Very general formulae giving one loop corrections to fermion masses and Higgs couplings in the MSSM are summarized in the article [10.1] by Carena and Haber.

[10.5]　H.E. Haber in G.L. Kane (ed). *Perspectives on Higgs Physics, op. cit., Bibl.*

[10.6]　H.E. Haber and R. Hempfling, Phys. Rev. **D48** (1993) 4280. H.E. Haber, R. Hempfling and A.H. Hoang, Z. Phys. **C75** (1997) 537.

[10.7]　T.P. Cheng and L.F. Li, *op. cit., Bibl.*

[10.8]　M.E. Peskin and D.V. Schroeder, *op. cit., Bibl.*

[10.9]　C. Itzykson and J-B. Zuber, *op. cit., Bibl.*

[10.10]　A. Djouadi, J. Kalinowski and P.M. Zerwas, Z. Phys. **C57** (1993) 569.

[10.11]　R. Tarrach, Nucl. Phys. **B183** (1980) 384.

[10.12]　F. Abe et al. Phys. Rev. Lett. **80** (1998) 2767. B. Abbot et al. Phys. Rev. **D58** (1998) 052001.

[10.13]　M. Carena, H.E. Haber, S. Heinemeyer, W. Hollik, C.E.M. Wagner and G. Weiglein, Nucl. Phys. **B580** (2000) 29.

[10.14]　G. Degrassi, S. Heinemeyer, W. Hollik, P. Slavich and G. Weiglein, Eur. Phys. J **C28** (2003) 133.

[10.15]　S.Y. Choi, M. Drees and J.S. Lee, Phys. Lett. **B481** (2000) 57.

[10.16]　V. Barger, M. Berger, A. Stange and R. Phillips, Phys. Rev. **D45** (1992) 4128. Z. Kunszt and F. Zwirner, Nucl. Phys. **B385** (1992) 3. S. Heinemeyer and W. Hollik, Nucl. Phys. **B474** (1996) 32.

[10.17]　K.S. Babu and C. Kolda, Phys. Rev. Lett. **84**, (2000) 228.

Chapter 11

Evolution from Very High Energies

11.1 The Need for a High Scale

We have seen in Ch.9 that supersymmetry cannot be broken spontaneously by MSSM fields alone. Faced with the lack of a satisfactory theory of the dynamical breakdown of supersymmetry, we resorted there to parameterization. We added a set of explicit "soft breaking terms" to the supersymmetric Lagrangian density. Such a model independent approach suffices for most phenomenological applications. However, it is neither very satisfying from a theoretical perspective, nor always practical. First of all, the number of parameters to describe the most general soft supersymmetry breaking Lagrangian density is very large. As mentioned in Ch.9, including CP violating complex phases, it is precisely 124 in the MSSM with exact R-parity. Usually, at the tree level, only a handful of parameters influence the cross section for a given sparticle production process. Nevertheless, decay widths and branching ratios of heavier sparticles are often sensitive to a much greater number of parameters. Take, for example, the branching ratios for three body decays of the chargino $\tilde{\chi}^+ \to f\bar{f}'\tilde{\chi}^0$, where f, f' are SM matter fermions and $\tilde{\chi}^0$ the LSP neutralino. These depend on the masses and mixing angles of nearly *all* sfermions. (Only $\tilde{t}$ and $\tilde{b}$ squarks are exempt if $M_{\tilde{\chi}^+} \leq m_t + m_b + M_{\tilde{\chi}^0}$ so that the decay channel $\tilde{\chi}^+ \to t\bar{b}\tilde{\chi}^0$ is kinematically inaccessible). Moreover, experiments often sum over many production channels. This is particularly true of measurements made at hadron colliders, where there are no sharp kinematical thresholds after integrating over parton distribution functions. Analyzing sparticle production at pp or $\bar{p}p$ colliders in a completely general softly broken MSSM is, therefore, all but impossible.

Another problem is the following. Only a tiny fraction of the 124-dimensional parameter space is phenomenologically allowed. Suppose one picked those parameters randomly, subject only to the naturalness constraint that sparticle masses should not (greatly) exceed 1 TeV. Then one would be almost certain to violate some of the experimental bounds on FCNC processes and CP violating amplitudes mentioned in Ch.9. The true theory of supersymmetry breaking must, of course, respect these bounds. Hence it is tempting to make an ansatz that avoids or at least greatly ameliorates the problems of unwanted FCNC and CP violating amplitudes. In particular, as discussed in Ch.9, supersymmetric contributions to FCNC processes vanish if there is a "universality" of sfermion masses across generations. In other words, identical masses for sfermions of the same gauge quantum numbers imply the (approximate) flavor diagonality of the soft breaking parameters. As a result, the number of free parameters is vastly reduced.

Notwithstanding the absence of a generally accepted theory of supersymmetry breaking (several candidates will be discussed in the next two chapters), we can abstract some of its properties. In particular, this theory is almost certain to involve an energy scale well above the weak scale. This is suggested by the physical impossibility of generating a spontaneous supersymmetry breakdown only with MSSM superfields at the weak scale. Indeed, there have been attempts postulating various new, yet unseen, superfields at around that scale to try to effect a spontaneous breaking of supersymmetry, but all these attempts have failed. Such numerous failed attempts at model building have shown the difficulty of constructing a realistic theory wherein the fields, whose VEVs break supersymmetry, couple to MSSM fields directly at the tree level. The problem is the supertrace sum rule (9.2), with its disastrous phenomenological consequences. It tends to survive in most of such models. As already mentioned in Ch.9, all nearly successful models so far make use of a hidden or secluded sector where supersymmetry is broken spontaneously. The construction of such a hidden sector is not difficult by itself. It is postulated to couple only very weakly to the observable sector containing the MSSM. Such a coupling is effected either (1) through loops (usually involving a "messenger sector"), or (2) gravitationally, where gravity itself acts as the messenger. The resulting soft supersymmetry breaking operators, written in terms of the MSSM fields, have dimensional coefficients much smaller than the original VEV's that break supersymmetry in the hidden sector. This suppression involves either of the following two things: (1) at least one loop factor $\alpha/\pi \sim 1/100$ in models with supersymmetry breaking mediated by gauge interactions, (2) inverse powers of the (reduced) Planck mass $M_{Pl} \simeq 2.4 \times 10^{18}$ GeV in models with gravity mediated supersymmetry breaking. Various such models, "still in business", will be described in more detail in the next two chapters.

The upshot of the above discussion is the following. The scale, not only of spontaneous supersymmetry breaking in the hidden sector, but also of the transmission of supersymmetry breaking to the visible sector, has to be significantly larger than the sparticle masses themselves. In other words, the MSSM, with soft explicit supersymmetry breaking, is expected to remain a viable description of Nature (in the sense of an effective Lagrangian) for energy scales that are considerably above sparticle masses. The ultimate theory of supersymmetry breaking will then impose some "boundary conditions" on the soft breaking parameters, valid at some very high energy scale, where the degrees of freedom responsible for the transmission of supersymmetry breaking to the MSSM sector are integrated out. We are thus faced with a piquaint situation. At least two very different energy scales are playing a role: the relatively low scale of MSSM sparticle masses $\leq \mathcal{O}(1)$ TeV and the much higher "messenger" scale where supersymmetry breaking begins to be felt by the MSSM fields. In order to either test the predictions of some theory of the breakdown of supersymmetry at this high scale, or translate the knowledge of experimental constraints on soft supersymmetry breaking parameters into restrictions on this putative theory, we need to connect these two scales. This is best achieved by means of equations following from [11.1] Renormalization Group Evolution (RGE) already discussed for a supersymmetric gauge theory in §6.9.

11.2 The Running of Gauge Couplings in the SM and the MSSM

The RGE equation for the gauge coupling strength g in a general nonabelian gauge theory without supersymmetry was given in (6.108). We rewrite here the corresponding expression for $b_{g^2}^{(1)}$, namely the (one-loop) coefficient of g^4 in the RHS of (6.108):

$$b_{g^2}^{(1)} = -\frac{11}{3}C_2(G) + \frac{2}{3}\sum_i T_i(\mathcal{R}) + \frac{1}{3}\sum_\alpha T_\alpha(\mathcal{R}) \ . \tag{11.1}$$

Recall that $C_2(G)$ is the quadratic Casimir constant of the gauge group G. Moreover, $T_i(\mathcal{R})$ is the representation constant of $\mathcal{R}$ which is the representation according to which the i-th left chiral fermion in the loop transforms and ditto for $T_\alpha(\mathcal{R})$ vis-a-vis the αth complex scalar. Let us apply (11.1) specifically to the gauge group $G_{SM} = SU(3)_C \times SU(2)_L \times U(1)_Y$ of the Standard Model. For an $SU(N)$ group, such as the $SU(3)_C$ or the $SU(2)_L$ factor of G_{SM}, $C_2(G) = N$ while $T(\mathcal{R}) = 1/2$ for the fundamental representation. On the other hand, for the hypercharge gauge group $U(1)_Y$, the $C_2(G)$ term is absent and $T(\mathcal{R})$ is simply $Y^2/4$. Note that, while summing over contributions from different fermions, one needs to count the two chirality states of each fermion separately. A massive quark contains two chirality states, both of which contribute to the running of the QCD $SU(3)_C$ coupling strength in the SM. In contrast, only left chiral fermions contribute to the running of the weak isospin $SU(2)_L$ coupling strength. Also, quarks and antiquarks contribute with a color factor of 3 to the running of the $SU(2)_L$ and $U(1)_Y$ gauge couplings; the same is true of squarks and antisquarks. Finally, as explained in §6.9, each Cartesian gaugino component, on account of its Majorana nature, counts only as a single chiral fermion in the adjoint representation.

Consider first the hypercharge coupling strength g_Y in the SM. Its β-function only receives contributions from the second and third RHS terms of (11.1). For $l = e, \mu, \tau$, $q_\uparrow = u, c, t$ and $q_\downarrow = d, s, b$, we have $(Y^2/4)_{l_L} = (Y^2/4)_{\nu_L} = 1/4$, $(Y^2/4)_{l_R} = 1$, $\sum_{\text{color}}(Y^2/4)_{q_{\uparrow L}} = \sum_{\text{color}}(Y^2/4)_{q_{\downarrow L}} = 1/12$, $\sum_{\text{color}}(Y^2/4)_{q_{\uparrow R}} = 4/3$, $\sum_{\text{color}}(Y^2/4)_{q_{\downarrow R}} = 1/3$. Summing all over three generations and multiplying by 2/3 (cf. 11.1), we then have $(b_{g_Y^2}^{(1)})_{\text{fermions}} = 20/3$. For the Higgs scalars, $(Y^2/4)_{\phi^+} = (Y^2/4)_{\phi^0} = 1/4$ so that, from the RHS of (11.1), $(b_{g_Y^2}^{(1)})_{\text{scalars}} = 1/6$. Finally, adding the two,

$$b_{g_Y^2}^{(1)}\Big|_{SM} = 41/6 \ . \tag{11.2}$$

The same procedure can be repeated for the $SU(2)_L$ and $SU(3)_C$ gauge couplings. For the former, $(\beta_{g_2^2}^{(1)})_{\text{gauge}} = -22/3$. Then, for the SM, the twelve matter fermion doublets and one Higgs doublet contribute $4g_2^4$ and $\frac{1}{6}g_2^4$ to $(b_{g_2^2}^{(1)})_{\text{fermions}}$ and $(b_{g_2^2}^{(1)})_{\text{scalars}}$ respectively. Thus

$$b_{g_2^2}^{(1)}\Big|_{SM} = -19/6 \ . \tag{11.3}$$

Turning to the QCD coupling[1] in the SM, $(b_{g_3^2}^{(1)})_{\text{gauge}} = -11$ from (11.1). Moreover, six quarks (of two chiralities) each contribute $4g_3^4$ to $(b_{g_3^2}^{(1)})_{\text{fermions}}$ and there is no scalar contribution. In

---

[1] For similarity with g_2, this is called g_3 in Ch.11 rather than g_s, as done elsewhere.

total, therefore, one has

$$b_{g_3^2}^{(1)}\Big|_{SM} = -7 \ . \tag{11.4}$$

Let us now repeat the above exercise for the MSSM. From (6.107) we can take

$$b_{g^2}^{(1)} = -3C_2(G) + \sum_i T_i(\mathcal{R}) \tag{11.5}$$

for a supersymmetric nonabelian gauge theory, with i specified as a type index of the chiral superfield contributing to the superloop (cf. Ch.6). Reconsider the evolution of the hypercharge coupling strength g_Y in the MSSM. We just need to sum $Y^2/4$ over all left chiral superfields. In addition to the value 10, obtained for this quantity from all chiral (matter) fermions of the SM (each of which gets elevated to a chiral superfield, cf. Ch.8), we need to include the contributions, $1/4$ each, to $Y^2/4$ from the four Higgs superfields of the MSSM. We therefore have from (11.5)

$$b_{g_Y^2}^{(1)}\Big|_{MSSM} = 11 \ . \tag{11.6}$$

For the $SU(2)_L$ gauge coupling strength g_2, the gauge superfield contribution to $b_{g_2^2}^{(1)}$ in (11.5) is -6. This needs to be added to $\sum_i T_i(\mathcal{R})$ which is half times the number of chiral weak doublets in the MSSM, to yield $-6 + \dfrac{1}{2} \times 14 = 1$. Thus

$$b_{g_2^2}^{(1)}\Big|_{MSSM} = 1 \ . \tag{11.7}$$

Finally, the QCD coupling evolution is straightforward. The gauge superfield contribution to $b_{g_2^2}^{(1)}$, namely -9, gets added to the sum of the color triplet and antitriplet chiral superfield contributions to $\sum_i T_i(\mathcal{R})$ which is $(1/2) \times 12$, to yield -3. Therefore

$$b_{g_3^2}^{(1)}\Big|_{MSSM} = -3 \ . \tag{11.8}$$

Altogether, referring back to (6.107), one can write the one loop evolution equations evolution equations [11.1]

$$\frac{dg_Y^2}{dt} = \frac{g_Y^4}{8\pi^2} \cdot \begin{cases} 41/6 \ , & SM \\ 11 \ , & MSSM \end{cases} , \tag{11.9a}$$

$$\frac{dg_2^2}{dt} = \frac{g_2^4}{8\pi^2} \cdot \begin{cases} -19/6 \ , & SM \\ 1 \ , & MSSM \end{cases} , \tag{11.9b}$$

$$\frac{dg_3^2}{dt} = \frac{g_3^4}{8\pi^2} \cdot \begin{cases} -7 \ , & SM \\ -3 \ , & MSSM \end{cases} , \tag{11.9c}$$

for the hypercharge, weak isospin and QCD couplings respectively. Eqs.(11.9) are valid for energy scales much larger than the mass of the heaviest particle whose contribution has been included in the loop calculations of the β-functions. In the SM this particle is the top quark or possibly the Higgs boson. In the MSSM it is most likely to be a sparticle or one of the heavy Higgs bosons. It is easy to derive β-functions for scales where some, but not all, sparticles contribute, using (11.1) and (11.4).

To one loop order, the β-functions of the three subgroups of the (MS)SM gauge group do not mix. Eqs.(11.9) are thus easily solved analytically. Using the shorter notation $b^{(1)}_{g^2_\alpha} \equiv b^{(1)}_\alpha$, we can write the solution as

$$g_\alpha^2(\mu) = g_\alpha^2(\mu_0) \left[1 - (8\pi^2)^{-1} b_\alpha^{(1)} g_\alpha^2(\mu_0) \ln (\mu/\mu_0)\right]^{-1} , \quad \alpha = Y, 2, 3 . \tag{11.10}$$

As with any first order differential equation, the solution depends on one unknown parameter. Here that is taken to be the value of g_α^2 at some "input scale" μ_0. Eq.(11.2) then allows one to compute g_α^2 at all other scales μ. Measurements at the Large Electron-Positron (LEP) collider at CERN, the Stanford Linear Collider and elsewhere have determined [11.2] the strengths of the three SM gauge couplings at the Z scale with quite small errors:

$$g_Y^2(M_Z) = 0.1277 \pm 0.0004 , \quad g_2^2(M_Z) = 0.424 \pm 0.001 , \quad g_3^2(M_Z) = 1.495 \pm 0.025 .$$

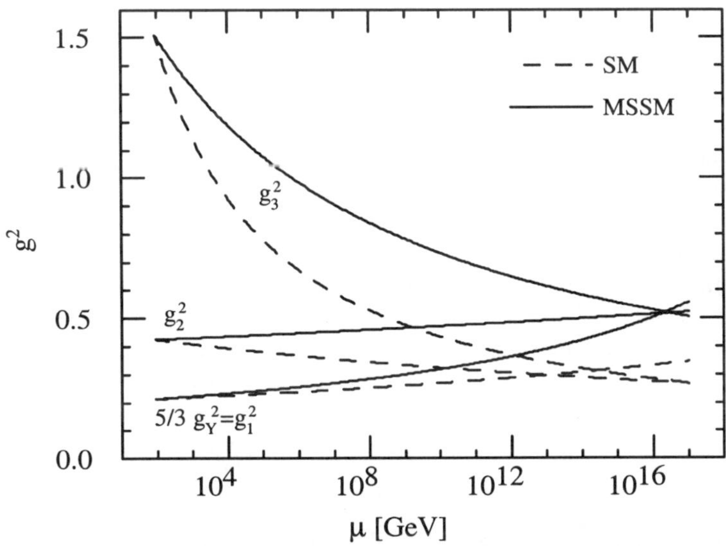

Fig. 11.1. The running of the $U(1)$, $SU(2)_L$ and $SU(3)_C$ squared gauge coupling strengths above the Z mass scale in the SM (dashed) and in the MSSM (solid).

Given these input numbers, (11.9) and (11.10) can be used to predict values of the running gauge couplings at higher energies. The results are shown in Fig. 11.1, where dashed curves have been used for the SM and solid lines for the MSSM. The $U(1)$ coupling, which appears in Fig. 11.1, is $g_1 = \sqrt{5/3}\, g_Y$. In effect, this means that all hypercharges have been reduced by a factor $\sqrt{3/5}$. Such a normalization is necessary if the hypercharge generator is to be interpreted as a generator of a larger nonabelian **simple gauge group**. The point is this. Tr T^2, taken over one complete generation of left chiral fermions in the SM, is the same (2) for T being any of the weak isospin or QCD generators. However, such is not the case for the weak hypercharge generator $Y/2$, Tr$(Y^2/4)$ being 10/3 i.e. 5/3 times too big. Thus a "scaling" $Y \to Y_1 = \sqrt{3/5}\, Y$ puts Y_1 on par with all other generators, as required in a [11.3] Grand Unified Theory (GUT). So long as there is also the scaling $g_Y \to g_1 = \sqrt{5/3}\, g_Y$, physical quantities remain unaffected since all weak hypercharge vertices involve only the product $g_Y(Y/2) = g_1(Y_1/2)$. Note also that $b_1^{(1)} = \frac{3}{5} b_Y^{(1)}$.

Observe that the gauge couplings, evolved according to the SM β-functions, do not meet at a point. The three dashed curves do come fairly close to one another. However, the errors on the input numbers for the gauge coupling strengths at the weak scale are so small that the central values would need to be changed by many standard deviations to make those curves meet. In contrast, the three solid lines, showing the MSSM prediction for the evolution of the couplings, meet at $\mu = M_U \simeq 2 \times 10^{16}$ GeV, where

$$g_\alpha^2(M_U^2) = g_U^2 = 0.51, \quad (\alpha = 1, 2, 3) \ .$$

We have so far ignored supersymmetry breaking and the role of M_s, characterizing the particle-sparticle mass differences. The fact, however, is that the above conclusion remains unaltered [11.4] so long as M_s lies in the range $M_Z < M_s < \mathcal{O}(1 \text{ TeV})$. This allows the embedding of the three separate gauge groups of the MSSM into a GUT simple group [11.3]. The latter symmetry then is supposed to be somehow broken at the energy scale M_U. Of course, the results, shown in Fig. 11.1, are based only on one loop β-functions. Moreover, we have assumed that all sparticles and Higgs bosons contribute fully to the MSSM β-functions already at the scale $\mu = M_Z$. Nevertheless, the main conclusion, reached by us, survives the inclusion of two loop corrections and a more careful treatment of thresholds [11.4]. Let us restate it. The $U(1)$, $SU(2)_L$ and $SU(3)_C$ gauge groups can be unified in the MSSM, but not in the SM, unless additional particles are introduced at an "intermediate scale" between M_Z and M_U. In the latter case, however, the SM is valid only for energies only up to this intermediate scale. In the MSSM no such intermediate scale needs to be postulated. There is, of course, the intra-supermultiplet splitting mass M_s in the latter scenario. But the MSSM can be valid beyond M_s – indeed, all the way upto M_U. In that case the MSSM coupling evolution equations (11.9) can be integrated and expressed as a single equation.

$$g_i^{-2}(M_Z) = g_U^{-2} + \frac{1}{2\pi} b_i^{(1)}\bigg|_{\text{MSSM}} \ln \frac{M_Z}{M_U} \ , \tag{11.11}$$

with $i = 1, 2, 3$. Eq. (11.11) can, in fact, be derived from (11.10) by setting $\mu_0 = M_U$. Moreover, the following amusing result follows from (11.11):

$$[g_3^{-2}(M_Z) - g_2^{-2}(M_Z)]/[g_2^{-2}(M_Z) - g_1^{-2}(M_Z)] = [b_2^{(1)} - b_3^{(1)}]/[b_1^{(1)} - b_2^{(2)}] \ .$$

It is remarkable that the LHS of the above equation, as determined [11.2] from experimental measurements, is $0.717 \pm 0.008 \pm 0.03$, while the RHS equals $5/7 = 0.714$.

The observations, made in the above discussion, strengthen the motivation for near weak scale supersymmetry. At the same time, they reinforce our earlier argument for the existence of a very high energy scale – beyond that of sparticle masses. Notice also that g_3^2 changes by about a factor of 3 between M_Z and M_U. This fact clearly illustrates a general feature of the running of squared gauge coupling strengths between such vastly different scales, namely that it can indeed lead to an effect of order unity, rather than a small radiative correction.

Specifically, the need to resum all powers of $(g_3^2/8\pi^2)\ln(M_Z/M_U)$ – as done by the solution (11.10), with $\alpha = 3$, to (11.9c) – is thereby demonstrated. Of course, the MSSM Lagrangian contains many more (as yet unknown) parameters describing supersymmetry breaking. We made a statement earlier that the unification conclusion in the MSSM is insensitive to these parameters so long as $M_s < O$ (1 TeV). In order to justify this statement, we need to investigate the evolution of those with the energy scale, i.e. compute their β-functions. That will be the topic of the following section.

11.3 Derivation of the Remaining RGE Equations

The first calculation of all MSSM β-functions to one loop order was performed in the early eighties [11.1]. It used explicit diagrammatic techniques based on conventional Feynman diagrams. Owing to the large number of (component) fields in the MSSM, there are many (seemingly) independent β-functions. Each of these usually receives contributions from several different diagrams. Consequently, the computation becomes quite nontrivial. Indeed, the original calculation contained errors which were corrected more than one year later.

A somewhat simpler and more versatile method for calculating β-functions is based on the effective potential. Introducing a momentum cutoff Λ, the effective potential density can be written to one-loop order [11.5] as

$$V_{\text{eff}}^{(1)} = V^{(0)} + \frac{1}{32\pi^2}\left[\Lambda^2 \,\text{STr}\,\mathcal{M}^2 - (\ln\Lambda)\,\text{STr}\,\mathcal{M}^4 + \text{finite terms}\right], \qquad (11.12)$$

where $V^{(0)}$ is the tree–level term. The supertrace is again (cf. 5.10) given, for any positive even integer n, by $\text{STr}\,\mathcal{M}^n \equiv \sum_J (2J+1)(-1)^{2J}\text{Tr}\,\mathcal{M}_J^n$, $\mathcal{M}_J$ being the field dependent mass matrix for particles with spin J. In general, the latter mass matrix is given by

$$(\mathcal{M}^p)_{mn} = -\frac{\partial^2 \mathcal{L}(\psi_i)}{\partial\psi_m \partial\psi_n}, \qquad (11.13)$$

ψ_i standing for any bosonic or fermionic field with the corresponding power p being 2 or 1 respectively. The first term within the square bracket in the RHS of (11.12) is quadratically divergent. It vanishes if supersymmetry is exact. Moreover, in softly broken supersymmetry, it can be shown [11.6] to contribute at most a field independent constant, which does not affect[2] renormalizable particle interactions.

The second term within the square bracket in the RHS of (11.12) is logarithmically divergent. It does give field dependent contributions to $V_{\text{eff}}^{(1)}$. In general, its coefficient has the same structure as the tree level potential density $V^{(0)}$. One is thereby able to directly read off all the β-functions of the coefficients of terms appearing in $V^{(0)}$ including the gauge couplings which occur in the D-term contributions. The problem has now been reduced to a computation of tree level mass matrices and taking appropriate powers of them. The only one loop β-functions, that cannot be obtained in this fashion, are those of the gaugino masses. But the latter can be computed in a straightforward manner using diagrammatic

---

[2]It might, in that event, contribute to the cosmological constant, however.

techniques, as shown below. This 'nondiagrammatic' derivation of the remaining β-functions makes use of the following fact. The renormalized effective potential must be independent of the renormalization scale. Consequently, as already emphasized in Ch.10, the dependence on μ of the tree level potential with running parameters is cancelled to one loop order by corrections proportional to $(\text{STr } \mathcal{M}^4) \ln(\mu/\mu_0)$.

The simplest and most elegant way to compute the β-functions of the parameters of the scalar potential is to use [11.7] the superfield formalism. Recall from the discussion in §7.7 that soft supersymmetry breaking can be described in terms of "**spurion**" superfields $\eta = \theta\theta$ and $\bar{\eta} = \bar{\theta}\bar{\theta}$ of mass dimension -1, as expressed in terms of the two component Grassmann spinorial coordinates $\theta, \bar{\theta}$ (cf. Ch.3). Let us dispense with the linear chiral superfield contribution $h_i\Phi_i$ to the superpotential $\mathcal{W}$ in (5.1b) since no gauge singlet chiral superfield is present in the MSSM. With the latter in mind, we can – in the manner of the discussion at the end of §7.7 – take the gauge group G to be $\prod_\alpha G_\alpha$, each factor G_α being either a simple Lie group or a $U(1)$ with an associated gauge coupling g_α. Let us further generalize the complex type subspace formalism, introduced in §5.1 for scalar fields, to chiral superfields. Thus, with i being a type index, $\bar{\Phi}^i \equiv \Phi_i^\dagger$ and $\partial\Phi_i/\partial\Phi_j = \delta_i{}^j$. Now the supersymmetric and the soft supersymmetry-breaking parts of the Lagrangian density respectively read:

$$\mathcal{L}_{\text{SUSY}} = \int d^4\theta \, \bar{\Phi}^i \left(\exp 2\sum_\alpha g_\alpha V^\alpha\right)^j_i \Phi_j + \frac{1}{4}\sum_\alpha \left(\int d^2\theta \, W^{\alpha A}W^\alpha_A + h.c.\right)$$

$$+ \left[\int d^2\theta \left(\frac{1}{3!}f^{ijk}\Phi_i\Phi_j\Phi_k + \frac{1}{2}\mu^{ij}\Phi_i\Phi_j\right) + h.c.\right], \qquad (11.14a)$$

$$\mathcal{L}_{\text{SOFT}} = \int d^2\theta \, \eta \left(\frac{1}{6}\mathcal{A}^{ijk}\Phi_i\Phi_j\Phi_k - \frac{1}{2}\mathcal{B}^{ij}\Phi_i\Phi_j - \frac{1}{2}\sum_\alpha M_\alpha W^{\alpha A}W^\alpha_A\right) + h.c.$$

$$- \int d^4\theta \, \bar{\eta}\eta \, \bar{\Phi}^i(m^2)_i{}^j \exp\left(2\sum_\alpha g_\alpha V^\alpha\right)^k_j \Phi_k . \qquad (11.14b)$$

Once again, we have dropped the singlet $C_i\phi_i$ term from (7.42) since there is no pure gauge singlet scalar in the MSSM. It is noteworthy that the type indices $i, j, k, \cdots$ in the above equations label both different representations of the gauge group and members[3] of a given representation. For instance, Φ_i might stand for the left chiral top superfield of a certain color, say "red", etc.

The sum of $\mathcal{L}_{\text{SUSY}}$ and $\mathcal{L}_{\text{SOFT}}$, as given in (11.14), represents the tree level Lagrangian density with bare couplings and parameters. On the incorporation of loop induced quantum corrections, the superfields in (11.14) will become unrenormalized superfields containing divergences. Apart from gauge fixing and Fadeev-Popov terms, that need to be added to

---

[3] We shall follow the convention that summation over α subsumes summation over adjoint indices within each factor group for which M_α is invariant. Note also that our $\mathcal{A}^{ijk}$ are opposite in sign to those of Ref. [11.7].

ensure gauge invariance, explicit counterterms have to be introduced, cf. §6.8, to renormalize the superfields and their couplings. This has to be done in such a way that the 'spurious' divergences, induced in association with the spurion superfield(s) and mentioned in §7.7, get completely excluded. Our only recourse is to introduce such terms in the original tree level Lagrangian density and cancel them through renormalization. Thus we modify [11.7] the starting $\mathcal{L}_{\text{SOFT}}$ of (11.14b) to $\mathcal{L}'_{\text{SOFT}}$, where

$$
\mathcal{L}'_{\text{SOFT}} = \int d^2\theta\, \eta \left(\frac{1}{6}\hat{\mathcal{A}}^{ijk}\Phi'_i\Phi'_j\Phi'_k - \frac{1}{2}\hat{\mathcal{B}}^{ij}\Phi'_i\Phi'_j - \frac{1}{2}\sum_\alpha M_\alpha W'^{\alpha A}W^{\prime\alpha}_A \right) + h.c.
$$

$$
+ \int d^4\theta \left[-\bar{\eta}\eta\, \bar{\Phi}'^i(\hat{m}^2)_i{}^j (\exp 2\sum_\alpha g_\alpha V'^\alpha)_j{}^k \bar{\Phi}'_k \right.
$$

$$
\left. + \bar{\Phi}'^i(\eta\kappa_i^j + \bar{\eta}\kappa_i^{*j})(\exp 2\sum_\alpha g_\alpha V'^\alpha)_j^k \Phi'_k + \frac{1}{4}\rho^\alpha(\bar{\mathcal{D}}\bar{\mathcal{D}}\bar{\eta})(\mathcal{D}\mathcal{D}\eta)V'^{\alpha U(1)} \right].(11.15)
$$

Here we have introduced new couplings κ, ρ^α with mass dimensions 1 and 2 respectively and have written the corresponding interactions in a gauge invariant form. In (11.15) primes have been associated with chiral superfields $\bar{\Phi}'$, Φ' as well as gauge and gauge field strength superfields V'^α, $W'^{\alpha A}$ and carets have been put on $\hat{\mathcal{A}}^{ijk}$, $\hat{\mathcal{B}}^{ij}$ and $(\hat{m}^2)_i{}^j$ to distinguish them from the corresponding quantities appearing without carets in (11.14). Furthermore, $V'^{\alpha U(1)}$ is a gauge superfield only when α is a $U(1)$ factor and vanishes otherwise. With $W'^{\alpha U(1)}$ understood in the same way, the following identity is obeyed:

$$
\frac{1}{4}\rho^\alpha \int d^4\theta(\bar{\mathcal{D}}\bar{\mathcal{D}}\bar{\eta})(\mathcal{D}\mathcal{D}\eta)V'^{\alpha U(1)} = \rho^\alpha \int d^4\theta\, \eta\bar{\eta}\, \mathcal{D}^A W'^{\alpha U(1)}_A.
$$

Eq. (11.15) can be transformed into the conventional form (11.14b) by the following θ-dependent superfield redefinitions [11.7]:

$$
\Phi_i = \Phi'_i - \eta\kappa_i^j\,\Phi'_j\,,
$$

$$
V^{\alpha U(1)} = V'^{\alpha U(1)} - \bar{\eta}\bar{\eta}\,\rho^\alpha\,.
$$

The second of the above equations implies a redefinition only of the $U(1)$ components (and not of others) of the gauge superfield, corresponding to $U(1)$ factors in the gauge group, if any. Then the coupling strengths in $\mathcal{L}'_{\text{SOFT}}$ and those in $\mathcal{L}_{\text{SOFT}}$ get related as follows:

$$
\mathcal{A}^{ijk} = \hat{\mathcal{A}}^{ijk} - f^{ljk}\kappa_l^i - f^{ilk}\kappa_l^j - f^{ijl}\kappa_l^k\,, \tag{11.16a}
$$

$$
\mathcal{B}^{ij} = \hat{\mathcal{B}}^{ij} + \mu^{lj}\kappa_l^i + \mu^{il}\kappa_l^j\,, \tag{11.16b}
$$

$$
(m^2)_i{}^j = (\hat{m}^2)_i{}^j + \kappa_i^{*k}\kappa_k^j + 2g_\alpha(T^\alpha)_i{}^j\rho^\alpha\,. \tag{11.16c}
$$

The renormalization of interactions in $\mathcal{L}_{\text{SOFT}}$ of (11.14b) proceeds under two conditions. (1) All couplings in (11.15) are taken to be independent of one another. (2) The renormalized couplings κ_i^j are set equal to zero. Thus the bare couplings $\hat{\mathcal{A}}$, $\hat{\mathcal{B}}$, $\hat{m}^2$, κ in (11.15) get expressed in terms of the renormalized couplings $\mathcal{A}$, $\mathcal{B}$ and m^2 of (11.14b). Since the bare couplings $\mathcal{A}$, $\mathcal{B}$ and m^2 are related to the former set by (11.16), they can then be expressed in terms of their own renormalized couplings. The effect of ρ^α comes only through the third RHS term of (11.16c).

We turn next to the relevant divergent supergraphs. On account of the nonrenormalization theorem (cf. §6.7) and the fact (cf. §6.3) that any nontrivial superloop must involve both left and right chiral vertices, it is sufficient for our purpose to study the divergent part of only the chiral superfield propagator $\langle \bar{\Phi}\Phi \rangle$. In the exactly supersymmetric case, the divergent part of this two point function can be written in terms of renormalized chiral superfields Φ_i as $\int d^4\theta\, \bar{\Phi}^i T_i^j \Phi_j$. Here T is related to the wavefunction renormalization constant Z and the anomalous dimension γ of the superfield Φ by[4] (cf. 6.92)

$$T_i^j = Z_i^j - \delta_i^{\ j} = 2\gamma_i^{\ j}(4 - 2\omega)^{-1} \ , \tag{11.17}$$

$2\omega = d$ being the number of spacetime dimension (cf. Ch.6). The one loop expression for $\gamma_i^{\ j}$ is given in (6.100a). Now the nonrenormalization (cf. §6.7) of the superpotential interactions $(3!)^{-1} f^{ijk}\Phi_i\Phi_j\Phi_k$ and $\mu^{ij}\Phi_i\Phi_j$ ensures that the RGE of the coupling strengths f^{ijk} and μ^{ij} is given by the β-functions of (6.101b) and (6.101a) respectively, i.e.

$$\frac{d}{dt} f^{ijk} = \gamma_l^i f^{ljk} + \gamma_l^j f^{jlk} + \gamma_l^k f^{ijl} \ , \tag{11.18a}$$

$$\frac{d}{dt} \mu^{ij} = \gamma_l^i \mu^{lj} + \gamma_l^j \mu^{il} \ . \tag{11.18b}$$

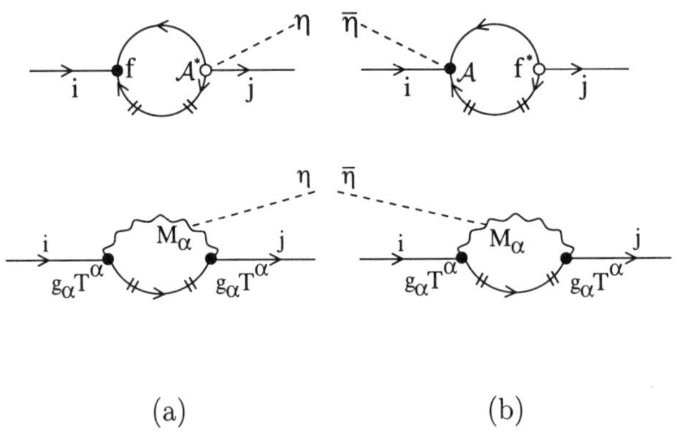

(a)   (b)

Fig. 11.2. One loop supergraphs generating (a) $T_i^{(1)j}$ and (b) $T_i^{\dagger(1)j}$.

When soft, explicit supersymmetry breaking is introduced via the spurion interactions of (11.15), the chiral superpropagator $\langle \bar{\Phi}\Phi \rangle$ can be expanded to include three and four point functions with one and two spurion superfields respectively. The inclusion of these 'vertex' functions, which revert to two point functions when the spurions are replaced by their VEVs,

---

[4] We have dropped the explicit reference to a one loop calculation in the form of the superscript (1) since (11.17) is more general.

enables one to write the total divergent part as [11.7] $\int d^4\theta\; \bar\Phi^i\, T_i^{(\eta)j}\Phi_j$ with $T_i^{(\eta)j}$ expanded in $\eta, \bar\eta$ as[5]

$$T_i^{(\eta)j} = T_i^j + \eta T_i^{(1)j} + \bar\eta T_i^{(1)\dagger j} + \eta\bar\eta\; T_i^{(2)j} \; . \tag{11.19}$$

The second and third RHS terms in (11.19) arise from new three legged supergraphs whose one loop incarnations are shown in Fig. 11.2. On the other hand, the last RHS term originates from new four legged supergraphs, shown in Fig. 11.3 to one loop. We can also

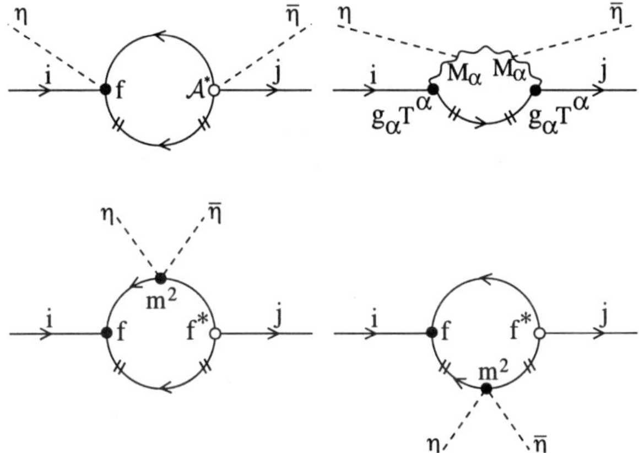

Fig. 11.3. One loop diagrams for $T_i^{(2)j}$.

introduce generalized anomalous dimensions $\gamma_i^{(1,2)j}$, in association with $T_i^{(1,2)j}$, and generalize (11.19) to be consistent with (11.17).

$$\gamma_i^{(\eta)j} = \gamma_i^j + \eta\gamma_i^{(1)j} + \bar\eta\gamma_i^{(1)\dagger j} + \eta\bar\eta\gamma_i^{(2)j} \; , \tag{11.20a}$$

$$T_i^{(1,2)j} = 2\gamma_i^{(1,2)j}(4 - 2\omega)^{-1} \; , \tag{11.20b}$$

From a comparison of the above with (11.15), we can infer the condition for the vanishing of the renormalized coupling κ_i^j, namely

$$\kappa_i^j(\text{bare}) = -T_i^{(1)j} \equiv -2\gamma_i^{(1)j}(4 - 2\omega)^{-1} \; , \tag{11.21}$$

to leading order in $1/(4 - 2\omega)$. On substituting (11.21) in (11.16a,b) and using the standard nonrenormalization argument for the couplings $\hat{A}^{ijk}\Phi_i\Phi_j\Phi_k\eta$ and $\hat{B}^{ij}\Phi_i\Phi_j\eta$, we can derive the RGE equations for $\mathcal{A}^{ijk}$ and $\mathcal{B}^{ij}$ in a manner similar to the way that (11.18) was obtained. They are:

$$\frac{d}{dt}\mathcal{A}^{ijk} = \gamma_l^i\mathcal{A}^{ljk} + \gamma_l^j\mathcal{A}^{ilk} + \gamma_l^k\mathcal{A}^{ijl} + 2(\gamma_l^{(1)i}f^{ljk} + \gamma_l^{(1)j}f^{ilk} + \gamma_l^{(1)k}f^{ijl}) \; , \tag{11.22a}$$

$$\frac{d}{dt}\mathcal{B}^{ij} = \gamma_l^i\mathcal{B}^{lj} + \gamma_l^j\mathcal{B}^{il} + 2(\gamma_l^{(1)i}\mu^{lj} + \gamma_l^{(1)j}\mu^{il}) \; . \tag{11.22b}$$

---

[5]The exclusion of terms with factors like $\mathcal{D}\eta$ and $\bar{\mathcal{D}}\bar\eta$ can be seen from power counting.

Turning to the scalar squared mass matrices $(m^2)_i{}^j$ in (11.16c), we note that there are three additive sources contributing to its t-evolution: (1) the regular anomalous dimensions γ_i^j from the standard wave function renormalization of the term $\bar{\Phi}^i(\hat{m}^2)_i^j\Phi_j$, (2) the counterterm containing $T_i^{(2)j}$ of (11.19) and (3) the contribution from the last term in the RHS of (11.16c) containing ρ^Y, Y being the hypercharge[6]. In fact, we have [11.7]

$$\frac{d}{dt}(m^2)_i{}^j = \gamma_i^{l\star}(m^2)_l{}^j + \gamma_l^j(m^2)_i{}^l + 2\gamma_i^{(2)j} + \frac{2g_Y^2}{16\pi^2}\delta_i^j\frac{Y}{2}\operatorname{Tr}\left(\frac{Y}{2}m^2\right), \tag{11.23}$$

The computation of the β-functions of the soft parameters of the MSSM has now been reduced to a calculation of the generalized anomalous dimensions $\gamma^{(\eta)}$ appearing in (11.22) and (11.23). The key observation here is the derivability of the matrices $\gamma_i^{(1)j}$ and $\gamma_i^{(2)j}$ from the exactly supersymmetric anomalous dimensions γ_i^j by simple algebraic manipulations. This becomes possible on account of the three following simplifications:

- At every step of the calculation, the spurion superfield η can be commuted with the covariant chiral derivative $\mathcal{D}_A$. This follows from our knowledge that $[\mathcal{D}_A, \eta] = (\mathcal{D}_A\eta)$ does not appear in the final result (11.19). The Grassmann algebra and momentum integrations are just the same as those without the η- or $\bar{\eta}$-insertions. The insertion of η and/or $\bar{\eta}$ at the relevant supergraph vertices then becomes a virtually trivial exercise since that can be made *after* the completion of all the θ-algebra and when all $\theta, \bar{\theta}$ integrations in the supergraph have been reduced to a single $\int d^4\theta$.

- One can account for $m^2\ \bar{\eta}\eta$ operator insertions on chiral superpropagators by utilizing the following form for the latter:

$$\int d^4x\ e^{ip\cdot x}\langle\Phi_i(x,\theta_1)\bar{\Phi}^j(x,\theta_2)\rangle = \delta^{(4)}(\theta_1 - \theta_2)\frac{i}{p^2}\left[\delta_i^j + (m^2)_i^j\ \bar{\eta}\eta\right];$$

  this has the same power dependence on the momentum p as the chiral superpropagator in the limit of exact sypersymmetry.

- Gaugino mass insertions on a vector superpropagator can also be done in an analogous way. One can find a gauge where the said superpropagator can be written in the presence of soft supersymmetry-breaking as its supersymmetric limit times a momentum independent factor that is quadratic in the gaugino mass M_α, viz. $1 + M_\alpha\eta + M_\alpha^\star\bar{\eta} + 2|M_\alpha|^2\bar{\eta}\eta$.

The consequence of the above three simplifications is that no new momentum integrals need to be performed when explicit soft supersymmetry breaking is introduced. Altogether, the supergraphs for $\gamma_i^{(\eta)j}$ obtain from those for γ_i^j by the following simple recipe [11.7]:

(1) Replace all $\Phi_i\Phi_j\Phi_k$ vertices f^{ijk} by $f^{ijk} + \eta\mathcal{A}^{ijk}$.

---

[6]Note that the second RHS term in (11.16c) is $O\left[(4 - 2\omega)^{-2}\right]$ and hence does not contribute to the β-functions [11.7].

(2) In gauge interaction vertices $\bar{\Phi}^i[(V^\alpha)^n]_i^j\Phi_j$, n being a positive integer (≤ 2 in the Wess-Zumino gauge), replace the coefficients $[(T^\alpha)^n]_i^j$ by $[(T^\alpha)^n]_i^k[\delta_k^j - (m^2)_k^j\bar{\eta}\eta]$.

(3) Replace the factors δ_i^j stemming from $\langle\Phi_i\bar{\Phi}^j\rangle$ propagators by $\delta_i^j + (m^2)_i^j\bar{\eta}\eta$.

(4) Multiply every $\langle V^\alpha V^\alpha\rangle$ propagator by $1 + M_\alpha\eta + M_\alpha^\star\bar{\eta} + 2|M_\alpha|^2\bar{\eta}\eta$.

(5) Multiply[7] vector superfield products $(V^\alpha)^n$, $n \geq 3$, by $1 - M_\alpha\eta - M_\alpha^\star\bar{\eta}$.

The first, second and fifth rules above are immediately evident by comparison of the softly broken supersymmetric Lagrangian density (11.14) with supersymmetric gauge theory vertices given in Ch.5. Rules (3) and (4) follow from the modified chiral and vector superpropagators given in the previous paragraph.

The application of the above recipe leads to the following substitution rules for the calculation of $\gamma^{(\eta)}$ from γ:

(A) $f^{ijk} \to f^{ijk} + \eta A^{ijk}$;

(B) $g_\alpha^2 \to g_\alpha^2(1 + M_\alpha\eta + M_\alpha^\star\bar{\eta} + 2|M_\alpha|^2\bar{\eta}\eta)$;

(C) insertion of the factor $\delta_i^l + \bar{\eta}\eta(m^2)_i^l$ between contracted indices i and l in $ff^\star$ products, i.e. $f^{ijk}f_{imn}^\star \to f^{ijk}f_{imn}^\star + \bar{\eta}\eta f^{ijk}(m^2)_i^l f_{lmn}^\star$;

(D) modification of Kronecker delta symbols coming from pure gauge interactions as per $\delta_i^{\ j} \to \delta_i^{\ j} - \bar{\eta}\eta(m^2)_i^j$.

(A) follows from rule (1) above, (C) from (3) and (D) from (2), while (B) follows from rules (4) and (5) on account of the identity

$$(1 + M_\alpha\eta + M_\alpha^\star\bar{\eta} + 2|M_\alpha|^2\bar{\eta}\eta)(1 - M_\alpha\eta - M_\alpha^\star\bar{\eta}) = 1 .$$

The actual computation of $\gamma_i^{(\eta)j}$ is now straightforward. The usual anomalous dimensions were calculated in Ch.6. From (6.100a), which was valid for a simple gauge group, we can generalize to the case of the group $G = \prod_\alpha G_\alpha$. Defining $C_\alpha(\Phi_i)$ by

$$C_\alpha(\Phi_i)\mathbf{1} = \sum_a (T^a T^a)_{\mathcal{R}_i} ,$$

with $\mathbf{1}$ as the unit matrix, $\mathcal{R}_i$ the representation of G that Φ_i belongs to and the summation over the generators in the factor group G_α, we can rewrite (6.100a) as

$$\gamma_i^j = \frac{1}{16\pi^2}\left[\frac{1}{2}f_{ikl}^\star f^{jkl} - 2\sum_\alpha g_\alpha^2 C_\alpha(\Phi_i)\delta_i^j\right] . \tag{11.24}$$

---

[7]This rule is relevant to the evaluation of the one loop diagrams of Figs. 11.2 and 11.3.

Note that, for the fundamental representation of $SU(N)$, $C_\alpha = (N^2 - 1)/(2N)$ and, for $U(1)_Y$, $C_\alpha = Y_{\Phi_i}^2/4$. Now the application of substitution rules (A) and (B) immediately leads to the result

$$\gamma_i^{(1)j} = \frac{1}{16\pi^2} \left[\frac{1}{2} f_{ikl}^* \mathcal{A}^{jkl} - 2 \sum_\alpha g_\alpha^2 C_\alpha(\Phi_i) M_\alpha \delta_i^j \right]. \tag{11.25}$$

The calculation of $\gamma_i^{(2)j}$ involves rules (C) and (D) also:

$$\gamma_i^{(2)j} = \frac{1}{16\pi^2} \left[f_{ikl}^* \left(m^2\right)_n^l f^{jkn} + \frac{1}{2} \mathcal{A}_{ikl}^* \mathcal{A}^{jkl} \right.$$
$$\left. - 2 \sum_\alpha g_\alpha^2 C_\alpha(\Phi_i) \left(2 \left| M_\alpha^2 \right| \delta_i^j - \left(m^2\right)_i^j \right) \right]. \tag{11.26}$$

For a general softly broken sypersymmetric gauge theory without any gauge singlet chiral superfield, (11.22)–(11.26) allow us to compute the one loop β-functions of couplings and masses in the superpotential as well as of scalar soft supersymmetry breaking operators. Before applying these expressions to the MSSM, let us make some remarks about further generalizations. First, in the presence of gauge singlet superfields, one has to allow linear terms in both the superpotential and $\mathcal{L}_{\text{SOFT}}$ of (11.14). Moreover, the divergences of "super–tadpoles" $\langle \Phi_i \rangle$ need to be investigated and these can be readily computed from the results for $\gamma^{(n)}$ described above [11.7]. Second, the true power of the superfield formalism becomes apparent when one goes to two (or higher) loops. The basic expressions (11.22)–(11.23) are in fact valid to all orders. However, if a scheme respecting supersymmetry such as $\overline{\text{DR}}$ (cf. Ch.6) is used to regularize infinities, new terms need to be added to $\mathcal{L}_{\text{SOFT}}$ of (11.14). Recall that in $\overline{\text{DR}}$ [11.8] *only* the indices of space time derivatives (or momenta) are continued to $d < 4$ dimensions with all four components of vector fields v_μ and also four Dirac gamma matrices retained. The fields v_ρ with ρ running between d and 4 then act as scalar fields in the adjoint representation of the gauge group. In general, soft supersymmetry breaking masses for these "ϵ-scalars", cf. §6.1, have to be added. Since a loop involving these scalars is suppressed by a factor $\epsilon \equiv (4 - 2\omega)/2$, they affect the β-functions only from the two loop level onwards. They also have some effect on finite one loop corrections. Fortunately a variant of the $\overline{\text{DR}}$ scheme has been found where these ϵ-scalars decouple completely from the physical degrees of freedom [11.9].

The use of the $\overline{\text{DR}}$ scheme allows the derivation of some results that are valid to all orders of perturbation theory. In particular, the β−functions for the gaugino masses, not treated so far, can be computed from the running of the gauge couplings via [11.10]

$$\frac{dM_\alpha}{dt} = \left(\sum_\beta M_\beta g_\beta^2 \frac{\partial}{\partial g_\beta^2} + \mathcal{A}^{ijk} \frac{\partial}{\partial f^{ijk}} \right) \left(g_\alpha^{-2} \frac{dg_\alpha^2}{dt} \right). \tag{11.27}$$

In (11.27) the factor $g_\alpha^{-2} dg_\alpha^2/dt$ should be considered a function of the gauge and Yukawa couplings of the theory. This result is quite useful, since the β-functions for the gauge couplings are known to three loops. To one loop order, dg_α^2/dt is proportional to g_α^4, vide (11.1). This implies that

$$\frac{d(M_\alpha/g_\alpha^2)}{dt} = 0 \qquad (11.28)$$

to one loop order, i.e., to this order, gaugino masses run the same way as squared gauge couplings do! This result is perhaps not so surprising. The one loop self energy diagrams, contributing to the running of gaugino masses, consist of a gaugino-gauge boson loop and antifermion-sfermion plus fermion-antisfermion loops. These are obtainable by supersymmetry transformations from the antifermion-fermion loop and gauge boson-gauge boson loop vacuum polarization diagrams which determine the running of gauge coupling strengths at one loop, i.e. by changing gauge bosons into gauginos, fermions into sfermions etc. However, starting at the two loop level, gaugino masses and squared gauge couplings run differently.

11.4 Application to the MSSM

We now specialize to the MSSM, as defined in Chs. 8 and 9. The results of the previous section allow an easy derivation of the RGE equations in the MSSM (or in any other softly broken supersymmetric theory). We shall omit the tedious algebra of counting coefficients and just list the one loop RGE equations for the supersymmetric couplings f^{ijk} and μ^{ij} as well as the soft supersymmetry breaking parameters $\mathcal{A}^{ijk}$, $\mathcal{B}^{ij}$, $(m^2)_i^j$ and M_a in the limit of negligible generation mixing of sfermions. One loop RGE equations for the completely general MSSM can be found in Ref. [11.11] along with extensions to the two loop level.

The β-functions for gauge couplings were already given in (11.9). These also determine the running of gaugino masses, cf. (11.27). The RGE equations for the parameters in the superpotential follow from (11.18) with the anomalous dimensions γ_i^j given by (11.24). We keep[8] only third generation Yukawa couplings $f_{t,b,\tau}$. Now the Yukawa coupling evolution equations can be given as

$$\frac{df_t}{dt} = \frac{f_t}{16\pi^2}\left(6f_t^2 + f_b^2 - \frac{16}{3}g_3^2 - 3g_2^2 - \frac{13}{9}g_Y^2\right), \qquad (11.29a)$$

$$\frac{df_b}{dt} = \frac{f_b}{16\pi^2}\left(6f_b^2 + f_t^2 + f_\tau^2 - \frac{16}{3}g_3^2 - 3g_2^2 - \frac{7}{9}g_Y^2\right), \qquad (11.29b)$$

$$\frac{df_\tau}{dt} = \frac{f_\tau}{16\pi^2}\left(3f_b^2 + 4f_\tau^2 - 3g_2^2 - 3g_Y^2\right), \qquad (11.29c)$$

The Higgsino mass parameter μ obeys the evolution equation

$$\frac{d\mu}{dt} = \frac{\mu}{16\pi^2}\left(3f_t^2 + 3f_b^2 + f_\tau^2 - 3g_2^2 - g_Y^2\right). \qquad (11.30)$$

The equations for the evolution of the corresponding trilinear and bilinear soft supersymmetry breaking parameters have a quite similar structure. Note that the soft $\mathcal{A}$-parameters are

---

[8] Since Yukawa couplings are proportional to fermion masses, those of the two lightest families are quite negligible in comparison with third generation ones.

also restricted in the same way as the Yukawa couplings, so that we can neglect them for the two lightest generations. Indeed, we define $A^{t,b,\tau} \equiv (A^{u,d,e})_{33}$, cf. (9.45a). From (11.22a) the third generation A-parameters are seen to obey[9]

$$\frac{dA^t}{dt} = \frac{1}{8\pi^2}\left(6f_t^2 A^t + f_b^2 A^b - \frac{16}{3}g_3^2 M_3 - 3g_2^2 M_2 - \frac{13}{9}g_Y^2 M_1\right), \tag{11.31a}$$

$$\frac{dA^b}{dt} = \frac{1}{8\pi^2}\left(6f_b^2 A^b + f_t^2 A^t + f_\tau^2 A^\tau - \frac{16}{3}g_3^2 M_3 - 3g_2^2 M_2 - \frac{7}{9}g_Y^2 M_1\right), \tag{11.31b}$$

$$\frac{dA^\tau}{dt} = \frac{1}{8\pi^2}\left(3f_b^2 A^b + 4f_\tau^2 A^\tau - 3g_2^2 M_2 - 3g_Y^2 M_1\right). \tag{11.31c}$$

There is only one bilinear soft supersymmetry breaking parameter in the MSSM (cf. Ch.8) and we define $B \equiv \mathcal{B}/\mu$. The evolution equation for the B-parameter follows from (11.22b) and is

$$\frac{dB}{dt} = \frac{1}{8\pi^2}\left(-3f_t^2 A^t - 3f_b^2 A^b - f_\tau^2 A^\tau + 3g_2^2 M_2 + g_Y^2 M_1\right). \tag{11.32}$$

Next, we come to scalar masses. These are conveniently expressed in terms of the quantities $S_{t,b,\tau}$ and S_Y, defined as linear combinations of sfermion and Higgs squared masses and the modulus squared of the A-parameters as follows:

$$S_t \equiv m_2^2 + m_{\tilde{Q}_3}^2 + m_{\tilde{t}_R}^2 + |A^t|^2, \tag{11.33a}$$

$$S_b \equiv m_1^2 + m_{\tilde{Q}_3}^2 + m_{\tilde{b}_R}^2 + |A^b|^2, \tag{11.33b}$$

$$S_\tau \equiv m_1^2 + m_{\tilde{\ell}_3}^2 + m_{\tilde{\tau}_R}^2 + |A^\tau|^2 \tag{11.33c}$$

and

$$S_Y \equiv \frac{1}{2}\sum_i Y_i m_i^2, \tag{11.34}$$

where Y_i is the hypercharge of the scalar i. In (11.34) the summmation $\sum_Y$ runs over all scalar fields of the MSSM. The RGE equations for the scalar soft supersymmetry breaking Higgs boson masses can now be written as

$$\frac{dm_1^2}{dt} = \frac{1}{8\pi^2}\left(3f_b^2 S_b + f_\tau^2 S_\tau - 3g_2^2\left|M_2^2\right| - g_Y^2\left|M_1^2\right| - \frac{1}{2}g_Y^2 S_Y\right); \tag{11.35a}$$

$$\frac{dm_2^2}{dt} = \frac{1}{8\pi^2}\left(3f_t^2 S_t - 3g_2^2\left|M_2^2\right| - g_Y^2\left|M_1^2\right| + \frac{1}{2}g_Y^2 S_Y\right). \tag{11.35b}$$

The corresponding evolution equations for soft breaking contributions to third generation squark masses read:

$$\frac{dm_{\tilde{Q}_3}^2}{dt} = \frac{1}{8\pi^2}\left(f_t^2 S_t + f_b^2 S_b - \frac{16}{3}g_3^2\left|M_3^2\right| - 3g_2^2\left|M_2^2\right| - \frac{1}{9}g_Y^2\left|M_1^2\right| + \frac{1}{6}g_Y^2 S_Y\right), \tag{11.36a}$$

---

[9]**Caution:** our A-parameters are negatives of those defined by Kazakov [11.11].

$$\frac{dm_{\tilde{t}_R}^2}{dt} = \frac{1}{8\pi^2}\left(2f_t^2 S_t - \frac{16}{3}g_3^2\left|M_3^2\right| - \frac{16}{9}g_Y^2\left|M_1^2\right| - \frac{2}{3}g_Y^2 S_Y\right), \tag{11.36b}$$

$$\frac{dm_{\tilde{b}_R}^2}{dt} = \frac{1}{8\pi^2}\left(2f_b^2 S_b - \frac{16}{3}g_3^2\left|M_3^2\right| - \frac{4}{9}g_Y^2\left|M_1^2\right| + \frac{1}{3}g_Y^2 S_Y\right). \tag{11.36c}$$

Finally, the evolution equations for the soft supersymmetry breaking contributions to third generation slepton masses are

$$\frac{dm_{\tilde{\ell}_3}^2}{dt} = \frac{1}{8\pi^2}\left(f_\tau^2 S_\tau - 3g_2^2\left|M_2^2\right| - g_Y^2\left|M_1^2\right| - \frac{1}{2}g_Y^2 S_Y\right), \tag{11.37a}$$

$$\frac{dm_{\tilde{\tau}_R}^2}{dt} = \frac{1}{8\pi^2}\left(2f_\tau^2 S_\tau - 4g_Y^2\left|M_1^2\right| + g_Y^2 S_Y\right). \tag{11.37b}$$

Now consider situations where sfermionic generation mixing is negligible. Then RGE equations for sfermions of the first and second generation can be obtained to a good approximation simply by the omission of all terms in (11.36) and (11.37) which contain Yukawa couplings. The RGE equations for the first and second generation A-parameters are

$$\frac{dA^{u,c}}{dt} = \frac{1}{8\pi^2}\left(3f_t^2 A^t - \frac{16}{3}g_3^2 M_3 - 3g_2^2 M_2 - \frac{13}{9}g_Y^2 M_1\right),$$

$$\frac{dA^{d,s}}{dt} = \frac{1}{8\pi^2}\left(3f_b^2 A^b + f_\tau^2 A^\tau - \frac{16}{3}g_3^2 M_3 - 3g_2^2 M_2 - \frac{7}{9}g_Y^2 M_1\right),$$

$$\frac{dA^{e,\mu}}{dt} = \frac{1}{8\pi^2}\left(3f_b^2 A^b + f_\tau^2 A^\tau - 3g_2^2 M_2 - 3g_Y^2 M_1\right).$$

We observe that every superpotential parameter renormalizes purely multiplicatively. This is a consequence of the nonrenormalization theorem (§6.7). Of course, in the presence of generation mixing, the former statement is true only in the sense of matrix multiplication. To the given one loop order, this is true also of gaugino masses[10] on account of chirality. RGE equations for all other soft breaking parameters are inhomogeneous already at the one loop level. In particular, nonvanishing A- and B-parameters as well as scalar soft supersymmetry breaking masses can all be generated radiatively from gaugino masses. It is not possible to give analytical solutions to these coupled RGE equations in general. Nevertheless, their general structure allows several conclusions of immediate phenomenological interest to be drawn. We discuss these in turn in three separate subsections that follow.

Infrared (Quasi-)Fixed Points and Upper Bounds on f_t

The advertised features arise from the β-functions of the Yukawa couplings, cf. (6.101b), (11.24) and (11.29) . The positive signs of the Yukawa coupling terms in the latter are particularly noteworthy. They make the Yukawa couplings develop Landau poles at some high energy scale if the contributions of the gauge couplings to the RGE equations are

---

[10]At the two loop level, chirality breaking products of A-parameters and Yukawa couplings also appear in the RGE equations for the gaugino masses, vide (11.27).

sufficiently small. However, we know that even the top Yukawa coupling is not much larger than the strong QCD gauge coupling[11]. Now, it is known that gauge couplings contribute with negative sign to the running of the Yukawa couplings. Therefore, they can either prevent the occurrence of such Landau poles or can at least push these beyond the grand unifying or Planck scale where the MSSM is most likely invalid. Eq.(11.29a) can be solved analytically [11.12] if only terms proportional to f_t^2 or g_3^2 are retained. The solution is :

$$f_t^2(\mu) = 7g_3^2(\mu) \left[18 - \left\{g_3^2(\mu_0)/g_3^2(\mu)\right\}^{7/9} \left\{18 - 7g_3^2(\mu_0)/f_t^2(\mu_0)\right\}\right]^{-1} , \qquad (11.38)$$

where $g_3^2(\mu)$ is given by (11.10) with $\beta_3 = -3$. First, note that the ansatz $f_t^2 = 7g_3^2/18$ is scale invariant; the ratio f_t^2/g_3^2 is then said to be at a fixed point. Second, this yields $f_t(m_t) \simeq 0.73$. For the top mass, one can write (cf. Ch.8)

$$m_t(m_t) = f_t(m_t) \, \sin \beta \, (174 \text{ GeV}) . \qquad (11.39)$$

The experimental lower bound $m_t(m_t) > 160$ GeV, corresponding to a pole mass of 168 GeV, substituted in (11.39), implies $f_t(m_t) > 0.92$. The fixed point solution for f_t/g_3 is therefore unphysical.

However, (11.38) also has a "quasi-fixed point" in the infrared region (i.e., for "low" scales μ). Suppose that at some high energy scale μ_0 one has $f_t^2(\mu_0) \gg g_3^2(\mu_0)$. The last term in the denominator of (11.38) is then negligible for scales $\mu \ll \mu_0$. Hence one becomes faced with the remarkable consequence that $f_t(\mu \ll \mu_0)$ is nearly *independent* of the "input" value $f_t(\mu_0)$. In this situation, f_t is said to have reached a "quasi-fixed point". However, it still does depend on the scale μ_0 where f_t is assumed to be large. We have seen earlier in this chapter that the MSSM gauge couplings appear to unify at the scale $M_U \simeq 2 \times 10^{16}$ GeV. If M_U is identified with the scale μ_0, one obtains $f_t(m_t) = 1.09$, including the contributions from electroweak gauge couplings as well as two loop effects [11.13]. In the same limit, several soft breaking parameters, e.g. A_t and some combinations of third generation squark and Higgs boson masses, also have infrared quasi-fixed points [11.13], i.e. they are not sensitive to their values at scale μ_0. However, the large "quasi-fixed" value of $f_t(m_t)$ implies a rather small value of $\tan \beta$, vide (11.39). For $m_t(m_t) = 166$ GeV, say, this implies that $\tan \beta = 1.8$. As discussed in detail in Ch.10, such a small value of $\tan\beta$ would correspond to a rather light neutral Higgs boson h. One then needs large one loop corrections to m_h in order to satisfy present LEP search limits. This requires large stop masses, and/or substantial $\tilde{t}_L$-$\tilde{t}_R$ mixing. Also, significant supersymmetric corrections are now induced from stop-gluino loops to the relation between the running and pole masses of the top quark; these tend to decrease the required $m_t(m_t)$ for a fixed pole mass[12]. For $m_{\tilde{t}}$ near 1 TeV, one typically finds $m_t(m_t) \simeq 160$ GeV; $f_t(m_t) = 1.09$ then implies $\tan\beta \simeq 1.6$, which in turn means $m_h \le 100$ GeV even for optimal $\tilde{t}_L$-$\tilde{t}_R$ mixing. This bound is now on the point of being excluded by Higgs searches

---

[11]The only way that f_t could exceed g_3 significantly is by $\tan \beta$ being less than unity. Higgs searches at the LEP accelerator then require $\tan \beta < 0.6$ in which case f_t would have a Landau pole, i.e. the theory would become nonperturbative just above the weak scale.

[12]Note that these corrections do not decouple for large sparticle masses; the resulting term, proportional to $\ln(M_s/M_t)$, M_s being a generic sparticle mass, tells us that we should have switched to the RGE of the SM at the scale M_s.

at LEP. We can, nonetheless, derive an important conclusion from this analysis. Eq.(11.38) shows that $f_t(m_t)$ is maximized when $f_t(\mu_0)$ is large. The implication is that, if f_t is required to remain finite up to the scale μ_0, the quasi-fixed point value of $f_t(m_t)$ is in fact its upper bound. The MSSM thus predicts that $m_t(m_t) \leq 190$ GeV, or $m_t^{\text{pole}} \leq 200$ GeV, if its (perturbative) validity extends all the way up to the grand unifying scale. The fact that the experimentally measured top quark mass lies well below this upper bound can therefore be regarded as comforting for the MSSM.

Sfermion to Gaugino Mass Ratios

Significant one loop contributions to the running of first and second generation sfermion masses come only from gauge interactions. We want to return to the quantity S_Y, defined in (11.34). S_Y renormalizes purely multiplicatively and therefore evolves homogeneously:

$$\frac{dS_Y}{dt} = \frac{g_Y^2}{8\pi^2} \sum_i \left(\frac{Y_i}{2}\right)^2 S_Y \ . \tag{11.40}$$

Eq. (11.40) has the same form as the RGE equation for the $U(1)_Y$ gaugino mass. This equation is solved by

$$S_Y(\mu) = S_Y(\mu_0) g_Y^2(\mu)/g_Y^2(\mu_0) \ .$$

The RGE equations for the first and second generation sfermion masses can therefore be written as

$$\frac{dm_i^2}{dt} = -\frac{1}{2\pi^2} \sum_{\alpha=1}^{3} C_{\alpha,i} \left|M_\alpha^2(t_0)\right| \frac{g_\alpha^6(t)}{g_\alpha^4(t_0)} + \frac{Y_i}{16\pi^2} S_Y(t_0) \frac{g_Y^4(t)}{g_Y^2(t_0)} \ , \tag{11.41}$$

Here $t_0 = 0$ and we have made use of the solution to the RGE equation (11.28) for the gaugino masses:

$$M_\alpha(t) = M_\alpha(t_0) g_\alpha^2(t)/g_\alpha^2(t_0) \ . \tag{11.42}$$

Moreover, $C_{\alpha i} \equiv C_\alpha(\Phi_i)$ are the quadratic Casimir coefficients already introduced in (11.24). The RGE equation (11.41) is easily solved, in terms of running coupling constants, by using (11.10). The result is

$$m_i^2(\mu) = m_i^2(\mu_0) - 2 \sum_{\alpha=1}^{3} \frac{C_{\alpha,i}}{b_\alpha^{(1)}} \left|M_\alpha^2(\mu_0)\right| \left[\frac{g_\alpha^4(\mu)}{g_\alpha^4(\mu_0)} - 1\right] + \frac{Y_i}{22} S_Y(\mu_0) \left[\frac{g_Y^2(\mu)}{g_Y^2(\mu_0)} - 1\right] . \tag{11.43}$$

In (11.43) $b_\alpha^{(1)} = -3, 1$ and 11 for the $SU(3)_C$, $SU(2)_L$ and $U(1)_Y$ factors of the MSSM gauge group respectively, cf. (11.6–11.8). Notice that the second RHS term in (11.43) is always negative for $\mu > \mu_0$. This is since $g_\alpha^2(\mu)$ is greater or less than $g_\alpha^2(\mu_0)$ for positive or negative β_α respectively. The third term, proportional to $S_Y(\mu_0)$, can have either sign, but is usually much smaller in magnitude than the second term. In fact, in the most widely studied models of supersymmetry breaking, to be discussed in more detail in the next two chapters, this third term vanishes identically. Therefore, we can neglect this term and will do so in most of the subsequent discussions.

The main significance of (11.43) is that it allows the derivation of lower bounds on sfermion to gaugino mass ratios at the weak scale. In case the gaugino masses at some scale

μ_0 are much larger than the sfermion masses at the same scale, the squared sfermion masses will become negative at some higher scale[13] $\mu > \mu_0$. This is undesirable, since scalar fields with negative squared masses develop nonvanishing VEVs that break $U(1)_{em}$ and/or $SU(3)_C$, in complete analogy with the Higgs mechanism discussed in Ch.10. There we showed that the proper scale to discuss the Higgs potential is approximately set by the mass of the heaviest particle that contributes to the one loop corrections to that potential. In the case at hand, large VEVs of some slepton and/or squark fields will generate gauge boson masses of the order of these VEVs. In most cases it is possible to find field configurations (involving at least two nonzero VEVs) such that the D-term contributions to the scalar potential vanish identically. The relevant quartic scalar coupling that stabilizes the potential at large values of the fields is then a Yukawa coupling f_i, giving a vev $\langle \phi_i \rangle \sim \sqrt{-m_i^2/f_i}$. For the first and second generation sfermions, $1/f_i \gg 1$. Since the heaviest particles will have masses of the order of the VEVs themselves, the proper condition for avoiding the existence of these unwanted minima is [11.14] $m_i^2(\langle \phi_i \rangle) > 0$. Note that these minima of the scalar potential are usually far deeper than the minimum discussed in Ch.10, where only neutral Higgs fields develop nonvanishing VEVs.

For first generation sfermions ϕ_i, with $|m_i|$ roughly of the order of the weak scale, a negative m_i^2 would induce VEVs of order 10^8 GeV since $f_i(\langle \phi_i \rangle) \simeq 3 \cdot 10^{-6}$. Thus, a necessary condition for avoiding minima of the potential that break charge and/or color is that all first generation squared sfermion masses remain nonnegative for scales up to about 10^8 GeV. For $\mu_0 \simeq 300$ GeV, (11.43) then yields numerically the expression

$$
\begin{aligned}
m_i^2(10^8 \text{ GeV}) &= m_i^2(\mu_0) - 0.13 \left(\frac{Y_i}{2} \right)^2 \left| M_1^2(\mu_0) \right| - 0.22 \frac{C_{2,i}}{3/4} \left| M_2^2(\mu_0) \right| \\
&\quad - 0.55 \frac{C_{3,i}}{4/3} \left| M_3^2(\mu_0) \right| + 0.027 \frac{Y_i}{2} S_Y(\mu_0) \, .
\end{aligned}
\tag{11.44}
$$

Recall that $C_{2,i} = 3/4$ if ϕ_i is an $SU(2)_L$ doublet, while $C_{3,i} = 4/3$ for $SU(3)_C$ triplets. In contrast, these coefficients vanish if ϕ_i is a singlet under the respective gauge group. Again, the coefficient in front of S_Y in the RHS of (11.44) is small. If we ignore this small contribution, the requirement $m_i^2(10^8 \text{ GeV}) \geq 0$ then implies that

$$
\begin{aligned}
m_{\tilde{e}_R} &\geq 0.36|M_1| \, , &\quad &(11.45a) \\
m_{\tilde{e}_L} &\geq 0.47|M_2| \, , &\quad &(11.45b) \\
m_{\tilde{q}} &\geq 0.74|M_3| \, . &\quad &(11.45c)
\end{aligned}
$$

In (11.45) all masses have been taken at the scale $\mu_0 = 300$ GeV. In particular, first generation squarks cannot be much lighter than gluinos[14] [11.15]. Conditions (11.45a,b) are somewhat less restrictive. They still allow substantial mass differences between sleptons and electroweak gauginos. Moreover, the masses of physical neutralino and chargino states can be further reduced due to mixing between gaugino and higgsino current states. These conditions,

---

[13]Unlike in the discussion of quasi-fixed points, we are now choosing the "input scale" μ_0 to be the weak scale and the variable scale μ to be well above that.

[14]The physical gluino mass differs from $|M_3(M_3)|$ by finite threshold corrections [11.16]; however, for $m_{\tilde{q}} \lesssim M_{\tilde{g}}$, these only amount to 10% or so.

nonetheless, significantly constrain the allowed MSSM parameter space at the weak scale, as outlined above. The assumptions, which have gone into making this statement, are that the MSSM remain valid up to a scale $\mu \simeq 10^8$ GeV, and that we are living in the absolute minimum of the scalar potential.

Radiative Electroweak Symmetry Breaking

It has been shown in the previous chapter that the Higgs sector of the MSSM can only break electroweak symmetry if the two Higgs doublets get *different* soft supersymmetry–breaking contributions to their squared masses. Otherwise the Higgs potential has its minimum either at the origin in which case the $SU(2)_L \times U(1)_Y$ symmetry remains unbroken, or it is at infinity i.e. the potential is unbounded from below. The RGE equations (11.35a,b) tell us that such a difference can be created *dynamically*. Even if $m_1^2 = m_2^2$ at some (high) scale, the Yukawa contributions to the RGE will drive these masses apart at lower scales. Since we know that the top Yukawa coupling strength f_t is of the order of one at the weak scale, this effect will in general be quite large.

Note that the Yukawa couplings appear with the positive sign here. Their contributions therefore tend to decrease the Higgs squared masses as one goes down in energy scale. Unless $\tan\beta$ is really very large, one has an inequality between top and bottom Yukawa coupling strengths, namely $f_t > f_b$. In this case, m_2^2 is reduced more than m_1^2; it might even be driven to a negative value though that is not necessary for electroweak symmetry breaking to occur if the Higgs mixing term $|B\mu|$ is sizable. Models, where $m_1^2 = m_2^2$ at some high scale, arc therefore said to exhibit[15] radiative symmetry breaking [11.17]. This then is a **dynamical explanation** for the negative determinant of the squared Higgs mass matrix (for vanishing VEVs). In the SM, a negative Higgs mass squared has to be introduced "by hand", so that this is a definite theoretical improvement. Finally, the Yukawa terms in the RGE equations (11.36) for squark masses have smaller coefficients than those in the same equations for Higgs masses. Moreover, the large contributions, proportional to $g_3^2 |M_3^2|$ tend to increase squark masses as one reduces the energy scale. Let us assume that the squark masses are positive at the scale where $m_1^2 = m_2^2$. Then they do remain positive at lower energy scales and color as well as charge remains conserved. Strictly speaking, the requirement $m_{\tilde{q}}^2 > 0$ is only a necessary condition for color to remain unbroken. In some cases one can find minima of the scalar potential where some squarks develop VEVs even if their squared masses are positive. Sufficient conditions for the absence of charge and color breaking minima have been discussed in the literature [11.18].

An interesting point is that $m_2^2 < m_1^2$ at the weak scale implies $\langle H_2^0 \rangle > \langle H_1^0 \rangle$ and vice versa. Assume that $m_1^2 = m_2^2$ at some high scale below which the MSSM is valid. Radiative electroweak symmetry breaking is then possible only if $\tan \beta \leq [m_t(m_t)S_t^{1/2}]/[m_b(m_t)S_b^{1/2}] \simeq 60(S_t/S_b)^{1/2}$. This can be demonstrated via the *reductio ad absurdum* procedure. Assume that this bound was violated, i.e. $f_b(m_t)S_b^{1/2} > f_t(M_t)S_t^{1/2}$. If $m_1^2 = m_2^2$ at some scale, we then find $m_1^2 < m_2^2$ at the weak scale. This in turn would imply $\langle H_1^0 \rangle > \langle H_2^0 \rangle$, i.e. $\tan \beta < 1$, which contradicts our original assumption since S_t/S_b is of order unity. In fact, most models of broken supersymmetry actually have $S_t \simeq S_b$, so that radiative EW symmetry breaking

---

[15] A more elaborate discussion of this phenomenon will be given in Ch.12, cf. Fig. 12.1, where the high scale boundary conditions are treated in detail.

in these models requires $\tan\beta \leq 60$.

Let us close this discussion with a general observation. Within a particular model of supersymmetry breaking, the study of RGE in the MSSM in general and the requirement of radiative symmetry breaking, in particular, lead to further predictions on (relations between) MSSM parameters at the weak scale. However, these depend on model assumptions about physics at some very high energy scale where supersymmetry breaking is transmitted to the visible sector. The currently most promising models of supersymmetry breaking will be described in the following two chapters, where we can be more specific.

References

[11.1] K. Inoue, A. Kakuto, H. Komatsu and H. Takeshita, Prog. Theor. Phys. **67** (1982) 1989; *ibid.* **68** (1982) 927.

[11.2] J. Erler and P. Langacker, in *Reviews of Particle Physics*, Particle Data Group, *loc. cit.*, *Bibl.*

[11.3] G.G. Ross, *op. cit.*, *Bibl.* R.N. Mohapatra, *op. cit.*, *Bibl.*

[11.4] P. Langacker and N. Polonsky, Phys. Rev. **D52** (1995) 3081. D.M. Pierce in *Proc. 1997 SLAC Summer Inst.* (eds. A. Breaux et al., SLAC, 1998).

[11.5] M.E. Peskin and D.V. Schroeder, *op. cit.*, *Bibl.*

[11.6] L. Girardello and M. Grisaru, Nucl. Phys. **B174** (1980) 193.

[11.7] Y. Yamada, Phys. Rev. **D50** (1994) 3537.

[11.8] I. Jack, D.R.T. Jones and K.L. Roberts, Z. Phys. **C63** (1994) 151.

[11.9] I. Jack, D.R.T. Jones, S.P. Martin, M.T. Vaughn and Y. Yamada, Phys. Rev. **D50** (1994) 5481.

[11.10] I. Jack and D.R.T. Jones, Phys. Lett. B415 (1997) 383.

[11.11] S.P. Martin and M.T. Vaughn, Phys. Rev. **D50** (1994) 2282. D.I. Kazakhov, *loc. cit.*, *Bibl.*

[11.12] M. Lanzagorta and G.G. Ross, Phys. Lett. **B349** (1995) 319; *ibid.* **B364** (1995) 163.

[11.13] S.A. Abel and B.C. Allanach, Phys. Lett. **B415** (1997) 371; *ibid.* **B431** (1998) 339.

[11.14] G. Gamberini, G. Ridolfi and F. Zwirner, Nucl. Phys. **B331** (1990) 331.

[11.15] U. Ellwanger, Phys. Lett. **B141** (1984) 435.

[11.16] S.P. Martin and M.T. Vaughn, Phys. Lett. **B318** (1993) 331.

[11.17] L.E. Ibáñez and G.G. Ross in *Perspectives on Higgs Physics*, G.L. Kane (ed.), *op. cit.*, *Bibl*, p229 and references therein.

[11.18] J.A. Casas, A. Lleyda and C. Munoz, Nucl. Phys. **B471** (1996) 3.

Chapter 12

GRAVITY MEDIATED SUPERSYMMETRY BREAKING

12.1 General Remarks

It has been demonstrated in §9.1 that MSSM fields alone cannot break supersymmetry spontaneously at the weak scale. Certain soft supersymmetry breaking operators involving some of those fields are explicitly added to the Lagrangian density to make sparticles much heavier than particles. The coefficients of these operators are treated as unknown parameters in the MSSM. Arguments have also been given at the end of Ch.11 that this description, while appropriate for many phenomenological applications, is unlikely to be complete. One would prefer a mechanism of spontaneous, rather than explicit, supersymmetry breaking – much like the Higgs mechanism which is used to implement the breakdown of electroweak symmetry. This analogy is not perfect, however. The Higgs mechanism offers the only known way to break electroweak symmetry, while keeping the theory unitary and renormalizable. In contrast, models with explicit and soft supersymmetry breaking make perfectly consistent quantum field theories satisfying the criterion of naturalness, cf. Ch.1. Our preference, nonetheless, is for a theory that actually explains supersymmetry breaking instead of merely parameterizing it. Such a theory should also lead to more predictions on the values of the numerous soft supersymmetry breaking parameters of the MSSM and to relations among them.

As argued earlier, by far the easiest way to satisfy the sum rule (7.49), while accommodating a spontaneous supersymmetry breakdown and without violating experimental constraints, is to follow the procedure mentioned in §9.1. Supersymmetry can be broken spontaneously (cf. Ch.7) in a new sector of the theory, taken to consist entirely of SM gauge singlet superfields. One can then assume that this sector couples only very weakly to MSSM superfields. Because of the very weak coupling, details of this "hidden" or "secluded" sector are of little phenomenological interest, though they may have ramifications for Big Bang cosmology (cf. Ch.16). However, it is this very weak coupling which transmits the supersymmetry breaking to the observable sector of MSSM fields and interactions. In this chapter and the next, we shall illustrate two broad ways how supersymmetry, once broken spontaneously in the hidden sector, might be communicated to the visible sector.

Unlike the details of the spontaneous breaking of supersymmetry in the hidden sector, the mechanism of its transmission from the latter to the MSSM fields does have an immediate impact on the observable sparticle spectrum and hence on supersymmetry phenomenology. The most economical mechanism of this kind uses [12.1] gravitational strength interactions. Models of this type, along with some of their problems, will be discussed in this chapter. They are based on *local* supersymmetry a.k.a. **supergravity**. Their chief advantage is the automatic presence in a locally supersymmetric Lagrangian of terms that can mediate supersymmetry breaking. Their main disadvantage is the necessity to appeal to Planck scale physics, which is still poorly understood. In particular, the requirement of a sufficient suppression of sparticle induced FCNC amplitudes (cf. §9.5) imposes severe constraints on this theory. These are constraints which cannot easily be given a dynamical explanation within this framework. One needs to appeal to (1) extra (e.g. generation) symmetries or (2) to the absence (for some reason) of tree level interactions between the hidden and observable sectors and the loop induced transmission of supersymmetry breaking from one to the other, say via the superconformal anomaly. We shall discuss all these questions in this chapter. In the next chapter we shall consider a second scenario in which supersymmetry breaking is mediated from the hidden sector to MSSM superfields through gauge interactions so that scalars with the same gauge quantum numbers *automatically* get the same soft supersymmetry breaking mass. As seen in Ch.9, this particular feature guarantees the absence of sparticle induced FCNC amplitudes. This is not to suggest that such models are free of problems. They are not and some of these problems are quite serious. All those issues will be addressed in Ch.13. Our emphasis in both Chs. 12 and 13 will be on the simplest, and hence the most predictive, potentially realistic models. However, some extensions and generalizations will also be mentioned.

Let us now focus on the transmission of supersymmetry breaking from the hidden to the observable sector via gravitational strength interactions. The main advantage of this approach has already been mentioned. The required interactions are a necessary ingredient of any Lagrangian density yielding an action that is invariant under *local* supersymmetry transformations. A sketch of the origin of this Lagrangian is given in the introduction to $N=1$ supergravity included in the *Annex* to this chapter. In the next section §12.2 we give a pedestrian treatment of a theory of spontaneously broken $N=1$ supergravity with matter and gauge couplings by listing the pieces in its Lagrangian density that are required for an understanding of the (transmission of) supersymmetry breaking. Then, in the following section §12.3, we consider the so-called minimal supergravity model (mSUGRA [12.1]), where an imposed global symmetry greatly reduces the number of independent soft supersymmetry breaking parameters. We discuss the phenomenological implications of mSUGRA in the succeeding section §12.4. Then, in section §12.5, we discuss some extensions of this minimal model; these include Grand Unified Theories (GUTs) and also attempts to find a common origin of the large intergeneration hierarchy observed in quark and lepton masses on one hand and of the suppression of sparticle mediated FCNC amplitudes on the other. Finally, in section §12.6, we briefly discuss the effects of quantum corrections to the supergravity interactions between the hidden and observable sector superfields as well as the role of possible extra dimensions in this context. Models like those with Anomaly Mediated Supersymmetry Breaking (AMSB) are also dealt with here. The *Annex* §12.7 contains a brief presentation of the essential features of the $N=1$ supergravity theory, as well as matter and gauge couplings therein, that are relevant to our considerations.

12.2 $N=1$ Supergravity Broken in the Hidden Sector

We have, so far, dealt with global $N=1$ supersymmetry (cf. §3.1) defined by an invariance under coordinate independent supersymmetry transformations. Significant changes are generally expected when a global symmetry is elevated to a local one in which the transformations are coordinate dependent. When a global bosonic symmetry – characterized by current density operators J_μ^a – gets elevated to a local symmetry, new spin one gauge boson fields A_μ^a need to be introduced with predetermined interactions. Since the supercurrent density K^μ, discussed in §5.1 and §7.3, is fermionic in character, the elevation of supersymmetry from a global to a local version should require a new (gauge) fermion.

We take (3.9b), i.e. the relation

$$[Q_a, \bar{Q}_b]_+ = 2(\gamma^\mu)_{ab} P_\mu$$

as the starting point of our discussion. This means that an invariance under local supersymmetry transformations necessarily implies an invariance under local coordinate shifts, generated by the four momentum operator P_μ. Eqs. (3.1) then imply local invariance under the full Poincaré algebra. That, of course, is the underlying principle of General Relativity on which Einstein's theory of gravity is based. In other words, the Hilbert action [12.2] of this theory is subsumed in what can be derived from the postulate of local supersymmetry, though – of course – without any information on the magnitude of Newton's constant G_N. It is for this reason that local supersymmetry is usually called **supergravity** and we shall be concerned only with $N=1$, cf. §3.1, supergravity here. Supergravity theories with higher values of N, though theoretically researched [12.3] with vigor, have not had much of a phenomenological application.

One can construct [12.1] the $N=1$ supergravity Lagrangian starting from the corresponding globally supersymmetric one. The application of a local infinitesimal supersymmetry transformation on the latter, described by a parametric local fermionic (four component spinor) function $\epsilon(x)$, generates terms proportional to $K^\mu \partial_\mu \epsilon$. The cancellation of these terms requires the introduction of a new (four component) real fermionic field Ψ_μ with[1] the infinitesimal supersymmetry transformation

$$\delta \Psi_\mu = 2 M_{Pl} \partial_\mu \epsilon + \cdots \ . \tag{12.1}$$

The addition to the Lagrangian density of a term proportional to $\Psi_\mu K^\mu$ with a suitable coefficient can now effect the cancellation. The dimensions match in (12.1) since $\epsilon(x)$, a four spinor Grassmann function of spacetime coordinates (cf. Ch.2), has the mass dimension of $-1/2$ while Ψ_μ, like any other fermionic field, has the mass dimension $3/2$. Now (12.1), with M_{Pl} interpreted as the inverse of a coupling strength, shows a complete analogy with

---

[1] The factor of 2 has been put in the RHS of (12.1) in order to make this $\epsilon(x)$ a spacetime dependent four spinor generalization of the ϵ defined in (4.25a). See also (12.86) in the *Annex* to this chapter.

the transformation property of vector bosons of bosonic gauge symmetries. The field Ψ_μ, apart from being fermionic, also carries the spacetime index μ. It must therefore represent a fermion with a spin of (at least) 3/2. In fact, it is the superpartner of the spin two graviton, cf. Table 3.1, i.e. the **gravitino**: the 'gauge fermion' of local supersymmetry. The coefficient in the RHS of (12.1) has been chosen to be the *reduced* Planck mass $M_{Pl} = (8\pi G_N)^{-1/2} \simeq 2.4 \times 10^{18}$ GeV in order to reproduce standard Einstein gravity in the nongravitino sector. This fact will become evident only after the complete supergravity Lagrangian density has been constructed.

The simplest consistent supergravity theory consists only of the graviton-gravitino supermultiplet described by a single superfield. However, we are interested in models containing chiral (matter) and vector (gauge) superfields as well; those will couple to the previous superfield. A detailed discussion of the construction [12.1, 12.4] of the corresponding Lagrangian density is beyond the scope of this book, but a brief summary appears in the *Annex* to this chapter. Here in the text we concentrate on the parts that are most important for phenomenological purposes. These can be described in terms of two quantities: (1) the Kähler potential $\mathcal{G}$, transforming as a gauge singlet real function of complex scalar fields, and (2) the "gauge kinetic function" f_{ab}, a and b being gauge group indices, transforming as the symmetric product of two adjoint representations of the gauge group, for instance containing a gauge singlet part proportional to δ_{ab}. The Kähler potential can be written as

$$\mathcal{G} = M_{Pl}^2 \left[K\left(\frac{\phi_i}{M_{Pl}}, \frac{\bar{\phi}^i}{M_{Pl}} \right) - \ln \frac{|\mathcal{W}(\phi_i)|^2}{M_{Pl}^6} \right]. \tag{12.2}$$

In (12.2) i is a type index with ϕ_i denoting scalar components of left chiral superfields Φ_i, while $\bar{\phi}^i \equiv \phi_i^\star$ (cf. §5.1) are the complex conjugates of the former and are scalar components of the right chiral superfield $\Phi_i^\dagger$. The superpotential is $\mathcal{W}(\Phi_i)$ with $\overline{\mathcal{W}} \equiv \mathcal{W}^\dagger$ and we have used the notation $\mathcal{W}(\phi_i) = \mathcal{W}(\Phi_i)|$, cf. §4.3. Note that the function K is real, cf. (12.85) in the *Annex*. In contrast, $\mathcal{W}$ – as in the case of global supersymmetry (§5.1) – is analytic; it depends only on left chiral superfields and not on their hermitian conjugates. The gauge kinetic function f_{ab} must also be an analytic function of the scalar components of left chiral superfields since $f_{ab}(\phi_i)W^{Aa}W_A^b$ must be analytic, cf. (12.84). The mass dimension of $\mathcal{G}$ is two, while f_{ab} is dimensionless. It may also be noted that $\mathcal{G}$ is invariant under the transformation

$$K \to K + h(\phi_i) + h^\star(\bar{\phi}^i) \,,$$
$$\mathcal{W} \to e^{-h(\phi_i)}\mathcal{W} \,,$$

for an arbitrary function h of ϕ_i.

The matrix of the second derivatives of $\mathcal{G}$ determines the form of the kinetic energy terms of the scalar members of chiral superfields while f_{ab} appears as the coefficient of those of gauge superfields. Thus the generally covariant kinetic energy terms in the Lagrangian density can be written in a flat[2] spacetime as [12.4]

$$\mathcal{L}_{\text{kin}} = -\mathcal{G}_j^i \left(\tilde{\mathcal{D}}_\mu \phi_i \tilde{\mathcal{D}}^\mu \bar{\phi}^j + i\overline{\psi_{iL}}\tilde{\slashed{\mathcal{P}}}\psi_L^j \right) + \Re e f_{ab} \left(\frac{i}{2}\overline{\lambda_M^a}\tilde{\slashed{\mathcal{P}}}\lambda_M^b - \frac{1}{4}F_{\mu\nu}^a F^{b\mu\nu} \right). \tag{12.3}$$

---

[2]Otherwise, there would be an overall factor of the tetrad determinant (cf. *Annex*) multiplying the RHS.

In (12.3) $\tilde{\mathcal{D}}_\mu$ is a general covariant derivative which is the gauge covariant derivative Δ_μ of (5.56a) plus the affine connection term of General Relativity [12.2]. Moreover, $\tilde{\mathcal{D}} \equiv \gamma^\mu \tilde{\mathcal{D}}_\mu$ and ψ_{iL} is the fermionic component of left chiral superfield Φ_i while λ^a_M denote Majorana gaugino fields. Summations over i, j and a, b are implied. We have employed the four component formalism for fermionic fields here. Furthermore, we have defined

$$\mathcal{G}^i_j \equiv \frac{\partial^2 \mathcal{G}}{\partial \phi_i \partial \bar{\phi}^j} \; . \tag{12.4}$$

The LHS of (12.4), known as[3] the Kähler metric, is independent of the superpotential $\mathcal{W}$ since the second term in the RHS of (12.2) drops out of $\mathcal{G}^i_j$. Canonical kinetic energy terms correspond to $\mathcal{G}^i_j = -\delta^i_j$ and $f_{ab} = \delta_{ab}$. If $\mathcal{G}^i_j$ and/or f_{ab} have any nontrivial dependence on some fields, the Lagrangian density of (12.3) will not, in general, be renormalizable. However, the complete (super-) gravity Lagrangian describes a nonrenormalizable effective field theory anyway[4]. Thus there is no rationale in requiring the renormalizability of (12.3). More reasonably, we can require that in the effective Lagrangian, describing physics at energies much below the Planck scale, nonrenormalizable terms should be suppressed by inverse powers of M_{Pl}.

As with global supersymmetry (cf. Ch.7), a spontaneous breakdown of local supersymmetry takes place if and only if the auxiliary component of (at least) one superfield develops a nonzero VEV. In the two component notation of §3.2, the equation of motion for the auxiliary component of a left chiral superfield can be worked out. The terms, that are relevant to our consideration, read

$$F_i = M_{Pl} e^{-\mathcal{G}/(2M_{Pl}^2)} (\mathcal{G}^{-1})^j_i \mathcal{G}_j + \frac{1}{4} f^\star_{ab,k} (\mathcal{G}^{-1})^k_i \bar{\lambda}^a \bar{\lambda}^b - (\mathcal{G}^{-1})^k_i \mathcal{G}^{jl}_k \xi_j \xi_l - \frac{1}{2} (M_{Pl})^{-2} \mathcal{G}_j \bar{\xi}^j \xi_i \; . \tag{12.5}$$

In (12.5) ξ_i is the two spinor fermionic component of the left chiral superfield mentioned earlier, $f^\star_{ab,k} \equiv \partial f^\star_{ab}/\partial \bar{\phi}^k$, $\mathcal{G}_j = \partial \mathcal{G}/\partial \bar{\phi}^j$ and $\mathcal{G}^{jl}_k \equiv \partial \mathcal{G}^j_k/\partial \phi_l$. The first RHS term of (12.5) might be nonzero if some elementary scalar field developed a VEV, as in an O'Raifeartaigh type of spontaneous supersymmetry breaking mechanism (cf. §7.4). Similarly, had some chiral fermions formed a condensate as in the dynamical supersymmetry breaking model of §7.6, either (or both) of the third and fourth RHS terms would have a nonzero value. Such a condensate, if formed by gauginos, could contribute to the second RHS term of (12.5). The latter possibility, which usually does not lead to supersymmetry breaking in globally supersymmetric models, has become quite popular [12.5] in the wake of superstring theories.

Let us now consider the phenomenon of the gravitino becoming massive. We can again draw an analogy with bosonic gauge interactions. If supersymmetry is broken spontaneously, the gravitino acquires a mass $m_{3/2}$ by 'eating' the goldstino (cf. §7.3). This is called the **super-Higgs effect**: a generalization of the Higgs mechanism in bosonic gauge theories. The expression for the gravitino mass, as explained after (12.90) in the *Annex*, turns out to be the coefficient of the first RHS operator in (12.5) with the Kähler potential replaced by its VEV:

$$m_{3/2} = M_{Pl} e^{-\langle \mathcal{G} \rangle/(2M_{Pl}^2)} \; . \tag{12.6}$$

---

[3]Our Kähler manifold is a manifold of complex scalar fields ϕ_i, $\bar{\phi}^j$ which is imbued with a metric of the form of (12.4).

[4]There is as yet no renormalizable quantum field theory of gravity.

This acquisition of mass increases the number of degrees of freedom of the gravitino field from two to four. The two original states have helicity $\pm 3/2$ and there are two 'longitudinal' states with helicity $\pm 1/2$. Indeed, the above mentioned analogy can be taken even further. According to the 'equivalence theorem' [12.6], at energies much larger than the mass of a massive gauge boson, the interactions of its longitudinal component approach those of the Goldstone 'eaten' by this gauge boson. Similarly, at energies much larger than the gravitino mass, the interactions of the helicity $\pm 1/2$ components of the gravitino approach those of the corresponding goldstino [12.7]. In particular, the interaction strength of a longitudinal gravitino scales like $1/(M_{Pl}m_{3/2})$, as can be seen from the following argument. For phenomenological purposes, the most important interactions of the gravitino in a flat spacetime are given by[5] (cf. 12.90a in the *Annex*)

$$\tilde{\mathcal{L}}_{\tilde{G}}^{int} = \frac{1}{M_{Pl}}\left(-\frac{1}{\sqrt{2}}\mathcal{G}_i^j\overline{\Psi}_\mu\tilde{\slashed{D}}\phi^i\gamma^\mu\psi_{jL} + \frac{1}{4}\Re e f_{ab}\overline{\lambda_M^a}\gamma^\mu\Sigma^{\rho\sigma}\Psi_\mu F_{\rho\sigma}^b\right) + \text{h.c.} .\qquad (12.7)$$

In (12.7) a,b (μ,ρ,σ) are gauge group (spacetime) indices and evidently we are using the four component form of spinorial fields. The gravitino wave function can, in general, be written as a Majorana four spinor times a polarization four vector. The supersymmetric equivalence theorem, mentioned above, follows from the form of the polarization four vector for the helicity $\pm 1/2$ gravitino components, i.e.

$$e_{1/2}^\mu = \frac{1}{m_{3/2}}(|\vec{p}|,\ |\vec{p}|^{-1}E\vec{p}) ,\qquad (12.8)$$

E and $\vec{p}$ being the energy and the three momentum of the gravitino respectively. One then recognizes that the terms listed in eq.(12.7) correspond[6] to two of the four terms appearing in the goldstino interactions of (7.29) and (7.30). The other two RHS terms in (7.29) do not contribute to enhanced interactions of the gravitino since the corresponding terms in the goldstino interaction (7.30) vanish for an on-shell goldstino.

Let us return to those terms in the supergravity Lagrangian density that give rise to renormalizable interactions at energies much lower than the Planck mass. This last specification is very important. Indeed, we shall eventually treat low energy supergravity as an effective theory by taking the limit $M_{Pl} \to \infty$ with $m_{3/2}$ held fixed[7]. We first consider the scalar potential which is contained in the following interaction terms, cf. (12.90c):

$$\mathcal{L}_{\text{SUGRA}}^I = F_i\mathcal{G}_j^i\bar{F}^j + 3M_{Pl}^4 e^{-\mathcal{G}/M_{Pl}^2} - \sum_\alpha \frac{g_\alpha^2}{2}f_{ab}^{-1}\tilde{D}^{\alpha a}\tilde{D}^{\alpha b} ,\qquad (12.9)$$

with[8]

$$\tilde{D}^{\alpha a} = \mathcal{G}^i(T^{\alpha a})_i^j\phi_j .\qquad (12.10)$$

---

[5]Other, less important, gravitino interactions terms have been omitted here, cf. (12.90), *Annex*.

[6]The precise correspondence lies in the replacements $\Psi^\mu \to \sqrt{\frac{2}{3}}e_{1/2}^\mu\lambda_{\tilde{G}M}$, and $F \to \sqrt{6}m_{3/2}M_{Pl}$, where $\lambda_{\tilde{G}M}$ is the properly normalized goldstino Majorana spinor.

[7]That this can be a sensible limiting procedure will shortly be illustrated by a simple model.

[8]Due to the gauge invariance of the superpotential only the term proportional to K^i contributes to $\mathcal{G}^i$ here. The correspondence to the D-term in global supersymmetry is now manifest, cf. (5.56c).

In (12.9) we have allowed for the fact that the gauge group is a direct product of factor groups: $G = \prod_\alpha G_\alpha$, cf. §11.3. So the index α labels the factor group while the index a labels generators within a factor. The first and the last terms in the RHS of (12.9) are rather similar to the F- and D-term contributions to the scalar potential, cf. (5.56b). The new and different feature is the appearance of the respective tensors $\mathcal{G}^i_j$ and f^{-1}_{ab} in them. Moreover, unlike in Ch.5, the auxiliary component F_i here contains fermionic fields too – as shown in (12.5). On the other hand, the second RHS term in (12.9) is entirely new. Note also that the requirement of positivity of the kinetic energy term in a physically well defined theory forces $\mathcal{G}^i_j$ to be negative (cf. 12.3). This implies opposite signs for the first two RHS terms of (12.9) allowing us to find solutions[9] with broken supersymmetry, where some $\langle F_i \rangle \neq 0$, but with a vanishing vacuum energy, i.e. $\langle \mathcal{H} \rangle = 0$.

In order to see how one set of nonvanishing $\langle F_i \rangle$ affects the other superfields, let us assign the chiral superfields Z_i, $\bar{Z}^i$ (with scalar components z_i, $\bar{z}^i$) to the observable sector and characterize the hidden sector by the generic[10] chiral superfield Σ with a scalar component σ, $\bar{\sigma}$ being σ^*. In other words,

$$\Phi_i \equiv \{Z_i, \Sigma\} ,$$

while for the scalar components

$$\phi_i = \{z_i, \sigma\} ,$$
$$\bar{\phi}^i = \{\bar{z}^i, \bar{\sigma}\} .$$

We also make the following ansätze

$$\mathcal{G} = -\sum_i z_i \bar{z}^i - H(\sigma, \bar{\sigma}) - M_{Pl}^2 \ln \frac{|\mathcal{W}(z_i, \sigma)|^2}{M_{Pl}^6} , \qquad (12.11a)$$

$$\mathcal{W}(\Phi_i) = \mathcal{W}_o(Z_i) + \mathcal{W}_h(\Sigma) . \qquad (12.11b)$$

In (12.11a) the part of the Kähler potential, that does not involve the superpotential, is taken to be the sum of a simple bilinear of observable sector scalar fields z_i, $\bar{z}^i$ and a term $H(\sigma, \bar{\sigma})$ which depends only on the scalar component of the hidden sector superfield Σ and its complex conjugate. Moreover, $\mathcal{W}_o$ and $\mathcal{W}_h$ in (12.11b) are the superpotential parts for the observable and the hidden sector respectively, i.e. *the total superpotential is assumed to be additively split between observable sector and hidden sector contributions*. The latter is chosen in such a way that $\langle F_\Sigma \rangle \neq 0$. We allow for the possibilities of σ, which is the scalar component of Σ, being a composite field and also of its developing a large VEV $\langle \sigma \rangle \lesssim M_{Pl}$. A further implication of (12.11) is that there are no renormalizable couplings between the components of Σ and those of Z_i.

Let us specifically consider the F-term contribution to the scalar potential[11], which can be derived from (12.5) and (12.9):

$$V_F = -e^{-\mathcal{G}/M_{Pl}^2} \left[M_{Pl}^2 \mathcal{G}^i (\mathcal{G}^{-1})^j_i \mathcal{G}_j + 3M_{Pl}^4 \right] . \qquad (12.12)$$

---

[9]Most models, though, are unable to achieve this naturally and need explicit fine tuning to make the vacuum energy vanish.

[10]The generalization from one Σ to several such hidden sector superfields is straightforward.

[11]Including the D-term, $V = -e^{-\mathcal{G}/M_{Pl}^2}[M_{Pl}^2 \mathcal{G}^i (\mathcal{G}^{-1})^j_i \mathcal{G}_j + 3M_{Pl}^4] + \sum_\alpha \frac{1}{2} g_\alpha^2 f^{-1}_{ab} \tilde{D}^{\alpha a} \tilde{D}^{\alpha b}$.

The ansätze (12.11) imply that

$$\mathcal{G}_i = -z_i - \frac{M_{Pl}^2}{|\mathcal{W}(z_i,\sigma)|}\frac{\partial \mathcal{W}_o^\dagger}{\partial Z_i^\dagger}\bigg|, \qquad (12.13a)$$

$$\mathcal{G}_{\bar\sigma} = -H_{\bar\sigma} - \frac{M_{Pl}^2}{|\mathcal{W}(z_i,\sigma)|}\frac{\partial \mathcal{W}_h^\dagger}{\partial \Sigma^\dagger}\bigg|, \qquad (12.13b)$$

$$\mathcal{G}_i^j = -\delta_i^j, \qquad (12.13c)$$

$$\mathcal{G}_{\bar\sigma}^\sigma = -H_{\sigma\bar\sigma}, \qquad (12.13d)$$

$$\mathcal{G}_i^\sigma = \mathcal{G}_{\bar\sigma}^i = 0, \qquad (12.13e)$$

as well as the hermitian conjugates of (12.13a,b). The subscript or superscript σ and/or $\bar\sigma$ in (12.13) means that the corresponding partial derivative has been taken in the space of scalar fields. Now, on using (12.13), the scalar potential of (12.12), with the notation $\mathcal{W}_o(z_i) = \mathcal{W}_o(Z_i)|$, becomes

$$V_F \simeq e^{H/M_{Pl}^2}\left\{ \left[\left|\frac{\partial \mathcal{W}_o(z_i)}{\partial z_i}\right|^2 + \left(z_i\frac{\partial \mathcal{W}_o(z_i)}{\partial z_i}\frac{|\mathcal{W}(z_i,\sigma)|}{M_{Pl}^2} + \text{h.c.}\right) + |z_i|^2\frac{|\mathcal{W}(z_i,\sigma)|^2}{M_{Pl}^4}\right] \right.$$
$$\left. + \frac{1}{H_{\sigma\bar\sigma}}\left(\mathcal{W}_o(z_i)\frac{\partial \mathcal{W}_h}{\partial \Sigma}\bigg|\frac{H_\sigma}{M_{Pl}^2} + \text{h.c.}\right) - 3\left(\mathcal{W}_o(z_i)\frac{\overline{\mathcal{W}_h(\bar\sigma)}}{M_{Pl}^2} + \text{h.c.}\right) + \cdots \right\}, \quad (12.14)$$

where the ellipsis stands for terms that are unimportant at energies well below M_{Pl}.

In order to bring (12.14) into a more easily recognizable form, we introduce the rescaled superpotential of the observable sector $\widehat{\mathcal{W}} \equiv e^{H/(2M_{Pl}^2)}\mathcal{W}_o$. The rescaling $\mathcal{W}_o \to \widehat{\mathcal{W}}$ partly absorbs the overall prefactor in (12.14). The first RHS term in it is then nothing but the F-term contribution to the scalar potential in a globally supersymmetric theory with a superpotential $\widehat{\mathcal{W}}$. Moreover, at low energies, we can replace σ by its VEV. The remaining terms in (12.14) then take the form of soft supersymmetry breaking operators:

$$V_F = \sum_i \left[\left|\frac{\partial \widehat{\mathcal{W}}(z_i)}{\partial z_i}\right|^2 + m_{3/2}^2|z_i|^2 + m_{3/2}\left(z_i\frac{\partial \widehat{\mathcal{W}}(z_i)}{\partial z_i} + \text{h.c.}\right)\right.$$
$$\left. + m_{3/2}\left[\widehat{\mathcal{W}}(z_i)\left(-3 + \left\langle\frac{\partial \mathcal{W}_h}{\partial \Sigma}\bigg|\frac{H_\sigma}{H_{\sigma\bar\sigma}\mathcal{W}_h(\sigma)}\right\rangle\right)\right) + \text{h.c.}\right] + \cdots \right]. \quad (12.15)$$

The dots in (12.15) are analogous to those in (12.14) and $m_{3/2}$ again is the gravitino mass of (12.6):

$$m_{3/2} = M_{Pl}\left\langle e^{H/(2M_{Pl}^2)}\left|\frac{\mathcal{W}_h(\sigma)}{M_{Pl}^3}\right|\right\rangle = M_{Pl}e^{-\langle\mathcal{G}\rangle/(2M_{Pl}^2)}. \qquad (12.16)$$

All the scalar masses have been given the universal value $m_{3/2}$ in the simple example leading to (12.15). This **universality**, imposed at a very high scale and with a value $\mathcal{O}(m_{3/2})$, will be seen later to be an important ingredient of the mSUGRA model [12.8]. A more general statement is that the magnitude of $m_{3/2}$ is the key factor determining the size of soft supersymmetry breaking terms in the scalar sector. For naturalness reasons (cf. Chs. 1 and 7), the latter should not exceed the weak scale by much. Under the natural assumption that $\langle H(\sigma)\rangle \lesssim M_{Pl}^2$, we thus need

$$\langle \mathcal{W}_h(\sigma)\rangle \simeq m_{3/2} M_{Pl}^2 \simeq M_Z M_{Pl}^2 \simeq (10^{13} \text{ GeV})^3 \,. \tag{12.17}$$

As discussed in Ch.7, this scale of 10^{13} GeV can be introduced into the hidden sector either 'explicitly' or 'dynamically' through dimensional transmutation.

A very important property of the low energy scalar potential V_F of (12.15) is that **all** scalar fields get the **same** soft supersymmetry breaking mass, $m_{3/2}$ independent of the index i. In addition, all A-parameters, cf. (7.44), also turn out to be identical, with the value

$$A_0 = m_{3/2} \left\langle \frac{\partial \mathcal{W}_h}{\partial \Sigma} \bigg| \frac{H_\sigma}{H_{\sigma\bar\sigma} \mathcal{W}_h(\sigma)} \right\rangle. \tag{12.18}$$

In (12.18) the VEV, appearing as the coefficient of $m_{3/2}$, is naturally $O(1)$. This property of soft supersymmetry breaking scalar masses and of the A-parameters, namely that they are **universal**, is valid at least at the classical level. It has been shown in §9.5 that this is one of the possible solutions of the FCNC problem in the MSSM. Eq. (12.15) also predicts a universal bilinear soft supersymmetry breaking parameter B_0, cf. (7.44), and a simple relation between it and A_0 namely $B_0 = A_0 - m_{3/2}$. However, this relation does not survive [12.9] if a high mass grand unified theoretic (GUT) sector exists in the theory or if the ansätze (12.11) are generalized even slightly. *So we do not assume this relation.*

This brings us to an important question. What would have been the change in the form of (12.15), which has desirable phenomenological properties, had we chosen a more general real function $K(z_i, \bar z^i)$ instead of the simple bilinear $\sum_i z_i \bar z^i$ in the RHS of the equation (12.11a) for the Kähler potential $\mathcal{G}$? The answer [12.9] is that our main results on the size of the soft supersymmetry breaking scalar masses and on the universality obeyed by them as well as by the A-parameters hold so long as the K of (12.2) depends on observable sector scalar fields only through the combination $u = \sum_i |z_i|^2$. This, in turn, is a consequence of imposing a global $U(n)$ symmetry on K, where n is the total number of independent chiral superfields in the observable sector[12]. In such a scenario the soft supersymmetry breaking parameters are still universal since the Kähler potential $\mathcal{G}$ depends on all scalar fields in the same way. In general, the scalar soft supersymmetry breaking masses will remain universal but not equal[13] to the gravitino mass $m_{3/2}$. Furthermore, the A and B parameters will also have universal values A_0 and B_0 respectively though the simple relation $B_0 = A_0 - m_{3/2}$, discussed above, will be invalidated. Contrariwise, if this $U(n)$ symmetry is broken by a choice such as

$$K = \sum_i X_i(\sigma, \bar\sigma) z_i \bar z^i - H(\sigma, \bar\sigma) \,,$$

---

[12]For the MSSM, we have $n = 17$.

[13]These two quantities will be of the same order of magnitude in most cases, but this cannot be strictly guaranteed.

with different real functions X_i for different values of i, flavor universality of the scalar soft supersymmetry breaking masses will be lost. Such a choice would, in general, predict unacceptably large FCNC strengths. Therefore, we shall for the most part assume that the Kähler potential does indeed possess a $U(n)$ global symmetry of the above type.

Finally, the supergravity Lagrangian density now also contains a term of the form

$$\mathcal{L}^{M_\lambda}_{\text{SUGRA}} = \frac{1}{4} M_{Pl} e^{-\langle \mathcal{G} \rangle/(2M_{Pl}^2)} \langle \mathcal{G}^\ell (\mathcal{G}^{-1})^k_\ell f^\star_{ab,k} \rangle \lambda^a \lambda^b + \text{h.c.} , \qquad (12.19)$$

coming from the cross product between the first and second RHS terms of (12.5) contributing to $F_i \bar{F}^i$. This term also occurs in four component notation in (12.90d) of the *Annex*. It will imply nonzero Majorana masses for the gauginos given by

$$M_\lambda = \frac{1}{2} m_{3/2} \Re e \langle \mathcal{G}^\ell (\mathcal{G}^{-1})^k_\ell f^\star_{ab,k} \rangle . \qquad (12.20)$$

These masses will have the same order of magnitude as the soft supersymmetry breaking dimensional parameters in (12.15) and (12.18) if the VEV, appearing as the coefficient of $m_{3/2}$ in (12.20), is of order unity. One could argue in favor of a canonical form for the gaugino kinetic energy terms in (12.7), i.e. $\Re e f_{ab} = \delta_{ab}$, on the grounds that any nontrivial dependence of $\Re e f_{ab}$ on the scalar fields z_i would make these terms nonrenormalizable. However, since the RHS of (12.20) involves a derivative of $f^\star_{ab}$, one would then have vanishing gaugino masses in conflict with experiment. The point, as already mentioned once, is that supergravity is a nonrenormalizable effective field theory and, in consequence, canonical gaugino kinetic energy terms can in no way be singled out. With the requirement that nonrenormalizable terms be suppressed by inverse powers of the Planck mass, the theory will become renormalizable in the 'flat limit' $M_{Pl} \to \infty$ for a fixed $m_{3/2}$. The latter is really the limit that is relevant at energies far below M_{Pl}.

The statement that the mass $m_{3/2}$ of the gravitino is expected in the gravity mediated supersymmetry breaking scheme to be of the same order[14] as the gaugino mass M_λ of (12.20), can be put on a stronger foundation. One can, in fact, derive [12.10] an upper bound on the ratio $M_\lambda/m_{3/2}$ from the requirement of the validity of perturbative unitarity till near-Planckian energies. The latter is a reasonable demand since generally one does not expect any nonperturbative behavior in the observable sector till those energies[15]. Consider the scattering of two massless gauge bosons with helicities λ_1 and λ_2 into two longitudinally polarized gravitinos of helicities σ_1 and σ_2. The relevant diagrams at the tree level involve [12.11] exchanges of the gravitino and the graviton as well as of light scalar and pseudoscalar fields which occur in the $N{=}1$ supergravity theory, broken spontaneously by the super-Higgs mechanism. The helicity amplitudes for this process have been computed from the vertex Feynman rules [12.11] of that theory. Let us specifically consider the Jth partial wave projected helicity amplitude $T^J_{\lambda_1,\lambda_2;\sigma_1,\sigma_2}$. For $J = 2$, $\lambda_1 = -\lambda_2 = 1$ and $\sigma_1 = -\sigma_2 = 1/2$, we have [12.10]

$$T^2_{1,-1;\frac{1}{2},-\frac{1}{2}} = \frac{E_{CM}^2}{288\pi M_{Pl}^2} \left(\frac{M_\lambda^2}{m_{3/2}^2} - \frac{6}{5} \right) , \qquad (12.21)$$

---

[14]This is also of the order of the soft supersymmetry breaking masses in the scalar sector.

[15]For instance, tree unitarity remains valid for a reaction such as $\gamma\gamma \to ZZ$, proceeding via s channel graviton exchange, till $E_{CM} = \mathcal{O}(M_{Pl})$.

E_{CM} being the CM energy. The partial wave unitarity requirement

$$\Re e \; T^2_{1,-1;\frac{1}{2},-\frac{1}{2}} < \frac{1}{2} \tag{12.22}$$

is then violated at the critical energy scale

$$E_{cr.} = 12\sqrt{\pi}(M_\lambda^2 m_{3/2}^{-2} - 6/5)^{-1/2} M_{Pl} \; .$$

Equivalently, the requirement of the validity of (12.21) upto $E_{CM} = M_{Pl}$ implies the upper bound[16]

$$M_\lambda/m_{3/2} < (144\pi + 6/5)^{1/2} \simeq 21 \; . \tag{12.23}$$

This completes our pedestrian treatment of the spontaneously broken $N{=}1$ supergravity theory in which we have highlighted the transmission of supersymmetry breaking from the hidden to the observable sector via gravitational interactions. We are now ready to explore the phenomenology of models constructed with this idea.

12.3 mSUGRA and Its Parameters

We first consider to the minimal supergravity model (with the acronym mSUGRA) [12.1] of supersymmetry breaking. The simplest ansatz for the Kähler metric, treating all chiral superfields symmetrically, was seen in the last subsection to lead to **universal** soft supersymmetry breaking parameters in the scalar sector:

$$m_{ij}^2 = m_0^2 \, \delta_{ij} \; , \tag{12.24a}$$

$$A_{ijk} = A_0, \; \forall i, j, k \; , \tag{12.24b}$$

$$B_{ij} = B_0 \; \forall \; i, j \; . \tag{12.24c}$$

Eq.(12.24c) is trivially true in the MSSM with conserved R-parity since that theory allows only one soft supersymmetry breaking B parameter. However, (12.24a) and (12.24b) are very restrictive indeed. Allowing A_0 and B_0 to be complex, they reduce the extra one hundred and five free parameters of the MSSM (cf. §9.1) to just five.

Let us recall the discussion in §11.4 of the renormalization group evolution of these parameters . The RGE is nonidentical for different soft supersymmetry breaking masses and A parameters. Therefore, (12.24) cannot hold at all scales once quantum effects, arising from the interactions of the MSSM, are included. Equations like (12.24) were derived in the previous section by taking M_{Pl} to infinity with $m_{3/2}$ (and hence the size of every soft supersymmetry breaking term) held fixed. This amounts to 'integrating out' from the Lagrangian density the terms that are suppressed by inverse powers of the Planck mass. The resulting effective Lagrangian density should thus become valid at energies somewhat below

---

[16]No such upper bound exists for scalar soft supersymmetry breaking masses. However, for phenomenological reasons, one generally wants weak scale sfermion and gaugino masses to be of roughly the same order of magnitude.

M_{Pl}. Thus (12.24) should be interpreted as "boundary conditions" for the RGE equations of those parameters *valid at some very high scale* Λ_{HS} which is $\lesssim M_{Pl}$.

Eqs.(12.24) do not cover the complete set of soft supersymmetry breaking parameters. We need to know, in addition, the gaugino masses at the high scale mentioned above to act as boundary conditions for their RGE equations. Not only do these determine the measurable gaugino masses at experimentally accessible energies, they also appear in the RGE equations of *all* scalar soft supersymmetry breaking parameters, cf. §11.4. The gaugino masses are related by (12.20) to the derivative of the gauge kinetic function f_{ab} whose real part appears as a coefficient of the term proportional to $F^a_{\mu\nu}F^{b\mu\nu}$ in the Lagrangian density (12.3). For a simple gauge group, the expectation value of $\Re e f_{ab}$ will be inversely proportional to the square of the gauge coupling. Fig. 11.2 illustrated the seeming unification (to a good approximation) of the three separate gauge couplings of the MSSM into a single one of presumably some simple Grand Unified group at the scale $M_U \simeq 2 \times 10^{16}$ GeV. It is thus reasonable to posit that, at that scale, $\langle f_{ab} \rangle$ is proportional[17] to δ_{ab}. This, in turn, suggests (but does not strictly imply) that the derivatives of f_{ab}, with respect to the fields responsible for the spontaneous breakdown of supersymmetry, should also be proportional to δ_{ab}. In case the latter is true, one can add to (12.24) a final additional boundary condition to the RGE equations. With α referring to the factor gauge group G_α, that can be stated as

$$M_\alpha(M_U) = M_{1/2} \,, \tag{12.25}$$

for all the three MSSM gaugino masses.

The scale M_U, where (12.25) should be imposed, is the scale of Grand Unification. This is, in fact, quite well defined vis-a-vis the MSSM. At the same time the (apparent) unification of couplings at that scale indicates that new degrees of freedom might have to be introduced there, viz. the new gauge and Higgs superfields of the operative Grand Unified Theory (GUT). Given the lack of our knowledge of the correct theory for energy scales higher than M_U, it is tempting to assume that the boundary conditions (12.24) also hold at the scale M_U. This is certainly the simplest assumption that we can make. From a supergravity point of view, M_U is really a bit too small, being about two orders of magnitude below the reduced Planck scale M_{Pl}. It may, nonetheless, be reasonable to assume that (hitherto unknown) deviations from universality will still be small at that scale. We shall discuss this question later in a more quantitative manner. For the time being, let us assume both (12.24) and (12.25) to hold at the scale $M_U = 2 \times 10^{16}$ GeV. Let us further assume that the MSSM remains valid upto that scale. This package of assumptions leads to the model known as mSUGRA or minimal supergravity. We shall now describe the spectrum of this model.

This discussion will closely mirror our general treatment of RGE in the MSSM given earlier in §11.4. In particular, the boundary condition (12.25) immediately implies the RGE invariant relation

$$M_3 : M_2 : M_1 = g_3^2 : g_2^2 : \frac{5}{3}g_Y^2 \,, \tag{12.26}$$

---

[17]We take a "GUT normalization" for the $U(1)_Y$ coupling, cf. §11.2. Note also that, though the GUT scale gaugino mass unification condition (12.25) has been included here as part of mSUGRA, it may be more generally valid than the universality assumptions (12.24) of that model on the soft supersymmetry breaking parameters associated with scalars.

where g_Y has been defined with the SM normalization (cf. Chs. 8 and 11). Eq. (11.28) shows that (12.26) holds at all energy scales between the weak scale and M_U upto two loop corrections. Evolving down to presently accessible energy scales, one can write the numerical values, cf. (9.21):

$$M_1(100 \text{ GeV}) \simeq 0.41 M_{1/2} , \tag{12.27a}$$

$$M_2(100 \text{ GeV}) \simeq 0.82 M_{1/2} , \tag{12.27b}$$

$$M_3(500 \text{ GeV}) \simeq 2.6 M_{1/2} , \tag{12.27c}$$

$$M_1 : M_2 : M_3 \simeq 1 : 2 : 7 . \tag{12.27d}$$

The last proportionality (12.27d) is valid at the weak scale[18] where the coefficient in the RHS of (12.27c) would be 2.8. RGE significantly reduces the mass of the $U(1)_Y$ gaugino and increases that of the gluino. Recall from discussions in Ch.9 that electroweak gauginos do mix with higgsinos. Thus M_1 and M_2 do not (exactly) correspond to physical masses, though they are usually close to some eigenvalues of the neutralino and chargino mass matrices. Note also that the coefficient in the RHS of (12.27c) depends rather significantly on the energy scale. The value 2.8, quoted above, can be treated as an approximate upper bound on the ratio, gluino mass divided by $M_{1/2}$, in mSUGRA. On the other hand, when Q increases to 2 TeV, the said coefficient decreases to 2.45 which can be taken as a fairly conservative lower bound on the same ratio from considerations of naturalness. Moreover, the difference between the running mass $M_3(M_3)$ and the physical (pole) mass of the gluino can be significant [12.12]:

$$M_{\tilde{g}}^{\text{pole}} \simeq M_3(M_3) \left[1 + \frac{\alpha_3(M_3)}{4\pi} \left(15 + \sum_{\tilde{q}} \ln \frac{m_{\tilde{q}}}{M_3} \right) \right] , \tag{12.28}$$

the summation being over all squarks. Despite such complications, one can definitely infer from (12.27) a significant mass splitting between the gluino on one hand and the (lighter) chargino/neutralino states on the other.

Eq.(11.36) shows that gaugino masses, in general, contribute significantly to the running of sfermion masses. Note also the implication of (12.24a) that the quantity S_Y, defined earlier in (11.34), vanishes. From (12.24), (12.25) as well as (11.43), we can now write expressions for the masses of the first generation sfermions that are linear in $m_0, M_{1/2}$ and $D \equiv -M_Z^2 \cos 2\beta$.

$$m_{\tilde{e}_R}^2(100 \text{ GeV}) = m_0^2 + 0.15 M_{1/2}^2 + \sin^2\theta_W D , \tag{12.29a}$$

$$m_{\tilde{e}_L}^2(100 \text{ GeV}) = m_0^2 + 0.53 M_{1/2}^2 + \left(\frac{1}{2} - \sin^2\theta_W \right) D , \tag{12.29b}$$

$$m_{\tilde{\nu}}^2(100 \text{ GeV}) = m_0^2 + 0.53 M_{1/2}^2 - \frac{1}{2} D , \tag{12.29c}$$

---

[18]It is worth pointing out that if, in addition, $|\mu| \gg |M_{1,2}|$ (cf. §9.2), one obtains the approximate mass equalities $2M_{\tilde{\chi}_1^0} \simeq M_{\tilde{\chi}_1^\pm} \simeq M_{\tilde{\chi}_2^0} \simeq \frac{1}{3} M_3$ at the weak scale.

$$m_{\tilde{u}_L}^2(500\text{ GeV}) \;=\; m_0^2 + 5.6 M_{1/2}^2 - \left(\frac{1}{2} - \frac{2}{3}\sin^2\theta_W\right) D \;, \tag{12.29d}$$

$$m_{\tilde{d}_L}^2(500\text{ GeV}) \;=\; m_0^2 + 5.6 M_{1/2}^2 + \left(\frac{1}{2} - \frac{1}{3}\sin^2\theta_W\right) D \;, \tag{12.29e}$$

$$m_{\tilde{u}_R}^2(500\text{ GeV}) \;=\; m_0^2 + 5.2 M_{1/2}^2 - \frac{2}{3}\sin^2\theta_W D \;, \tag{12.29f}$$

$$m_{\tilde{d}_R}^2(500\text{ GeV}) \;=\; m_0^2 + 5.1 M_{1/2}^2 + \frac{1}{3}\sin^2\theta_W D \;. \tag{12.29g}$$

Not surprisingly, the corrections proportional to $M_{1/2}^2$ are much larger for squarks than for sleptons. This is why we have listed the running masses of the former at a scale 500 GeV, higher than 100 GeV, at which the latter have been taken. As with the gluino mass, the numerical coefficients of $M_{1/2}^2$ in (12.29d-g) depend significantly on the mass scale chosen. For instance, the coefficient 5.1 in (12.29g) becomes 5.9 at the scale $Q = 200$ GeV and decreases to 4.5 at scale[19] $Q = 2$ TeV. However, unlike in the gluino case, the differences between the running masses $m_{\tilde{q}}(m_{\tilde{q}})$ and the physical pole masses are quite small for these squarks.

Since $D < M_Z^2$, the D-term contributions to (12.29) are subdominant, though they may not be negligible for sleptons. Nevertheless, the ratio of squark and slepton masses can vary significantly, depending on the ratio of m_0 and $M_{1/2}$. For instance, one can write

$$1 \le m_{\tilde{u}_L}/m_{\tilde{e}_R} \le 6 \;, \tag{12.30}$$

where the lower (upper) bound gets saturated for $m_0^2 \gg M_{1/2}^2$ $(M_{1/2}^2 \gg m_0^2)$. Only lower bounds can be given for sfermion to gaugino mass ratios, which become very large for $m_0^2 \gg M_{1/2}^2$:

$$0.95 \;\le\; \frac{m_{\tilde{e}_R}(100\text{ GeV})}{|M_1(100\text{ GeV})|} \;, \tag{12.31a}$$

$$0.9 \;\le\; \frac{m_{\tilde{e}_L,\tilde{\nu}}(100\text{ GeV})}{|M_2(100\text{ GeV})|} \;, \tag{12.31b}$$

$$0.8 \;\le\; \frac{m_{\tilde{u}_L,\tilde{d}_L}(500\text{ GeV})}{|M_3(500\text{ GeV})|} \;, \tag{12.31c}$$

$$0.77 \;\le\; \frac{m_{\tilde{u}_R}(500\text{ GeV})}{|M_3(500\text{ GeV})|} \;, \tag{12.31d}$$

$$0.75 \;\le\; \frac{m_{\tilde{d}_R}(500\text{ GeV})}{|M_3(500\text{ GeV})|} \;. \tag{12.31e}$$

These constraints are significantly stronger than the bounds (11.45), following merely from the required positivity of squared sfermion masses at least upto an energy scale of $\sim 10^8$

<hr>

[19]The corresponding mass values can be regarded as approximate lower and upper bounds on squark masses of the first two generations, as predicted by mSUGRA.

GeV. Specifically, (12.27a) and (12.31a) together suggest that the Lightest Supersymmetric Particle[20] (LSP) should be the lightest neutralino $\tilde{\chi}_1^0$ whose mass is expected to be less than $|M_1|$ on account of gaugino-higgsino mixing. Were all the sparticles of (12.29) light, the sneutrino could have been the LSP, but this possibility is now all but excluded by the negative outcome of sparticle searches at the LEP collider.

The magnitudes of second generation Yukawa coupling strengths, though larger than those of the first, are still small enough to be ignored in sfermion mass formulae. Consequently, sfermions of the second generation are practically mass-degenerate with their first generation cousins: $m_{\tilde{\mu}} = m_{\tilde{e}}$, $m_{\tilde{c}} = m_{\tilde{u}}$ etc. However, (12.29) do not apply to the third generation, the Yukawa couplings of which cannot be ignored. Indeed, in some regions of the parameter space, the combination of Yukawa contributions to the RGE (which reduce sfermion masses at low energies) and $\tilde{f}_L$-$\tilde{f}_R$ mixing can lower the mass of the lighter stop or (for $\tan\beta \gg 1$) of the lighter stau mass eigenstate below that of $\tilde{\chi}_1^0$. Such scenarios with a charged LSP can be excluded by cosmological considerations (cf. Ch.16).

Yukawa coupling contributions to the running of squared scalar masses also play a crucial role in the phenomenon of radiative electroweak symmetry breaking, cf. §11.4. The large hierarchy between the weak and Grand Unified scales, quantified by the number $\ln(M_U/M_Z) = 33$, and the large value of the top quark mass are the key factors here. They make it easy to find parametric values leading to a negative squared soft supersymmetry breaking mass m_2^2 (cf. Ch.10) at the weak scale. In fact, m_2^2 could (but should not) become 'too negative'. Were the parameter m_{2h}^2, appearing in the Higgs potential (10.5), equal to m_2^2, the resulting VEV v_2 of the Higgs field h_2^0 would be too large. Fortunately, m_{2h}^2 also receives the positive supersymmetric contribution $|\mu|^2$, cf. (10.7). The requirement of M_Z having the experimentally measured value then allows the *determination* of $|\mu|$ at the weak scale, as a function of the soft supersymmetry breaking parameters m_0, $M_{1/2}$, A_0 and B_0. In practice, one usually 'trades' $\tan\beta$ as a free parameter for B_0, using (10.18a); (10.18b) can then be used[21] to determine $|\mu|$. Notice that this procedure leaves the sign (or more generally the phase) of μ undetermined. The constraint of correct $SU(2)_L \times U(1)_Y$ symmetry breaking then completely determines the mass spectrum of sparticles in mSUGRA, assuming the absence of extra CP violation, in terms of the values of one discrete and four continuous real parameters. If we denote this set of parameters as $\{p\}$, then

$$\{p\} = \{\mathrm{sgn.}\ \mu, m_0, M_{1/2}, A_0, \tan\beta\}\ . \tag{12.32}$$

12.4 Phenomenology with mSUGRA

We now discuss the phenomenological consequences of mSUGRA in the parameter space defined by (12.32). Consider first how the RGE of some soft supersymmetry breaking masses can be affected by Yukawa coupling contributions. Fig. 12.1 shows an example with details explained in the caption. We have, in addition, chosen $m_0 = M_{1/2} = 200$ GeV, $A_0 = 0$,

---

[20]The LSP may not lie in the observable sector at all. Such a possibility will be discussed in Ch.13 in connection with gauge mediated supersymmetry breaking.

[21]These equations are modified somewhat by radiative corrections, as discussed in §10.6.

$\mu < 0$. For $\tan\beta = 3$ the contributions of the b and τ Yukawa couplings are practically

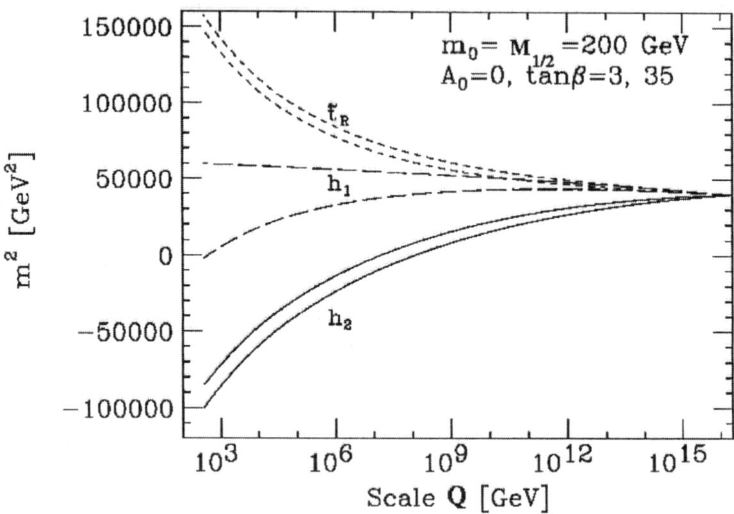

Fig.12.1. Squares of scalar masses running with the energy scale. The upper (lower) curve refers to $\tan\beta$ equal to 3 (35) for h_1 and to 35 (3) for the others.

negligible. Gauge contributions then increase m_1^2 slightly as the energy scale is reduced. The RGE equation (11.36b) for $m_{\tilde{t}_R}^2$ receives sizable contributions proportional to the square of the top quark coupling strength f_t. However, for $m_0 \simeq M_{1/2}$, the negative (strong) gauge contributions proportional to $M_{1/2}^2$ are even larger. As a result, $m_{\tilde{t}_R}^2$ increases with decreasing energy scale. An exactly opposite behavior is exhibited by m_2^2, which decreases rapidly as the energy scale comes down, turning negative at around $Q \sim 10^9$ GeV. Note, however, that for $\tan\beta = 3$, $m_{2h}^2 = m_2^2 + |\mu|^2$ remains positive at all energy scales. This smaller value of $\tan\beta$ corresponds to the lower solid and short-dashed as well as to the upper long-dashed curves. Increasing $\tan\beta$ to 35 has relatively little effect on m_2^2 and $m_{\tilde{t}_R}^2$. The observed increase of the two latter parameters occurs since terms, proportional to the square of the b-quark coupling strength f_b, reduce $m_{Q_3}^2$ at low energies. That, in turn, diminishes the importance of the terms proportional to f_t^2 in the RGE equations (11.35b), (11.36b). However, this large a value of $\tan\beta$ and the correspondingly larger strength of the bottom Yukawa coupling suffice to ensure the positivity of the β-function for a positive m_1^2. Now m_1^2 also decreases with a decreasing energy scale, eventually even turning (slightly) negative. Yet, $m_{1h}^2 = m_1^2 + |\mu|^2$ remains positive at all scales, while m_{2h}^2 becomes negative very near the weak scale for large $\tan\beta$.

The parameters, whose RGE is depicted in Fig.12.1, do not correspond directly to physical quantities. In order to impart to the reader a feeling for the dependence of the sparticle and Higgs mass spectra on the values of the input parameters, we have drawn Figs. 12.2–12.5. These figures show the behavior of various quantities of interest as functions of one of the continuous parameters in the set $\{p\}$ of (12.32). Shown in these figures are the absolute value of the μ-parameter, the mass of the CP odd Higgs boson A (cf. Ch.10) as well as masses of three sparticles: the lighter stop mass eigenstate $\tilde{t}_1$, the lighter stau $\tilde{\tau}_1$ and the lightest neutralino $\tilde{\chi}_1^0$. What we have plotted are the running masses at the scale Q_0, defined by the requirement of smallness of one loop corrections to the Higgs potential (cf. §10.6). These running masses are, in fact, quite close to the respective pole masses. All parameters from $\{p\}$, except one, are fixed in each of these figures to the values chosen for Fig.12.1, while the remaining parameter is varied on the horizontal axis. Specific comments follow on each of these figures.

(a) Fig.12.2 explores the dependence of the said masses on the gaugino mass parameter $M_{1/2}$. For $M_{1,2}^2 \gg m_0^2$, there is a large splitting between the lightest squark $\tilde{t}_1$ and the lightest slepton $\tilde{\tau}_1$. Note, furthermore, that $|\mu|$ is now much larger than $M_{\tilde{\chi}_1^0}$. This indicates that $\tilde{\chi}_1^0$ is gauginolike. Eq. (12.27a) and the behavior of $M_{\tilde{\chi}_1^0}$ in Fig.12.2 imply that $M_{\tilde{\chi}_1^0} \sim M_1$ (100 GeV), i.e. the bino or $U(1)_Y$ component of $\tilde{\chi}_1^0$ is larger than its $SU(2)_L$ wino component. In the case at hand, $|\mu|$ is also significantly larger than the $SU(2)$ gaugino mass M_2. Given a top quark mass $m_t \simeq 175$ GeV, one typically has in mSUGRA gauginolike $\tilde{\chi}_1^0$, $\tilde{\chi}_2^0$ and $\tilde{\chi}_1^\pm$, while the heavier $\tilde{\chi}_3^0$, $\tilde{\chi}_4^0$ and $\tilde{\chi}_2^\pm$ states are higgsinolike with masses $\simeq |\mu|$. We should add the caveat, however, that scenarios with a higgsinolike $\tilde{\chi}_1^0$ are still possible if one takes $m_0^2 \gg M_{1/2}^2$ and $\tan\beta > 5$.

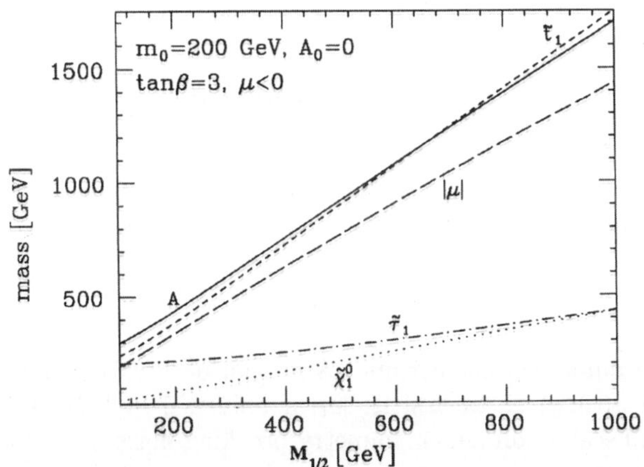

Fig.12.2. The masses of A (solid), $\tilde{t}_1$ (short-dashed), $\tilde{\tau}_1$ (dot-dashed) and $\tilde{\chi}_1^0$ (dotted) as functions of $M_{1/2}$, the common gaugino mass at the GUT scale. The dependence of $|\mu|$ on $M_{1/2}$, as determined from radiative electroweak symmetry breaking, is also shown.

(b) The dependence on the universal scalar mass parameter m_0 of the same quantities as before is illustrated in Fig.12.3. In this case, while $m_{\tilde{\tau}_1}$ and m_A (the latter for $m_0^2 \gg M_{1/2}^2$) show an almost linear dependence, $M_{\tilde{\chi}_1^0}$ remains essentially constant. In fact, $m_{\tilde{t}_1}$ and $|\mu|$ too vary relatively little, since increasing m_0 also increases the terms proportional to f_t^2 in the RGE equations. As a result, $m_{\tilde{\tau}_1}$ exceeds $m_{\tilde{t}_1}$ for $m_0^2 > 4M_{1/2}^2$, though the masses of the first and second generation squarks always remain above $m_{\tilde{\tau}_1}$.

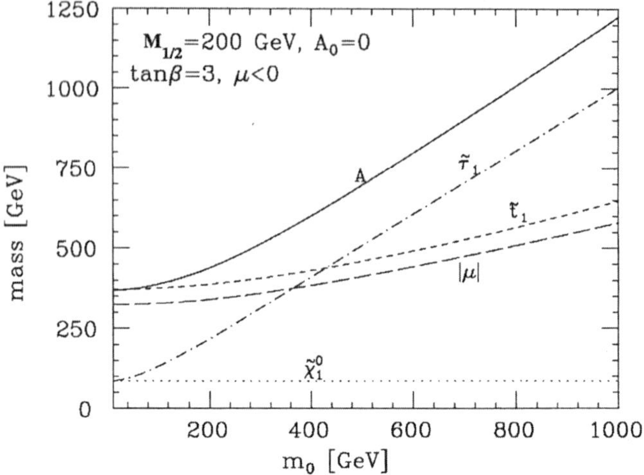

Fig.12.3. The dependence of the spectrum on the universal soft supersymmetry breaking parameter m_0. The notation is as in Fig.12.2.

Let us point out an interesting fact concerning the large m_0 region. Eq. (10.18b), which is one of the EW symmetry breaking conditions, can be rewritten as

$$\frac{1}{2}M_Z^2 = \frac{m_1^2 - m_2^2 \tan^2\beta}{\tan^2\beta - 1} - |\mu|^2 .$$

A numerical investigation of this equation in mSUGRA in terms of the parameters (12.32), assuming A_0 to be zero, allows it to be recast as

$$\frac{1}{2}M_Z^2 = c_0 m_0^2 + c_1 M_{1/2}^2 - |\mu|^2 ,$$

where $c_{0,1}$ are numerical coefficients. A crucial observation now is that[22], for $m_t \simeq$ 175 GeV and $g_3^2(M_Z) \simeq 1.5$, $|c_0| \ll |c_1|$ if universal high scale boundary values are assumed for all scalar soft supersymmetry breaking masses. In comparison, c_1 ($\simeq 2.0$) has a relatively insensitive dependence on the scale and the magnitudes of the input parameters. This small magnitude of c_0 means that fairly large values of m_0, even in excess of a TeV, do not result in much of a fine tuning in the Higgs potential. A **focus point** [12.13] in the RGE trajectories of m_2^2, as shown in Fig. 12.4, is the explanation

---

[22] As argued earlier, when discussing EW symmetry breakdown, the proper scale at which to terminate the RGE should be near the masses of third generation sfermions. If $m_0 \gg |M_{1/2}|$, those masses are determined by m_0 alone. For an on-shell top mass of 175 GeV, the coefficient c_0 at the scale m_0 will change from a small positive value when m_0 is a few hundred GeV to a small negative value if m_0 is in the multi-TeV range, the precise magnitude depending sensitively on m_t, the gauge coupling strengths and two loop terms in the RGE equation.

for this small coefficient. For a large range of initial values of m_0, these trajectories focus to a point with $m_2^2 \simeq 0$ at the weak scale. In other words, naturalness bounds on m_2 at the weak scale *cannot* constrain m_0 tightly at all. Nevertheless, despite its insensitivity to m_2, m_0 does set the scale for the masses of all sfermions. Thus a large value of m_0, which is possible now, can make the latter very massive in mSUGRA without increasing the values of the mass parameters in the Higgs potential (10.6). This is useful in alleviating the flavor problem, c.f. §9.5, and the CP problem discussed later in this section. In the focus point region ($m_0 \gtrsim 1$ TeV and $|M_{1/2}| \ll m_0$), $|\mu|$ is forced to be small by the above equation and $M_{1/2}$ can be comparable to $|\mu|$. Thus, while squarks and sleptons can lie in the multi-TeV mass range, the masses of the electroweak charginos and neutralinos as well as the strongly interacting gluino can be within a few hundred GeV.

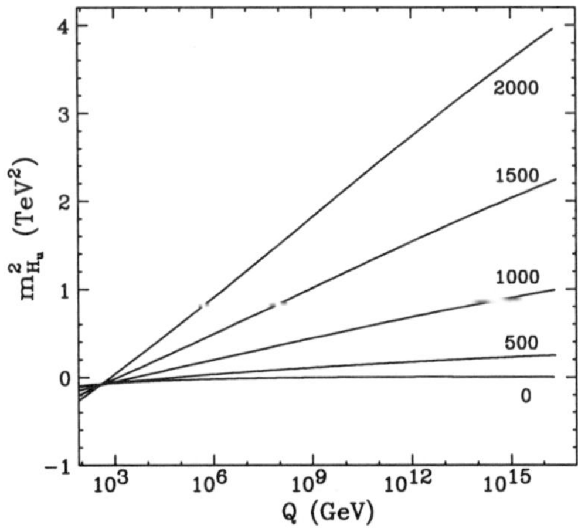

Fig. 12.4. RGE trajectories of m_2^2 in mSUGRA for various values of m_0 labelled in GeV and $M_{1/2} = 300$ GeV, $A_0 = 0$, $\tan\beta = 10$.

(c) The variation of the quantities that have been considered in Figs. 12.2 and 12.3, with the third dimensional input parameter A_0 (scaled by m_0) is explored in Fig.12.5. Both $m_{\tilde{\tau}_1}$ and $M_{\tilde{\chi}_1^0}$ are seen to be nearly independent of this parameter. For the given choice $\tan\beta = 3$, the tau Yukawa coupling is still rather small, so that $m_{\tilde{\tau}_1} \simeq m_{\tilde{e}_R}$ is essentially given by (12.29a), while $M_{\tilde{\chi}_1^0}$ remains nearly equal to M_1, as discussed in (a). On the other hand, m_A and $|\mu|$ show a quadratic dependence on A_0 since $|A_0|^2$ appears in the quantity S_t (cf. 11.33a) multiplying the f_t^2 terms in the RGE equations for the squared scalar soft supersymmetry breaking masses. However, the extrema of the parabolas get shifted from $A_0 = 0$ to $A_0 \simeq -M_{1/2}$ due to an interplay[23] between the gauge and the Yukawa contributions to (11.31a), which is the RGE equation for A_t. Finally, $m_{\tilde{t}_1}$ depends on A_0 both through the RGE equations for the diagonal entries in the stop mass matrix and through the offdiagonal entries proportional to $A_t m_t$. In fact, both

---

[23]Moving A_0 away from $A_0 \simeq -M_{1/2}$ decreases m_2^2 at the weak scale, i.e. makes it *more negative*. The value of $|\mu|$ consequently increases in accordance with (10.18b).

effects decrease $m_{\tilde{t}_1}$ as A_0 moves away from $-M_{1/2}$.

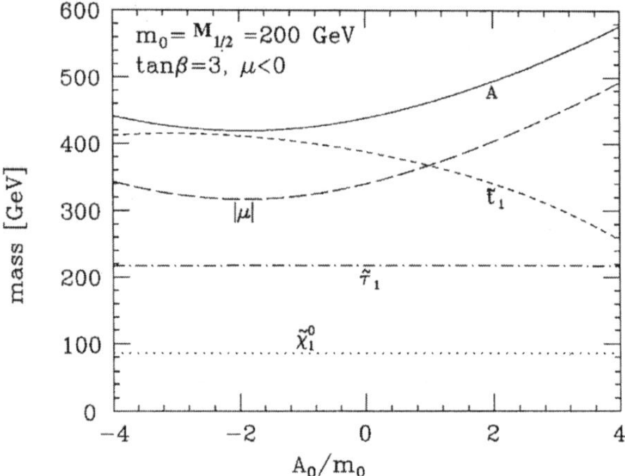

Fig.12.5. Dependence of the sparticle spectrum and SUSY parameters on the soft supersymmetry breaking parameter A_0 divided by m_0. The notation is as in Fig.12.2.

(d) Fig.12.6 illustrates the dependence of the spectrum on $\tan\beta$. The mass of the lightest neutralino $\tilde{\chi}_1^0$ is almost unaffected since $|\mu| \gg M_1$ and hence $M_{\tilde{\chi}_1^0} \simeq M_1$ throughout. The mass of $\tilde{t}_1$ also changes only slightly. This, in part, is due to the tendency of the two contributions to the offdiagonal entries of the stop mass matrix to cancel each other for the given choice of signs. This cancellation is quite pronounced at small $\tan\beta$,

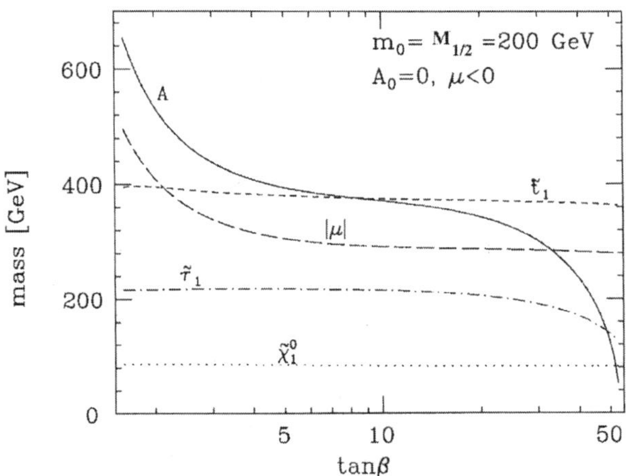

Fig.12.6. Sparticle spectrum and the parameters of Fig.12.2 as varying with $\tan\beta$.

where $|A_t|$ and $|\mu|\cot\beta$ are of comparable magnitude. Increasing $\tan\beta$ has little effect on A_t, but reduces $|\mu|\cot\beta$ quickly. At the same time, a larger $\tan\beta$ means a smaller

f_t and hence larger diagonal entries in the stop mass matrix. Altogether, then, an increase in $\tan\beta$ (for $\mu A_t < 0$) not only increases the absolute size of the off-diagonal entries, mentioned above, but also enhances the diagonal elements, leaving $m_{\tilde{t}_1}$ almost unchanged. Finally, for $\tan\beta \gtrsim 30$, the τ Yukawa contributions to the RGE equations become significant. The decrease of $m_{\tilde{\tau}_1}$, with increasing $\tan\beta$, is caused partly by the reduction of the diagonal elements of the stau mass matrix due to these RGE effects, and partly by the increase of the contribution proportional to $m_\tau \mu \tan\beta$ to the off-diagonal entries. Similar effects, associated with the rise of the magnitude of the b Yukawa coupling strength, lead to a decrease in the mass of the lighter $\tilde{b}$ eigenstate with increasing $\tan\beta$.

We shall now make some general comments on Figs. 12.2–12.6 for $m_A^2 \gg M_Z^2$ when the CP even neutral Higgs H^0 and the charged Higgs $H^\pm$ are nearly degenerate with A. First, let us focus on the mass m_A of the CP odd Higgs boson A. Recall from §10.4 that $m_A \to \infty$ is the decoupling limit of the Higgs sector of the MSSM. One striking common feature of Figs. 12.2, 12.3 and 12.5 is that A is one of the heaviest of the new MSSM particles. Thus, for not too large a value of $\tan\beta$, a heavy nearly degenerate triplet of Higgs bosons $A, H^\pm, H^0$ is a generic feature of mSUGRA. In fact, one can establish a quantitative link between $m_A^2 = m_{1h}^2 + m_{2h}^2 = m_1^2 + |\mu|^2 + m_{2h}^2$ (cf. 10.17 and 10.24b) and the squared mass of the sneutrino. If $\tan\beta$ is not large, the Yukawa contributions to (11.35a), the RGE equation for m_1^2, are negligible. Then m_1^2 runs like the squared mass of the $SU(2)_L$ doublet sleptons which have the same gauge quantum numbers as the h_1 doublet. Solving (10.18b) for m_{2h}^2, rather than for $|\mu|$, one finds[24] [12.14] that

$$m_A^2 = \frac{m_{\tilde{\nu}}^2 + |\mu|^2}{\sin^2\beta} - O(f_b^2, f_\tau^2) \ . \tag{12.33}$$

The negative sign of the terms in (12.33) that are proportional to the squared b and τ Yukawa couplings, means that the latter tend to reduce m_1^2, as is clear from Fig.12.1. Though (12.33) shows the dependence of m_A on $\tan\beta$, it is applicable only for small to moderate values of $\tan\beta$. The variation of m_A in that regime parallels that of $|\mu|$, which diverges as $\tan\beta \to 1$ i.e. when $\cos 2\beta \to 0$, cf. (10.18b). For $\tan\beta \geq 5$, $|\mu|$ becomes nearly independent of $\tan\beta$. However, when $\tan\beta$ exceeds 20, the corrections to (12.33), that are proportional to f_b^2, become sizable and m_A starts decreasing again, cf. Fig. 12.6. This decrease in m_A for large $\tan\beta$ can be put to a good phenomenological use by obtaining an upper bound on $\tan\beta$ from the experimental lower bound of 75 GeV on m_A from the LEP collider. Also, since S_t and S_b of mSUGRA are equal at the unification scale M_U, the discussion of §11.4 implies that

$$\tan\beta < \frac{m_t(Q_0)}{m_b(Q_0)} \simeq 60 \ , \tag{12.34}$$

where Q_0 is the scale where the running due to RGE is terminated. However this is weaker than the bound obtained from m_A.

---

[24]Eq.(12.30) has been derived from the renormalization group improved tree level potential. Loop corrections to this relation remain small, provided the RGE is stopped at the energy scale Q_0 which minimizes loop induced shifts in Higgs VEVs.

Both the lower and the upper ends of the abscissa in Fig.12.6 are quite well defined, the former through the Landau pole of the top quark Yukawa coupling strength, as discussed in §11.4, and the latter through the experimental lower bound on m_A. Such is not always the case for Figs. 12.2, 12.3 and 12.5, however. The lower limit on $M_{1/2}$ in Fig.12.2 is from the LEP lower bound on the mass of the lighter chargino $M_{\tilde{\chi}_1^\pm}$. All other abscissa limits are based on more theoretical considerations. For instance, the corresponding upper limits in Figs.12.3 and 12.5 are inspired by the naturalness arguments of Ch.1 which suggest that sparticles should not be (much) heavier than 1 TeV. The lower limit $m_0 > 10$ GeV, used in Fig.12.3, follows from the cosmological (vide Ch.16) requirement that the LSP be neutral, implying $M_{\tilde{\chi}_1^0} < m_{\tilde{\tau}_1}$. In this case, further constraints can be derived, from the requirement that supersymmetric Big Bang relics do not 'overclose' the Universe. A more detailed discussion of these issues is postponed to Ch.16. Let us just state here that this requirement forces the upper bounds $M_{1/2} \leq 740$ GeV in Fig.12.2 and $m_0 \leq 260$ GeV in Fig.12.3.

Finally, the bound $A_0 > -4m_0$ in Fig. 12.5 follows from the constraint that the scalar potential should not have its absolute minimum at a field configuration that is charged or colored so that $U(1)_{\rm em}$ or $SU(3)_C$ does not get spontaneously broken. Recall, from §9.1, that this potential contains terms proportional to $f_t A^t \tilde{t}_L \tilde{t}_R^\star h_2^0$ etc. These terms can always be made negative by an appropriate choice of signs (or phases) of the scalar fields involved. For a sufficiently large $|A^t|$, the potential then develops a minimum where $\langle \tilde{t}_L \rangle$ and $\langle \tilde{t}_R \rangle$ are nonzero. Other dangerous directions in field space can be found by setting $\mu h_2^0 + f_{d_i} \tilde{q}_{iL} \tilde{d}_{iR}^\star = 0$ or $\mu h_2^0 + f_{\ell_i} \tilde{\ell}_{iL} \tilde{\ell}_{iR}^\star = 0$. In these two cases the potential receives a sizably negative contribution from the term $m_2^2 |h_2^0|^2$. Minima of this kind can be avoided by demanding [12.15] that

$$m_2^2 + m_{\tilde{\ell}_i}^2 > 0, \ \forall i \ .$$

The requirement of the absence of such minima **excludes scenarios where $M_{1/2}$ is much greater than** m_0. In particular, $M_{1/2}$ (cf. Fig.12.2) and m_0 (cf. Fig.12.3) are constrained to be less than 600 GeV and greater than 100 GeV respectively. However, one cannot strictly exclude scenarios where charge and/or color are broken in the absolute minimum of the scalar potential so long as the tunnelling time from our vacuum to this 'false vacuum' is (much) greater than the age of the universe. Such is usually the case since the barrier between the two minima of the potential is quite high. For these reasons, we have chosen quite generous limits for the abscissae in Figs. 12.2, 12.3 and 12.5. It is worth mentioning here that cosmological constraints can be avoided by introducing a small amount of R-parity violation. The latter might be too small to produce any visible effects at collider experiments but sufficiently large to make the LSP decay on cosmological time scales. Alternatively, one might postulate that the true LSP resides in the hidden sector.

We shall now briefly comment on constraints from FCNC processes. In mSUGRA, as described above, all three lepton type numbers are separately conserved. So a radiative process such as the decay $\mu \to e\gamma$ cannot take place[25]. However, flavor mixing does occur in the quark sector, described by the CKM matrix $\mathbf{V}$ introduced in Ch.8. For our purpose, it is most convenient to work in a 'current basis' with the up quark mass matrix chosen as

---

[25]Recent evidence for neutrino flavor oscillations [12.16] indicates that lepton type numbers have to be violated in some sector of the theory. We will see in §14.6 that in some supersymmetry scenarios this can indeed lead to flavor violating charged lepton decays at an observable level.

diagonal[26]. The quark Yukawa terms in the superpotential of (8.33) can now be written, in the superfield notation of §8.3, as [12.17]

$$\mathcal{W}_q = H_2 \cdot Q \mathbf{f^u} \bar{U} + Q \cdot H_1 \mathbf{V}^\dagger \mathbf{f^d} \bar{D} \ . \tag{12.35}$$

In (12.35) the left chiral superfields $Q, \bar{U}$ and $\bar{D}$ are vectors in generation space and $\mathbf{f^{u,d}}$ refers to the diagonal Yukawa coupling matrix for u, d quarks. Since f^t is large, it reduces the mass of $\tilde{b}_L$ relative to those of $\tilde{d}_L$ and $\tilde{s}_L$ through the RGE of those soft supersymmetry breaking parameters. At the weak scale, the mass matrix for the $SU(2)_L$ doublet charge $-1/3$ squarks, while still diagonal, is therefore no longer proportional to the unit matrix. As explained in §9.7, the couplings of neutral gauginos to charge $-1/3$ quark and squark mass eigenstates, therefore, do have a nontrivial family structure specified completely by the CKM matrix $\mathbf{V}$. In particular, for the gluino, we can write

$$\mathcal{L}_{\tilde{g}\overline{\tilde{d}_L}\tilde{d}_L} = \sqrt{2}g_3 \overline{\tilde{d}_L} T^a \mathbf{V}^\dagger \tilde{d}_L \tilde{g}^a \ , \tag{12.36}$$

T^a being an $SU(3)_C$ generator.

Let us take values for squark and gaugino masses which are above current lower limits established by direct searches. Then their contributions to FCNC processes, induced from the interaction (12.36), are found to be well below present experimental bounds. The reason for this is something that has already been pointed out, namely that the mass splitting between the first and the second generation squarks is very small in mSUGRA. Therefore, supersymmetric contributions to FCNC processes involving these sparticles are strongly suppressed through an analog of the GIM mechanism, cf. §9.5. There is, of course, significant splitting between second and third generation squark masses. However, the corresponding CKM elements are small and experimental bounds on FCNC processes involving third generation quarks are generally weak. One class of FCNC processes can nonetheless get significant new contributions even in mSUGRA: radiative decays of the type $b \to s\gamma$ [12.18]. The dominant SM contribution to this decay comes from loop diagrams involving an up type quark and a W-boson. In mSUGRA there are additional gluino-squark loops from (12.36), though these usually yield smaller contributions than tH^- and $\tilde{t}\tilde{\chi}^-$ loops. The charged Higgs loop always contributes with the same sign as the tW^- loop of the SM, while the chargino loop can have either sign, depending on mSUGRA parameters. This feature makes it impossible to derive strict lower limits on sparticle or Higgs masses from the observed branching ratio of this decay, which is very close to the SM prediction. The measurement does exclude some regions of the mSUGRA parameter space, though. These constraints actually become rather stringent for large $\tan\beta$, where the chargino-stop loop contribution can be large even for quite massive sparticles, since the $\tilde{\chi}^-\overline{\tilde{t}b}$ vertex factor contains a term proportional to f_b which is itself proportional to $1/\cos\beta$. There could also be significant new contributions to yet unobserved decays such as $b \to s\ell^+\ell^-$ [12.18] in this region of the mSUGRA parameter space. For relatively light sparticle masses and not a very large $|A_0|$, $\mu > 0$ is preferred, since, for $\mu \le 0$, the chargino-stop loop contribution tends to add constructively with that of the $H^\pm$-top loop, leading to too large a value for $BR\,(b \to s\gamma)$.

---

[26]In a 'current basis' all gauge interactions are diagonal in generation space (cf. Ch.9), allowing one to diagonalize either the up or the down quark mass matrix, but not both.

We need to add the caveat that a certain amount of caution is generally called for in interpreting bounds from such loop induced FCNC processes. The resulting constraints depend rather sensitively on the assumption of **exact universality** of squark masses at the input scale M_U. It is quite possible that such a universality holds only approximately even at this high scale. The deviations might be negligible for computations of sparticle production cross sections and decay branching ratios. Yet, they might totally change the pattern of supersymmetric loop contributions to FCNC processes. Later discussions will, in fact, make it clear that one should not expect the boundary conditions (12.24) to hold exactly.

Less model dependent constraints on the mSUGRA parameter space (as well as on other supersymmetry breaking schemes) come from flavor *conserving* amplitudes involving 'static' photons. We have in mind specifically those leading to the anomalous muon magnetic moment $g_\mu - 2$ [12.19] and the electron electric dipole moment d_e [12.20], the latter being practically zero in the SM. The contributing loop diagrams here have a structure similar to those of $b \to s\gamma$ decay, with loops containing a charged slepton-neutralino or a chargino-sneutrino combination. The corresponding effective interactions require muon/electron chirality violation, the source of which in the MSSM matter sector can only be Yukawa couplings involving Higgs bosons. As a result of this, the supersymmetry contributions to both $g_\mu - 2$ and d_e turn out to be proportional to $\tan\beta$ if $\tan\beta \gg 1$. Of course, a measurable d_e can only arise via nontrivial CP-violating phases in the chargino-neutralino-slepton sectors. Another feature is that the supersymmetry contributions to both quantities decrease quadratically with increasing sparticle mass scale M_s. The level of present precision in the $g_\mu - 2$ measurement [12.21] is nonetheless able to exclude[27] certain regions of the m_0–$M_{1/2}$ plane that are otherwise allowed by present sparticle and Higgs search limits. Turning to d_e, the strong experimental upper limit [12.24] on its magnitude implies very tight constraints on the additional CP violating phases in the MSSM. That is unless the various contributions 'happen' to cancel[28] in which case a significant phases would be allowed between A_0 and the gaugino masses and more generally between the $U(1)_Y$ and $SU(2)_L$ gaugino masses. Otherwise, phases of order unity are allowed only if all sfermion masses lie well above 1 TeV, clearly violating (at least for mSUGRA) our notion of naturalness. Similar remarks also apply to the yet unseen [12.24] electric dipole moment of the neutron which however receives contributions from gluonic operators and in any case involves poorly understood nonperturbative physics in computing neutron matrix elements of partonic operators.

This brings us to the end of the phenomenological discusions of the mSUGRA model.

---

[27]As of this writing, there is a dichotomy in the SM 'prediction' for $g_\mu - 2$. The latter has a nonperturbative hadronic component, contributing at two loops, which is not a priori computable but needs to be extracted from data on hadron production in e^+e^- collision. Two extracted values using two sets of data, namely hadronic τ decays at LEP and direct hadron production at low energy e^+e^- colliders, do not quite agree [12.22]. The use of the latter suggests a deviation from the SM prediction, but its significance is hard to assess, given the aforementioned lack of agreement. It may be noted that mSUGRA can easily generate [12.23] an additional contribution to $g_\mu - 2$ of the required size and sign, provided $\mu > 0$.

[28]Generically, the experimental upper limit requires $(2 \text{ TeV}/M_s)^2 |\tan\beta \sin\phi_{\rm CP}| \leq 1$, where M_s is the biggest relevant (slepton or chargino) mass scale. In other words, one needs unconfortably heavy selectrons/charginos or small CP violating phases $\phi_{\rm CP} \ll 1$. It has been suggested that cancellations between different loop contributions might relax this constraint. But, as has been pointed out [12.25], such a cancellation is unlikely to work for the (also yet unseen) electric dipole moment of the mercury atom [12.24].

A very recent analysis [12.26] has updated all constraints on mSUGRA combining inputs from (1) LEP lower limits on sparticle and Higgs masses, (2) relic densities of neutralinos originating from the Big Bang, (3) $BR\ (b \to s\gamma)$, (4) $g_\mu - 2$ and (5) the experimental upper bound on $BR\ (B_s \to \mu^+\mu^-)$.

12.5 Beyond mSUGRA

We shall now briefly present two well motivated extensions of the mSUGRA model: Grand Unification and Flavor Symmetry. The effects of Grand Unification (of the MSSM gauge interactions) will be described first. We shall then discuss attempts to understand the hierarchy of quark and lepton masses in terms of a broken flavor symmetry. An important emergent feature will be the following: the universality of the boundary conditions (12.24) and (12.25), imposed at the unification scale $M_U \simeq 2 \times 10^{16}$ GeV, is generally not sacrosanct beyond mSUGRA. In fact, specific instances of **nonuniversality** will be given.

Grand unification with baryon and lepton (flavor) nonconservation

Our choice of the input scale M_U, at which (12.24) and (12.25) are supposedly valid, has been motivated by the meeting of the three evolving gauge couplings of the MSSM there, cf. Fig.11.1. However, in the discussion following (12.24), it was suggested that these, i.e. the boundary conditions for the scalar soft supersymmetry breaking parameters, should really be imposed at a high scale Λ_{HS} close to the reduced Planck mass $M_{Pl} = 2.4 \times 10^{18}$ GeV. If the MSSM continues to describe Nature at energies beyond M_U, the difference between imposing (12.24) at M_U or at M_{Pl} would be quite small. The observed meeting of the gauge couplings at M_U certainly hints at a Grand Unification of all MSSM gauge interactions there. If there is a significant range of energy scales between Λ_{HS} and M_U, where Nature is described [12.27] by a Grand Unified Theory (GUT), the scalar soft supersymmetry breaking parameters will quite likely differ significantly from the simple universal boundary conditions (12.24).

Let us take for our GUT a field theory formulated in $(3+1)$ spacetime dimensions[29] and a simple model, more for illustration than as a realistic candidate. We consider the simplest GUT model based on the rank four [12.27] simple gauge group $SU(5)$ which is broken at the scale M_U by appropriate VEVs in a single **24** dimensional representation Σ of Higgs scalars. Some of the left chiral matter superfields, namely the quark doublets Q, the u-type antiquark singlets $\bar{U}$ and the antilepton singlets $\bar{E}$, are combined in **10** dimensional representations T_i of $SU(5)$. The remaining left chiral matter superfields, i.e. the d-type antiquark singlets $\bar{D}$ and the lepton doublets L, reside in $\bar{\mathbf{5}}$ representations $\bar{F}_i$. Finally, the MSSM Higgs doublet superfields are placed in separate $\bar{\mathbf{5}}$ and **5** representations $\mathcal{H}_1$ and $\mathcal{H}_2$ respectively. The superpotential is given by

$$\mathcal{W}_5 = \lambda \mathrm{Tr}\ \Sigma^3 - M \mathrm{Tr}\ \Sigma^2 + \mathcal{H}_2(\lambda'\Sigma + M')\mathcal{H}_1 + T\mathbf{f}^u T \mathcal{H}_2 + T\mathbf{f}^d \bar{F}\mathcal{H}_1\ . \tag{12.37}$$

In (12.37) λ, λ' are coupling strengths while M, M' are mass parameters. In particular, M is a high scale mass, introduced "by hand". The breaking of $SU(5)$ symmetry is caused by

---

[29]Other, more exotic, possibilities exist, e.g. superstring theories or those formulated on different three-branes in a higher dimensional bulk. But it gets much harder to extract sound quantitative predictions from these.

a nonzero[30] $\langle \Sigma \rangle$, of the order of M_U, which must not break supersymmetry. We want the scale of spontaneous supersymmetry breaking not to exceed 10^{13} GeV and that is much less than M_U. Therefore $\langle \Sigma \rangle$ must be computable from $\partial \mathcal{W}_5 / \partial \Sigma = 0$. This VEV is practically determined by the first two RHS terms of (12.37).

Turning to the rest of the superpotential $\mathcal{W}_5$, we see that the last two RHS terms of (12.37) lead to the Yukawa interactions of the MSSM. On the other hand, the third term generates masses for the $SU(3)_C$ triplet, $SU(2)_L$ singlet GUT partners of the MSSM Higgs superfields. These must be made very heavy due to the following reason. The exchange of their fermionic components gives rise to baryon and lepton number violating operators in the effective low energy Lagrangian. Those operators are of mass dimension five, i.e. are only suppressed by a single power of the heavy triplet 'higgsino' masses. As will be shown in more detail later, such operators contribute to baryon and lepton nonconserving nucleon decay processes at the one loop level. At the same time, we want the bilinear $\mu H_1 \cdot H_2$ term of the $SU(2)_L$ doublet MSSM Higgs superfields to be small since $|\mu|$ is expected to be of the order of the electroweak scale on account of (10.18b). In (12.37) this 'doublet-triplet splitting' has to be achieved by finetuning the parameters to one part in 10^{14}. This is what makes this supersymmetric $SU(5)$ GUT model rather unattractive. It is nonetheless a useful paradigm inasmuch as it illustrates the possible effects of a superheavy GUT sector on the MSSM sparticle spectrum.

The new GUT interactions leave their imprint on the low energy spectrum through the RGE between $\Lambda_{HS} \simeq M_{Pl}$ and M_U. The first such effect is due to gauge interactions: gaugino loops contribute differently to the masses of the scalars in the **5** or $\bar{\mathbf{5}}$ representations and to those in the **10** on account of the different Casimir coefficients $C_5(\bar{F}, \mathcal{H}_1, \mathcal{H}_2) = 24/5$, $C_5(T) = 36/5$. The one loop β-function coefficient of the gauge coupling in minimal $SU(5)$ is $\beta_5 = -3$. On using (11.43), this makes the additional contributions $+0.24 M_{1/2}^2$ to $m_{\tilde{\ell}_L}^2$, $m_{\tilde{d}_R}^2$ and $+0.36 M_{1/2}^2$ to $m_{\tilde{q}_L}^2, m_{\tilde{u}_R}^2$ and $m_{\tilde{\ell}_R}^2$. Here $M_{1/2}$ is the gaugino mass at the scale M_U. These additional contributions significantly increase the lower bound on the ratio between the $SU(2)_L$ singlet slepton and bino masses. In fact, (12.31a) now becomes

$$1.75 < \frac{m_{\tilde{e}_R}(100 \text{ GeV})}{|M_1(100 \text{ GeV})|} , \tag{12.38}$$

implying a significant mass splitting between sleptons (of the first and second generations) and the lightest neutralino whose mass is bounded from above by $|M_1|$.

The couplings that appear in the superpotential of (12.37) also change the low energy mass spectrum [12.28]. As mentioned above, the color triplet GUT partners of the MSSM higgsinos must be very heavy. In order to arrange this, the coupling strength λ' is forced to be quite large, $\lambda'(M_U) > 0.7$. It was shown in §11.3 that Yukawa couplings always appear with positive sign in the RGE equations for the squared scalar soft supersymmetry breaking masses. The RGE from Λ_{HS} to M_U thus reduces the soft supersymmetry breaking contributions $m_{1,2}^2$ to the Higgs squared masses. The latter are significantly affected by the coupling λ', unlike squark and slepton masses which remain independent of λ' at the one-loop level. There is further reduction of m_2^2 by the contribution from the top quark

---

[30] We use the notation $\langle \Sigma \rangle$ to mean the VEV of the scalar component of Σ.

Yukawa coupling to the GUT RGE equations. In total, $m_{1,2}^2$ get substantially more reduced at the weak scale, as compared to the scenario where the boundary conditions (12.24) are imposed at M_U. The electroweak symmetry breaking requirement (10.13) then forces one to **significantly increase the size of the μ-parameter**. The net effect is that not only are the lighter charginos and neutralinos more gauginolike, the heavier higgsinolike neutralinos and charginos get even heavier[31]. However, (12.33) survives to a good approximation. Yet another important effect comes from the fourth RHS term of (12.37), namely that the third generation $SU(2)_L$ singlet $\tilde{\tau}_R$ is influenced by the top Yukawa coupling strength f_t at scales above M_U in that $m_{\tilde{\tau}_R}^2(M_U)$ is reduced relative to $m_{\tilde{e}_R}^2(M_U) \simeq m_{\tilde{\mu}_R}^2(M_U)$. For small or moderate values of $\tan\beta$, the corresponding mass splitting is typically $\sim 15\%$ at the weak scale but can be[32] as high as 50%.

In the light of the above effects, one often chooses a supergravity induced supersymmetry breaking framework [12.29] that is somewhat more general than mSUGRA, though a price is paid in the proliferation of parameters. Let us describe a top-down scheme in this genre with parameters specified at the unification scale M_U and having reasonable predictive power. We shall call it $\tilde{\text{C}}$MSSM. Eq. (12.25) is assumed here, but not all of (12.24); (12.24c) is retained but (12.24a,b) are modified, i.e. **nonuniversality** is introduced into the **scalar sector**. A common left/right squark (slepton) mass $m_{\tilde{q}_L}/m_{\tilde{u}_R,\tilde{d}_R}$ $(m_{\tilde{\ell}_L}/m_{\tilde{\ell}_R})$ is taken[33] at M_U. Furthermore, among the trilinear scalar interaction coefficients A_{ijk}, only three diagonal third generation ones are kept nonzero as free parameters A_t, A_b, A_τ defined at the scale M_U. Additional parameters are $m_{H_{1,2}}$ at M_U and $\tan\beta$ which together describe the Higgs(-ino) mass spectrum, at least at the tree level. The relations (12.29) get modified inasmuch as m_0 in the RHS gets replaced by the appropriate GUT-scale sfermion mass, the contributions from the RGE and the electroweak D-terms remaining unchanged. A major difference with mSUGRA is that the masses of the MSSM Higgs scalars at M_U are allowed to be different from those of the sfermions. The rationale for this is that the former and the latter transform in general as different representations of the grand unifying symmetry group and evolve differently in the passage from M_{Pl} to M_U. The magnitude of μ can now be quite different from that in mSUGRA but adjusts itself to be compatible with radiative EW symmetry breaking. As a result of this, the mixing patterns for charginos and neutralinos can be quite unlike those in mSUGRA with important consequences for their collider signals. A departure from mSUGRA, that is even more radical than $\tilde{\text{C}}$MSSM, is effected [12.30] by giving up (12.25), i.e. gaugino mass universality at M_U. As mentioned immediately before (12.25), even if the gauge kinetic function f_{ab} is proportional to δ_{ab}, such need not be the case for its derivatives with respect to hidden sector scalar fields responsible for causing spontaneous supersymmetry breakdown in that sector. With appropriate GUT representations chosen for the said fields[34], (12.25) can be violated in which case low energy gaugino mass values

---

[31]Recall that, for most of the parameter space of mSUGRA, these states lie much above the corresponding gauginolike states.

[32]This corresponds to the 'quasi-fixed point' scenario, cf. §11.4.

[33]Universal sfermion mass values at the scale M_U, assumed separately with each left (and right down type as well as the allowed right up type) chiral fermion set for the first two generations, suffice in ensuring the validity of experimental upper bounds on FCNC amplitudes (cf. §9.5). Third generation sfermion and Higgs masses are the least constrained by FCNC considerations.

[34]If these fields are nonsinglets under the GUT gauge group, they also have SM gauge interactions and

will no longer be given by (12.27).

The lack of degeneracy of slepton masses at M_U, mentioned earlier, leads to a nontrivial slepton mixing. To see this, we can work in a basis where the Yukawa coupling $\mathbf{f^u}$ of (12.37) is diagonal at the scale Λ_{HS}. If $\tan\beta$ is not very large, the entries of the matrix $\mathbf{f^d}$ would all be fairly small so that $\mathbf{f^u}$ would remain essentially diagonal at all scales down to M_U. But $\mathbf{f^d}$ has to have some off-diagonal entries in order to account for the well-known flavor mixing in the quark sector. Once the GUT symmetry is broken at M_U, one is free to rotate quark and lepton superfields independently. In particular, one can rotate the lepton basis so as to diagonalize the matrix of charged lepton Yukawa couplings $\mathbf{f^e}$. On account of $SU(5)$ invariance, the latter also equals $\mathbf{f^d}$ at M_U. Therefore, at this high scale, the required rotation is just the CKM one. The $\tilde{\ell}_R$ mass matrix remains diagonal. However, RGE running proportional to f_t^2 leads to nonuniversality at the scale M_U, so that $\tilde{\ell}_R$ fields can no longer be rotated arbitrarily. This results in the appearance of lepton flavor violating lepton-slepton-neutralino interactions whose size is controlled by the CKM matrix.

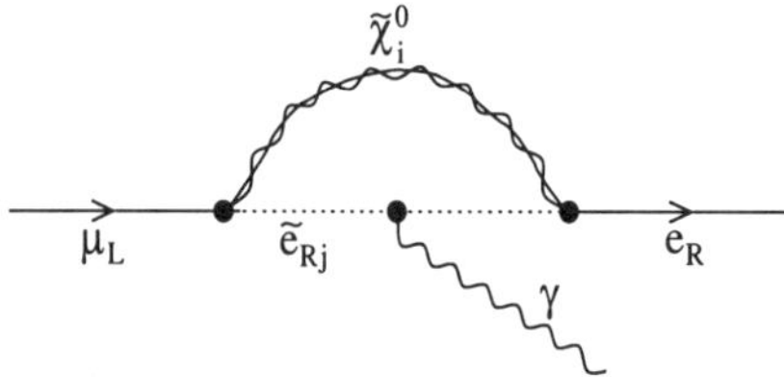

Fig.12.7. A typical $\mu \to e\gamma$ diagram in the presence of slepton flavor mixing.

The most sensitive probes of this kind of leptonic flavor violation are the decay $\mu \to e\gamma$ and the closely related $\mu \to e$ conversion in muonic atoms [12.31]. In the case at hand, the former decay proceeds through diagrams of the type shown in Fig.12.7. Let us make a couple of comments on the calculation of the decay amplitude. (1) The photon in the diagram could be attached to any charged line. (2) The effective operator leading to this decay is $\bar{e}\Sigma_{\alpha\beta}\mu F^{\alpha\beta}$ which flips the lepton chirality; the latter can occur on either of the lepton lines through a mass insertion or at one of the vertices involving the neutralino through its higgsino component. The resulting branching ratio can be estimated to be

$$
B(\mu \to e\gamma) \;\sim\; \frac{6\alpha_{em}}{\pi} \tan^4\theta_W |V_{ts}V_{td}|^2 \left(\frac{M_W}{m_{\tilde{\tau}_R}}\right)^4 \left(\frac{m_{\tilde{e}_R}^2 - m_{\tilde{\tau}_R}^2}{m_{\tilde{e}_R}^2}\right)^2
$$

$$
\sim\; 5 \times 10^{-11} \left(\frac{M_W}{m_{\tilde{\tau}_R}}\right)^4 \left(\frac{m_{\tilde{e}_R}^2 - m_{\tilde{\tau}_R}^2}{m_{\tilde{e}_R}^2}\right)^2 , \tag{12.39}
$$

V_{ij} being the appropriate element of the CKM matrix of Chs. 8 and 9. This is a very rough estimate since we have ignored contributions of similar size involving $\tilde{\mu}_L$-$\tilde{\mu}_R$ and $\tilde{e}_L$

---

hence are no longer 'hidden'. However, it is true that such fields often decouple from physics at energy scales far below M_U.

$-\tilde{e}_R$ mixing. The point, however, has been made that in a supersymmetric GUT model one generically expects the branching ratio for the $\mu \to e\gamma$ decay to be within a few orders of magnitude of the current experimental limit $B_{\exp}(\mu \to e\gamma) \leq 4.9 \times 10^{-11}$. Though major cancellations between different contributions could reduce the calculated answer, this prediction is at least as robust as that on proton decay to which we now turn.

As already mentioned, the dominant contribution to baryon nonconservation in minimal supersymmetric $SU(5)$ comes from the exchange [12.32] of the GUT partners of the MSSM higgsinos. A typical diagram for the subprocess $ud \to \bar{s}\bar{\nu}$, contributing to proton decay, is shown in Fig. 12.8. It involves the Yukawa coupling strengths of the higgsino triplets $\tilde{h}_1^{(3)}$, $\tilde{h}_2^{(3)}$ to light (s)quarks and (s)leptons. The arrows indicate the flow of left chiral superfield components, while the cross on the higgsino internal line marks a mass insertion. The internal sfermions as well as the external neutrino can be of any generation. The outgoing charge $-1/3$ antiquark must be an $\bar{s}$, since the total effective four fermion operator must be antisymmetric in the quark indices to produce a color singlet. This diagram thus contributes to $p \to K^+ \bar{\nu}_i$.

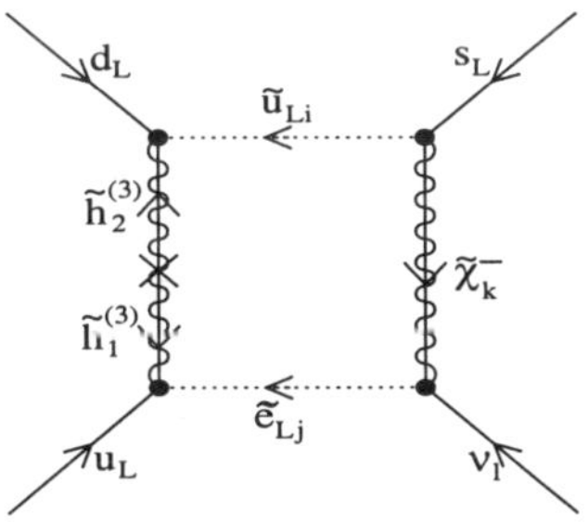

Fig.12.8. A typical diagram for the subprocess $ud \to \bar{s}\bar{\nu}$ contributing to the decay $p \to K^+\bar{\nu}_i$ through the exchange of the GUT partners of the MSSM higgsinos.

Some additional comments on the decays $\mu \to e\gamma$ and $p \to K^+\bar{\nu}_i$ are in order. First, the superpotential (12.37), describing the minimal model, is unlikely to be completely right since it suffers from the standard $SU(5)$ problem [12.27] of making wrong predictions for relations between the masses of light SM fermions. One example is $m_s(M_U) = m_\mu(M_U)$. After accounting for the RGE between M_U and the weak scale, this relation is found to be wrong by a factor of three. Some extra terms need to be added to (12.37) to correct this. These will change the Yukawa couplings appearing in Figs.12.7 and 12.8, affecting the corresponding branching ratio predictions. Second, all contributions to nucleon decay, involving the exchange of higgsino triplets, require a mass insertion to turn the $\tilde{h}_1^{(3)}$ into $\tilde{h}_2^{(3)}$. In the minimal $SU(5)$ model this insertion is that of the higgsino mass itself, and this does not involve any suppression. There exist models, nonetheless, where these triplets acquire masses by pairing up with other triplets which are not directly related by GUT symmetry to the MSSM Higgs fields [12.33]. In such models the higgsino triplet exchange contribution can be strongly suppressed by additional mixing angles. This insertion is given in the most extreme case by the supersymmetric higgsino mass parameter μ of the MSSM. In that situation the higgsino contributions become completely negligible. Nucleon decay then takes place predominantly from the exchange of superheavy X and Y gauge bosons just as in nonsupersymmetric GUTs [12.27]. However the corresponding lifetime [12.32] is very long:

$$\tau(p \to e^+\pi^0)_{X-\text{exch.}} \simeq 10^{36} \text{ yrs.} \left(\frac{M_U}{10^{16} \text{ GeV}}\right)^4 \left(\frac{0.003 \text{ GeV}^3}{\mathcal{A}}\right)^2. \qquad (12.40)$$

In (12.40) $\mathcal{A}$ is the size of the relevant hadronic matrix element and the dimensional numerical factor 0.003 GeV^3 was estimated using chiral Lagrangian techniques.

Flavor symmetries

The existence of a GUT would indeed revolutionize our overall perception of Nature. Nevertheless, its impact on the sparticle spectrum at low energies would be relatively minor so long as the boundary conditions (12.24) are valid at a high scale near M_{Pl}. Recall from §12.2 that these conditions were derived by imposing a global symmetry on the Kähler manifold: a $U(n)$ invariance, where n is the number of independent chiral superfields in the observable sector. This was motivated mostly by the requirement of the suppression of sparticle loop induced FCNC amplitudes – implemented by the universality of scalar masses and of A parameters. However, this global $U(n)$ symmetry is badly broken in other sectors of the theory, by both gauge and Yukawa interactions that differ for different superfields. There would, therefore, be at least some quantum corrections to the boundary conditions (12.24), derived from this $U(n)$ symmetry. Unfortunately, one cannot estimate the size of these corrections with any degree of confidence owing to the absence of a reliable quantum theory of all interactions, including gravity, at Planckian energies.

What one can do is construct models on the basis of the following fact. The suppression of supersymmetric FCNC amplitudes requires only the near mass degeneracy of *scalars with identical gauge quantum numbers*. One needs $m_{\tilde{q}_{1L}} \simeq m_{\tilde{q}_{2L}}$, $m_{\tilde{u}_{1R}} \simeq m_{\tilde{u}_{2R}}$ etc. (1, 2 being generation indices); the argument does *not* require $m_{\tilde{q}_{1L}} \simeq m_{\tilde{u}_{1R}}$ etc. The near mass degeneracy of scalars with identical gauge quantum numbers can be derived from much smaller 'family' or 'flavor' symmetry groups. The generators of the latter commute with the gauge symmetry generators of the MSSM. Even this smaller symmetry must be violated by the Yukawa couplings of the latter, since all charged SM fermions have different masses. However, the first and second generation Yukawa couplings are quite small. Note, moreover, that there is a certain regularity between ratios of running charge 2/3 quark and lepton masses, taken at some energy scale above the top quark mass:

$$m_t : m_c : m_u \simeq 1 : \epsilon^2 : \epsilon^4, \quad m_\tau : m_\mu : m_e \simeq 1 : \epsilon : \epsilon^3, \qquad (12.41)$$

with $\epsilon^2 \simeq 1/200$. Though the mass hierarchy of charge $-1/3$ quarks does not quite fit into such a pattern, (12.41) does suggest a symmetry ansatz with ϵ as a small symmetry breaking parameter. In the limit of unbroken symmetry when $\epsilon \to 0$, the first and the second generation fermions would be massless and their superpartners would be exactly degenerate. This means $m_{\tilde{u}_R} \simeq m_{\tilde{c}_R}$ etc., but $m_{\tilde{u}_R}$, $m_{\tilde{d}_R}$, $m_{\tilde{q}_L}$, $m_{\tilde{e}_R}$ and $m_{\tilde{\ell}_L}$ could all be different as in the $\tilde{\text{C}}$MSSM scheme, described earlier. The need then is to construct a model where the degeneracy between the first and the second generation sfermions is broken only at order ϵ^3 or higher. Referring back to the discussion of the flavor problem in §9.5, we can see that

$O(\epsilon)$ corrections to universality are not acceptable. Indeed, even $O(\epsilon^2)$ corrections would be at best marginally allowed unless the first generation sfermions lie well above a TeV.

The small parameter ϵ can be interpreted as the ratio of a VEV $\langle\phi\rangle$, breaking the flavor symmetry, and some large mass scale M which could be the Planck mass or the string scale of superstring theory. The scalar field ϕ can be presumed to be a component of some generic 'flavon' superfield Φ which must be a singlet under the MSSM gauge group. The family symmetry is then chosen in such a way that renormalizable Yukawa couplings are allowed for the third generation, but not for the first and the second. Fermion masses for the latter can only come from nonrenormalizable superpotential terms of mass dimensions four and five respectively:

$$\mathcal{W}_{\text{Yuk}} \sim \tilde{h}_3 HR_3L_3 + \tilde{h}_2 \frac{\Phi}{M}HR_2L_2 + \tilde{h}_1 \frac{\Phi^2}{M^2}HR_1L_1 \ . \tag{12.42}$$

In (12.42) H stands for one of the MSSM Higgs superfields while R_i and L_i are the $SU(2)_L$ singlet and doublet matter superfields respectively and $\tilde{h}_{1,2,3}$ are dimensionless coefficients. We need not specify the gauge quantum numbers of the matter and MSSM Higgs superfields in this and the following discussions. Thus (12.42) should be understood to contain separate up and down type quark doublet and singlet as well as lepton superfields. At energy scales much less than $\langle\phi\rangle$, the effective second and first generation Yukawa couplings are then given respectively by $\tilde{h}_2\epsilon$ and $\tilde{h}_1\epsilon^2$ with $\epsilon = \langle\phi\rangle/M$. Sizable fermion mass hierarchies can thus be generated between generations while keeping the magnitudes of the dimensionless coefficients $\tilde{h}_3$, $\tilde{h}_2$ and $\tilde{h}_1$ to be of the same order[35].

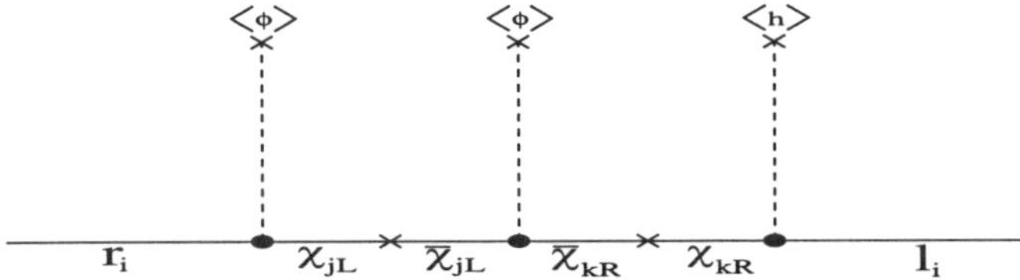

Fig.12.9. A diagram producing an entry of order $\langle H\rangle\langle\phi\rangle^2/(M_jM_k)$ in the SM fermion mass matrix.

An especially elegant mechanism for the origin of Yukawa couplings of the form of (12.42) has been suggested by Frogatt and Nielsen [12.34]. In that the scale M of (12.42) is interpreted as the mass of some left- and right-chiral superfields X_{iL}, X_{iR}, $\bar{X}_{iL}$, $\bar{X}_{iR}$ that form vectorlike representations of the gauge group. Unprotected by any chiral symmetry, M can be quite large. In fact, the flavor symmetry is chosen in such a way that the superpotential does contain large mass terms of the form $\overline{\chi_R}\mathbf{M}_L\chi_L + (L \leftrightarrow R)$, where we have written χ_L as a column vector of fermionic components χ_{iL} of the superfields X_{iL} etc. The nonvanishing entries of the matrix $\mathbf{M}_L$ are all assumed to be similar in magnitude. However, one has to either allow the X_i to transform differently under the family symmetry or introduce several VEVs that break this symmetry or do both. The mass matrices of the SM fermions are then

---

[35]The hierarchy $m_t \gg m_b$ cannot be explained in this manner, however.

generated through the kind of diagram shown[36] in Fig.12.9. Depending on the symmetry and the field content chosen, the resulting low-energy superpotential might well be more restrictive than the general ansatz (12.42).

A model of family symmetry

It is instructive to illustrate the previous general arguments with a concrete example. Following Barbieri et al. [12.35], let us choose a global $U(2)$ family symmetry. The first two generations of MSSM matter supermultiplets are assumed to transform as doublets L_a, R_a, $a = 1, 2$ while the third generation superfields R_3, L_3 as well as the MSSM Higgs superfields are taken to be $U(2)$ singlets. This choice, singling out the third generation, ensures that only fermions of that generation can acquire masses in the limit of unbroken $U(2)$. Moreover, sfermions of the first two generations with identical gauge quantum numbers are exactly degenerate in this limit and cannot mix with third generation ones. The family symmetry in this model is broken spontaneously by the VEVs of two flavon superfields: a doublet Φ^a and an antisymmetric tensor Φ^{ab}, $a, b = 1, 2$ being $U(2)$ indices. Both these superfields are taken to be singlets under the MSSM gauge group. Without loss of generality, their VEVs can be written, with lower case fields signifying scalar components, as

$$\langle \phi^a \rangle = \begin{pmatrix} 0 \\ V \end{pmatrix} ; \quad \langle \phi^{ab} \rangle = v\epsilon^{ab} = \begin{pmatrix} 0 & v \\ -v & 0 \end{pmatrix} . \tag{12.43}$$

For phenomenological reasons, which will become clear shortly, one needs to assume that $V \gg v$. The breaking of the $U(2)$ symmetry can then be envisaged as a two step process:

$$U(2) \underset{V}{\to} U(1) \underset{v}{\to} \text{nothing} . \tag{12.44}$$

In addition to the above, one introduces two heavy families $X^a_{L,R}$, $\bar{X}_{aL,R}$ of Frogatt-Nielsen (FN) chiral superfields which are vectorlike under the MSSM gauge group and transform like doublets under the family $U(2)$. The most general renormalizable superpotential for the model can then be written, with summation implied for repeated indices, as

$$\begin{aligned} \mathcal{W} &= \tilde{h}_3 H L_3 R_3 + H(\lambda_1 L_a X^a_R + \lambda_1' R_a X^a_L) + \Phi^{ab}(\lambda_2 L_a \bar{X}_{bL} + \lambda_2' R_a \bar{X}_{bR}) \\ &\quad + \Phi^a(\lambda_3 L_3 \bar{X}_{aL} + \lambda_3' R_3 \bar{X}_{aR}) + X^a_L \mathbf{M}_L \bar{X}_{aL} + X^a_R \mathbf{M}_R \bar{X}_{aR} . \end{aligned} \tag{12.45}$$

The effective low-energy superpotential can be derived by integrating out the heavy superfields under the assumption that $(M_L)_{ij} \simeq (M_R)_{ij} \gg V \gg v$:

$$\begin{aligned} \mathcal{W}_{eff} &= H \left[\tilde{h}_3 L_3 R_3 + \Phi^a \left(\frac{\lambda_1 \lambda_3'}{M_R} L_a R_3 + \frac{\lambda_1' \lambda_3}{M_L} R_a L_3 \right) \right. \\ &\quad \left. + \Phi^{ab} \left(\frac{\lambda_1 \lambda_2'}{M_R} L_a R_b + \frac{\lambda_1' \lambda_2}{M_L} R_a L_b \right) \right] , \end{aligned} \tag{12.46}$$

---

[36]The order in which the three scalar VEVs appear in Fig.12.9 depends on details of the model.

where we have used the notation $M_{L,R}$ to denote the largest element of the matrix $\mathbf{M}_{L,R}$.

It is noteworthy that terms of the form $(\Phi^a \Phi^b / M^2) H L_a R_b$, though allowed by all the symmetries of the model, do not emerge from the superpotential (12.45). The structure of the Yukawa coupling matrices f_{ij} (cf. 8.11) of the SM fermions can now be obtained by replacing ϕ^a and ϕ^{ab} by their VEVs (12.40), to wit:

$$\mathbf{f} = \begin{pmatrix} 0 & d & 0 \\ -d & 0 & b \\ 0 & c & a \end{pmatrix}. \tag{12.47}$$

The elements of $\mathbf{f}$ have the following orders of magnitude: $a \sim O(1), b, c \sim O(V/M) \equiv \epsilon \ll 1$ and $d \sim O(v/M) \equiv \epsilon' \ll \epsilon$. The last strong inequality becomes necessary since (12.47) implies that

$$\epsilon^2 \sim \frac{M_{f_2}}{M_{f_3}}, \quad \epsilon'^2 \sim \frac{M_{f_1} M_{f_2}}{M_{f_3}^2} \sim \epsilon^2 \frac{M_{f_1}}{M_{f_3}}, \tag{12.48}$$

M_{f_i} standing for an i-th generation fermion mass. A successful description of the known quark mixing phenomenology is allowed by the structure of (12.47) within current errors. It also predicts that

$$\left| \frac{V_{ub}}{V_{cb}} \right| = \sqrt{\frac{M_b}{M_c}} = 0.061 \pm 0.009 \,, \tag{12.49a}$$

$$\left| \frac{V_{td}}{V_{ts}} \right| = \sqrt{\frac{M_d}{M_s}} = 0.226 \pm 0.009 \,. \tag{12.49b}$$

What is more interesting from a supersymmetry standpoint is the impact of the assumed family symmetry on the sparticle spectrum. Let us assume[37] that the energy scale Λ_S of the transmission of supersymmetry breaking from the hidden to the observable sector is at least as large as the mass M of the FN fields. The soft supersymmetry breaking operators at the scale M must then respect the $U(2)$ family symmetry. Any deviation from an exact mass degeneracy between first and second generation sfermions and any mixing between them and those of the third generation are then immediately required to be a consequence of integrating out the superheavy fields. It is then easy to see that the trilinear soft supersymmetry breaking 'A-terms' develop the same family structure as the Yukawa coupling matrix (12.47). This result can be understood by means of the spurion formalism of §11.3, for instance. Our assumption $\Lambda_S \geq M$ means that the spurion fields are $U(2)$ singlets. Just like the Yukawa couplings, the trilinear scalar interactions originate from an F-term contribution to the Lagrangian. The only difference is that the former contributions are linear in the spurion field, cf. (11.14b).

In contrast to the above, the squared scalar soft supersymmetry breaking masses originate from D-term contributions to the Lagrangian in the spurion formalism. Within the effective low energy theory, the mass degeneracy between the first two generation sfermions may then be expected to be lifted by operators such as

$$\hat{O} = \int d^4\theta \, \eta \bar{\eta} \, m^2 \left(\frac{\Phi^a \Phi^b}{M^2} \right) \Psi_a^\dagger \Psi_b \,,$$

---

[37]This assumption is very natural in models with gravity mediated supersymmetry breaking.

where $\Psi_{a,b}$ stands for any first or second generation matter (chiral) superfield. The said degeneracy would then be broken at $O(\epsilon^2)$. As discussed earlier, that is marginally allowed by FCNC constraints. This potentially troublesome operator $\widehat{O}$ is, however, not generated in the renormalizable model [12.35] with FN fields. Instead, the mass-degeneracy is lifted at $O(\epsilon^2\epsilon'^2)$, leading – on account of (12.48) – to

$$\frac{m_{\tilde{f}_1}^2 - m_{\tilde{f}_2}^2}{m_{\tilde{f}_1}^2 + m_{\tilde{f}_2}^2} \sim \frac{M_{f_1} M_{f_2}^2}{M_{f_3}^3} \; . \tag{12.50}$$

Note that the mass-splitting must contain at least one factor of ϵ'. This is since first and second generation matter superfields couple to the flavon sector that breaks the $U(2)$ family symmetry only through Φ^{ab}, cf. (12.45).

The precise magnitude of the splitting, discussed above, obtains from the supersymmetry-breaking part of the scalar potential, namely

$$V = m_1^2 \sum_{a=1}^{2} \left|\tilde{\ell}_a\right|^2 + m_2^2 \left|\tilde{\ell}_3\right|^2 + m_3^2 \sum_a |\tilde{x}_L^a|^2 + m_4^2 \sum_a |\tilde{\bar{x}}_{aL}|^2$$

$$+ \left(B_L M_L \tilde{x}_L^a \tilde{\bar{x}}_{aL} + A_2 \lambda_2 \phi^{ab} \tilde{\ell}_a \tilde{\bar{x}}_{bL} + A_3 \lambda_3 \phi^a \tilde{\ell}_3 \tilde{\bar{x}}_{aL} + \text{h.c.} \right)$$

$$+ \text{ analogous terms for right handed fields.} \tag{12.51}$$

In (12.51) $\tilde{\ell}_{a,3}$ is the scalar component of the chiral superfield $L_{a,3}$ while the same goes for $\tilde{x}_L^a$ ($\tilde{\bar{x}}_{aL}$) with respect to the chiral superfield X_L^a ($\bar{X}_{aL}$). Note that we can write

$$m_3^2 \simeq m_4^2 = M_L^2 + O(M_s^2) \; , \tag{12.52}$$

M_s being the typical intra-supermultiplet mass-split of Ch.1. In fact, every other parameter in (12.48) with the dimension of mass is $O(M_s) \lesssim$ TeV, cf. §1.2. The heavy fields can thus all be integrated out by requiring

$$\frac{\partial V}{\partial \tilde{x}_L^a} = \frac{\partial V}{\partial \tilde{\bar{x}}_{aL}} = 0$$

and replacing the Higgs fields ϕ^a, ϕ^{ab} by their VEVs. The supersymmetry breaking contribution to the squared mass matrix of the light sfermions then develops the form

$$\underset{\sim}{m}_{\text{light}}^2 = \begin{pmatrix} \tilde{m}_1^2 & 0 & c\epsilon\epsilon' \\ 0 & \tilde{m}_1^2 & 0 \\ c^\star\epsilon\epsilon' & 0 & \tilde{m}_2^2 \end{pmatrix} , \tag{12.53}$$

with

$$\tilde{m}_1^2 = m_1^2 + |A_2\lambda_2|^2\epsilon'^2 \; , \tag{12.54a}$$

$$\tilde{m}_2^2 = m_2^2 - |A_3\lambda_3|^2\epsilon^2 \; , \tag{12.54b}$$

$$c = -\lambda_2\lambda_3^\star A_2 A_3^\star \; . \tag{12.54c}$$

Eq. (12.53) immediately implies (12.50). Notice, however, that this scalar squared mass matrix has been written in the same basis as the Yukawa coupling matrix of (12.47). A transformation to the fermion mass basis will thus produce a significant mixing between third generation sfermions and those of the first two generations. The mass splitting between these sfermions is expected to be sizable in this model since $\tilde{m}_1^2$ and $\tilde{m}_2^2$ are independent parameters. As a result, one expects the decay $\mu \to e\gamma$ to occur with a rate close to the current experimental upper bound. Supersymmetric contributions to K^0-$\bar{K}^0$ mixing will also now be expected to be large. Finally, it is reassuring that the $U(2)$ family symmetry can also be combined [12.36] with a GUT gauge group $SU(5)$ or $SO(10)$.

The above model is based on a global, continuous family symmetry. Such an ansatz is subject to the criticism [12.37] that gravitational interactions may not respect any global symmetry. The latter argument is based on the observation that a black hole cannot carry any global charge without an associated gauge invariance. Gravitational quantum corrections might be quite dangerous in the present context since these considerations are at energies not far from the Planck scale. This argument can, of course, be countered by elevating the global family symmetry to a local one. However, one will then have additional D-term contributions, cf. (5.22), to the scalar potential. These will contribute to scalar masses once the scalar fields in them develop VEVs at the time of symmetry breaking, cf. Ch.7. Such D-term contributions do not respect the family symmetry and can therefore introduce large mass splittings between the first and the second generation sfermions. The final possibility is to postulate a gauged discrete family symmetry. Unfortunately, the existing implementations [12.38, 12.39] of this idea seem quite baroque. Let us just remark that the model of Ref. [12.38], based on the group $(S_3)^3$, does establish a connection among fermion mass hierarchies, sfermion degeneracy and approximate R-parity conservation.

The main lesson from our discussions in this section can be summed up as follows. Even if gravitational strength interactions do transmit supersymmetry breaking from the hidden to the observable sector, the mSUGRA boundary conditions (12.24) do not necessarily hold at $M_U \simeq 2 \times 10^{16}$ GeV. The correct and precise form of these boundary conditions is unfortunately not known with any degree of reliability. Moreover, corrections to (12.24) are quite possibly rather small. For these reasons we shall continue to use mSUGRA as one of our benchmark models whenever we need to specify the spectrum of sparticles and Higgs bosons with a managable number of parameters.

12.6 Quantum Effects and Extra Dimensions

We have so far treated supergravity interactions at the classical level and only in four space-time dimensions. Quantum corrections to classical supergravity interactions should certainly exist. However, since the supergravity Lagrangian is not renormalizable, the calculation of such corrections is in general neither straightforward nor unambiguous. Quantum corrections to soft supersymmetry breaking parameters in certain cases can, nonetheless, be computed unambiguously. Such is particularly the case for models with Anomaly Mediated Supersymmetry Breaking (**AMSB**) [12.39–12.42] which we discuss below.

In the **AMSB** scenario the size of the soft supersymmetry breaking parameters is determined by the loop induced superconformal (Weyl) anomaly because there are no direct tree level couplings between the superfields of the hidden and observable sectors. The latter is attained by the following ansätze for the Kähler potential and the superpotential:

$$K(z_i, \sigma_k) = -3M_{Pl}^2 \ln\left[1 - f_o(z_i M_{Pl}^{-1}, \bar{z}^i M_{Pl}^{-1}) - f_h(\sigma_k, \bar{\sigma}^k)\right], \qquad (12.55a)$$

$$\mathcal{W}(Z_i, \Sigma_k) = \mathcal{W}_o(Z_i) + \mathcal{W}_h(\Sigma_k) . \qquad (12.55b)$$

In (12.55) the Z_i are observable sector left chiral superfields, while the Σ_k stand for hidden sector ones and the corresponding lower case symbols refer to their scalar components. Eq. (12.55b) is analogous to (12.11b). But (12.55a) differs from (12.11a) in that we have taken a logarithmic form for the Kähler potential and allowed observable sector fields to contribute via the general dimensionless real function[38] f_o. As a result of the above ansätze, all tree level contributions to soft supersymmetry breaking terms vanish though gaugino masses and A-terms get generated at one loop:

$$M_a = \frac{\beta(g_a)}{g_a} M , \qquad (12.56a)$$

$$A_{ijk} = -\frac{1}{2}(\gamma_i + \gamma_j + \gamma_k)M , \qquad (12.56b)$$

where M is a mass parameter[39], $\beta(g_a)$ is the β-function of the running coupling strength g_a and γ_i is the anomalous dimension of the ith matter superfield, the last two quantities having been computed in §6.8. We assume here that all anomalous dimensions are diagonal in field space, i.e.

$$\gamma_{ij} = \gamma_i \, \delta_{ij} .$$

Note that (12.56a) is compatible with (11.28) to one loop order, but is totally *incompatible* with the assumption (12.25) of high scale gaugino mass unification. Now suppose we make the *additional* assumption [12.43] that

$$\sum_k F^{\Sigma_k}| \, K_{,\sigma_k} = 0 , \qquad (12.57)$$

where F^{Σ_k} is the auxiliary component of the left chiral superfield Σ_k. Then the mass parameter M in (12.56) turns out to equal the gravitino mass $m_{3/2}$. More importantly, the one loop contributions to squared scalar soft breaking masses m_i^2 vanish in this case. This is important since, if nonvanishing, these would have led to scalar masses that are much larger than gaugino masses. Given the strong experimental lower bounds on gaugino masses, scalars would then need to be uncomfortably heavy. Nonvanishing squared scalar masses do get generated at the two loop level:

$$m_i^2(Q) = -\frac{1}{4}\frac{d\gamma_i(Q)}{d\ln Q}m_{3/2}^2 = -\frac{1}{4}\left[\beta(g_a)\frac{\partial\gamma_i}{\partial g_\alpha} + \beta_Y\frac{\partial\gamma_i}{\partial Y}\right]m_{3/2}^2 , \qquad (12.58)$$

---

[38] A consistent low energy theory requires that the Taylor expansion of f_o starts with the bilinear term $z_i \bar{z}_i M_{Pl}^{-2}$.

[39] This arises from the VEV of the auxiliary component of a conformal compensator superfield [12.39].

where Q is the renormalization scale and Y a generic Yukawa coupling strength with β_Y being the corresponding β-function. Repeated indices are summed as usual.

One interesting feature of the results (12.56a), (12.56b) and (12.58) is that, when $M = m_{3/2}$, they are controlled by one mass parameter, namely $m_{3/2}$. Another remarkable property is that they hold at all scales Q in the range $M_s \leq Q \leq M_{Pl}$. In other words, these relations are renormalization group invariant[40]. This means that the spectrum of light sparticles is almost completely insensitive to the details of Planck scale physics, apart from the overall size, which is governed by the gravitino mass of (12.6). It is important to bear in mind, though, that this spectrum results only under specific assumptions about Planck scale physics, as described by (12.55) and (12.57). We see, moreover, that M_s – a typical observable sector supersymmetry breaking mass – is smaller than the gravitino mass by a loop factor $\mathcal{O}(g^2/16\pi^2)$, where g is a generic coupling. *The gravitino in this scenario thus needs to be fairly heavy*, typically, $m_{3/2} \sim$ 20–100 TeV. This eases certain problems of gravitino cosmology, as will be discussed in §16.4.

From a phenomenological viewpoint, (12.56a) elicits maximum interest, since it predicts that the winolike states are lighter than the binolike neutralino. Numerically, at the weak scale, one has

$$M_2 = \frac{\cot^2 \theta_W}{11} M_1 \simeq 0.30 M_1 \; .$$

But this is ignoring NLO corrections. In fact, including NLO corrections, the AMSB prediction is

$$M_1 : M_2 : M_3 = 2.8 : 1 : 7.1 \tag{12.59}$$

at the weak scale, grossly different from (12.27d) of mSUGRA. Thus AMSB is incompatible with high scale gaugino mass unification. In fact, the tree level relation $M_2 \simeq 0.30 M_1$ contradicts (9.21), namely $M_2 \simeq 2M_1$, which was obtained just by assuming that the two gaugino masses become equal at the unification scale. As a result, for wide regions of the AMSB parameter space, the LSP is mostly a neutral wino, and is nearly degenerate in mass with the lighter chargino, also a predominant wino. Generally, because of a relatively large coannihilation cross section, such a winolike LSP does not make a good thermal Dark Matter candidate, as will be shown in §16.3. However, the decay of the heavy gravitino and hidden sector 'moduli' fields in the AMSB scenario can lead to copious LSP production counterbalancing the large coannihilation [12.45].

Unfortunately, the AMSB model, as described so far, is not realistic. Eq. (12.58) leads to renormalization group invariant negative squared masses for sleptons, making them tachyonic. This is a catastrophe for charged sleptons since it would lead to the spontaneous nonconservation of electric charge. Recall, however, that (12.58) is (even) more model-dependent than (12.56). This suggests that it should be possible to modify the former, while leaving the latter unchanged. The simplest solution to the $m_{\tilde{l}}^2 < 0$ problem is just to add [12.42] a universal constant m_0^2 to all squared scalar masses. The RGE invariance of (12.58) is lost and the model is then characterized by four parameters which control sparticle mass

---

[40]These relations were first found in searches for an RGE invariant ansatz for the soft supersymmetry breaking terms [12.44].

spectra: $M, \tan\beta, m_0$ and sgn.(μ). Squares of weak scale scalar masses are now given by

$$m_i^2 = C_i(16\pi^2)^{-2}m_{3/2}^2 + m_0^2 \,. \tag{12.60}$$

Keeping only third generation Yukawa coupling strengths $f_{t,b,\tau}$, the C_i's of (12.60) can be written in terms of those and the gauge coupling strengths $g_{1,2,3}$, cf. Table 12.1. Moreover, the three surviving A-terms can be written as

$$A_{t.b,\tau} = (16\pi^2)^{-1}m_{3/2}f_{t,b,\tau}^{-1}\hat{\beta}_{f_{t,b,\tau}}, \tag{12.61}$$

where the coefficients $\hat{\beta}_{f_{t,b,\tau}}$ have also been listed in Table 12.1.

$$\hat{\beta}_{f_t} = f_t\left(-\frac{13}{15}g_1^2 - 3g_2^2 - \frac{16}{3}g_3^2 + 6f_t^2 + f_b^2\right),$$

$$\hat{\beta}_{f_b} = f_t\left(-\frac{7}{15}g_1^2 - 3g_2^2 - \frac{16}{3}g_3^2 + f_t^2 + 6f_b^2 + f_\tau^2\right),$$

$$\hat{\beta}_{f_\tau} = f_\tau\left(-\frac{9}{5}g_1^2 - 3g_2^2 + 3f_b^2 + 4h^2\tau\right),$$

$$C_Q = -\frac{11}{50}g_1^4 - \frac{3}{2}g_2^4 + 8g_3^4 + f_t\hat{\beta}_{f_t} + f_b\hat{\beta}_{f_b},$$

$$C_{\bar{U}} = -\frac{88}{25}g_1^4 + 8g_3^4 + 2f_t\hat{\beta}_{f_t},$$

$$C_{\bar{D}} = -\frac{22}{25}g_1^4 + 8g_3^4 + 2f_b\hat{\beta}_{f_b},$$

$$C_L = -\frac{99}{50}g_1^4 - \frac{3}{2}g_2^4 + f_\tau\hat{\beta}_{f_\tau},$$

$$C_{\bar{E}} = -\frac{198}{25}g_1^4 + 2f_\tau\hat{\beta}_{f_\tau},$$

$$C_{H_2} = -\frac{99}{50}g_1^4 - \frac{3}{2}g_2^4 + 3f_t\hat{\beta}_{f_t},$$

$$C_{H_1} = -\frac{99}{50}g_1^4 - \frac{3}{2}g_2^4 + 3f_b\hat{\beta}_{f_b} + f_\tau\hat{\beta}_{f_\tau}$$

Table 12.1. Expressions for C_i's and $\hat{\beta}$'s for third generation sfermions; those for the first two generations follow simply by setting the Yukawa couplings to zero.

A distinctive prediction of this model is that the $SU(2)_L$ singlet and doublet charged sleptons are highly mixed and nearly mass degenerate, i.e. $m_{\tilde{e}_L}^2 \simeq m_{\tilde{e}_R}^2$: a relation which holds at the percent level independently of the values of $m_{3/2}$ and m_0. The third generation sleptons are rather light here [12.42]. However, the relative ordering of slepton and gaugino masses can no longer be predicted. Nevertheless, some interesting statements can be made on the charginos and neutralinos in this model. While $\tilde{\chi}_2^0$ is a $U(1)_Y$ gaugino described by the field $\tilde{\lambda}_0$ of (9.30b), the LSP $\tilde{\chi}_1^0$ is a nearly pure wino described by the field λ_3 of (9.30a). Moreover, $\tilde{\chi}_1^{\pm}$ (also winos) are highly degenerate with $\tilde{\chi}_1^0$, the mass difference being of the order of hundreds of MeV. Thus a produced $\tilde{\chi}_1^+/\tilde{\chi}_1^-$ will decay into a $\tilde{\chi}_1^0$ and a soft charged pion with a large decay length – leading to distinct and unique signatures for collider experiments. As usual, squarks cannot be lighter than gluinos; in fact, the masses of the former have rather high values because the RHS of (12.60) contains a term proportional to g_3^4 for squarks. A very interesting prediction [12.46] of this model is that the CP odd Higgs boson A (cf. Ch.10) is quite heavy – typically over 500 GeV in mass – and the CP even lighter neutral Higgs scalar h has a mass less than or equal to 118 GeV.

Adding a universal contribution to the RHS of (12.60) means that the equation is no longer form invariant under RGE, i.e. one has to specify a scale Q_U (typically taken $\simeq M_U$) where m_0^2 is supposed to be universal. The invariance under RGE can be saved if, instead of putting in m_0 by hand, one postulates the existence of extra D-term contributions to squared scalar masses. In the simplest case [12.47] one introduces a D-term from a gauged $U(1)_{B-L}$. This symmetry might be broken at an arbitrarily high scale, so long as it leaves *global* $U(1)_{B-L}$ invariance unbroken in the observable sector. The Lagrangian density can then also contain the operator which mixes the field strengths for $U(1)_Y$ and $U(1)_{B-L}$, namely

$$c \int d^2\theta\, W_{\alpha,B-L} W_Y^\alpha \ ,$$

with an arbitrary coefficient c. This leads to a Fayet-Illiopoulos term ξ_Y for the hypercharge $U(1)_Y$, again with an unknown coefficient[41]. One altogether gets two extra contributions to squared scalar masses [12.47], summed as

$$\Delta_D m_i^2 = -(B_i - L_i)D_{B-L} + g_Y^2 Y_i \xi_Y \ , \tag{12.62}$$

which is again invariant under RGE. For a judicious choice of the ratios D_{B-L}/M and ξ_Y/M, where M is the mass parameter of (12.56), all squared scalar masses can be made positive[42]. In this case, however, the left-right mass degeneracy of the charged sleptons gets lifted. Nevertheless, other features remain more or less the same, so that most of the earlier mentioned signatures survive.

Returning to the theoretical basis of the AMSB scenario, the ansätze (12.55), (12.57) may not seem particularly natural from a four dimensional standpoint. However, they emerge naturally in higher dimensional "braneworld" scenarios [12.48] in which our world is taken to be embedded in a bulk with a higher number of spatial dimensions with the extra dimensions curled up. Suppose one takes the hidden and the observable sector superfields to be localized on two parallel but distinct spacelike hypersurfaces known as[43] three-branes, located $\mathcal{O}(r_c)$

---

[41]This term is not to be confused with the usual MSSM D-term contribution, proportional to $M_Z^2 \cos 2\beta$, which is also present.

[42]Other solutions of the $m_i^2 < 0$ problem can be found in the papers cited in Refs. [12.47].

[43]These emerge as solitonic solutions in String Theory.

apart in the higher dimensional bulk, r_c being the radius of compactification, cf. Fig. 12.10.

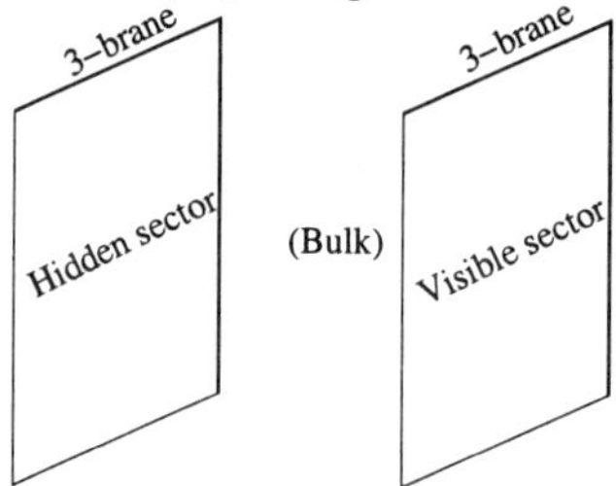

Fig. 12.10. Supersymmetry breaking across extra dimension(s) in a braneworld scenario.

Now any tree level exchange with bulk fields of mass m ($> r_c^{-1}$) will be suppressed by the factor e^{-mr_c}. Supergravity fields propagate in the bulk, but the supergravity mediated tree level couplings can now be eliminated by a rescaling transformation. The problem can be considered in the background of a compensator left chiral superfield S whose VEV is given by [12.40] $\langle S \rangle = 1 + m_{3/2}\,\theta\theta$. The said rescaling transformation for a generic observable sector superfield Z is then given by $Z \to ZS$. However, this rescaling symmetry is anomalous at the quantum level and the communication of supersymmetry breaking from the hidden to the observable sector takes place through the loop generated superconformal anomaly. The situation is schematically depicted in Fig. 12.11.

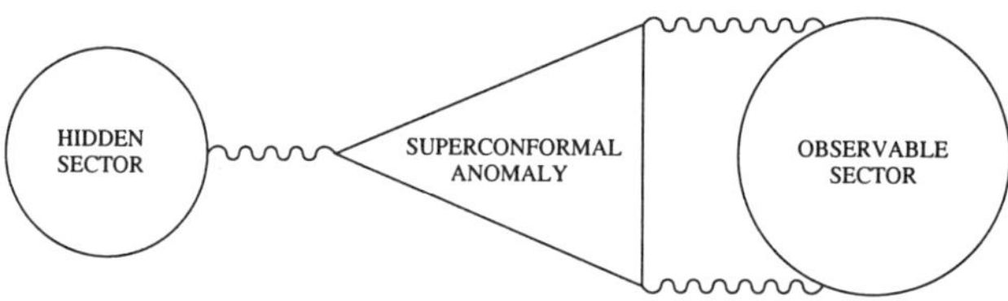

Fig. 12.11. Schematic depiction of anomaly mediation.

The superconformal anomaly is topological in origin and naturally conserves flavor. Thus no new FCNC amplitudes are introduced (apart from calculable Yukawa corrections) from supersymmetry breaking terms in an AMSB scenario. This is an advantage that the latter has over usual schemes of gravity mediated supersymmetry breaking from tree level exchanges between the hidden and the observable sectors. Though gravity is flavor blind, the supergravity invariance of the Lagrangian in the latter case cannot prevent the occurrence of tree level terms like

$$\mathcal{L}_{eff} \simeq \int d^4\theta \; h M_{Pl}^{-2} \Sigma^+ \Sigma Z^+ Z \,, \tag{12.63}$$

$\Sigma(Z)$ being a generic hidden (observable) sector superfield and h being a dimensionless coupling strength of order unity. In ordinary tree level gravity mediated supersymmetry breaking models, there is no a priori symmetry (unless postulated in an ad hoc manner) to keep h diagonal in flavor space. As a result, when soft sfermion mass matrices get induced from supersymmetry breaking (say through a nonzero VEV $\langle F_\Sigma \rangle$) in the hidden sector, they develop significant off diagonal terms in flavor space, leading to a major conflict with strong experimental upper bounds on FCNC amplitudes (cf. §9.5). On the other hand, a term like (12.63) is naturally absent (both at the tree level and in the loop induced contribution) in the AMSB scenario which thereby claims to have overcome the flavor problem in broken supersymmetry.

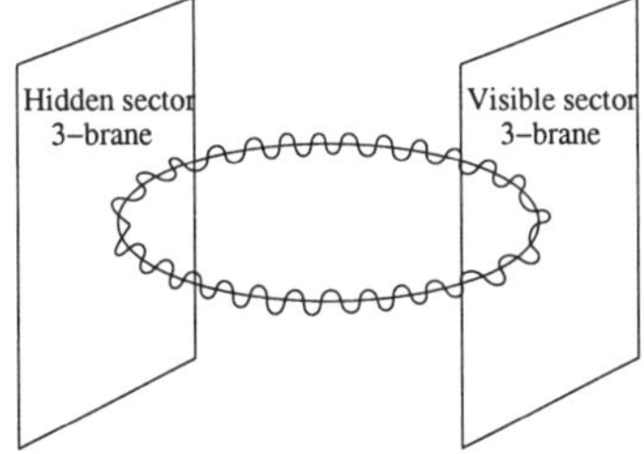

Fig. 12.12. An interbrane gaugino/higgsino loop.

There have been several braneworld extensions of the AMSB scenario. A particular one [12.49], that is worth mentioning, goes by the name of gaugino/higgsino mediated supersymmetry breaking. In this scheme once again there are two separated 3-branes in a higher dimensional bulk. But now only observable matter superfields live on the corresponding brane, while gauge and Higgs superfields can propagate in the bulk. In this situation, an interbrane gaugino/higgsino loop (cf. Fig. 12.12), in addition to the superconformal anomaly, can transmit supersymmetry breaking from the hidden to the visible sector. This kind of a scenario is characterized by the following features

- $M_{1/2} \sim m_{3/2} \sim |M_{H_1}| \sim |m_{H_2}| \sim |\mu B|$.

- Sleptons are never tachyonic.

- The μ problem can be tackled.

- The near mass degeneracies $M_{\tilde{\chi}_1^0} \sim M_{\tilde{\chi}_1^\pm}$, $m_{\tilde{e}_L} \sim m_{\tilde{e}_R}$ of minimal AMSB are lost.

In fact, one can construct even more general mechanisms for the transmission of supersymmetry breaking with two separated and parallel 3-branes in a higher dimensional bulk. This research is currently (2003) being pursued vigorously, but the field has not matured enough to be covered in a book. The reader is referred to some interesting papers [12.50] in this genre.

There have been other approaches that introduce supersymmetry breaking in the four dimensional spacetime via mechanisms enacted in compact extra dimensions. A review of these may be found in Ref. [12.51]. We can mention one of the first such mechanisms that is due to Scherk and Schwarz [12.52]. Bosons and fermions, belonging to the same supermultiplet in four dimensions, are here given different boundary conditions or topological properties in the compact space – thereby breaking Bose-Fermi degeneracy. We can, in fact, make one remark in this general context. Dynamical supersymmetry breakdown in a higher dimensional spacetime and the compactification of the extra dimensions within a size r_c can naturally induce soft supersymmetry breaking terms of order r_c^{-1} [12.53]. However, in this case one would go directly from a four dimensional nonsupersymmetric theory at energies well below r_c^{-1} to a higher dimensional supersymmetric theory at high energies. It is not *a priori* clear whether there exists a range of energies where physics can be usefully described by a softly broken supersymmetric field theory in four dimensions, which is the theme of this book. Moreover, if generic nonrenormalizable operators in the low energy theory are only suppressed by powers of r_c, electroweak precision experiments and bounds from rare processes would force r_c^{-1} to lie well above 1 TeV. This would imply sparticle masses that are uncomfortably high in terms of solving the naturalness problem in four dimensions. We therefore do not pursue this idea any further.

12.7 Annex to Chapter 12

A brief introduction to $N{=}1$ supergravity

We give here a short and pedestrian presentation (without detailed derivations) of those essential features of the $N{=}1$ supergravity theory that are relevant to the contents of this chapter. In this context, the first point to be noted is the utility of the tetrad formalism [12.2]. The use of this formalism is especially convenient owing to the presence of a gauge fermion, namely the gravitino, in supergravity theory. So we develop it first.

Tetrad formulation of gravity and supergravity

While labelling spacetime coordinates as x^μ, with μ (= 0, 1, 2, 3) as the world index, we can locally introduce at each spacetime point a tangent frame known as the vierbein or tetrad frame. The latter consists of a sufficiently differentiable field of four vectors $e_m(x) \equiv e_m{}^\mu \, \partial_\mu$ with $m = 0,1,2,3$ and a constant Minkowski metric $\eta_{mn} = \eta^{mn} = (1, -1, -1, -1)$. This is in contrast with the variable metric tensor $g_{\mu\nu}(x)$ for the spacetime manifold. The tetrads obey the orthonormality condition $e_m \cdot e_n = \eta_{mn}$. The dual frame is defined by $e^m \equiv e^m{}_\mu dx^\mu$ with the conditions $e^n{}_\mu e_m{}^\mu = \delta^n_m, e^m{}_\mu e_m{}^\nu = \delta^\nu_\mu$. Components of a vector can be defined in the tetrad basis, e.g. $A^m \equiv e^m{}_\mu A^\mu$ and the same is true of a tensor of any rank. We thus have $\eta_{mn} = e_m{}^\mu e_n{}^\nu g_{\mu\nu}$, $A^m B_m = e^m{}_\mu A^\mu e_m{}^\nu B_\nu = A^\mu B_\mu$. General coordinate transformations $x^\mu \to x'^\mu$ leave the tangent frame invariant while local Lorentz transformations $e^m \to \Lambda^m{}_n e^n$ in the latter take place at every spacetime point independently from those at any other. In order to introduce the covariant derivative of an object such as $e^m{}_\nu$ with mixed types of indices, one needs two types of connections: (1) the **affine connection** [12.2] $\Gamma^\lambda{}_{\mu\nu}$ for differentiating w.r.t. world indices and (2) the **spin connection** [12.54] $\omega^{mn}{}_\mu$ for differentiating w.r.t. local tangent frame indices. Thus the action of a general covariant derivative on a tetrad can be written as

$$D_\mu e^m{}_\nu \equiv \partial_\mu e^m{}_\nu + \Gamma^\lambda{}_{\mu\nu} e^m{}_\lambda - \omega^{mn}{}_\mu e_{n\nu} \; . \tag{12.64}$$

For supergravity, one needs to work in an Einstein-Cartan spacetime which has a nonzero torsion tensor

$$S^\lambda{}_{\mu\nu} = \frac{1}{2}\left(\Gamma^\lambda{}_{\mu\nu} - \Gamma^\lambda{}_{\nu\mu}\right). \tag{12.65}$$

In fact, the affine connection here can be split into two additive parts, namely the standard Christoffel-Riemann connection $\overset{\circ}{\Gamma}{}^\lambda{}_{\mu\nu}$, definable for any manifold, plus a contorsion tensor which is linear in the components of torsion:

$$\Gamma^\lambda{}_{\mu\nu} = \overset{\circ}{\Gamma}{}^\lambda{}_{\mu\nu} - K^\lambda{}_{\mu\nu}\,, \tag{12.66a}$$

$$\overset{\circ}{\Gamma}{}^\lambda{}_{\mu\nu} = \overset{\circ}{\Gamma}{}^\lambda{}_{\nu\mu} = -\frac{1}{2} g^{\lambda\alpha} \left(\partial_\mu g_{\nu\alpha} + \partial_\nu g_{\mu\alpha} - \partial_\alpha g_{\mu\nu}\right), \tag{12.66b}$$

$$K^\lambda{}_{\mu\nu} = -g^{\lambda\alpha} \left(g_{\alpha\beta} S^\beta{}_{\mu\nu} + g_{\mu\beta} S^\beta{}_{\alpha\nu} + g_{\nu\beta} S^\beta{}_{\alpha\mu}\right). \tag{12.66c}$$

A similar additive split occurs [12.54] for the spin connection in an Einstein-Cartan space-time, namely

$$\omega^{mn}{}_\mu = \overset{\circ}{\omega}{}^{mn}{}_\mu + K^{mn}{}_\mu\,, \tag{12.67}$$

where

$$K^{mn}{}_\mu = -K^{nm}{}_\mu = K^\lambda{}_{\mu\nu} e^{m\nu} e^n{}_\lambda\,, \tag{12.68a}$$

$$\overset{\circ}{\omega}{}^{mn}{}_\mu = \frac{1}{2}\left[e^{m\nu}(\partial_\nu e^n{}_\mu - \partial_\mu e^n{}_\nu) + \frac{1}{2}\, e^{m\alpha} e^{n\beta}(\partial_\alpha e_{l\beta} - \partial_\beta e_{l\alpha}) e^l{}_\mu\right] - (m \leftrightarrow n)\,. \tag{12.68b}$$

The nonminimal covariant derivative $\hat{\mathcal{D}}_\mu$ acting on a tetrad can be defined as

$$\hat{\mathcal{D}}_\mu e^m{}_\nu \equiv \partial_\mu e^m{}_\nu - \omega^{mn}{}_\mu e_{n\nu}\,. \tag{12.69}$$

One can then derive that

$$\hat{\mathcal{D}}_\nu e^m{}_\mu - \hat{\mathcal{D}}_\mu e^m{}_\nu = 2S^\lambda{}_{\mu\nu} e^m{}_\lambda\,. \tag{12.70}$$

The metricity constraint [12.54] on the tetrad relates the affine connection Γ to the spin connection ω and the tetrad e as follows:

$$\Gamma^\mu{}_{\nu\lambda} = e_m{}^\mu(-\partial_\nu e^m{}_\lambda + \omega^{mn}{}_\nu e_{n\lambda}). \tag{12.71}$$

The Riemann curvature tensor can be given as a function of the affine connection as

$$-R^\mu{}_{\nu\rho\lambda}(\Gamma) = \partial_\rho \Gamma^\mu{}_{\lambda\nu} - \partial_\lambda \Gamma^\mu{}_{\rho\nu} - \Gamma^\mu{}_{\rho\alpha}\Gamma^\alpha{}_{\lambda\nu} + \Gamma^\mu{}_{\lambda\alpha}\Gamma^\alpha{}_{\rho\nu}\,. \tag{12.72}$$

The same tensor in the tetrad basis is defined by

$$R^{mn}{}_{\rho\lambda}(\omega) = e^{m\nu} e^n{}_\mu R^\mu{}_{\nu\rho\lambda}(\Gamma). \tag{12.73}$$

In terms of the spin connection, the LHS of (12.73) can be expressed as

$$R^{mn}{}_{\rho\lambda}(\omega) = \partial_\rho \omega^{mn}{}_\lambda - \partial_\lambda \omega^{mn}{}_\rho + \omega^{ml}{}_\rho \omega_l{}^n{}_\lambda - \omega^{ml}{}_\lambda \omega_l{}^n{}_\rho\,. \tag{12.74}$$

The scalar curvature R is then given by

$$R = \frac{1}{2}(e_m{}^\rho e_n{}^\lambda - e_m{}^\lambda e_n{}^\rho) R^{mn}{}_{\rho\lambda}(\omega). \tag{12.75}$$

It can also be cast in the determinant form

$$R = -\frac{1}{4e} e^p{}_\mu e^q{}_\nu \, \epsilon_{pqmn} \, \epsilon^{\mu\nu\lambda\rho} \, R^{mn}{}_{\lambda\rho} \, , \tag{12.76}$$

where the tetrad determinant[44] e is defined by

$$e \equiv \det(e^m{}_\mu) = 1/\det(e_m{}^\mu) = [-\det(g_{\mu\nu})]^{1/2} \, . \tag{12.77}$$

The Lagrangian density $M_{Pl}^2 \, (-g)^{1/2} R(g,\Gamma)$ for the Hilbert action [12.2] can then be written (in the first order form) as

$$\mathcal{L}_G = -\frac{M_{Pl}^2}{2} \, eR(e,\omega) = \frac{M_{Pl}^2}{8} \, e^p{}_\mu e^q{}_\nu \epsilon_{pqmn} \, \epsilon^{\mu\nu\lambda\rho} \, R^{mn}{}_{\lambda\rho}(\omega) \, . \tag{12.78}$$

The equations of motion are obtained from (12.78) by treating $e^m{}_\mu$ and $\omega^{mn}{}_\mu$ as independent sets of variables. The variation with respect to the tetrad[45] yields the Einstein equations

$$e\left(R^m{}_\sigma - \frac{1}{2} e^m{}_\sigma R\right) = \frac{1}{M_{Pl}^2}(\Theta_M)^m{}_\sigma \, , \tag{12.79}$$

where

$$R^m{}_\sigma = R^{mn}{}_{\rho\sigma}(\omega) e_n{}^\rho \tag{12.80}$$

and the energy momentum tensor $(\Theta_M)^m{}_\sigma$ in the tetrad basis is defined through the change in the matter action S_M for an infinitesimal tetrad variation:

$$\delta S_M = \int d^4x (\Theta_M)^m{}_\sigma \delta e_m{}^\sigma \, . \tag{12.81}$$

While extending this theory to $N=1$ supergravity [12.55], we have to consider the multiplet of supergravity fields consisting of (apart from auxiliary fields) the set $\{e^m{}_\mu, \, \Psi_\mu\}$ where Ψ_μ is a Rarita-Schwinger spin 3/2 field describing the gravitino. Eq.(12.78) now generalizes to

$$\mathcal{L}_{SG} = -\frac{M_{Pl}^2}{2} \, eR\left(e,\omega(e,\Psi)\right) - \frac{1}{2} \, \epsilon^{\mu\nu\rho\sigma} \bar{\Psi}_\mu \gamma_5 \gamma_\nu \hat{D}_\rho \Psi_\sigma \, . \tag{12.82}$$

The second RHS term in (12.82) corresponds to the kinetic energy of the gravitino field and here

$$\hat{D}_\rho = \partial_\rho - \frac{i}{4} \, \omega^{mn}{}_\rho \Sigma_{mn} \, .$$

Furthermore, one has

$$K^{mn}{}_\mu = \frac{i}{4M_{Pl}^2}(\bar{\Psi}_\mu \gamma^m \Psi^n - \bar{\Psi}_\mu \gamma^n \Psi^m + \bar{\Psi}^m \gamma_\mu \Psi^n) \, ,$$

---

[44]Some useful relations with e are : $e\epsilon_{\mu\nu\rho\lambda} = e^m{}_\mu e^n{}_\nu e^p{}_\rho e^q{}_\lambda \epsilon_{mnpq}$, $e^{-1}\epsilon^{\mu\nu\rho\lambda} = e_m{}^\mu e_n{}^\nu e_p{}^\rho e_q{}^\lambda \epsilon^{mnpq}$, $e\epsilon_{\mu\nu\rho\lambda}e^\rho{}_p = e^m{}_\mu e^n{}_\nu e^q{}_\lambda \, \epsilon_{mnpq}$, $e^{-1} \, \epsilon^{\mu\nu\rho\lambda}e^n{}_\nu = e_m{}^\mu e_p{}^\rho e_q{}^\lambda \epsilon^{mnpq}$.

[45]The variation with respect to the spin connection yields the algebraic equation that relates the spin density and the torsion tensors [12.54].

i.e. the contorsion is a bilinear in the gravitino field. This fact provides an *a posteriori* justification of the need to set up this theory in an Einstein-Cartan spacetime.

General matter and gauge couplings in N=1 supergravity

We shall only sketch the procedure here stating the final results; details may be found in Refs. [12.3] and [12.56]. Recall first the most general renormalizable and globally (N=1) supersymmetric Lagrangian density in a flat spacetime with matter and nonabelian gauge fields, namely (5.54). Utilizing (4.12c,d), the corresponding action may be written in the notation of (5.54) as

$$S = \int d^6z \left[-\frac{1}{8}\bar{\mathcal{D}}\bar{\mathcal{D}}\Phi_i^\dagger (e^V)_{ij}\Phi_j + \mathcal{W}(\Phi_i) + \frac{1}{4}W^{aA}W_A^a \right] + h.c. \tag{12.83}$$

Any extension of (12.83) to curved spacetime will lose renormalizability and hence will have to allow nonrenormalizable kinetic energy terms for chiral (matter) as well as vector (gauge) superfields so long as they are compatible with global supersymmetry. This means two extensions in practice: (1) the expression to the right of the $(-\bar{\mathcal{D}}\bar{\mathcal{D}}/8)$ factor in the integrand of the first RHS term in (12.83) gets changed to a general **real** function of the form $\mathcal{K}[(\Phi^\dagger e^V)_i, \Phi_i]$ with mass dimension two and (2) the coefficient $1/4$ of the gauge kinetic energy term gets multiplied by $f_{ab}(\Phi_i)$ which is an arbitrary dimensionless analytic function of the chiral superfields, symmetric in the gauge group indices a, b. Thus a general extension of (12.83) to curved spacetime, while retaining global ($N = 1$) supersymmetry, would be

$$S_{\text{GLOBAL}} = \int d^6z \left[-\frac{1}{8}\bar{\mathcal{D}}\bar{\mathcal{D}}\mathcal{K}\left[(\Phi^\dagger e^V)_i, \Phi_i\right] + \mathcal{W}(\Phi_i) + \frac{1}{4}f_{ab}(\Phi_i)W^{aA}W_A^a \right] + h.c. \tag{12.84}$$

It is convenient to redefine the function $\mathcal{K}$ in terms of another function K:

$$K \equiv -3\ln\left[-\frac{1}{3}\frac{\mathcal{K}}{M_{Pl}^2}\right]. \tag{12.85}$$

The generalization of (12.84), obeying *local* N=1 supersymmetry or supergravity, comes across two new features. First, the tetrad $e^m{}_\mu$ has to be extended to a supertetrad [12.4] which contains $e^m{}_\mu$ and the gravitino field Ψ_μ, apart from auxiliary fields, as components. Specifically, the tetrad determinant e generalizes to the superspace determinant E so that the superspace coordinate invariant integral measure has the factor E, as compared with the spacetime coordinate invariant measure which has the factor e. Second, the scalar curvature R needs to be generalized [12.4] to a left chiral superfield $\mathcal{R}$; indeed, the action corresponding to the superspace generalization of (12.78) turns out [12.4] to be

$$S_{SG} = -3M_{Pl}\int d^6z E\mathcal{R} + h.c. \tag{12.86}$$

Now (12.84) generalizes in the supergravity case to [12.4]

$$S_{\text{LOCAL}} = \int d^6z E\left[\frac{3}{8}M_{Pl}(M_{Pl}\bar{\mathcal{D}}\bar{\mathcal{D}} - 8\mathcal{R})\exp\left[-\frac{1}{3}K\{(\Phi^\dagger e^V)_i, \Phi_i\}\right]\right.$$

$$\left. + \mathcal{W}(\Phi_i) + \frac{1}{4}f_{ab}(\Phi_i)W^{aA}W_A^a \right] + h.c. \tag{12.87}$$

The final Lagrangian density obtains after expanding all the superfields of (12.87) into their components and doing very lengthy algebra. The real function K of superfields contributes $K| \equiv K(\phi_i, \bar{\phi}^i)$ of (12.2), ϕ_i and $\bar{\phi}^i$ being the scalar components of Φ_i and $\Phi_i^\dagger$ respectively. Then a Weyl rescaling of the tetrad, the fermionic two spinor component ξ of Φ and the gravitino field Ψ_μ, namely

$$e^m{}_\mu \to e^m{}_\mu \exp\left[\frac{1}{6}\frac{K}{M_{Pl}^2}\right], \, \xi_i \to \xi_i \exp\left[-\frac{1}{12}\frac{K}{M_{Pl}^2}\right], \, \Psi_\mu \to \Psi_\mu \exp\left[\frac{1}{12}\frac{K}{M_{Pl}^2}\right], \qquad (12.88)$$

has to be done to bring the Lagrangian to the desired form. One finds that $K(\phi_i, \bar{\phi}_i)$ combines with $\mathcal{W}(\phi_i, \bar{\phi}_i)$ in the manner of (12.2) yielding the Kähler potential $\mathcal{G}$, i.e. K enters only through $\mathcal{G}$. For our purpose, the Lagrangian density for an on-shell description [12.55–12.57] of $N=1$ supergravity, coupled to a general system of gauge fields and gauge variant matter fields, can be written as

$$\mathcal{L} = \mathcal{L}_{FK} + \mathcal{L}_P + \mathcal{L}_{BK} + \mathcal{L}_{FM} + \mathcal{L}_{(4)F}\,. \qquad (12.89)$$

The subscripts BK, P, FK, FM and $(4)F$ in (12.89) respectively stand for boson kinetic energy, potential, fermion kinetic energy, fermion masslike and four fermion interaction terms. The last RHS term in (12.89) is of no relevance to this chapter and we ignore it.

We shall now explicitly display the first four RHS terms of (12.89) with the notation which has been explained in §12.2. Recall from (12.3) that $\tilde{\mathcal{D}}_\mu$ is a general covariant derivative including both the affine connection and the gauge field terms. Furthermore, we employ the four component spinor formalism. Gauginos are described by Majorana fermionic fields λ_M^a, while chiral matter fermions are described by left chiral components ψ_{iL} of four spinor fields[46] ψ_i. Note that the right chiral components ψ_i do not appear in any of the following equations. If ψ_i were to be a Dirac field, ψ_{iR} would need to arise from the charge conjugate of some other left chiral antifermion field, cf. Ch.5. The said terms can be given as

$$
\begin{aligned}
e^{-1}\mathcal{L}_{FK} = {}& \frac{i}{2}\Re e\, f_{ab}\bar{\lambda}_M^a \slashed{\tilde{\mathcal{D}}}\lambda_M^b - \frac{1}{2}e^{-1}\epsilon^{\mu\nu\alpha\beta}\bar{\Psi}_\mu\gamma_5\gamma_\nu\hat{\mathcal{D}}_\alpha\Psi_\beta - i\mathcal{G}_j^i\overline{\psi_L^j}\slashed{\tilde{\mathcal{D}}}\psi_{iL} \\[4pt]
& -\frac{i}{4}\Im m\, f_{ab}\tilde{\mathcal{D}}^\mu(e\bar{\lambda}_M^a\gamma_5\gamma_\mu\lambda_M^b) - \frac{1}{4}\frac{e^{-1}}{M_{Pl}^2}\epsilon^{\mu\nu\alpha\beta}\bar{\Psi}_\mu\gamma_\nu\Psi_\alpha(\mathcal{G}^i\tilde{\mathcal{D}}_\beta\phi_i - \mathcal{G}_i\tilde{\mathcal{D}}_\beta\bar{\phi}^i) \\[4pt]
& +\left[\frac{i}{2}\overline{\psi_L^i}\slashed{\tilde{\mathcal{D}}}\phi_j\psi_{kL}(\mathcal{G}_i^{kj} + \frac{1}{2M_{Pl}^2}\,\mathcal{G}_i^k\mathcal{G}^j) - \frac{1}{\sqrt{2}}\,\frac{1}{M_{Pl}}\mathcal{G}_i^j\bar{\Psi}_\mu\slashed{\tilde{\mathcal{D}}}\bar{\phi}^i\gamma^\mu\psi_{jL} + h.c.\right] \\[4pt]
& +\frac{1}{M_{Pl}}\left(\frac{1}{4}\Re e\, f_{ab}\bar{\lambda}_M^a\gamma^\mu\Sigma^{\alpha\beta}\Psi_\mu F_{\alpha\beta}^b - \frac{\Re e\, f_{ab}}{4M_{Pl}}\mathcal{G}^i\tilde{\mathcal{D}}^\mu\phi_i\bar{\lambda}_M^a\gamma_\mu\lambda_{ML}^b \right. \\[4pt]
& \left. -\frac{M_{Pl}}{2\sqrt{2}}f_{ab},{}^i\overline{\psi_{iL}{}^C}\Sigma^{\mu\nu}F_{\mu\nu}^a\lambda_{ML}^b + h.c.\right),
\end{aligned}
\qquad (12.90a)
$$

---

[46] We follow the normalization convention of Ref. [12.4] for these ψ_i fields. Therefore these are $\sqrt{2}$ times the χ_i's of Ref. [12.56]. Note also that our $\mathcal{G}$ is the negative of that of Ref. [12.56]. Moreover, our $\epsilon(x)$, cf. (12.1), is half the infinitesimal parameter of local $N=1$ supersymmetric transformations in Ref. [12.56].

$$e^{-1}\mathcal{L}_{BK} = -\frac{M_{Pl}^2}{2}R - \mathcal{G}_j^i\tilde{\mathcal{D}}_\mu\phi_i\tilde{\mathcal{D}}^\mu\bar{\phi}^j - \frac{1}{4}\Re e f_{ab}F_{\mu\nu}^a F^{b\mu\nu} + \frac{1}{4}\Im m\, f_{ab}\tilde{F}_{\mu\nu}^a F^{b\mu\nu}, \tag{12.90b}$$

$$e^{-1}\mathcal{L}_P = -\frac{1}{2}\sum_\alpha g_\alpha^2 \Re e f_{ab}^{-1}\mathcal{G}^i(T^{\alpha a})_i^j\phi_j\mathcal{G}^k(T^{\alpha b})_k^l\phi_l$$

$$+e^{-\mathcal{G}/M_{Pl}^2}\left[\mathcal{G}^i(\mathcal{G}^{-1})_i^j\mathcal{G}_j + 3M_{Pl}^2\right]M_{Pl}^2, \tag{12.90c}$$

$$e^{-1}\mathcal{L}_{FM} = \frac{i}{2}M_{Pl}e^{-\mathcal{G}/(2M_{Pl}^2)}\bar{\Psi}_\mu\Sigma^{\mu\nu}\Psi_\nu + \left[-\frac{i}{\sqrt{2}}e^{-\mathcal{G}/(2M_{Pl}^2)}\mathcal{G}^i\bar{\Psi}_\mu\gamma^\mu\psi_{iL}\right.$$

$$+\frac{M_{Pl}}{2}e^{-\mathcal{G}/(2M_{Pl}^2)}\left\{\mathcal{G}^{ij} - M_{Pl}^{-2}\mathcal{G}^i\mathcal{G}^j - \mathcal{G}_k^{ij}(\mathcal{G}^{-1})_l^k\mathcal{G}^l\right\}\overline{\psi_{iL}{}^C}\psi_{jL} + h.c.\bigg]$$

$$+\frac{M_{Pl}}{4}e^{-\mathcal{G}/(2M_{Pl}^2)}\left\{f_{ab,k}^*(\mathcal{G}^{-1})_i^k\mathcal{G}^i\bar{\lambda}_M^a P_R\lambda_M^b + h.c.\right\}$$

$$+\left[\frac{i}{2M_{Pl}}\sum_\alpha g_\alpha\{\mathcal{G}^i(T^{\alpha a})_i^j\phi_j\bar{\Psi}_\mu\gamma^\mu\lambda_{MR}^a - 2i\sqrt{2}\mathcal{G}_j^i(T^{\alpha a})_k^j\bar{\phi}^k\bar{\lambda}_M^a\psi_{iL}\right.$$

$$\left.-\frac{1}{\sqrt{2}}\,\Re e(f^{-1})_a^b f_{bc,}^{\star}{}^k\mathcal{G}_i(T^{\alpha a})_j^i\bar{\phi}^j\bar{\lambda}_M^c\psi_{kL}\} + h.c.\right]. \tag{12.90d}$$

When $\mathcal{G}$ acquires a VEV, owing to the spontaneous breakdown of local supersymmetry, the coefficient multiplying $\frac{1}{2}i\bar{\Psi}_\mu\Sigma^{\mu\nu}\Psi_\nu$ in the first RHS term of (12.90d) becomes the Majorana mass of the gravitino, cf. (12.6).

The full action from the Lagrangian density of (12.89) is invariant under the following infinitesimal $N=1$ local supersymmetric transformations characterized by the coordinate dependent Grassmann four spinor function $\epsilon(x)$:

$$\delta\phi_i = \sqrt{2}\,\bar{\epsilon}\psi_{iL}, \tag{12.91a}$$

$$\delta e^m{}_\mu = -\frac{i}{M_{Pl}}\bar{\epsilon}\gamma^m\Psi_\mu, \tag{12.91b}$$

$$\delta\Psi_\mu = 2M_{Pl}\tilde{\mathcal{D}}_\mu\epsilon - \frac{1}{M_{Pl}}\,\epsilon(\mathcal{G}^i\tilde{\mathcal{D}}_\mu\phi_i - \mathcal{G}_i\tilde{\mathcal{D}}_\mu\bar{\phi}^i) + ie^{-\mathcal{G}/(2M_{Pl}^2)}\gamma_\mu\epsilon + \text{two fermion terms}, \tag{12.91c}$$

$$\delta\psi_{iL} = -\frac{i}{\sqrt{2}}\tilde{\slashed{\mathcal{D}}}\phi_i\epsilon_R - \frac{\sqrt{2}}{M_{Pl}}\,e^{-\mathcal{G}/(2M_{Pl}^2)}(\mathcal{G}^{-1})_i^j\mathcal{G}_j\epsilon_L + \text{two fermion terms}, \tag{12.91d}$$

$$\delta A_\mu^a = -\bar{\epsilon}\gamma_\mu\lambda_{MR}^a + h.c., \tag{12.91e}$$

$$\delta\lambda_{ML}^a = \Sigma^{\mu\nu}F_{\mu\nu}^a\epsilon_L - \frac{i}{4}\sum_\alpha g_\alpha\Re e(f^{-1})_b^a\mathcal{G}^i(T^{\alpha b})_i^j\phi_j\epsilon_L + \text{two fermion terms}. \tag{12.91f}$$

It may be checked that in the limit of a flat spacetime ($M_{Pl} \to \infty$), these transformations match with those of $N=1$ global supersymmetry, cf. (4.8).

References

[12.1] H.P. Nilles, *loc. cit.*, *Bibl.* R. Arnowitt, A.H. Chamseddine and P. Nath, *op. cit.*, *Bibl.*

[12.2] R.M. Wald, *op. cit.*, *Bibl.* S. Weinberg #1, *op. cit.*, *Bibl.*

[12.3] P. van Nieuwenhuizen, *loc. cit.*, *Bibl.* L. Castellini, R. D'Auria and P. Fré, *op. cit.*, *Bibl.*

[12.4] J. Wess and J. Bagger, *op. cit.*, *Bibl.*

[12.5] H.P. Nilles, Int. J. Mod. Phys. **A5** (1990) 4199.

[12.6] M.E. Peskin and D.V. Schroeder, *op. cit.*, *Bibl.*

[12.7] P. Fayet, Phys. Lett. **B70** (1977) 40. R. Casalbuoni, S. De Curtis, D. Dominici, F. Feruglio and R. Gatto, Phys. Rev. **D39** (1989) 2281.

[12.8] A.H. Chamseddine, R. Arnowitt and P. Nath, Phys. Rev. Lett. **49** (1982) 970. R. Barbieri, S. Ferrara and C. Savoy, Phys. Lett. **B119** (1982) 343.

[12.9] L.J. Hall, J. Lykken and S. Weinberg, Phys. Rev. **D27** (1982) 2359.

[12.10] T. Bhattacharya and P. Roy, Phys. Lett. **B206** (1988) 655; Nucl. Phys. **B328** (1989) 469; *ibid.* **B328** (1989) 481.

[12.11] T. Bhattacharya and P. Roy, Phys. Rev. **D38** (1988) 2284.

[12.12] S.P. Martin and M. Vaughn, Phys. Lett. **B318** (1993) 331.

[12.13] J.L. Feng, K.T. Matchev and F. Wilczek, Phys. Lett. **B482** (2000) 388. J.L. Feng and K.T. Matchev, Phys. Rev. Lett. **84** (2000) 2322; Phys. Rev. **D61** (2000) 075005, *ibid* **D63** (2001) 095003.

[12.14] M. Drees and M.M. Nojiri, Phys. Rev. **D45** (1992) 2482.

[12.15] J.A. Casas, A. Lleyda and C. Muñoz, Nucl. Phys. **B471** (1996) 3.

[12.16] S. Fukuda et. al., Phys. Rev. Lett. **86** (2001) 5651. M.H. Ahn et. al., Phys. Rev. Lett. **87** (2001) 071301. Q.R. Ahmad et. al., Phys. Rev. Lett. **90** (2003) 041801. K. Eguchi et. al., Phys. Rev. Lett. **90** (2003) 021802.

[12.17] R. Barbieri, L.J. Hall and A. Strumia, Nucl. Phys. **B449** (1995) 437.

[12.18] S. Bertolini, F. Borzumati, A. Masiero and G. Ridolfi, Nucl. Phys. **B353** (1991) 591.

[12.19] T. Moroi, Phys. Rev. **D53** (1996) 6565; *errtm. ibid.* **D56** (1997) 4424.

[12.20] T. Ibrahim and P. Nath, Phys. Rev. **D57** (1998) 478; *errtm. ibid.* **D58** (1998) 019901; *ibid.* **D60** (1999) 079903; *ibid.* **D60** (1999) 19901.

[12.21] G.W. Bennett et al., Phys. Rev. Lett. **89** (2002) 101804; *errtm. ibid.* **89** (2002) 129903.

[12.22] M. Davier, S. Eidelman, A. Höcker and Z. Zhang, Eur. Phys. J. **C27** (2003) 497. K. Hagiwara, A.D. Martin, D. Nomura and T. Teubner, Phys. Lett. **B557** (2003) 69.

[12.23] R. Arnowitt, B. Dutta, B. Hu and Y. Santoso, Phys. Lett. **505** (2001) 177. A. Djouodi, M. Drees and J.L. Kneur, JHEP **0108** (2001) 055. J.R. Ellis, K.A. Olive and Y. Santoso, New J. Phys. **4** (2002) 32.

[12.24] Particle Data Group, *loc. cit., Bibl.*

[12.25] S. Abel, S. Khalil and O. Lebedev, Nucl. Phys. **B606** (2001) 151.

[12.26] H. Baer, C. Balazs, A. Belayev, J.K. Mizukoshi, X. Tata and Y. Wang, hep-ph/0210441.

[12.27] G.G. Ross, *op. cit., Bibl.* R.N. Mohapatra, *op. cit., Bibl.*

[12.28] T. Goto, Y. Okada and Y. Shimizu, Phys. Rev. **D58** (1998) 094006. N. Polonsky and A. Pomarol, Phys. Rev. **D51** (1995) 6532.

[12.29] D. Matalliotakis and H.P. Nilles, Nucl. Phys. **B435** (1995) 115. M. Olechowski and S. Pokorski, Phys. Lett. **B344** (1995) 201. P. Nath and R. Arnowitt, Phys. Rev. **D56** (1997) 2820.

[12.30] J. Ellis, K. Enqvist, D. Nanopoulos and K. Tamvakis, Phys. Lett. **155B** (1985) 381. K. Huitu, Y. Kawamura, T. Kobayashi and K. Pulomaki, Phys. Rev. **D61** (2000) 035001.

[12.31] R. Barbieri, L.J. Hall and A. Strumia, Nucl. Phys. **B445** (1995) 219.

[12.32] J. Hisano, H. Murayama and T. Yanagida, Nucl. Phys. **402** (1993) 46.

[12.33] K.S. Babu and S.M. Barr, Phys. Rev. **D48** (1993) 5354; S.M. Barr and S. Raby, Phys. Rev. Lett. **79** (1997) 4748.

[12.34] C. Froggatt and H.B. Nielsen, Nucl. Phys. **B147** (1979) 277.

[12.35] R. Barbieri, G. Dvali and L.J. Hall, Phys. Lett. **B377** (1996) 76.

[12.36] R. Barbieri, L.J. Hall, S. Raby and A. Romanino, Nucl. Phys. **B493** (1997) 3.

[12.37] R. Kallosh, A. Linde, D. Linde and L. Susskind, Phys. Rev. **D52** (1995) 912.

[12.38] N. Arkani-Hamed, C. Carone, L.J. Hall and H. Murayama, Phys. Rev. **D54** (1996) 7032.

[12.39] D. Kaplan and M. Schmaltz, Phys. Rev. **D49** (1994) 3741.

[12.40] L. Randall and R. Sundrum, Nucl. Phys. **B557** (1999) 79.

[12.41] G.F. Giudice, M.A. Luty, H. Murayama and R. Rattazzi, JHEP **9812** (1998) 27.

[12.42] T. Gherghetta, G.F. Giudice and J.D. Wells, Nucl. Phys. **B559** (1999) 27.

[12.43] M.K. Gaillard and B. Nelson, Nucl. Phys. **B588** (2000) 197.

[12.44] I. Jack, D.R.T. Jones and A. Pickering, Phys. Lett. **B426** (1998) 73.

[12.45] T. Moroi and L. Randall, Nucl. Phys. **B570** (2000) 455.

[12.46] S. Su, Nucl. Phys. **B75** (2000) 87.

[12.47] I. Jack and D.R.T. Jones, Phys. Lett. **B482** (2000) 167; N. Arkani-Hamed, D.E. Kaplan, H. Murayama and Y. Nomura, JHEP **0102** (2001) 041.

[12.48] A. Giveon and D. Kutasov, *loc. cit., Bibl.*

[12.49] D.E. Kaplan, G. Kribs and M. Schmaltz, Phys. Rev. **D62** (2000) 035010. M. Schmaltz and W. Skiba, *ibid* 095004. Z. Chacko, M. Luty, A. Nelson and E. Ponton, JHEP **0001** (2000) 003.

[12.50] T. Gherghetta and A. Pomarol, Nucl. Phys. **B586** (2000) 141; *ibid.* **B602** (2001) 3. E. Bergshoeff, R. Kallosh and A. van Proeyen, Fortsch Phys. **49** (2001) 625.

[12.51] K.R. Dienes, *loc. cit., Bibl.*

[12.52] J. Scherk and J.H. Schwarz, Phys. Lett. **B82** (1979) 60; Nucl. Phys. **B153** (1979) 61.

[12.53] I. Antoniadis, C. Muñoz and M. Quiros, Nucl. Phys. **B397** (1993) 515.

[12.54] P.P. Srivastava, *op. cit., Bibl.*

[12.55] D. Friedman, P. van Nieuwenhuizen and W. Siegel, Phys. Rev. **D13** (1976) B214. S. Deser and B. Zumino, Phys. Lett. **B62** (1977) 335.

[12.56] E. Cremmer, S. Ferrara, L. Girardello and P. van Nieuwenhuizen, Nucl. Phys. **B212** (1983) 413. S. Ferrara in *Supersymmetry and Supergravity/Nonperturbative QCD* (P. Roy and V. Singh eds.), *op. cit., Bibl.*

[12.57] D. Bailin and A. Love, *op. cit., Bibl.*

Chapter 13

GAUGE MEDIATED SUPERSYMMETRY BREAKING

13.1   The Basic Ingredients

We start with some brief remarks to motivate these scenarios which were alluded to in §9.1 and §12.1. In the previous chapter we have touched upon the difficulties faced by generic gravity mediated supersymmetry breaking models in naturally respecting experimental bounds on FCNC and CP violating processes. In order to ameliorate these, one needs to postulate some generation symmetry there – beyond the gauge symmetries of the MSSM or of Grand Unified Theories. In sharp contrast, the main advantage of Gauge Mediated Supersymmetry Breaking (GMSB) models [13.1, 13.2] is the automatic creation of identical soft supersymmetry breaking masses for scalars with the same gauge quantum numbers but different flavors. Consequently, there is a priori no problem with the FCNC or CP violation constraints. This is not to claim that the GMSB scenario is free of any trouble since such is far from the case. Specifically, it is more difficult here than in gravity mediated supersymmetry breaking models to naturally generate the MSSM higgsino mass parameter μ which is required (cf. Ch.10) to be of the order of the EW symmetry breaking scale. We shall discuss such questions later in the chapter. Our aim, in this section, is to set up the basic GMSB framework and broadly discuss the main results that are needed to construct these models. The simplest potentially realistic (i.e. minimal) model mGMSB is then described in the next section §13.2. The subsequent section §13.3 is devoted to a discussion of nonminimality in the messenger sector. Two important issues of the GMSB option, namely the μ problem and superpotential couplings between messenger and visible matter superfields, are discussed in §13.4. The final section §13.5 explores the implications of flavor symmetries in this class of models.

As in gravity mediated supersymmetry breaking, here also supersymmetry is taken to be spontaneously broken in some 'hidden' or 'secluded' sector of the theory which is a singlet under MSSM gauge transformations. Details of this sector are again of little phenomenological interest. All that are needed are nonzero VEVs for both the scalar and auxiliary components s and F_S respectively of some (SM) gauge singlet chiral superfield S. There is a coupling between S and messenger left chiral superfields[1] Φ_i and $\bar{\Phi}_i$. The latter are nonsinglets vis-à-vis the SM gauge group $SU(3)_C \times SU(2)_L \times U(1)_Y$. Indeed, Φ_i and $\bar{\Phi}_i$ are chosen to form **vectorlike** representations[2] of it so that they can acquire masses that are much larger than the weak scale. The messenger superfields enter the scene via the superpotential

$$\mathcal{W}_{\mathrm{mess}} = \sum_i \lambda_i S \Phi_i \bar{\Phi}_i \ . \tag{13.1}$$

Since both $\langle s \rangle$ and $\langle F_S \rangle$ are nonzero, (13.1) leads to the following squared mass matrix, cf. (5.8), for the scalar components of the messenger superfields:

$$m^2_{\sim i} = \begin{pmatrix} |\lambda_i \langle s \rangle|^2 & \lambda_i \langle F_S \rangle \\ \lambda_i^\star \langle F_S^\star \rangle & |\lambda_i \langle s \rangle|^2 \end{pmatrix} . \tag{13.2}$$

Without loss of generality, we have chosen a basis in which the couplings λ_i and hence the squared mass matrix (13.2) are diagonal in the type index i labelling the messenger superfields. Now the requirement, that the smaller eigenvalue of this matrix be positive, yields the important constraint

$$x_i \equiv \left| \frac{\langle F_S \rangle}{\lambda_i \langle s \rangle^2} \right| < 1 \ . \tag{13.3}$$

The masses of the messenger superfield components still satisfy the supertrace sum rule (5.10) for each pair of left chiral superfields $(\Phi_i, \bar{\Phi}_i)$. Nevertheless, the splitting between the masses of the scalar and fermionic members of Φ_i, induced by $\langle F_S \rangle$, signifies supersymmetry breaking. In the final step, this is communicated to the MSSM fields through quantum corrections, cf. Fig. 9.2. Specifically, the gauginos acquire masses through the one loop diagram of Fig. 13.1a which yields [13.2]

$$M_\alpha = \frac{g_\alpha^2}{16\pi^2} M_s \sum_i 2 T_\alpha(\mathcal{R}_i) g(x_i) \ . \tag{13.4}$$

Here x_i are as in (13.3), while M_s is a scale determining the overall size of the soft supersymmetry breaking masses in the observable sector, being given by

$$M_s \equiv \left| \frac{\langle F_S \rangle}{\langle s \rangle} \right| , \tag{13.5}$$

so that $|\lambda_i \langle s \rangle| = M_s / x_i$. Furthermore, $T_\alpha(\mathcal{R}_i)$ is the representation constant of the representation $\mathcal{R}_i$ by which ϕ_i transforms under the gauge group G_α which is a factor of the SM gauge group[3]: $\mathrm{Tr}\, T^a(\phi_i) T^b(\phi_i) = T_\alpha(\mathcal{R}_i) \delta^{ab}$. Recall that $T_\alpha(\mathcal{R}_i)$ appeared in the β-functions for gauge couplings, cf. §6.8 and §11.2.

---

[1] $\bar{\Phi}_i$ is a left chiral superfield whose components transform as conjugates of those of Φ_i under the SM gauge group.

[2] Mass terms of chiral fermions, transforming as complex irreducible representations (irreps) of a gauge group, are not gauge invariant. Even after spontaneous symmetry breakdown, their masses remain low, being protected [13.3] by chiral symmetry. Fermions which transform as real, generally reducible, representations (called vectorlike, e.g. the direct sum of a complex irrep and its conjugate), have gauge invariant Dirac mass terms and can be very heavy.

[3] Here g_α is the gauge coupling strength corresponding to the factor group G_α and $g_\alpha^2/(16\pi^2)$ is the 'proper' loop prefactor.

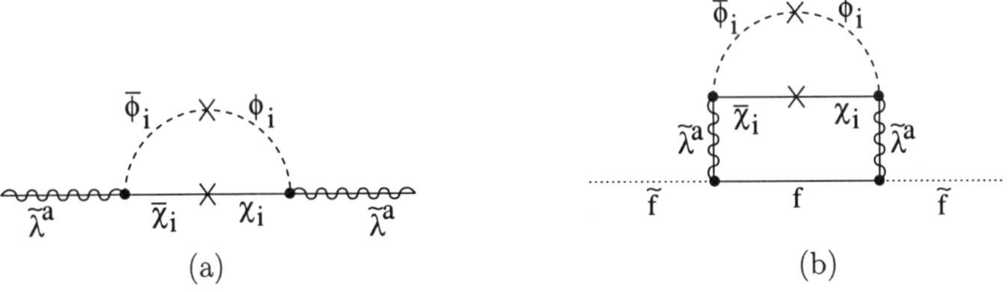

Fig.13.1. (a) One loop diagram yielding a mass for the gaugino $\tilde{\lambda}^a$ and (b) two loop diagram generating a squared mass for the sfermion $\tilde{f}$. The crosses denote mass insertions, while ϕ_i and χ_i are scalar and fermionic components respectively of the messenger superfield Φ_i.

Thus[4] $T_\alpha(\mathcal{R}_i) = 1/2$ if $\mathcal{R}_i$ is the fundamental representation of an $SU(N)$ gauge group, while $T_Y(\mathcal{R}_i) = (Y_i/2)^2$ for $U(1)_Y$, Y_i being the hypercharge carried by ϕ_i. Finally, the function $g(x)$ is given by

$$g(x) = \frac{1}{x^2}\left[(1+x)\ln(1+x) + (1-x)\ln(1-x)\right] \simeq 1 + \frac{x^2}{6} + \frac{x^4}{15} + \cdots . \tag{13.6}$$

Since, from (13.3), $x_i < 1$, $g(x)$ ranges between 1 and $2\ln 2 \simeq 1.39$. However, it remains [13.4] very close to unity in the range $0 < x < 0.5$ and deviates significantly from that value only for $0.5 < x < 1$.

Supersymmetry breaking contributions to *squared* scalar masses are induced only at the two loop level. This implies sfermion masses of the same order as gaugino masses in the MSSM: a feature that is rather important for the viability of this scenario. A typical two loop diagram is shown in Fig. 13.1b. Summing the complete set of such diagrams, one obtains the following finite and gauge invariant result [13.2]:

$$m^2_{\tilde{f},H} = 2M_s^2 \sum_\alpha \left(\frac{g_\alpha^2}{16\pi^2}\right)^2 C_\alpha \sum_i 2T_\alpha(\mathcal{R}_i) f(x_i) . \tag{13.7}$$

In (13.7) M_s is as defined in (13.5), x_i are as in (13.3) and C_α is[5] the Casimir $\sum_\alpha (T^a T^a)_\alpha$ of the sfermion representation $\mathcal{R}_i$ under G_α, the summation over α covering all the factors of the SM gauge group. The function $f(x)$ is given by

$$f(x) = \frac{1+x}{x^2}\left[\ln(1+x) - 2\mathrm{Li}_2\left(\frac{x}{1+x}\right) + \frac{1}{2}\mathrm{Li}_2\left(\frac{2x}{1+x}\right)\right] + (x \to -x)$$

$$\simeq 1 + \frac{x^2}{16} - \frac{11}{450}x^4 + \cdots , \tag{13.8}$$

where $\mathrm{Li}_2(y) \equiv -\int_0^1 dt\, t^{-1}\ln(1-yt)$ is the dilogarithm, i.e. the Spence function. The behavior of $f(x)$ is noteworthy; it remains close to unity, nearly independent of x, for $0 < x < 0.8$ and then sharply drops to $f(1) \simeq 0.702$.

---

[4]It is because of this normalization that we have written $2T_\alpha$ inside the sum in (13.4).
[5]$C_{SU(n)} = (n^2 - 1)(2n)^{-1}$ and $C_{U(1)_Y} = (Y/2)^2$.

Finally, trilinear soft supersymmetry breaking terms are generated at the two loop level. Note that the corresponding A-parameters have mass dimension one, as opposed to two loop generated scalar mass terms which have mass dimension two. The A-parameters (cf. 9.3) are thus much smaller in magnitude than the masses of scalars and also of gauginos which have the same order of magnitude. Therefore, for most phenomenological applications, it is sufficient to simply set them to zero:

$$A_{ijk} \to 0 \; . \tag{13.9}$$

Now (13.3)–(13.9) represent the completely specified boundary conditions for soft supersymmetry breaking parameters of the MSSM in the GMSB option. The pertinent question to ask then is about the scale where these boundary conditions should be imposed. A clue towards answering this question comes from the origin of the messenger mass-scale. Since the messenger superfields Φ_i, $\bar{\Phi}_i$ together form a vectorlike (as opposed[2] to chiral) representation of the gauge group, they can and do acquire large masses

$$m_i = |\lambda_i \langle s \rangle| = \frac{M_s}{x_i} \; . \tag{13.10}$$

This is the ballpark messenger scale where the boundary conditions (13.3), (13.7) and (13.9) can be naturally taken to be valid. We therefore impose these relations at the messenger mass scale M_M defined by

$$M_M \equiv |\lambda \langle s \rangle| = \frac{M_s}{x} \; , \tag{13.11}$$

where λ is some sort of an average over the couplings appearing in the messenger superpotential[6] (13.1) and x is the corresponding average over x_i. The mass spectrum of scalars and gauginos at experimentally accessible energies can then be computed by using the RGE equations of §11.4.

A noteworthy feature of (13.7) is its implication of the degeneracy of the soft supersymmetry breaking Higgs mass parameters, $m_1^2 = m_2^2$, at the scale M_M. Electroweak symmetry then gets broken radiatively with m_2^2 running down to a negative value at a lower energy scale – pretty much the same way as in mSUGRA. There is one significant difference, though. The scale M_M here need not be as high as the GUT scale where boundary conditions on the soft supersymmetry breaking parameters are imposed in mSUGRA. This is because (13.7) generates much larger soft supersymmetry breaking masses for squarks than for Higgs bosons, simply because $g_3^4 \gg g_2^4$. Consequently, at the scale M_M, the parameter S_t – defined in (11.33a) – is much larger than m_2^2 itself. Radiative symmetry breaking is thus possible in the GMSB scenario for quite low values of M_M, such as $M_M \simeq M_s \sim \mathcal{O}$ (50 TeV). There is, in fact, an upper bound on M_M. As emphasized in §12.2, any supersymmetric theory incorporating gravity will produce gravity mediated contributions to the soft supersymmetry breaking parameters in the observable sector, once supersymmetry gets broken in *some* sector of the theory. In general, such contributions will spoil the mass-degeneracy between

---

[6]In principle, there could be large hierarchies among the λ_i. In this case, the various terms in the summation of over i in (13.4) and (13.7) have to be added one by one while going down in scale from the largest to the smallest $|\lambda_i \langle x \rangle|$ with the appropriate RGE performed in between. No reason is known, however, for the existence of such large hierarchies.

sfermions with identical gauge quantum numbers. Recall from §9.5 that, unless generation mixing is suppressed by some mechanism, the occurrence of this degeneracy for the first two generations is dictated by experimental upper limits on FCNC amplitudes holding up at the permille level. One demands that these gravity mediated contributions, proportional to $\langle F_S \rangle / M_{Pl}$ (cf. §12.7), amount at most to 0.1% of the value given by (13.7) from gauge-mediation. That then leads – with the couplings $|\lambda_i|$ taken to be $\lesssim \mathcal{O}(1)$ – to the upper bound

$$M_M \leq \langle s \rangle < 2 \times 10^{-3} \sqrt{n} \left(\frac{\alpha}{\pi} \right) M_{Pl} \simeq \sqrt{n} \times 10^{14} \text{ GeV} \qquad (13.12)$$

on the messenger mass scale. In (13.12) n is the effective number of messenger superfields and α is a generic gauge fine structure coupling. Altogether, therefore, the messenger scale can lie between a few tens of TeV and about 10^{14} GeV.

If $|\langle F_S \rangle|$ lies near the lower end of the abovementioned range, gravitino interactions might become sufficiently strong to produce effects that are observable in the laboratory. We know from §12.2 that the effective interaction strength of the longitudinal gravitino involves $(m_{3/2} M_{Pl})^{-1}$. It is evident from (12.15) that the relevant quantity here is the sum (in quadrature) of all auxiliary components. Requiring $\langle V_F \rangle$ to vanish in order to avoid a disastrously large cosmological constant and keeping in mind[7] that, at energies well below M_{Pl}, $\mathcal{G}_j^i \to -\delta_i^j$ and $m_{3/2} \mathcal{G}_i \to F_i$, cf. (12.5), we can derive from (12.5), (12.6) and (12.12) an expression for the gravitino mass that is applicable in the present context, namely

$$m_{3/2} = \sqrt{\frac{1}{3} \frac{1}{M_{Pl}}} \left(\sum_i |\langle F_i \rangle|^2 \right)^{1/2} . \qquad (13.13)$$

The assumption that all relevant VEVs are small compared to M_{Pl} has gone into the derivation of (13.13). The hidden sector is, in general, expected to have nonzero VEVs of the auxiliary components of some superfields. These superfields need not be directly involved with the transmission of supersymmetry breaking to the observable sector, but could yet contribute to the RHS of (13.13). Such contributions could make the gravitino heavier. Nevertheless, a light gravitino is more natural in the GMSB option than in most gravity mediated supersymmetry breaking models.

In particular, the gravitino is likely to be the LSP in GMSB scenarios. The lightest sparticle, which participates in MSSM gauge and Yukawa interactions, will now be called[8] the Next Lightest Superpartner (NLSP). Being unstable, the latter is not cosmologically constrained to be electrically neutral. It will decay predominantly into a gravitino and an SM particle. This might be a fast decay, before the NLSP crosses the detector system, once an observable sparticle has been produced in a collider experiment. Some of these partial widths have been computed [13.2, 13.5]; the corresponding expressions are given in Table 13.1. For

---

[7]Recall that i, j are type indices for hidden sector superfields, F_i are auxiliary components and $\mathcal{G}_j^i$ is the Kähler metric.

[8]We are considering only the visible sector here. In principle, there could be lighter sparticles in the hidden sector. However, these would have even weaker couplings to the observable sector than the gravitino does. These would, therefore, be produced only very rarely in decays of observable sector sparticles. So we ignore them.

Mode	Partial width Γ
$\tilde{\chi}_i^0 \to \gamma \tilde{G}$	$\dfrac{1}{48\pi} \lvert N_{i1}c_W + N_{i2}s_W \rvert^2 M_{Pl}^{-2} m_{3/2}^{-2} M_{\tilde{\chi}}^5$
$\tilde{\chi}_i^0 \to Z \tilde{G}$	$\dfrac{1}{48\pi} \left[\lvert N_{i1}s_W - N_{i2}c_W \rvert^2 + \dfrac{1}{2}\lvert N_{i4}c_\beta - N_{i3}s_\beta \rvert^2 \right] M_{Pl}^{-2} m_{3/2}^{-2} M_{\tilde{\chi}}^5 (1 - M_Z^2 M_{\tilde{\chi}}^{-2})$
$\tilde{\chi}_i^0 \to h \tilde{G}$	$\dfrac{1}{96\pi} \lvert N_{i4}s_\alpha - N_{i3}c_\alpha \rvert^2 M_{Pl}^{-2} m_{3/2}^{-2} M_{\tilde{\chi}}^5 (1 - m_h^2 M_{\tilde{\chi}}^{-2})$
$\tilde{\chi}_i^0 \to H \tilde{G}$	$\dfrac{1}{96\pi} \lvert N_{i4}c_\alpha + N_{i3}s_\alpha \rvert^2 M_{Pl}^{-2} m_{3/2}^{-2} M_{\tilde{\chi}}^5 (1 - m_H^2 M_{\tilde{\chi}}^{-2})$
$\tilde{\chi}_i^0 \to A \tilde{G}$	$\dfrac{1}{96\pi} \lvert N_{i4}s_\beta - N_{i3}c_\beta \rvert^2 M_{Pl}^{-2} m_{3/2}^{-2} (1 - m_A^2 M_{\tilde{\chi}}^{-2}) M_{\tilde{\chi}}^5$
$\tilde{f} \to f \tilde{G}$	$\dfrac{1}{48\pi} M_{Pl}^{-2} m_{3/2}^{-2} m_{\tilde{f}}^5$

Table 13.1. Expressions for the partial widths of various gravitino decay modes of $\tilde{\chi}_i^0$ and $\tilde{f}$ in terms of the elements N_{ij} of the neutralino mixing matrix (cf. §9.2), $c_{W,\beta,\alpha} \equiv \cos(\theta_W, \beta, \alpha)$ and $s_{W,\beta,\alpha} \equiv \sin(\theta_W, \beta, \alpha)$, α and β being the angles defined in Fig.10.1.

the lifetime of the NLSP, one typically finds

$$\tau_{\rm NLSP} \simeq 6 \times 10^{-14} \left(\frac{100 \text{ GeV}}{m_{\rm NLSP}} \right)^5 \left(\frac{m_{3/2}}{\text{eV}} \right)^2 \text{ sec. ,} \tag{13.14a}$$

$$\gtrsim 6 \times 10^{-14} \left(\frac{100 \text{ GeV}}{m_{\rm NLSP}} \right)^5 \left[\frac{M_s M_M}{(\lambda \cdot 65 \text{ TeV})^2} \right]^2 \text{ sec. ,} \tag{13.14b}$$

where we have used (13.5) and (13.11) as well as the inequality $m_{3/2} \geq \lvert \langle F_S \rangle \rvert / (\sqrt{3} M_{Pl})$. This latter inequality allows for additional contributions to the gravitino mass. Let us again state that λ in (13.14b) is an 'average' coupling strength from the messenger superpotential (13.1). Furthermore, by using naturalness arguments (cf. Ch.1), we can argue that the scale M_s should be roughly in the region of tens of TeV. Eqs. (13.14) then tell us that the NLSP decay into a gravitino will be prompt (i.e. not outside the detector system) only if the scale M_M of the transmission of supersymmetry breaking to the observable sector lies near the lower end of its allowed range. If M_M is much larger, the average flight path will be significantly longer. However, even if the latter is in the km range, a small number of NLSPs will still decay inside the detector. The signature for such a decay, with the secondary decay vertex $O(1\ m)$ away from the primary production vertex, is quite unique. Therefore, if the total event sample is huge in a machine such as the LHC, NLSP decays may be detectable even for a large M_M. A search, already conducted at the TEVATRON for the $\gamma \not{E}_T$ signal in $p\bar{p}$ collisions with $\sqrt{s} = 1.8$ TeV by the CDF collaboration, has yielded [13.6] the lower bound $\left(\sum_i |\langle F_i \rangle|^2\right)^{1/4} > 221$ GeV, i.e. from (13.13), $m_{3/2} > 1.17 \times 10^{-5}$ eV.

Let us finally remark on the expected large value of the gaugino-gravitino mass ratio $M_\alpha/m_{3/2}$ in GMSB models. It may be recalled from the discussion at the end of §12.2 that this ratio is related to a critical energy $E_{cr.}$ above which tree level partial wave unitarity is violated:

$$M_\alpha/m_{3/2} = \left[\left(\frac{12\sqrt{\pi}M_{Pl}}{E_{cr.}}\right)^2 + \frac{6}{5}\right]^{1/2}. \tag{13.15}$$

The very large expected value of the above LHS in the GMSB scenario will imply a breakdown in perturbative unitarity at energies much lower than M_{Pl}. This, however, does not mean that the gravitino becomes strongly coupled at $E_{cr.} \ll M_{Pl}$. The gaugino mass M_α arises (cf. 13.4) in GMSB scenarios from a dynamical supersymmetry breakdown transmitted to the observable sector at the scale M_M, cf. (13.4). So long as

$$M_M \ll E_{cr.} ,$$

there is no problem. This is since, for energies above M_M, all observable sector supersymmetry breaking parameters disappear and, in particular, the gaugino mass may be taken to be zero. This is in contrast with the gravitino mass $m_{3/2}$ which remains a "hard" parameter at those energies. It can be checked that the above strong inequality is satisfied in GMSB models, owing to loop factors in (13.4) and numerical factors in (13.15), even if only F_S contributes to $m_{3/2}$ in (13.13).

13.2 The Minimal Model mGMSB

The minimal GMSB model, with the acronym mGMSB, is the simplest potentially realistic one that has been formulated. It is based on the assumption that the messenger superfields form **complete**[9] representations of $SU(5)$. The latter can be viewed as the overlying lowest rank simple group containing all the SM gauge groups. This will, of course, change the predicted values of the gauge coupling strengths at the scale of Grand Unification. However, the apparent unification of these couplings, observed in the MSSM (cf. Fig. 11.1), remains unaffected[10]. As will be shown in Ch.16, models with a stable gravitino heavier than the keV scale (1 keV $< m_{3/2}$), are severely restricted since such gravitinos would 'overclose' the Universe. The requirement $|\langle F_S \rangle| \leq (2 \times 10^6 \text{ GeV})^2$ ensures safety in this respect. With the natural assumption that the couplings in the messenger superpotential are $\mathcal{O}(1)$ and the

---

[9]If the messengers are put in incomplete $SU(5)$ representations, the gauge coupling unification condition gets spoilt. We have already seen that, in order to be heavy, messenger superfields need to be put in vectorlike representations of the SM gauge group. The conclusion, therefore, is that the messenger superfields need to be put in complete, vectorlike representations of $SU(5)$.

[10]Note that this does not necessarily imply any obligatory commitment to a supersymmetric $SU(5)$ Grand Unified Theory at the energy scale of this unification.

experimental fact that the lighter chargino weighs more than a 100 GeV, one is then led to
the inequalities

$$M_M \lesssim \mathcal{O}(10^8) \text{ GeV} , \tag{13.16a}$$

$$\mathcal{O}(10^{-4}) < \frac{M_s}{M_M} = x < 1 , \tag{13.16b}$$

for the messenger mass scale M_M. The rather low value of the messenger scale (relative
to a GUT scale, say), as required by (13.16a), has implications for the maximal number of
messenger superfields that can be introduced. Recall the discussion in §11.2 that matter
fields always contribute with positive coefficients to gauge β-functions. Renormalization
Group Evolution then leads to higher magnitudes of the gauge coupling strengths at energy
scales above M_M. We know, however, that the condition $g_\alpha^2(M_U)/(4\pi) < 1$, M_U being the
GUT scale, is necessary for a perturbative unification of the MSSM gauge interactions. The
imposition of this condition yields a constraint on the number of allowed vectorlike messenger
representations. At most five $\mathbf{5} \oplus \bar{\mathbf{5}}$ pairs or alternatively one $\mathbf{10} \oplus \bar{\mathbf{10}}$ and two $\mathbf{5} \oplus \bar{\mathbf{5}}$ pairs
are allowed. However, one $\mathbf{10} \oplus \bar{\mathbf{10}}$ pair contributes to the boundary conditions (13.14) and
(13.7) what three $\mathbf{5} \oplus \bar{\mathbf{5}}$ pairs do. Thus we need to consider only models with $n_5 \leq 5$, n_5
being the number of messenger superfields in the $\mathbf{5} \oplus \bar{\mathbf{5}}$ representation. In addition, (13.16b)
suggests that x be considered on a logarithmic scale. Then, for 'most of the range of x', the
functions $f(x)$ and $g(x)$ can be replaced by unity, as is clear from the discussions after (13.6)
and (13.8). The minimal GMSB model **assumes** that x is such that such a replacement can
be made[11]. Now the couplings λ_i of the messenger superpotential need not be counted as
separate parameters but can be subsumed as a whole in the messenger scale M_M.

Strictly speaking, the minimal model should have a very small value of the bilinear soft
supersymmetry breaking parameter B (cf. 9.3) at the messenger scale. The latter is produced
at the two loop level just like the A parameters. In other words, it is suppressed by a loop
factor α/π relative to sfermion and gaugino masses. However, we keep B and the higgsino
mass μ as free parameters here. Ways to generate these two quantities dynamically will be
discussed in §13.6. As already mentioned, identical masses get attributed to the two Higgs
doublets of the MSSM, meaning that electroweak symmetry has to be broken radiatively.
This allows one to 'trade' $|\mu|$ and B for M_Z and $\tan\beta$, exactly as in mSUGRA. The free
parameters of the minimal GMSB model are thus [13.4]

$$\{p_{\text{mGMSB}}\} = M_s, M_M, \tan\beta, n_5, \text{sgn.}(\mu) , \tag{13.17}$$

subject to the constraints (13.16) as well as the bound $n_5 \leq 5$.

The expression (13.4) for the gaugino masses now reduces, with $g(x)$ set equal to unity,
to

$$M_\alpha \simeq \frac{g_\alpha^2}{16\pi^2} n_5 M_s , \tag{13.18}$$

for the subscript $\alpha = 1, 2, 3$. We can write α_1 for the 'fine structure' coupling of the $U(1)$
factor group with the GUT normalization, i.e. $\alpha_1 \equiv g_1^2/(4\pi)$ and similarly $\alpha_2 \equiv g_2^2/(4\pi)$
and $\alpha_3 = g_3^2/(4\pi)$ for the $SU(2)_L$ and $SU(3)_C$ factor groups respectively. In terms of the

---

[11]The somewhat pathological limit of x tending to 1, i.e. when M_M comes close to its lower bound M_s
and the separate λ_i become relevant, will be dealt with separately.

corresponding coupling α_Y of the hypercharge gauge group $U(1)_Y$, $\alpha_1 = 5/3\ \alpha_Y$ as per the discussion after[12] (11.10). To one loop order, (13.18) holds at all energy scales lower than M_M. This is since, to that order, gaugino masses run in the same way as squared gauge couplings, cf. (11.28). Eq.(13.18) shows that the gluino, wino and bino masses are in the same proportion as the corresponding 'fine structure' couplings α_3, α_2 and $(5/3)\alpha_Y$, cf. (12.25). This part of the sparticle spectrum thus looks just like that of the mSUGRA model[13]; the lightest and heaviest charginos are mostly winos and higgsinos respectively, while the two heaviest neutralinos are dominantly higgsinos and the lightest (next-to-lightest) is dominantly a bino (wino).

This is not at all the picture in the scalar sector [13.7], however. In the case at hand, the sfermion mass squared formula (13.7) can be rewritten as

$$m^2_{\tilde{f},H}(M_M) = 2\sum_\alpha \frac{C_\alpha M^2_\alpha}{\sum_i 2T_\alpha(\mathcal{R}_i)} = \frac{2}{n_5}\left[C_3 M^2_3(M_M) + C_2 M^2_2(M_M) + \frac{3}{5}\left(\frac{Y}{2}\right)^2 M^2_1(M_M)\right],$$

(13.19)

with $C_3 = 4/3$ (0) for $SU(3)_C$ triplets (singlets) and $C_2 = 3/4$ (0) for $SU(2)_L$ doublets (singlets)[14]. Eq.(13.19) disallows sfermions from being much heavier than gauginos whereas such a hierarchy is quite possible in mSUGRA if $m_0 \gg M_{1/2}$. It may also be noted that the sfermion to gaugino mass ratio scales like $1/\sqrt{n_5}$. Of course, (13.19) is not scale invariant; the corresponding sparticle masses at experimentally accessible energies have to be obtained by using Renormalization Group Evolution equations. For first and second generation sfermions, the relevant RGE equations which can be solved analytically, cf. (11.34). The constraint (13.16a) implies that these masses need to be evolved over two to six orders of magnitude, as compared with fourteen orders of magnitude in mSUGRA. For a given value of M_s, any increase in the messenger scale M_M has two effects: (1) the range of the RGE is increased, thereby raising the sfermion mass values at the weak scale, (2) the input values (13.19) are changed. These two effects tend to cancel for squarks since M_3, like α_3, decreases with increasing energy. On the other hand, slepton masses are only subject to electroweak corrections. Their final dependence on the ratio M_M/M_s is quite mild [13.1]. The functional dependence of all sfermion squared masses on M_M/M_s is, of course, logarithmic on account of RGE. We can express it in terms of $t_s \equiv \ln(M_M/M_s) = -\ln x$. The exact expressions, that can be obtained from (13.19) and (11.43), are rather cumbersome. A more practically useful procedure is to provide numerical fits, that are linear in t_s, to those expressions. If the constraint (13.16b) is satisfied, these fits can be given after linear interpolation, cf. 12.29 for mSUGRA, as

$$m^2_{\tilde{e}_R}(100\text{ GeV}) = M^2_1(100\text{ GeV})\left[1.54n_5^{-1} + 0.05 + (0.072n_5^{-1} + 0.01)t_s\right] + s^2_W D\ , \quad (13.20\text{a})$$

---

[12]In detail, $M_1 = (16\pi^2)^{-1}g^2_Y M_s \Sigma_i(Y^2_i/2)$, where i runs over all the messenger fields. For every $\mathbf{5} \oplus \mathbf{\bar{5}}$ of $SU(5)$, $\sum_i(Y^2_i/2) = 5/3$. Hence $M_1 = (16\pi^2)^{-1}(5g^2_Y/3)M_s n_5 = (16\pi^2)^{-1}g^2_1 M_s n_5$.

[13]If $M_s \simeq M_M$ and the messenger couplings λ_i differ significantly from one another, there could be deviations of upto 40% from the 'unification ratio relation' (12.26).

[14]Once again, $\sum_i 2T_\alpha(\mathcal{R}_i)$ is n_5 for the $SU(3)_C$ or $SU(2)_L$ factor of $SU(5)$, but equals $\sum_i(Y^2_i/2) = \frac{5}{3}n_5$ for the $U(1)_Y$ factor.

$$m_{\tilde{\ell}_L}^2(100\text{ GeV}) \;=\; M_2^2(100\text{ GeV})\left[1.71n_5^{-1} + 0.11 + (0.023n_5^{-1} + 0.02)t_s\right] + (0.5 - s_W^2)D \;,$$

$$(13.20b)$$

$$m_{\tilde{\nu}}^2(100\text{ GeV}) \;=\; M_2^2(100\text{ GeV})\left[1.71n_5^{-1} + 0.11 + (0.023n_5^{-1} + 0.02)t_s\right] - 0.5D \;, \qquad (13.20c)$$

$$m_{\tilde{u}_L}^2(500\text{ GeV}) \;=\; M_3^2(500\text{ GeV})\left[1.96n_5^{-1} + 0.31 - (0.102n_5^{-1} - 0.037)t_s\right]$$
$$-(0.5 - 0.66s_W^2)D \;, \qquad (13.20d)$$

$$m_{\tilde{d}_L}^2(500\text{ GeV}) \;=\; M_3^2(500\text{ GeV})\left[1.96n_5^{-1} + 0.31 - (0.102n_5^{-1} - 0.037)t_s\right]$$
$$+(0.5 - 0.66s_W^2)D \;, \qquad (13.20e)$$

$$m_{\tilde{u}_R}^2(500\text{ GeV}) \;=\; M_3^2(500\text{ GeV})\left[1.78n_5^{-1} + 0.30 - (0.103n_5^{-1} - 0.035)t_s\right] - 0.66s_W^2D \;,$$

$$(13.20f)$$

$$m_{\tilde{d}_R}^2(500\text{ GeV}) \;=\; M_3^2(500\text{ GeV})\left[1.77n_5^{-1} + 0.30 - \left(0.103n_5^{-1} - 0.034\right)t_s\right] + 0.33s_W^2D \;,$$

$$(13.20g)$$

where $D = -M_Z^2\cos 2\beta$ and we have written s_W^2 for $\sin^2\theta_W$.

Let us now comment on (13.20). Experimental lower bounds on squark masses are already above 200 GeV. That is why we have evaluated them at 500 GeV. Slepton masses have been taken at 100 GeV since the corresponding lower bounds are in that ballpark. Note further the implication of the constraint (13.16b) that $t_s \lesssim 9$. The terms in the RHS of (13.20), that are proportional to n_5^{-1}, describe[15] the t_s-dependence of the boundary conditions. On the other hand, the n_5-independent terms are due to evolution from the scale M_M to the sparticle mass scale. Also, one should not forget that, in writing (13.20), one has set $f(x) = g(x) = 1$. In practice, this is justified if $M_M > 1.5M_s$. As remarked in ftnt. 8, the case when the scales M_M and M_s are really close has to be handled separately. Without going into the details of that situation, let us just observe that the M_M-independent terms in the RHS of (13.20) can then be reduced by as much as a factor of 2.6. Barring this possibility, in the $n_5 = 1$ case squarks may be seen to be somewhat heavier than gluinos and ditto for sleptons vis-a-vis the corresponding electroweak gauginos. In either case the ordering gets reversed for $n_5 = 3$. For $n_5 = 1$, the gross features of the spectrum (13.20) resemble those of a typical mSUGRA sfermion spectrum with $m_0 \ll M_{1/2}$. This is hardly surprising; even in mSUGRA it is the gauge interactions which transmit supersymmetry breaking to the scalar sector in the limit of small $m_0/M_{1/2}$. However, there are differences in detail between the sfermion mass spectra of the two scenarios. In particular, *the squark to slepton and the slepton doublet to slepton singlet mass ratios are larger in the GMSB case*. For a relatively light mass spectrum, these ratios can and do get reduced by the EW D-terms. Thus a GMSB sfermion mass spectrum at *large* $\tan\beta$ can be made to resemble that in mSUGRA with $m_0 \ll M_{1/2}$ at *small* $\tan\beta$. Significant differences do persist, however, in the Higgs and higgsino sectors so that an experimental distinction between the two scenarios is still feasible.

---

[15]Note that the expansion parameter, relevant to the linear fits (13.20), is $\alpha t_s/\pi$ and hence quite small.

Eqs.(13.20) hold for first and second generation sfermions. Owing to the shorter range here of the energy scale of evolution via the renormalization group, as compared with mSUGRA, Yukawa effects on sfermion masses are less important in the present case. Moreover, the A-parameters being fairly small, the amount of $\tilde{t}_L$-$\tilde{t}_R$ mixing is also less. The picture is quite different in the Higgs sector, though. While the boundary condition on the supersymmetry breaking contributions to the Higgs masses is the same as that for slepton doublets, the Yukawa contributions to the RGE equations for the former are proportional to squark masses. As a result, as the energy scale decreases, m_2^2 is driven rapidly to large, negative values. EW symmetry breaking, à la (10.18b), can then be achieved only by a sizeable value of $|\mu|$, typically

$$|\mu| \gtrsim \frac{2}{3}(n_5)^{-1/2}|M_3| \ .$$

The lighter chargino/neutralinos are then perforce more gauginolike, while their heavier partners have to be mostly higgsinos. Significant gaugino-higgsino mixing can still occur, but only for a positive sgn. μ. A second implication of a large $|\mu|$ is substantial $\tilde{b}_L$-$\tilde{b}_R$ and $\tilde{\tau}_L$-$\tilde{\tau}_R$ mixing, provided $\tan\beta$ is not too small. The lighter mass eigenstates $\tilde{b}_1$ and $\tilde{\tau}_1$ can thus be about 20% lighter than $\tilde{d}_R$ and $\tilde{e}_R$ respectively. In particular, for $n_5 \gtrsim 2$, $\tilde{\tau}_1$ is often the NLSP.

Finally, the large value of $|\mu|$ makes the Higgs particles $A, H, H^\pm$ (cf. Ch.10) quite heavy. In particular, at the energy scale equal to M_M, one has the boundary condition $m_1^2 = m_2^2 = m_\nu^2$. Consequently, the mass relation (12.33) holds in the minimal GMSB model too. Moreover, the b and τ Yukawa contributions to the RGE equations again reduce m_A. As a result, the upper bound on $\tan\beta$ is oftentimes given by the experimental lower bound on m_A, just as in mSUGRA. Note, in this connection, that the upper bound

$$\tan\beta < \frac{m_t(m_t)}{m_b(m_t)} \simeq 60$$

is valid in this minimal GMSB scenario as well. Turning to the mass of the lightest Higgs scalar h, the numerical upper bound $m_h \lesssim 132$ GeV, derived in Ch.10 for the MSSM in general, can be sharpened here to [13.4]

$$m_h \lesssim 120 \text{ GeV} \ ,$$

since a large $\tilde{t}_L$-$\tilde{t}_R$ mixing is not allowed in this picture. Thus the minimal GMSB model has a distinctly tighter constraint on the lightest Higgs mass than the MSSM.

13.3 Nonminimal Messenger Sector

In this section we begin to explore beyond the mGMSB model by considering generalizations of the messenger sector. In the said model, the messenger superfields are taken to form complete, vectorlike representations of $SU(5)$. Moreover, nonzero VEVs are assumed to accrue to the lowest and highest components of a single chiral superfield S in the hidden sector. Both of these assumptions can be relaxed. As pointed out in Ref. [13.4], even if one insists on introducing new chiral superfields at the messenger scale M_M in complete representations of $SU(5)$, not all of them need act as messengers. In practice, this implies that some fields have to acquire masses $\mathcal{O}(M_M)$ either through explicit mass terms in the messenger superpotential or through the VEV of a second singlet field that does not contribute to supersymmetry breaking. In this case the unification proportionality (12.25) between the gaugino masses can be badly violated. For instance, suppose the messenger sector contains many more $SU(2)_L$ doublets than $SU(3)_C$ triplets. Then the $SU(2)_L$ gauginos might become heavier than gluinos. However, (13.19) still remains valid if M_M is not very close to M_s. There could nonetheless be large deviations from the scalar mass spectrum (13.20) of the minimal model. For instance, a model with $|M_2| > |M_3|$ might have $SU(2)_L$ singlet squarks close in mass to $SU(2)_L$ doublet sleptons. However, some constraints on ratios of sparticle masses still survive in such models with a generalized messenger sector, if the perturbative unification of the MSSM gauge interactions is insisted upon. The latter condition limits the number of messenger fields that can be introduced. For instance, one finds[16] [13.4] that $m_{\tilde{d}_R} > m_{\tilde{e}_L}$ and $|M_3| > |M_1|$.

This kind of generalization of the minimal GMSB model suffers from the following problem. If only some, but not all, members of an $SU(5)$ multiplet couple to the superfield S, then the scheme cannot be easily integrated into a GUT model. This problem does not occur in the *second* straightforward generalization of the minimal model, where the lone singlet superfield S is replaced by a set $\{S_i\}$ of such fields. We shall now give an extended discussion this possibility. The messenger superpotential (13.1) changes in this case to

$$\mathcal{W}_{\text{mess}} = \sum_{i,j,k} \lambda_{ijk} S_i \Phi_j \bar{\Phi}_k \ . \tag{13.21}$$

A caveat on $\mathcal{W}$ is that it can no longer be assumed to be diagonal in the messenger indices j, k. A simplification, however, is that for phenomenological purposes it suffices to write the VEVs of S_i in their place. This step gives rise to the supersymmetric mass terms

$$\mathcal{W}_m = \sum_{j,k} M_{jk} \Phi_j \bar{\Phi}_k \ , \tag{13.22}$$

with

$$M_{jk} = \sum_i \lambda_{ijk} \langle s_i \rangle \ . \tag{13.23}$$

Along with these mass terms, a nonsupersymmetric contribution to the scalar potential is also produced, namely

$$\delta V_{\text{mess}} = \sum_{j,k} F_{jk} \phi_j \bar{\phi}_k + \text{h.c.} \ , \tag{13.24}$$

with

$$F_{jk} = \sum_i \lambda_{ijk} \langle F_{S_i} \rangle \ . \tag{13.25}$$

---

[16]These conclusions can be avoided if the messenger sector contains no $SU(3)$ triplets at all. But in such a case the gluinos become massless.

Without loss of generality, a basis can be chosen where M_{jk} is diagonal. In this basis, however, F_{jk} will in general have nonvanishing entries with $j \neq k$.

Let us introduce a new basis for the scalar messenger fields in terms of $\phi_{i\pm}$ defined as

$$\phi_{i\pm} \equiv \frac{1}{\sqrt{2}}(\phi_i \pm \bar{\phi}_i^\star) \,. \tag{13.26}$$

The squared mass matrix of the scalar messenger fields in this (ϕ_+, ϕ_-) basis is

$$\mathcal{M}_\phi^2 = \begin{pmatrix} M^2 + \frac{1}{2}(F + F^\dagger) & \frac{1}{2}(F^\dagger - F) \\ \frac{1}{2}(F - F^\dagger) & M^2 - \frac{1}{2}(F + F^\dagger) \end{pmatrix} \,. \tag{13.27}$$

In (13.27) each entry is a matrix by itself, cf. (13.23) and (13.25). This squared mass matrix will, in general, lead to a $U(1)_Y$ D-term at the one loop level. The latter is quite dangerous in that it would make negative contributions to some squared sfermion masses. Fortunately, the generation of such a term can be postponed to the three loop level [13.8] by requiring hermiticity for the matrix (13.25):

$$F = F^\dagger \,. \tag{13.28}$$

What is the main significance of introducing several $\langle s_i \rangle$ and $\langle F_{S_i} \rangle$? The answer is that the scales of gaugino and scalar masses get separated in the process. The former are proportional to Tr F, while nonsupersymmetric contributions to the latter are proportional to Tr F^2. One can thus envision models where Tr F is suppressed or even is vanishing, e.g. due to an (approximate) R-symmetry, while Tr F^2 is not. Such an extension[17] of the discrete R-parity to a continuous symmetry forbids the generation of operators in $\mathcal{L}_{\text{SOFT}}$ of §9.1 with mass dimension three, i.e. gaugino mass terms and trilinear scalar interactions with A-parameters. Even in this case, though, Tr F^2 remains nonzero so long as supersymmetry is broken in the hidden sector. *Ratios* of sfermion masses at the messenger scale M_M are, however, unaffected by this change so long as the messenger fields continue to form complete multiplets of $SU(5)$. The boundary conditions (13.4) and (13.7) continue to hold, except that they now depend on separate energy scales $M_{\tilde{g}}$ and $M_{\tilde{f}}$ respectively. In particular, the ratio of squark to slepton masses still tends to be larger than in mSUGRA. One can try to mimic here an mSUGRA type of sfermion spectrum with $m_0 > M_{1/2}$, i.e. with only moderate mass splittings between squarks and sleptons and $m_{\tilde{q}}$ significantly above $M_{\tilde{g}}$. But then one would have to introduce several hidden sector superfields S_i *and* messenger fields in incomplete representations of $SU(5)$. As mentioned earlier, such a procedure would lead to a violation of the unification proportionality (12.26) for the gaugino masses. One is thus led to conclude that it is essentially impossible to completely mimic an mSUGRA type of sparticle mass spectrum with $m_0 \gtrsim M_{1/2}$ within the framework of gauge mediated supersymmetry breaking.

13.4 The μ and $B\mu$ Problems

So far, we have not put any restrictions on the supersymmetric higgsino mass parameter μ and the associated soft supersymmetry breaking parameter B. We have assumed that they

---

[17]As mentioned in §4.5, a continuous R-symmetry would lead to an unwanted anomaly to cancel which additional fields would need to be postulated.

can be chosen freely, respecting (10.18), which is required for electroweak symmetry breaking. Such an approach can indeed be justified for models with gravity mediated supersymmetry breaking. Several methods are known, for the latter, that lead to a μ-term of the same size as the supersymmetry breaking parameters in the observable sector [13.9] and B is automatically of the right order of magnitude. Possibilities of solving the μ and $B\mu$ problems within the framework of the MSSM with gravity mediated supersymmetry breaking will be discussed at the beginning of §14.2. Creating a similarly sized μ-parameter in GMSB models may not seem difficult to start with. The simplest possibility is to add a term $\lambda_\mu S H_1 \cdot H_2$ to the messenger superpotential (13.1). Suppose we require the higgsino mass μ to be of the order of $\alpha M_s/\pi$, M_s being as in (13.5). Now the following logical conclusion becomes binding on us:

$$\mu \sim \frac{\alpha}{\pi} M_s \Rightarrow \lambda_\mu \sim \frac{\alpha}{\pi} \frac{M_s}{|\langle s \rangle|} \ll 1 \ .$$

In other words, the coupling λ_μ of this new messenger superpotential term is obliged to be unappealingly small. What is worse is that the same term also generates the soft supersymmetry breaking term (cf. 9.3) with a coefficient $B\mu = \lambda_\mu \langle F_S \rangle$ such that

$$|B| = \left| \frac{\lambda_\mu \langle F_S \rangle}{\mu} \right| = M_s > 30 \text{ TeV} \ . \tag{13.29}$$

Such a minimum magnitude for the B-parameter is far too large, in fact by an inverse loop factor π/α, α again being a generic gauge fine structure coupling. Eq. (13.29) represents the heart of the μ-problem in the GMSB scenario. More generic GMSB models are still saddled with (13.29), including those in which μ itself is created at the one loop level. Though there is no need to postulate very small couplings in the superpotential in the latter case, usually $B\mu$ is also created at the one loop level and (13.29) still holds at least approximately.

So the question remains: how to naturally generate a B parameter with an order of magnitude comparable to that of the other soft supersymmetry breaking parameters in the observable sector? One possibility is to construct a model where B obtains only at the two loop level. The simplest scheme of this kind, proposed [13.8] so far, uses the messenger superpotential

$$\mathcal{W}_{\text{mess}} = \lambda S \Phi \bar{\Phi} + T \left(\lambda_1 H_1 \cdot H_2 + \frac{1}{2} \lambda_2 N^2 + \lambda_3 \Phi \bar{\Phi} - M_N^2 \right) \ . \tag{13.30}$$

In (13.30) T and N are additional superfields that are singlets with respect to the SM gauge group, while $\lambda_{1,2,3}$ represent new couplings. Furthermore, by assumption, the hidden sector gives rise[18] to an auxiliary field VEV $\langle F_S \rangle \sim M_N^2$ and certain discrete symmetries are imposed to forbid undesirable terms[19] in the RHS of (13.30) that would otherwise be allowed. At the tree level, (13.30) leads to a scalar potential that is minimized by $\langle n \rangle^2 = 2\lambda_2^{-1} M_N^2$, n being the scalar component of N, with vanishing VEVs for all scalar fields other than s. A

---

[18]Explicit examples effecting this are known.

[19]A term like $S^2 T$ in the RHS of (13.30) will spoil the mechanism and needs to be forbidden. Such a crucial dependence on some discrete symmetry, with only *a posteriori* justification, is a major weakness of this proposal.

tadpole, though, is created at the one loop level for the scalar component t of the superfield T through its coupling to the messenger superfields. One then finds [13.8] that

$$\mu = \lambda_1 \langle t \rangle = -\frac{5}{32\pi^2} \frac{\lambda\lambda_1\lambda_3}{\lambda_2} \frac{\langle F_S \rangle}{M_N^2} M_s \ . \tag{13.31}$$

The key point here is that $\langle F_T \rangle$ remains zero at this order so that no $B\mu$ term is generated at one loop. One can easily see how this happens. With the scalar component VEVs $\langle h_1 \rangle = \langle h_2 \rangle = \langle \phi \rangle = \langle \bar\phi \rangle = 0$ at the one loop level, F_T can only be a function of n. Now since the superfield N does not couple directly to the messenger superfields Φ and $\bar\Phi$, no tadpole for n gets generated at one loop. Consequently, the one loop effective potential depends on n only through $|F_T|^2$. In the minimum of the potential, $\langle n \rangle$ then automatically adjusts itself such that $\langle F_T \rangle = 0$. This mechanism is known as 'dynamical relaxation'.

At the two loop level, all fields begin to feel the effect of supersymmetry breaking parametrized by $\langle F_S \rangle$. Now all the remaining soft supersymmetry breaking parameters get generated. In particular, one finds that

$$B\mu = \lambda_1 \langle F_T \rangle = -\frac{10\lambda_3^2\lambda_1\lambda_2}{(16\pi^2)^2} \left(1 + \frac{5\lambda^2 \langle F_S \rangle^2}{8\lambda_2^2 M_N^4} \right) M_s^2. \tag{13.32}$$

Eq. (13.32) implies a B-parameter of the same order of magnitude as the other soft supersymmetry breaking parameters. Owing to different dependences on the different couplings that occur in the messenger superpotential (13.30), μ and B can indeed be considered as independent parameters. However, (13.30) also generates additional soft supersymmetry breaking contributions [13.8] to the Higgs mass squared parameters:

$$\delta m_1^2 = \delta m_2^2 = 10 \left(\frac{\lambda_1\lambda_3}{16\pi^2} \right)^2 M_s^2. \tag{13.33}$$

These are generally of the same order of magnitude as the gauge contribution (13.7). This result is fairly generic in that the sector producing μ and $B\mu$ is also expected to contribute to the Higgs mass parameters. However, a degree of uncertainty is now introduced on the boundary conditions for m_1^2 and m_2^2. The additional contributions (13.33) destroy the equality of m_1^2 and $m_{\bar\nu}^2$ at the messenger scale, thereby invalidating the sum rule (12.33), even for small to moderate values of $\tan\beta$. Note, moreover, that the contribution (13.33) is positive. The soft supersymmetry breaking parameter m_2^2 will therefore be less negative at the weak scale than in the minimal model. As a result, the value of $|\mu|$ required to generate correct W and Z masses will be smaller. It might thus be possible in this model to have a strongly mixed, even higgsinolike, lightest neutralino. In contrast and as noted earlier, the lightest neutralino in the minimal GMSB model is always a nearly pure gaugino.

It is interesting to compare the above attempt to solve the μ problem in the GMSB scenario with that in the gravity mediated context. We shall discuss in more detail in Ch.14 a possible method to generate an effective μ-term in gravity mediated supersymmetry breaking models by going beyond the MSSM, but let us mention it briefly here. That is done via the introduction to the MSSM spectrum of an additional singlet superfield T somewhat like that above. This T couples to $H_1 \cdot H_2$ as in (13.20) and has an extra T^3 term in the superpotential. For appropriate choices of the soft supersymmetry breaking parameters, the scalar component t will acquire a weak scale VEV. In this scenario T remains at the weak scale, whereas in the scheme for GMSB models, discussed above, $O(M_N)$ masses get attributed to both T and N via (13.30). In fact, a blind imitation of this mechanism for the simplest GMSB models [13.10] runs into a further problem: the A-parameters turn out to be too small. This difficulty can be overcome if there are at least two generations (1,2) of messenger superfields in the superpotential. Suppose the latter is

$$\mathcal{W}_{\mathrm{mess}} = \lambda S(\Phi_1 \bar{\Phi}_1 + \Phi_2 \bar{\Phi}_2) + \lambda_t T \bar{\Phi}_1 \Phi_2 \,, \tag{13.34}$$

with some discrete symmetry postulated to forbid terms that would mix S and T. Any mixing of the latter kind would destroy the mechanism and this is why at least two generations of messengers are needed, transforming differently under a discrete symmetry, so that diagonal superpotential terms like $T\bar{\Phi}_i\Phi_i$ (inducing S-T mixing) are forbidden while the off-diagonal term $T\bar{\Phi}_1\Phi_2$ is allowed. Eq. (13.34) facilitates the generation of sizable A-terms at the one-loop level for the $t h_1 \cdot h_2$ and t^3 couplings. It also leads to additional, possibly negative, two loop contributions to the soft supersymmetry breaking singlet Higgs squared mass, allowing for a consistent mechanism to induce electroweak symmetry breakdown.

Our bottomline on the μ-problem in GMSB models is the following. Schemes can be invented to generate desirable terms with μ, μB and A_{ijk} as coefficients but only at the cost of using ad hoc discrete symmetries.

13.5   Direct Messenger-Matter Couplings

Yet another difficulty with the GMSB superpotential (13.1) is that the messenger sector is totally isolated from the observable sector. As a result, the lightest messenger particle is stable. This can be cosmologically problematic: the contribution of Big Bang relics of the above species to the overall mass density of the Universe tends to be too high, cf. Ch.16 for details. One solution is to make a provision for some interaction between the two sectors. Even in the simplest model of §13.2, where the messengers only consist of $\mathbf{5} \oplus \bar{\mathbf{5}}$ representations of $SU(5)$, a large number of such couplings are allowed [13.11] within the required invariance under SM gauge interactions:

$$\begin{aligned}
\mathcal{W}_{\mathrm{mix}} &= H_1 \cdot L_m \bar{E} + H_1 \cdot Q \bar{D}_m + L \cdot L_m \bar{E} + Q \cdot L_m \bar{E} + Q \cdot L_m \bar{D} + Q \cdot \bar{L}_m \bar{U} \\
&\quad + \bar{E}\bar{U} D_m + L \cdot Q \bar{D}_m + Q \cdot Q D_m + \bar{U}\bar{D}\bar{D}_m + Q \cdot L_m \bar{D}_m \\
&\quad + \mu_{\mathrm{mix}}(H_2 \cdot L_m + H_1 \cdot \bar{L}_m + L \cdot \bar{L}_m + \bar{D} D_m) \,.
\end{aligned} \tag{13.35}$$

We have suppressed dimensionless coefficients as well as generation indices in the RHS of (13.35). The messenger superfields, each carrying the label m, transform under the SM gauge group $SU(3)_C \times SU(2)_L \times U(1)_Y$ as : $L_m = (\mathbf{1}, \mathbf{2}, -1)$, $\bar{L}_m = (\mathbf{1}, \mathbf{2}, 1)$, $D_m = \left(\mathbf{3}, \mathbf{1}, -\frac{2}{3}\right)$, $\bar{D}_m = \left(\bar{\mathbf{3}}, \mathbf{1}, \frac{2}{3}\right)$. Not all terms in (13.35) are allowed to be simultaneously present. For instance, with both baryon and lepton number violation, the proton would decay catastrophically rapidly since D_m and $\bar{D}_m$ transform exactly like the $SU(5)$ partner triplets of the MSSM Higgs doublets, cf. §12.5. This constraint of avoiding rapid proton decay will be particularly severe if one wishes to embed (13.35) into a simple group GUT. Weaker, but still significant, constraints on various products of coupling strengths appearing in (13.35) come from [13.11] bounds on $K_L \to \mu^\mp e^\pm$ decays, K^0-$\bar{K}^0$ mixing etc.

There is, in fact, another possible problem with sizeable superpotential couplings between the messenger and observable sectors: negative contributions to squared scalar masses generated at the one loop level. These contributions are in general neither universal nor diagonal in flavor space. A coupling of the form $\lambda_{ijk}\Psi_i\Psi'_j\Phi_k$, where Ψ, Ψ' on one hand and Φ on the other are observable sector and messenger sector superfields respectively, yields:

$$\delta m^2_{\psi_{il}} \;=\; \frac{1}{16\pi^2}\sum_{j,k}\lambda_{ijk}\lambda^\star_{ljk}N_{jk}M_s^2 u(x_k) \,, \tag{13.36a}$$

$$\delta m^2_{\psi'_{jl}} \;=\; \frac{1}{16\pi^2}\sum_{i,k}\lambda_{ijk}\lambda^\star_{ilk}N'_{ik}M_s^2 u(x_k) \,. \tag{13.36b}$$

N_{jk} and N'_{ik} in the above equations are multiplicity factors that count the number of degrees of freedom circulating inside the relevant loop. Moreover,

$$x_k \equiv \frac{M_s}{M_k} \,,$$

with M_k being the supersymmetric mass of the messenger superfield Φ_k while M_s is as in (13.5). The function $u(x)$ is negative in the range $0 < x < 1$, being given by

$$u(x) = \frac{1}{x^2}\left[\ln(1 - x^2) + \frac{x}{2}\ln\frac{1+x}{1-x}\right]. \tag{13.37}$$

For x close to zero, $u(x) \simeq -x^2/6 + O(x^4)$ and thus becomes very small if the messenger scale $M_M \simeq M_k \gg M_s$. However, for $M_s \simeq M_k$, (13.36) constrain some couplings in (13.35) quite severely. Since $u < 0$, these one loop contributions could lead to some negative squared sfermion masses unless the relevant couplings are sufficiently small. This requirement can be quantified to [13.11]

$$\sum |\lambda|^2 < 10^{-3}$$

for couplings λ that contribute to m_E^2 for $x_k \simeq 1$. Bounds on FCNC processes lead to even stronger constraints on some products of flavor off-diagonal couplings (for $x_k \simeq 1$). In fact, the lack of observation of $\mu \to e\gamma$ decay and of $\mu \to e$ conversion in atoms requires some of these coupling products to lie [13.12] below the 10^{-5} level.

13.6 Flavor Symmetries for the GMSB Scenario

The viability of gravity mediated supersymmetry breaking models, constructed with flavor symmetries in conjunction with suitably chosen Frogatt-Nielsen (FN) fields [12.34], was discussed in §12.5. These are characterized both by sufficiently small sparticle loop contributions to FCNC amplitudes and by realistic textures for Yukawa coupling matrices. Finding such symmetries becomes much simpler in GMSB models *if* the energy scale Λ_F, where the flavor symmetry is broken, lies above the messenger scale M_M. In this case the (broken) flavor symmetry leaves no imprint at all on the soft supersymmetry breaking masses and the supersymmetric contributions to FCNC amplitudes are all very small. The latter situation is in sharp contrast with that discussed in §12.5 where finding the right flavor symmetry was argued to be quite a nontrivial exercise. On the other hand, for reasons of economy, one might be tempted to identify the two energy scales, at least approximately: $\Lambda_F \sim M_M$. To date, such a possibility has been discussed only within the limited framework of a particular type of models. These are the ones where supersymmetry breaking is communicated to the hidden sector superfield S appearing in the messenger superpotential (13.1), through a $U(1)_m$ messenger symmetry group. In such a scenario the auxiliary component VEV $\langle F_S \rangle$ is suppressed by a loop factor $\sim \alpha/\pi$ relative to the supersymmetry breaking scale in the hidden sector. One can then follow either of two routes, as elaborated below, in the construction of potentially realistic models.

The first route [13.13] is, in fact, the conceptually simpler one. Here one identifies the messenger $U(1)_m$ symmetry with the flavor symmetry and messenger superfields with the FN superfields. Some MSSM matter superfields must consequently carry nonvanishing $U(1)_m$ charges $q^{(m)}$. This means that they communicate *directly* with the hidden sector, which also carries nonzero $U(1)_m$ charges. At energy scales below M_M, the scalar potential is then of the form

$$V_{\text{eff}} = \sum_i \left| \frac{\partial \mathcal{W}}{\partial \Psi_i} \right|^2 + \frac{g_m^2}{2} \left(\xi^2 + \sum_i q_i^{(m)} |\psi_i|^2 \right)^2$$

$$+ \tilde{m}^2 \sum_i q_i^{(m)^2} |\psi_i|^2 + (\text{two loop terms}) + (\text{MSSM } D-\text{terms}) . \qquad (13.38)$$

The sizes of the $U(1)_m$ Fayet-Iliopoules term proportional to ξ^2 as well as of the soft supersymmetry breaking parameter $\tilde{m}^2$ in (13.38) depend on the details of the hidden sector. One can define ξ^2 to be positive, thereby fixing the signs of the $U(1)_m$ charges. However, $\tilde{m}^2$ can have either sign. Now the leading contributions to the masses of scalars, carrying nonzero messenger charges, depend either linearly or quadratically on those charges and are therefore *nonuniversal* for different generations. Then the only solution, consistent with both naturalness arguments (Ch.1) and FCNC constraints, is to make the following assumption. Third generation[20] superfields as well as Higgs superfields are taken to carry no $U(1)_m$ charge, while first and second generation superfields are made to carry them. Thus the scalar components of the latter acquire masses in the range of tens of TeV. This then is an explicit realization of the solution to the supersymmetric FCNC problem that relies on very large masses of first and second generation sfermions, cf. §9.5. The particular model of Ref. [13.13] yields supersymmetric contributions to K^0-$\bar{K}^0$ mixing that are, in magnitude, lower than but close to the present experimental limit. It also predicts large supersymmetric contributions to CP-violating asymmetries in B-decays.

The second route is to postulate the existence of a 'flavor sector', in addition to and essentially decoupled from the messenger sector [13.14]. This flavor sector consists of Frogatt-

---

[20]This condition on third generation superfields may be relaxed sometimes.

Nielsen superfields as well as 'flavon' superfields whose scalar VEVs break the flavor symmetry. Just as the messenger superfields acquire their masses, so do the FN superfields in a similar fashion – through their couplings to some superfield(s) X_f that are singlets under the SM gauge group and whose scalars acquire large VEVs. However, the need to avoid large nonuniversal contributions to sfermion masses forces the requirement

$$|\langle F_{X_f} \rangle| \ll |\langle F_S \rangle| .$$

In other words, it has to be required that the dynamics of the hidden sector distinguish between the superfields S and X_f. An explicit example of this route is given by the model of Ref. [13.14]. This model assumes that

$$\Lambda_F \lesssim M_M \simeq M_s \sim \mathcal{O}(50) \text{ TeV}$$

and consequently becomes tightly constrained[21]. In particular, a global, continuous flavor symmetry is excluded. This is since the couplings of the corresponding Goldstone boson(s) ('familons'), which are proportional to inverse powers of Λ_F, would get too large. Such large couplings for them would be in conflict with experimental upper bounds on unobserved $\mu \to e+$ familon and $K \to \pi+$ familon decay rates. Gauged continuous symmetries are also dangerous because of both nonuniversal D-term contributions to scalar masses and gauge boson exchange contributions to FCNC processes. This scenario thus favors discrete flavor symmetries.

The superpotential of the flavor sector has to contain terms like

$$\mathcal{W}_{\text{flavor}} = X_f \bar{\Psi}\Psi + \bar{\Psi}\Phi F + H\Psi F + H F_3 \bar{F}_3 . \tag{13.39}$$

A symbolic notation has been used in (13.39) above: Ψ and $\bar{\Psi}$ stand for Frogatt-Nielsen superfields and Φ for flavon superfields, while F and $\bar{F}$ stand for MSSM matter superfields and H for Higgs superfields. Note that Φ, like X_f, is a singlet with respect to SM gauge transformations. A distinct choice of F, $\bar{F}$ in the third generation has been made in the last RHS term of (13.39). Terms containing additional MSSM superfields are needed to stabilize the VEVs[22] of x_f and ϕ. The superpotential (13.39) generates masses for first and second generation fermions through diagrams like that shown in Fig. 12.9.

It may be recalled that X_f couples to the hidden sector. Consequently, the scalar flavons ϕ, the MSSM scalars $\tilde{f}$ and the Higgs bosons h get negative, nonuniversal two loop corrections to their squared soft supersymmetry breaking masses. These corrections are proportional to products of superpotential couplings which are supposed to appear in (13.39). In case of flavons, this is a desirable feature; it automatically generates a hierarchy

$$\frac{\langle \phi \rangle}{\Lambda_F} \sim \frac{\sqrt{-m_\phi^2}}{\Lambda_F} \sim \frac{1}{16\pi^2} . \tag{13.40}$$

Eq.(13.40) can be used to produce realistic quark and lepton mass matrices through the Frogatt-Nielsen mechanism [12.34]. However, these contributions must not be too large in

---

[21]If $M_M \gg M_s$, many of the constraints discussed below become far weaker. However, once one is allowing a hierarchy of scales, one might as well assume that $\Lambda_F \gg M_M$.

[22]Note that $|\langle F_{X_f} \rangle| \sim (\alpha/\pi)|\langle x_f \rangle|^2$.

comparison with the usual gauge mediated contribution (13.7). Otherwise, squared Higgs scalar masses would become too negative, to counterbalance which an excessive finetuning in $|\mu|$ would be required. Furthermore, the supersymmetric FCNC problem might resurface through nonuniversal contributions to sfermion masses. On the other hand, the scale Λ_F cannot be too small, either. That is since the exchange of FN fields, with masses $\mathcal{O}(\Lambda_F)$, mediates *tree-level* FCNC amplitudes. In the model of Ref. [13.14], the authors choose $\langle x_f \rangle \equiv \Lambda_F \simeq M_M/5 \sim 10$ TeV. Flavon fields, which are singlets with respect to the gauge transformations of the MSSM, are then expected to acquire masses of a few hundered GeV. Even for this optimal choice of Λ_f, the agreement with extant FCNC bounds is rather marginal. Once again, one expects new contributions at a level not far beyond the present bounds. Indeed, a significant experimental tightening of the latter would pose a severe challenge to this scenario. We shall have more to say on the future exploration of GMSB scenarios at colliders in Ch.15.

References

[13.1] M. Dine and A.E. Nelson, Phys. Rev. **D48** (1993) 1277. M. Dine, A.E. Nelson and Y. Shirman, Phys. Rev. **D51** (1995) 1362. M. Dine, A. Nelson, Y. Nir and Y. Shirman, Phys. Rev. **D53** (1996) 2658.

[13.2] G.F. Giudice and R. Rattazzi, *loc. cit.*, *Bibl.* C.F. Kolda, *loc. cit.*, *Bibl.*

[13.3] H. Georgi, *op. cit.*, *Bibl.* G. Ross, *op. cit.*, *Bibl.*

[13.4] S.P. Martin, Phys. Rev. **D55** (1997) 3177.

[13.5] S. Ambrosanio, G.L. Kane, G.D. Kribs, S.P. Martin and S. Mrenna, Phys. Rev. **D54** (1996) 5395, *ibid.* **D55** (1997) 3188.

[13.6] D. Acosta et al. (CDF collaboration), Phys. Rev. Lett. **89** (2002) 281801-1.

[13.7] S. Dimopoulos, S. Thomas and J.D. Wells, Nucl. Phys. **B488** (1997) 39.

[13.8] G. Dvali, G.F. Giudice and A. Pomarol, Nucl. Phys. **B478** (1996) 31.

[13.9] G.F. Giudice and A. Masiero, Phys. Lett. **B206** (1988) 480. J. Casas and C. Muñoz, Phys. Lett. **B306** (1993) 208. G.F. Giudice and E. Roulet, Phys. Lett. **B315** (1993) 107.

[13.10] A. de Gouvea, A. Friedland and H. Murayama, Phys. Rev. **D57** (1998) 5676.

[13.11] T. Han and R-J. Zhang, Phys. Lett. **B428** (1998) 120.

[13.12] S.L. Dubovsky and D.S. Gorbunov, Phys. Lett. **B419** (1998) 223.

[13.13] D.E. Kaplan, F. Lepeintre, A. Masiero, A.E. Nelson and A. Riotto, Phys. Rev. **D60** (1999) 055003.

[13.14] N. Arkani-Hamed, C.D. Carone, L.J. Hall and H. Murayama, Phys. Rev. **D54** (1996) 7032.

Chapter 14

BEYOND THE MSSM

14.1 Motivation and Outline

Certain extensions of the MSSM are discussed in this chapter. We consider in §14.2 the so-called Next-to-the-Minimal Supersymmetric Standard Model (NMSSM). This is a model where just the Higgs sector of the MSSM is extended by the addition of a new Higgs superfield which is a singlet under SM gauge transformations. This step enables one to construct a realistic scheme without introducing dimensional quantities in the superpotential. Recall from discussions in Ch.10 that the higgsino mass parameter μ is a bit of an embarassment in the MSSM. The μ term is supersymmetric, being part of the superpotential, and μ can a priori have any value. However, in order to implement EW symmetry breaking, the magnitude of μ needs to be (cf. §10.2) of the same order as the EW scale or comparable to the magnitudes of the soft supersymmetry breaking parameters which supposedly induce the latter. The MSSM, by itself, does not provide an explanation for this puzzle. In the NMSSM μ is replaced by the VEV of the singlet Higgs scalar. In this way some rationale emerges for μ turning out to have a magnitude of the order of the electroweak scale. After this, we turn to a discussion of the violation of R-parity, the assumed discrete symmetry of the MSSM, for which a really convincing reason has never been given. We show in §14.3 that certain types of R_p interactions may well exist in nature, leading to interesting new signals, though strong phenomenological bounds exist on many R_p coupling strengths. These bounds are enumerated in §14.4 along with descriptions of the processes from which they emerge. A specific model of bilinear R_p violation is described in §14.5. Finally, the question of introducing neutrino masses to the MSSM is taken up. We have repeatedly alluded earlier to the necessity for this move. There is now overwhelming experimental evidence in favor of tiny nonzero neutrino masses and these need to be included in a broken supersymmetric theory of the world. Various ways of doing that are discussed in §14.6. Of course, many other extensions of the MSSM are possible, involving new superfields and/or new interactions. Quite a few of these can be included in models based on the Grand Unified gauge group $E(6)$. We refer the reader to the relevant literature for further details [14.1].

343

14.2 The Next-to-the-Minimal Supersymmetric Standard Model

As a minimal addition to the MSSM field content, one can introduce a chiral Higgs superfield which is a singlet under the SM gauge group. The resulting model is usually described[1] as the Next-to-the-Minimal Supersymmetric Standard Model (NMSSM) [14.2]. Let us first explain the main motivation for introducing such a Higgs singlet superfield. The real advantage of the latter is that it can lead to a purely weak scale solution of the so-called μ problem. More recent studies of the NMSSM have also been motivated by other observations. For instance, Higgs search limits from LEP can be easily accommodated in this model even if $\tan\beta$ is near unity and all sparticles are fairly light. Recall from our discussions in Ch.10 that this goal is hard to attain within the MSSM on account of quite sizable radiative corrections needed there in the Higgs sector. Those, in turn, require large soft supersymmetry breaking masses for third generation squarks, thereby aggravating the fine tuning problem. The evasion of this difficultly is a plus point for the NMSSM, as will be made clear when we discuss the Higgs sector of the NMSSM in detail. Another motivation is that the said limits from LEP are compatible with electroweak baryogenesis within the aegis of the NMSSM, but not with the MSSM alone unless $m_{\tilde{t}} > 1$ TeV. This issue will be elaborated upon in §16.4. Here we give a brief account of the μ problem as well as of the theoretical issues arising from the introduction of a Higgs singlet superfield.

We saw earlier (§10.2) the need for a nonvanishing mixed Higgs squared mass m_{12}^2 $(= \mu B)$ in order to give VEVs to both neutral Higgs fields $h_{1,2}^0$. A vanishing m_{12}^2 would have further implied the presence in the Higgs potential of a global $U(1)$ symmetry under which both Higgs doublets transform in the same way. The breakdown of electroweak symmetry would also mean the breaking of this global $U(1)$ symmetry, leading to an unwanted massless Goldstone boson, cf. (10.24b). We thus need a nonzero m_{12}^2 and hence nonvanishing values for both the soft supersymmetry breaking parameter B and the higgsino mass parameter μ. Moreover, as will be discussed later in §15.2, unsuccessful chargino searches at LEP already tell us directly that $|\mu| \gtrsim M_W$. In other words, in the MSSM one is forced to choose a value of μ, which is a supersymmetry invariant parameter, to be of the same order of magnitude as the soft supersymmetry breaking parameters, i.e. $|\mu| = \mathcal{O}(M_s)$. This is the infamous "μ problem" [14.3]. Evidently, this is "only" a question of naturalness since nothing prevents us from choosing these two a priori independent scales to be close to each other. However, since the main motivation for weak scale supersymmetry comes from naturalness arguments (cf. Ch. 1), such a choice cannot be justified within just the MSSM.

Several resolutions of the μ problem have been suggested in the literature. In all cases one starts with a purely cubic superpotential in the observable sector by postulating that $\mu = 0$ in the fundamental theory, via the imposition of some symmetry on the starting superpotential, say. A nonvanishing μ can then be generated radiatively, e.g. in a GUT sector [14.4]. Such a proposal would require the soft supersymmetry breaking terms to violate the symmetry which protects the superpotential from acquiring a (very large) μ term. Another problem with this approach is that the generated μ would be smaller in

---

[1]References to the original work on the NMSSM may be found in the review articles cited in Ref. [14.2].

magnitude than typical soft supersymmetry breaking parameters by a loop factor. In order to have a weak scale μ, i.e. $\mu = \mathcal{O}(M_W)$, one is then forced to choose an uncomfortably large $M_s \gg M_W$. An alternative procedure would be to introduce a nonrenormalizable coupling between the bilinear superfield product $H_1 \cdot H_2$ and hidden sector superfields. This can be done either in the 'kinetic' part [14.5] of the Kähler potential (12.2) or in the superpotential [14.6]. In both scenarios an effective μ term appears in the low energy superpotential once the hidden sector scalar fields acquire the VEVs needed to break supersymmetry. A direct connection is thereby established to supersymmetry breaking. In the former option, known as the *Giudice-Masiero mechanism*, one can extend the symmetry, ensuring the absence of a μ term in the original superpotential, into the hidden sector. This symmetry can now be identified with the Peccei-Quinn (PQ) symmetry [14.7], used to alleviate the strong CP problem. However, if elementary scalar fields of the hidden sector have nontrivial charges under this symmetry, the natural PQ symmetry breaking scale would be $\mathcal{O}(M_{Pl})$, well above the range around 10^{10} GeV favored by cosmological arguments [14.7]. While these arguments are not completely airtight [14.5], it might be more natural [14.8] to instead couple $H_1 \cdot H_2$ to a hidden sector "quark" condensate $\langle \bar{Q}_h Q_h \rangle$. This procedure exploits the (approximate) numerical coincidence between the desired PQ symmetry breaking scale and the scale $\sqrt{m_{3/2} M_{Pl}}$ characterizing the VEVs of hidden sector F-terms, cf. (12.5). Finally, an effective μ parameter of the right size also emerges [14.9] upon integrating out heavy fields in a supergravity GUT, provided there is a renormalizable coupling of the bilinear superfield product $H_1 \cdot H_2$ to some heavy GUT superfield.

All these mechanisms [14.5, 14.6, 14.8, 14.9] produce an effective μ parameter $\mathcal{O}(M_s)$, proving that such a coincidence can indeed be natural. There is an adverse feature, though. They all operate at very high energies, leaving essentially no imprint on the weak scale spectrum[2]. This means that they cannot be tested experimentally. The NMSSM is very different in this respect, since here the form of the MSSM superpotential (8.33) is changed. The term $\mu H_1 \cdot H_2$ is replaced by two new terms with dimensionless coupling strengths λ and κ, namely

$$\mathcal{W}_N = \lambda N H_1 \cdot H_2 + \frac{1}{3} \kappa N^3, \tag{14.1}$$

the Yukawa coupling terms in (8.33) remaining the same. Evidently, (14.1) is gauge invariant only if the new chiral superfield N is a gauge singlet. One needs to arrange the parameters of the theory in a way such that the scalar component n of N obtains a VEV of the order of M_s. In that situation $\lambda \langle n \rangle$ plays the role of an effective μ parameter. This is usually not too difficult to achieve. The NMSSM, as defined above, has the appealing property that all dimensionful parameters break supersymmetry. Therefore, if $x \equiv \sqrt{2} \langle n \rangle$ is nonzero, it is expected to be naturally $\mathcal{O}(M_s)$. This model[3] then is manifestly free of the μ problem. However, the existence of a 'light' chiral gauge singlet superfield in the observable sector can

---

[2]Even the axion, produced in the spontaneous breakdown of the PQ symmetry, is "invisible" since its couplings are suppressed by inverse powers of the PQ symmetry breaking scale. Besides, one can easily construct [14.7] "invisible" axion models which have nothing to do with the μ problem.

[3]The terms $\mu H_1 \cdot H_2 + \frac{1}{2} \mu' N^2$ could also be introduced in the superpotential if one were merely interested in studying the phenomenological consequences of introducing a Higgs singlet superfield. That would, of course, "double" the μ problem. However, apart from the possibility of spontaneous CP violation, to be discussed later, the phenomenology would remain qualitatively the same [14.10].

cause other difficulties.

The first of these difficulties is related to the stability of the gauge hierarchy. Let us work within the framework of a supersymmetric grand unified field theory. It *can* admit superpotential terms of the form $\tilde{\lambda} N X_1 X_2$, where $X_{1,2}$ are very heavy superfields from the GUT sector. Such a term will, in turn, induce [14.11] in the scalar potential an $n^2 + n^{\star 2}$ term with a coefficient $\sim (8\pi^2)^{-1}\tilde{\lambda}\kappa M_s M_U \gg M_s^2$ as well as a linear (tadpole) term with a coefficient $\sim (8\pi^2)^{-1}\tilde{\lambda}M_s^2 M_U \gg M_s^3$, M_U being a characteristic GUT mass scale. Such terms would induce a VEV $x \gg M_s \sim \mathcal{O}(M_W)$, thereby destabilizing the hierarchy; they need to be forbidden by excluding any $\tilde{\lambda} N X_1 X_2$ term from the superpotential. On the other hand, if the superfield N is a singlet under the GUT gauge group, such an exclusion becomes untenable since $\tilde{\lambda}$ gets related by gauge invariance to the coupling strength λ in the superpotential, taking $X_{1,2}$ to be the GUT partners of the two Higgs superfields $H_{1,2}$. Therefore, the NMSSM can thus be embedded into a field theoretic GUT *only if* N is a nonsinglet under the GUT group. This restriction considerably complicates the construction of realistic models.

There is another problematic feature of the NMSSM, as defined by the superpotential (14.1). It possesses a discrete Z_3 symmetry under which all chiral superfields are multiplied by the factor $\exp(2\pi i/3)$. This symmetry is also respected by soft supersymmetry breaking scalar mass terms and the trilinear A-terms[4]. There is consequently a severe cosmological problem for the scenario in which the Universe underwent the electroweak phase transition or crossover after the end of inflation. Causally disconnected parts of the Universe would have randomly "chosen" one of the three Z_3-equivalent minima of the scalar potential, leading to the formation of disastrous domain walls [14.12]. The solution of this problem requires the addition of some terms to the scalar potential that explicitly break the Z_3 symmetry; at least one of these terms should have, at the minimum of the potential, a size [14.12] $\gtrsim 10^{-6}\tilde{v}^3 M_W^2/M_{Pl}$, $\tilde{v}$ being the largest VEV that breaks the Z_3 symmetry spontaneously. Within the framework of gravity mediated supersymmetry breaking, the required amount of violation of the Z_3 symmetry can be introduced through nonrenormalizable operators. The dominant contribution comes not from these operators themselves but radiatively [14.13] from the n-tadpole (a term in the scalar potential that is linear in n), induced by them. The imposition of another discrete symmetry (often a discrete R-symmetry) on these non-renormalizable terms can ensure [14.14] that this tadpole is sufficiently suppressed so as not to be able to destabilize the hierarchy, and yet remain large enough to solve the domain wall problem. Indeed, the lower and upper bounds on the tadpole coefficient differ by some twenty orders of magnitude[5]. We shall shortly see that this tadpole can also have important phenomenological implications [14.15] if its size is close to the upper bound of $\mathcal{O}(M_s^3)$.

---

[4]Though the superpotential is purely trilinear, one could in principle introduce bilinear soft supersymmetry breaking terms proportional to $H_1 H_2$ or n^2 which break the Z_3 symmetry explicitly. In most schemes of supersymmetry breaking, however, it is difficult to generate such terms in the absence of corresponding terms in the superpotential.

[5]The arguments of Refs. [14.13, 14.14] are based on estimates of radiative corrections in a nonrenormalizable field theory. A momentum cutoff equal to the Planck mass is used there to derive finite results. Moreover, the additional discrete symmetry can be cleverly chosen [14.14, 14.15] so as to forbid the κ-term in (14.1). This more simplified (with $\kappa = 0$) version of the NMSSM has been given the name Minimally Nonminimal Supersymmetric Standard Model (MNMSSM).

These problems might lead one to conclude that the NMSSM is no more attractive or natural a solution to the μ problem than the mechanisms suggested in Refs. [14.5, 14.6, 14.8, 14.9]. One strength of the NMSSM, nonetheless, is that it is a concrete proposal that is amenable to experimental tests. It is therefore worthwhile to discuss the phenomenology of the NMSSM to which we turn, starting with the Higgs sector.

The Higgs sector of the NMSSM

We start with the superpotential (14.1), but supplement it with an additional 'tadpole' term $T^2 N$, generated through the nonrenormalizable operators mentioned above. Similarly, we add a true scalar tadpole term $t^3 n + $ h.c. to the scalar potential. The latter also needs to include trilinear A-terms involving $nh_1 \cdot h_2$ and n^3. Here n is the scalar component of the chiral superfield N, while T and t are dimensional constants. Clearly, a stable hierarchy requires the magnitudes of the dimensional parameters T and t to be $\lesssim M_s$. Altogether, the Higgs potential of the NMSSM is then given by [14.10, 14.15],

$$
\begin{aligned}
V_H \;=\; & |\lambda|^2 \Big\{ |n|^2 \left(|h_1|^2 + |h_2^0|^2 \right) + |h_1 \cdot h_2|^2 \Big\} + |\kappa|^2 |n|^4 \\[2mm]
& + \Big\{ \lambda h_1 \cdot h_2 \left(\kappa^\star n^{\star^2} + T^{\star^2} \right) - \lambda A_\lambda n h_1 \cdot h_2 - \frac{\kappa}{3} A_\kappa n^3 + t^3 n + \text{h.c.} \Big\} \\[2mm]
& + m_n^2 |n|^2 + m_1^2 |h_1|^2 + m_2^2 |h_2|^2 + D-\text{terms} .
\end{aligned}
\tag{14.2}
$$

The D-terms in (14.2) are identical to those in the MSSM potential (10.5). Let us assume for the moment that CP is conserved in the Higgs sector, so that all parameters in the scalar potential (14.2) as well as the VEVs $x \equiv \sqrt{2}\langle n \rangle$, $v_1 \equiv \sqrt{2}\langle h_1^0 \rangle$ and $v_2 \equiv \sqrt{2}\langle h_2^0 \rangle$ (with $\tan\beta = v_2/v_1$ as usual) are real. The five physical neutral Higgs particles can then be grouped into two CP odd and three CP even states. Together with the charged Higgs bosons, they form the seven Higgs particles of the NMSSM, cf. Table 14.1, where all the notation to be used by us is explained.

Symbol	Description	Decomposition
$H^\pm$	Charged Higgs particles	$h_1^\pm \sin\beta + h_2^\pm \cos\beta$
A_1, A_2 with $m_{A_1} \le m_{A_2}$	Neutral CP odd Higgs particles	$A_1 = a^0 \cos\alpha_{PS} + \sqrt{2}\Im m\, n \sin\alpha_{PS}$ $A_2 = -a^0 \sin\alpha_{PS} + \sqrt{2}\Im m\, n \cos\alpha_{PS}$
h_1, h_2, h_3 with $m_{h_1} \le m_{h_2} \le m_{h_3}$	Neutral CP even Higgs particles	$h_i = \sqrt{2}\Re e\left[\mathcal{O}_{i1}(h_1^0 - v_1) + \right.$ $\left. \mathcal{O}_{i2}(h_2^0 - v_2) + \mathcal{O}_{i3}(n - x) \right]$

Table 14.1. The physical Higgs states of the NMSSM under the assumption of CP conservation. The angle α_{PS} defines the rotation needed to diagonalize the mass matrix (14.3), where the state a^0 is defined. The orthogonal matrix $\mathcal{O}$ diagonalizes the mass matrix (14.5).

Under the stated assumptions, the tree level mass matrices for the CP even and CP odd neutral Higgs bosons can be computed by taking second derivatives of the scalar potential

(14.2) w.r.t. the real and imaginary parts, respectively, of the fields h_1^0, h_2^0 and n. The soft supersymmetry breaking squared masses m_1^2, m_2^2 and m_n^2 can be eliminated from these matrices by use of the equations of constraint $\partial V_H/\partial h_1^0 = \partial V_H/\partial h_2^0 = \partial V_H/\partial n = 0$. The latter have to be satisfied in the minimum of the potential. As in the MSSM, one finds $G^0 = \sqrt{2}(\Im m\, h_2^0 \sin\beta - \Im m\, h_1^0 \cos\beta)$ to be the massless Goldstone boson. The mass matrix for the other two CP odd states in the basis $(a^0, \sqrt{2}\Im m\, n)$, with $a^0 = \sqrt{2}(\Im m\, h_1^0 \sin\beta + \Im m\, h_2^0 \cos\beta)$, is

$$
m^2_{\Im m\, h^0} = \begin{pmatrix} m_a^2 & \lambda v(\kappa x - A_\lambda/\sqrt{2}) \\ \lambda v(\kappa x - A_\lambda/\sqrt{2}) & m_{n,I}^2 \end{pmatrix}. \tag{14.3}
$$

In (14.3)

$$
m_a^2 = \lambda(\sin 2\beta)^{-1}(2T^2 + \kappa x^2 - \sqrt{2}A_\lambda x)\,, \tag{14.4a}
$$

$$
m_{n,I}^2 = \lambda v^2 \sin 2\beta \left\{ \kappa - A_\lambda(2\sqrt{2}x)^{-1} \right\} + 3\kappa A_\kappa x/\sqrt{2} - \sqrt{2}t^3 x^{-1} \tag{14.4b}
$$

and $v^2 = v_1^2 + v_2^2 = 1/(\sqrt{2}G_F)$. The superpotential 'tadpole' term $T^2 N$ affects the 'mass' of the $SU(2)_L$ doublet field a^0, while the scalar tadpole term $t^3 n$ only affects $m_{n,I}^2$, the latter effect stemming solely from the equation of constraint for x. The diagonalization of the matrix (14.3) is straightforward. Since its smaller eigenvalue has to be less than the smallest of the diagonal elements, both m_a^2 and $m_{n,I}^2$ have to be positive for a physical solution. This requirement imposes stringent constraints on the allowed parameter space. Yet, no nontrivial upper or lower bounds on the eigenvalues can be given. This is just as in the MSSM, where the mass m_A of the single CP odd Higgs boson can a priori have any value.

The squared mass matrix of the CP even Higgs scalars can be given in the basis $\{\sqrt{2}\Re e\,(h_1^0 - v_1), \sqrt{2}\Re e\,(h_2^0 - v_2), \sqrt{2}\Re e\,(n - x)\}$ by

$$
m^2_{\Re e\, h^0} = \begin{pmatrix} m_a^2 \sin^2\beta + M_Z \cos^2\beta & \frac{1}{2}\sin 2\beta(\lambda v^2 - m_a^2 - M_Z^2) & C\cos\beta + C'\sin\beta \\ \frac{1}{2}\sin 2\beta(\lambda v^2 - m_a^2 - M_Z^2) & m_a^2 \cos^2\beta + M_Z^2 \sin^2\beta & C\sin\beta + C'\cos\beta \\ C\cos\beta + C'\sin\beta & C\sin\beta + C'\cos\beta & m_{n,R}^2 \end{pmatrix},
\tag{14.5}
$$

with m_a^2 given by (14.4a). In (14.5) we have introduced the quantities

$$
C = \lambda^2 v x\,, \tag{14.6a}
$$

$$
C' = \lambda v(A_\lambda/\sqrt{2} - \kappa x)\,, \tag{14.6b}
$$

$$
m_{n,R}^2 = 2\kappa^2 x^2 - \lambda A_\lambda v^2 \sin 2\beta(2\sqrt{2}x)^{-1} - \kappa A_\kappa x/\sqrt{2} - \sqrt{2}t^3 x^{-1}. \tag{14.6c}
$$

It may be noted that the (1,1) and (2,2) entries of the matrix (14.5) are identical to the corresponding entries of the analogous MSSM matrix (10.26), if one replaces m_A^2 by m_a^2. However, each of the off-diagonal (1,2) and (2,1) entries receives an additional contribution proportional to λv^2. This makes a crucial difference for the upper bound on the smallest eigenvalue of (14.5) which determines the squared mass of the lightest Higgs scalar. Again, this eigenvalue has to be smaller than any diagonal element, i.e. $m_{n,R}^2$ of (14.6c) has to be positive. A much more useful bound follows from the condition that the smallest eigenvalue of the matrix (14.5) must be less than the lower eigenvalue of any 2×2 submatrix that is centered on the diagonal. The application of this theorem to the top left submatrix describing the $SU(2)_L$ doublet fields leads [14.10] to the upper bound

$$m_{h_1}^2 \leq \frac{1}{2}\left[m_a^2 + M_Z^2 - \left\{(m_a^2 + M_Z^2)^2 - 4m_a^2 M_Z^2 \cos^2\beta + \lambda v^2(\lambda v^2 - 2m_a^2 - 2M_Z^2)\sin^2 2\beta\right\}^{1/2}\right],$$

$$(14.7)$$

where h_1 is the lightest CP even physical Higgs scalar of the NMSSM.

The terms, proportional to λv^2 within the square root in (14.7), have two consequences. First, m_{h_1} no longer vanishes in the limit $m_a \to 0$, provided $0 < \lambda < (g_2^2 + g_Y^2)/2$. This means that, if the lightest CP even and CP odd mass eigenstates have small singlet components, the former can be heavier than the latter even at the tree level[6]. Eq. (10.30c) shows that this is not possible in the MSSM. The opposite limit $m_a^2 \gg M_Z^2$ is of even greater phenomenological relevance since it maximizes the upper bound in (14.7). In this limit one has [14.10]

$$m_{h_1}^2 \leq M_Z^2\left[\cos^2 2\beta + 2\lambda^2(g_2^2 + g_Y^2)^{-1}\sin^2 2\beta\right]. \tag{14.8}$$

The new quartic coupling strength λ between the Higgs doublets in the potential (14.2) has introduced an additional term to the RHS of (14.8), as compared with MSSM result (10.30c). Note that the upper bound in (14.8) can indeed be saturated. In order to attain saturation, one has to choose $m_a^2 \gg M_Z^2$ as well as $m_{n,R}^2 \gg |C|, |C'|$, cf. (14.6), implying that the Higgs singlet decouples from the doublets. In this 'decoupling limit', the tree level couplings of h_1 to SM particles approach those of the SM Higgs – just as in the MSSM with $m_A^2 \gg M_Z^2$. Furthermore, the appearance of λ^2 in the RHS of (14.8) means that no upper bound on m_{h_1} can be given unless we find a way to limit $|\lambda|$ from above. This is rather similar to the Higgs sector of the nonsupersymmetric SM. The squared Higgs mass there is also proportional to the unknown quartic Higgs self-coupling strength. *Without assumptions about physics at energies well above the weak scale*, one can then derive only the mild upper [14.16] bound $m_{h_1} \lesssim 700$ GeV from the requirement of perturbative unitary for S-matrix elements of the elastic scattering process $V_L V_L \to V_L V_L$, V_L standing for a massive longitudinal weak boson, W or Z.

On the other hand, unlike in the SM, the appearance of large hierarchies is at least technically natural with supersymmetry. Moreover, as discussed in §11.2, the three gauge coupling strengths appear to meet a scale $\sim 2 \times 10^{16}$ GeV (with $M_s \sim 1$ TeV), if the MSSM gauge β-functions, calculated perturbatively, remain valid upto that scale. The new NMSSM coupling strengths λ and κ actually do affect the running of the gauge coupling strengths but only at two and three loops respectively. Therefore, the coupling unification result continues to hold in the NMSSM provided the NMSSM remains amenable to a perturbative treatment upto such energies. In particular, the coupling strengths λ and κ should remain in the perturbative region ($\lesssim \pi$ in magnitude, say) upto the scale of Grand Unification. The one loop renormalization group evolution of these coupling strengths is connected to that of the Yukawa coupling strengths $h_{t,b}$ [14.17]. With $t = \ln(\mu/\mu_0)$ and $g_{Y,2,3}$ as the concerned gauge coupling strengths, as in Ch.11, the relevant evolution equations can be written as

$$\frac{d\lambda}{dt} = \frac{\lambda}{16\pi^2}\left[4\lambda^2 + 2\kappa^2 + 3(f_t^2 + f_b^2) - 3g_2^2 - g_Y^2\right], \tag{14.9a}$$

---

[6]Eqs. (14.4b) and (14.6c) show that the mass ordering of the Higgs fields, that are mostly $SU(2)_L$ singlets, is also unconstrained.

$$\frac{d\kappa}{dt} = \frac{3\kappa}{8\pi^2}(\lambda^2 + \kappa^2) \,, \tag{14.9b}$$

$$\frac{df_t}{dt} = \frac{f_t}{16\pi^2}\left(6f_t^2 + f_b^2 + \lambda^2 - \frac{16}{3}g_3^2 - 3g_2^2 - \frac{13}{9}g_Y^2\right), \tag{14.9c}$$

$$\frac{df_b}{dt} = \frac{f_b}{16\pi^2}\left(6f_b^2 + f_t^2 + \lambda^2 - \frac{16}{3}g_3^2 - 3g_2^2 - \frac{7}{9}g_Y^2\right). \tag{14.9d}$$

The changes in the β-functions of the top and bottom Yukawa coupling strengths, as compared to their MSSM versions (11.29a,b), are not too important. However, (14.9a) shows that $|\lambda|$ will grow faster with energy, and hence the upper bound on the weak scale value of $|\lambda|$ will be stronger, when the other superpotential coupling strengths $|\kappa|$, $|f_t|$ and $|f_b|$ are increased. In particular, an increase in the mass of the top quark (for fixed $\tan\beta$) would reduce the upper bound on $|\lambda|$ at the weak scale. The latter would, in turn, *reduce* the tree level upper bound (14.8) on the mass of the lightest CP even Higgs h_1.

Of course, the squared mass matrices (14.3) and (14.5) are subject to radiative corrections. As in the MSSM, the most important contributions come from the top-stop sector (and, if $\tan\beta \gg 1$, from the $b\text{-}\tilde{b}$ sector). In fact, the corrections to m_a^2 and to the top left 2×2 submatrix of (14.5) are identical to the corresponding corrections in the MSSM given by (10.63)–(10.67), with the replacement $\mu \to \lambda x/\sqrt{2}$. There are also $t\text{-}\tilde{t}$ and $b\text{-}\tilde{b}$ corrections to the other entries of the matrices in (14.3) and (14.5), but these do not affect the upper bound on m_{h_1}. There are also corrections [14.18] involving the new coupling strengths λ and κ themselves, but these are of lesser importance. To a pretty good approximation, then, one has

$$m_{h_1}^2 \leq M_Z^2\left[\cos^2 2\beta + 2\lambda_{\max}^2(g_2^2 + g_Y^2)^{-1}\sin^2 2\beta\right]$$
$$+\frac{3G_F}{\sqrt{2}\pi^2\sin^2\beta}\left[m_t^4(M_i)\ln\frac{M_s^2}{m_t^2} + \frac{(A^t)^2 m_t^4(M_s)}{M_s^2}\left(1 - \frac{(A^t)^2}{12M_s^2}\right)\right], \tag{14.10}$$

with $M_s \equiv \sqrt{m_{\tilde{t}_1}m_{\tilde{t}_2}}$ and $M_i \equiv \sqrt{M_s m_t}$. Here $\lambda_{\max}$ is the maximum value (at the scale M_s) of λ that is compatible with grand unification. The upper bound (14.10) reduces to that for the MSSM, namely (10.72), when $\lambda_{\max} \to 0$ and $\sin^2\beta \to 1$.

The following two features of the upper bound (14.10) are noteworthy.

- As the top mass increases, $\lambda_{\max}$ decreases while the correction terms increase. Consequently, the upper bound on m_{h_1} is much less sensitive to m_t than the corresponding bound on m_h in the MSSM. In fact, by varying m_t, one finds [14.18] that the former reaches a minimum around 170 GeV, the precise value depending on stop mass parameters. The occurrence of this minimum so close to the experimentally measured value of m_t implies that the bound (14.10) is insensitive to variations in the value of the top mass within its experimental errors.

- The tree level upper bound (14.8) gets maximized for $\tan\beta = 1$ if $\lambda^2 > (g_2^2 + g_Y^2)/2$. In fact, $\lambda_{\max}^2$ is above this value so long as [14.18] $h_t(M_s) \leq 1.05$. On the other hand, the requirement that the top Yukawa coupling strength itself remain perturbative upto the

GUT scale forces $\tan\beta$ to be above 1.2–1.4, the precise lower bound depending on the exact values of m_t, α_s and on the size of the supersymmetric loop corrections to the relation between the running and on-shell masses of the top quark [14.19]. Moreover, if $\tan\beta$ is very close to this lower bound, f_t is very large so that $\lambda_{\max}$ becomes quite small. On the other hand, the top-stop loop corrections term in (14.10) decreases with increasing $\tan\beta$ on account of the $1/\sin^2\beta$ factor in front. The highest value of m_{h_1} is then typically found for $\tan\beta$ between 1.5 and 2. For instance, with $\lambda_{\max} = 0.7$ [14.18] and $\tan\beta = 1.5$, the tree level upper bound (14.8) on m_{h_1} already amounts to 117 GeV, above the limit established by LEP searches. Those searches do exclude large chunks of the NMSSM parameter space, but the resulting constraints are much weaker than in the MSSM; specifically, no constraints are imposed on $\tan\beta$ or M_s.

Numerically, for $M_s < 1$ TeV and $m_t = 175$ GeV, one finds [14.18] that

$$m_{h_1} < 145 \text{ GeV} . \tag{14.11}$$

Eq. (14.11) is a reasonable numerical upper bound on the lightest Higgs mass in the NMSSM. Note that this cannot be further increased [14.10] by introducing additional $SU(2)_L$ doublet or singlet Higgs fields. One can always choose a basis, even in such extended models, where only two $SU(2)_L$ doublets (with opposite hypercharge) and one singlet have nonvanishing VEVs. In that basis the top left 2×2 submatrix of the squared mass matrix of the CP even Higgs bosons will have the same form as in (14.5). Furthermore, the top-stop radiative corrections to this mass matrix will also be as in the NMSSM. Since the upper bounds (14.8) and (14.10) depend *only* on the form of this submatrix, they will not be changed at all. On the other hand, the numerical bound (14.11) *does depend on the assumption of a grand desert*. By introducing judiciously chosen additional fields in vectorlike representations of $SU(5)$, so that the gauge coupling strengths $g_{1,2,3}$ unify, this upper bound can be raised [14.20] to about 155 GeV. One can raise it further [14.20] to about 200 GeV by introducing $SU(2)_L$ triplet Higgs superfields, but taking care that the weak ρ-parameter, $M_W/(M_Z \cos\theta_W)$, is preserved within a few tenths of a percent of unity, as required by the electroweak precision data. To our knowledge, **200 GeV is the absolute limit to which the upper bound on the lightest Higgs mass can be raised in any perturbatively treatable model with weak scale supersymmetry**. That is the basis of the statement, sometimes made, that killing the entire idea of weak scale supersymmetry (as opposed to just the MSSM or NMSSM) will require the experimental exclusion of any Higgs particle below 200 GeV in mass.

The last members of the NMSSM Higgs spectrum that we need to discuss are the charged Higgs bosons. As in the MSSM, the physical spectrum only contains a single complex charged Higgs field $H^+ = (h_1^-)^\star \sin\beta + h_2^+ \cos\beta$, with a tree level mass m_{H^+} given [14.10, 14.15] by

$$m_{H^+}^2 = m_a^2 + M_W^2(1 - 2\lambda^2 g_2^{-2}) . \tag{14.12}$$

Eq. (14.12) shows that, unlike the MSSM, the NMSSM permits the charged Higgs to be lighter than the W. However, once the LEP search limits (cf. Ch.15) for neutral Higgs bosons as well as charginos (which implies $|\lambda x|/\sqrt{2} \gtrsim M_W$, see below) are imposed, such a light charged Higgs particle can be realized [14.15] only if the tadpole parameters T and/or t are roughly of the order of the soft supersymmetry breaking masses. It may be recalled here that values of the said parameters well below 1 MeV would be sufficient to solve the cosmological domain wall problem.

Of course, the phenomenology of the NMSSM Higgs sector depends also on the couplings of the Higgs bosons – especially on their couplings to the SM particles. The tree level couplings of the charged Higgs boson to matter fermions and gauge bosons are identical to those in the MSSM. The couplings of the CP odd Higgs bosons $A_{1,2}$ to the SM particles are given by the corresponding couplings of the Higgs boson A, listed in §10.4, but multiplied by the 'dilution factor' $\cos \alpha_{PS}$ describing the size of the $SU(2)_L$ singlet components of $A_{1,2}$, cf. Table 14.1. Similarly, the couplings of the CP even NMSSM Higgs boson h_i ($i = 1, 2, 3$) can be read off from the analogous couplings of the MSSM Higgs boson H, after replacing $\cos \alpha$ and $\sin \alpha$ respectively by the first and second components of the ith eigenvector of the matrix (14.5). An important fact is that, regardless of its CP properties, a pure $SU(2)_L$ singlet Higgs boson has vanishing tree level couplings to matter fermions and gauge bosons. Thus the unsuccessful Higgs searches at LEP cannot exclude the possibility that h_1 and/or A_1 are very light in case they are mostly $SU(2)_L$ singlets. To be sure, pure singlets cannot be directly involved in creating nonzero masses for the SM weak bosons and matter fermions. In this sense they are not really Higgs fields at all. LEP searches do impose severe constraints on the NMSSM Higgs bosons which are mostly $SU(2)_L$ doublets, though these do not translate into significant constraints on any one single parameter, as already mentioned. Recall also that the upper bound (14.11) on m_{h_1} can be saturated only if h_1 has a vanishing singlet component. Therefore, at an e^+e^- collider with an accumulated data set of a few fb^{-1} and CM energy $\sqrt{s} \gtrsim 300$ GeV, at least one NMSSM Higgs particle must be found [14.21] if the model is correct. Thus, in a way, the NMSSM is as readily testable[7] as the MSSM. Finally, without going into any detail, let us mention that the couplings of NMSSM Higgs bosons to sparticles often involve [14.22] the new coupling strengths λ and κ. In general, therefore, they cannot be simply read off from the corresponding MSSM couplings given in §10.5.

We have so far treated all soft supersymmetry breaking parameters at the weak scale as independent parameters as in our discussion of the Higgs sector of the MSSM in Ch.10. This raises the following question. Which characteristic features of the NMSSM would survive in a more restrictive scenario with universal boundary conditions (cf. §12.3) imposed on the said parameters at a high energy scale? The issue has been discussed in the literature within the Z_3 symmetric version of the NMSSM where $t = T = 0$ [14.23] as well as within a version with a $\mu H_1 \cdot H_2$ term in the superpotential in addition to the tadpole [14.24]. In the Z_3 symmetric version, the imposition of universal conditions forces one into a region of parameter space where the NMSSM closely resembles the MSSM [14.23]. The absolute values of the new coupling strengths λ and κ are both small here (typically < 0.1 at the weak scale), but the singlet VEV is large so that both λx and κx are of the size of generic soft supersymmetry breaking parameters. The allowed parameter space is, in fact, more constrained in the NMSSM scenario than in the MSSM. This is since one needs a large $|A_0|$ to achieve the required large value of x. Care must then be taken to avoid the occurrence

---

[7]On the other hand, an experimental discrimination between these two models might become a daunting task.

of deeper minima of the complete scalar potential where some colored or charged sfermion fields would acquire VEVs. Indeed, but for the RGE of the soft supersymmetry breaking parameters, no acceptable solution would exist[8]. Eqs. (14.3) and (14.5) show that there will be little mixing between doublet and singlet Higgs fields in this scenario. Thus the latter effectively decouple. Moreover, a small $|\lambda|$ implies that the upper bound on the mass of the lightest Higgs scalar is essentially the same as in the MSSM. Thus, in this case, the only experimental signal distinguishing between the NMSSM from the MSSM would be the existence of a light 'singlino' as the LSP. We shall discuss this later in connection with the NMSSM neutralinos. On the other hand, the NMSSM without the Z_3 symmetry allows more general solutions. These include scenarios with a relatively small $|x|$, a sizable $|\lambda|$ and significant amounts of mixing between Higgs doublets and singlets, even with universal [14.24] boundary conditions at a high scale.

There is another distinct feature of the NMSSM, as compared with the MSSM. The former permits spontaneous CP violation but only in the version without the Z_3 symmetry [14.25]. Parameter combinations can be found in this case with CP violated at the absolute minimum of the potential, though all the parameters in the latter are real. Moreover, one usually finds an unacceptably light Higgs particle unless the CP-violating phase of the singlet VEV is quite large. There is an explanation [14.24] for this. CP being a discrete symmetry, any CP-violating minimum implies the existence of an equally deep minimum where all phases have the opposite sign. If all phases are small, these two degenerate minima lie very close to each other in field space, implying that the curvature of the potential in the direction containing these two minima is small. That implication, in turn, leads to (at least) one light Higgs particle. The authors of Ref. [14.25] explore the possibility of this spontaneous CP violation being the only significant CP violation in the model[9]. It is indeed possible to explain the observed pattern of CP violation in the neutral K or B systems via trilinear A-terms with a nontrivial flavor structure. However, the yet unseen electric dipole moments of the electron and the neutron come out too large in this scenario. The only solution seems to be in arranging cancellations between different contributions to these electric dipole moments by the fine tuning of parameters.

Charginos and neutralinos in the NMSSM

The charged field content of the NMSSM is the same as that of the MSSM. Therefore, the NMSSM mass matrices for charginos and charged sfermions are still given by (9.7) and (9.50) respectively with the replacement $\mu \to \lambda x/\sqrt{2}$. In particular, the LEP chargino limit $M_{\tilde{\chi}_1^+} \gtrsim 100$ GeV implies that $|\lambda x| \gtrsim \sqrt{2}M_W$. This constraint means that $|x|$ has to be large if $|\lambda|$ is small.

A different situation prevails in the neutralino sector. The NMSSM contains a fifth neutralino field in the form of a singlino $\tilde{n}$ which acts as an extra current eigenstate. The MSSM

---

[8]Unlike in the MSSM, in the Z_3 symmetric NMSSM there can exist unwanted minima where only a single Higgs field has a nonvanishing VEV. Requiring these to be shallower than the desired minimum with nonzero VEVs for all three neutral Higgs fields leads to nontrivial constraints on the parameter space [14.17, 14.23].

[9]A small amount of explicit CP violation is probably required somewhere to avoid cosmological problems with domain walls. However, this explicit CP violating phase could be too small to be phenomenologically relevant [14.24].

neutralino mass matrix (9.24) now extends to a 5×5 form. In the basis $(\lambda_0, \lambda_3, \tilde{h}_1^1, \tilde{h}_2^2, \tilde{n})$, it reads

$$\mathcal{M} = \begin{pmatrix} M_1 & 0 & -M_Z s_W c_\beta & M_Z s_W s_\beta & 0 \\ 0 & M_2 & M_Z c_W c_\beta & -M_Z c_W s_\beta & 0 \\ -M_Z s_W c_\beta & M_Z c_W c_\beta & 0 & -\lambda x/\sqrt{2} & -\lambda v_1/\sqrt{2} \\ M_Z s_W s_\beta & -M_Z c_W s_\beta & -\lambda x/\sqrt{2} & 0 & -\lambda v_2/\sqrt{2} \\ 0 & 0 & -\lambda v_1/\sqrt{2} & -\lambda v_2/\sqrt{2} & \sqrt{2}\kappa x \end{pmatrix}. \tag{14.13}$$

It is noteworthy that the singlino does not mix directly with the gauginos. Hence two of the neutralino mass eigenstates are still dominantly gauginolike, if the absolute values of the differences between the soft supersymmetry breaking masses $M_{1,2}$ and the effective μ parameter $\lambda x/\sqrt{2}$ are significantly larger than M_Z. The singlet and doublet higgsinos do mix directly. This mixing is numerically important only if $|x|$ is not much larger than the doublet VEVs $v_{1,2}$. Scenarios where $|x|$ is of the same order as $v_{1,2}$ are, in fact, singled out [14.24] if one wants to produce the baryon asymmetry of the Universe (cf. Ch.16) during the electroweak phase transition. In this case three neutralino mass eigenstates are required to be mixed higgsino-singlino states. However, we have also seen that one often has $|x| \gg v_{1,2}$. In this situation, the singlino effectively decouples from the other four neutralinos, which will be MSSM-like. If $|\kappa|$ is very small so that $|\kappa| < |\lambda|/2$ and $|\sqrt{2}\kappa x| < |M_{1,2}|$, this singlinolike state will become the LSP. This happens quite frequently in the Z_3 symmetric version of the NMSSM with universal GUT-scale boundary conditions for the soft supersymmetry breaking parameters [14.23].

The singlet superfield N does not couple to other gauge or matter superfields. Thus the couplings of $\tilde{\chi}$ pairs to gauge bosons as well as the neutralino/chargino-matter fermion-sfermion couplings are still as given in §9.3 and §9.7 respectively. Of course, the neutralino mixing matrix $\mathbf{Z}$ of (9.26) is now a 5×5 matrix. But the $Z_{\ell 5}$ components of the neutralino mass eigenstates $\tilde{\chi}_l^0$, $l = 1, \cdots, 5$ do not appear in any of these couplings. The total cross section for $e^+ e^-$ pair production of higgsinolike neutralinos at beam energies sufficiently above the mass of the heaviest higgsinolike state will thus be the same as in the MSSM. However, in scenarios with a significant amount of mixing between the singlet and the doublet higgsinos, this cross section will be distributed over three neutralinos i.e. over six final states, as opposed to two neutralinos i.e. over three final states of the MSSM[10]. A careful analysis of the production thresholds and/or of the decay patterns of the heavier neutralino states should then allow a discrimination between the NMSSM and the MSSM.

On the other hand, the pair production cross section will be very small for a neutralino mass eigenstate which is mostly a singlino. Moreover, in case the decays of the heavier neutralinos and charginos proceed largely via the exchange of (real or virtual) gauge bosons or sfermions, no singlinolike state will be produced (unless it is the LSP, see below) either in cascade decays of heavier neutralinos/charginos or in sfermion decays. The decay of heavy $SU(2)_L$ doublet Higgs particles into a doublet higgsino and the singlino may be kinematically allowed in some regions of the parameter space. Similarly, a heavy doublet higgsino

---

[10] Analogous statements also hold for higgsino pair production at hadron colliders, including the production of one charged and one neutral higgsino. However, this will probably be difficult to detect and impossible to analyze in detail, given the large backgrounds, including those from other supersymmetric processes.

might decay into the singlino and a lighter doublet Higgs particle. Both kinds of decays are controlled by the magnitude of the coupling strength λ. However, it is not clear whether the branching ratios for these decays can be sizable.

Large deviations from MSSM phenomenology can occur, nonetheless, in the NMSSM even if one of the neutralinos is nearly a pure singlino, provided it is the LSP [14.26]. Owing to the feeble couplings of such a singlinolike state, the next-to-the-lightest sparticle might be long-lived enough to be able to travel over a macroscopic distance before decaying. Barring some extreme fine tuning of parameters, such a scenario can arise in the MSSM only if the LSP is essentially a pure neutral wino[11]. In this case the mass splitting between it and the lightest chargino could be $\mathcal{O}(100)$ MeV, so that the decay width of the lighter chargino $\tilde{\chi}_1^{\pm}$ is kinematically suppressed. In contrast, in the NMSSM with a singlinolike LSP, the next-to-the-lightest sparticle would be expected to be just another neutralino or perhaps a sneutrino. In general, the mass difference between the NLSP and the LSP will amount to several GeV at least. This should allow an experimental discrimination between the MSSM and the NMSSM plus singlino LSP options.

14.3 Introduction to R-parity Violation

The discrete symmetry R-parity was briefly discussed in §4.5. It is characterized by the multiplicative quantum number R_p which is $+1$ for particles and -1 for sparticles, cf. (4.52b). Equivalently, it is matter parity for superfields, being odd for quark and lepton superfields and even for all others. In supersymmetric theories with baryon and lepton conservation, R_p can be expressed as $(-1)^{3B+L+2S}$. The assumption of an exact conservation law for R_p is one of the key ingredients of the MSSM, as enunciated in §8.3. However, the absence of any really convincing theoretical motivation behind this assumption was mentioned already in §4.5. The basic point is that baryon number B and lepton number L, including lepton type number L_i for each generation, get conserved quite accidentally in the SM. These conservation laws follow from the choice of the specific representations needed for the SM quark and lepton fields with respect to the gauge group $SU(3)_C \times SU(2)_L \times U(1)_Y$. They do not come about in consequence of any gauge symmetry or a similar deep principle and are anyway expected to be violated in very high scale theories such as grand unified ones. One can then speculate that some combinations[12] of these charges may not be conserved at much lower (such as $\sim$ TeV) energies. Indeed, neutrino oscillations show that the L_i are not separately conserved, though $L = \sum_i L_i$ might be. Thus it is not *a priori* clear why B and L should at all be maintained as good quantum numbers in a supersymmetric extension of the SM operative at energies beyond the weak scale. The possibility of R_p violation and a consequent new phenomenology at such energies is therefore something not to be dismissed out of hand.

---

[11]Gaugino masses will not then unify at any high mass scale. A winolike LSP arises, for instance, in the minimal AMSB model, cf. §12.6.

[12]Phenomenological constraints disallow all these charges from being violated together at those energies. For example, as discussed later, the simultaneous violation of B and L at laboratory energies would be forbidden by the longevity of the proton.

A supersymmetric extension of the SM with broken R_p invariance was first proposed by Aulakh and Mohapatra [14.27]. In this proposal R_p invariance gets broken spontaneously, but the implementation of this process requires the presence of extra chiral superfields that are singlets under the SM gauge group. A minimal model of this kind, constructed later, needed at least three SM-singlet chiral superfields – apart from containing a new Goldstone boson, viz. the Majoron. The CP odd Majoron is massless but has an associated light CP even scalar, both being components of the same chiral superfield. Since Z decay into such a Majoron and its light CP even scalar partner is not observed, the occurrence of spontaneous R_p violation is unlikely. A more likely possibility is the explicit violation of R_p and we choose to confine ourselves to scenarios with such explicit violation. It became clear some time ago [14.28] that perfectly reasonable gauge invariant terms, violating R-parity explicitly, could be present in the superpotential. To be specific, we can define an 'extended' model as follows. Keep everything else the same as in the MSSM, but just add some new terms to the superpotential $\mathcal{W}_{MSSM}$ of (8.33): $\mathcal{W}_{MSSM} \to \mathcal{W}_{MSSM} + \mathcal{W}_{\not R_p}$, where

$$\mathcal{W}_{\not R_p} = -\epsilon_i L_i \cdot H_2 + \frac{1}{2}\lambda_{ijk} L_i \cdot L_j \bar{E}_k + \lambda'_{ijk} L_i \cdot Q_j \bar{D}_k + \frac{1}{2}\lambda''_{ijk} \bar{U}_i \bar{D}_j \bar{D}_k \ . \qquad (14.14)$$

In (14.14) i, j, k are flavor indices while $\epsilon_i, \lambda_{ijk}, \lambda'_{ijk}$ and λ''_{ijk} are coupling strengths, ϵ_i with the dimension of mass and the other three being dimensionless. The ϵ_i, λ_{ijk} and λ'_{ijk} couplings violate lepton number L (including lepton type number L_i) while the λ''_{ijk} couplings violate baryon number B. Respective invariances under the $SU(2)_L$ and $SU(3)_C$ symmetries of the SM require the antisymmetry of λ_{ijk} in i, j and of λ''_{ijk} in j, k. The Lagrangian density, arising out of the $\mathcal{W}_{\not R_p}$ of (14.14), can be written in terms of component fields. For instance, the R_p violating supersymmetric trilinear interaction terms read:

$$
\begin{aligned}
\mathcal{L}_{\text{trilinear } \not R_p} &= \lambda_{[ij]k}\left[\tilde{\nu}_i\overline{e_{kR}}e_{jL} + \tilde{e}_{jL}\overline{e_{kR}}\nu_i + \tilde{e}^\star_{kR}\overline{\nu^C_i}e_{jL} - \tilde{\nu}_j\overline{e_{kR}}e_{iL} - \tilde{e}_{iL}\overline{e_{kR}}\nu_j \right.\\
&\quad \left. -\tilde{e}^\star_{kR}\overline{\nu^C_j}e_{iL}\right] + \lambda'_{ijk}\left[\tilde{\nu}_i\overline{d_{kR}}d_{jL} + \tilde{d}_{jL}\overline{d_{kR}}\nu_i + \tilde{d}^\star_{kR}\overline{\nu^C_i}d_{jL} - \tilde{e}_{iL}\overline{d_{kR}}u_{jL}\right.\\
&\quad \left. -\tilde{u}_{jL}\overline{d_{kR}}e_{jL} - \tilde{d}^\star_{kR}\overline{(e_{iL})^C}u_{jL}\right] + \lambda''_{i[jk]}\epsilon_{\alpha\beta\gamma}\left[\tilde{u}^\star_{iR\alpha}\overline{d_{kR\beta}}d^C_{jR\gamma}\right.\\
&\quad \left. +\tilde{d}^\star_{jR\beta}\overline{d_{kR\gamma}}u^C_{iR\alpha} + \tilde{d}^\star_{kR\gamma}\overline{(d_{jR\beta})}u^C_{iR\alpha}\right] + \text{h.c.} \qquad (14.15)
\end{aligned}
$$

Here, in an otherwise transparent notation, e_i ($\tilde{e}_i$) is a charged lepton (slepton) of generation i etc., while L, R are chiral and α, β, γ are color indices. The square brackets in the generation subscripts means antisymmetrization with a factor $1/2$. The $\not R_p$ vertices, following from the interactions of (14.15), are shown in Fig. 14.1; the corresponding hermitian conjugate vertices obtain by flipping the arrows and putting stars on the couplings. A similar exericse can be performed with the bilinear terms in $\mathcal{W}_{\not R_p}$, but we postpone a detailed discussion of bilinear R_p violation to §14.5. We shall later give a justification for such a separate treatment of the bilinear and trilinear R_p violating terms.

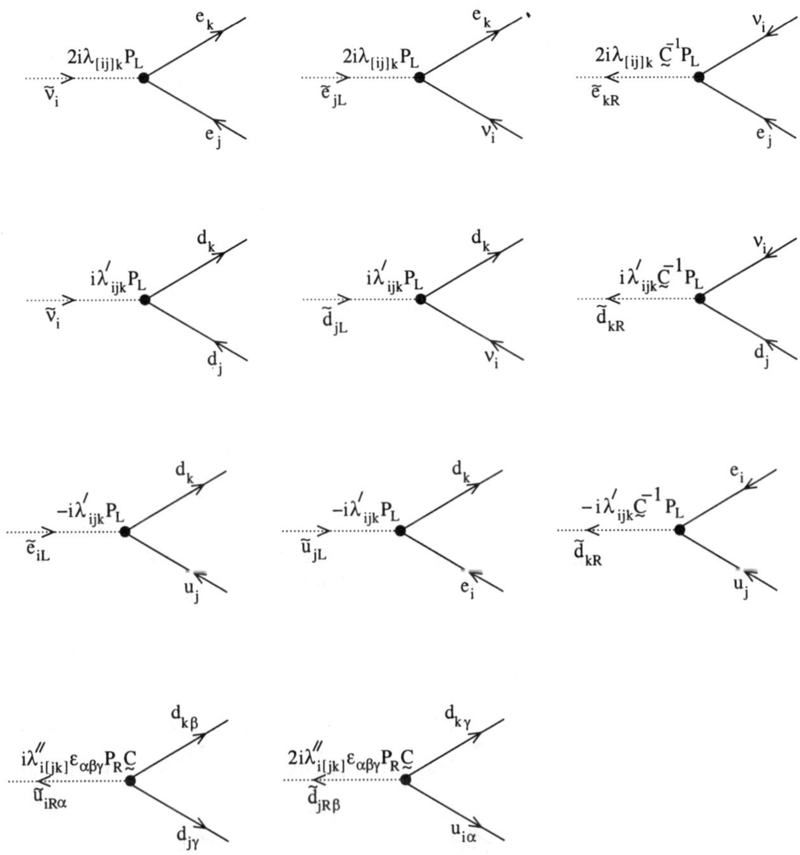

Fig. 14.1. Vertices with R_p violation; square brackets in the generation subscripts mean antisymmetrization with a weight factor $1/2$. The corresponding Hermitian conjugate vertices can be obtained by reversing the arrows and putting stars on the couplings.

A law of conservation of R_p would require vanishing values for all the coupling strengths of (14.14). The pertinent question is: why make such a law? Let us go into the history of it a bit. The R_p invariance assumption was really introduced to provide 'easy' explanations of two observed facts on an *a posteriori* basis: (1) the longevity of the proton and (2) the inferred existence (cf. Ch.16) of pervasive Cold Dark Matter (CDM) in the Universe. Consider first the question of the total lifetime of the proton which is experimentally known to have a lower bound $\sim 10^{32}$ years. Yet, the superpotential of (14.14) allows a squark-mediated mechanism for catastrophically rapid proton decay, say into $\ell^+\pi^0$, as shown in Fig. 14.2. The amplitude

for this process can be estimated to be

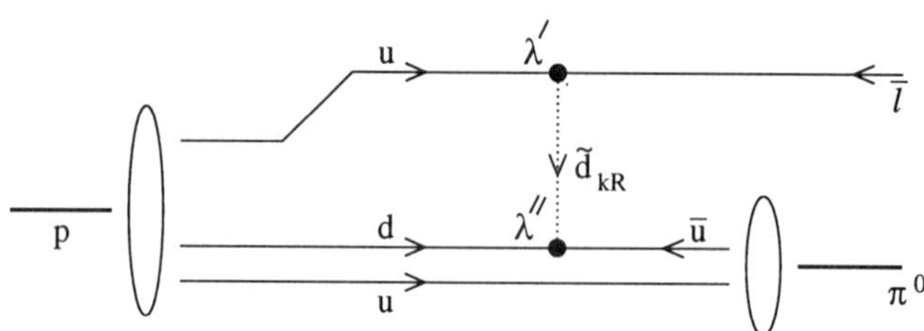

Fig. 14.2. A typical diagram for $p \to \ell^+\pi^0$ with $\not{R}_p$ couplings λ' *and* λ''. Here $\tilde{d}_k = \tilde{s}, \tilde{b}$.

$$\mathcal{M}(p \to \ell_i^+ \pi^0) \sim \sum_{k=2,3} \frac{\lambda'_{i1k} \lambda''^*_{11k}}{m_{\tilde{k}}^2} \,, \qquad i = 1, 2 \,, \tag{14.16}$$

$m_{\tilde{k}}$ standing for the mass of a right squark with down type flavor of a generation higher than the lowest. This is, of course, not the only diagram responsible for this decay and there are others, but all have similar amplitudes. In the assumed absence of some magical cancellations between different coherent diagrams, it is reasonable to estimate the lifetime τ_p of the proton from (14.16). Taking λ', λ'' to be of semiweak strength, i.e. each being $\mathcal{O}(10^{-1})$, and $m_{\tilde{k}} = \mathcal{O}(\text{TeV})$, τ_p is estimated to be $\simeq 10^{-9}$ s. On the other hand, the use of the factual information that $\tau_p > 10^{32}$ years yields the upper bound

$$|\lambda'_{11k} \lambda''_{11k}| \lesssim 10^{-24} (m_{\tilde{k}}/100 \text{ GeV})^2, \; k = 2, 3 \,. \tag{14.17}$$

Though (14.17) does not cover certain other product combinations of λ' and λ'' couplings, the latter do come into the picture through one loop corrections to the diagram of Fig. 14.2. A careful analysis of all such corrections leads to [14.29] a bound on each product combination of λ' and λ'' couplings which is not included in (14.17). Indicating the indices of such couplings by dots, one has the result

$$|\lambda'_{...} \lambda''_{...}| \lesssim 10^{-9} \, (\tilde{m}/100 \text{ GeV})^2 \,, \tag{14.18}$$

$\tilde{m}$ being a generic squark mass. Eqs. (14.17) and (14.18) provide severe constraints on all products of couplings λ' and λ''. Comparable bounds on all product combinations of λ and λ'' couplings have also been obtained from the proton longevity constraint and are given in the second paper cited in Ref.[14.29]. Needless to say, this problem does not arise in the MSSM where R_p is conserved [14.30]. However, that is quite an overkill since, in reality, either B or L conservation implies $\lambda'_{...} \lambda''_{...} = 0 = \lambda_{...} \lambda''_{...}$ and thus trivially satisfies (14.17) and (14.18). It is, therefore, quite possible to have R_p violating supersymmetric theories while ensuring the longevity of the proton through the conservation of *either B or L*.

We turn next to the second issue concerning cold dark matter (CDM). The cosmological need for a stable *weakly interacting massive particle* (WIMP) to produce a CDM that pervades the Universe will be discussed in §16.3. It will be seen there that an R_p conserving supersymmetric scenario does provide a very good candidate for this purpose in the form of a stable LSP with a mass in the range of tens to hundreds of GeV. However, this is not the only possibility that has been thought of. There exist viable models with other stable particles as candidates for CDM: the invisible axion [14.31] or new types of WIMPs [14.32]. Therefore, the inferred existence of CDM from cosmological observations (cf. §16.3) does not necessarily preclude an unstable LSP decaying via R_p violating interactions. Of course, while on the issue of cosmological constraints, one could raise the question of the preservation of a GUT-generated baryon asymmetry of the Universe or of the occurrence of electroweak baryogenesis (cf. Ch.16) which might not be feasible in the presence of R_p-violating interactions. Indeed, very strong constraints on the corresponding coupling strengths had been claimed on this count. The general wisdom [14.33] in the subject, nonetheless, is that these constraints are model dependent and can be evaded by utilizing loopholes in arguments deriving them. We shall have more to say on this in §16.5.

From a theoretical standpoint, R_p invariance or matter parity as a superfield symmetry is the relic (cf. §4.5) of a global $U(1)_R$ symmetry broken down to its discrete Z_2 subgroup. But something like baryon parity $(-1)^{3B}$, under which every quark superfield changes sign but all other superfields remain invariant, can also be viewed as a Z_2 symmetry broken down from a global baryonic charge invariance $U(1)_B$. The same goes for lepton parity $(-1)^L$, under which only lepton superfields change sign while all others remain the same, in relation to $U(1)_L$. The way these transformations act on the superfields of §8.2 is shown in Table 14.2. An invariance under either B parity or L parity would ensure proton longevity and yet allow R_p-violating interactions. No one has so far found a really convincing dynamical reason for selecting a particular one out of these three Z_2 symmetries.

Matter parity	Baryon parity	Lepton parity
$(Q_i, \bar{U}_i, \bar{D}_i, L_i, \bar{E}_i)$ $\to -(Q_i, \bar{U}_i, \bar{D}_i, L_i, \bar{E}_i)$	$(Q_i, \bar{U}_i, \bar{D}_i) \to -(\bar{Q}_i, \bar{U}_i, \bar{D}_i)$	$(L_i, \bar{E}_i) \to -(L_i, \bar{E}_i)$
$(H_1, H_2, V_\gamma, V_Z, V_{W\pm})$ $\to (H_1, H_2, V_\gamma, V_Z, V_{W\pm})$	$(L_i, \bar{E}_i, H_1, H_2, V_\gamma, V_Z, V_{W\pm})$ $\to (L_i, \bar{E}_i, H_1, H_2, V_\gamma, V_Z, V_{W\pm})$	$(Q_i, \bar{U}_i, \bar{D}_i, H_1, H_2, V_\gamma, V_Z, V_{W\pm})$ $\to$ $(Q_i, \bar{U}_i, \bar{D}_i, H_1, H_2, V_\gamma, V_Z, V_{W\pm})$

Table 14.2. Transformations of superfields under various Z_2 symmetry operations

The speculation that either baryon parity or lepton parity, as opposed to matter parity, might be a true symmetry of nature raises an interesting issue. Suppose the supersymmetric extension of the SM is a low energy effective description of a supergrand unified theory defined at the energy scale M_{GUT} (cf. §12.5). In such a theory quarks and leptons would be unified in one gauge multiplet[13]. That, however, militates against a discrimination between the matter parities of their superfields at a lower energy. Does not grand unification therefore argue against R_p violation? One can offer several answers to this poser. First of all, there

---

[13]The nonconservation of baryon and lepton numbers in such theories does not necessarily imply that R_p is violated. The one to one correspondence between the breaking of the former and that of the latter, discussed in §4.5, does not hold in certain extensions of the MSSM, such as the super-GUTs of §12.5.

exists a specific supersymmetric grand unified model [14.34] with R_p violation via only the $-\epsilon_i L_i{\cdot}H_2$ term in the low energy superpotential $\mathcal{W}_{\slashed{R}_p}$. This term can be 'rotated away' by the Higgs and lepton superfield redefinitions

$$H_1 \;\rightarrow\; H_1' = (\mu^2 + \epsilon_j \epsilon_j)^{-1/2}(\mu H_1 - \epsilon_i L_i) \,, \tag{14.19a}$$

$$L_i \;\rightarrow\; L_i' = (\mu^2 + \epsilon_j \epsilon_j)^{-1/2}(\epsilon_i H_1 + \mu L_i) \,, \tag{14.19b}$$

μ being the higgsino mass parameter of $(8.33)^{14}$. However, R_p violation then reappears in trilinear lepton number violating terms in the superpotential as well as in the soft supersymmetry breaking part of the scalar potential [14.34, 14.35] which get generated in consequence of this redefinition. Moreover, even if such an elimination of this bilinear $\slashed{R}_p$ term is effected at a certain energy scale, it shows up at a different energy scale through RGE [14.36]. Second [14.37], in a general grand unified theory, certain $\slashed{R}_p$ operators, containing new matter superfields added to the superpotential, can give rise to *specific* couplings of λ-, λ'- or λ''-type, once the new superfields acquire masses $\mathcal{O}(M_{GUT})$. In general, the other couplings in the low energy superpotential (14.14) also get generated, but their strengths are suppressed by powers of M_W/M_{GUT} so that the constraint (14.18) is satisfied. It is possible in such models to have R_p violating interactions without shortening the proton lifetime which remains at a characteristic super-GUT value (cf. 12.40). Third, unification can be achieved in string theories without a simple gauge group so that quark and lepton superfields could in principle have different matter parities. In practice, string-unified theories with R_p violating couplings have been constructed [14.38].

Turning to energies at or near the Planck scale, some arguments have been given to favor the conservation of baryon parity a little bit over that of lepton parity. These are within the framework of a putative anomaly free gauge theory, defined on a higher dimensional manifold, whose gauge symmetry can break down to a discrete subgroup in the process of compactification to four dimensional spacetime. A systematic analysis of all Z_N symmetries revealed [14.39] that the only permitted discrete symmetries are baryon parity and matter (i.e. R-) parity and not lepton parity. Indeed, an invariance under baryon parity does prohibit certain dimension five operators which are embarassing for GUT-induced proton decay. Therefore, B parity is preferred over matter parity which does not have this virtue. Such arguments, however, have been found [14.40] to depend sensitively on various questionable assumptions/approximations; the former do not go through if the latter are removed. Another issue has been the consistency of R-parity violation with coupling strength unification via RGE (cf. Ch.11) which is supposed to be a triumph of the MSSM. The answer seems to depend on the region chosen in the parameter space of the latter theory. The fact is that there do exist low energy solutions [14.41] of the RGE equations for gauge couplings, unified at M_{GUT}, in which the evolution is driven partially by R_p violating couplings and is also influenced by perturbative unitarity constraints. Indeed, these have been used to put (rather loose) bounds [14.42] on some of the λ'' couplings of (14.14). The bottomline after all this is that any argument against R_p violation, based on unification, does not really hold.

---

[14]The analogous rotation of the $\bar{D}_i$ fields, the partners of the L_i fields in an $SU(5)$ super-GUT (cf. §12.5), is suppressed by the very large triplet mass M_t of the $SU(5)$ partners of the Higgs superfield, i.e. the corresponding mixing angle is $\mathcal{O}(M_s/M_t)$.

An appraisal of the above discussions should convince one that R_p violation, in the form of *either L or B* violation, is quite a feasible possibility in the supersymmetric extension of the SM. On the other hand, the joint occurrence of L and B violation is excluded phenomenologically from the lower bound on the lifetime of the proton. This conclusion motivates us to study the couplings of (14.15) in more detail. Of course, in $\not{R}_p$ scenarios the lack of a stable LSP, produced from sparticle decay and escaping the detectors in a collider process, will mean the absence of the main source of large missing energy signalling the production of sparticles. But, in contrast, these theories will be shown to lead to unusual multilepton and multijet events instead[15], uniquely characterizing R_p violating supersymmetric processes. In any event, one should be able to detect the decay products of the unstable LSP produced directly or created in cascade decays of other sparticles directly produced in collider machines. A neutralino[16] LSP with a mass M_{LSP}, for instance, could decay into $e_i\bar{e}_j\bar{\nu}_k$ and $\bar{e}_i e_j \nu_k$ via the λ_{kji} coupling as well as into $d_i\bar{d}_j\bar{\nu}_k$ and $\bar{d}_i d_j \nu_k$, $d_i\bar{u}_j\bar{e}_k$ and $\bar{d}_i u_j e_k$ via the λ'_{kji} coupling and also into $u_i d_j d_k$ and $\bar{u}_i\bar{d}_j\bar{d}_k$ via the λ''_{ijk} coupling, with i,j,k being generation indices. All these processes can occur via diagrams shown in Fig. 14.3. Suppose $m_{\tilde{\ell},\tilde{q}}$ is the mass of the exchanged slepton, squark. The corresponding LSP mean free path can be estimated [14.43] to be typically

$$c\tau_{\tilde{\chi}_1^0} \simeq 3 \times 10^{-13}(m_{\tilde{\ell},\tilde{q}}/100 \text{ GeV})^4(100 \text{ GeV}/M_{\tilde{\chi}_1^0})^5(\lambda,\lambda',\lambda'')^{-2} \text{ meters.}$$

Thus the mean free path of $\tilde{\chi}_i^0$ would be less than a meter and its decay products

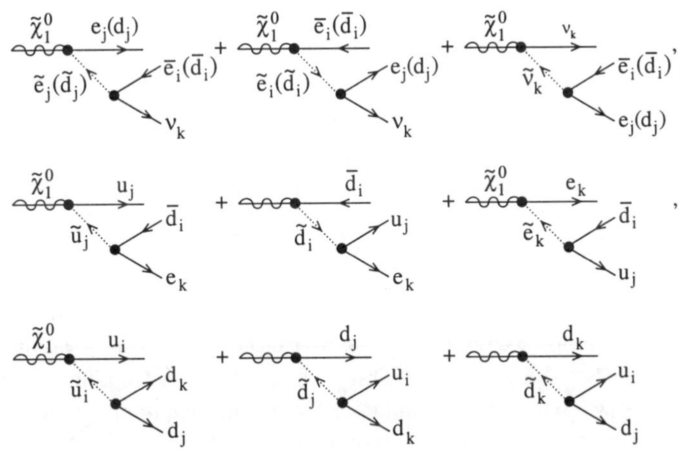

Fig. 14.3. Diagrams for *LSP* decay: $\tilde{\chi}_1^0 \to \bar{e}_i e_j \nu_k$, $\bar{d}_i d_j \nu_k$, $\bar{d}_i u_j e_k$, $u_i d_j d_k$. Diagrams for the charge conjugate processes obtain by reversing the arrows.

---

[15]This occurs since (1) sleptons behave as *dileptons* in the presence of λ-type couplings, (2) squarks (sleptons) behave as *leptoquarks (diquarks)* in the presence of λ'-type couplings and (3) squarks behave as *diquarks* in the presence of λ''-type couplings.

[16]The lightest neutralino $\tilde{\chi}_1^0$ need not be the LSP in an R_p violating scenario (e.g. the stau could be lighter), though most authors assume that it is. We take the $\tilde{\chi}_1^0$ to be the LSP here; the expression for the mean free path would be different if the latter were a scalar.

would be detectable if the appropriate λ or λ' or λ'' coupling strength obeys the condition[17]

$$|\lambda, \lambda', \lambda''| > 5 \times 10^{-7} (m_{\tilde{\ell}, \tilde{q}}/100 \text{ GeV})^2 (100 \text{ GeV}/M_{LSP})^{5/2} \,. \qquad (14.20)$$

Any violation of the lower bound of (14.20) will ensure that the LSP escapes undetected.

14.4   Phenomenological Limits on Trilinear $\not{R}_p$ Couplings

The number of extra parameters, introduced in the $\not{R}_p$MSSM, can be counted by utilizing the symmetry properties of the $\not{R}_p$ couplings in (14.14). Thus there could be a maximum of twenty seven λ'-type and nine each of the λ- and λ''-types of coupling strengths (each complex) that are all different from one another, leading to ninety additional Yukawalike real parameters in the theory. In addition, there would be the corresponding A-terms plus bilinear $\not{R}_p$ terms in the superpotential and corresponding B-terms. Recall from the discussion in §9.1 that the MSSM itself introduced one hundred and five new parameters, added to the nineteen of the SM. Indeed, we have invoked various models of supersymmetry breaking in Chs. 9, 12 and 13 in order to drastically reduce this number so that once supersymmetry is discovered, meaningful phenomenology becomes possible. The further addition of so many new parameters would be a nightmare to any serious investigator of R-parity violating phenomena. One thus needs to adopt some kind of a simplifying philosophy. The latter could be based on the assumption that just one of the $\not{R}_p$ couplings is dominant at a time – taking the others to vanish for all practical purposes.

There is some phenomenological basis for the last-mentioned assumption. Wherever one has been able to relate products of pairs of different $\not{R}_p$ couplings to measurable effects, one has found [14.44] that the upper bounds[18] on such products are far more severe than products of individual upper bounds on those two coupling strengths. This statement is true of products not only of different trilinear R_p violating couplings but also of bilinear and trilinear ones. It is therefore reasonable to make the single coupling dominance assumption for any process and this is the justification for our separate discussions of bilinear and trilinear R-parity violation. The usual approach is to implement the said assumption in the physical mass basis of the fermions, specifically quarks. This is since such a pattern has already been seen in the Yukawa coupling strengths of the SM with $|f_t| \gg |f_{i \neq t}|$. An alternative procedure [14.45] might choose to take one $\not{R}_p$ coupling strength to be dominant in the gauge basis prior to the EW symmetry breakdown. The latter would, however, mean the generation of other $\not{R}_p$ couplings in the physical basis on account of the fermion mixing caused by the unitary transformation from that to the gauge basis. In principle, this sounds like a more rational assumption; the previously proposed one might seem to require some cancellations and/or small mixing angles in the physical basis so as to finally keep one coupling as dominant. Some such mechanism may nonetheless be at work since precisely such a thing happens in the Yukawa sector of the SM. Without further ado and largely for convenience, then, we

---

[17]When the coupling strength λ''_{3jk} is involved, (14.20) will not hold unless $\tilde{\chi}_1^0$ is much heavier than the top. This point will be discussed more thoroughly in relation to $\not{R}_p$ signals at colliders in §15.6.

[18]These [14.44] have been derived not only from searches for flavor changing decays (or conversions) of quarks and leptons, but also from neutrino mass considerations.

assume the dominance of one $\not{R}_p$ coupling strength at a time in the physical basis, leaving the explanation to a putative deeper theory.

Bounds on the magnitudes of $\not{R}_p$ coupling strengths.

We come now to indirect phenomenological limits that can be derived on various $\not{R}_p$ coupling strengths. These emerge from the consideration of yet unobserved processes. Alternatively, they can come from unmeasured additional contributions, involving virtual sparticles exchanged in tree diagrams or loops, to observed SM processes. We do not consider constraints on the ϵ_i parameters of (14.14) here but postpone their discussion to the next section. Moreover, for simplicity, we take all trilinear couplings to be real. Starting with the interaction Lagrangian density of (14.15), one can calculate suitably chosen amplitudes and their rates which can be compared [14.46, 14.47] with experimental constraints. We enumerate below different relevant processes and at the end show in a consolidated way all the bounds on various $\not{R}_p$ coupling strengths that emerge from them.

- **Neutron-antineutron oscillation**

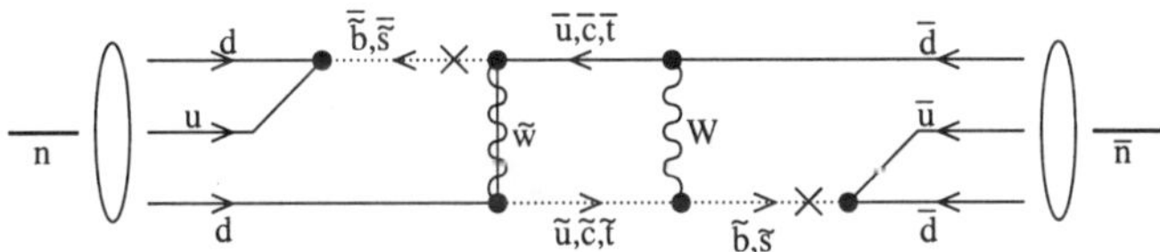

Fig. 14.4. Box graph for n-$\bar{n}$ oscillation via $\not{R}_p$ vertices; each cross on a squark line represents an L-R insertion.

This process can occur via the box graph[19], shown in Fig. 14.4, involving the coupling strengths λ''_{113} and λ''_{112} on which limits can be obtained following Ref. [14.42]. The limit on $|\lambda''_{112}|$ gets diluted relative to $|\lambda''_{113}|$ by the quark mass ratio m_b/m_s. However, a stronger constraint on the former coupling strength comes from the lack of observation of the $\Delta B = \Delta S = -2$ double nucleon annihilation into two kaons, namely $N + N \to KK$. A typical diagram for this process is shown in Fig. 14.5.

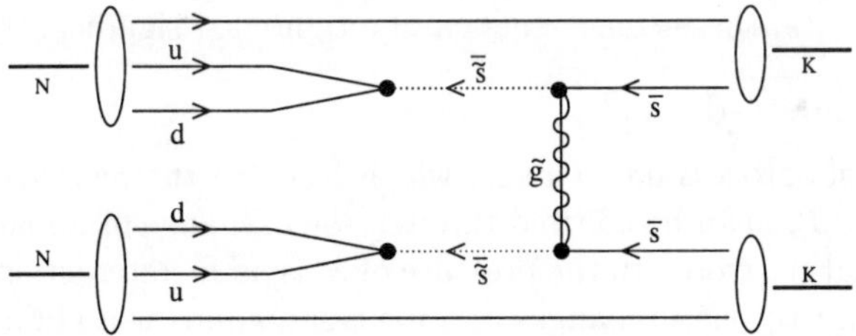

Fig. 14.5. A typical diagram for double nucleon annihilation into a kaon pair via $\not{R}_p$ vertices.

---

[19]There could also be an additional tree level graph involving the exchange of a single gluino provided the sflavor violating d-$\tilde{g}$-$(\tilde{b}, \tilde{s})$ vertex is allowed.

• Majorana masses of neutrinos

Couplings both of the λ- and λ'-types can induce Majorana masses for neutrinos via self energy loops of the kind shown in Fig. 14.6 in the case of ν_e. This type of a mass can be bounded from above by considering unobserved neutrinoless nuclear double β-decay (which is separately discussed next). The typical expression for the contribution [14.48] from a diagram[20] such as that in Fig. 14.6 is

$$\delta m_{\nu_k} \sim \frac{1}{16\pi^2} \left(\lambda_{kii} \text{ or } \lambda'_{kii}\right)^2 N_c \frac{M_s m_i^2}{\tilde{m}_i^2}. \tag{14.21}$$

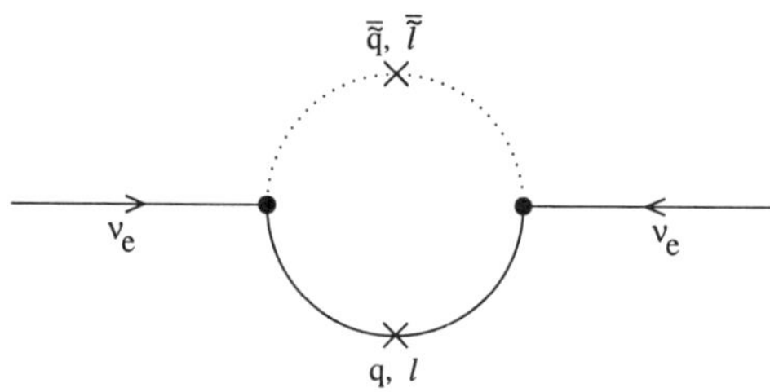

Fig. 14.6. A self-energy loop generating a ν_e Majorana mass with crosses representing chirality flips.

In (14.21) N_c, the number of colors, is three or one depending on the intermediate state being a quark-squark pair with a λ'-type coupling or a lepton-slepton pair with a λ-type coupling. In the numerator one power of m_i, the mass of the exchanged fermion, comes from its chirality flip while the factor $m_i M_s$ comes from the left-right mixing of the sfermion. The experimental requirement $\delta m_{\nu_e} < 1$ eV and the assumption $M_s \sim \tilde{m}_i$ lead, for a $\tau\bar{\tau}$ or $b\bar{b}$ loop ($N_c = 1$ or 3 respectively), to constraints on $|\lambda_{133}|$ and $|\lambda'_{133}|$. It has recently [14.50] been pointed out that there are ways of avoiding such constraints at the one loop level but not at the two loop level and the latter are significant. One might expect looser constraints on the analogous coupling strengths of the higher generations since the upper bounds on the corresponding Majorana masses are weaker. However, using the presently known information on squared mass differences between neutrinos of different generations from the oscillation data, these constraints can be tightened significantly [14.51].

• Neutrinoless double β-decay

The L violating subprocess $dd \to uuee$, which leads to the neutrinoless nuclear double beta decay process $(\beta\beta)_{o\nu}$ can be effected through the exchange of a massive Majorana neutrino between the final electrons. In the presence of λ'-type $\not{R}_p$ interactions, however, the said subprocess can go directly via exchanges of a neutralino and/or a gluino plus selectrons/u-squarks. There is a plethora of possible diagrams, a typical one being that of Fig. 14.7. A

---

[20]There exist [14.49] other one loop diagrams contributing to δm_ν with gauge couplings at the neutrino vertices and additional λ-couplings, but the bounds on the λ-couplings from these are weaker. Loop diagrams can yield stronger constraints, if coupled with tree level bilinear R_p violation to be covered in §14.5.

comparison of the calculated rate with the experimental lower bound on the corresponding lifetime leads to [14.52] an upper bound on $|\lambda'_{111}|$. Currently, there is controversy about whether neutrinoless double β–decay has been observed or not.

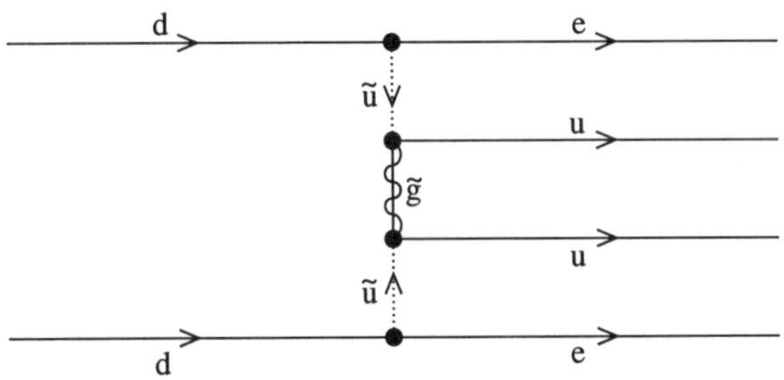

Fig. 14.7. A typical diagram with $\not{R}_p$ couplings for the subprocess $dd \to uuee$.

• Charged Current universality

Quark-lepton universality in Charged Current (CC) interactions is violated if λ- and/or λ'-type couplings are present. In effect, the new sfermion mediated contributions involving $\not{R}_p$ vertices to any two-particle to two-particle amplitude can, after a Fierz transformation, be written in the $(V - A) \otimes (V - A)$ form of the usual W exchange terms of the SM. There is nonetheless a difference with the latter in that the CKM factors in the former amplitude get multiplied by factors containing $\not{R}_p$ couplings. A careful analysis [14.47, 14.53], utilizing experimental constraints on nonuniversal terms in CC scattering and decay processes allowed in the SM, leads to upper limits on $|\lambda_{12k}|$ as well as on $|\lambda'_{11k}|$, $|\lambda'_{12k}|$ and $|\lambda'_{13k}|$. More specifically[21], the charm semileptonic decay subprocess $c \to s\ell^+\nu_\ell$ is mediated at the tree level by W-exchange in the SM. However, λ'-type couplings – if present – will generate an additional term from a sfermion exchange diagram. Again, by a Fierz transformation, this term can be recast in the $(V - A) \otimes (V - A)$ form of the former. Experiments, searching for departures from μ-e universality in semileptonic D decay can now be utilized. In particular, data on the ratio of rates for the semileptonic decays $\bar{D}^0 \to \bar{K}_L \mu^+ \nu_\mu$ and $\bar{D}^0 \to \bar{K}_L e^+ \nu_e$ can be used to constrain $|\lambda'_{12k}|$ and $|\lambda'_{22k}|$. A similar treatment of the ratio of rates for the leptonic decays of D_s mesons, namely $\Gamma(D_s \to \tau\nu_\tau)/\Gamma(D_s \to \mu\nu_\mu)$, leads to upper bounds on $|\lambda'_{32k}|$.

• $\nu_\mu(\bar{\nu}_\mu)$ scattering from electron and nucleon

Consider the amplitudes for the processes $\nu_\mu(\bar{\nu}_\mu)e \to \nu_\mu(\bar{\nu}_\mu)e$. These acquire additional contributions from the exchange of charged sleptons with λ-type couplings to the external leptons; ditto for the subprocesses $\nu_\mu(\bar{\nu}_\mu)q \to \nu_\mu(\bar{\nu}_\mu)q$ and $\nu_\mu(\bar{\nu}_\mu)\bar{q} \to \nu_\mu(\bar{\nu}_\mu)\bar{q}$ contributing to deep inelastic Neutral Current (NC) scattering of muon neutrinos/antineutrinos from nucleons, except that here the new contributions involve squark exchange and λ'-type couplings.

---

[21] A consideration of new terms in K-$\bar{K}$ or D-$\bar{D}$ mixing from λ'-type couplings, in conjunction with strong limits on FCNC processes with strange and charmed mesons, can lead to limits on $|\lambda'_{ijk}|$; but these are basis dependent.

Upper bounds [14.47, 14.53] on $|\lambda_{12k}|$, $|\lambda_{231}|$, $|\lambda'_{21k}|$ and on $|\lambda'_{2j1}|$ emerge from the success of the SM in explaining all experimental results on these processes.

• Precision measurements on Z-decays

Precision data on the fermion antifermion decays of the Z lead to interesting constraints on various $\not{R}_p$ couplings [14.54]. Heavy virtual quarks can participate in quark-squark mediated triangle graph vertex corrections (induced by λ'-type couplings) to the decay $Z \to \ell\bar{\ell}$ for any charged lepton ℓ. Various branching ratios for Z-decays falling in this category have been measured very precisely at LEP and are in brilliant agreement with SM predictions. Despite the suppression from heavy masses circulating in the loop, reasonably severe constraints follow on the previously mentioned corrections and get translated to upper bounds on the corresponding λ'-couplings. Especially interesting are the new $\not{R}_p$ corrections involving the top quark (cf. Fig. 14.8). These lead to upper bounds on $|\lambda'_{13k}|$, $|\lambda'_{23k}|$ and $|\lambda'_{33k}|$. If one assumes quark lepton universality and imposes the related constraints on $Z \to q\bar{q}$ (on which the data are less precise) from the leptonic Z-decay data, one can constrain $|\lambda'_{i3k}|$ and $|\lambda''_{3jk}|$ also.

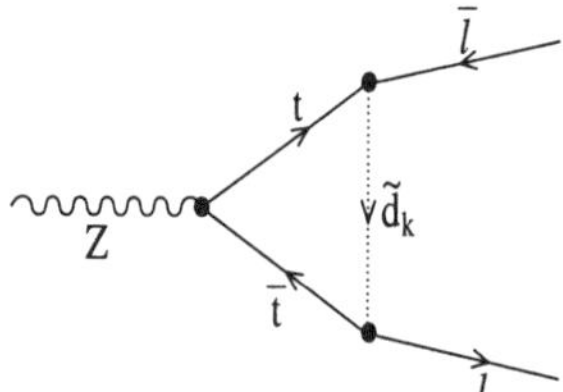

Fig. 14.8. An $\not{R}_p$ vertex correction to the $Z \to \ell\bar{\ell}$ decay amplitude involving the t-quark.

• Atomic parity violation

The parity violating part of the four fermion electron-quark interaction Lagrangian density can be written in standard notation as

$$\mathcal{L}_{PV}^{eq} = \frac{G_F}{\sqrt{2}} \sum_i (C_1^i \, \bar{e}\gamma_\mu\gamma_5 e \, \bar{q}_i\gamma^\mu q_i + C_2^i \, \bar{e}\gamma_\mu e \, \bar{q}_i\gamma^\mu\gamma_5 q_i) \,, \tag{14.22}$$

where the SM values for C_1^i and C_2^i are T_{3L}^i and $-T_{3L}^i$ respectively, T_{3L}^i being the third component of the weak isospin $SU(2)_L$ of the ith quark. The occurrence of $\not{R}_p$ interactions leads to additional terms in $C_{1,2}^i$ involving $|\lambda'_{11k}|^2$, $|\lambda'_{1j1}|^2$ as well as $|\lambda'_{12k}|^2$. A comparison of the calculated weak charge Q_W with the expression in the SM and the experimentally measured value leads to significant constraints on these coupling strengths.

Upper bounds from most of the processes, discussed above, are summarized in Fig. 14.9. An examination of this figure reveals the pattern of strongest bounds for the first generation, weakening as one moves through the second to the third. This is most likely due to the increasing difficulty of measuring R_p violating effects for the latter. There is nonetheless the exciting possibility that evidence for R_p violation may show up in experiments studying processes with heavier, especially third generation, fermions. A further comment concerns bounds based on processes that do not violate baryon or lepton number. These bounds have been derived assuming the absence of any other possible loop induced supersymmetric R_p-conserving contribution to the process(es) under consideration. This procedure would have been realistic if the sparticle exchanged in an $\not{R}_p$ interaction were to be the lightest. Such is far from the case in practice, though! The question then naturally arises: how realistic are these bounds? It is not easy to satisfactorily answer this query since a complete study will involve all the parameters of the MSSM and will make the analysis extremely messy. Preliminary investigations [14.55] in this direction do show, however, that many of these bounds are robust.

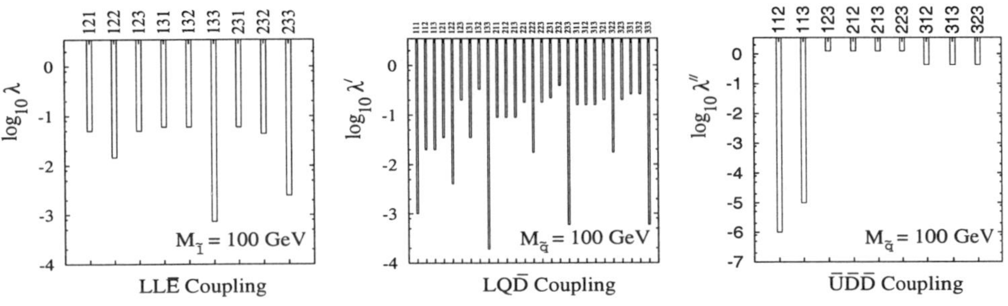

Fig. 14.9. Bounds on $\not{R}_p$ coupling strengths [14.46, 14.47] with vertical bars denoting ruled out regions; these are proportional to the mass of the relevant exchanged sfermion, taken to be 100 GeV here.

Additionally, if complex phases are allowed in $\not{R}_p$ couplings, these can lead to electric dipole moments (EDMs) of the neutron and of the electron. The lack of observation of such EDMs has led to limits [14.56] on the imaginary parts of certain λ, λ' as well as λ'' couplings.

14.5 Bilinear R_p Violation

Let us now consider a model [14.57, 14.58] in which the explicit violation of R-parity is effectively parametrized by only the bilinear superpotential terms $\epsilon_i L_i \cdot H_2$. Two reasons for the unavoidability of such interactions were given earlier in §14.3 in connection with (14.19). A couple of additional arguments can be put forward for the presence of such bilinear $\not{R}_p$ terms in the superpotential. First, the trilinear L violating λ- and λ'-terms, cf. (14.14), themselves effectively give rise to such bilinear terms at the one loop level [14.58]. Second, the model, in which R_p violation is effected only through such bilinear superpotential terms, has been shown [14.59] to be consistent with minimal $N = 1$ supergravity unification incorporating radiative electroweak symmetry breaking and universal scalar as well as gaugino masses. The superpotential for this model can be written, cf. (8.33) and (14.14), as

$$\mathcal{W} = -\epsilon_i L_i \cdot H_2 + \mu H_1 \cdot H_2 - f^e_{ij} H_1 L_i \bar{E}_j - f^d_{ij} H_1 Q_i \bar{D}_j - f^u_{ij} Q_i H_2 \bar{D}_j \ . \tag{14.23}$$

Note that one can always rotate to a basis in which only one of the ϵ_i, say ϵ_3, is nonzero. This only requires a rotation among the three L_i's; no mixing between the L_i's and H_1 is needed. We shall therefore take the superpotential as

$$W = W_{\text{MSSM}} - \epsilon L_3 \cdot H_2 \; . \tag{14.24}$$

Once this extra term with the coefficient ϵ is added, one is obliged for the sake of consistency to include the correspondingly allowed soft supersymmetry breaking term in the Lagrangian density. This is analogous to incorporating the $B\mu h_1 \cdot h_2$ term in $\mathcal{L}_{\text{SOFT}}$ on account of the Higgs superfield bilinear $\mu H_1 \cdot H_2$ present in the superpotential, cf. (9.3). However, the above mentioned rotation, when applied to the left slepton fields $\tilde{\ell}_{iL}$, also removes two of the three $B_\epsilon \epsilon_i$ terms if the soft supersymmetry breaking parameters are flavor universal. Such will be assumed to be the case. Thus two quantities $\epsilon_3 \equiv \epsilon$ and B_ϵ are sufficient to parametrize the extension of the MSSM that we consider in this section. We shall see later that the values of these parameters are severly limited by upper bounds on neutrino masses. Cosmological arguments indicate that the largest neutrino mass can at most be a fraction of an electron volt. This means that $|\epsilon| \lesssim \mathcal{O}$ (MeV). In other words, while B_ϵ can be of the same size as other soft supersymmetry breaking parameters, $|\epsilon|$ has to be orders of magnitude less. So we now have a new soft supersymmetry breaking part of the Lagrangian density and extend (9.3) to

$$-\mathcal{L}_{\text{SOFT}} = -(\mathcal{L}_{\text{SOFT}})_{\text{MSSM}} - (B_\epsilon \epsilon \tilde{\ell}_{3L} \cdot h_2 + \text{h.c.}) \; . \tag{14.25}$$

In (14.25), $-(\mathcal{L}_{\text{SOFT}})_{\text{MSSM}}$ stands for the RHS of (9.3), B_ϵ is a constant with the dimension of mass and $\tilde{\ell}_{3L}$ is the $SU(2)_L$ doublet slepton field which is the scalar component of L_3. The B_ϵ term in (14.25) is important in the following way. Even if the $\epsilon_3 L_3 \cdot H_2$ term is 'formally' eliminated form the superpotential by a superfield redefinition of the type of (14.19), the latter stays and shows the effect of bilinear R_p violation in the scalar sector.

An immediate consequence of the new terms in (14.24) and (14.25) can be pointed out. A nonzero VEV can be acquired by the sneutrino field $\tilde{\nu}_3$ in conjunction with those accruing to the neutral scalar fields $h^0_{1,2}$. In fact, we can define $\langle h^0_{1,2} \rangle \equiv v_{1,2}/\sqrt{2}$, $\langle \tilde{\nu}_3 \rangle \equiv v_3/\sqrt{2}$ in analogy with the notation introduced in §10.2. We thus have three scalar VEVs $v_{1,2,3}$, constrained by the weak boson masses, as

$$M_W^2 = \frac{1}{4} g_2^2 v^2, \;\; M_Z^2 = \frac{1}{4}(g_2^2 + g_Y^2)v^2, \;\; v^2 = v_1^2 + v_2^2 + v_3^2 \simeq (246 \text{ GeV})^2 \; . \tag{14.26}$$

These three VEVs can be parametrized in spherical polar coordinates by v, the MSSM angle β (with $\tan\beta = v_2/v_1$) and a new angle θ as

$$v_1 = v \sin\theta \cos\beta, \;\; v_2 = v \sin\theta \sin\beta, \;\; v_3 = v \cos\theta \; . \tag{14.27}$$

Evidently, this new angle θ tends to $\pi/2$ in the limit of the MSSM when v_3 vanishes.

We turn next to the tree level scalar potential

$$V_{\text{SCALAR}} = \sum_i |\partial W/\partial \Phi_i||^2 + V_D + V_{\text{SOFT}} \; . \tag{14.28}$$

Here Φ_i a generic chiral superfield of the theory, V_D consists of the usual D-terms and V_{SOFT} is the part of $-\mathcal{L}_{\text{SOFT}}$ of (14.25) that contains only scalar fields. The superpotential W in

(14.28) is that of (14.24). We can now concentrate on that part V_{H,L_3} of V_{SCALAR} which contains only the Higgs and the third generation slepton fields. That is since these are now the only scalars directly influencing the breakdown of electroweak symmetry.

In detail, V_{H,L_3} reads

$$
\begin{aligned}
V_{H,L_3} =\ & |f_\tau|^2 \left\{ h_1^\dagger h_1(\tilde{\ell}_3^\dagger \tilde{\ell}_3 + |\tilde{\tau}_R|^2) + \tilde{\ell}_3^\dagger \tilde{\ell}_3 |\tilde{\tau}_R|^2 - |h_1^\dagger \tilde{\ell}_3|^2 \right\} \\[2mm]
& + \frac{g_Y^2}{8}(\tilde{\ell}_3^\dagger \tilde{\ell}_3 + h_1^\dagger h_1 - h_2^\dagger h_2 + 2|\tilde{\tau}_R|^2)^2 \\[2mm]
& + \frac{g_2^2}{8}\left(|h_1^0|^2 - |h_1^-|^2 + |h_2^+|^2 - |h_2^0|^2 + |\tilde{\nu}_\tau|^2 - |\tilde{\tau}_L|^2 \right)^2 \\[2mm]
& + \frac{g_2^2}{2}\left\{ \left[(h_1^- h_2^+)^\star h_1^0 h_2^0 + (h_1^- \tilde{\nu}_\tau)^\star h_1^0 \tilde{\tau}_L + (h_2^0 \tilde{\nu}_\tau)^\star h_2^+ \tilde{\tau}_L + \text{h.c.} \right] \right. \\[2mm]
& \left. \quad + |h_1^-|^2|h_1^0|^2 + |h_2^0|^2|h_2^+|^2 + |\tilde{\tau}_L|^2|\tilde{\nu}_\tau|^2 \right\} \\[2mm]
& - \left\{ f_\tau A^\tau(\tilde{\tau}_L h_1^0 - \tilde{\nu}_\tau h_1^-)\tilde{\tau}_R^\star + \text{h.c.} \right\} + \left\{ B_\epsilon \epsilon(\tilde{\nu}_\tau h_2^0 - \tilde{\tau}_L h_2^+) + \text{h.c.} \right\} \\[2mm]
& - \left\{ B\mu(h_1^0 h_2^0 - h_1^- h_2^+) + \text{h.c.} \right\} - (\mu\epsilon^\star \tilde{\ell}_3^\dagger h_1 + \text{h.c.}) - (f_\tau^\star \epsilon h_1^\dagger h_2 \tilde{\tau}_R + \text{h.c.}) \\[2mm]
& - (\mu f_\tau h_2^\dagger \tilde{\ell}_3 \tilde{\tau}_R^\star + \text{h.c.}) + m_{1h}^2 h_1^\dagger h_1 + (m_{2h}^2 + |\epsilon|^2)h_2^\dagger h_2 + (m_{\tilde{\ell}_3}^2 + |\epsilon|^2)\tilde{\ell}_3^\dagger \tilde{\ell}_3 + m_{\tilde{\tau}}^2 |\tilde{\tau}_R|^2 .
\end{aligned}
$$

$$(14.29)$$

At its minimum, it becomes[22] a generalization of (10.16):

$$
\begin{aligned}
V_{H,L_3}^{\min} =\ & \frac{1}{32}(g_Y^2 + g_2^2)(v_1^2 - v_2^2 + v_3^2)^2 + \frac{1}{2}m_{1h}^2 v_1^2 + \frac{1}{2}m_{2h}^2 v_2^2 - B\mu v_1 v_2 - \mu\epsilon v_1 v_3 \\[2mm]
& + \frac{1}{2}m_{\tilde{\ell}}^2 v_3^2 + \frac{1}{2}\epsilon^2(v_2^2 + v_3^2) + B_\epsilon \epsilon v_2 v_3 .
\end{aligned}
$$

$$(14.30)$$

In (14.30) $m_{1,2h}$ are as defined in (10.7) and $m_{\tilde{\ell}_3}$ is the mass associated with the third generation $SU(2)_L$ slepton doublet. One can crosscheck that the above RHS reduces to that of (10.16) when $v_3, \epsilon \to 0$. Now the requirements that $\partial V_{H,L_3}^{\min}/\partial v_1$, $\partial V_{H,L_3}^{\min}/\partial v_2$ and $\partial V_{H,L_3}^{\min}/\partial v_3$ vanish imply the respective relations

$$
0 = m_{1h}^2 v_1 - B\mu v_2 - \mu\epsilon v_3 + \frac{1}{8}(g_Y^2 + g_2^2)v_1(v_1^2 - v_2^2 + v_3^2) , \tag{14.31a}
$$

$$
0 = (m_{2h}^2 + \epsilon^2)v_2 - B\mu v_1 + B_\epsilon \epsilon v_3 - \frac{1}{8}(g_Y^2 + g_2^2)v_2(v_1^2 - v_2^2 + v_3^2) , \tag{14.31b}
$$

$$
0 = (m_{\tilde{\ell}}^2 + \epsilon^2)v_3 - \mu\epsilon v_1 + B_\epsilon \epsilon v_2 + \frac{1}{8}(g_Y^2 + g_2^2)v_3(v_1^2 - v_2^2 + v_3^2) . \tag{14.31c}
$$

---

[22]We are now taking all the three VEVs $v_{1,2,3}$ to be real and this is a nontrivial assumption. In the MSSM at the tree level, one can always choose the relative phase between h_1 and h_2 in such a way that the phase of μB gets absorbed so that all parameters and both VEVs $v_{1,2}$ are real. In contrast, there are three possible complex parameters in V_{SCALAR} here, namely $\mu\epsilon$, μB and $B_\epsilon\epsilon$, while there are only two relative phases to play with. In general, one of the three combinations of those parameters could be complex, making one combination of VEVs complex. But we are **assuming** that such is not the case.

Note also that the condition for EW symmetry breakdown at the minimum of the potential (14.32) now is that the minimum eigenvalue of the matrix

$$
\begin{pmatrix}
m_{1h}^2 & -B\mu & -\epsilon\mu \\
-B\mu & m_{2h}^2 + \epsilon^2 & \epsilon B_\epsilon \\
-\epsilon\mu & \epsilon B_\epsilon & m_{\tilde{\ell}}^2 + \epsilon^2
\end{pmatrix}
$$

has to be negative, cf. (10.15). However, since $|\epsilon| \ll |\mu|$, for all practical purposes condition (10.15) still applies.

By appropriate double differentitation of (14.29) and going to the minimum, one can compute the mass squared matrices of the charged as well as neutral scalars and pseudoscalars, exactly as was done for the MSSM in §10.3. The charged scalar mass term can be written, after the use of (14.31), as

$$
V_{\text{Quadr.}}^{\text{ch.}} = (h_1^+ \ h_2^+ \ \tilde{\tau}_L^+ \ \tilde{\tau}_R^+)
\begin{pmatrix}
\underset{\sim}{M}{}^2_{hh} & \underset{\sim}{M}{}^2_{h\tilde{\tau}} \\
\underset{\sim}{M}{}^{2\dagger}_{h\tilde{\tau}} & \underset{\sim}{M}{}^2_{\tilde{\tau}\tilde{\tau}}
\end{pmatrix}
\begin{pmatrix}
h_1^- \\ h_2^- \\ \tilde{\tau}_L^- \\ \tilde{\tau}_R^-
\end{pmatrix}
+ \text{h.c.} , \tag{14.32}
$$

where

$$
\underset{\sim}{M}{}^2_{hh} =
\begin{pmatrix}
B\mu\dfrac{v_2}{v_1} + \dfrac{1}{4}g_2^2(v_2^2 - v_3^2) + \mu\epsilon\dfrac{v_3}{v_1} + \dfrac{1}{2}|f_\tau|^2 v_3^2 & B\mu + \dfrac{1}{4}g_2^2 v_1 v_2 \\
B\mu + \dfrac{1}{4}g_2^2 v_1 v_2 & B\mu\dfrac{v_1}{v_2} + \dfrac{1}{4}g_2^2(v_1^2 + v_3^2) - B_\epsilon\epsilon\dfrac{v_3}{v_2}
\end{pmatrix} ,
\tag{14.33a}
$$

$$
\underset{\sim}{M}{}^2_{\tilde{\tau}\tilde{\tau}} =
\begin{pmatrix}
a & b^\star \\
b & c
\end{pmatrix}
\tag{14.33b}
$$

with

$$
a = \frac{1}{2}|f_\tau|^2 v_1^2 - \frac{1}{4}g_2^2(v_1^2 - v_2^2) + \mu\epsilon\frac{v_1}{v_3} - B_\epsilon\epsilon\frac{v_2}{v_3} + m_{\tilde{\ell}}^2 ,
$$

$$
b = -\frac{1}{\sqrt{2}}f_\tau\epsilon v_2(A^\tau + \mu^\star \tan\beta) ,
$$

$$
c = m_{\tilde{R}}^2 + \frac{1}{2}|f_\tau|^2(v_1^2 + v_3^2) - \frac{1}{4}g_Y^2(v_1^2 - v_2^2 + v_3^2)
$$

and

$$
\underset{\sim}{M}{}^2_{h\tilde{\tau}} =
\begin{pmatrix}
-\mu\epsilon - \dfrac{1}{2}|f_\tau|^2 v_1 v_3 + \dfrac{1}{4}g_2^2 v_1 v_3 & -\dfrac{1}{\sqrt{2}}(f_\tau\epsilon v_2 - f_\tau^\star A^{\tau\star} v_3) \\
-B_\epsilon\epsilon + \dfrac{1}{4}g_2^2 v_2 v_3 & -\dfrac{1}{\sqrt{2}}f_\tau^\star(\mu v_3 + \epsilon v_1)
\end{pmatrix} .
\tag{14.33c}
$$

The off-diagonal block $\underset{\sim}{M}{}^2_{h\tilde\tau}$, which induces the mixing between the Higgs and stau sectors, vanishes in the limit when v_3, $\epsilon \to 0$. In the same limit $\underset{\sim}{M}{}^2_{hh}$ and $\underset{\sim}{M}{}^2_{\tilde\tau\tilde\tau}$ also reduce to the corresponding MSSM expressions given in (10.20) and (9.62a) respectively. One consequence of (14.32) is that the tree level physical charged Higgs mass $m_{H^\pm}$ is different from that predicted by the MSSM on account of stau-Higgs mixing, the difference being controlled by the parameter ϵ.

The neutral scalar and pseudoscalar squared mass matrices can be obtained similarly. The 3×3 generalizations of (10.23) and (10.26) respectively are

$$\underset{\sim}{m}{}^2_P = \begin{pmatrix} B\mu\dfrac{v_2}{v_1} + \mu\epsilon\dfrac{v_3}{v_1} & B\mu & -\mu\epsilon \\[2ex] B\mu & B\mu\dfrac{v_1}{v_2} - B_\epsilon\epsilon\dfrac{v_3}{v_2} & -B_\epsilon\epsilon \\[2ex] -\mu\epsilon & -B_\epsilon\epsilon & \mu\epsilon\dfrac{v_1}{v_3} - B_\epsilon\epsilon\dfrac{v_2}{v_3} \end{pmatrix} \tag{14.34a}$$

and

$$\underset{\sim}{m}{}^2_S =$$

$$\begin{pmatrix} B\mu\dfrac{v_2}{v_1} + \mu\epsilon\dfrac{v_3}{v_1} + (g_2^2 + g_Y^2)\dfrac{v_1^2}{4} & -B\mu - (g_2^2 + g_Y^2)\dfrac{v_1 v_2}{4} & -\mu\epsilon + (g_2^2 + g_Y^2)\dfrac{v_1 v_3}{4} \\[2ex] -B\mu - (g_2^2 + g_Y^2)\dfrac{v_1 v_2}{4} & B\mu\dfrac{v_1}{v_2} - B_\epsilon\epsilon\dfrac{v_3}{v_2} + (g_2^2 + g_Y^2)\dfrac{v_2^2}{4} & B_\epsilon\epsilon - (g_2^2 + g_Y^2)\dfrac{v_2 v_3}{4} \\[2ex] -\mu\epsilon + (g_2^2 + g_Y^2)\dfrac{v_1 v_3}{4} & B_\epsilon\epsilon - (g_2^2 + g_Y^2)\dfrac{v_2 v_3}{4} & \mu\epsilon\dfrac{v_1}{v_3} - B_\epsilon\epsilon\dfrac{v_2}{v_3} + (g_2^2 + g_Y^2)\dfrac{v_3^2}{4} \end{pmatrix} .$$

$$\tag{14.34b}$$

We note that the Higgs spectrum now contains two pairs of charged scalars described by orthogonal linear combinations of the fields $H^\pm$ and $\tilde\tau^\pm$. In addition, there are now three neutral CP even scalars as orthogonal linear combinations of the fields h, H and $\Re\,\tilde\nu_\tau$. Finally, there are two CP odd neutral Higgs bosons.

Another consequence of a nonzero ϵ in (14.24) and the resulting VEV of $\tilde\nu_\tau$ is the mixing induced between charged SM leptons and charginos on one hand and between SM neutrinos and neutralinos on the other. For the sake of simplicity, let us assume that $\epsilon_1 = \epsilon_2 = 0$ in the basis where the leptonic Yukawa couplings are diagonal. The said mixing then involves only the leptons of the third generation. In the chargino sector, for instance, there are additional bilinear masslike terms mixing the τ field with gaugino and higgsino fields. Indeed, one can extend the definitions $\psi^\pm$ and $(\psi^\mp)^T$ of (9.5) to $\psi^{\pm\prime}$ and $(\psi'^{\pm})^T$ defined by

$$(\psi'^+)^T \equiv \begin{pmatrix} \lambda^+ & \tilde h_2^1 & \overline{\tau_R} \end{pmatrix}, \text{ and } (\psi'^-)^T \equiv \begin{pmatrix} \lambda^- & \tilde h_1^2 & \tau_L^- \end{pmatrix},$$

i.e. including the two component parts of the τ field as a third element. So the chargino mass term of (9.6) now becomes

$$-\mathcal{L}^c_{\text{MASS}} = (\psi'^-)^T \mathbf{X}' \psi^{+\prime} + \text{h.c.}$$

with

$$\mathbf{X}' = \begin{pmatrix} M_2 & g_2 \dfrac{v_2}{\sqrt{2}} & 0 \\[2ex] g_2 \dfrac{v_1}{\sqrt{2}} & \mu & -f_\tau \dfrac{v_3}{\sqrt{2}} \\[2ex] g_2 \dfrac{v_3}{\sqrt{2}} & -\epsilon & f_\tau \dfrac{v_1}{\sqrt{2}} \end{pmatrix}. \tag{14.35}$$

The physical m_τ^2 is given by the smallest eigenvalue of $\mathbf{X}'\mathbf{X}'^\dagger$, i.e. the relation between m_τ and $f_\tau \equiv f_{33}$ will be modified slightly. However, since $v_3, |\epsilon| \ll |\mu|$, this modification can in practice be ignored, i.e. we can set $\sin\theta = 1$ in (14.27).

Similarly, the definition of ψ^0 and $(\psi^0)^T$ of (9.22) get extended to $\psi^{0'}$ and $(\psi^0)'^T$ with

$$(\psi'^0)^T \equiv (\lambda_0 \quad \lambda_3 \quad \tilde{h}_1^1 \quad \tilde{h}_1^2 \quad \nu_\tau),$$

where ν_τ denotes a two component (Weyl) spinor. The neutralino mass term of (9.23) now becomes

$$-\mathcal{L}^n_{\mathrm{MASS}} = \frac{1}{2}(\psi'^0)^T \mathcal{M}'^n \psi'^0 + \text{h.c.},$$

with

$$\mathcal{M}'^n = \begin{pmatrix} M_1 & 0 & -\dfrac{g_Y}{2}v_1 & \dfrac{g_Y}{2}v_2 & -\dfrac{g_Y}{2}v_3 \\[2ex] 0 & M_2 & \dfrac{g_2}{2}v_1 & -\dfrac{g_2}{2}v_2 & \dfrac{g_2}{2}v_3 \\[2ex] -\dfrac{g_Y}{2}v_1 & \dfrac{g_2}{2}v_1 & 0 & -\mu & 0 \\[2ex] \dfrac{g_Y}{2}v_2 & -\dfrac{g_2}{2}v_2 & -\mu & 0 & \epsilon \\[2ex] -\dfrac{g_Y}{2}v_3 & \dfrac{g_2}{2}v_3 & 0 & \epsilon & 0 \end{pmatrix}. \tag{14.36}$$

The 3×3 matrix $\mathbf{X}'$ of (14.35) can be put into a diagonal form in the same way as $\mathbf{X}$ was put in (9.8), i.e. by a biunitary transformation. Now $\mathcal{U}'$ $(\mathcal{V}')$, which is the extended version of $\mathcal{U}$ $(\mathcal{V})$ of (9.9), will be a 3×3 unitary matrix acting not only on the right chiral (left chiral) components of the charged gaugino and higgsino fields, but also on those components of the τ-field. Similarly, the 5×5 matrix $\mathcal{M}'^n$ gets diagonalized like $\mathcal{M}^n$ in (9.26), except that the extension of $\mathbf{Z}$, namely $\mathbf{Z}'$, is now a 5×5 matrix acting on ν_τ as well as on two component neutral gaugino and higgsino fields.

The entire discussion above has been conducted with the idea that the deviation from the MSSM, controlled by the parameters ϵ and B_ϵ, is not very large. Thus physical spin zero states, that are dominantly charged sleptons or sneutrinos, have small Higgs components in them. Similarly, there are some gaugino/higgsino admixtures in the states that are mostly τ or ν_τ. The couplings of the physical fields are what get related to experimental observables and hence are subject to phenomenological constraints. For instance, trilinear lepton nonconserving interaction terms involving only leptonic and sleptonic fields arise from the Higgs and gaugino interactions of leptons in the current eigenstate basis. L violating quark-quark-slepton or quark-squark-lepton vertices also emerge in like fashion. The strengths of these vertices can be worked out in terms of MSSM couplings and appropriate elements of the $\mathcal{U}, \mathcal{V}, \mathbf{Z}$ and $\mathbf{W}$ matrices, introduced in Ch.9, after tedious but straightforward algebra. We do not catalog these vertices here, but refer the reader to the relevant literature [14.60]. Most of these vertices are already contained in the set given in Fig. 14.1. An interesting difference here is that interaction terms like

$$\overline{\tau_L^C}\nu_\tau \tilde{\tau}_L^\star, \ \ \overline{\tau}_R\tau_L\tilde{\nu}_\tau, \ \ \overline{\nu_\tau^C}\tau_L\tilde{\tau}_L^\star, \ \ \overline{\nu_\tau^C}\nu_\tau\tilde{\nu}_\tau^\star, \ \ \overline{\tau}_R\nu_\tau\tilde{\tau}_R, \ \ \overline{\tau_R^C}\nu_\tau\tilde{\tau}_R^\star, \ \ \overline{\nu_\tau^C}\tau_L\tilde{\tau}_R^\star \ ,$$

are forbidden to originate from an explicit R_p superpotential on account of gauge invariance, cf. (14.14); but these can arise here after mass diagonalization.

The LSP, which is usually taken to be the neutralino $\tilde{\chi}_1^0$, is unstable in this model of R_p violation. The different types of mixing, that have been discussed above, make several channels of two body decay open up for the $\tilde{\chi}_1^0$ provided they are kinematically allowed. These are: $\tilde{\chi}_1^0 \to \tau^\pm W^\mp, \nu_\tau Z, \nu_\tau h$ etc. In addition, there are three body decay channels which become available through mediation by virtual gauge/Higgs bosons and sfermions. While the channels $\tilde{\chi}_1^0 \to e_i^+\bar{u}_j d_k, \nu_i \bar{d}_j d_k, e_i^+\nu_j e_k^-$ are available both in this bilinear R_p model as well as with direct trilinear superpotential violations of R_p, decays such as $\tilde{\chi}_1^0 \to u_i\bar{u}_i\nu_\tau$, $\tilde{\chi}_i^0 \to \nu_\ell\bar{\nu}_\ell\nu_\tau$ can only take place in the bilinear R_p model. One can easily find [14.61] experimentally allowed values of the parameters of this model for which the LSP will decay quite rapidly before travelling from the vertex of production to the detector system. The radiative decay $\tilde{\chi}_1^0 \to \nu_\tau \gamma$ is also significant and can be 5–10% in the BR [14.60] and can, in principle, be used as a means to tag the $\tilde{\chi}_1^0$.

A final topic of concern in this section is the interrelation between the parameters of bilinear R_p violation, sneutrino VEVs and neutrino masses. We have already seen in §14.4 how λ- and λ'-type trilinear L violating terms can give rise to Majorana masses for neutrinos. Turning to the bilinear R_p model, let us return to the superpotential of (14.23) and rewrite it keeping only the superfields of the heaviest generation. This reads

$$\mathcal{W} = -\epsilon L_3 \cdot H_2 + \mu H_1 \cdot H_2 - f_t Q_3 \cdot H_2 \bar{U}_3 - f_b H_1 Q_3 \bar{D}_3 - f_\tau H_1 L_3 \bar{E}_3 \ . \tag{14.37}$$

We now rotate the superfields H_1, L_3 to a new basis H_1', L_3' by a linear transformation, which is different from (14.19), and is given by

$$H_1' = \frac{\mu H_1 - \epsilon L_3}{\sqrt{\mu^2 + \epsilon^2}} \ , \tag{14.38a}$$

$$L_3' = \frac{\epsilon H_1 + \mu L_3}{\sqrt{\mu^2 + \epsilon^2}} \ . \tag{14.38b}$$

In this new basis, the superpotential $\mathcal{W}$ of (14.37) takes the form

$$\mathcal{W} = -f_t Q_3 \cdot H_2 \bar{U}_3 + f_b \frac{\mu}{\mu'} H_1' Q_3 \bar{D}_3 - f_\tau H_1' \cdot L_3' \bar{E}_3 + \mu' H_1' \cdot H_2 + f_b \frac{\epsilon}{\mu'} L_3' Q_3 \bar{D}_3 \ , \tag{14.39}$$

where $\mu'^2 = \mu^2 + \epsilon^2$. In this new basis, the first four terms in the RHS of (14.37) look like MSSM terms while the last term violates R-parity invariance with a trilinear coupling.

We also have to consider the relevant parts of the soft supersymmetry breaking and supersymmetry invariant terms in the scalar potential, cf. (14.23) and (14.25):

$$
\begin{aligned}
V_{\text{SOFT}} \;=\;& m_{h_1}^2 |h_1|^2 + m_{\tilde{\ell}_3}^2 |\tilde{\ell}_3|^2 + (B\mu h_1 \!\cdot\! h_2 - B_\epsilon \epsilon \tilde{\ell}_3 h_2 + \text{h.c.}) + \cdots \\[2mm]
=\;& \mu'^{-2}(m_{h_1}^2 \mu^2 + m_{\tilde{\ell}_3}^2 \epsilon^2)|h_1'|^2 + \mu'^{-2}(m_{h_1}^2 \epsilon^2 + m_{\tilde{\ell}_3}^2 \mu^2)|\tilde{\ell}_3'|^2 \\[2mm]
& + \frac{B\mu^2 + B_\epsilon \epsilon^2}{\mu'} h_1' \!\cdot\! h_2 + \frac{\epsilon\mu}{\mu'}(m_{\tilde{\ell}_3}^2 - m_{h_1}^2)\tilde{\ell}_3'^{\dagger} h_1' \\[2mm]
& + \frac{\epsilon\mu}{\mu'}(B - B_\epsilon)\tilde{\ell}_3' \!\cdot\! h_2 + \text{h.c.} + \cdots \, .
\end{aligned}
\tag{14.40}
$$

In the second step of (14.40), we have rewritten the expression for V_{SOFT} in the rotated basis. The first three terms in its RHS are MSSM-like. Indeed, we can define the coefficients of $|h_1'|^2$ and $|\tilde{\ell}_3'|^2$ as $m_{h_1'}^2$ and $m_{\tilde{\ell}_3'}^2$ in the rotated basis. Similarly, the coefficient of the $h_1' \!\cdot\! h_2$ term can be identified as the new soft bilinear coefficient $B'\mu'$. On the other hand, the last two RHS terms violate R-parity and, being linear in the slepton field, induce a nonzero sneutrino VEV.

We shall not derive in detail the conditions of formation of the three nonzero VEVs in the rotated basis which proceeds in analogy with (14.30) and (14.31). Suffice it to say that v_2 remains the same while we have $\langle h_1'^0 \rangle = v_1'/\sqrt{2}$ and $\langle \tilde{\nu}_\tau' \rangle = v_3'/\sqrt{2}$ with

$$
v_1' \;=\; \mu'^{-1}(\mu v_1 - \epsilon v_3) \, ,
\tag{14.41a}
$$

$$
v_3' \;=\; \mu'^{-1}(\epsilon v_1 + \mu v_3) \, .
\tag{14.41b}
$$

We can now reconsider the neutralino mass matrix (14.36) in the rotated basis. This can be easily done by doing the substitution $(v_1, v_3, \epsilon, \mu) \to (v_1', v_3', 0, \mu')$. Thus v_2 remains the same and in this basis the ϵ term is absent. As a consequence of the latter, the only source of the τ-neutrino mass is the VEV v_3'. In the approximation that the latter is small, one can solve for the smallest eigenvalue of the neutralino mass matrix yielding the ν_τ mass. The answer is

$$
m_{\nu_\tau} \approx -\frac{(g_2^2 M_2 + g_Y^2 M_1)(\mu' v_3')^2}{4 M_1 M_2 \mu'^2 - 2(g_2^2 M_2 + g_Y^2 M_1)v_1' v_2 \mu'} \, .
\tag{14.42}
$$

We see that the τ-neutrino mass is controlled by the numerator factor $(\mu' v_3')^2 = (\epsilon v_1 + \mu v_3)^2$. Any experimental constraint on m_{ν_τ} will therefore be a constraint on $|\epsilon v_1 + \mu v_3|$. Eq. (14.42) shows that an upper bound of a fraction of an eV on the mass of the heaviest neutrino implies an upper bound of the order of $\sqrt{m_\nu M_s} \simeq$ MeV on v_3'. Barring accidental cancellations, a similar upper bound can then be derived on $|\epsilon|$.

Eq. (14.42) can, in fact, be used as a starting point for the construction of realistic neutrino mass models. It should be clear that, even in a general basis, bilinear R_p violation produces only one nonvanishing eigenvalue of the tree level neutrino mass matrix since the eigenvalues are basis independent. However, it also generates new lepton nonconserving couplings, which in turn give rise through loop corrections [14.61] to additional entries in the neutrino mass matrix, thereby leading to (hierarchically smaller) masses for the other two neutrinos. Due to the constrained nature of the model, LSP decay patterns get correlated [14.62] with neutrino mixing patterns.

14.6   Neutrino Masses in Supersymmetric Theories

The evidence [14.63] for nonvanishing neutrino masses comes from the experimental observation that certain kinds of neutrinos seem to "disappear" on their way from the source to the detector. There is a deficit of muon (anti-)neutrinos produced in cosmic ray induced air showers near the top of the atmosphere [14.64] ("atmospheric neutrino problem"), and of electron neutrinos produced in nuclear reactions in the Sun ("solar neutrino problem") [14.65] as well as in nuclear reactors [14.66]. These observations can most easily be explained in terms of oscillations between mass eigenstates that are linear combinations of several flavor eigenstates. The oscillation length L is then proportional to the neutrino energy E and also inversely proportional to the difference of squared masses Δm^2 between the oscillating neutrinos. Depending on the zenith angle, the observation of atmospheric neutrinos can probe oscillation lengths between ~ 10 km (for neutrinos coming from above) to about 10^4 km (for neutrinos coming from below). The agreement of the data with the predicted behavior of L being proportional to $E/\Delta m^2$ is very good, providing strong support for the explanation of the observed anomaly in terms of oscillations. The atmospheric neutrino data require the relevant difference of squared masses (in the ν_μ-ν_τ sector) to be $\sim 3 \times 10^{-3}$ eV^2, and the corresponding mixing angle to be nearly maximal, i.e. $\sim \pi/4$.

The explanation of the solar neutrino deficit in terms of oscillations is complicated by the possibility of matter induced resonant flavor conversion: what is called the MSW effect which is reviewed in Bahcall's article [14.65]. The most favored solution is now with $\Delta m^2 \sim 7 \times 10^{-5}$ eV^2 and a large mixing angle $\sim \pi/6$ (of ν_e, in this case). Note also that the distance between source and detector is essentially fixed here (to $\sim 1.5 \times 10^8$ km). The fact that the flux of ν_e's with energy around 1 MeV seems to be more strongly suppressed than it is at either higher or lower energies nonetheless favors an interpretation in terms of neutrino oscillations. It is important to emphasize that oscillation phenomena depend only on the *differences* of squared masses. The masses themselves could in principle be much larger than their differences, if some neutrinos are nearly degenerate in mass [14.67]. However, any stable neutrino with a mass $\gtrsim 1$ eV would produce too much Hot Dark Matter. In fact, a recent limit [14.68], derived from cosmological observations, constrains the fractional contribution of stable massive neutrinos to the total mass density of the Universe – thereby requiring an upper bound on the sum of neutrino masses $\sum_i m_{\nu_i} < 0.71$ eV. The model builder's task is then to explain why neutrino masses, while nonvanishing, are so much smaller than the masses of the charged leptons and quarks.

We have already discussed in §14.4 how neutrino masses might originate at the loop level in R-parity violating scenarios. Here we shall discuss how neutrino mass operators can arise in certain supersymmetric theories at the tree level itself, provided one is willing to introduce very heavy "sterile neutrinos"[23]. The latter are fermions that are singlets under the SM gauge group. Defining the vector (in flavor space) of two component neutrinos $\xi^T = (\xi_\nu^T, \xi_N^T)$, where ξ_ν are left handed $SU(2)_L$ doublets and ξ_N are the singlets, the neutrino mass term can in

---

[23]They are often called "right handed neutrinos". However, they need not be the right chiral partners of the standard left chiral neutrinos.

general be written as

$$\mathcal{L}_{\nu\;\text{mass}} \;=\; \frac{1}{2}\xi^T \mathcal{M}_\nu \xi + \text{h.c.} ,\qquad\qquad (14.43a)$$

$$\mathcal{M}_\nu \;=\; \begin{pmatrix} \underset{\sim}{m_1} & \underset{\sim}{m_2} \\[4pt] \underset{\sim}{m_2^T} & \underset{\sim}{M} \end{pmatrix} . \qquad\qquad (14.43b)$$

Here $\underset{\sim}{m_1}$ and $\underset{\sim}{M}$ are Majorana mass matrices involving only standard and only sterile neutrinos, respectively, while the Dirac mass matrix $\underset{\sim}{m_2}$ mixes these types of neutrinos. Note that both the m_1 and m_2 terms violate electroweak gauge symmetry. The latter can be written as a matrix of Yukawa couplings $\underset{\sim}{f_\nu}$ times the VEV of a Higgs doublet (h_2, in simple extensions of the MSSM), just like the other Dirac masses in the (MS)SM:

$$\underset{\sim}{m_2} \;=\; \underset{\sim}{f_\nu} \langle h_2^0 \rangle . \qquad\qquad (14.44)$$

However, the m_1 term transforms like a triplet under $SU(2)_L$. In contrast to the $m_{1,2}$ terms, the M term does not break electroweak gauge symmetry and could therefore be very large. However, it does violate lepton number conservation and, for $\underset{\sim}{m_2} \neq 0$, lets that enter the sector of electroweak active neutrinos. In models with an extended gauge group, e.g. those derived from the Grand Unified group $SO(10)$, $\underset{\sim}{M}$ can be generated by a VEV that breaks the $B\text{–}L$ symmetry which is part of $SO(10)$.

In general, the mass matrix (14.43b) allows three different kinds of ansätze leading to light neutrinos:

- $\underset{\sim}{m_1} = \underset{\sim}{M} = 0,\; \underset{\sim}{m_2} \neq 0$: This is the least attractive possibility. Since the m_2 term has the same structure as the mass matrices of charged leptons and quarks, this possibility offers no a priori explanation why neutrinos are light except for merely introducing Yukawa couplings $f_\nu \lesssim 10^{-10} f_{q,\ell}$. There have been speculations on the origin of such a tiny Yukawa coupling from a higher dimensional nonrenormalizable term in the superpotential which, in the SUGRA context, can arise as an effective coupling, duly suppressed by M_{Pl}^{-1}. These are reviewed in Ref.[14.69].

- $\underset{\sim}{m_1} \neq 0,\; \underset{\sim}{m_2}\,\underset{\sim}{M}^{-1} \to 0$: In this limit the sterile neutrinos decouple completely from physics at the electroweak scale. The model is then equivalent to a model without any singlets. Since the m_1 term transforms as a triplet under $SU(2)_L$, in models without fundamental Higgs triplets it can only be generated through nonrenormalizable operators, e.g. of the form of the RHS term in (16.84), cf. Ch.16, or through radiative corrections. A point to note is that the m_1 term also breaks lepton number by two units. In models with MSSM field content it can therefore be created radiatively (from renormalizable interactions) only if R-parity is broken[24]. In this scenario neutrino masses are naturally small since they are suppressed by loop factors. We have already given a discussion of this possibility in §14.4 and §14.5.

---

[24] One should remember that the said operator in (14.43a) does not break R-parity. The 1:1 correspondence between the violation of lepton number and R-parity holds only for renormalizable Lagrangians.

- $\underset{\sim}{M} \gg \underset{\sim}{m_2}$, $\underline{m_2^T \underset{\sim}{M}^{-1} \underset{\sim}{m_2} \gtrsim \underset{\sim}{m_1}}$: In this case[25] the sterile neutrinos can again be integrated out, but they leave an imprint on the mass matrix of the light neutrinos, which becomes[26]

$$\underset{\sim}{\mathcal{M}}_{\text{light}} = \underset{\sim}{m_1} - \underset{\sim}{m_2^T} \underset{\sim}{M}^{-1} \underset{\sim}{m_2} . \tag{14.45}$$

The suppression of the second term in the RHS by the inverse of $\underset{\sim}{M}$ is the celebrated "seesaw" mechanism [14.70]. The latter naturally allows one to explain the smallness of neutrino masses even if the eigenvalues of $\underset{\sim}{m_2}$ are of the order of the known quark and lepton masses. For the remainder of this section we shall focus on models of this type.

As noted above, the heaviest light neutrino probably has a mass between a few times 10^{-2} eV and $\lesssim 1$ eV, while the largest eigenvalue of $\underset{\sim}{m_2}$ can be expected to lie roughly between m_τ and m_t. Barring cancellations between the two terms in the RHS of (14.45), the corresponding eigenvalue of $\underset{\sim}{M}$ should then lie between $\sim 10^9$ and $\sim 10^{15}$ GeV. Of course, this already quite large range applies only to the mass of that sterile neutrino which is most strongly coupled to the heaviest of the light neutrinos. For simplicity, one often assumes that all eigenvalues of $\underset{\sim}{M}$ are of the same order of magnitude, but such need not be the case. The scale of $\underset{\sim}{M}$ is often associated with the scale where an extended gauge symmetry containing a gauged $U(1)_{B-L}$ is broken with the heavy neutrinos being nonsinglets under this extended symmetry. Indeed, the very existence of $SU(2)_L \times U(1)_Y$ singlet superfields hints towards $SO(10)$ unification [14.71], where one such singlet per generation is predicted as a member of the **16** representation which also contains all MSSM matter superfields. Since the group $SO(10)$ has the rank five, $B - L$ can be gauged in $SO(10)$-based models. On the other hand, in supergravity models (cf. Ch.12) the scale of M can also be identified [14.72] with the intermediate scale $M_I \simeq \sqrt{M_s M_{Pl}} \sim 10^{11}$ GeV that characterizes supersymmetry breaking in the hidden sector, where M_s is a typical visible sector soft supersymmetry breaking parameter.

We shall see later in §16.5 that the existence of very heavy sterile (s)neutrinos can explain the baryon asymmetry of the Universe through the leptogenesis mechanism. Though there are quantitative differences, this mechanism can operate in both supersymmetric and nonsupersymmetric theories. However, only supersymmetric models allow one to test the ansätze for the neutrino mass matrix (14.43b) in processes that do not involve neutrinos. Moreover, in such models, there are effects that are not suppressed by the small neutrino masses, *provided* that the energy scale of transmission of supersymmetry breaking to the visible sector exceeds the masses of the sterile neutrinos. In this case the Yukawa couplings giving rise to m_2 will also affect the running of the soft supersymmetry breaking parameters $\mathcal{M}_{\tilde{l}}^2$ and A_e in the slepton sector. In particular, off-diagonal entries in m_2 will generate flavor mixing between $SU(2)_L$ doublet sleptons. This is quite analogous to radiatively induced slepton flavor mixing which occurs in supersymmetric Grand Unified theories, cf. §12.5. In

---

[25]These "matrix inequalities" are meant to hold for the eigenvalues.

[26]We are assuming that $\underset{\sim}{M}$ is a nonsingular matrix. If one or more eigenvalues of $\underset{\sim}{M}$ vanish, the spectrum of light neutrinos will contain one or more sterile states. Eqs.(14.43) and (14.45) can still be used, but $\underset{\sim}{m_1}$ is then no longer a 3×3 matrix.

fact, a recent analysis [14.73] finds that constraints from the unobserved radiative $\mu \to e\gamma$ decay disfavor scenarios with large ν_e-ν_μ mixing in models with mSUGRA type of boundary conditions and a seesaw mechanism, if at least one of the f_ν couplings is $\mathcal{O}(1)$. Finally, in the framework of Grand Unified theories, the couplings f_ν will also affect the RGE of the masses of $SU(2)_L$ singlet sleptons as well as of $SU(2)_L$ doublet and charge $2/3$ $SU(2)_L$ singlet squarks. Consequently, new contributions may result in both leptonic and hadronic flavor violating processes [14.74].

The presence of a generic gauge singlet superfield $\bar{N}$ and of the new superpotential term $H_2 \cdot L\bar{N}$ will also lead to extra contributions to the RGE equations for the squared soft supersymmetry breaking Higgs mass m_2^2 and the Yukawa coupling of the top quark. In models with radiative EW symmetry breaking this leads to slightly different predictions on $|\mu|$ for given boundary conditions on the soft supersymmetry breaking parameters at a high scale. However, these effects are not large and therefore difficult to distinguish from minor modifications of the boundary conditions, due for instance to a Grand Unified Theory. Suppose we consider a number of gauge singlet superfields and use the symbol $\tilde{N}$ to denote a column vector with those. Including the gauge singlets, the sneutrino mass terms can be written as

$$\mathcal{L}_{\tilde{\nu}-\text{mass}} = -\tilde{\nu}^\dagger \mathcal{M}_{\tilde{\nu}}^2 \tilde{\nu} - \left(\tilde{n}^\dagger \underset{\sim}{B_N}\, \underset{\sim}{M} \tilde{n}^* + \tilde{\nu}_L^T \underset{\sim}{m_2}\, \underset{\sim}{M^*}\, \tilde{n} + h.c. \right), \tag{14.46}$$

where $\tilde{n}^*$ is a column vector of the scalar components of the superfields $\tilde{N}$, the vector $\tilde{\nu}^T \equiv (\tilde{\nu}_L^T,\ \tilde{n}^T)$ and[27]

$$\mathcal{M}_{\tilde{\nu}}^2 = \begin{pmatrix} \mathcal{M}_{\tilde{l}}^2 + D \cdot \mathbb{1} + \left(\underset{\sim}{m_2}\, \underset{\sim}{m_2^\dagger} \right)^T & -\underset{\sim}{m_2} \left(\underset{\sim}{A_\nu^\dagger} + \mu \cot\beta \cdot \mathbb{1} - \right) \\ -\left(\underset{\sim}{A_\nu} + \mu^* \cot\beta \cdot \mathbb{1} \right) \underset{\sim}{m_2^\dagger} & \mathcal{M}_{\tilde{N}}^2 + \left(\underset{\sim}{M^\dagger}\, \underset{\sim}{M} + \underset{\sim}{m_2^\dagger} - \underset{\sim}{m_2} \right)^T \end{pmatrix}. \tag{14.47}$$

Here $D = (M_Z^2 \cos 2\beta)/2$ and $\underset{\sim}{A_\nu}$ is the matrix coefficient of the trilinear soft supersymmetry breaking term corresponding to the superpotential term giving rise to $\underset{\sim}{m_2}$ and in (14.46) $\underset{\sim}{B_N}$ is a similar coefficient for the $\tilde{n}$-sector. We have set $\underset{\sim}{m_1} = 0$ since its effect on sneutrino masses is negligible. The mixing between $SU(2)_L$ singlet and doublet sneutrinos is dominated by the last term in (14.46). The latter contributes with opposite signs to the mass matrices of the real and imaginary parts of the complex sneutrino fields, leading to a mass-splitting between 'scalar' and 'pseudoscalar' sneutrinos. However, for light sneutrinos, this splitting is of the order of the light sneutrino mass and is hence negligible, unless [14.49] $|B_N| \gg 1$ TeV. Flavor mixing between $SU(2)_L$ doublet sneutrinos is therefore essentially determined by $\mathcal{M}_{\tilde{l}}^2$. As noted above, the latter also contributes to mixing between charged $SU(2)_L$ doublet sleptons.

Though neutrino flavor mixing, as described by the m_2 term, affects $\mathcal{M}_{\tilde{l}}^2$ only at the one loop level, it can lead to significant effects through terms $\mathcal{O}(\underset{\sim}{f_\nu}\underset{\sim}{f_\nu^\dagger}/16\pi^2)$. The latter show up not only in the lepton flavor violating rare decays mentioned above but also in collider experiments. This has been demonstrated in Ref. [14.75], where a minimal supergravity model is analyzed with universal boundary conditions for soft supersymmetry breaking masses at

---

[27]We denote the scalar component of $\bar{N}_i$ by n_i^*.

the Planck scale, cf. §12.3. Moreover, it is assumed that the near-maximal ν_μ-ν_τ mixing angle, which one needs to explain the atmospheric neutrino anomaly in models with only three light neutrinos, is produced by off-diagonal terms in the matrix $\underset{\sim}{m}_2$ (in the basis where the mass matrix of charged leptons is diagonal). In addition, $\underset{\sim}{M}$ is taken to be essentially proportional to the unit matrix. In this case the model allows large $\tilde{\mu}_L$-$\tilde{\tau}_L$ and $\tilde{\nu}_\mu$-$\tilde{\nu}_\tau$ mixing, with mass differences roughly between 0.1 and 10 GeV. One should not forget that the flavor diagonal τ Yukawa coupling also contributes to the mass difference. The size of flavor violating effects diminishes with a decreasing sterile neutrino mass scale, since for a fixed light neutrino mass,

$$\underset{\sim}{f}_\nu \underset{\sim}{f}_\nu^\dagger \; \propto \; \underset{\sim}{m}_2 \underset{\sim}{m}_2^\dagger \; \propto \; \underset{\sim}{m}_\nu \underset{\sim}{M} \,.$$

This overcompensates the increase of logarithms like $\ln(M_{Pl}/M)$ which appear due to the running of $\mathcal{M}_{\tilde{l}}^2$ from the Planck scale to energy scales $\sim M$. In this model the loop induced difference between sneutrino or slepton masses is often comparable to, or even smaller than, the decay widths of the corresponding particles. The occurrence of signals of slepton flavor violation can then be understood as being due to oscillations between $\tilde{\mu}_L$ and $\tilde{\tau}_L$, and/or between $\tilde{\nu}_\mu$ and $\tilde{\nu}_\tau$ [14.76]. In particular, for a very small mass splitting Δm, the size of the signal decreases as $\left(\Delta m/\overline{\Gamma}\right)^2$, where $\overline{\Gamma}$ is the average of the decay widths of the two oscillating particles[28]. The signal diminishes for very large values of Δm as well, since the latter correspond to relatively small mixing angles[29]. However, there is an interesting range of parameters for which observable signals are expected.

The simplest signal of slepton flavor violation at lepton colliders can be found in the production of intermediate states like $\mu^- \tau^+ \tilde{\chi}_i^- \tilde{\chi}_j^+$ $(i,j = 1,2)$ or $\mu^- \tau^+ \tilde{\chi}_i^0 \tilde{\chi}_j^0$ $(i,j = 1,\ldots,4)$ (plus their charge conjugates) which can arise through slepton pairs produced [14.74] either directly or in cascades of decaying pair-produced charginos. (Recall that the charginos $\tilde{\chi}_{k,m}^\pm$ and all neutralinos $\tilde{\chi}_{l,n}^0$, except the lightest, are unstable.) In order to avoid possible confusion between the primary leptons and those from $\tilde{\chi}$ decays, it is easiest to focus on purely hadronic decay modes of the latter. Since the most significant mixing is expected to be between the second and the third generations, $\mu^+\mu^-$ colliders will give significantly larger signals of this type than do e^+e^- colliders of equal luminosity. The reason is that at muon colliders smuons and muon sneutrinos can be pair-produced through the t channel exchange of neutralinos and charginos, respectively, leading to enhanced cross sections, cf. §15.3. Similar signals are also possible at hadron colliders, but there the detection of τ leptons is far more challenging.

In case slepton mass matrices contain a nontrivial phase, one can even hope to find signals of CP violation [14.77], e.g. a difference between the branching ratios for the decays $\tilde{\chi}_2^0 \to e^+\mu^- \tilde{\chi}_1^0$ and $\tilde{\chi}_2^0 \to e^-\mu^+\tilde{\chi}_1^0$. However, if $\mathcal{M}_{\tilde{l}}^2$ is the only relevant source of slepton flavor violation, CP noninvariance will require a nontrivial mixing between all three generations. In models with hierarchical neutrino masses, the mixing of sleptons of the first generation with those of the second and the third, and/or the corresponding mass differences, will be small. Moreover, independently of the mixing angles, CP violating rate asymmetries can

---

[28]Unlike in the familiar cases of the K and B meson systems, slepton oscillations are expected to be much too rapid to be experimentally resolvable, having time scales $\lesssim 10^{-24}$ sec. Only time-integrated observables can be used to study this phenomenon.

[29]Note that Δm also receives contributions from the flavor conserving τ Yukawa coupling term.

only be sizable if $\Delta m \simeq \bar{\Gamma}$, i.e. if they are suppressed for both small and large Δm.

One should keep in mind that neutrino mixing depends in general on both $\underset{\sim}{m_2}$ and $\underset{\sim}{M}$ (and $\underset{\sim}{m_1}$, if nonzero), cf. (14.45). In comparison, the size of loop induced slepton flavor violation depends essentially only on $\underset{\sim}{m_2}$ (through $\underset{\sim}{f_\nu}$, see 14.44), except for the logarithmic dependence on the mass of the sterile neutrinos mentioned above. Consequently, one cannot make model independent predictions for slepton flavor violation even if the mass matrix of light neutrinos were known exactly. On the other hand, studies of slepton flavor violation, when combined with analyses of neutrino oscillations, could determine, or at least considerably constrain, both $\underset{\sim}{m_2}$ and $\underset{\sim}{M}$. In contrast, in nonsupersymmetric theories no particle physics experiment is sensitive to $\underset{\sim}{m_2}$ and $\underset{\sim}{M}$ separately at energy scales much below the masses of the heavy sterile neutrinos. Even in supersymmetric theories a separate determination of $\underset{\sim}{m_2}$ and $\underset{\sim}{M}$ is possible only if one has an accepted theory of supersymmetry breaking. If an arbitrary $\mathcal{M}_{\tilde{l}}^2$ is allowed already at the Planck scale, it will be impossible to extract the radiatively induced flavor violation from the "primordial" one.

References

[14.1] J.L. Hewett and T.G. Rizzo, Phys. Rept. **183**, (1989) 193.

[14.2] B. Ananthanarayan and P.N. Pandita, Int. J. Mod. Phys. **A12** (1997) 2321. U. Ellwanger and C. Hugonie, hep-ph/9901309, *ibid.*/0006222.

[14.3] J.E. Kim and H.P. Nilles, Phys. Lett. **B138** (1984) 150.

[14.4] L.J. Hall, J. Lykken and S. Weinberg, Phys. Rev. **D27** (1983) 2359.

[14.5] G.F. Giudice and A. Masiero, Phys. Lett. **B206** (1988) 480.

[14.6] J.A. Casas and C. Muñoz, Phys. Lett. **B306** (1993) 288.

[14.7] J.E. Kim, Phys. Rept. **150** (1987) 1.

[14.8] J.E. Kim and H.P. Nilles, Phys. Lett. **B315** (1993) 107.

[14.9] G.F. Giudice and E. Roulet, Phys. Lett. **B315** (1993) 107.

[14.10] M. Drees, Int. J. Mod. Phys. **A4** (1989) 3635.

[14.11] H.P. Nilles, M. Srednicki and D. Wyler, Phys. Lett. **124B** (1983) 337. K. Tamvakis, Phys. Lett. **124B** (1983) 341.

[14.12] S.A. Abel, S. Sarkar and P.L. White, Nucl. Phys. **B454** (1995) 663.

[14.13] S.A. Abel, Nucl. Phys. **B480** (1996) 55.

[14.14] C. Panagiotakopoulos and K. Tamvakis, Phys. Lett. **B446** (1999) 224; *ibid.* **B469** (1999) 145.

[14.15] C. Panagiotakopoulos and A. Pilaftsis, Phys. Rev. **D63** (2001) 055003; Phys. Lett. **B505** (2001) 184.

[14.16] B.W. Lee, C. Quigg and H.B. Thacker, Phys. Rev. **D16** (1977) 1519.

[14.17] J.-P. Derendinger and C.A. Savoy, Nucl. Phys. **B237** (1984) 307.

[14.18] T. Elliot, S.F. King and P.L. White, Phys. Rev. **D49** (1994) 2435.

[14.19] D.M. Pierce, J. Bagger, K.T. Matchev and R.-J. Zhang, Nucl. Phys. **B491** (1997) 3.

[14.20] J.R. Espinosa and M. Quiros, Phys. Rev. Lett. **81** (1998) 516. P.N. Pandita, Pramana J. Phys. **51** (1998) 169: *Proc.* WHEPP-5, Bangalore.

[14.21] J. Kamoshita, Y. Okada and M. Tanaka, Phys. Lett. **B328** (1994) 67.

[14.22] J.F. Gunion, H.E. Haber, G.L. Kane and S. Dawson, *op. cit., Bibl.*

[14.23] U. Ellwanger, M. Rausch de Traubenberg and C.A. Savoy, Nucl. Phys. **B492** (1997) 21.

[14.24] S.J. Huber and M.G. Schmidt, Nucl. Phys. **B606** (2001) 183.

[14.25] G.C. Branco, F. Krüger, J.C. Romão and A. Texeria, JHEP **0107** (2001) 027.

[14.26] U. Ellwanger and C. Hugonie, Eur. Phys. J. **C13** (2000) 681.

[14.27] C.S. Aulakh and R.N. Mohapatra, Phys. Lett. **B119** (1982) 136. A. Santamaria and J.W.F. Valle, Phys. Lett. **B195** (1987) 423; Phys. Rev. Lett. **60** (1988) 397; Phys. Rev. **D39** (1989) 1780.

[14.28] L.J. Hall and M. Suzuki, Nucl. Phys. **B231** (1984) 419. J. Ellis, G. Gelmini, C. Jarlskog, G.G. Ross and J.W.F. Valle, Phys. Lett. **150** (1985) 142.

[14.29] A.Yu. Smirnov and F. Vissani, Phys. Lett. **B380** (1996) 317. G. Bhattacharyya and P.B. Pal, Phys. Rev. **D59** (1999) 097701.

[14.30] G.R. Farrar and P. Fayet, Phys. Lett. **B76** (1978) 575.

[14.31] M. Turner, Phys. Rep. **190** (1990) 67.

[14.32] B. Sadoulet, *American Institute of Physics Conf. Proc.* **47B** (1999) 363.

[14.33] H.K. Dreiner and G.G. Ross, Nucl. Phys. **B410** (1993) 183.

[14.34] L.J. Hall and M. Suzuki, Nucl. Phys. **B231** (1984) 419.

[14.35] I.H. Lee, Phys. Lett. **B138** (1984) 121; Nucl. Phys. **B246** (1984) 120.

[14.36] B. de Carlos and P.L. White, Phys. Rev. **D54** (1996) 3427.

[14.37] A.Yu. Smirnov and F. Vissani, Nucl. Phys. **B460** (1996) 37.

[14.38] M.C. Bento, L.J. Hall and G.G. Ross, Nucl. Phys. **B292** (1987) 400. N. Ganoulis, G. Lazarides and Q. Shafi, Nucl. Phys. **B323** (1989) 374.

[14.39] L.E. Ibañez and G.G. Ross, Nucl. Phys. **B368** (1992) 3.

[14.40] T. Banks and M. Dine, Phys. Rev. **D45** (1992) 1424.

[14.41] B.C. Allanach, A. Dedes and H.K. Dreiner, Phys. Rev. **D60** (1999) 056002.

[14.42] M. Bastero-Gil and B. Brahmachari, Phys. Rev. **D56** (1997) 6912. M. Bastero-Gil, B. Brahmachari and R.N. Mohapatra, hep-ph/9606447. B. Brahmachari and P. Roy, Phys. Rev. **D50** (1994) 39; *ibid.* **D51** (1995) 3974 (E). J.L. Goity and M. Sher, Phys. Lett. **B346** (1995) 69.

[14.43] S. Dawson, Nucl. Phys. **B261** (1985) 297. H.K. Dreiner, P. Richardson and M.H. Seymour, JHEP **0004** (2000) 008. F. Borzumati, R.M. Godbole, J.L. Kneur and F. Takayama, JHEP **07** (2002) 037. E.A. Baltz and P. Gondolo, Phys. Rev. **D57** (1998) 2969.

[14.44] C.E. Carlson, P. Roy and M. Sher, Phys. Lett. **B357** (1995) 99. D. Choudhury and P. Roy, Phys. Lett. **B378** (1996) 153. J.H. Jang, J.K. Kim and J.S. Lee, Phys. Rev. **D55** (1997) 7296. K. Huitu, J. Maalampi, M. Raidal and A. Santamaria, Phys. Lett. **B430** (1998) 335. G. Bhattacharya and P.B. Pal, Phys. Lett. **B439** (1998) 81.

[14.45] K. Agashe and M. Graesser, Phys. Rev. **D54** (1996) 4445.

[14.46] G. Bhattacharyya, Nucl. Phys. Proc. Suppl. **52A** (1997) 83; also in *Tegernsee 1997, Beyond the Desert 1997*, hep-ph/9709395. H.K. Dreiner in *Pespectives on Superysymmetry* (ed. G.L. Kane, World Scientific, Singapore, 1998), p462. O.C. Kong, Nucl. Phys. Proc. Suppl. **62** (1998) 266. S. Raychaudhuri, hep-ph/9905576, in *Mumbai 1999: Hadron Collider Physics*, p350. B.C. Allanach, A. Dedes and H.K. Dreiner, Phys. Rev. **D60** (1999) 075014. P. Roy in *Pacific Particle Physics Phenomenology*, eds. M. Drees, C.S. Kim and S.K. Kim (World Scientific, Singapore 1999), p15. J.S. Lee, *ibid.*, p307. D.K. Ghosh, K-G. He, B.H.J. McKellar and J-Q. Shi, JHEP **0207** (2002) 067. H.K. Dreiner, G. Polesollo and M. Thormeier, Phys. Rev. **D65** (2002) 115006. J.P. Saha and A. Kundu, *ibid.* **D66** (2002) 054021. S. Bar-Shalom, G. Eilam and Y-D. Yang, *ibid.* **D67** (2003) 014007. G. Bhattacharyya, H. Päs, L. Song and T.J. Weiler, hep-ph/0302191.

[14.47] R. Barbier et al, hep-ph/9810232, *ibid.* /0406039. F. Borzumati et al., *loc. cit.* [14.43].

[14.48] S. Dimopoulos and L.J. Hall, Phys. Lett. **B207** (1988) 210. R.M. Godbole, P. Roy and X. Tata, Nucl. Phys. **B401** (1993) 67. M. Drees, S. Pakvasa, X. Tata and T. der Veldhuis, Phys. Rev. **D57** (1998) 5335. S. Rakshit, G. Bhattacharyya and A. Raychaudhuri, Phys. Rev. **D59** (1999) 091701.

[14.49] Y. Grossman and H. Haber, Phys. Rev. Lett. **78** (1997) 3438; Phys. Rev. **D59** (1999) 093008; *ibid.* **D63** (2001) 075011.

[14.50] F. Borzumati and J.S. Lee, Phys. Rev. **D66** (2002) 115012.

[14.51] G. Bhattacharyya, H.V. Klapdor-Kleingrothaus and H. Päs, Phys. Lett. **B463** (1999) 77.

[14.52] M. Hirsch, H.V. Klapdor-Kleingrothaus and S.G. Kovalenko, Phys. Rev. Lett. **75** (1995) 17; Phys. Rev. **D53** (1996) 1329.

[14.53] V. Barger, G.F. Giudice and T. Han, Phys. Rev. **D40** (1989) 2987. F. Ledroit and G. Sajor, available in http://cdfinfo.in2p3.fr/store/Gdrsusy/lim.html.

[14.54] G. Bhattacharyya, D. Choudhury and K. Sridhar, Phys. Lett. **B355** (1995) 193. O. Lebedev, W. Loinaz and T. Takeuchi, Phys. Rev. **D62** (2000) 015003; in Osaka 2000, *High Energy Physics*, Vol. 2, p.1042.

[14.55] A. Kundu and S. Raychaudhuri, private communication.

[14.56] R.M. Godbole, S. Pakvasa, S.D. Rindani and X. Tata, Phys. Rev. **D61** (2000) 113003. S.A. Abel, A. Dedes and H.K. Dreiner, JHEP **0005** (2000) 023. Y.Y. Keum and O.C. Kong, Phys. Rev. Lett. **86** (2001) 393. K. Choi, E. J. Chun and K. Hwang, Phys. Rev. **D63** (2001) 013002.

[14.57] F.de Campos, M.A. Garcia-Jareño, A.S. Joshipura, J. Rosiek and J.W.F. Valle, Nucl. Phys. **B451** (1995) 3. A.S. Joshipura and M. Nowakowski, Phys. Rev. **D51** (1995) 2421; *ibid* **D51** (1995) 5271. M. Nowakowski and A. Pilaftsis, Nucl. Phys. **B461** (1996) 19. R. Hempfling, Nucl. Phys. **B478** (1996) 37. S. Roy and B. Mukhopadhyaya, Phys. Rev. D55 (1997) 7020; *ibid.* **D59** (1999) 091701. A.G. Akeroyd, M.A. Díaz, J. Ferrandis and J.W.F. Valle, Nucl. Phys. **B529** (1998) 3. C.H. Chang and T.F. Feng, Eur. Phys. J. **C12** (2000) 137.

[14.58] B.de Carlos and P.L. White, *loc. cit.* Ref. [14.36]; Phys. Rev. **D55** (1997) 4222.

[14.59] M.A. Díaz, J.C. Romão and J.W.F. Valle, Nucl. Phys. **B524** (1998) 23.

[14.60] B. Mukhopadhyaya and S. Roy, Phys. Rev. **D60** (1999) 115012. S. Roy, Ph.D. thesis, Mehta Research Institute (1999).

[14.61] Y. Grossmann and H.E. Haber, *loc. cit.*, Ref. [14.49]. S. Davidson and M. Losada, Phys. Rev. **D65** (2002) 075025.

[14.62] W. Porod, M. Hirsch, J. Romão and J.W.F. Valle, Phys. Rev. **D3** (2001) 115004. E.J. Chun, D.-W. Jung and J.D. Park, Phys. Lett. **B557** (2003) 233.

[14.63] Particle Data Group, *loc. cit.*, *Bibl.*

[14.64] T. Kajita and Y. Totsuka, Rev. Mod. Phys. **73** (2001) 85.

[14.65] J.N. Bahcall, Phys. Rept. **333** (2000) 47. Q.R. Ahmad et al., SNO collaboration, Phys. Rev. Lett. **89** (2002) 011301; *ibid.* **89** (2002) 011302. S. Fukuda et al., Super-Kamiokande collaboration, Phys. Lett. **B539** (2002) 179.

[14.66] K. Eguchi et al., KamLAND collaboration, Phys. Rev. Lett. **90** (2003) 021802.

[14.67] D. Caldwell and R.N. Mohapatra, Phys. Rev. **D48** (1993) 3259. A.S. Joshipura, Z. Phys. **C64** (1994) 31; Phys. Rev. **D55** (1995) 1321. A. Ioannisian and J.W.F. Valle, Phys. Lett. **B332** (1994) 93.

[14.68] D.N. Spergel et al., WMAP collaboration, Astrophys. J. Suppl., **148** (2003) 175. Also, M. Colles *et al.* (2df Galactic Redshift Survey), Month. Not. Roy. Astr. Soc. **328** (2001) 1039 and M. Tegmark et al., (Sloan Digital Sky Survey), astro-ph/0310723.

[14.69] B. Mukhopadhyaya, hep-ph/0301278, to appear in Proc. Ind. Nat. Sci. Acad., special volume on neutrinos.

[14.70] M. Gell-Mann, P. Ramond, R. Slansky in *Supergravity*, ed. P. van Nieuwemhuizen and D.Z. Freedman, North Holland (Amsterdam 1979). T. Yanagida, Proc. Wkshp. *Unified Theory and the Baryon Number in the Universe*, ed. O. Sawada and A. Sugamoto, KEK Report No. 79-18 (Tsukuba, 1979). See also S.F. King, *loc. cit., Bibl.*

[14.71] T. Blazek, S. Raby and K. Tobe, Phys. Rev. **D60** (1999) 113001; C.H. Albright and S.M. Barr, Phys. Rev. **D62** (2000) 093008.

[14.72] N. Arkani-Hamed, L.J. Hall, H. Murayama, D.R. Smith and N. Weiner, Phys. Rev. **D64** (2001) 115011.

[14.73] J.A. Casas and A. Ibarra, Nucl. Phys. **B618** (2001) 171.

[14.74] S. Baek, T. Goto, Y. Okada and K. Okumura, Phys. Rev. **D64** (2001) 095001. N. Akama, Y. Kiyo, S. Komine and T. Moroi, *ibid.* **D64** (2001) 095012.

[14.75] J. Hisano, M.M. Nojiri, Y. Shimizu and M. Tanaka, Phys. Rev. **D60** (1999) 055008. M. Guchait, J. Kalinowski and P. Roy, Eur. Phys. J. **C21** (2001) 163.

[14.76] N. Arkani-Hamed, H-C. Cheng, J.L. Feng and L.J. Hall, Phys. Rev. Lett. **77** (1996) 1937.

[14.77] N. Arkani-Hamed, H-C. Cheng, J.L. Feng and L.J. Hall, Nucl. Phys. **B505** (1997) 3.

Chapter 15

SUPERSYMMETRY AT COLLIDERS

15.1 Introduction

The fascinating nature of phenomenological supersymmetry notwithstanding, the results of all efforts to find any direct experimental evidence of sparticles have so far been negative [15.1]. We have only lower bounds on sparticle masses and upper bounds on the extra phases and flavor mixing angles (cf. §9.5) present, say in the MSSM. Nevertheless, sparticles are expected (cf. Ch.1) to lie in the sub-TeV to a few TeV mass range. Their detection would constitute direct, hard experimental evidence of weak scale supersymmetry. The search for such sparticles via their production and decay processes, therefore, continues to be an important part of the physics programs of all currently running high energy colliders as well as of those under construction or planning. Our objective in this chapter is to describe and discuss these search strategies, but we first present an overview in this introductory section.

The masses of various sparticles and Higgs bosons are determined to a large extent by soft supersymmetry breaking parameters (cf. §9.2), which in turn are controlled (cf. Chs. 12 and 13) by as yet unknown higher scale physics. An experimental determination of the sparticle and Higgs mass spectra will doubtless help unravel the nature of supersymmetry breaking and shed light on the role of such higher scale physics. Yet, this dependence on unknown physics makes a general description of sparticle search prospects at colliders rather complicated. Moreover, as seen in Ch.9, electroweak symmetry breaking causes gaugino-higgsino mixing which is also affected by supersymmetry breaking. Consequently, not only the masses of the physical charginos and neutralinos, but also their couplings, become functions of supersymmetry model parameters. Search strategies for different sparticles depend quite critically on their relative masses as well as on the mass and couplings of the lightest supersymmetric particle, which again are determined by the soft supersymmetry breaking terms. The former are thus best discussed within the framework of some specific models for the latter.

In Ch.9 we had taken all the supersymmetry breaking parameters to be completely arbitrary. Commitment to a specific model of supersymmetry breaking, cf. Chs. 12, 13, usually leads to a drastic reduction in the number of these parameters. The general class of supersymmetry breaking models, which boast of such simplifications in the number of parameters involved, goes under the name Constrained MSSM or CMSSM, cf. §9.1. The overall sparticle mass scale remains unknown even here. However, a given model makes predictions on ratios of sparticle and/or Higgs boson masses (unfortunately often in the form of inequalities only), as well as on the amounts of mixing in the chargino and neutralino sectors. The most thoroughly studied member of this class of models is mSUGRA, cf. §12.3, §12.4. The mass spectra of gauginos and sfermions of the first two generations in this model are given in (12.27)–(12.31). But we shall also consider extensions of mSUGRA, such as $\tilde{\mathrm{C}}$MSSM, as discussed in §12.5. Another case of interest is that of the minimal AMSB model described in §12.6, see in particular (12.56)–(12.60). A third constrained model that will frequently come in our discussions is mGMSB, cf. §13.2. Eqs. (13.18)–(13.20) describe the mass spectra of gauginos and sfermions of the first two generations in the latter. Generalizations of these models will also be mentioned occasionally.

We largely ignore flavor mixing in this chapter. Squark flavor mixing is often immaterial in so far as collider physics is concerned. This is since it is experimentally hard to identify the flavor of a first or second generation quark/antiquark produced in squark decay; these flavors need to be summed over. The effects of slepton flavor mixing can be minimized by simply adding final states containing electrons and muons, though this may cause a loss of valuable information. Specific effects of slepton flavor mixing may be looked for in slepton oscillations [9.16]. Our neglect of flavor mixing amounts to putting all the $\mathbf{U}^{\tilde{f}}$ and $\mathbf{W}^{\tilde{f}}$ matrices of Ch.9 to unity. We also neglect $L\text{-}R$ mixing for the first two generations of sfermions throughout our collider search discussions. Measuring additional CP violating phases, which occur in supersymmetric models, at colliders is difficult but not impossible [15.2]. Without going into the details of such measurements, we generally take these phases to be zero in our discussions.

Definitive signatures are needed to discriminate between sparticle production reactions and Standard Model backgrounds. Let us elaborate on an important ingredient of this signature present in almost all such processes. This is the famous **missing transverse energy** $\not{E}_T$ which is a characteristic diagnostic for models with conserved R-parity, e.g. for the general MSSM as defined in Ch. 9. As originally mentioned in §4.5, R_p conservation implies the stability of the LSP in which case cosmological arguments (cf. §16.3) indicate that it must have no electromagnetic charge. The conservation of R-parity also means that sparticles can only be produced in even numbers in any reaction, the minimal configuration being a pair of sparticles which are sometimes (but not mostly) identical. The production of two sparticles will be possible in a collider process only when the invariant energy in the production channel exceeds the summed masses of the concerned sparticles. The existing lower bounds on sparticle masses make the direct production of two or more such pairs quite unlikely in current or forthcoming colliders. Therefore, if R_p is conserved, sparticle production processes can be taken to generally involve, among other things, exactly two sparticles, each decaying in a *cascade which always ends with an* LSP. The LSP will essentially behave like a heavy neutrino since it can interact with ordinary matter only by exchanging other (heavy) sparticles. Therefore, like a neutrino, it escapes the detectors used at any collider[1]. The two LSPs in the final state configuration are thus expected to carry away a

---

[1]Special experiments have been designed to detect LSP's that are expected to be ubiquitously present

substantial amount of energy which is called the 'missing energy'. The latter is the difference between the total initial energy of collision and the sum of the energies of the particles detected by the experiment. This missing energy plus leptons (photons) and/or jets in the cascades from the decays of produced sparticles will characterize a supersymmetry event.

At hadron colliders, the total energy of the parton subsystem producing the sparticle pair in a 'hard' subprocess is not known exactly. On the other hand, the net momentum (and hence energy) in the transverse direction is clearly zero for the initial system. Hence the presence of two neutral and stable (or quasistable) but undetectable objects will be signalled by an imbalance in the momentum (energy) in the transverse direction for the final state. This is the famous $\not{E}_T$ or 'missing transverse energy' signal mentioned above. At an e^+e^- collider experiment with full calorimetric coverage, one often does not have to make the restriction to the transverse direction and can use $\not{E}$ or missing total energy in sparticle searches [15.3]. However, some SM processes, e.g. those mediated by two photons in the t channel with the collinear $e^\pm$ vanishing into the beam pipe, can give rise to events with a large amount of missing longitudinal energy and thus sometimes provide important backgrounds [15.3]. In such a situation, $\not{E}_T$ again has to be employed in looking for sparticles.

Of course, neutrinos contribute to $\not{E}_T$ too. A significant amount of $\not{E}_T$ can be generated in mundane SM processes like the production and subsequent decay into (anti)neutrinos of W/Z bosons or heavy quarks. In addition to this, incomplete solid angle coverage as well as measuring errors, such as the finite energy resolution of the detector and the mismeasurement of jet energies, can also give rise[2] to 'fake' $\not{E}_T$. Nevertheless, the amount of $\not{E}_T$ generated by the production of sparticles which then decay in an R_p conserving supersymmetric scenario is usually *significantly greater* than that produced by these SM processes. Therefore, a suitable lower cut on the $\not{E}_T$ in any candidate event will usually increase the signal to background ratio significantly. The magnitude of this $\not{E}_T$ cut will depend on the beam CM energy $\sqrt{s}$ under consideration. While 40 GeV may be good for analyses of $\bar{p}p$ annihilation at the TEVATRON [15.5] with $\sqrt{s} \simeq 2$ TeV, a cut more like 150 GeV may be reasonable [15.6, 15.7] for pp collision at the LHC operating at $\sqrt{s} = 14$ TeV. In addition to such a large $\not{E}_T$, there would be additional model dependent signatures that vary for different processes. Examples are leptons/jets for mSUGRA, isolated energetic photons for GMSB, a heavily ionizing track with a displaced vertex and/or a soft $\pi^\pm$ with a characteristic impact parameter distribution for AMSB etc.

We shall also discuss signals for $\not{R}_p$ supersymmetry. If R_p is violated, the unstable LSP need not be neutral. It is nonetheless common in phenomenological studies of models with broken R_p to take the lightest neutralino $\tilde{\chi}_1^0$ as the LSP, as is the case at least in mSUGRA type of models. We shall focus on the production of sparticle pairs via R-parity conserving (RPC) gauge vertices first; that of a single sparticle – though possible through an $\not{R}_p$ amplitude – will generally have a low rate on account of the small strength of the

---

as Cold Dark Matter in the Universe, by looking for the very rare interactions between LSPs and ordinary matter, cf. §16.3.

[2]In fact, the first ever claimed 'signal' for supersymmetry [15.4] turned out to be a damp squib owing to an inaccurate estimate of the SM background from the decays $W \to \tau\bar{\nu}_\tau, \bar{\tau}\nu_\tau$ as well as the effects of cracks in the detector.

relevant couplings[3], cf. §14.4. However, sparticle decay chains will now be different from those in RPC scenarios. Specifically, the final state (containing the decay products of the LSP pair) will have an unusual flavor structure and quite distinct multilepton and/or multijet configurations. At a first glance the viability of the $\not{E}_T$ signature could be in doubt in this case since this signature seems to depend crucially on the stability of the charge neutral LSP, caused by R-parity conservation. However, even with unstable LSPs in a $\not{R}_p$ scenario, the final state will contain energetic neutrinos. They will come not only from leptonic decays of the LSPs but also from their hadronic decays leading to heavy quarks[4] that then decay semileptonically. These will be in addition to other energetic neutrinos produced earlier in sparticle decay cascades. All such neutrinos will cause an increase in the total $\not{E}_T$. The upshot of this discussion is that $\not{E}_T$ will remain the basic signature of supersymmetry even with R_p violation, though now there will be additional leptons and/or jets in the final state.

Let us comment on the types of colliders relevant to sparticle search, namely $p\bar{p}$ or pp, e^+e^- and ep machines. The most stringent current bounds on the masses of sparticles and Higgs bosons come from LEP and the TEVATRON, while the world's only ep collider HERA has produced nontrivial constraints only for the $\not{R}_p$ scenario. Hadron colliders generally have the highest energy reach and can produce large numbers of strongly interacting (s)particles. Yet, the necessary integration over parton distribution functions introduces uncertainties, both in the prediction of a cross section and in the kinematics of an event. Moreover, a hard scattering event with a large momentum transfer is always on top of an 'underlying event' with (usually) soft hadrons from beam remnants as well as QCD radiation, generally clubbed under the letter X. The diagnostic cuts required to isolate a signal at a hadron machine are often more severe, on account of the very large QCD background, than at a lepton collider. During the next decade, hadron colliders will occupy center stage: TEV-II ($p\bar{p}$ collisions at $\sqrt{s} = 2$ TeV) has started and the LHC (pp collisions at $\sqrt{s} = 14$ TeV) will commence operations in a few years. The production cross sections are smaller at e^+e^- colliders, but any weakly interacting (s)particle with a mass less than half the beam energy will be pair-produced at a comparable rate. The final state here is simpler than at hadron colliders because of the paucity of soft hadron stragglers which are present copiously in hadronic collisions. Moreover, one can scan across new thresholds in e^+e^- collision events and extract chirality information by using longitudinally polarized beams. The measurement of masses and couplings, crucial to a test of the supersymmetric nature of any observed new interactions, is hence relatively straightforward at e^+e^- colliders. LEP has been decommissioned, but various e^+e^- collider proposals, such as JLC, NLC and TESLA, are being discussed seriously [15.8] and there are plans for a Global Linear Collider. The options [15.9] of studying γe and $\gamma\gamma$ collisions are also available with a linear e^+e^- collider with the photon(s) produced by the Compton back-scattering of laser light off the incoming e beam(s). Finally,

---

[3]A possible exception is resonant sparticle production via $\not{R}_p$ couplings.

[4]Heavy quark production is likely to be more common in supersymmetry models than in the SM. In the latter case, heavy quarks are mostly produced from gluon fusion, the cross section for which falls sharply with the quark mass. In contrast, all (s)quark flavors are probably produced more or less democratically. In fact, third generation squarks could even be produced preferentially, since they might well be lighter than those in the first and second generations. In the absence of flavor mixing, third generation squarks will decay into third generation quarks. For $\not{R}_p$ processes, the $\not{R}_p$ couplings involving t and b quarks are the least constrained, thus allowing their possible production in the final states.

a muon collider [15.10] is a new and interesting possibility – especially with respect to the supersymmetric Higgs sector.

We first present in §15.2 a detailed consideration of the possible production and detection of charginos and neutralinos at different colliders. Our focus is specifically on e^+e^- pair production of the lightest charginos and the light neutralinos of the MSSM and their subsequent decays. These are treated as prototypes for all supersymmetric processes to be directly studied in accelerators. We then extend these discussions to their hadroproduction. §15.3 contains a similar but less detailed discussion of sleptons. Squarks and gluinos are treated in like fashion in §15.4, while §15.5 is devoted to a treatment of the signals expected at colliders from the production of supersymmetric Higgs bosons. The final §15.6 contains a discussion of collider processes with R_p violation and the characteristic signals expected from them.

15.2 Signals of charginos and neutralinos

As enumerated in §9.2, the electroweak charginos $\tilde{\chi}^{\pm}_{1,2}$ and neutralinos $\tilde{\chi}^0_{1,2,3,4}$ obey the mass hierarchy

$$M_{\tilde{\chi}^{\pm}_1} < M_{\tilde{\chi}^{\pm}_2} \,, \; M_{\tilde{\chi}^0_1} < M_{\tilde{\chi}^0_2} < M_{\tilde{\chi}^0_3} < M_{\tilde{\chi}^0_4} \,,$$

with $\tilde{\chi}^0_1$ often being the LSP in the visible sector. In most scenarios $\tilde{\chi}^{\pm}_2$ and $\tilde{\chi}^0_{3,4}$ are much heavier than $\tilde{\chi}^{\pm}_1$, $\tilde{\chi}^0_{1,2}$ and we concentrate on the latter lighter states. While their masses and coupling strengths depend on the MSSM parameters $M_{1,2}, \mu$ and $\tan\beta$, the decay branching fractions for those that are unstable involve sfermion masses and mixing parameters too. A study of their direct production and decays can yield information on a large subset of the entire gamut of new parameters in the MSSM.

Pair production

Chargino and neutralino pair production are typical of sparticle pair production in general and deserve to be discussed in some detail. The first reactions [15.11] of interest to us are $e^+e^- \to \tilde{\chi}^+_m\tilde{\chi}^-_k/\tilde{\chi}^0_l\tilde{\chi}^0_n$, the subscripts being the same as in §9.2. The relevant lowest order diagrams are shown in Figs. 15.1 and 15.2 and one needs to apply the Feynman rules given in Figs. 9.4, 9.13 and 9.15. In our convention[5] e^-, $\tilde{\chi}^+$ are particles and e^+, $\tilde{\chi}^-$ antiparticles,

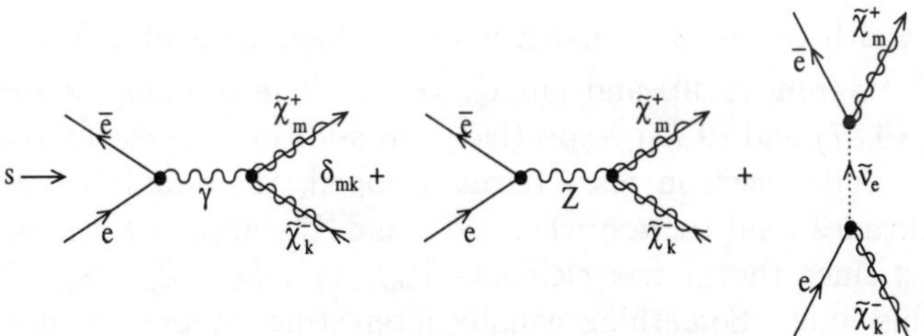

Fig.15.1. Tree diagrams for the process $e^+e^- \to \tilde{\chi}^+_m\tilde{\chi}^-_k$

---

[5] A reverse choice for charginos would have put $\bar{u}(\tilde{\chi}_k)$ on the left and $v(\tilde{\chi}_m)$ on the right of $\gamma^\mu P_\beta$ in (15.1) instead of what we have. But the identity $\bar{u}(\tilde{\chi}_k)\gamma_\mu P_{L,R} v(\tilde{\chi}_m) = \bar{u}(\tilde{\chi}_m)\gamma_\mu P_{R,L} v(\tilde{\chi}_k)$ shows that the two conventions yield the same result.

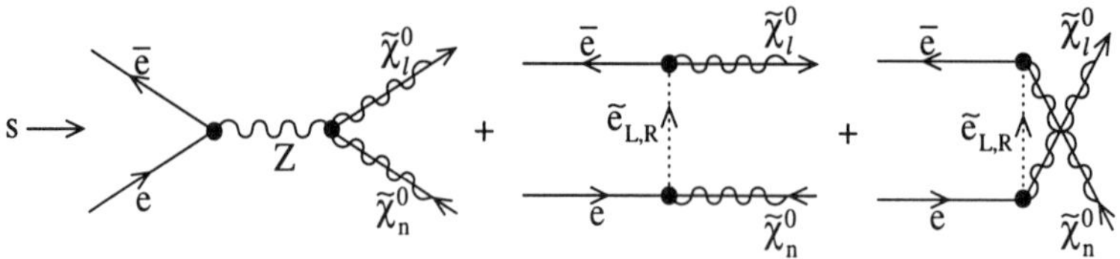

Fig.15.2. Tree diagrams for the process $e^+e^- \to \tilde{\chi}^0_l \tilde{\chi}^0_n$

while[6] $|\tilde{\chi}^0\rangle = |\bar{\tilde{\chi}}^0\rangle$. Clashing arrows in the t channel or u channel exchange diagrams signify the presence of a $\underset{\sim}{C}$ or $\underset{\sim}{C}^{-1}$ matrix in the relevant vertex factor, cf. Fig.9.13. The neglect of generation mixing in the slepton sector allows one to equate $U^{\tilde{\nu},e}_{km}$ of §9.4 to δ_{km}. Also, with the electron mass neglected, the $\tilde{e}_L$-$\tilde{e}_R$ mixing vanishes whereas the sneutrino couples only to the left handed electron and the right handed positron. The amplitude for chargino (neutralino) pair production involves the additional parameter(s) $m_{\tilde{\nu}_e}$ ($m_{\tilde{e}_{L,R}}$). The total amplitude for either process can be written after appropriate Fierz transformation(s) as

$$\mathcal{M}^{pq}_{\alpha\beta} = \frac{ie^2}{s} Q^{pq}_{\alpha\beta}\, \bar{v}(e)\gamma_\mu P_\alpha u(e)\, \bar{u}(\tilde{\chi}_p)\gamma^\mu P_\beta v(\tilde{\chi}_q)\ . \tag{15.1}$$

Here the subscripts p, q are generalized from m, k for charginos and n, l for neutralinos, while α and β refer to the chirality L/R of the electron and the neutralino or positive chargino currents respectively.

Detailed expressions for the $Q^{pq}_{\alpha\beta}$ of (15.1) in the two processes are given in Table 15.1. The various Δ's, used in the table, are given by

$$\Delta_Z = s(s - M_Z^2 + iM_Z\Gamma_Z)^{-1}\ , \tag{15.2a}$$

$$\Delta_t^{\tilde{\nu}_e} = s(t - m_{\tilde{\nu}_e}^2)^{-1}\ , \tag{15.2b}$$

$$\Delta_t^{\tilde{e}_{L,R}} = s(t - m_{\tilde{e}_{L,R}}^2)^{-1}\ , \tag{15.2c}$$

$$\Delta_u^{\tilde{e}_{L,R}} = s(u - m_{\tilde{e}_{L,R}}^2)^{-1}\ , \tag{15.2d}$$

where Γ_Z is the total width of the Z. Furrhermore, we have used $g^e_{L,R}$ from Fig. 8.1, $\mathcal{V}$ from (9.8) plus the $O^{L,R}$'s from (9.40) and put $Q_e = -1$. The coupling factors $N^{L,R}_{ln}$ and $G^{\tilde{e}L,R}_l$ are as defined in (9.37) and (9.73) respectively. In addition, $\cos\theta_W$ ($\sin\theta_W$) has been shortened to c_W (s_W). With electron mass terms dropped, the t and (where present) u channel exchange amplitudes tend to zero when $\tilde{\chi}^\pm_{m,k}$ and $\tilde{\chi}^0_{l,n}$ approach the limit of being pure higgsinos. This is since the matrix elements $\mathcal{V}_{m1}$, $\mathcal{V}_{k1}$, Z_{n1}, Z_{n2}, Z_{l1}, Z_{l2} of (9.18) and (9.31) vanish in this limit. Something equally interesting happens in neutralino pair production when $\tilde{\chi}^0_{l,n}$ tend to become pure gauginos. In this limit the s channel exchange

---

[6]A $\langle\tilde{\chi}^0|$ in the final state can be contracted either with the $\bar{\tilde{\chi}}^0_l$ field leading to a $\bar{u}$-spinor or with the $\tilde{\chi}^0_n$ field yielding a v-spinor. The relative signs between the different diagrams of Fig. 15.2 in the neutralino case are negative between the t and u channel exchanges as well as between the t and s channel ones. Some explicit computations appear in Appendix D of Haber and Kane [9.10].

amplitude vanishes since Z_{n1}, Z_{n2}, Z_{l1}, Z_{l2} of (9.31) all tend to zero. Hence any interference between s channel and t plus u channel exchange amplitudes in the neutralino case can only arise from a nonzero gaugino-higgsino mixing in that sector.

Chargino pair production : $p = m, q = k$	Neutralino pair production : $p = l, q = n$
$Q_{LL}^{mk} = Q_e \delta_{mk} - \dfrac{\Delta_Z}{c_W^2 s_W^2} g_L^e O_{mk}^L - \dfrac{1}{2 s_W^2} \Delta_t^{\tilde{\nu}_e} \mathcal{V}_{m1} \mathcal{V}_{k1}^\star$	$Q_{LL}^{ln} = -\dfrac{\Delta_Z}{c_W^2 s_W^2} g_L^e N_{ln}^L + \Delta_u^{\tilde{e}_L} e^{-2} G_l^{e_L} (G_n^{e_L})^\star$
$Q_{LR}^{mk} = Q_e \delta_{mk} - \dfrac{\Delta_Z}{c_W^2 s_W^2} g_L^e O_{mk}^R$	$Q_{LR}^{ln} = -\dfrac{\Delta_Z}{c_W^2 s_W^2} g_L^e N_{ln}^R + \Delta_t^{\tilde{e}_L} e^{-2} G_l^{e_L} (G_n^{e_L})^\star$
$Q_{RL}^{mk} = Q_e \delta_{mk} + \dfrac{\Delta_Z}{c_W^2 s_W^2} g_R^e O_{mk}^L$	$Q_{RL}^{ln} = \dfrac{\Delta_Z}{c_W^2 s_W^2} g_R^e N_{ln}^L + \Delta_t^{\tilde{e}_R} e^{-2} G_l^{e_R} (G_n^{e_R})^\star$
$Q_{RR}^{mk} = Q_e \delta_{mk} + \dfrac{\Delta_Z}{c_W^2 s_W^2} g_R^e O_{mk}^R$	$Q_{RR}^{ln} = \dfrac{\Delta_Z}{c_W^2 s_W^2} g_R^e N_{ln}^R + \Delta_u^{\tilde{e}_R} e^{-2} G_l^{e_R} (G_n^{e_R})^\star$

$$Q_1^{pq} = \frac{1}{4} \left(|Q_{RR}^{pq}|^2 + |Q_{LL}^{pq}|^2 + |Q_{RL}^{pq}|^2 + |Q_{LR}^{pq}|^2 \right)$$

$$Q_2^{pq} = \frac{1}{2} \Re e \left(Q_{RR}^{pq} Q_{RL}^{pq\star} + Q_{LL}^{pq} Q_{LR}^{pq\star} \right)$$

$$Q_3^{pq} = \frac{1}{4} \left(|Q_{RR}^{pq}|^2 + |Q_{LL}^{pq}|^2 - |Q_{RL}^{pq}|^2 - |Q_{LR}^{pq}|^2 \right)$$

$$Q_1'^{pq} = \frac{1}{4} \left(|Q_{RR}^{pq}|^2 + |Q_{RL}^{pq}|^2 - |Q_{LR}^{pq}|^2 - |Q_{LL}^{pq}|^2 \right)$$

$$Q_2'^{pq} = \frac{1}{2} \Re e \left(Q_{RR}^{pq} Q_{RL}^{pq\star} - Q_{LL}^{pq} Q_{LR}^{pq\star} \right)$$

$$Q_3'^{pq} = \frac{1}{4} \left(|Q_{RR}^{pq}|^2 + |Q_{LR}^{pq}|^2 - |Q_{RL}^{pq}|^2 - |Q_{LL}^{pq}|^2 \right)$$

Table 15.1. Expressions for the quantities Q appearing in (15.1) and (15.4) and Q' appearing in (15.6). All other symbols are explained in the text.

The kinematic variables s, t, u have the usual definitions, namely $s \equiv (p_{e^-} + p_{e^+})^2$, $t \equiv (p_{e^+} - p_{m,l})^2$ and $u \equiv (p_{e^+} - p_{k,n})^2$, with p standing for the four-momentum of the particle (sparticle) whose label it carries. Let us also introduce dimensionless reduced squared masses $\mu_{p,q}^2 = M_{p,q}^2 / s$ and a kinematic function of them $\lambda(1, \mu_p^2, \mu_q^2)$, where

$$\lambda(a^2, b^2, c^2) \equiv [a^2 - (b - c)^2][a^2 - (b + c)^2]. \tag{15.3}$$

The differential cross section for both chargino and neutralino pair production with unpolarized e^+, e^- beams can now be given in terms of the CM scattering angle θ between the incoming e^- and the outgoing $\tilde{\chi}_p$ as[7]

---

[7]For chargino pair production, Choi, Djouadi, Song and Zerwas [15.11] chose $\tilde{\chi}_1^-$ as a particle and $\tilde{\chi}_1^+$ as its antiparticle – the opposite of our convention. As a result, their combinations $\tilde{Q}_{\alpha\beta}^{mk}$ are related to ours by $\tilde{Q}_{LL}^{mk} = -Q_{LR}^{mk}$, $\tilde{Q}_{LR}^{mk} = -Q_{LL}^{mk}$, $\tilde{Q}_{RL}^{mk} = -Q_{RR}^{mk}$, $\tilde{Q}_{RR}^{mk} = -Q_{RL}^{mk}$, i.e. $\tilde{Q}_1^{mk} = Q_1^{mk}$, $\tilde{Q}_2^{mk} = Q_2^{mk}$, but $\tilde{Q}_3^{mk} = -Q_3^{mk}$. The differential cross sections match since their θ is our $\pi - \theta$.

$$\frac{d\sigma_{pq}}{d\cos\theta} = \frac{\pi\alpha_{EM}^2}{2s}\lambda^{1/2}(1,\mu_p^2,\mu_q^2)\Big\{\big[1 - (\mu_p - \mu_q)^2 + \lambda(1,\mu_p^2,\mu_q^2)\cos^2\theta\big]Q_1^{pq}$$

$$+4\mu_p\mu_q Q_2^{pq} + 2\lambda^{1/2}(1,\mu_p^2,\mu_q^2)Q_3^{pq}\cos\theta\Big\}. \qquad (15.4)$$

The quantities $Q_{1,2,3}^{pq}$ above are as listed in Table 15.1. The dependence of the differential cross section as well as its integrated version σ_{pq} on all couplings and slepton masses comes only through these quantities. The third RHS term in (15.4), giving rise to a forward-backward ($\cos\theta \leftrightarrow -\cos\theta$) asymmetry, is especially interesting. For the pair production of two identical neutralinos ($l = n$), $t \leftrightarrow u$ when $\cos\theta \leftrightarrow -\cos\theta$, i.e. $Q_3^{ll} \leftrightarrow -Q_3^{ll}$, so that the differential cross section is forward-backward symmetric. This is a consequence of the antisymmetrization in the final state containing two identical fermions which incidentially implies an additional factor of $1/2$ in the total cross section σ_{ll}^{00}. If $s \gtrsim m_{\tilde{e},\tilde{\nu}}^2$, the angular distribution (15.4) peaks in the forward (forward and backward) direction(s) on account of $\tilde{\nu}_e$ ($\tilde{e}_{L,R}$) exchange in the t (t,u) channel(s) for chargino (neutralino) pair production. These peaks will become visible if $\tilde{\chi}_{p,q}$ are gauginolike ($|\mu| \gg |M_{1,2}|$) or also in the mixed ($|\mu| \sim |M_{1,2}|$) regime. However, if $\tilde{\chi}_{p,q}$ are higgsinolike ($|\mu| \ll |M_{1,2}|$), the strengths of their couplings to an electron and a sneutrino (selectron) become insignificant so that these peaks disappear. A quantitative study of this angular distribution will hence provide a measure of the gaugino/higgsino content of $\tilde{\chi}_{p,q}$. Note finally that higher order corrections to chargino pair production are known [15.12] and will need to be included in the analysis when highly accurate measurements are made at a linear collider.

The interactions of gauginos/higgsinos depend on the chiralities of the fermions that they couple to. Hence beam polarization[8] becomes a sensitive probe of the gaugino/higgsino contents of $\tilde{\chi}_p$, $\tilde{\chi}_q$. Let us illustrate this by an example. The cross section for producing a pure wino pair with an e_R^- beam tends to zero in proportion to M_Z^4/s^3 as $s \to \infty$ since e_R^- is an $SU(2)_L$ singlet. In contrast, that for producing a pure higgsino pair scales like $1/s$ for both e_L^- and e_R^- beams. However, only hypercharge interactions survive at high energies in the last case. Specifically, for a pure e_R^- beam, $\tilde{\nu}_e$ exchange does not contribute to the $\tilde{\chi}_m^+\tilde{\chi}_k^-$ production amplitude which then becomes independent of $m_{\tilde{\nu}_e}$ and gets determined by chargino parameters alone. For a polarized e^- beam, one can define the total and differential left-right asymmetries respectively by

$$A_{LR}^{pq} \equiv \frac{\sigma_{pq}^L - \sigma_{pq}^R}{\sigma_{pq}^L + \sigma_{pq}^R}, \qquad (15.5a)$$

$$A_{LR}^{pq}(\cos\theta) \equiv \frac{d\sigma_{pq}^L/d\cos\theta - d\sigma_{pq}^R/d\cos\theta}{d\sigma_{pq}^L/d\cos\theta + d\sigma_{pq}^R/d\cos\theta}, \qquad (15.5b)$$

where $d\sigma_{pq}^{L,R} = d\sigma_{pq}(e^+e_{L,R}^- \to \tilde{\chi}_p\tilde{\chi}_q)$ etc. The LHS of (15.5b) can be explicitly computed to

---

[8]A similar statement holds vis-à-vis the polarizations of the final state sparticles, but we do not go into that aspect.

be

$$A_{LR}^{pq}(\cos\theta) = -\frac{[1 - (\mu_p^2 - \mu_q^2)^2 + \lambda\cos^2\theta]Q_1'^{pq} + 4\mu_p\mu_q Q_2'^{pq} + 2\lambda^{1/2}Q_3'^{pq}\cos\theta}{[1 - (\mu_p^2 - \mu_q^2)^2 + \lambda\cos^2\theta]Q_1^{pq} + 4\mu_p\mu_q Q_2^{pq} + 2\lambda^{1/2}Q_3^{pq}\cos\theta}, \qquad (15.6)$$

λ being $\lambda(1, \mu_p^2, \mu_q^2)$ of (15.3) and the quantities[9] $Q_{1,2,3}'^{pq}$, listed in the last three rows of Table 15.1, being different linear combinations of $|Q_{\alpha\beta}^{pq}|^2$ as compared with $Q_{1,2,3}^{pq}$.

We shall now make some specific remarks on the total cross section σ_{pq} obtained by integrating (15.4) over the entire range of the scattering angle θ. Consider the chargino version σ_{mk}^{+-} first. If the t channel $\tilde\nu_e$-exchange contribution is present in strength in this process, σ_{mk}^{+-} becomes dependent on the sneutrino mass $m_{\tilde\nu_e}$ and so does $d\sigma_{mk}^{+-}/d\cos\theta$. Indeed, this dependence is rather strong since the t channel exchange contribution interferes destructively with those from s channel exchanges of γ and Z. Fig.15.3 shows σ_{11}^{+-} at $\sqrt{s} = 500$ GeV as a

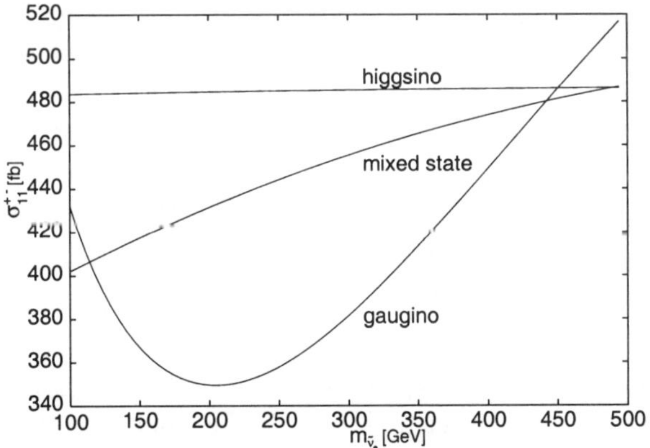

Fig.15.3. The total cross section σ_{11}^{+-} in femtobarns for chargino pair production from e^+e^- annihilation at $\sqrt{s} = 500$ GeV plotted against the sneutrino mass for higgsinolike, typically mixed and gauginolike charginos.

function of $m_{\tilde\nu_e}$ for three different cases with $M_{\tilde\chi_1^\pm}$ kept the same[10] at approximately 120 GeV. First, note that the highest value of the cross section exceeds half a picobarn. Next, the curve for the case when $\tilde\chi_1^\pm$ is a pure wino clearly displays a big dip due to the said destructive interference. For a pure higgsino $\tilde\chi_1^\pm$, the near absence of sneutrino exchange is shown by the insensitivity of the cross section to the value of $m_{\tilde\nu_e}$. In a mixed situation, the destructive interference tends to pull down the cross section, but not enough to cause a minimum. Moreover, as $m_{\tilde\nu_e}$ increases, σ_{mk}^{+-} for $m \neq k$ (not shown in Fig.15.3) falls off more strongly than σ_{mm}^{+-}, reflecting the domination of the former by this t channel contribution. Near threshold, the chargino pair production cross section σ_{mk}^{+-} become proportional to $\beta \equiv \lambda^{1/2}(1, \mu_m^2, \mu_k^2)$; for the equal mass ($\mu_m = \mu_k$) case, the latter reduces to the three-velocity of the $\tilde\chi$ in the CM frame. However, in the neutralino case, the threshold behavior

---

[9]Once again, as compared with the primed quantitites of Choi, Djoudi, Song and Zerwas [15.11], now marked with tilde, we have $\tilde Q_1'^{mk} = Q_1'^{mk}$, $\tilde Q_2'^{mk} = Q_2'^{mk}$, but $\tilde Q_3'^{mk} = -Q_3'^{mk}$.

[10]This cross section is insensitive to variations in $M_{\tilde\chi_1^\pm}$ except very near threshold.

of σ_{ln}^{00} is controlled by the signature factor[11] $\cos[2\arg(Z_{lk}Z_{nk})]$. The threshold dependence is S-wave (P-wave) if the signature factor is -1 ($+1$).

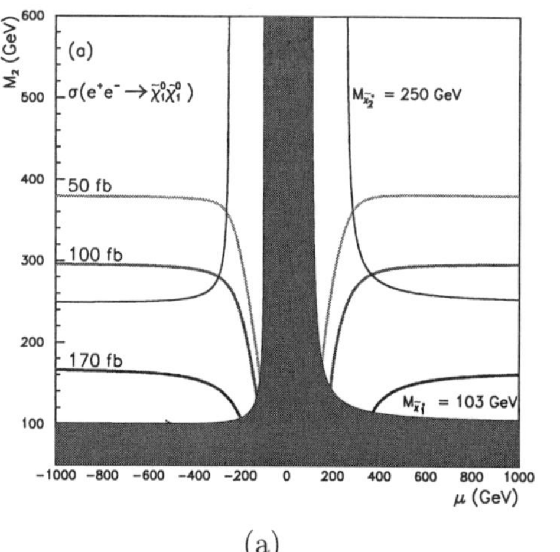

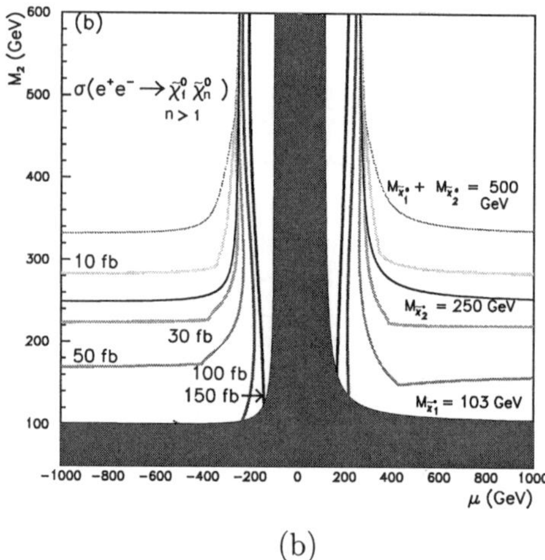

(a)   (b)

Fig. 15.4. Plots of (a) σ_{11}^{00} and (b) σ_{1n}^{00} ($n > 1$) in the $M_2-\mu$ plane for $\sqrt{s} = 500$ GeV, $\tan\beta = 5$ and $m_{\tilde{e}_{L,R}} = 200$ GeV at the scale of grand unification M_U.

Fig.15.4a (15.4b) shows illustrative contour plots in the $M_2-\mu$ plane for the cross section σ_{11}^{00} ($\sigma_{1n}^{00}, n > 1$) at $\sqrt{s} = 500$ GeV within a $\tilde{\text{C}}$MSSM framework, cf. §12.5, assuming a universal sfermion mass of 200 GeV for $\tilde{e}_L$ and $\tilde{e}_R$ at the unification scale M_U. The central shaded domains are the exclusion regions from LEP 2. The boundaries in Fig.15.4, indicated by the labels $M_{\tilde{\chi}_2^0} = 250$ GeV and $M_{\tilde{\chi}_1^0} + M_{\tilde{\chi}_2^0} = 500$ GeV, correspond to the kinematic reach for a given $\sqrt{s} = 500$ GeV LC. A comparison of Figs. 15.3 and 15.4 forcefully brings home the smallness of the cross section for the pair production of gauginolike neutralinos relative to that for similar charginos. This is essentially due to the vanishing of the tree level $Z\tilde{\chi}^0\tilde{\chi}^{0\prime}$ coupling for gauginolike $\tilde{\chi}_0, \tilde{\chi}_0'$. In this situation, the pertinent σ^{00} scales like $m_{\tilde{e}}^{-4}$ when $m_{\tilde{e}} \gg \sqrt{s}/2$, the selectron mass value getting determined by the choice $m_{\tilde{e}_{L,R}} = 200$ GeV (at the high scale) in $\tilde{\text{C}}$MSSM. On the other hand, for higgsinolike lighter $\tilde{\chi}$ states, $\tilde{\chi}_1^+\tilde{\chi}_1^-$ and $\tilde{\chi}_1^0\tilde{\chi}_2^0$ production in e^+e^- collisions have comparable cross sections. However, strong cancellations between the two contributions to the coupling (9.37a) make the cross sections for $\tilde{\chi}_1^0\tilde{\chi}_1^0$ and $\tilde{\chi}_2^0\tilde{\chi}_2^0$ production very small for higgsinolike light neutralinos.

We now turn to the direct pair production[12] of charginos and neutralinos at hadron colliders. These are inclusive processes with the underlying event containing soft unspecified particles. The rates for hadroproduction are calculated by convoluting the cross sections for the hard partonic [15.14] subprocesses with parton luminosities in a standard way [15.15].

---

[11]See also the discussion after (9.37).

[12]The associated production [15.13] of an electroweak $\tilde{\chi}$ together with a gluino or a squark, though present, is not sizable and we do not discuss it. However, standard supersymmetry event generators such as ISAJET, HERWIG and SPYTHIA do take it into account.

Suffice it to say that our production subprocess always comprises two partons (P_1 and P_2, say) going into two sparticles with a subprocess invariant CM energy squared $\hat{s}$ and invariant momentum transfers squared $\hat{t}$ and $\hat{u}$. The link between these subprocesses and the pair production processes in e^+e^- collision is established through the correspondence $s \leftrightarrow \hat{s}$, $t \leftrightarrow \hat{t}$ and $u \leftrightarrow \hat{u}$. The main subprocess reactions are $q_i\bar{q}_i \rightarrow \tilde{\chi}_m^+\tilde{\chi}_k^-$ or $\tilde{\chi}_l^0\tilde{\chi}_n^0$, $q_i\bar{q}_i' \rightarrow \tilde{\chi}_m^\pm\tilde{\chi}_n^0$, where it is sufficient to restrict the generation index i to $i = 1, 2$. The tree level diagrams describing $\tilde{\chi}_m^+\tilde{\chi}_k^-$ and $\tilde{\chi}_l^0\tilde{\chi}_n^0$ are exactly analogous to those in Figs.15.1 and 15.2 respectively and the amplitudes can be obtained from (15.1) and Table 15.1 with the replacements $Q_e \rightarrow Q_q$, $g_{L,R}^e \rightarrow g_{L,R}^q$. Furthermore, there is a color averaging factor of $1/3$ in the cross section. The characteristic features of the cross sections, except for the overall magnitudes, are similar to those in the corresponding e^+e^- production cases. But there is one additional point. If $|M_3| \simeq 3.5\,|M_2|$ (cf. 12.27d), the first and second generation squarks would be significantly heavier[13] than winolike charginos and neutralinos, suppressing the t and u channel graphs. On the other hand, since most events have $\hat{s}$ not much above threshold, the s channel exchange diagrams would dominate. With the Z coupling only to the higgsino components of the neutralinos, the s channel contribution for gauginolike neutralinos is drastically reduced. Thus the pair production cross section for gauginolike neutralinos is much smaller than that for similar charginos. In contrast, higgsinolike charginos and neutralinos would be pair-produced in a hadron collider with comparable rates.

The really new and distinct subprocess, specific to a hadron collider, is $q_i\bar{q}_i', q_i'\bar{q}_i \rightarrow \tilde{\chi}_l^0\tilde{\chi}_k^\pm$ ($i = 1, 2$). The corresponding tree diagrams are shown in Fig.15.5. Ignoring flavor mixing,

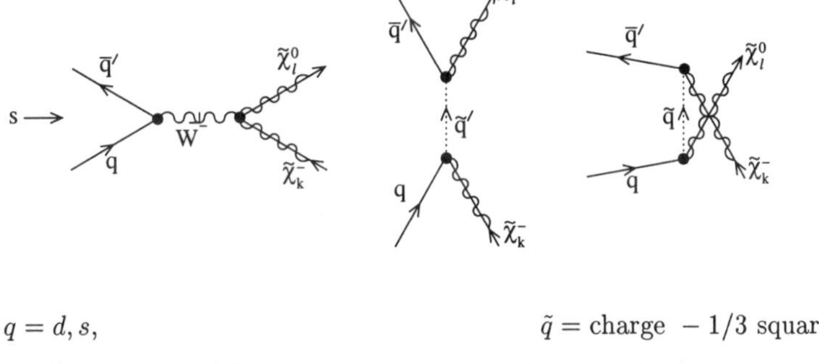

Legend

$q = d, s,$ $\qquad\qquad\qquad\qquad\qquad$ $\tilde{q} = $ charge $-1/3$ squark,

$\bar{q}' = \bar{u}, \bar{c},$ $\qquad\qquad\qquad\qquad$ $\tilde{q}' = $ charge $2/3$ squark,

$l = 1, 2, 3, 4,$ $\quad$ $k = 1, 2$

Fig.15.5. Tree diagrams for the subprocess of the production of a neutralino and a negative chargino by quark–antiquark annihilation

the partonic differential cross section $d\sigma_{lk}^{0-}$ $(d_i\bar{u}_i \rightarrow \tilde{\chi}_l^0\tilde{\chi}_k^-)/d\cos\theta^\star$ is again given [15.16] by (15.4) with the replacements $\theta \rightarrow \theta^\star$ (the scattering angle in the subprocess CM frame), $p\,(q) \rightarrow l\,(k)$ and $Q_{\alpha\beta}^{pq} \rightarrow Q_{\alpha\beta}^{lk}$, as listed in Table 15.2. Note that the Δ's for the W and the squarks are given,

---

[13]Since sleptons need not be much heavier than winolike charginos/neutralinos, t and u channel contributions to $\tilde{\chi}$ pair production at e^+e^- colliders may still be significant, cf. (11.45a-b).

in analogy with those in (15.2), by the expressions

$$\Delta_W = \hat{s}(\hat{s} - M_W^2 + i\Gamma_W M_W)^{-1} , \tag{15.7a}$$

$$\Delta_{\hat{t}}^{\tilde{u}_L} = \hat{s}(\hat{t} - m_{\tilde{u}_L}^2)^{-1} , \tag{15.7b}$$

$$\Delta_{\hat{u}}^{\tilde{d}_L} = \hat{s}(\hat{u} - m_{\tilde{d}_L}^2)^{-1} . \tag{15.7c}$$

$$
\begin{aligned}
Q_{LL}^{lk} &= \frac{\Delta_W}{\sqrt{2}s_W^2}C_{lk}^R + \frac{\Delta_{\hat{u}}^{\tilde{d}_{iL}}}{\sqrt{2}s_W^2}\mathcal{U}_{k1}\left(\frac{1}{2}Z_{l2}^\star - \frac{1}{6}Z_{l1}^\star \tan\theta_W\right) \\[2ex]
Q_{LR}^{lk} &= \frac{\Delta_W}{\sqrt{2}s_W^2}C_{lk}^L + \frac{\Delta_{\hat{t}}^{\tilde{u}_{iL}}}{\sqrt{2}s_W^2}(\mathcal{V}_{k1})^\star\left(\frac{1}{2}Z_{l2} + \frac{1}{6}Z_{l1} \tan\theta_W\right) \\[2ex]
Q_{RL}^{lk} &= 0 \\[2ex]
Q_{RR}^{lk} &= 0
\end{aligned}
$$

Table 15.2. $Q_{\alpha\beta}^{lk}$ expressions for $d_i \bar{u}_i \to \tilde{\chi}_l^0 \tilde{\chi}_k^-$.

In Table 15.2 $C_{lk}^{L,R}$ are as defined in (9.34) while $\mathcal{U}_{k1}, \mathcal{V}_{k1}$ and Z_{l1}, Z_{l2} are as per (9.18a,b) and (9.31a–d) respectively. As in chargino pair production, the $\hat{s}$ and $\hat{t}$ channel amplitudes interfere destructively. However, for models with high scale gaugino mass unification (12.25), squark exchange contributions are much smaller than W exchange ones, thereby weakening the effect of that interference. Let us also mention for completeness that the helicity amplitudes for $\tilde{\chi}$ hadroproduction, keeping complete spin correlations, are available in the literature [15.16].

Decay patterns

Consider first tree level two body decays of charginos and neutralinos into (s)particles of the observable sector. Whenever allowed kinematically, these will always dominate. The possible chargino decay modes[14] are:

$$\tilde{\chi}_k^+ \to \ell_i^+ \tilde{\nu}_j, \ k = 1, 2, \ i, j = 1, 2, 3 ; \tag{15.8a}$$

$$\tilde{\chi}_k^+ \to \nu_i \tilde{\ell}_s^+, \ \bar{d}_i \tilde{u}_s, \ u_i \tilde{d}_s, \ k = 1, 2, \ i = 1, 2, 3, \ s, = 1, \ldots, 6 ; \tag{15.8b}$$

$$\tilde{\chi}_k^+ \to W^+ \tilde{\chi}_l^0, \ k = 1, 2, \ l = 1, \ldots, 4 ; \tag{15.8c}$$

$$\tilde{\chi}_k^+ \to H^+ \tilde{\chi}_l^0, \ k = 1, 2, \ l = 1, \ldots, 4 ; \tag{15.8d}$$

$$\tilde{\chi}_2^+ \to \tilde{\chi}_1^+ Z ; \tag{15.8e}$$

$$\tilde{\chi}_2^+ \to \Phi \tilde{\chi}_1^+, \ \Phi = h, H, A . \tag{15.8f}$$

The analogous neutralino two body decay modes, without necessarily taking $\tilde{\chi}_1^0$ to be the stable LSP, are (the charge conjugate modes need always to be added):

$$\tilde{\chi}_l^0 \to \tilde{\nu}_i \bar{\nu}_j, \ l = 1, 2, 3, 4, \ i, j = 1, 2, 3 ; \tag{15.9a}$$

$$\tilde{\chi}_l^0 \to \ell_i^+ \tilde{\ell}_s^-, \ \bar{u}_i \tilde{u}_s, \ \bar{d}_i \tilde{d}_s, \ l = 1, 2, 3, 4, \ i = 1, 2, 3, \ s = 1, \ldots, 6 ; \tag{15.9b}$$

---

[14]Corresponding $\tilde{\chi}^-$ decays follow by taking charge conjugation.

$$\tilde{\chi}_l^0 \;\rightarrow\; W^+\tilde{\chi}_k^-, \; l = 1,2,3,4, \; k = 1,2 \; ; \qquad (15.9c)$$

$$\tilde{\chi}_l^0 \;\rightarrow\; H^+\tilde{\chi}_k^-, \; l = 1,2,3,4, \; k = 1,2 \; ; \qquad (15.9d)$$

$$\tilde{\chi}_l^0 \;\rightarrow\; Z\tilde{\chi}_n^0, \; l = 2,3,4, \; n = 1,\ldots,l-1 \; ; \qquad (15.9e)$$

$$\tilde{\chi}_l^0 \;\rightarrow\; \Phi\tilde{\chi}_n^0, \; l = 2,3,4, \; n = 1,\ldots,l-1, \; \Phi = h, H, A \; . \qquad (15.9f)$$

Given a generic interaction of the form $\bar{\psi}(a + b\gamma_5)\tilde{\chi}\phi$, where ϕ and ψ stand for any scalar and spin half fermion respectively, the partial width for the two body decay $\tilde{\chi} \rightarrow \psi\phi$ is given by

$$\Gamma(\tilde{\chi} \rightarrow \psi\phi) = \frac{M_{\tilde{\chi}}}{16\pi}\lambda^{1/2}(1, \mu_\psi^2, \mu_\phi^2)\left[(|a|^2 + |b|^2)(1 + \mu_\psi^2 - \mu_\phi^2) + 2\mu_\psi(|a|^2 - |b|^2)\right]. \qquad (15.10)$$

In (15.10) $\mu_{\phi,\psi} = m_{\phi,\psi}/M_{\tilde{\chi}}$, $m_{\phi,\psi}$ being the mass of ϕ, ψ. Moreover, $\lambda^{1/2}$ is the kinematic function of (15.3) with (ψ, ϕ) standing for a (matter fermion, sfermion) combination in (15.8a,b) as well as in (15.9a,b) and for a (chargino/neutralino, Higgs) combination in (15.8d,f) as well as in (15.9d,f). The coupling strengths a, b for the two body decay, are enumerated in Ch.9 and Appendix A for the matter fermion and matter sfermion cases, while the relevant Higgs boson couplings are listed in Fig. 10.6 in[15] Appendix B. Note that for decays into a quark-squark combination, one needs to multiply the RHS of (15.10) by the color factor $N_c = 3$. With the neglect of sfermionic generation mixing, neutralino and chargino decays can be taken as diagonal in generation space. This is since the CKM matrix for quark mixing is 'almost' diagonal and since final states involving first or second generation quarks are not easy to distinguish experimentally. Turning to a generic gauge boson V, its general coupling with a pair of $\tilde{\chi}$'s is $q_2 V^\mu \bar{\tilde{\chi}}_l \gamma^\mu(c + d\gamma_5)\tilde{\chi}_k$. The partial width for the decay $\tilde{\chi}_k \rightarrow V\tilde{\chi}_l$ is then given by

$$\Gamma(\tilde{\chi}_k \rightarrow V\tilde{\chi}_l) = \frac{g_2^2}{16\pi}\frac{M_V}{(\mu_{\tilde{\chi}_k})^3}\,\lambda^{1/2}(\mu_{\tilde{\chi}_k}^2, 1, \mu_{\tilde{\chi}_l}^2)$$

$$\cdot \left[(|c|^2 + |d|^2)\{\mu_{\tilde{\chi}_k}^2 + \mu_{\tilde{\chi}_l}^2 - 2 + (\mu_{\tilde{\chi}_k}^2 - \mu_{\tilde{\chi}_l}^2)^2\} - 6(|c|^2 - |d|^2)\mu_{\tilde{\chi}_k}\mu_{\tilde{\chi}_l}\right],$$

$$(15.11)$$

where $\mu_{\tilde{\chi}_{k,l}} = M_{\tilde{\chi}_{k,l}}/M_V$. Eq.(15.11) describes the decays (15.8c,e) as well as (15.9c,e). The relevant couplings are listed in Figs. 9.3 and 9.4 in Appendix A.

Only gauginolike charginos and neutralinos will decay into first or second generation matter (s)fermions with large partial widths. Dominantly $SU(2)_L$ gauginos will have full strength decays only into $SU(2)_L$ doublet (s)fermions. In contrast, both gauginos and higgsinos can decay into (s)tops and, for $\tan\beta \gg 1$, (s)taus and (s)bottoms. In fact, $\tilde{\chi}_1^+ \rightarrow \nu_\tau\tilde{\tau}_1^+$ can be the dominant $\tilde{\chi}_1^+$ decay mode even for small $\tan\beta$, since the analogous decays with $\tilde{e}_1 \simeq \tilde{e}_R$ and $\tilde{\mu}_1 \simeq \tilde{\mu}_R$ in the final state are suppressed by the small e, μ Yukawa couplings and correspondingly small L-R mixing angles. Such a '$\tilde{\tau}$ dominance' can occur in $\tilde{\chi}_2^0$ decays only if $\tilde{\tau}_1$ is significantly lighter than $\tilde{e}_R, \tilde{\mu}_R$, or if $\tilde{\chi}_2^0$ is an almost pure higgsino. For small $\tan\beta$, the decay (15.8c), if kinematically allowed, will dominate over decays into $\tilde{\tau}_1$. Moreover, both will be superceded by decays into $SU(2)_L$ doublet sleptons once those open up. Decays with Higgs bosons in the final state will have full gauge strength only if one of the $\tilde{\chi}$ involved is largely a higgsino while the other is mostly a gaugino. Note also that decays with massive gauge bosons in the final state can have large partial widths if the $\tilde{\chi}$

---

[15]The $c_{L,R}$ couplings, listed there, are related to a, b by $2a = c_L + c_R$ and $2b = c_L - c_R$. Recall that the couplings of the neutral Higgs bosons will either be purely scalar or purely pseudoscalar if CP is conserved in both the Higgs and $\tilde{\chi}$ sectors.

states involved have either sizable higgsino components or − in case at least one of the $\tilde{\chi}$ states is charged − large $SU(2)_L$ gaugino components. The last, but not the least feature to be highlighted, is the occurrence[16] of unstable sparticles and Higgs/gauge bosons in most of the decays (15.8, 15.9). These will undergo further decays that need to be taken into account.

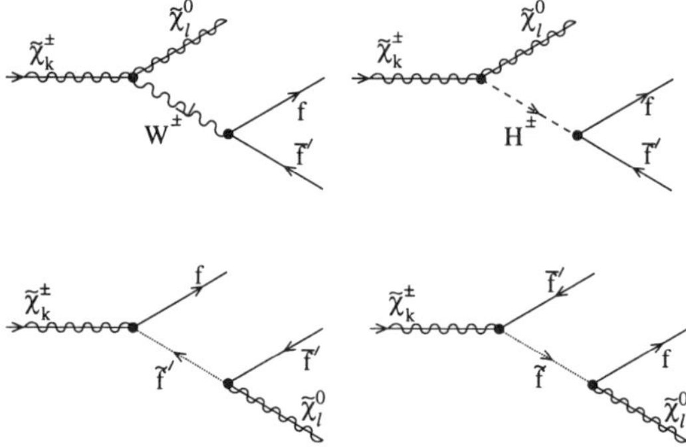

Fig.15.6. Tree diagrams for three body chargino decay

The kinematic inaccessibility of two body modes may force some $\tilde{\chi}$'s to decay into three body final states through the exchange of gauge or Higgs bosons or matter sfermions. The lowest order diagrams describing the decays

$$\tilde{\chi}_k^\pm \to \tilde{\chi}_l^0 f \bar{f}', \ k = 1, 2, \ l = 1, \dots, 4,$$
(15.12)

are drawn in Fig. 15.6, while Fig.15.7 shows the ones for the analogous neutralino decays:

$$\tilde{\chi}_l^0 \ \to \ f\bar{f}\tilde{\chi}_n^0, \ l = 2, 3, 4, \ n = 1, \dots, l - 1,$$
(15.13a)
$$\tilde{\chi}_l^0 \ \to \ f\bar{f}'\tilde{\chi}_k^+, \ l = 1, 2, 3, 4, \ k = 1, 2.$$
(15.13b)

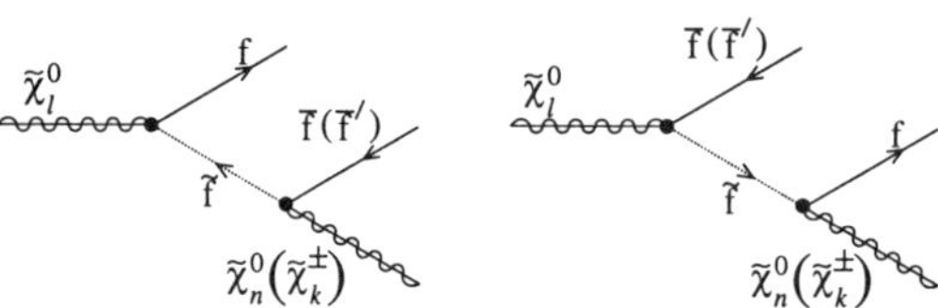

Fig.15.7. Tree diagrams for three body neutralino decay

[16]A notable exception could be the final states $\ell\bar{\tilde{\nu}}_\ell$, $\bar{\ell}\tilde{\nu}_\ell$ in case $\tilde{\nu}_\ell$ is the LSP; the latter possibility is, however, disfavored by cosmological arguments, cf. §16.3

In models incorporating high scale gaugino mass unification (12.25), the decay (15.13b) is subdominant on account of the small $\tilde{\chi}_2^0$-$\tilde{\chi}_1^\pm$ mass difference. Heavier neutralinos usually undergo two body decays. The Higgs exchange diagrams are significant only for third generation fermions in the final state. Furthermore, the $W^\pm$ couple to both the gaugino and the higgsino components of charginos, while the Z only couples to the higgsino components of neutralinos. Three body chargino decays are therefore dominated by the W exchange diagram, unless the difference between the chargino mass and the mass of some sfermion with significant $SU(2)_L$ doublet component is comparable to, or smaller than, M_W. The Z exchange diagram in Fig. 15.7 is often suppressed by small higgsino components of the neutralino in the initial and/or final state. Sfermion exchange can therefore play a prominent role in three body neutralino decays even if the sfermion-neutralino mass difference exceeds M_Z significantly. In particular, there can be strong destructive interference between sfermion and Z exchanges in certain channels. Three body neutralino decay branching ratios therefore typically vary much more rapidly over the allowed parameter space than do corresponding chargino branching fractions. Take the situation when the final state contains a τ or b and $\tan\beta$ is sizable. Now, because of the lightness of h and the frequent suppression of Z exchange, the Higgs exchange diagrams can make nontrivial contributions – especially in neutralino decays. These can interfere with sfermion exchange contributions, but not with those having W or Z exchange, provided fermion masses are neglected. Note that three body decays with a top quark in the final state are never very important, even if kinematically allowed. This is since two body decays involving W, Z or h bosons will then also be allowed and will dominate. Expressions for chargino and neutralino three body partial widths can be found in Refs. [15.17–15.19].

Complications arise when the mass difference $\Delta M_{\tilde{\chi}}$ between the decaying $\tilde{\chi}$ and the final $\tilde{\chi}_1^0$ is less than a few GeV. Such a situation occurs for $\tilde{\chi}_1^\pm$ decays if $\tilde{\chi}_1^0$ and $\tilde{\chi}_1^\pm$ are $SU(2)_L$ gauginolike, as in AMSB models (cf. §12.6), or for both $\tilde{\chi}_1^\pm$ and $\tilde{\chi}_2^0$ decays if $|\mu| \ll |M_{1,2}|$ in which case $\tilde{\chi}_{1,2}^0$ and $\tilde{\chi}_1^\pm$ are almost pure higgsinos. Accuracy now demands the inclusion [15.20] of loop corrections to the masses of $\tilde{\chi}_{1,2}^0$ and of $\tilde{\chi}_1^\pm$. Three body decays, with partial widths scaling like $(\Delta M)^5$, are suppressed. Moreover, hadronic final states now have small multiplicity even after hadronization and can thus no longer be described by the parton model. Instead, one has to treat final states with one or a few pions explicitly [15.21]. For example,

$$\tilde{\chi}_1^+ \to \tilde{\chi}_1^0 + \pi^+ \tag{15.14}$$

is the dominant chargino decay mode if $m_\pi < \Delta M \lesssim 1$ GeV. The strong phase space suppression of tree level decays also implies that the loop induced radiative decay [15.22]

$$\tilde{\chi}_2^0 \to \tilde{\chi}_1^0 \gamma \tag{15.15}$$

can have a relatively large branching ratio. In fact, this decay can compete with tree level decays even for quite a substantial mass splitting, provided $\tilde{\chi}_2^0$ is higgsinolike and $\tilde{\chi}_1^0$ is gauginolike or vice versa and if $\tan\beta$ is not large. This is because both sfermion and Z exchange contributions are then suppressed by small neutralino mixing angles.

Finally, despite being the lightest sparticle in the observable sector, $\tilde{\chi}_1^0$ might be unstable. One possibility leading to this, namely that of R-parity violation, will be discussed in the collider context in §15.6. $\tilde{\chi}_1^0$ will also decay if it is the NLSP, as in GMSB models (cf. Ch.13), where an ultralight gravitino $\tilde{G}$ is the LSP. The two body decay modes of (15.9a-d), with $l = 1$, will be inaccessible in this case since $\tilde{\chi}_1^0$ is the NLSP. If $\tilde{\chi}_1^0$ is gauginolike (higgsinolike), the dominant decay in such models will be $\tilde{\chi}_1^0 \to \gamma\tilde{G}$ ($Z\tilde{G}$), cf. Table 3.1. But, for a small $\tan\beta$, the final state $h\tilde{G}$ may become prominent in the decay of a higgsinolike $\tilde{\chi}_1^0$ with a mass well above m_h. Relevant expressions for partial widths may be found in Ref. [13.2].

Diagnostics, bounds and reach prospects

<u>LEP and future LC</u>

Consider charginos first. The most stringent constraints on the MSSM parameter space have come largely from LEP 2 studies (upto $\sqrt{s} \simeq 209$ GeV) of e^+e^- pair production of $\tilde{\chi}^\pm$. The negative outcome of simultaneous sfermion and Higgs searches leaves the three body modes of (15.12) as the only allowable decays for both charginos in the final state. These lead to four jets or two jets plus one charged lepton or two leptons with opposite charges, in all cases accompanied by large missing energy. The production of $W^\pm$ pairs, followed by their decays, constitute the main SM background. These events are reconstructable and can be eliminated. The lack of a signal then establishes a lower bound of 104 GeV [15.23] on $M_{\tilde{\chi}^\pm}$ for either a higgsinolike or a gauginolike $\tilde{\chi}^0_1$ with the assumptions (1)[17] $m_{\tilde{\nu}} > 300$ GeV and (2) $\Delta M_{\tilde{\chi}} \equiv M_{\tilde{\chi}^\pm} - M_{\tilde{\chi}^0_1}$ for the decay $\tilde{\chi}^\pm \to \tilde{\chi}^0_1 f \bar{f}'$ exceeds 10 GeV. A relaxation of assumption (1) marginally reduces the said bound to ~ 100 GeV for a gauginolike chargino. On the other hand, if $\Delta M_{\tilde{\chi}}$ is as small as 3 GeV and the sneutrino is heavy, the bound on the gauginolike chargino goes down to ~ 97 GeV. Two photon backgrounds force a change of strategy in case $\Delta M_{\tilde{\chi}} < 3$ GeV. When $\Delta M_{\tilde{\chi}} \lesssim 200$ MeV, the chargino lives sufficiently long to leave a 'heavy' ionizing track in the detector. The negative result of the search for a metastable charged particle has led to a lower bound, again right at the kinematical limit. However, if 200 MeV $< \Delta M_{\tilde{\chi}} \lesssim 3$ GeV, the presence of an additional (ISR or FSR) hard photon has to be required [15.21] to establish a bound. The lack of observation of such events has led to a lower mass bound ~ 90 GeV for a gauginolike $\tilde{\chi}^\pm_1$, assuming the sneutrino to be heavy.

While pair-produced charginos will be easily seen in a future linear e^+e^- collider (except possibly for 'beamstrahlung'[18] complications [15.24]), the extraction of the MSSM parameters controlling their masses will be more involved. The three parameters of the chargino mass matrix (9.7) as well as the sneutrino mass $m_{\tilde{\nu}_e}$ enter the expression for $\sigma(\tilde{\chi}^+_1 \tilde{\chi}^-_1)$. The decay kinematics also involve $M_{\tilde{\chi}^0_1}$ and possibly some sfermion masses in case two body decays[19] (15.8) are allowed. The detailed procedure[20] for kinematic reconstruction along with the modelling of effects like the broadening of energy endpoints by ISR and beamsstrahlung, are outlined in Refs. [15.8] and [15.25]. The conclusion from those studies is that 50 fb^{-1} of data should suffice for the determination of $M_{\tilde{\chi}^\pm_1}$ and $M_{\tilde{\chi}^0_1}$ at the 5% statistical accuracy level. Note, though, that the most precise determination of chargino masses in a future LC is likely to come from threshold scans [15.25], facilitated by the steep onset of the $S-$wave pair production cross section.

Further information about MSSM parameters can be obtained from heavier chargino pair $(\tilde{\chi}^+_m \tilde{\chi}^-_k)$ production cross sections as well as from the polarization of the $\tilde{\chi}^\pm$ and spin correlations between the two charginos [15.26]. There are two independent cross sections σ_L and σ_R

---

[17]Recall that a light sneutrino reduces the $\tilde{\chi}^\pm$ pair production cross section.

[18]This consists of real photons, emitted by an $e^\pm$ in the beam on being accelerated by the EM fields of the opposite bunch, which collide to produce a large number of soft e^+e^- pairs in near forward directions.

[19]The archetypal case is (15.8c) with $k = 1, l = 1$, i.e. $\tilde{\chi}^\pm_1 \to W^\pm \tilde{\chi}^0_1$. Another interesting decay $\tilde{\chi}^\pm_1 \to \tilde{\nu}_\ell \ell^+$, $\ell = e/\mu$, if allowed, may enable the determination of $M^\pm_{\tilde{\chi}_1}$ and $m_{\tilde{\nu}}$.

[20]Consider the sequential processes $e^+e^- \to P_1 P_2$, $P_2 \to P_3 P_4$, P_i standing for a generic particle with mass m_i. The energy $E_{1,2}$ of $P_{1,2}$ is given by $E_{1,2} = (s + m^2_{1,2} - m^2_{2,1})/(2\sqrt{s})$, $\sqrt{s}$ being the CM energy. The energy E_3 of P_3 lies in the range $\gamma(E^\star_3 - \beta p^\star_3) \leq E_3 \leq \gamma(E^\star_3 + \beta p^\star_3)$, where $E^\star_3$ and $p^\star_3$ are respectively the energy and the absolute value of the momentum of P_3 in the rest frame of P_2 : $E^\star_3 = (m^2_2 + m^2_3 - m^2_4)/(2m_2)$, $p^\star_3 = \sqrt{E^{\star 2}_3 - m^2_3}$, $\gamma = E_2/m_2$, $\beta = \sqrt{1 - \gamma^{-2}}$. When $m_1 = m_2$ and m_4 is known, $m_{1,3}$ can be determined from the spectrum endpoints of E_3. The analysis is extendable to the three body decay $\tilde{\chi}^\pm_1 \to q\bar{q}'\tilde{\chi}^0_1$ if the $q\bar{q}'$ pair is restricted to have a fixed invariant mass which serves as m_4.

for a left handed and right handed e^- respectively for every combination of m, k. The quantity σ_R is independent of the sneutrino mass $m_{\tilde{\nu}_e}$ and measures the higgsino contents of the produced charginos in the high energy limit. In contrast, σ_L is nonzero for both higgsinolike and gauginolike charginos but depends on $m_{\tilde{\nu}_e}$ only in the latter case. Once the masses of the chargino decay products get measured, $\tilde{\chi}^\pm$ production events should be completely reconstructable (often with a discrete ambiguity). This would allow a direct determination of the $\tilde{\chi}^\pm$ production angle, the distribution of which is sensitive to $m_{\tilde{\nu}_e}$. With polarized beams, both $\cos 2\phi_u$ and $\cos 2\phi_v$ of (9.15c) can be determined from $\tilde{\chi}_1^+ \tilde{\chi}_1^-$ pair production alone. In case CP is conserved in the chargino mass matrix, a knowledge of these two angles and of $M_{\tilde{\chi}_1^\pm}$ allows the determination of M_2, μ and $\tan\beta$. If there is CP violation, the relative phase between M_2 and μ becomes an additional parameter to be determined via a measurement of the $\tilde{\chi}_1^\pm$ polarization vector. The latter, however, requires a detailed analysis of $\tilde{\chi}_1^\pm$ decays which can become quite complicated, as discussed above. Any large hierarchy between $|\mu|$ and $|M_2|$ will make $\tilde{\chi}_1^\pm$ a nearly pure wino or higgsino enabling this kind of measurement to establish a lower bound on max. $(|M_2|, |\mu|)$. A more accurate determination of this parameter will be possible only if the $\tilde{\chi}_1^\pm \tilde{\chi}_2^\mp$ state is accessible. Even so, the determination of $\tan\beta$ from chargino production and decay will be difficult in the large $\tan\beta$ limit. This is since the relevant quantity $\cos 2\beta$ $(\simeq -1)$ is insensitive to the precise value of $\tan\beta$ in such a limit.

Turning to neutralino pair production, the $\tilde{\chi}_1^0 \tilde{\chi}_1^0$ state is invisible for a stable $\tilde{\chi}_1^0$. Furthermore, the cross section to produce this pair with an additional hard ISR photon for detection is too small and in any event there is a large $\nu\bar{\nu}\gamma$ SM background. Instead, $e^+e^- \to \tilde{\chi}_1^0 \tilde{\chi}_2^0$, leading to 'one-sided events' with either two acoplanar jets or an acoplanar charged lepton pair, is the first useful neutralino pair production process. However, its cross section is sizable only if the selectrons are light or if $\tilde{\chi}_1^0$ has a significant higgsino component. Assuming the latter to be the case, negative LEP searches have generated an exclusion region in the MSSM parameter space that is only marginally larger than that from the nonobservation of higgsinolike charginos. A combination of all information from negative chargino, neutralino and slepton searches at LEP, together with the assumption of high scale gaugino mass unification (12.25), has yielded [15.23] the lower bound (see the table on p xviii) $\mathbf{M_{\tilde{\chi}_1^0} > 40~GeV}$ for a stable or quasistable $\tilde{\chi}_1^0$. It is, however, important to emphasize that there is no assumption independent[21] lower bound on the mass of a stable or quasistable $\tilde{\chi}_1^0$. An arbitrarily light stable $\tilde{\chi}_1^0$ could still have escaped collider searches if $|M_1| \ll |M_2|$ and either $M_{\tilde{\chi}_1^0} + M_{\tilde{\chi}_2^0} > 200$ GeV or selectrons are very heavy. On the other hand, the question what happens if the $\tilde{\chi}_1^0$ decays inside the detector can be asked. One consequence is that chargino mass bounds approach the kinematic limit independent of the sneutrino mass or the chargino-neutralino mass difference. But there are additional effects depending on the decay mode. Consider the mode $\tilde{\chi}_1^0 \to \gamma\tilde{G}$ with a low mass gravitino: a characteristic feature of GMSB models. The pair-produced $\tilde{\chi}_1^0$'s now become visible through two hard acoplanar photons with $\not{E}$. For a binolike $\tilde{\chi}_1^0$, a natural expectation in the mGMSB model, the pair production amplitude is dominated by t or u channel selectron exchange. This has enabled one to exclude a region in the $(M_{\tilde{\chi}_1^0}, m_{\tilde{e}})$ plane for such models, using the negative outcome of the search for this signal at LEP. Assuming left-right selectron mass degeneracy $(m_{\tilde{e}_L} \simeq m_{\tilde{e}_R})$ for simplicity, one can conclude [15.23] that $M_{\tilde{\chi}_1^0} > 100$ GeV for $m_{\tilde{e}} < 200$ GeV. In case the gravitino mass is quite a bit larger, a significant fraction of these events will have only one hard photon in the final state, not pointing back to the interaction vertex. Though such a signal is essentially free of background, no published search results for it are available.

---

[21] A different set of assumptions, including the requirement of LSPs not overclosing the Universe, has yielded a weaker lower bound of 20 GeV on $M_{\tilde{\chi}_1^0}$ [15.27].

The final state $\tilde{\chi}_1^0\tilde{\chi}_2^0$, produced in e^+e^- collision, is expected to play a greater role [15.28] in a future high luminosity LC – specifically determining its gaugino mass reach within the framework of the electroweak gaugino mass unification condition (9.21). Given sufficient CM energy and statistics, both the neutralino masses $M_{\tilde{\chi}_{1,2}^0}$ and the selectron masses $m_{\tilde{e}_{L,R}}$ should be measurable in such a machine from the angular distributions of the $\tilde{\chi}_2^0$ decay products in the process $e^+e^- \to \tilde{\chi}_1^0\tilde{\chi}_2^0$ with left handed and right handed e^- beams. However, the MSSM parameters determined from this final state will have (on account of the smaller cross section) larger errors than from $\tilde{\chi}_1^+\tilde{\chi}_1^-$ production if the lighter $\tilde{\chi}$ states are gauginolike. When the latter are higgsinolike, the event rates in the two channels become comparable, but then $\tilde{\chi}_1^0\tilde{\chi}_2^0$ production has an edge. This is since the $\tilde{\chi}_1^0$-$\tilde{\chi}_2^0$ mass difference will now be significantly larger[22] than the $\tilde{\chi}_1^\pm$-$\tilde{\chi}_1^0$ one, leading to more energetic final state particles and hence a smaller two photon background. In the GMSB scenario, the NLSP $\tilde{\chi}_1^0$, decaying into an ultralight $\tilde{G}$ and a visible photon, makes $\tilde{\chi}_1^0\tilde{\chi}_1^0$ production a detectable process. The latter then will set the gaugino mass reach of a future LC if $\tilde{\chi}_1^0$ is gauginolike. Moreover, in such a case, the NLSP mass $M_{\tilde{\chi}_1^0}$ (though not the gravitino mass $m_{3/2}$) can be determined kinematically, cf. ftnt. 20. A binolike $\tilde{\chi}_1^0$ in a GMSB scenario, with its mass and the gravitino mass in the respective ranges 50 GeV $\leq M_{\tilde{\chi}_1^0} \leq$ 300 GeV and 10^{-5} eV $< m_{3/2} < 1$ keV, will have a decay length in the interval 1 fm $\lesssim c\tau_{\tilde{\chi}_1^0} \lesssim 10^3$ m. A part of this interval will admit[23] direct measurements [15.29] in a future LC.

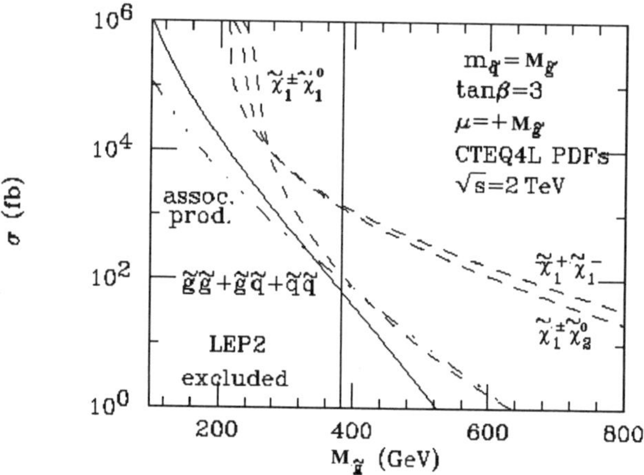

Fig.15.8. Expected sparticle production cross sections at the TEVATRON in mSUGRA, adapted from Barger et al. [15.5]. Note that the assumptions are high scale gaugino mass unification and large $|\mu|$. Only $\tilde{\chi}_1^\pm\tilde{\chi}_1^0$ production depends sensitively on sgn. μ.

---

[22]The ratio of these mass differences is about a factor of two if gaugino mass unification at a high scale holds even approximately.

[23]Decay lengths down to $\sim 10^{-3}$ cm can be measured with a (micro)vertex detector, using photon conversion to e^+e^- pairs and/or Dalitz decays $\tilde{\chi}_1^0 \to e^+e^-\tilde{G}$. In the opposite extreme, values in the interval 1m $\lesssim c\tau_{\tilde{\chi}_1^0} \lesssim 50$ m are measurable through the statistical distribution of decay lengths in the detector – using photons that do not point back to the interaction vertex. In fact, the ability to identify such nonpointing photons defines one of the requirements of the electromagnetic calometer in a future detector. The aforementioned interval should be extendable upwards by considering events with only one nonpointing photon or by counting the fraction of other supersymmetry events (e.g. $\tilde{\chi}_1^\pm$ pair production) containing one hard, isolated photon.

<u>TEVATRON</u>

For a hadron collider such as the TEVATRON (with p and $\bar{p}$ beams and a CM energy of 2 TeV), chargino and neutralino production are best considered together in an integrated way. Indeed, if kinematically allowed, these should be the most copious of all sparticle production processes in that machine, provided the gaugino mass unification condition (12.25) holds. The latter implies that the pole masses of gluinos are at least thrice as much as those of $\tilde{\chi}_1^{\pm}$ and $\tilde{\chi}_2^0$ and moreover first and second generation squarks cannot be much lighter than gluinos, cf. (11.45c). Cross section suppression by large masses overcompensates[24] the factor $(g_s/g_2)^4$ enhancing the weak cross section for producing charginos and neutralinos over the strong cross section, as illustrated in Fig. 15.8. Here, with the assumption $\mu = M_{\tilde{g}} = m_{\tilde{q}}$, $m_{\tilde{q}}$ being the average mass of the squarks of the first two generations, $\tilde{\chi}_{1,2}^0$ and $\tilde{\chi}_1^{\pm}$ become gauginolike. We see that the rates of hadroproduction of a $\tilde{\chi}_1^{\pm}$ pair and $\tilde{\chi}_1^{\pm}\tilde{\chi}_1^0$ as well as $\tilde{\chi}_1^{\pm}\tilde{\chi}_2^0$ are then larger than those of strong sparticles. On the other hand, because of the near absence of an s channel electroweak gauge boson exchange the cross sections for producing $\tilde{\chi}_1^0\tilde{\chi}_1^0$, $\tilde{\chi}_2^0\tilde{\chi}_2^0$ as well as $\tilde{\chi}_1^{\pm}\tilde{\chi}_1^0$ (for the last, depending on how binolike $\tilde{\chi}_1^0$ is) are significantly smaller than those for $\tilde{\chi}_1^+\tilde{\chi}_1^-$ and $\tilde{\chi}_1^{\pm}\tilde{\chi}_2^0$ production. With the lighter $\tilde{\chi}$ states being higgsinolike in the opposite limit $|\mu| \ll |M_{1,2}|$, the hadroproduction rates for $\tilde{\chi}_1^+\tilde{\chi}_1^-$, $\tilde{\chi}_{1,2}^0\tilde{\chi}_1^{\pm}$ and $\tilde{\chi}_1^0\tilde{\chi}_2^0$ will be quite close to each other and at about a third of the value shown for $\tilde{\chi}_1^+\tilde{\chi}_1^-$ in Fig.15.8.

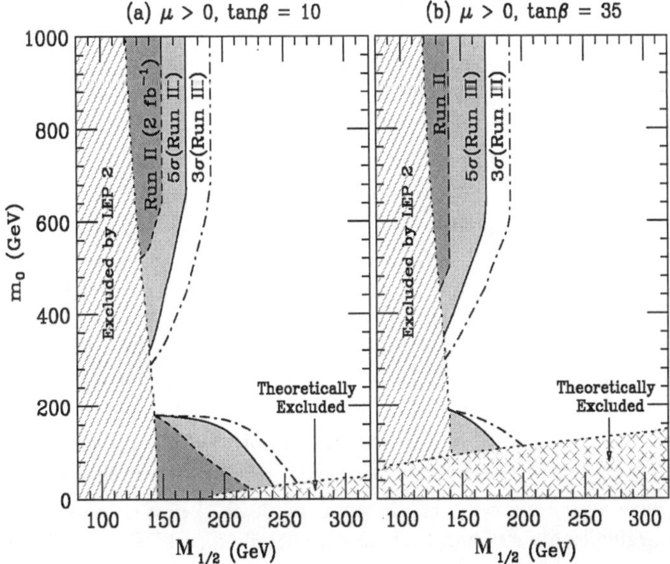

Fig.15.9. Expected reach of TEV-II in the trilepton $+$ $\not{E}_T$ channel in the mSUGRA parameter space; adapted from Barger et al. [15.5].

The most promising channels, from among the plethora of final states possible in the hadroproduction of a $\tilde{\chi}$ pair, are $\tilde{\chi}_2^0 (\to \tilde{\chi}_1^0\ell^+\ell^-)\,\tilde{\chi}_1^+ (\to \tilde{\chi}_1^0\ell^+\nu_e)$ and $\tilde{\chi}_2^0 (\to \tilde{\chi}_1^0\ell^+\ell^-)\,\tilde{\chi}_1^- (\to \tilde{\chi}_1^0\ell^-\bar{\nu}_e)$ in case $\tilde{\chi}_1^0$ is invisible. These lead to the hadronically quiet (i.e. with no significant large E_T hadronic activity in the event) **trilepton $+$ large $\not{E}_T$** signal with all the three leptons being *isolated* in

---

[24]The cross section for $\tilde{t}_1$ pair production might exceed that for a $\tilde{\chi}$ pair.

phase space. The main SM backgrounds to this signal come from the production of virtual gauge boson pairs $W^\star\gamma^\star$, $W^\star Z^\star$, followed by their decays into leptons with $\not{E}_T$ due to the neutrino. This background can be restricted to the $O(\mathrm{fb})$ range by cuts on the $\ell^+\ell^-$ invariant mass, on $\not{E}_T$ and on the angular acceptance of each charged lepton. The likelihood of the occurrence of a signal event that will pass all these cuts is determined by the $\tilde\chi$ pair production cross section as well as by the leptonic branching fractions and the decay kinematics. All of these are controlled by far too many MSSM parameters in the gaugino, sfermion and Higgs sector. An appeal to specific CMSSM scenarios thus becomes obligatory. Now mSUGRA, cf. §12.3 and §12.4, is a particularly convenient model in which to analyze [15.5] the reach from the trilepton + large $\not{E}_T$ channel of upcoming TEVATRON runs. The results, shown in Fig. 15.9, imply that the highest reach in $M_{1/2}$ is expected at small m_0 and not too large $\tan\beta$ which is where B.R. $(\tilde\chi_2^0 \to \tilde\ell_{L,R}^{\pm}\ell^{\mp})$, $\ell = e/\mu$, is large. If $\tan\beta$ is large (and even if m_0 is small), the decay $\tilde\chi_2^0 \to \tau^+\tau^-\tilde\chi_1^0$ will dominate through a real or virtual $\tilde\tau$, degrading the signal[25]. As m_0 increases, the destructive interference between slepton and Z exchanges reduces the leptonic BR of a decaying $\tilde\chi_2^0$ when $\mu > 0$. However, once $m_0 \geq 600$ GeV, slepton exchange can be neglected and the BR's of $\tilde\chi_2^0$ track those of the Z. Thus, even in mSUGRA, the lack of such a signal would not translate into a lower bound on any one mass parameter despite ruling out sizable chunks of the parameter space.

What about other models? Let us take AMSB, cf. §12.6, where M_2 at the weak scale is nearly a third of M_1 instead of being almost double the latter as is the case in mSUGRA. In consequence, $\tilde\chi_1^{\pm}$ and $\tilde\chi_1^0$ become nearly mass degenerate winos. The leading $\tilde\chi$ pair production cross sections can still be read off from Fig. 15.8 by simply interchanging $\tilde\chi_1^0 \leftrightarrow \tilde\chi_2^0$. Now the $\tilde\chi_1^{\pm}\tilde\chi_2^0$ production rate is quite small. Therefore the trilepton $+\not{E}_T$ signal does not work. Yet, if the $\tilde\chi_1^{\pm}$ is sufficiently long-lived (specifically $\Delta M_{\tilde\chi} \lesssim 250$ MeV), it will leave a macroscopic track in the detector [15.30]. However, the best case with $c\tau_{\tilde\chi_1^{\pm}} > 1$ m requires the inequality $\Delta M_{\tilde\chi_1} < m_\pi$. In this situation, the mass reach with[26] 2 (30) fb^{-1} of integrated TEVATRON luminosity extends to $M_{\tilde\chi_1^{\pm}} \simeq 320$ (450) GeV, the corresponding numbers being 120 (185) GeV for $\Delta M_{\tilde\chi_1} = 200$ MeV. Increasing $\Delta M_{\tilde\chi_1}$ means decreasing the $\tilde\chi_1^{\pm}$ lifetime, leading to a rapid decrease in the mass reach which here is somewhat worse than in mSUGRA. Turning to mGMSB, the $\tilde\chi$ pair production cross sections are more or less as in Fig. 15.8, but the mass reach depends crucially on the nature and lifetime of the NLSP. For a binolike $\tilde\chi_1^0$ NLSP with a long lifetime, the situation here parallels that of mSUGRA with a small m_0: search prospects are best in the trilepton + large $\not{E}_T$ channel and the mass reach falls with increasing $\tan\beta$ because of domination by τ leptonic final states. All these discussions become nearly irrelevant if the decay $\tilde\chi_1^0 \to \gamma\tilde{G}$ occurs promptly within the detector. Then, using the $\gamma\gamma\not{E}_T$ signal, $M_{\tilde\chi_1^{\pm}}$ upto 290 (350) GeV can be reached[27] [15.31] with 2 (30) fb^{-1} of TEVATRON data. For a $\tilde\tau_1$ NLSP with a prompt $\tilde\tau_1 \to \tau\tilde{G}$ decay, even 30 fb^{-1} of TEVATRON data will only yield a reach that is marginally higher than that already attained at LEP, the relevant constraints coming from searches for $\tilde\tau$ production. But a long-lived $\tilde\tau_1$ will extend the TEVATRON reach to $M_{\tilde\chi_1^{\pm}} \sim 350$ (420) GeV with 2 (30) fb^{-1} of data [15.31].

---

[25]This is since only a third of all τ's decay purely leptonically and even there the produced leptons are often too soft to pass the acceptance cuts. We recall, in this context, that $\tilde\chi_1^{\pm}$ decays are frequently dominated by the $\tilde\tau_1\nu_\tau$ mode even if $\tan\beta$ is not large. Therefore, at least one of the leptons of the signal should be allowed to have a fairly small p_T, not much above 7 GeV.

[26]The 30 fb^{-1} figure is notional since that is now known to be far beyond the limit of what will be realistically attainable at the TEVATRON.

[27]This degrades significantly if $c\tau_{\tilde\chi_1^0} \gtrsim 0.5$ m; but part (not all) of the sensitivity can be recovered by searching instead for $\gamma\not{E}_T$ events with the photon not pointing back to the interaction vertex.

LHC

The situation at the LHC, which will collide two proton beams at $\sqrt{s} = 14$ TeV, is very different on account of the absence of valence antiquarks in the initial state of any collision subprocess. Hence, cross sections for $\tilde{\chi}$ pair production reach only 3 (0.1) pb for two winolike states with $M_{\tilde{\chi}} \sim 100$ (300) GeV [15.32]. Two additional complications arise at the LHC. First, the above mentioned direct $\tilde{\chi}$ pair production gets swamped by the production of $\tilde{\chi}$ states in the decays of gluinos and squarks which are produced copiously. The former can be separated from the latter by selecting only the purely leptonic decay modes of the $\tilde{\chi}$'s and applying a strict jet veto. Second, the 'piling up' (i.e. occurrence in the same bunch crossing as the 'hard' signal event) of 'soft' or 'minimum bias' events will make one lose the signal event if the lepton isolation cuts are too stringent, calling for an optimized loosening of the latter or even for the use of a slightly reduced instantaneous luminosity.

In mSUGRA-like models, with an invisible $\tilde{\chi}_1^0$ and high scale gaugino mass unification (12.25), the trilepton + large $\not{E}_T$ signal from $\tilde{\chi}_1^\pm \tilde{\chi}_2^0$ production is still very viable at the LHC, though the mass reach will not be particularly superior to that at the TEVATRON if sfermion masses exceed $M_{\tilde{\chi}_1^\pm}$. For $M_{\tilde{\chi}_2^0} > 200$ GeV, the two body decays $\tilde{\chi}_2^0 \to \tilde{\chi}_1^0 + (Z, h)$ open up acting as *spoiler modes*, effectively reducing the signal to zero. On the other hand, if $\tilde{\chi}_2^0 \to \tilde{\ell}_L^\pm \ell^\mp$ decays are kinematically allowed, the $\tilde{\chi}_1^\pm$ mass reach can be extended to 360 GeV with 100 fb^{-1} of LHC data[28] [15.32]. Nevertheless, if the TEVATRON does detect the trilepton + $\not{E}_T$ signal, the LHC will be able to accumulate a much larger sample of such events. Indeed, the subset with $\ell^+\ell^-\ell'^\pm$ (and $\ell \neq \ell'$, both being charged leptons) is of particular interest. The dilepton invariant mass $m_{\ell'\ell}$ might be of considerable utility in deciding if $M_{\tilde{\chi}_2^0} < m_{\tilde{\ell}}$ or if $M_{\tilde{\chi}_2^0} > m_{\tilde{\ell}}$. In the former case, the relevant decay is $\tilde{\chi}_2^0 \to \tilde{\chi}_1^0 \ell^+\ell^-$ and $m_{\ell^+\ell^-}$ develops a sharp upper edge at

$$m_{\ell^+\ell^-}^{\max} = M_{\tilde{\chi}_2^0} - M_{\tilde{\chi}_1^0} \, .$$

In the latter case the relevant decay proceeds via $\tilde{\chi}_2^0 \to \tilde{\ell}^\pm \ell^\mp \to \ell^\pm \tilde{\chi}_1^0 \ell^\mp$ and the upper edge comes at[29]

$$m_{\ell^+\ell^-}^{\max} = M_{\tilde{\chi}_2^0} \left(1 - m_{\tilde{\ell}}^2/M_{\tilde{\chi}_2^0}^2\right)^{1/2} \left(1 - M_{\tilde{\chi}_1^0}^2/m_{\tilde{\ell}}^2\right)^{1/2} \, .$$

Even in the former case, the shape of the invariant $\ell^+\ell^-$ mass distribution can yield information [15.33] on slepton masses if slepton exchange is important.

We conclude this section by commenting on chargino/neutralino hadroproduction at the LHC in two specific non-mSUGRA scenarios. For AMSB with a very small $\Delta M_{\tilde{\chi}_1}$, the trilepton + large $\not{E}_T$ signal is feeble, but direct $\tilde{\chi}$ production can be detected as at the TEVATRON, if $\tilde{\chi}_1^\pm$ is sufficiently long-lived ($\Delta M_{\tilde{\chi}_1} < 200$ MeV) to leave a heavy ionizing track in the detector. One saving grace is that the cross section for producing $\tilde{\chi}_1^\pm \tilde{\chi}_2^0$ is somewhat larger. Thus it may be possible to accomplish the aforementioned kinematic reconstruction of the edge in $M_{\ell^+\ell^-}$ from the decay $\tilde{\chi}_2^0 \to \tilde{\chi}_1^0 \ell^+\ell^-$ despite the small cross section of this process. For GMSB with a short-lived $\tilde{\chi}_1^0$ NLSP, two hard isolated photons – each from the decay $\tilde{\chi}_1^0 \to \tilde{G}\gamma$ – will characterize any two sparticle state. The decay chain

$$\tilde{\chi}_2^0 \to \tilde{\ell}^\pm \ell^\mp \to \tilde{\chi}_1^0 \tilde{\ell}^+ \ell^- \to \tilde{G}\gamma \ell^+\ell^-$$

---

[28]This conclusion might be over-optimistic since backgrounds from virtual photon exchange were not included in the consideration.

[29]There will be two edges if $\tilde{\chi}_2^0$ decays into both $\tilde{\ell}_R$ and $\tilde{\ell}_L$ channels.

can provide precisely measurable $\ell^+\ell^-\gamma$ and $\ell^\pm\gamma$ invariant mass endpoints in addition to the $\ell^+\ell^-$ endpoint discussed above. These will yield more measurements of sparticle mass differences. The NLSP lifetime provides a good handle on the sparticle mass scale M_s. It should be measurable at the LHC for $c\tau_{\tilde{\chi}_1^0} < 100$ m [15.6]. Such an analysis will be best performed on an inclusive sparticle event sample involving a process with a much larger cross section as in squark/gluino production rather than on exclusive $\tilde{\chi}$ production. We shall return to this issue when we discuss squark and gluino searches in §15.4.

15.3 Signals of sleptons

Sleptons are likely to be among the lighter sparticles whose early discovery is anticipated. As already shown in the previous section, a knowledge of the mass parameters $m_{\tilde{\ell}_L}$, $m_{\tilde{\ell}_R}$ and $m_{\tilde{\nu}_\ell}$ will be of great use in studying signals of charginos and neutralinos. These features underscore the importance of the direct detection of the signals of sleptons and the determination of their properties at various colliders.

Pair production

Tree diagrams for e^+e^- pair production of charged sleptons are shown in Fig. 15.10. Given flavor diagonal fermion-neutralino-sfermion couplings, a t channel exchange of $\tilde{\chi}_l^0$ ($l = 1, 2, 3, 4$) can take place only for final state selectrons. When present, this can enhance $\tilde{e}\bar{\tilde{e}}$ production by an order of magnitude over that of $\tilde{\mu}\bar{\tilde{\mu}}$ or $\tilde{\tau}\bar{\tilde{\tau}}$. The possible final state selectron pairs are $\tilde{e}_L\bar{\tilde{e}}_L$, $\tilde{e}_R\bar{\tilde{e}}_R$, $\tilde{e}_L\bar{\tilde{e}}_R$ and $\tilde{e}_R\bar{\tilde{e}}_L$. With a negligible Yukawa coupling strength for the electron, only the gaugino components of the exchanged neutralinos can contribute significantly to selectron pair production. Therefore, the magnitudes of these cross sections away from threshold are determined largely by the gaugino masses $M_{1,2}$. Smuon and stau pairs are a different story. They are produced only

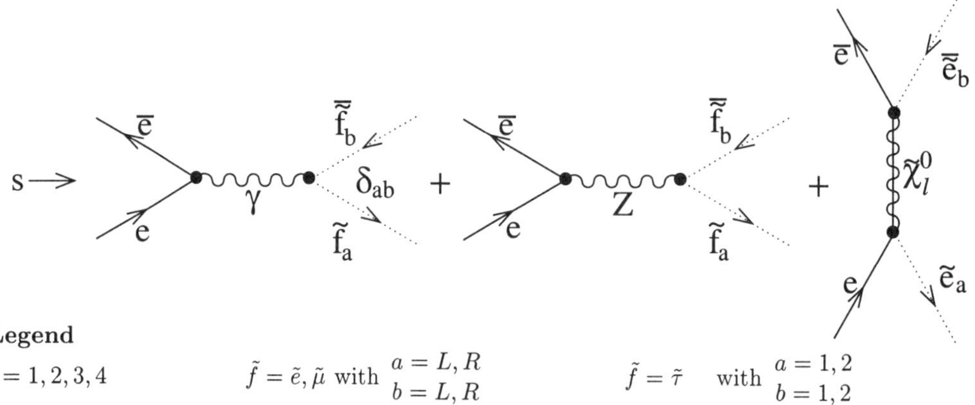

Legend

$l = 1, 2, 3, 4$ $\tilde{f} = \tilde{e}, \tilde{\mu}$ with $\begin{matrix} a = L, R \\ b = L, R \end{matrix}$ $\tilde{f} = \tilde{\tau}$ with $\begin{matrix} a = 1, 2 \\ b = 1, 2 \end{matrix}$

Fig.15.10. Tree diagrams for e^+e^- pair production of charged sleptons

by an off–shell $\gamma^\star$ or $Z^\star$ exchanged in the s channel. Smuons are simpler since the L-R mixing is small in this case. Thus $\tilde{\mu}_{L,R}$ are expected to be mass eigenstates and moreover the pairs $\tilde{\mu}_L\bar{\tilde{\mu}}_R$, $\tilde{\mu}_R\bar{\tilde{\mu}}_L$ cannot be realized in the final state. Gauge coupling strengths and smuon masses completely determine the cross sections for the two allowed processes $e^+e^- \to \tilde{\mu}_L\bar{\tilde{\mu}}_L, \tilde{\mu}_R\bar{\tilde{\mu}}_R$. The case of staus is more complicated on account of their nontrivial L-R mixing, the possible final states being $\tilde{\tau}_a\bar{\tilde{\tau}}_b$,

$a, b = 1, 2$. The combination with $a \neq b$ can be realized through an s channel $Z^\star$ exchange diagram and allows the creation and study of a $\tilde{\tau}_2$ below the $\tilde{\tau}_2\bar{\tilde{\tau}}_2$ production threshold. The cross section for stau pair production is sensitive to those supersymmetry parameters that determine L-R mixing in this sector in addition to the relevant gauge coupling strengths. There is, of course, no $\gamma^\star$ exchange for sneutrino pair production, the diagrams for which are shown in Fig. 15.11. The t channel exchange diagram now contributes only to the $\tilde{\nu}_e\bar{\tilde{\nu}}_e$ final state, with charginos $\tilde{\chi}_k^\pm$ ($k = 1, 2$) being exchanged. In contrast, $\tilde{\nu}_\mu\bar{\tilde{\nu}}_\mu$ and $\tilde{\nu}_\tau\bar{\tilde{\nu}}_\tau$ pairs are produced only via s channel $Z^\star$ exchange, with cross sections being completely determined by sneutrino masses and gauge coupling strengths. Thus, especially if gauginolike charginos are light, the production rate of $\tilde{\nu}_e\bar{\tilde{\nu}}_e$ will be substantially larger than those for the $\tilde{\nu}_\mu\bar{\tilde{\nu}}_\mu$ and $\tilde{\nu}_\tau\bar{\tilde{\nu}}_\tau$ pairs.

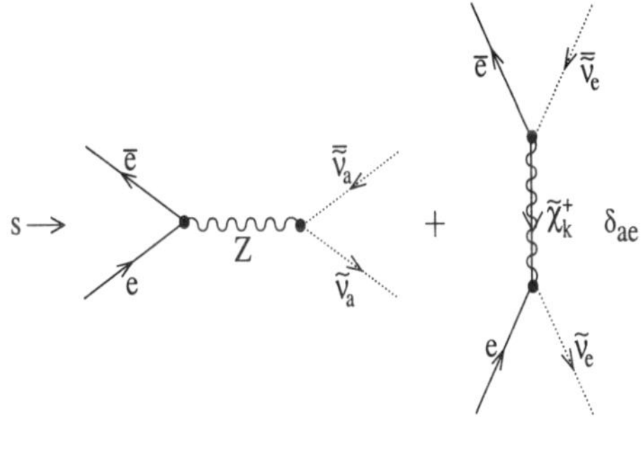

Fig.15.11. Tree level diagrams for e^+e^- pair production of sneutrinos

There exist many discussions in the literature [15.34–15.39] of cross sections of slepton pair production in e^+e^- collision. We follow the formalism of Ref. [15.37] which is particularly transparent in demonstrating how polarized e^+e^- beams will yield information on gaugino-higgsino mixing. Define the 'neutralino functions' $M_{ab}(t)$ and $N_{ab}(t)$, for the helicity flipping and helicity conserving cases respectively, with $a, b = L/R$, as

$$[M, N]_{ab}(t) = \sum_{l=1}^{4} V_{al} \frac{M_{\tilde{\chi}_1^0}\left[M_{\tilde{\chi}_l^0}, M_{\tilde{\chi}_1^0}\right]}{M_{\tilde{\chi}_l^0}^2 - t} V_{bl} . \tag{15.16}$$

The quantities V_{Ll} and V_{Rl} are the couplings G_l^{eL} and G_l^{eR} of (9.73) respectively, divided by the factor $\sqrt{2}e$. These multiply the t channel propagator along with neutralino mass factors. In the numerator of M_{ab}, the factor of the mass of the exchanged neutralino $\tilde{\chi}_l^0$ arises from fermionic helicity flip. The square (single power) of the mass of the lightest neutralino has been put in N_{ab} (M_{ab}) for dimensional reasons. Expressions for the differential cross section $d\sigma/d\cos\theta$ can now be given [15.37] for all polarization and chirality combinations that are possible in e^+e^- pair production of selectrons. These generally involve M_{ab} or N_{ab} and $\beta_{\tilde{e}} = 2k_{\tilde{e}}/\sqrt{s}$, $k_{\tilde{e}}$ being the selectron momentum in the CM frame. We can also consider the $e^+e^- \to \tilde{\nu}_e\bar{\tilde{\nu}}_e$ production cross sections with different beam polarizations. The case with the left handed electron beam involves the 'chargino function' $C_{LL}(t)$, defined by

$$C_{LL}(t) = \frac{1}{\sqrt{2}} \sum_{k=1}^{2} U_{kL} \frac{M_{\tilde{\chi}_1^\pm}^2}{M_{\tilde{\chi}_k^\pm}^2 - t} U_{kL}^\star, \tag{15.17}$$

U_{kL} being the coupling d^L_{11k} of (9.70) divided by g_2. One can also define $\beta_{\tilde{\nu}}$ in analogy with $\beta_{\tilde{e}}$. Expressions for all slepton pair production processes with nonvanishing cross sections are given in[30] Table 15.3. The differential cross section for producing any of the listed final states with unpolarized beams can be read off from this table by averaging over initial helicities.

Process	$d\sigma/d\cos\theta$
$e^-_R e^+_L \to \tilde{e}^-_R \tilde{e}^+_R$	$\pi\alpha^2_{EM}(2s)^{-1}\beta^3_{\tilde{e}}\sin^2\theta\|sM^{-2}_{\tilde{\chi}^0_1}N_{RR}(t) - 1 - s^2_W c^{-2}_W s(s - M^2_Z)^{-1}\|^2$
$e^-_R e^+_L \to \tilde{e}^-_L \tilde{e}^+_L$	$\pi\alpha^2_{EM}(2s)^{-1}\beta^3_{\tilde{e}}\sin^2\theta\|1 - c^{-2}_W(1/2 - s^2_W)s(s - M^2_Z)^{-1}\|^2$
$e^-_L e^+_L \to \tilde{e}^-_L \tilde{e}^+_R$	$2\pi\alpha^2_{EM}\beta_{\tilde{e}}M^{-2}_{\tilde{\chi}^0_1}\|M_{LR}(t)\|^2$
$e^-_R e^+_R \to \tilde{e}^-_R \tilde{e}^+_L$	$2\pi\alpha^2_{EM}\beta_{\tilde{e}}M^{-2}_{\tilde{\chi}^0_1}\|M_{RL}(t)\|^2$
$e^-_L e^+_R \to \tilde{e}^-_R \tilde{e}^+_R$	$\pi\alpha^2_{EM}(2s)^{-1}\beta^3_{\tilde{e}}\sin^2\theta\|1 - c^{-2}_W(1/2 - s^2_W)s(s - M^2_Z)^{-1}\|^2$
$e^-_L e^+_R \to \tilde{e}^-_L \tilde{e}^+_L$	$\pi\alpha^2_{EM}(2s)^{-1}\beta^3_{\tilde{e}}\sin^2\theta\|sM^{-2}_{\tilde{\chi}^0_1}N_{LL}(t) - 1 - c^{-2}_W s^{-2}_W(1/2 - s^2_W)^2 s(s - M^2_Z)^{-1}\|^2$
$e^-_R e^+_L \to \tilde{\nu}_e \bar{\tilde{\nu}}_e$	$\pi\alpha^2_{EM}(s/8)\beta^3_{\tilde{\nu}_e}\sin^2\theta c^{-4}_W(s - M^2_Z)^{-2}$
$e^-_L e^+_R \to \tilde{\nu}_e \bar{\tilde{\nu}}_e$	$\pi\alpha^2_{EM}(s/2)\beta^3_{\tilde{\nu}_e}\sin^2\theta\|M^{-2}_{\tilde{\chi}^\pm_1}C_{LL}(t) + c^{-2}_W s^{-2}_W(s - M^2_Z)^{-1}(1/2 - s^2_W)/2\|^2$

Table 15.3. Differential cross sections for e^+e^- production of selectron and sneutrino pairs

The following features about the formulae given in Table 15.3 are noteworthy. First, M_{ab} (N_{ab}) enter the expressions for processes with the same (opposite) sign polarizations of the electron and the positron and opposite (same) 'chiralities' of the selectrons. Second, each helicity amplitude involves at most one neutralino/chargino function. Third, production processes in different chiral channels can be distinguished not only by beam polarization but also by kinematic studies. Specifically, the cross section for the production of a slepton and its own antiparticle in the final state has a $\beta^3_{\tilde{\ell}}$ dependence on account of the production amplitude being P-wave near threshold. The production of the chirally mixed pairs $\tilde{e}^-_L \tilde{e}^+_R$ and $\tilde{e}^+_L \tilde{e}^-_R$ takes place[31] through an S-wave amplitude yielding a linear dependence on $\beta_{\tilde{e}}$. All other slepton pair production cross sections, possible in e^+e^- collisions, are $\beta^3_{\tilde{\ell}}$ suppressed near threshold. Explicit expressions for the total cross sections for producing various slepton pairs from unpolarized e^+e^- beams are given in the Appendix of Ref. [15.28] in a notation similar to ours. Smuon, μ-sneutrino and τ-sneutrino pair production cross sections are simply obtainable from Table 15.1 by discarding the t channel exchange terms. The case with the final state $\tilde{\tau}_a \bar{\tilde{\tau}}_a$ ($a = 1, 2$) is slightly more complicated due to L-R mixing, but the expression [15.28] for $e^+e^- \to \tilde{t}_a \bar{\tilde{t}}_b$ carries over with the replacement $t \to \tau$. A noteworthy point is that the $Z\tilde{\tau}_a \bar{\tilde{\tau}}_a$

---

[30] We use the notation where $\tilde{e}^+_R$, $\tilde{e}^+_L$ are $\bar{\tilde{e}}_R$, $\bar{\tilde{e}}_L$ respectively.

[31] Ditto for $e^-e^- \to \tilde{e}_L \tilde{e}_L, \tilde{e}_R \tilde{e}_R$, making these processes more suitable for a precision measurement of the mass of the left or right selectron via threshold scans [15.38]. The lack of any s channel exchange term, causing destructive interference, increases the sensitivity of these cross sections to supersymmetry model parameters.

coupling vanishes when $\cos^2\theta_{\tilde{\tau}}$ ($\sin^2\theta_{\tilde{\tau}}$) equals $2s_W^2$ for $a = 1$ (2). Fortunately, the nonzero value of the $\gamma\tilde{\tau}_a\overline{\tilde{\tau}}_a$ coupling guarantees the tree level production of $\tilde{\tau}_a\overline{\tilde{\tau}}_a$ even in this case. However, there is a characteristic minimum in the unpolarized cross section as a function of the stau mixing angle $\theta_{\tilde{\tau}}$, the position of the minimum depending on $\sqrt{s}$. Note also that $\gamma^\star$-$Z^\star$ interference is small in this case owing to the smallness of the Z vector coupling to charged leptons. The dependence of $\sigma(e^+e^- \to \tilde{\ell}^+\tilde{\ell}^-)$ on beam polarization is quite strong and very different for left and right chiral sleptons. For instance, with a completely right handed electron beam (and $\sqrt{s} \gg M_Z$), the $\tilde{\mu}_R\overline{\tilde{\mu}}_R$ production rate is four times that of $\tilde{\mu}_L\overline{\tilde{\mu}}_L$ for equal mass $\tilde{\mu}_{L,R}$. The difference reflects the bigger value of the hypercharge $Y_{\tilde{\mu}_R}$ of $\tilde{\mu}_R$. Contrariwise, a purely left handed electron beam will yield higher rates for the production of left chiral slepton pairs. Beam polarization also changes the dependence of the stau pair cross sections on the stau mixing angle. For a pure e_R^- beam, the cross section for $\tilde{\tau}_1\overline{\tilde{\tau}}_1$ ($\tilde{\tau}_2\overline{\tilde{\tau}}_2$) production decreases (increases) monotonically with increasing $\cos\theta_{\tilde{\tau}}$; this behavior is reversed for a pure e_L^- beam. The typical size of a stau or smuon pair production cross section at $\sqrt{s} = 500$ GeV, far away from threshold, is ~ 50 fb.

Direct hadroproduction of charged slepton pairs is dominated by the Drell-Yan subprocesses

$$q\bar{q} \to Z^\star/\gamma^\star \to \tilde{\ell}_a\overline{\tilde{\ell}_b}\ (a = b = L \text{ or } R \text{ for } \ell = e, \mu;\ a, b = 1, 2 \text{ for } \ell = \tau),$$

$$q\bar{q}' \to W^\star \to \tilde{\ell}_L\overline{\tilde{\nu}_\ell},$$

both of which proceed via s channel exchanges only. Electroweak gauge boson fusion can also lead to slepton pair production, but with a negligible (rather small) cross section at TEVATRON (LHC) energies. Higgs mediated 'direct' slepton pair production from gluon gluon fusion can be substantial [15.40] at the LHC but only at low $\tan\beta$ and at somewhat low values of the mass m_A of the CP odd Higgs boson. In case $M_{\tilde{\chi}_2^0}$ exceeds $m_{\tilde{\ell}}$, the decays of $\tilde{\chi}_2^0$, produced either 'directly' or in decay chains of gluinos and squarks, can also be slepton sources. The latter can even be the dominant modes of slepton production at the LHC. The region of parameter space in mSUGRA models that allow the decay $\tilde{\chi}_2^0 \to \tilde{\ell}_R\bar{\ell}$ corresponds to $m_0 < M_{1/2}$. If $m_0 \lesssim 0.45\, M_{1/2}$, $\tilde{\ell}_L$ and $\tilde{\nu}_e$ will also be produced in the decays of $\tilde{\chi}_2^0$ and $\tilde{\chi}_1^\pm$. All slepton pair hadroproduction cross sections are completely determined by the masses of those sleptons and their couplings with gauge bosons. As a result, the production of $\overline{\tilde{\nu}_\ell}\tilde{\ell}_L$ will have the highest cross section owing to the large $SU(2)_L$ coupling strength that controls it. Note that the differential cross sections for the subprocesses $q\bar{q} \to \tilde{\ell}_a\overline{\tilde{\ell}_a}$, with $a = L, R$, $\ell = e, \mu$, and $q\bar{q} \to \tilde{\tau}_a\overline{\tilde{\tau}_b}$ with $a, b = 1, 2$, can be simply obtained from the corresponding ones in the e^+e^- case given earlier. Only an additional factor of $1/3$ has to be inserted for averaging over the colors of the initial quark-antiquark pair. The explicit expressions for $d\hat{\sigma}/dt$, required for the calculation of the charged slepton pairs $\tilde{\ell}_a\overline{\tilde{\ell}_a}$, as well as of $\tilde{\ell}_L\overline{\tilde{\nu}_\ell}$ and $\tilde{\nu}_\ell\overline{\tilde{\ell}_L}$ in pp and $\bar{p}p$ collisions are available in Refs. [15.13, 15.41].

Decay patterns

Unlike charginos and neutralinos, sleptons can always undergo two body decays which become the dominant modes. The simplest situation prevails for sleptons of the first two generations which have practically no L-R mixing. Right chiral sleptons can only decay into final states containing $\tilde{\chi}_l^0$ via $U(1)_Y$ gauge interactions. In contrast, if kinematically allowed, $\tilde{\ell}_L$'s decay predominantly with the larger $SU(2)_L$ coupling strength and into final states involving both $\tilde{\chi}_k^\pm$ and $\tilde{\chi}_l^0$. Possible two body modes, assuming that they are kinematically allowed, are:

$$\tilde{\ell}_{L,R}^\pm \to \ell^\pm \tilde{\chi}_l^0,\ \tilde{\ell}_L^\mp \to \overset{(-)}{\nu_\ell} \tilde{\chi}_k^\mp,\ \overset{(-)}{\tilde{\nu}_\ell} \to \overset{(-)}{\nu_\ell} \tilde{\chi}_l^0,\ \overset{(-)}{\ell} \tilde{\chi}_k^\mp \tag{15.18}$$

The indices l and k can take values $1, \cdots, 4$ and $1, 2$ respectively, depending on the masses of the sparticles involved. Decays of a sneutrino into a neutrino and a neutralino will give rise to an invisible final state for the case $l = 1$ if $\tilde{\chi}_1^0$ is the LSP. In an mSUGRA type of a scenario, the decay of a slepton into a lepton plus $\tilde{\chi}_1^0$ via $U(1)_Y$ interactions is always allowed, but its decays into a lepton plus $SU(2)_L$ gauginos are allowed only if $m_0 \gtrsim 0.45 \, M_{1/2}$.

For those GMSB models which have $\tilde{\tau}_1$ as the NLSP, the relevant decay is $\tilde{\tau}_1 \to \tau \tilde{G}$, $\tilde{G}$ being the (ultralight) gravitino. The corresponding decay width (and hence the path length $c\tau_{\mathrm{NLSP}}$) is given by an expression similar to that in the first row of Table 13.1, with the replacements $\tilde{\chi}_1^0 \to \tilde{\tau}_1$ and $\gamma \to \tau$, but without the neutralino mixing factor in front and with an additional overall factor of $(1 - m_\tau^2/m_{\tilde{\tau}_1}^2)^4$. The general dependence of the path length L on the gravitino mass $m_{3/2}$ is thus very similar to that for the decay of a binolike neutralino NLSP into $\gamma\tilde{G}$. For small intergenerational mass-splitting, i.e. small to moderate values of $\tan\beta$, the three body decays $\tilde{\ell}_R \to \tilde{\tau}_1 \bar{\tau} \ell$ are kinematically inaccessible. Each of the three light charged sleptons will then decay into a charged lepton plus the gravitino. This happens for the co-NLSP scenario of GMSB models where

$$m_{\tilde{\ell}_R} < \min. \, [m_{\tilde{\tau}_1} + m_\tau, M_{\tilde{\chi}_1^0}]$$

for $\ell = e, \mu$.

Expressions for partial widths of the decay channels of (15.18) can be simply obtained from the corresponding expressions for chargino/neutralino decay by interchanging the masses of the scalars and sfermions involved and with an additional factor of two from summing over the spins of the final state fermion. For example, the partial width for $\tilde{\ell}_R \to \ell \tilde{\chi}_l^0$ is

$$\Gamma(\tilde{\ell}_R \to \ell \tilde{\chi}_l^0) = m_{\tilde{\ell}_R}(g_2^2/18\pi)\tan^2\theta_W |Z_{l1}|^2 (1 - M_{\tilde{\chi}_l^0}^2/m_{\tilde{\ell}_R}^2)^2. \tag{15.19}$$

Decays of the $\tilde{\tau}$ states are somewhat more involved. In particular, for $\tan\beta \gg 1$, the sizable τ Yukawa coupling strength induces not only a significant mixing between $\tilde{\tau}_L$ and $\tilde{\tau}_R$ but also significant couplings to the higgsino components of charginos and neutralinos. Specifically, both $\tilde{\tau}$ mass eigenstates can decay into final states containing charginos, if kinematically allowed. On the other hand, $\tilde{\tau}_a \to \tau \tilde{\chi}_l^0$ decays offer useful information on supersymmetry parameters not only through the corresponding branching ratios but also through the polarization [15.35] of the τ. The latter can be measured from the energy distribution of the decay products of the τ itself. Gauge interactions couple the $\tilde{\tau}_{L,R}$ component of $\tilde{\tau}_a$ to $\tau_{L,R}$, while Yukawa interactions couple $\tilde{\tau}_{L,R}$ to $\tau_{R,L}$. The average polarization of the τ coming from $\tilde{\tau}_a \to \tau \tilde{\chi}_l^0$ can be worked out easily by calculating the partial widths $\Gamma(\tilde{\tau}_a \to \tilde{\chi}_l^0 \tau_{L,R})$. On account of the inequality $m_\tau \ll m_{\tilde{\tau}_1}$, one can identify chiral τ states as polarized ones and define the average polarization of the emerging τ as

$$P_\tau^l \equiv \frac{\mathrm{BR}(\tilde{\tau}_a \to \tilde{\chi}_l^0 \tau_R) - \mathrm{BR}(\tilde{\tau}_a \to \tilde{\chi}_l^0 \tau_L)}{\mathrm{BR}(\tilde{\tau}_a \to \tilde{\chi}_l^0 \tau_R) + \mathrm{BR}(\tilde{\tau}_a \to \tilde{\chi}_l^0 \tau_L)} = \frac{|a_{al}^R|^2 - |a_{al}^L|^2}{|a_{al}^R|^2 + |a_{al}^L|^2}. \tag{15.20}$$

In (15.20) the quantities a_{al}^b, $b = L, R$, are simply the $\tilde{\tau}_a$-$\tilde{\chi}_l^0$-τ_b coupling strengths. These are given by $G_{3sl}^{e_{L(R)}}$, $s = 3, 6$ of (9.77) and involve the elements of the flavor rotation matrix $W_{is}^{\tilde{e}*}$, as given by (9.60). For instance, if the contribution to a_{al}^b proportional to m_τ can be ignored, as is generally the case unless $\tan\beta \gg 1$ or $\tilde{\chi}_1^0$ is an almost pure higgsino, (15.20) simplifies (for $l = 1$) to

$$P_\tau = \frac{1 - x \cot^2\theta_{\tilde{\tau}}}{1 + x \cot^2\theta_{\tilde{\tau}}}, \tag{15.21}$$

where

$$x = \frac{1}{4}\left|1 + \frac{Z_{12}^\star}{Z_{11}}\cot\theta_W\right|^2, \tag{15.22}$$

with $\theta_{\tilde{\tau}}$ being determinable from the $\tilde{\tau}$ pair production cross section. More generally, a measurement of P_τ will yield information not only on neutralino mixing but also on the size of the τ Yukawa coupling strength at large $\tan\beta$ and hence on $\tan\beta$ itself. Recall that it is often not possible to accurately determine $\tan\beta$ just using the chargino-neutralino sector.

Extant constraints and future strategies

Let us now discuss search strategies plus discovery prospects for sleptons at various colliders and comment on the extraction of MSSM model parameters subsequent to their discovery. However, we need to keep in mind constraints that have already emerged from negative searches in completed experiments.

<u>LEP</u>

Since sleptons have only electroweak interactions, a lepton collider is better suited to search for them. The best lower limits so far are from the LEP 2 e^+e^- annihilation data ranging from $\sqrt{s} = 184$ GeV to $\sqrt{s} = 208$ GeV. Signatures for slepton pair production would have been: (i) a pair of hard acoplanar leptons with missing energy, (ii) missing energy accompanied by pairs of acoplanar jets arising from the decay of a chargino produced in the second of the decays (15.18), (iii) some combination of leptons and jets together with a significant amount of missing energy. However, the absence of any signal for $\tilde{\chi}_1^+ \tilde{\chi}_1^-$ production at LEP, cf. §15.2, almost uniquely singles out case (i), produced by the sequence

$$
e^+e^- \rightarrow \quad \tilde{\ell}_a^- \qquad + \qquad \tilde{\ell}_a^+ \quad , \tag{15.23}
$$
$$
\qquad\qquad \ \ \ \ \raisebox{0.3ex}{$\llcorner$}\!\!\!\rightarrow \ \ell^- \tilde{\chi}_1^0 \qquad\quad \raisebox{0.3ex}{$\llcorner$}\!\!\!\rightarrow \ \ell^+ \tilde{\chi}_1^0
$$

with $a = L, R$ for $\ell = e, \mu$ and $a = 1, 2$ for $\ell = \tau$. For staus, the decaying $\tau^\mp$ pair would lead to a pair of charged leptons or a pair of low multiplicity hadronic jets or one of each in the acoplanar event.

The slepton search efficiency is largely controlled by the slepton-LSP mass difference $\Delta M = m_{\tilde{\ell}_a} - M_{\tilde{\chi}_1^0}$. As with chargino searches, the major background here is WW production ($\gamma\gamma$ processes) for large (small) values of ΔM. The absence of a signal with an acoplanar dilepton pair and missing energy can be converted into an upper limit on $\tilde{\ell}\tilde{\ell}$ production, assuming a 100% branching fraction for the $\tilde{\ell} \rightarrow \ell\tilde{\chi}_1^0$ decay channel. The latter can then be used to obtain exclusion regions in the $M_{\tilde{\chi}_1^0}$-$m_{\tilde{\ell}}$ plane. These depend on the kind of slepton being looked for and, in the case of selectrons, also on the neutralino spectrum. In most experimental analyses, the lack of a signal is used to yield a lower bound on the mass of the right chiral slepton, assuming its left chiral partner to be beyond the kinematic limit in mass. For staus, on the other hand, the possibility of a nontrivial amount of L-R mixing needs to be taken into account; limits are generally quoted for the two extreme cases of no mixing (i.e. $\tilde{\tau}_1 = \tilde{\tau}_R$, $\tilde{\tau}_2 = \tilde{\tau}_L$) and of a $\tilde{\tau}_1$ decoupled from the Z giving rise to the minimal cross section. The τ-detection efficiency is also reduced somewhat by the energy taken away by neutrinos among its decay products. Finally, the two photon background needs careful handling. Current slepton mass limits [15.23] from LEP, assuming a value of 40 GeV for the $\tilde{\chi}_1^0$ mass are given on p *xviii* at the beginning of the book. Fig. 15.12 shows exclusion regions in the $M_{\tilde{\chi}_1^0}$-$m_{\tilde{\ell}_R}$ plane, obtained from the combined data given by all the four LEP collaborations (the ADLO set) within the MSSM. The additional assumption of high scale gaugino mass unification (12.25), implying that $\tilde{\chi}_1^0$ is nearly a bino, has gone in. These limits are rather conservative since the chosen values of

$\tan\beta$ and μ correspond to the region where the limit on the $\tilde{\chi}_1^0$ mass from chargino and neutralino searches is at its weakest and the selectron cross section is relatively low.

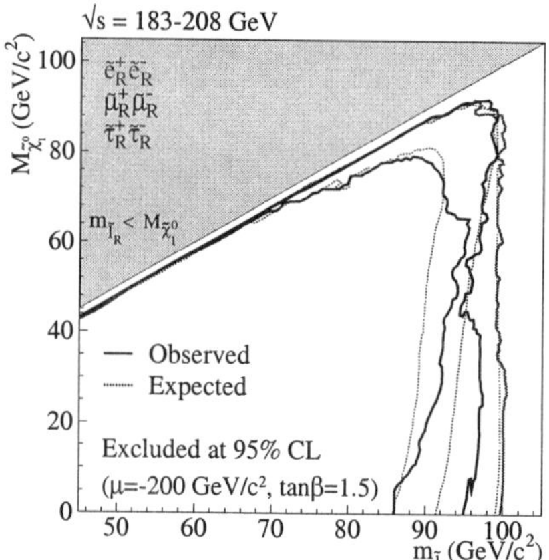

Fig.15.12. Exclusion regions in the $M_{\tilde{\chi}_1^0}$–$m_{\tilde{\ell}_R}$ plane, adapted from the ADLO analysis [15.23]. The leftmost (rightmost) pair of solid and dotted curves refers to staus (right selectrons) and the middle pair to right smuons.

In the GMSB class of models, with $\tilde{\tau}_1$ as the NLSP and $\tilde{G}$ as the ultralight gravitino LSP, the decay $\tilde{\tau}_1 \to \tilde{G}\tau$ from each of a directly produced $\tilde{\tau}_1$ pair gives rise to a final state containing $\tau^+\tau^-\tilde{G}\tilde{G}$. The decay $\tilde{\ell}^\pm \to \ell^\pm\tilde{G}$ for each of the three charged sleptons, mentioned earlier, will yield a signal comprising a pair of acoplanar prompt charged leptons plus a large amount of missing energy. If the decay length L of the slepton is as long as a meter, there will be two back-to-back heavy ionizing tracks as well as displaced vertices with large impact parameter kinks shown by the detected leptons. For small (large) values of L, the ultimate LEP search limit on $\tilde{\ell}^\pm$ in these models are as good as (better than) that in mSUGRA with a small value of $M_{\tilde{\chi}_1^0}$.

<u>Hadron colliders</u>

The hadroproduction of a charged slepton pair would be characterizable, after accounting for decays, by n_j jets and n_ℓ charged leptons. Here n_j can take values upto 4 and n_ℓ upto 6, depending on the masses of $\tilde{\chi}_2^0, \tilde{\chi}_1^\pm$ and of the concerned slepton. The simplest events would have $n_j = 0$ and $n_\ell = 2$, comprising hard, acollinear, isolated dileptons (and a lot of missing energy) with no central jet activity. Such an event would be generated from the production of an $\tilde{\ell}_a\tilde{\ell}_a$ ($a = L, R$) pair decaying via the first channel of (15.18) with $l = 1$. Sneutrino pair production could also give rise to a similar event. Despite its clean appearance, this signal has a low rate and high backgrounds from WW, $t\bar{t}$ and ZZ pair production. (In the last case one Z can decay into $\tau\bar{\tau}$ and the other into $\nu\bar{\nu}$). Detailed studies [15.41], in fact, find no TEVATRON reach in the dilepton channel in regions of mSUGRA parameter space that is allowed by LEP even after including other supersymmetric sources for this final state. In the GMSB scenario, the pair production of sleptons at the TEVATRON would be followed by the decay of each into $\ell\tilde{\chi}_1^0$, with the NLSP $\tilde{\chi}_1^0$ further decaying into $\tilde{G}\gamma$. This sequence would lead to single-photon/diphoton events along with high energy, isolated leptons. In the case of $\tilde{\tau}_1$, there are different possible decay channels involving the unstable $\tilde{\chi}_1^0$ – providing a distinct profile of additional jets and leptons. The search for sleptons would certainly be aided a lot by the occurrence of such decays, provided the mass differences involved are not too small. In case the $\tilde{\tau}_1$ itself is the NLSP, it could be searched for in two ways: (1) via the τ emerging from the decay $\tilde{\tau}_1 \to \tilde{G}\tau$ in terms of a track with a kink and a large impact parameter, (2) directly as a long-lived charged particle. For (2), the possible reach would depend crucially on the detector characteristics. Various detailed studies [15.5, 15.42] indicate, though, that the slepton mass reach at TEV-II, even with a 25 fb^{-1} integrated luminosity, will not be much beyond the LEP 2 limits quoted earlier. A similar statement can be made for the AMSB case.

Prospects for slepton discovery at the LHC are quite a bit brighter [15.6, 15.7, 15.43]. The cleanest signal is still the jet free $\ell^+\ell^- \not{E}_T$ channel. In order to suppress the very large top pair background of the LHC, a jet veto is essential. $SU(2)_L$ doublet slepton masses upto ~ 350 GeV can be probed [15.6, 15.7] in this channel for mSUGRA. Similar conclusions have been reached for the GMSB scenario [15.44]. An observation is that hard cuts required to suppress the W^+W^- pair background often remove most $\tilde{\ell}_R\tilde{\ell}_R$ as well as $\tilde{\chi}_1^+\tilde{\chi}_1^- \to \ell^+\ell^- X$ events, leaving a relatively pure, if small, sample of $\tilde{\ell}_L\tilde{\ell}_L$ events. The direct production of slepton pairs can give rise to spectacular signals for the GMSB scenario if the NLSP is either a rather long-lived $\tilde{\tau}_1$ or a short-lived $\tilde{\chi}_1^0$. Prompt $\tilde{\tau}_1 \to \tau\tilde{G}$ decays lead to signals similar to those from $\tilde{\tau}_1$ pair production in mSUGRA. The latter is difficult to detect at hadron colliders. On the other hand, in case $\tilde{\tau}_1$ is long-lived, it might be possible [15.45] to measure both its mass and lifetime, the latter yielding the gravitino mass, cf. Ch.13. In GMSB models, with $\tilde{\tau}_1$ as the NLSP, the large $\tilde{\ell}_L$-$\tilde{\ell}_R$ mass splitting makes $\tilde{\ell}_R$ pair production dominate the total slepton hadroproduction rate. This differs sharply from the situation in an mSUGRA type of a scenario where the dominant hadroproduction cross section is for the final state $\tilde{\ell}_L\bar{\tilde{\nu}}_\ell$. If a rather short-lived $\tilde{\chi}_1^0$ is the NLSP in the GMSB case, the production and decays of a pair of sleptons may yield the $\ell\ell\gamma\gamma + \not{E}_T$ signal. These spectacular events are almost free from any SM background and can be utilized not only towards slepton mass reach studies but also to extract information on $m_{3/2}$ and hence on $(\sum_i |\langle F_i \rangle|^2)^{1/2}$, cf. (13.13). In contrast, for a long-lived $\tilde{\chi}_1^0$ that decays mostly outside the detector, again only the 'canonical' dilepton plus $\not{E}_T$ signal is available. The latter yields an $\tilde{\ell}_L$ mass reach similar to that in mSUGRA. Finally, the LHC will have nonnegligible indirect slepton production [15.6, 15.7] from the decays of charginos and neutralinos; the latter in turn will often result from the decays of gluinos and squarks, cf. §15.4. Under favorable circumstances, even in those regions of the supersymmetry parameter space where direct slepton pair production no longer yields a viable signal, sleptons should be identifiable as decay products of heavier sparticles. Such is the situation in mSUGRA with $m_0 \lesssim M_{1/2} \lesssim 500$ GeV, so long as $\tan\beta$ is not too large.

<u>Future LC</u>

In a post-LHC regime, search strategies seeking both charged sleptons and sneutrinos in a linear collider would be extensions of those adopted at LEP with many of the same conclusions still valid. For instance, the mass reach for charged sleptons would be about half the CM energy. In the first round of LC's, being planned with $\sqrt{s} \leq 500$ GeV, such a reach would be comparable to that at the LHC. The focus of slepton studies at an LC would in fact be on (a) an accurate determination of slepton masses, either by kinematics or by threshold scan of the slepton pair production cross section, and (b) the establishment of their quantum numbers such as spin and hypercharge including the measurement of L-R mixing, if any. To start with, consider the simpler case of right chiral sleptons in the first two families. The utilization of partial information from LHC studies might enable one to tune the energy of an LC to produce pairs of different sleptons sequentially. The lightest charged slepton pair, produced first, would give rise via two body decays to events with a pair of acoplanar electrons/muons, as mentioned earlier. The scalar smuon would lead to a flat energy distribution of the muon produced from its two body decay $\tilde{\mu} \to \mu\tilde{\chi}_1^0$. An accurate measurement of its endpoints should yield the values of $m_{\tilde{\mu}_R}$ and $M_{\tilde{\chi}_1^0}$ precisely. The backgrounds, from W^+W^- and perhaps $\tilde{\chi}_1^0\tilde{\chi}_2^0$ production, could be handled by choosing a polarized e^-/e^+ beam. A simulation [15.8], performed for TESLA with $\sqrt{s} = 500$ GeV, $\int dt\mathcal{L} = 500$ fb^{-1} and **both beams polarized** ($P_{e^-} = 80\%$ and $P_{e^+} = 60\%$), suggested that the $\tilde{\mu}_R$ mass could be measured in this manner with an accuracy of 0.3%. For heavier charged left chiral sleptons, one might need to use the decay modes into the heavier neutralino, e.g. $\tilde{\mu}_L \to \mu\tilde{\chi}_2^0 \to \mu\tilde{\chi}_1^0\ell^+\ell^-$, because of the higher branching ratios in those channels. The clean environment of the LC should allow the reconstruction of the decay chains and the determination of the endpoints of the energy spectrum of the muon. As for $\tilde{e}_L^{\pm}$, the amplitude for its associated production with $\tilde{e}_R^{\mp}$ has only a t channel exchange contribution. An acoplanar e^+e^- pair (with missing energy) could be separated as a signal for $\tilde{e}_L^{\pm}\tilde{e}_R^{\mp}$ production from the dominant background of $\tilde{e}_R$ pairs by means of the observed $e^{\pm}$ energy distributions. Moreover, with both beams polarized, the production of LL or RR chiral selectron pairs could be suppressed by choosing *equal* polarizations for the initial e^- and e^+, as can be seen from Table 15.3.

The determination of the masses of third generation sleptons, in particular $\tilde{\nu}_\tau$ and $\tilde{\tau}_1$, by energy endpoint measurements becomes less precise because the τ's in the final state decay via $\tau \to \nu_\tau X$ ($X = \pi, \rho, a_1, \cdots$) and also because of the cuts that need to be imposed to control the 2γ background. The feasibility of such an analysis and of a consequent measurement of the $\tilde{\chi}_1^0, \tilde{\tau}_1$ masses with a 2–3% accuracy at a linear collider with an integrated luminosity of 100 fb^{-1} has been demonstrated [15.46]. It has been claimed, cf. Aguilar-Saavedra et al [15.8], that, even for third generation sleptons, mass determinations to per mille accuracy might be attained by the threshold scan method. A criticism is that the study did not include complications such as internal supersymmetry backgrounds and unknown branching ratios of the decaying sparticle. In fact, an analysis [15.47] for the sneutrinos $\tilde{\nu}_\tau, \tilde{\nu}_\mu$, addressing precisely these issues, reaches a less optimistic conclusion. It shows that the best possible mass determination comes not from threshold scans but from a measurement of the energy dependence of the cross section near threshold as well as in the continuum along with a judicious distribution of the total luminosity between the two measurements. However, the much larger pair production cross section for $\tilde{\nu}_e$ does allow the use of the threshold scan method in determining its mass. Since these mass measurements will play a key role in extracting the values of the soft supersymmetry breaking parameters, the question of their accuracy deserves more scrutiny. Coming to quantum number determinations, consider the production of a pair of right chiral smuons proceeding only via s channel exchange diagrams. Once the masses of the $\tilde{\chi}_1^0$ and the $\tilde{\mu}_R$ are determined kinematically, the three-momenta of the smuons can be reconstructed upto a twofold ambiguity. The distribution of the production angle of the $\tilde{\mu}_R$ with respect to the e^- beam direction can then be reconstructed modulo the same ambiguity. The wrong solution yields a flat background to this distribution so that the $\sin^2\theta$ distribution of the signal cross section can be clearly picked up above this background even for a low integrated luminosity of 20 fb^{-1} or so. Moreover, a measurement of the polarization dependence of the smuon pair production cross section, along with the above kinematic determination of the smuon mass, offers the best possibility of establishing the hypercharge of the produced smuon.

As for staus, a measurement of the $\tilde{\tau}_1$ pair production rate, coupled with the kinematic determination of the $\tilde{\tau}_1$ mass, allows a determination of the stau mixing angle $\theta_{\tilde{\tau}}$. Polarized beams substantially heighten the $\sin\theta_{\tilde{\tau}}$ dependence of the cross section. For instance, for $m_{\tilde{\tau}_1} = 100$ GeV, the $\tilde{\tau}_1$ pair production cross section at a $\sqrt{s} = 500$ GeV e^+e^- collider decreases (increases) by 13% (360%) for $P_L = 0$ (1) as $\tilde{\tau}_1$ ranges from a $\tilde{\tau}_L$ to a $\tilde{\tau}_R$. The error in the determination of $\sin\theta_{\tilde{\tau}}$, dominated by that in measuring $m_{\tilde{\tau}_1}$, can be reduced by using an accurate knowledge of the $\tilde{\chi}_1^0$ mass from some other studies. In the end, a determination of $\sin\theta_{\tilde{\tau}}$ to an accuracy of 3% has been shown [15.46] to be feasible for $m_{\tilde{\tau}_1} = 150$ GeV, $M_{\tilde{\chi}_1^0} = 100$ GeV at an LC with $\sqrt{s} = 500$ GeV and an integrated luminosity of 100 fb^{-1}. If the high scale gaugino mass unification condition (12.25) is assumed, the determination of $\sin\theta_{\tilde{\tau}}$ from the measurement of P_τ and of production cross sections of a stau pair as well as of a right chiral charged slepton pair with a right handed electron beam can be used to extract $\tan\beta$, M_1 and μ (upto a sign ambiguity [15.46]).

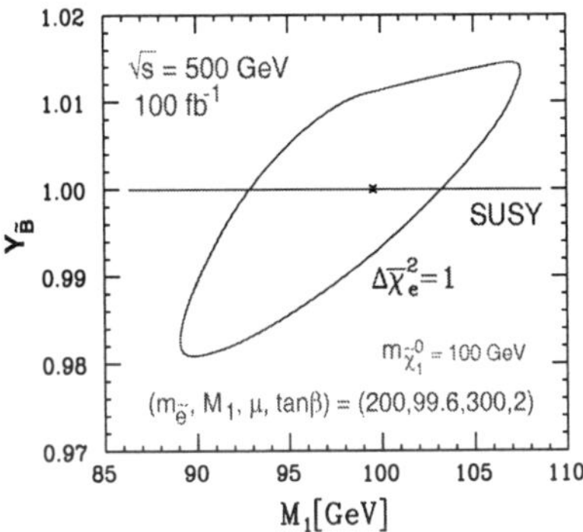

Fig.15.13. Simulated $\Delta\chi^2 = 1$ contour in the $Y_{\tilde{B}}$–M_1 plane from a right selectron pair produced at a linear collider, adapted from Ref. [15.25].

The production amplitude for $e^+e^- \to \tilde{e}_R^+\tilde{e}_R^-$, with a right handed electron beam, receives contributions from a t channel diagram involving the $\tilde{e}_R^+ e_R^- \tilde{B}$ coupling strength $g_{\tilde{e}_R^+ e_R^- \tilde{B}} = \sqrt{2}g_1 Y_{\tilde{B}}$. This bino coupling $Y_{\tilde{B}}$, which is unity at the tree level, can therefore be extracted (to within 1%) by a high statistics measurement of the corresponding differential cross section $d\sigma/d\cos\theta$, provided the $\tilde{\chi}_1^0$ mass is known from a separate kinematic determination [15.25, 15.46]. We show in Fig. 15.13 the $\Delta\chi^2 = 1$ contour in the $Y_{\tilde{B}}$–M_1 plane obtained [15.25] in a Monte-Carlo study of a pair of 100 GeV $\tilde{e}_R$'s produced by e^+e^- annihilation at $\sqrt{s} = 500$ GeV with 100 fb^{-1} of data. The input values, used in this study, are shown on the figure with dimensional quantities being in units of GeV. This is a case where, staying only within the slepton sector, one can precisely test [15.48] one of the basic tenets of supersymmetry, namely that *the lepton-slepton-gaugino and the lepton-lepton-gauge couplings are equal.* Furthermore, this equality receives radiative corrections in broken supersymmetry and can thus be used to extract indirect information on the masses of those squarks which may be sufficiently heavy to be beyond the reach of the LHC. These corrections can be [15.48, 15.49] of the order of 1–3% for $m_{\tilde{q}}/m_{\tilde{\ell}}$, $m_{\tilde{q}}/m_{\tilde{\nu}} \sim 10$ and may well be within the limits of measurability in the linear colliders being planned.

Finally, let us turn to the minimal AMSB scenario. There the pair-produced left chiral sleptons can lead [15.50] to a $\tilde{\chi}_1^\mp \tilde{\chi}_1^0 \nu_\ell \ell^\pm$ combination among their decay products providing a prompt lepton trigger and missing energy. The lighter chargino can have a long lifetime in this scenario, as discussed in the previous section, and will yield a heavily ionizing track and/or a displaced vertex with an impact parameter kink shown by the pion from the decay $\tilde{\chi}_1^\mp \to \pi^\mp \tilde{\chi}_1^0$. A doubling of this effect may be expected from the combination $\tilde{\chi}_1^\pm \tilde{\chi}_1^\mp \nu_\ell \overline{\nu}_\ell$. But, a distinction from the corresponding signal of direct $\tilde{\chi}_1^\pm$ pair production will now be possible only if the mass difference between the chargino and the (heavier) charged slepton is quite large. An inclusive analysis of data seeking evidence for one or more long-lived charged particles can nonetheless provide information on sleptons in this scenario.

15.4   Signals of gluinos and squarks

Though gluinos and squarks (of the first two generations[32]) are expected to be among the heaviest of sparticles, their large production cross sections have made them the main focus of supersymmetry studies at the TEVATRON and the LHC [15.5–15.7]. If the running gaugino masses unify at a high scale via (12.25), gluinos are predicted to be the heaviest of all MSSM gauginos. Such is also the case in many scenarios with nonuniversal high scale gaugino masses. On the other hand, the large top mass makes L-R mixing important for stops and indeed these effects are nonnegligible even for sbottoms at large $\tan\beta$. Thus $\tilde{t}_1$ could be the lightest strongly interacting sparticle, and might even be the NLSP. Squarks and gluinos mostly decay into quarks, antiquarks and charginos/neutralinos; the latter in turn decay as discussed in §15.2, resulting in long decay chains, generically described as **cascade decays**. A detailed analysis of the production and decay of squarks and gluinos is therefore often difficult, but does offer the prospect of probing many facets of the underlying model.

Hadroproduction

We choose hadronic colliders first since only these lead to the primary production of gluinos at the lowest order. Squark and gluino hadroproduction cross sections are determined only by their masses. The gluino mass $M_{\tilde{g}}$ affects not only the gluino but also some squark pair production cross sections, while *all* such cross sections depend on the squark masses. The basic subprocesses for gluino pair production are

$$q_i\bar{q}_i \quad\to\quad \tilde{g}\tilde{g}, \tag{15.24a}$$

$$gg \quad\to\quad \tilde{g}\tilde{g}. \tag{15.24b}$$

Squark pair production proceeds via

$$q_iq_j \quad\to\quad \tilde{q}_i\tilde{q}_j, \tag{15.25a}$$

$$q_i\bar{q}_j \quad\to\quad \tilde{q}_i\bar{\tilde{q}}_j,\ (i\neq j), \tag{15.25b}$$

$$q_i\bar{q}_i \quad\to\quad \tilde{q}_j\bar{\tilde{q}}_j, \tag{15.25c}$$

$$gg \quad\to\quad \tilde{q}_i\bar{\tilde{q}}_i. \tag{15.25d}$$

Finally,

$$gq_i \to \tilde{g}\tilde{q}_i \tag{15.26}$$

often makes the biggest contribution to the inclusive squark production cross section. As usual, charge conjugate modes should be added to (15.25a–b) and (15.26). Here i, j are flavor indices and q stands for either u or d. Chirality labels for squarks have been suppressed. Contributions from t channel gluino exchange to the reactions (15.25a–c) contain all possible left and right chiral squark combinations, while only left-left and right-right combinations are produced in s channel diagrams as well as in (15.25d). As mentioned in §15.1, squark flavor mixing, which is inconsequential here, has been neglected.

---

[32]Their masses are relatively constrained by the inequality (11.45c).

The tree level diagrams, corresponding only to the strong interaction contributions to the subprocesses (15.24), (15.25) and (15.26), are shown in Figs. 15.14, 15.15 and 15.16 respectively.

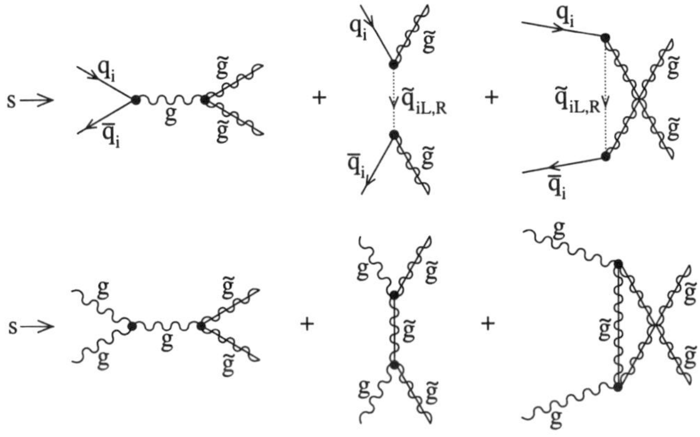

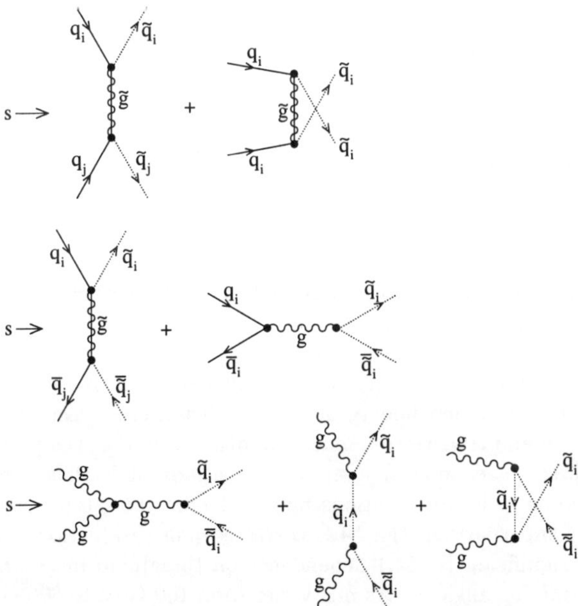

Fig. 15.14. Tree diagrams for gluino pair production in hadronic collisions.

Detailed expressions for the subprocess differential cross sections $d\hat{\sigma}/dt$ are available in the literature [15.13, 15.51].

Fig. 15.15. Tree diagrams for squark pair production in hadronic collisions with arrows indicating the flow of baryon number.

Sizable gluon fluxes and (for the $\tilde{g}\tilde{g}$ final state) large color factors make gg fusion contributions dominant at LHC energies energies if $M_{\tilde{g}}, m_{\tilde{q}} \lesssim 1$ TeV, while reactions involving valence quarks dominate squark and gluino production at the TEVATRON in the allowed mass range. The rate

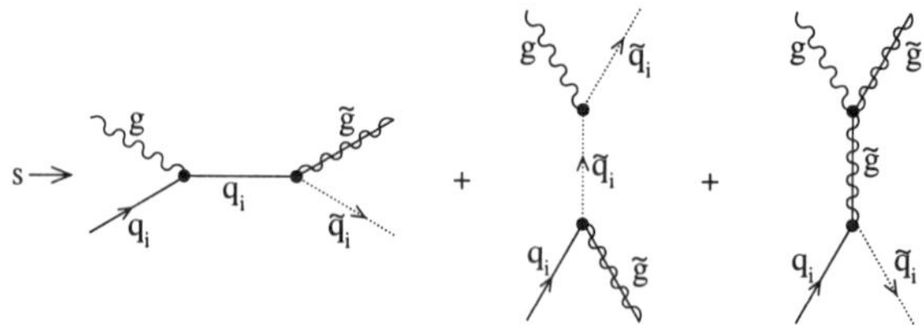

Fig. 15.16. Tree diagrams for the gluino squark production channel.

of pair production of all strongly interacting sparticles, taken together, reaches the highest value when $M_{\tilde{g}} \simeq m_{\tilde{q}}$. This is where none of the t/u channel $\tilde{g}/\tilde{q}$ exchange contributions is suppressed, while the contribution from $\tilde{g}\tilde{q}$ production is substantial. Note that stop pair production via strong interactions is special in that its leading order cross section depends only on the mass of the produced sparticle. The cross section for the direct pair production of the lightest stop $\tilde{t}_1$ is expected at about a tenth of that of top quarks of the same mass. The reduction is due to the β^3 suppression of stop production near threshold as well as to the fewer degrees of freedom contained in $\tilde{t}_1$ than in t. If $m_{\tilde{t}_1} < m_t$, a possibility not yet ruled out by experiment (see below), stop pairs can even be produced from the decays of a top and an antitop. The branching ratio for the decay $t \to \tilde{t}_1 \tilde{\chi}_1^0$ is constrained by the current information on $\sigma(p\bar{p} \to t\bar{t}X)$ at the TEVATRON and the kinematically measured value of m_t. These two together restrict any possible deviation of the top semileptonic decay branching ratio from its SM value and hence any nonstandard decay channel for the top. The phase space available for the $t \to \tilde{t}_1 \tilde{\chi}_1^0$ decay channel [15.52] is currently getting squeezed by increasing lower limits on $m_{\tilde{t}_1}$ from searches for $\tilde{t}_1$ pair production, as well as by (model dependent) lower limits on $M_{\tilde{\chi}_1^0}$. Thus, though the top pair production cross section $\sigma_{t\bar{t}}$ is fairly high at TEV-II/LHC, the contribution to stop production from top decay, if present at all, is likely to be quite limited.

The next-to-leading order (NLO) supersymmetric QCD corrections to all the subprocesses (15.24)–(15.26) have been found in general to be quite large [15.53]. For instance, on incorporating these corrections the corresponding cross sections, evaluated with a factorization and renormalization scale [15.14] near the average mass of the produced sparticles, increase typically by 10–50% at $\sqrt{s} = 2$ TeV. Moreover, the scale dependence of each predicted cross section is reduced to about 15%. However, the rapidity and transverse momentum distributions of the produced sparticles are not much affected by these corrections. For most applications, it is thus a good approximation to simply rescale the leading order distributions by a K factor. For stops, characterized by large L-R mixing and a heavy superpartner, the NLO-corrected pair production cross section acquires a (numerically not very significant [15.54]) dependence on the gluino mass $M_{\tilde{g}}$ and other squark masses as well as on the mixing angle $\theta_{\tilde{t}}$. As $m_{\tilde{t}_1}$ varies from 100 GeV to 300 GeV, the effective K factor varies between 1.4 and 1.1 at the TEVATRON and stays near 1.4 at the LHC. An important point is that the shape of the transverse momentum distribution *is altered* by NLO corrections in this case.

The expected cross sections [15.55, 15.56] for the pair production of different strongly interacting sparticles at the TEVATRON and the LHC are shown respectively in Figs. 15.17 and 15.18. The latter also includes electroweak bosino pair production cross sections for comparison. Five degenerate species of left and right chiral quarks have been taken for Fig. 15.17: an assumption expected to break down at large $\tan\beta$. The mass[33] M_X on the abcissa is different for different subprocesses. One has chosen $M_X = m_{\tilde{q}}$ and $M_{\tilde{g}} = 200$ GeV for $\tilde{q}\bar{\tilde{q}}$ and $\tilde{q}\tilde{q}$ production, whereas $M_X = M_{\tilde{g}}$ and $m_{\tilde{q}} = 200$ GeV have been taken for $\tilde{g}\tilde{g}$ production. For $\tilde{q}\tilde{g} + \bar{\tilde{q}}\tilde{g}$ production, the region hatched with vertical (horizontal) bars corresponds to $M_{\tilde{g}}$ $(m_{\tilde{q}}) = 200$ GeV and $M_X = m_{\tilde{q}}$ $(M_{\tilde{g}})$. The bands show the effect of changing the scale from $\langle\tilde{m}\rangle/2$ to $2\langle\tilde{m}\rangle$. The two panels of Fig. 15.18 show [15.56] leading order predictions for various sparticle pair production cross sections,

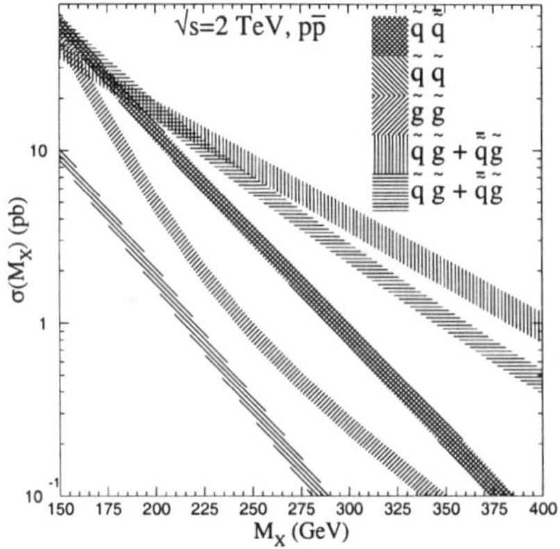

Fig. 15.17. Predicted squark and gluino production cross sections (adapted from Ref. [15.55]) at the TEVATRON with NLO corrections taken into account. The variable M_X has been defined in the text.

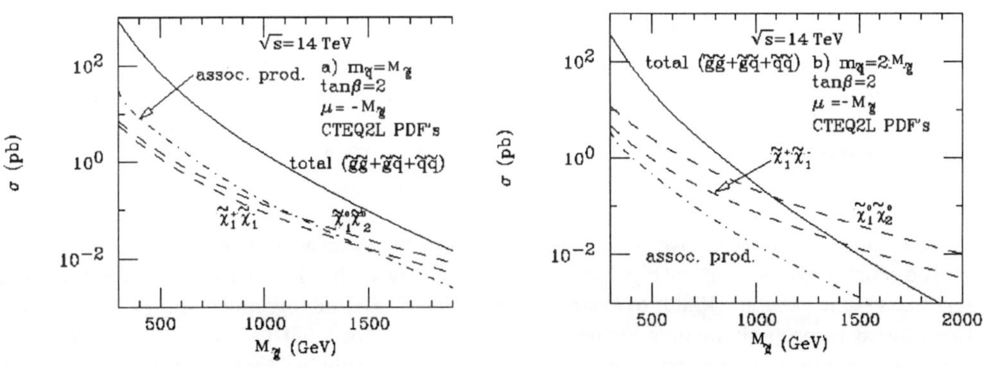

Fig. 15.18. LO production cross sections of sparticle pairs at the LHC, adapted from Ref. [15.56].

including the direct production of electroweak bosinos at the LHC, as functions of $M_{\tilde{g}}$ for two different choices of $m_{\tilde{q}}$, viz. $m_{\tilde{q}} = M_{\tilde{g}}$ and $m_{\tilde{q}} = 2M_{\tilde{g}}$. As in the previous figure, $\tilde{b}$ has been taken to be mass degenerate with the four other squark species of the first two generations; furthermore, high scale gaugino mass unification (12.25) has been assumed. The values of μ and $\tan\beta$ shown

---

[33]Some of the mass values shown in Figs. 15.17 and 15.18 may not be compatible with the current lower bounds [15.57] on $M_{\tilde{g}}$ and $m_{\tilde{q}}$.

on the figures are not relevant to the production of strongly interacting sparticles. For the case
$m_{\tilde{q}} = M_{\tilde{g}}$, this dominates the total sparticle production rate at the LHC all the way upto $M_{\tilde{g}} = 2$
TeV. Also shown is the rate of associated production of a gluino or squark with a neutralino
$\tilde{\chi}_l^0$, $l = 1, \cdots, 4$ or a chargino $\chi_k^{\pm}$, $k = 1, 2$, which is always subdominant. The relatively high
values of the cross sections for gluino and squark hadroproduction make it clear why sparticle
search at the LHC will be centered around them.

Let us comment separately on the production of the state $\tilde{t}_1\bar{\tilde{t}}_1$ [15.54]. For a $\tilde{t}_1$ mass between 80
and 200 GeV, this cross section at TEV-II varies from about 40 pb to 0.2 pb. On the other hand,
for $m_{\tilde{t}_1}$ between 200 and 400 GeV, it lies in the range 70–2.5 pb at LHC energies. In comparison,
the $t\bar{t}$ production cross section at TEV-II and at the LHC are expected to be ~ 8 pb and ~ 600
pb respectively, though $BR\,(t \to \tilde{t}_1\tilde{\chi}_1^0)$ is never high. For heavy gluinos and the mass hierarchy
$m_t + m_{\tilde{t}_1} < m_{\tilde{g}}$, the $\tilde{t}_1\bar{\tilde{t}}_1$ combination can arise from $\tilde{g}\tilde{g}$ pair production and the subsequent decays
[15.58] $\tilde{g} \to \bar{t} + \tilde{t}_1,\ t + \bar{\tilde{t}}_1$.

Production at an e^+e^- collider

All squarks get produced at the lowest order only through s channel $Z^\star$ and $\gamma^\star$ exchanges
in e^+e^- collision, while gluinos can emerge from such a collision only by radiation off final state
(s)quarks. The pair production of squarks of the first two generations is just like that of smuon pairs
with the rate completely decided by squark masses and EW couplings. But stop and sbottom pairs
are produced like stau pairs, controlled by the corresponding L-R mixing. Thus the expressions
[15.36] for tree level e^+e^- squark pair production cross sections can be simply derived from Table
15.1 by multiplying with a color factor of three, dropping the t channel contributions and inserting
the appropriate charges and weak isospins of the produced squarks.

For a sfermion $\tilde{f}_1$, the $Z\tilde{f}_1\bar{\tilde{f}}_1$ vertex vanishes when the L-R mixing angle $\theta_{\tilde{f}}$ obeys the relation
$|\cos\theta_{\tilde{f}}| = \sqrt{2|Q_f|}\sin\theta_W$, Q_f being the charge of f. For $f = t, b$, this leads to a minimum in the
$\tilde{t}_1\bar{\tilde{t}}_1$, $\tilde{b}_1\bar{\tilde{b}}_1$ production cross section as a function of $\cos\theta_{\tilde{t}/\tilde{b}}$, the location of the minimum depending
on the CM energy and on beam polarization. Just as with sleptons, QED ISR affects both the $Z^\star$
and $\gamma^\star$ exchange contributions strongly, producing a net effect of upto 10–15%. There is a new
feature here. Significant higher order effects come not only from Yukawa interactions [15.59], which
can be quite large for $\tilde{t}\bar{\tilde{t}}$ production, but also from QCD [15.60, 15.61] and SQCD [15.62] vertices.
Gluon QCD corrections are always positive and quite a bit bigger (by about a factor four at large
$\sqrt{s}$) than those for fermion partners of sfermions $\tilde{f}$. The gluino QCD corrections can be either
negative or positive, but are never more than a few percent. A measurement of the stop (sbottom)
pair production cross section with polarized beams – along with a kinematic determination of the
$\tilde{t}_1$ ($\tilde{b}_1$) mass – can be used [15.39] to extract information on the mixing angle $\theta_{\tilde{t}}$ ($\theta_{\tilde{b}}$). It is essential
in this context to properly include higher order corrections. For a left handed e^- beam, the $\tilde{t}_1$ ($\tilde{t}_2$)
pair production cross section increases (decreases) monotonically, as $\cos\theta_{\tilde{t}}$ increases. In contrast,
for a right handed e^- beam, the opposite dependence on $\cos\theta_{\tilde{t}}$ results. The $\tilde{t}_1\bar{\tilde{t}}_2/\bar{\tilde{t}}_1\tilde{t}_2$ production
process offers the possibility of studying a $\tilde{t}_2$ below its pair production threshold and has a cross
section upto 10 fb at $\sqrt{s} = 500$ GeV with a clear $\sin 2\theta_{\tilde{t}}$ dependence.

Decay characteristics

While a separate treatment for decaying $\tilde{t}$ and $\tilde{b}$ squarks is given later, we first discuss the
decay chains of gluinos and squarks in the first two generations. We assume the latter to be eight
degenerate states with a common mass $m_{\tilde{q}}$. On being produced, all these will have cascade decays.
Generally, strongly interacting particles and sparticles tend to come at the top end of the cascade,

while electroweak ones appear in the lower part of the chain. The sequences are different in detail for the two mass inequality regimes $m_{\tilde{q}} < M_{\tilde{g}}$ and $M_{\tilde{g}} < m_{\tilde{q}}$, as shown in the two halves of Table 15.4. Note also that the neutralino $\tilde{\chi}_l^0$ and the charginos $\tilde{\chi}_k^{\pm}$, generated along the way, will decay further (cf. §15.2).

$m_{\tilde{q}} < M_{\tilde{g}}$	$M_{\tilde{g}} < m_{\tilde{q}}$
$\tilde{g} \to \bar{q}\tilde{q}_{L,R} + c.c.$	$\tilde{q}_{L,R} \to q\tilde{g}$
	$\quad\to q\tilde{\chi}_l^0,\ l = 1,\cdots,4$
	$\tilde{q}_L \to q'\tilde{\chi}_k^{\pm},\ k = 1,2$
$\tilde{q}_{L,R} \to q\tilde{\chi}_l^0,\ l = 1,\cdots,4$	$\tilde{g} \to q\bar{q}\tilde{\chi}_l^0,\ l = 1,\cdots,4$
	$\quad\to q\bar{q}'\tilde{\chi}_k^{\pm},\ k = 1,2$
$\tilde{q}_L \to q'\tilde{\chi}_k^{\pm},\ k = 1,2$	$\quad\to \bar{t}\tilde{t}_1 + t\bar{\tilde{t}}_1$ if $M_{\tilde{g}} > m_{\tilde{t}_1} + m_t$
$\tilde{q} \to q\bar{b}\tilde{b}_1 + q b\bar{\tilde{b}}_1$ if $m_{\tilde{q}} > m_b + m_{\tilde{b}_1}$	$\quad\to \bar{b}\tilde{b}_1 + b\bar{\tilde{b}}_1$ if $M_{\tilde{g}} > m_{\tilde{b}_1} + m_b$

Table 15.4. Gluino and squark decay chains depending on which is heavier.

An examination of the entries in the above table makes it clear how to write an expression for[34] $\Gamma(\tilde{g} \to \bar{q}\tilde{q}_{L,R} + c.c.)$. One substitutes the couplings of (9.78) into (15.10) and multiplies by a color factor of $1/2$. The partial widths of the decays of squarks into different two body final states via electroweak interactions, determined by the gaugino/higgsino contents of the various $\tilde{\chi}_l^0$, $\tilde{\chi}_k^{\pm}$ and their masses, are given by expressions very similar to those occurring in (15.19). These are the only possible two body decays of squarks if $m_{\tilde{q}}$ is less than $M_{\tilde{g}}$. This is the case (left half of Table 15.4) we consider first. The right chiral squark $\tilde{q}_R$ has the strongest coupling to the $\tilde{\chi}_l^0$ with the highest bino content. Since the latter is the $\tilde{\chi}_1^0$ in most of the mSUGRA parameter space, $BR\,(\tilde{q}_R \to q\tilde{\chi}_1^0)$ is almost 98% there. On the other hand, the larger $SU(2)_L$ coupling strength favors the decay of the left chiral squark $\tilde{q}_L$ into a charged or neutral $SU(2)_L$ gaugino plus a quark. These decays are open in nearly all supersymmetry models; this is since $m_{\tilde{q}}$ exceeds M_2 so long as gaugino masses unify at least approximately at a high scale. On the other hand, $m_{\tilde{\ell}_R}$ is less than $M_{\tilde{\chi}_2^0}$ if $m_0 < M_{1/2}$ in mSUGRA. Thus, in that scenario, the decays of $\tilde{\chi}_2^0$ produced from those of $\tilde{q}_{L,R}$ will give rise to multilepton states with high branching ratios. Finally, the three body decays, shown in the last line of the left half of Table 15.4, are gluino mediated. These can have large BR's if the lighter charginos and neutralinos are higgsinolike and hence have small squark coupling [15.63] strengths.

We turn next to the region with $M_{\tilde{g}} < m_{\tilde{q}}$, considered in the right half of Table 15.4. Here the allowed strong two body decays of squarks dominate over their electroweak ones, e.g. $BR\,(\tilde{d}_R \to d\tilde{g})$ is about 98% unless the $\tilde{g}$ and the $\tilde{d}_R$ are nearly mass degenerate. However, even for a much smaller $M_{\tilde{g}}$, $BR\,(\tilde{q}_L \to q\tilde{g})$ varies between 80 and 90%; the remaining fraction goes into $SU(2)_L$ gauginolike charginos/neutralinos plus quarks. The gluino decays, shown in Table 15.4, are allowed at the tree level. Now $q, \bar{q}$ include third generation quarks and antiquarks which, together with the decays of the last two lines (if energetically allowed), will contribute to the t/b content of the final state. Each three body decay, with a final state neutralino (chargino), is mediated by a virtual $\tilde{q}_{L,R}$ ($\tilde{q}_L$), the amplitudes with virtual $\tilde{q}_L$ and $\tilde{q}_R$ not interfering for massless final state quarks. In case $M_{\tilde{g}} \ll m_{\tilde{q}}$,

---

[34]Analogous decays into third generation quarks and squarks may often occur with larger BR's.

gluino decay into a chargino/neutralino plus a first or second generation quark antiquark pair would roughly follow the decay pattern of $\tilde{q}_L$. This is since the larger $SU(2)_L$ coupling strength would give a large weight to the decay mediated by a virtual $\tilde{q}_L$. However, since the square of the $\tilde{q}_i$ exchange diagram scales like $(m_{\tilde{q}}^2 - M_{\tilde{g}}^2)^{-2}$, the $\tilde{q}_L$-$\tilde{q}_R$ mass splitting, even if small, is an important factor especially if $m_{\tilde{q}}$ is not too far above $M_{\tilde{g}}$. As a result, gluino decay into $q\bar{q}\tilde{\chi}_l^0$ can get a boost. Similarly, the reduced $\tilde{b}_1$ and $\tilde{t}_1$ masses can enhance the branching ratios into three body final states containing b and/or t quarks [15.64, 15.65]. There is also the loop induced gluino decay [15.64] $\tilde{g} \to g\tilde{\chi}_l^0$ which may become significant if the LSP is a pure higgsino or if $m_{\tilde{q}} \gg M_{\tilde{g}}$. For most regions of the mSUGRA parameter space, however, its branching fraction is below a few percent level. Tree level expressions for the partial widths, given in Refs. [15.19, 15.63, 15.66], are the most complete ones and can easily be transcribed to our notation.

Let us now discuss the decays of third generation squarks. Their special features are (1) the large value of m_t, (2) the large expected mass splitting between the top squark mass eigenstates on account of large L-R mixing and (3) the large Yukawa couplings of the stop. All of the kinematically allowed strong two body decays of $\tilde{t}_a, \tilde{b}_a$, i.e.

$$\tilde{t}_a \to t\tilde{g}, \qquad \tilde{b}_a \to b\tilde{g}, \quad a = 1, 2, \tag{15.27}$$

will be important, their partial widths being trivially obtainable from (15.19). If $\tilde{t}_a, \tilde{b}_a$ $(a = 1, 2)$, happen to be lighter than the gluino, several interesting possibilities arise. When $m_{\tilde{t}_a} - m_t > M_{\tilde{\chi}_l^0}$, $m_{\tilde{t}_a} - m_b > M_{\tilde{\chi}_k^+}$ and $m_{\tilde{b}_a} - m_t > M_{\tilde{\chi}_k^-}$ for $a = 1$ and/or 2, one can have the decays

$$\tilde{t}_a \to t\tilde{\chi}_l^0, \qquad b\tilde{\chi}_k^+, \quad l = 1, \cdots, 4, \quad k = 1, 2, \tag{15.28a}$$

$$\tilde{b}_a \to b\tilde{\chi}_l^0, \qquad t\tilde{\chi}_k^-, \quad l = 1, \cdots, 4, \quad k = 1, 2. \tag{15.28b}$$

Also, if the mass difference $|m_{\tilde{t}_a} - m_{\tilde{b}_b}|$ is large enough, the decays

$$(\tilde{t}, \tilde{b})_b \to (\tilde{b}, \tilde{t})_a W^{+,-}(H^{+,-}), \tag{15.29}$$

can occur. Moreover, decays into final states with neutral bosons, namely

$$(\tilde{t}, \tilde{b})_2 \to (\tilde{t}, \tilde{b})_1 Z(h, H, A). \tag{15.30}$$

will become significant for a sufficiently large $\tilde{Q}_2$-$\tilde{Q}_1$ mass splitting, where $\tilde{Q} = \tilde{t}, \tilde{b}$. Higher order corrections due to large Yukawa interactions [15.67] as well as supersymmetric QCD corrections [15.68] to the decays of $\tilde{t}_a, \tilde{b}_a$ $(a = 1, 2)$, have been calculated. These are generally 10–20% in most cases.

The lighter stop $\tilde{t}_1$ is in many ways the most interesting of all the squarks. For a wide range of supersymmetry model parameters, every two body decay of $\tilde{t}_1$ is forbidden at the tree level. This allows the loop induced decay [15.69] $\tilde{t}_a \to c\tilde{\chi}_1^0$ to take over. The amplitude of the latter is proportional to the mixing between $\tilde{t}$ and $\tilde{c}$ flavor eigenstates. Such a mixing is absent in mSUGRA or GMSB models at the high (input) scale, but gets induced in the $SU(2)_L$ doublet sector by weak radiative corrections proportional to the CKM element V_{cb}. Even in mSUGRA, where these corrections are enhanced by $\ln(M_{Pl}/m_{\tilde{t}})$, this loop decay will be small if there is no mixing in the stop sector. It will also be small if $\tilde{t}_L$ is much lighter than $\tilde{c}_L$ since the $\tilde{t}_L$-$\tilde{c}_L$ mixing angle, coming from a one loop diagram, scales like $(m_{\tilde{t}_L}^2 - m_{\tilde{c}_L}^2)^{-1}$. A heavy $\tilde{t}_1$ will further have competitive three body decays

$$\tilde{t}_1 \to \bar{\tilde{\ell}}\nu_\ell b, \quad \tilde{\ell}\bar{\nu}_\ell b, \quad b\tilde{\chi}_l^0 W^+(H^+). \tag{15.31}$$

For lighter sleptons, the occurrence of two heavy particles in the final state will disfavor the last channel of (15.31) as compared to the other two. Indeed, in mSUGRA, the first two channels of (15.31) will be favored (specially at low m_0) since sleptons are generally lighter there than squarks. Furthermore, at large $\tan\beta$, the $\tilde{\tau}_1$, being the lightest of all charged sleptons, helps the channel of (15.31) with $\ell = \tau$, eventually yielding a final state $b\tau^+\nu_\tau\tilde{\chi}_1^0$ on account of the decay $\tilde{\tau}_1 \to \tau\tilde{\chi}_1^0$. However, for large m_0, the mode with $c\tilde{\chi}_1^0$ as the final state has an appreciable branching ratio even at high $\tan\beta$. When the $\tilde{t}_1$ is so light that even the three body decays of (15.31) are forbidden, yet $m_{\tilde{t}_1} > m_b + M_{\tilde{\chi}_1^0}$, the four body decays $\tilde{t}_1 \to b\tilde{\chi}_1^0 f\bar{f}'$ could occur. The latter could have substantial branching fractions and could dominate over the $\tilde{t}_1 \to c\tilde{\chi}_1^0$ mode. Of course, this is provided the exchanged sparticles in the corresponding diagrams are not highly virtual, i.e. do not have masses much higher than that of $\tilde{t}_1$ [15.70]. Finally, if $\tilde{t}_1$ has no tree level two body decays, it should be so long-lived as to hadronize [15.71] before decaying.

We conclude this subsection with comments on two specific topics in relation to gluino and squark decays.

(a) The importance of contributions from $\tilde{g}/\tilde{q}$ decays to the production of the electroweak bosinos $\tilde{\chi}_l^0, \tilde{\chi}_k^\pm$ was mentioned earlier. In most models, either gluinos or $SU(2)_L$ doublet squarks almost always have large branching ratios into winolike charginos and neutralinos. This does not happen if $m_{\tilde{q}_L} > M_{\tilde{g}} > m_{\tilde{t}_1} + m_t$ or $|M_2| \gtrsim |M_3|$ at the weak scale, the latter requiring a violation of the assumption (12.25) of high scale gaugino mass unification by at least a factor of three. These $SU(2)_L$ gauginolike bosinos play important roles both in defining signals for gluino and squark production at hadron colliders and in attempts at event reconstruction, as will be discussed later.

(b) Decays of $\tilde{g}/\tilde{q}$ as well as of the heavier third generation squarks $\tilde{t}_2, \tilde{b}_2$ can be copious sources of Higgs particles. Moreover, we have just seen that $\tilde{g}$ or $\tilde{q}_L$ production often leads to final states containing winolike neutralinos whose decay products contain Higgs bosons. For instance, in mSUGRA, $\tilde{\chi}_2^0$ is often winolike and would decay predominantly into $\tilde{\chi}_1^0 h$, were this channel open and were $\tilde{\chi}_2^0$ lighter than $\tilde{\ell}$. ATLAS and CMS studies [15.6, 15.7] show the feasibility of signal isolation for $h \to b\bar{b}$ decays in inclusive SUSY event samples. Gluino and squark decays have been found to provide [15.72] an excellent source for the heavier charged Higgs as well when its mass exceeds that of the top quark.

Search results and futurology

<u>TEV-I and LEP</u>

The negative outcome of $\tilde{q}/\tilde{g}$ searches at the TEVATRON [15.57, 15.73] and $\tilde{q}$ searches at LEP [15.23] has established lower bounds on the masses $M_{\tilde{g}}, m_{\tilde{q}}$ and $m_{\tilde{t}_1/\tilde{b}_1}$. Any tree level decay of a produced gluino has to lead to at least a pair of jets along with substantial $\not{E}_T$ carried away by a stable $\tilde{\chi}_1^0$ produced at the end of a decay chain. For $m_{\tilde{q}} \leq M_{\tilde{g}}$, squark decay would produce at least one hard jet and a $\tilde{\chi}_1^0$, cf. Table 15.4. In contrast, for $M_{\tilde{g}} < m_{\tilde{q}}$, it would produce one jet via $\tilde{q} \to q\tilde{g}$ and the decay of the resultant $\tilde{g}$ would lead to additional jets. We have just seen that $\tilde{q}_L$ and $\tilde{g}$ decays often yield $SU(2)_L$ gauginolike charginos and neutralinos. The final decay products would involve, in addition to the hard jets mentioned above, hard isolated charged leptons, neutrinos and/or additional jets via the decays $\tilde{\chi}_1^+ \to f'\bar{f}\tilde{\chi}_1^0$ and $\tilde{\chi}_2^0 \to f\bar{f}\tilde{\chi}_1^0$. The production and decays of gluinos would, in consequence of their Majorana character, give rise to equal numbers of leptons of either sign. A clean signature [15.74] of $\tilde{g}\tilde{g}$ or $\tilde{q}\tilde{g}$ production is then *a large fraction of the generated lepton pairs being same sign (SS) dileptons*. SM backgrounds to this signal, as contrasted with those for opposite sign (OS) dileptons, are very small.

The final states, resulting from the production and decay of squarks and gluinos at hadron colliders, can be broadly classified as containing $n_j \geq 2$ jets, $n_\ell \geq 0$ charged leptons (electrons and muons) and large $\not{E}_T$. These constitute the 'canonical' signals for strong sparticle hadroproduction. Furthermore, for a higher gluino mass $M_{\tilde{g}}$, the decay products of $\tilde{\chi}_1^\pm/\tilde{\chi}_2^0$ may involve real W/Z's or Higgs bosons. In general, events from $\tilde{g}\tilde{g}$ production will have higher jet multiplicity. Again, for higher values of $M_{\tilde{g}}$, gluino decays will give rise to final states with higher t/b content. As already mentioned, a larger $\tan\beta$ (and hence lighter $\tilde{\tau}_1$'s) reduce the decay branching ratios into lighter leptons, thereby enhancing the relative number of τ's in the final state. All these features can be used to define additional subclasses of supersymmetry events and constitute signatures of sparticle hadroproduction. For $n_\ell = 2$, one has events with OS, SS dileptons. Multilepton events with $n_\ell = 2$ or $n_\ell = 3$ arise also from the 'direct' production of $\tilde{\chi}_2^0\tilde{\chi}_1^0$, $\tilde{\chi}_1^\pm\tilde{\chi}_2^0$, as discussed earlier. The two sources can be separated by looking for hadronically clean/quiet events. For the GMSB scenario, with a short-lived $\tilde{\chi}_1^0$ as the NLSP, the decay of the latter into $\tilde{G}\gamma$ will give rise to a characteristic final state: two energetic isolated hard photons and large $\not{E}_T$, carried by the gravitinos, along with hard jets and possibly e's and μ's coming from initial $\tilde{g}/\tilde{q}$ decays. For the case of a short-lived slepton NLSP, there will be additional leptons instead of photons. If the lifetime of the NLSP slepton is large, it may show up in the detector as a highly ionizing track. In the minimal AMSB model, on the other hand, the long-lived $\tilde{\chi}_1^\pm$ will produce its own characteristic signals, as discussed earlier.

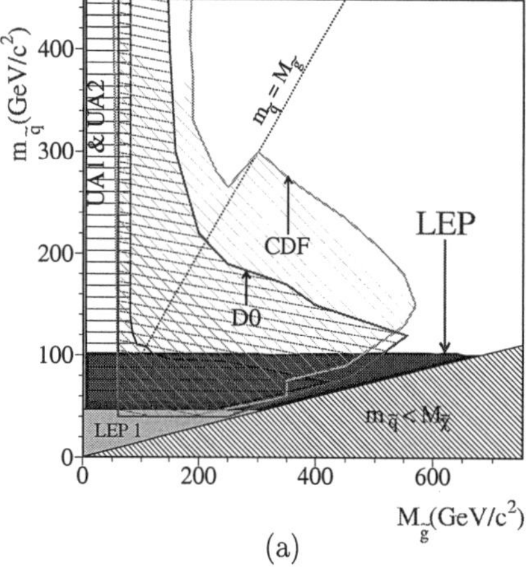

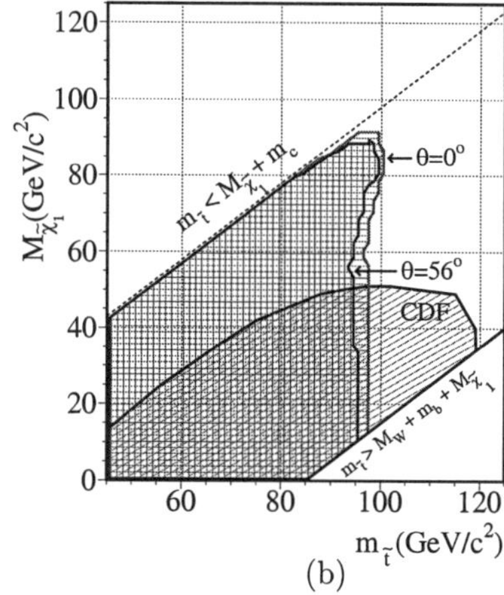

Fig. 15.19. Exclusion regions in the (a) $M_{\tilde{g}}$–$m_{\tilde{q}}$ and (b) $M_{\tilde{\chi}_1^0}$–$m_{\tilde{t}_1}$ planes from the TEVATRON and LEP 2, adapted from Ref. [15.23]. In (b) the lighter (darker) unlabelled curve corresponds to the LEP-generated exclusion region for $\theta \equiv \theta_{\tilde{t}} = 0^0$ (56^0).

The best limits from TEV-I come from an analysis of the multijets plus large $\not{E}_T$ events. The main SM backgrounds to these arise from (1) jets + W ($\to \tau\bar{\nu}_\tau$) with hadronic decays of the τ, (2) jets + Z ($\to \nu\bar{\nu}$) and (3) $t\bar{t}$ with W's from $t \to bW$ decaying into $\tau\nu$ and the τ's also decaying hadronically. In addition, the decay $W \to e(\mu)\bar{\nu}$ generates background events when the e/μ is misidentified as a jet or is lost. On the other hand, QCD jets tend to generate some $\not{E}_T$ due to the mismeasurement of jet energies or lost jets. Signal events can be distinguished from the background by their spherical nature as well as by the usually larger amount of E_T deposited. Pure QCD jet backgrounds can be suppressed by a cut on the azimuthal separation between the $\not{p}_T$ direction and the jets. The optimal values of the cuts would depend on the masses of the sparticles being searched for. The exclusion region, obtained in the $m_{\tilde{q}}$–$M_{\tilde{g}}$ plane from analyses of multijets +

large $\not{E}_T$ events by the CDF and DØ collaborations is shown in Fig. 15.19a. The effect of NLO corrections on the mass bounds is at the level of 5–10 GeV and has been included in the analysis. Ten exactly degenerate squarks were assumed as well as high scale gaugino mass unification (12.25). Moreover, the choices -800 GeV $< \mu < -400$ GeV and $2 < \tan\beta < 4$ have been made, though the dependence of the results on the values of μ and $\tan\beta$ is weak. The assumed high scale gaugino mass unification constrains the mass of a binolike LSP $\tilde{\chi}_1^0$ to be an increasing function of $M_{\tilde{g}}$. This leads to softer jets and less $\not{E}_T$ if $M_{\tilde{g}} \gg m_{\tilde{q}}$. As a result, the TEVATRON bound disappears in this regime. Note however, that such a regime is incompatible with the theoretical constraint (11.45c).

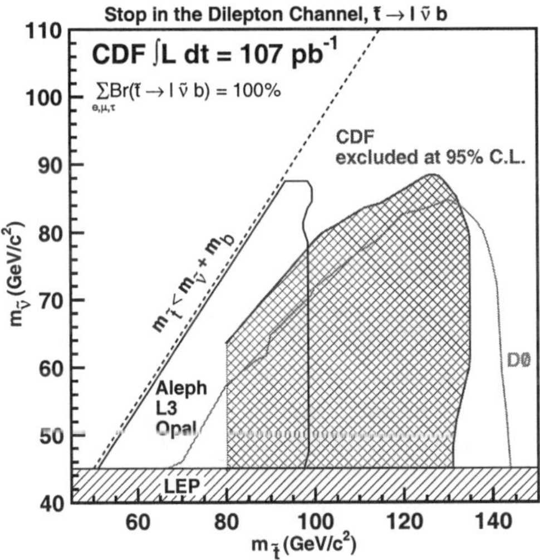

Fig. 15.20. Exclusion region in the stop-sneutrino mass plane from the TEVATRON and LEP experiments, adapted from the first paper of Ref. [15.73].

The lighter squark states $\tilde{t}_1, \tilde{b}_1$ need to be treated separately. In case $\tilde{t}_1$ is accessible at the TEVATRON, its possible decays – given the current mass limits on $\tilde{\chi}_1^0, \tilde{\chi}_1^\pm$ – are (i) the second channel of (15.28a), (ii) the loop decay $\tilde{t}_1 \to c\tilde{\chi}_1^0$, (iii) the first two three body channels of (15.31) and (iv) the four body decay $\tilde{t}_1 \to b\tilde{\chi}_1^0 f\bar{f}'$. Current searches have looked at the following four cases: (a) strong production of $\tilde{t}_1\bar{\tilde{t}}_1$, followed by the second decay of (15.28a), when allowed, or by the first two channels of (15.31) yielding a final state containing two b quarks, leptons/τ's and large $\not{E}_T$; (b) strong $\tilde{t}_1\bar{\tilde{t}}_1$ production, followed by the two body decay mentioned earlier, leading to the dijets + large $\not{E}_T$ signature; (c) strong $\tilde{t}_1\bar{\tilde{t}}_1$ production, followed by the second decay of (15.31), where the $\tilde{\nu}$ is stable; (d) production of a $\tilde{t}_1$ (from top decay) which then decays. In the last case the top quark event sample will contain fewer leptons and jets, with an $\not{E}_T$ distribution that is different from what is that expected for the SM $t\bar{t}$ sample. Thus $BR\,(t \to \tilde{t}_1\tilde{\chi}_1^0)$ can be restricted in a straightforward manner. If the $\tilde{t}_1$ decays produce b quarks in the final state, the analysis becomes much more complicated. No signal has yet been found in any of the current searches [15.73] ruling out significant regions in the $M_{\tilde{\chi}_1^0} - m_{\tilde{t}_1}$ or $m_{\tilde{\nu}} - m_{\tilde{t}_1}$ plane. An example is shown in Fig. 15.19b where the lower bound 135 GeV $< m_{\tilde{t}_1}$ has been established, assuming that $BR\,(\tilde{t}_1 \to c\tilde{\chi}_1^0) = 100\%$ and $M_{\tilde{\chi}_1^0} \gtrsim 40$ GeV. The results of a different analysis are shown in terms of exclusion regions in the $m_{\tilde{\nu}_\ell} - m_{\tilde{t}_1}$ plane in Fig. 15.20. Here the decays $\tilde{t}_1 \to \bar{\ell}\nu_\ell b$ have been considered, assuming equipartition into each ℓ type for the three body decay. This excludes the region 80 GeV

$< m_{\tilde{t}_1} < 135$ GeV. The four body decay of the stop can have a substantial branching ratio for this range and a reanalysis, taking due account of this fact, might be worthwhile.

The production from e^+e^- collision of a squark pair, decaying via the channels shown in Table 15.4, gives rise to acoplanar multijet events. Fig. 15.19a also shows the results of an analysis of the data from LEP 2. These were obtained assuming high scale gaugino mass unification and taking the same ranges of $\tan\beta$ and μ, as used in the TEVATRON analysis, with five flavors of degenerate left and right chiral squarks. The LEP results improve upon the bound established by the TEVATRON experiments only in the (theoretically disfavored) region where the mass of the LSP approaches the squark mass. Much more interesting at an e^+e^- collider would be the search of the lighter third generation squarks $\tilde{t}_1, \tilde{b}_1$. Recall that their production cross sections depend on their masses as well as on the L-R mixing angles. The weakest limits are obtained by choosing the mixing angles to be such that the Z is decoupled. In view of the kinematical limits on the masses that could be explored at LEP 2 and the limits given by experiments done there on chargino masses, the accountable decays of $\tilde{t}_1$ are the same as those considered at the TEVATRON: the loop decay $\tilde{t}_1 \to c\tilde{\chi}_1^0$, the first two channels of (15.31) and the four body decay $\tilde{t}_1 \to b\tilde{\chi}_1^0 f\bar{f}'$. On the other hand, the relevant decay for the $\tilde{b}_1$ is (15.28b) with $l = 1$. For $\tilde{b}_1\bar{\tilde{b}}_1$ and $\tilde{t}_1\bar{\tilde{t}}_1$ production, with the stops undergoing loop decays into $c(\bar{c})\chi_1^0$, the final states would consist of acoplanar jets with large missing energy. In contrast, the other decays would give rise to final states with two jets, two leptons and large $\not{E}$ as well. For Δm, the mass difference between the members of the sparticle pair involved, in excess of 5 GeV, values of $m_{\tilde{t}_1}$ upto 95 GeV are excluded independent of the mixing angle. LEP searches extend the exclusion region in the $M_{\tilde{\chi}_1^0}$ direction of the $M_{\tilde{\chi}_1^0}-m_{\tilde{t}_1}$ plane, as compared to efforts at the TEVATRON, but cannot do better in the large $m_{\tilde{t}_1}$ region. Assuming the branching fraction for the loop decay $\tilde{t}_1 \to c\tilde{\chi}_1^0$ to be 100%, the TEVATRON and LEP limits can be combined in this case as well and have been summarized in Fig. 15.19b.

TEV-II

The TEV-II reach studies of Ref. [15.5], performed in mSUGRA, come to the following conclusion. The $\not{E}_T$ + multijets channel should provide the greatest reach for gluinos, extending upto $M_{\tilde{g}} \simeq 400$ (475) GeV at low $\tan\beta$ with an integrated luminosity[35] of 2 (25) fb^{-1} so long as the squarks are not too heavy. The mass reach in more general models with high scale gaugino mass unification should be very similar so long as $\tilde{\chi}_1^0$ is the stable LSP. However, there will be no reach in these models at TEV-II in the LEP-allowed gluino mass range once $m_{\tilde{q}} \gtrsim 500$ GeV. In mSUGRA the trilepton signal from the 'direct' production of charginos/neutralinos is expected to provide a much better reach so long as $\tan\beta$ is not too high. Analyses of the gluino mass reach in specific non-mSUGRA models exist in the literature [15.75].

Turning to stop search, note that the stop production cross section should increase by about 40% as the TEVATRON CM energy is increased from 1.8 to 2 TeV. Possible signals have already been discussed. For an integrated luminosity[35] of 20 fb^{-1}, the reach is $\sim 200 - 220$ GeV, assuming $BR\,(\tilde{t}_1 \to b\tilde{\chi}_1^+)$ or $BR\,(\tilde{t}_1 \to b\tilde{\chi}_1^0 f\bar{f}')$ to be 100%. A parton level analysis [15.76] shows that a stop mass reach ~ 200 GeV should be feasible at TEV-II with an integrated luminosity[35] of 25 fb^{-1} even if $\tilde{t}_1$ dominantly decays into $\tau^+\nu_\tau \tilde{b}\tilde{\chi}_1^0$. For $\tilde{b}_1$, if only the first of the decays (15.28b) with $l = 1$ is allowed, the availability of very good b-tagging would imply that – even with an integrated luminosity of 2 fb^{-1} – the channel with two b's and large $\not{E}_T$ will provide a reach upto 200 GeV. The latter gets degraded by $\lesssim 30$ GeV, in case the second $\tilde{b}_1$ decay of (15.28b) with $k = 2$ also

---

[35]Again, 20 fb^{-1} and 25 fb^{-1} are notional and possibly unrealistic figures for the TeV-II integrated luminosity.

occurs [15.77].

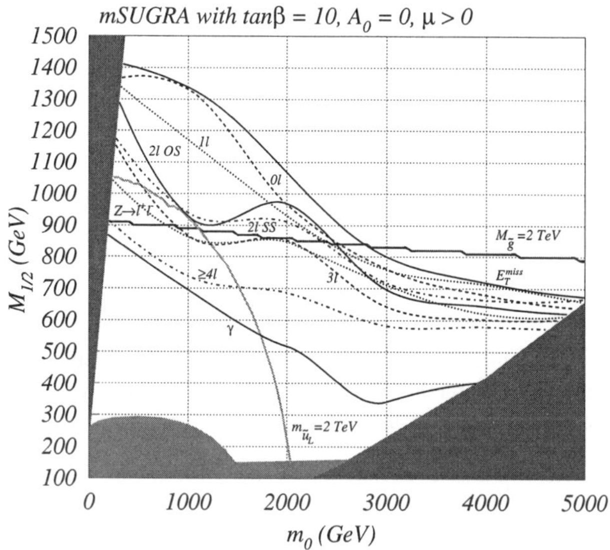

Fig. 15.21. LHC discovery reach, with an integrated luminosity of 100 fb^{-1} and selected $\tan\beta$ and μ, for inclusive supersymmetry channels in the m_0–$M_{1/2}$ plane of mSUGRA. The labels on different curves indicate the number of leptons in the final state and OS (SS) stands for opposite (same) sign dileptons. Adapted from Ref. [15.78].

LHC

Supersymmetry studies at the LHC will have two different goals. As Fig. 15.18 shows, even in the low luminosity option, the expected number of sparticle events is quite large. Hence the first objective would be to discover supersymmetry through an excess above the SM background processes in different types of inclusive events with large $\not{E}_T$, jets and leptons. The second objective would be to make detailed studies of these events to extract information on sparticle masses and other supersymmetry model parameters. The strategy for handling the SM background events would be to isolate the supersymmetry signal events in much the same way as discussed earlier for the TEVATRON. However, production cross sections (even for substantially heavier gluinos and squarks) would be much larger here. This means that more severe cuts can be employed in an inclusive sample with at least four hard jets and large $\not{E}_T$ where the supersymmetry signal should show up as a shoulder in the variable $M_{eff} = \not{E}_T + \sum_{jets} E_T$.

The mSUGRA sparticle reach in several channels at the LHC is summarized [15.78] in Fig. 15.21. It follows that with an integrated luminosity of 100 fb^{-1}, a substantial portion of the m_0–$M_{1/2}$ plane, upto 2 TeV in m_0 and 1.5 TeV in $M_{1/2}$, can be covered. Moreover, in the region of parameter space where $m_{\tilde{u}_L} \lesssim 2$ TeV, which is favored by naturalness arguments, at least three independent channels will yield a supersymmetry signal. The best LHC reach is offered by the jetty channel with no lepton (i.e. $n_\ell = 0$); the channel with $n_\ell = 1$, i.e. single e or μ, becomes comparable around $m_0 \simeq 2$ TeV and takes over for $m_0 > 3$ TeV. Since signal characteristics in different channels are related to one another in a model dependent way, a study of correlations between different signals would aid the determination of supersymmetry model parameters. For instance, a study of the $\not{E}_T$ sample could provide information on the mass ratio $M_{\tilde{g}}/m_{\tilde{q}}$ by use of the higher jet multiplicity expected for the gluino signal. The higher b-hadron multiplicity of events with sparticles could also be used to sharpen the supersymmetry signal. More direct measurements of (differences of) sparticle masses are possible by (partly) reconstructed SUSY events. Such analyses usually start by studying the dilepton invariant mass distribution in OS events. Decays such as $\tilde{\chi}_l^0 \to \tilde{\chi}_n^0 \ell^+ \ell^-$, possibly via on-shell slepton intermediate states, lead to edges in this distribution [15.6, 15.79]. The location of such an edge, as explained by the two equations on p 405, is expected to provide the first constraint on the masses of the involved sparticles. Additional information on neutralino, squark and possibly slepton masses should be obtainable from structures spotted in $j\ell^\pm$ and $j\ell^+ \ell^-$ invariant mass distributions, j standing for one of the two hardest jets in the event. Even more direct determinations of sparticle masses are possible in some cases. For example, the decay chain

$$\tilde{g} \to \tilde{b}_a \bar{b} \to \tilde{\chi}_2^0 b\bar{b} \to \tilde{\chi}_1^0 \ell^+ \ell^- b\bar{b} \tag{15.32}$$

could be reconstructed by selecting events with the $\ell^+ \ell^-$ invariant mass $m_{\ell\bar{\ell}}$ just below the edge in this distribution. The latter implies that the LSP $\tilde{\chi}_1^0$ is nearly static in the $\tilde{\chi}_2^0$ rest frame. The laboratory frame three momenta then satisfy

$$\vec{p}_{\tilde{\chi}_2^0} \simeq (1 + M_{\tilde{\chi}_1^0} m_{\ell\bar{\ell}}^{-1})(\vec{p}_{\ell^+} + \vec{p}_{\ell^-}), \tag{15.33}$$

where $\vec{p}_{\ell\pm}$ refers to the three momentum of the concerned lepton. The successive combination of the reconstructed $\tilde{\chi}_2^0$ with tagged b-jets then should lead to the determination of the $\tilde{b}_a$ and gluino masses. In particular, one should be able to pin down the $\tilde{g}$-$\tilde{b}_a$ mass difference within an error of ~ 2 GeV in the scenario studied in Ref. [15.6]. The partial reconstruction of events should also be possible in other decay chains [15.6] and, in particular, those which have $\tilde{t}_1$ [15.58].

While a majority of the studies [15.6, 15.7, 15.17, 15.56] of LHC sparticle reach has been carried out in mSUGRA, some of the investigations have explored the GMSB and AMSB options. These have tried to determine two things: (1) the reach yielded by the canonical analysis of events with large $\not{E}_T$, n_j jets and n_ℓ leptons in those options, without using the special characteristics that the final states will have in each case; (2) the utility of specific features of the final state in each case in improving on that reach or in extracting information on soft supersymmetry breaking parameters. GMSB studies, with different choices for the short-lived NLSP [15.80, 15.81], show some interesting features. For a short-lived NLSP $\tilde{\tau}_1$, the LHC reach turns out to be the highest in multilepton channels with $n_\ell \geq 2$. This is because of the additional hard charged lepton that gets produced in the decay of the τ emanating from $\tilde{\tau}_1$ decay. The reach in this case is upto $M_{\tilde{g}} \sim 2$ TeV. The selection of signals specific to τ's does not seem to improve the reach, but again that could depend on the sensitivity of the detector. A fast decaying binolike $\tilde{\chi}_1^0$ NLSP would enhance the gluino mass reach as compared to that in mSUGRA, namely upto $M_{\tilde{g}} = 2.8$ TeV, owing to the presence of a hard photon (along with leptons and jets) in the final state. Since the gravitino is ultralight in GMSB scenarios, the use of the prompt $\tilde{\chi}_1^0 \to \gamma\tilde{G}$ decay facilitates event reconstruction [15.81]. An analysis [15.45], using the details of the ATLAS detector, shows how to detect a long-lived $\tilde{\tau}_1$ NLSP at the LHC by employing strategies to look for a stable, charged particle. In the minimal AMSB case, with low values of m_0, the best gluino mass reach has been shown [15.82] to be provided by the $\not{E}_T$ + dileptons + jets channel and is upto $M_{\tilde{g}} \simeq 2.3$ TeV. In contrast, for high m_0 ($\simeq 2$ TeV), that reach becomes maximum for the channel with zero or one isolated lepton, large $\not{E}_T$ and jets, being upto $M_{\tilde{g}} \simeq 1.3$ TeV.

It has been demonstrated [15.79, 15.81] both for the gauge and gravity mediated supersymmetry breaking models that, once supersymmetry is observed at the LHC, studies of kinematic distributions in variables (such as the total E_T) would offer a rather good measure of the sparticle mass scale M_s. For GMSB, measurements of the NLSP $\tilde{\tau}_1$ lifetime and mass could actually reconstruct the supersymmetry breaking parameter $m_{3/2}$ and hence $(\sum_i |\langle F_i \rangle|^2)^{1/2}$. Furthermore, a combination of mass measurements of strongly interacting sparticles at the LHC, with additional accurate mass measurements of electroweak sparticles at a future LC, should enable one to test [15.83] the model hypothesis of a possible unification (or lack thereof) of common scalar and gaugino masses at a high scale. The focus of recent simulation studies has, however, been to estimate the accuracies of a model independent extraction of squark, gluino and LSP masses.

15.5　The quest for supersymmetric Higgs bosons

There is a plethora of new particles which are predicted to exist in the MSSM. Out of these, the lightest neutral Higgs boson h is the only one with an unambiguous upper bound on its mass, cf. (10.73). Whether or not h exists with a mass below this bound is therefore a critical test of the MSSM. Unfortunately, the final energy of the LEP collider at CERN was ~ 20 GeV too low to complete this test. We shall see that this test can be performed in a straightforward manner at future high energy e^+e^- colliders. In general, Higgs discovery prospects at hadron colliders are not that bright. Yet, after a great deal of effort, it has been established that at least one MSSM Higgs boson should be discovered at the LHC. A discovery of h, with a mass less than the predicted upper bound, would only mean consistency with the MSSM, not its confirmation. Indeed, an interesting feature was discussed in §10.4. For a large value of the mass m_A of the CP odd Higgs boson A, the couplings of the lighter CP even Higgs scalar h to the known particles approach those of the Standard Model Higgs particle[36]. Thus, while *not* finding h with any mass $\lesssim 132$ GeV **would clearly exclude the MSSM**, a positive signal could not necessarily be interpreted as evidence for physics beyond the SM. Much effort has therefore been devoted in recent years to the analysis of circumstances which would enable future collider experiments to distinguish between the Higgs sector of the MSSM and that of the SM.

A couple of general remarks need to be made before a detailed description of collider quests for MSSM Higgs particles is given. In the SM, the strength of the Higgs coupling to *any* particle is proportional to the mass of that particle, the constant of proportionality being determined uniquely by a single VEV, namely $v \simeq 246$ GeV. The situation in the MSSM is far more complex. Higgs couplings to SM particles here are still proportional to the masses of the latter, but the coefficients of proportionality now depend on the ratio $\tan\beta$ of the two concerned VEVs and, in the case of the neutral CP even Higgs particles, on the mixing angle α. Moreover, the strengths of couplings of Higgs bosons to sparticles and to other Higgs bosons are *not* proportional to the masses of the latter. The reason is that these masses are only partly due to the Higgs mechanism, the dominant contribution usually coming from gauge invariant soft supersymmetry breaking terms. As a result, already at the tree level of the MSSM, the branching ratios in the decays of Higgs bosons are *not* determined uniquely by their masses. They depend also on additional unknown parameters.

Further complications arise from large loop corrections discussed in §10.6. The masses and mixing angles of all MSSM Higgs bosons get determined at the tree level if the values of just two free parameters are fixed, the most common choice being the set $\{m_A, \tan\beta\}$. However, at the one loop level, all parameters appearing in the stop mass matrix also show up in the mass matrix (10.66) of the CP even neutral Higgs particles. Since these corrections increase the upper bound on

---

[36]This statement remains true even in the presence of (loop induced) CP violation in the Higgs sector. In this case m_A is no longer a physical mass, but it still sets the scale for the masses of the heavy MSSM Higgs bosons.

m_h, it is customary in phenomenological analyses of the MSSM Higgs sector to choose a large stop mass scale $\sqrt{m_{\tilde{t}_L} m_{\tilde{t}_R}} \simeq 1$ TeV. In case such a choice leads to measurable signals, scenarios with smaller (and more natural) stop masses should be testable even more easily. Since the $\tilde{t}_L$-$\tilde{t}_R$ mass splitting does not affect the MSSM Higgs sector by much[37], one usually takes $m_{\tilde{t}_L} = m_{\tilde{t}_R} \equiv m_{\tilde{t}} \simeq 1$ TeV. Finally, in order to investigate the impact of $\tilde{t}_L$-$\tilde{t}_R$ mixing, one usually considers two extreme scenarios: (1) "no mixing", where $\tilde{A}^t \equiv A^t + \mu \cot\beta = 0$, and (2) "maximal mixing" (or "m_h-max"), where $\tilde{A}^t \simeq \sqrt{6} m_{\tilde{t}}$, the latter choice maximizing the one loop corrections to m_h, provided $m_{\tilde{t}_L} = m_{\tilde{t}_R}$. Most of the parameter space can be adequately covered with these two choices, but two caveats need to be stated. First, nontrivial complex phases, if present in the stop mass matrix, will induce CP violation in the Higgs sector. In this case all three neutral MSSM Higgs bosons will mix and the phenomenology will become correspondingly more complicated [15.84]. Second, $b, \tilde{b}$ loops can also become important[38] for $\tan\beta \gg 1$. Moreover, $\tilde{g}$-$\tilde{b}$ and $\tilde{t}$-$\tilde{\chi}^-$ loop corrections to m_b (or, for a fixed physical bottom mass, to the bottom Yukawa coupling) can become large enough to significantly change Higgs decay branching ratios [15.85]. In what follows we shall mostly ignore these possible complications.

e^+e^- colliders

The two most important tree diagrams for the production of neutral Higgs bosons at e^+e^- colliders are shown in Fig. 15.22. The higgsstrahlung process of Fig. 15.22a is of particular importance, since the ZZ Higgs coupling exists only for a Higgs field with a nonvanishing VEV. Thus this process directly probes the mechanism which gives rise to the mass of the Z boson. The total cross section for this process, with $\phi = h$ or H, is given by [15.86]

$$\sigma(e^+e^- \to Z\phi) = \frac{\pi\alpha_{\rm em}^2}{192 s_W^4 c_W^4 s} \frac{a_e^2 + v_e^2}{(1 - M_Z^2 s^{-1})^2} \beta_{Z\phi}(\beta_{Z\phi}^2 + 12 M_Z^2 s^{-1}) c_{ZZ\phi}^2. \tag{15.34}$$

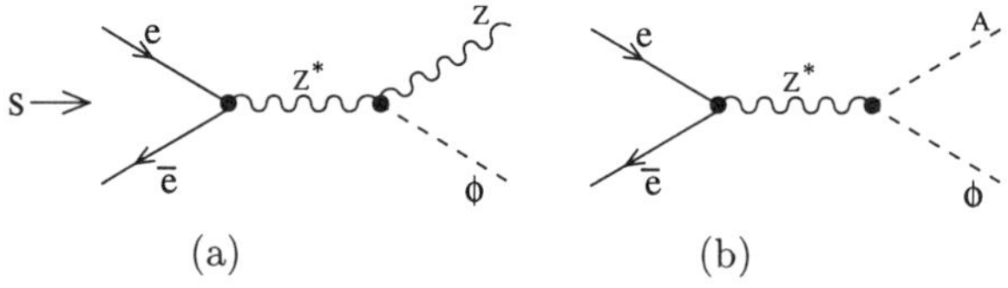

Fig. 15.22. Tree diagrams for (a) higgsstrahlung and (b) associated $A\phi$ production, ϕ standing for one of the two neutral CP even MSSM Higgs bosons.

In (15.34) $c_{ZZ\phi}$ determines the strength of the right vertex, while $a_e = -1$ and $v_e = -1 + 4s_W^2$ are the SM Ze^+e^- couplings. Moreover, $\beta_{ij} = \sqrt{\lambda(1, \mu_i^2, \mu_j^2)}$, cf.(15.3), determines the absolute value of the final state three-momentum in the CM frame. The associated production of CP even and CP odd states proceeds through the diagram of Fig. 15.22b for all scalar fields that transform nontrivially under $SU(2)_L \times U(1)_Y$. In the case of MSSM Higgs bosons, the total cross section can be written as [15.86]

$$\sigma(e^+e^- \to A\phi) = \frac{\pi\alpha_{\rm em}^2}{192 s_W^4 c_W^4 s} \frac{a_e^2 + v_e^2}{(1 - M_Z^2 s^{-1})^2} \beta_{A\phi}^3 c_{ZA\phi}^2. \tag{15.35}$$

---

[37]An important exception occurs in electroweak baryogenesis, which requires $m_{\tilde{t}_L}^2 \gg m_{\tilde{t}_R}^2$, cf. §16.5 for further details.

[38]They do not affect the upper bound on m_h significantly, but can change the mixing angle α.

Two crucial observations enable one to formulate "no lose" theorems pertaining to any MSSM Higgs search at a high energy e^+e^- collider. First, the coupling factors $c_{ZZ\phi}$, $c_{ZA\phi}$ are pairwise complementary to each other: $c_{ZZh}^2 = c_{ZAH}^2 = \sin^2(\alpha - \beta)$, while $c_{ZZH}^2 = c_{ZAh}^2 = \cos^2(\alpha - \beta)$, cf. Fig. 10.3, Appendix B. Second, (10.30e) implies that c_{ZZh}^2 is large if m_A is large and hence so is m_H. Thus, if A and H are heavy, the cross section for ZZh production is close to its maximum, as given by the value of the analogous SM cross section (for an SM Higgs mass equal to m_h). Conversely, if the magnitude of the ZZh coupling is too small for Zh production to be detectable, the ZZH and ZAh coupling strengths are forced to be large, and $m_{H,A}$ will have to be $\lesssim 200$ GeV. This means that either Zh or ZH production, as well as either Zh or Ah production, must have a sizable cross section at an e^+e^- collider operating at a CM energy $\sqrt{s} \gtrsim 250$ GeV. For a realistic luminosity, therefore, some kind of an MSSM Higgs signal in such a collider is inevitable.

The experimental lower bounds on sparticle masses, discussed earlier in this chapter, imply that h will predominantly decay into SM particles[39]. The channel $h \to b\bar{b}$ will be the dominant mode of decay of h for most of the parameter space, but $h \to WW^*$ (where W^* denotes an off-shell W boson) will become significant if $m_h \gtrsim 120$ GeV and $\sin^2(\alpha - \beta)$ is sizable. The branching ratio for h decaying into $\tau^+\tau^-$ is about ten times less than that for the $b\bar{b}$ mode, while decays into $c\bar{c}$ and gg are even rarer. All these h decay channels provide good signals for Zh production. For either of the decays $h \to b\bar{b}$ and $h \to \tau^+\tau^-$, all Z decay channels are useful (and have been used in Zh searches at LEP). Conversely, zeroing in on $Z \to \ell^+\ell^-$ ($\ell = e$ or μ), one can reconstruct the Higgs boson without having to detect its decay products. This is possible since the four momentum p_Z of the produced Z boson can be measured accurately as the sum of the four momenta of the ℓ^+ and ℓ^-, hence allowing the determination of the "missing mass"

$$M_{\text{miss}}^2 \equiv (p_{e^+} + p_{e^-} - p_Z)^2. \tag{15.36}$$

Here $p_{e^\pm}$ are the four-momenta of the incoming $e^\pm$. Signal events evidently correspond to $M_{\text{miss}} = m_\phi$, ϕ being h or (if $\cos^2(\alpha - \beta)$ is not too small) H, whereas many background processes would have a smoothly distributed M_{miss} or a peak at $M_{\text{miss}} = M_Z$. The signal should thus be identifiable as a peak at $M_{\text{miss}} = m_\phi$ over a smooth background, which can be determined by interpolation from the side bins. The advantage of this method is that it enables one to isolate a signal for Zh production in a way that is completely *independent* of the modes and mechanisms of Higgs decays.

Given a sufficiently small m_A, which makes c_{ZAh}^2 sizable, the CP odd Higgs A would also predominantly decay into SM particles. Since CP invariance forbids A from coupling to two electroweak gauge bosons at the tree level, by far the most dominant final states now would be $b\bar{b}$ and $\tau^+\tau^-$, produced in approximately the ratio 10:1. Associated Ah production would thus give rise to $b\bar{b}b\bar{b}$, $b\bar{b}\tau^+\tau^-$ and $\tau^+\tau^-\tau^+\tau^-$ final states with relative abundances $\sim 100 : 20 : 1$. The 4τ final state is therefore very rare, so that b tagging would be crucial to beat down multijet backgrounds from QCD or W pair events. At the LEP collider, which operated with CM energy upto $\sqrt{s} \le 209$ GeV, Zh and Ah production were the only potentially useful MSSM Higgs production channels. No unambiguous signal was observed. The combined LEP limits[40] [15.88] are shown on the m_A–$\tan\beta$ plane in Fig. 15.23. It has been assumed here that h and A do not decay into sparticles. We see that the "no mixing scenario" (left frame) is already severely constrained by the data, with only three small regions of the plane being still allowed. The area with $\tan\beta < 0.7$, $m_A < 40$ GeV is difficult to cover since the $h \to AA$ decay is possible there and the $A \to c\bar{c}$ decay can always compete with,

---

[39]If the unification condition (9.21) does not hold, the invisible mode $h \to \tilde{\chi}_1^0\tilde{\chi}_1^0$ can still be significant [15.87].

[40]As of this writing, these limits were still preliminary, but they are not expected to change significantly after the completion of the analysis.

or even dominate over the $A \to b\bar{b}$ decay. However, the condition $\tan\beta < 1$ is incompatible with radiative symmetry breaking and, in fact, $\tan\beta < 0.7$ implies that the top Yukawa coupling has a Landau pole well below the GUT scale, cf. §11.4. In the allowed region with $\tan\beta \gtrsim 11$ and m_A in the range 90 GeV $\lesssim m_A \lesssim$ 120 GeV, the factor $\sin^2(\alpha - \beta)$ is small enough to dynamically suppress Zh production, while the masses m_A and m_H are sufficiently large to kinematically suppress Ah and ZH production.

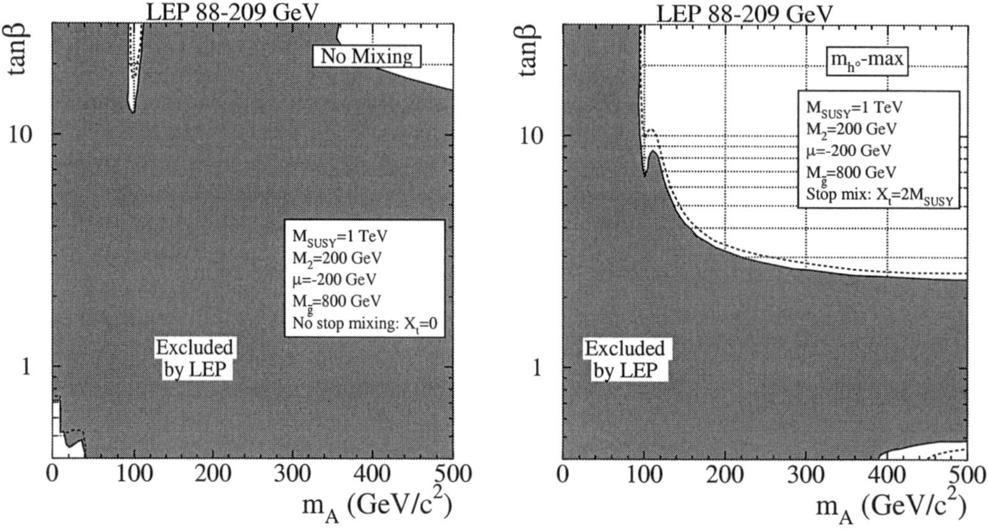

Fig. 15.23. The region in the $(m_A, \tan\beta)$ plane excluded by Higgs searches at LEP, for $m_t = $ 174.3 GeV and $m_{\tilde{t}} = 1$ TeV. The left and the right frames are for the "no mixing" and the "maximal mixing" scenarios respectively. The dotted lines show the limits that could have been expected in the absence of any signal. Adapted from Ref.[15.88].

Let us discuss the third region 'of high $\tan\beta$' next. The region $\tan\beta \gtrsim 15$, $m_A \gtrsim 350$ GeV, $\sin^2(\alpha - \beta) \simeq 1$ has m_h at a value slightly above the 95% c.l. lower bound of 114.1 GeV on the mass of the Higgs boson in the SM. The precise boundaries of this last allowed region depend very sensitively on the details of the calculation of m_h. The latter can be computed from m_A, $\tan\beta$ and the parameters of the stop sector (ignoring smaller electroweak gauge corrections). The allowance of a 2 GeV theoretical uncertainty in this calculation [15.89], i.e. requiring the calculated value of m_h to only exceed 112.1 GeV if $\sin^2(\alpha - \beta) \simeq 1$, shifts the lower bound on $\tan\beta$ to about 7.5 for large m_A. Moreover, the region with $m_h > 112.1$ GeV merges with the region where Zh production is dynamically suppressed. This means that all values of $m_A > 90$ GeV are allowed for large $\tan\beta$! The allowed region is similarly enlarged if the top mass is increased by about 4.5 GeV from its current central value of 174.3 GeV. In comparison, the current experimental error on m_t is 5.1 GeV. The reason that these uncertainties have such a large impact on the size of the allowed region is that, in the "no mixing" scenario, the value $m_{h,\mathrm{max}} = 114.3$ GeV for the given inputs [15.89] is extremely close to the lower bound on the mass of the SM Higgs particle. Moreover, for $\tan^2\beta \gg 1$, m_h depends only weakly on $\tan\beta$ and also becomes insensitive to m_A

once $m_A > m_{h,\text{max}}$. On the other hand, in the "maximal mixing" scenario shown in the right frame of Fig. 15.24, $m_{h,\text{max}} \simeq 126$ GeV is comfortably above the Higgs mass bound in the SM. In this case, the allowance of a theoretical uncertainty of 2 GeV in the calculation of m_h only reduces the lower bound on $\tan \beta$ from 2.4 to 2.2.

The production of heavy MSSM Higgs particles should eventually become feasible at a future LC [15.8]. For a value of m_A above 200 GeV (the precise value depending on $\tan \beta$), $\cos^2(\alpha - \beta) \ll 1$, making the cross sections for Ah and ZH production too small to be detectable. The heavy Higgs bosons will then have to be pair produced, i.e. one will need to look for signals from the processes $e^+e^- \to AH$ and $e^+e^- \to H^+H^-$, requiring $\sqrt{s} > 2m_A$. The decay characteristics of the Higgs particles, produced in such an environment, merit discussion. The decays of the heavy neutral ones H and A could be quite complicated if $\tan \beta$ were not large. Each might decay into a chargino pair or a neutralino pair, while H might in addition decay into either $\tilde{f} + \bar{\tilde{f}}$ or an hh pair. Decays into neutralinos would occur with full $U(1)_Y$ gauge coupling strength only if $m_A > |M_1| + |\mu|$, since each Higgs boson couples to the one gaugino and one higgsino state, cf. (10.35, 10.36). Similarly, decays into charginos and neutralinos will reach the full $SU(2)_L$ gauge coupling strength only if $m_A > |M_2| + |\mu|$. However, for $\tan \beta \gtrsim 20$, the only decay modes that might compete with the $b\bar{b}$ final state are those into $\tilde{t}$ or $\tilde{b}$ squarks. Note that A cannot decay into two light squarks, $\tilde{t}_1\bar{\tilde{t}}_1$ or $\tilde{b}_1\bar{\tilde{b}}_1$ (unless CP is violated). Moreover, (10.31) and (10.33) show that the couplings of A and (in the decoupling limit) of H to b and t quarks are proportional to $m_b \tan \beta$ and $m_t \cot \beta$, respectively. Hence decays into $b\bar{b}$ begin to dominate over decays into $t\bar{t}$ once $\tan \beta > \sqrt{m_t(m_A)/m_b(m_A)} \simeq 8$, even if $m_{A,H} > 2m_t$. Of course, $\tau^+\tau^-$ final states are again produced at about 10% of the rate for $b\bar{b}$ final states. Finally, any A decay into a W^+W^- or a ZZ pair is forbidden by CP invariance, while the analogous H decay modes are suppressed by the coupling factor $\cos^2(\alpha - \beta)$ which is proportional to $(M_Z^2/m_A^2) \cot^2 \beta$. Over much of the parameter space, then, one expects $4b$ and $b\bar{b}\tau^+\tau^-$ final states to remain the most promising signatures of AH production.

The cross section for the pair production of charged Higgs particles is given by (15.35) with the replacements $\beta_{A\phi} \to \beta_{H^+H^-}$, $v_e \to v_e - 8s_W^2 c_W^2(1 - M_Z^2/s)/\cos 2\theta_W$ and $c_{ZA\phi}^2 \to \cos^2 2\theta_W$. The charged Higgs boson would predominantly decay into $t\bar{b}$ if $M_{\tilde{t}_1} + m_{\tilde{b}_1} > m_{H^+} > m_t + m_b$. If $\tan \beta \gg 1$, $\tau^+\nu_\tau$ final states would occur at about 10% of the rate of $t\bar{b}$ final states. If $m_{H^+} < m_t + m_b$, the $\tau^+\nu_\tau$ final state will dominate over the $c\bar{s}$ final state in charged Higgs decay over the entire allowed parameter space with $\tan \beta > 1$. The only supersymmetric decay mode that might be able to compete with the $t\bar{b}$ final state then is $\tilde{t}_1\bar{\tilde{b}}_1$, if these squarks are sufficiently light. The most robust signature for H^+H^- pair production with $m_{H^+} > m_t + m_b$ thus comes from $t\bar{t}b\bar{b}$ final states, which can generate up to eight jets [15.8]. On the other hand, if $m_{H^+} < m_t - m_b$, one could also look for H^+ production in the decay of the top quark, $t \to bH^+ \to b\tau^+\nu_\tau$. This leads to an excess of τ's over electrons and muons in semileptonic top decays. Moreover, each of these τ leptons will have a helicity opposite to that of a τ produced in $W^\pm$ decay. The average helicity of τ leptons, determinable from the energy distribution of their decay products [15.90], would therefore be a useful diagnostic.

One can expect the pair production of heavy MSSM Higgs particles to lead to detectable signals at future e^+e^- colliders if the beam energy is at least a few GeV above the masses of these bosons[41]. Moreover, experiments at a high luminosity e^+e^- collider might be able to distinguish between the Higgs sectors of the SM and of the MSSM even if $\sqrt{s} < 2m_A$. To this end, one needs a large sample of $e^+e^- \to Zh$ events. Since m_A is large in this scenario, the ZZh coupling is guaranteed to

---

[41]Otherwise, a cross section very close to the threshold would get suppressed by the β^3 factor in (15.35) leading to an unacceptably low rate.

be close to its SM value. One can infer from (10.31–10.33) that a deviation from the SM only occurs at the order $(M_Z/m_A)^4 \cot^2 \beta$. Thus the Zh production cross section is not a sensitive probe of the MSSM Higgs sector. On the other hand, the large event sample will allow one to measure ratios of h branching ratios with high precision. We see from (10.32) and (10.33) that the $hf\bar{f}$ couplings deviate from their SM values already at the order $(M_Z/m_A)^2$. In fact, the deviation in the coupling to $I_3 = -1/2$ fermions is further enhanced by a factor $\tan\beta$. For $\tan\beta > 1$ the ratio of branching ratios into $WW^\star$ and $b\bar{b}$ therefore offers the best sensitivity. This sensitivity would be highest if the two branching ratios are approximately equal, since in this case they would be measured with comparable accuracies. Such a situation prevails in the SM for $m_{H,SM} \simeq 140$ GeV. Significantly smaller Higgs masses are less favorable since the branching ratio into $WW^\star$ then becomes small. However, already for $m_h \simeq 120$ GeV, this measurement should show a more than two standard deviation departure from the SM, provided $m_A \leq 600$ GeV [15.8]. Furthermore, if $m_A < 500$ GeV, its value would be extractable with an error of less than 100 GeV from the measured ratio of branching ratios.

It should be possible to extend the mass reach for directly produced neutral heavy MSSM Higgs particles by converting an e^+e^- collider into a $\gamma\gamma$ collider [15.9], cf. §15.1. The shape of the $\gamma\gamma$ luminosity spectrum as well as the photon helicities are expected to be controlled by the experimenter. Given some prior knowledge of m_A, e.g. through the indirect method described in the previous paragraph, one can tune the luminosity function in a way such that most events have the $\gamma\gamma$ CM energy nearly equal to m_A. Moreover, by choosing the two incident photons to have the same helicity, the Higgs production cross section can be maximized, while suppressing the QED pair production of light fermions[42]. For $\tan\beta \gtrsim 5$, a Higgs discovery in the $b\bar{b}$ channel should be possible, provided $m_A \lesssim 0.8\sqrt{s_{e^+e^-}}$, i.e. the mass reach gets extended by a factor ~ 1.6. This should cover [15.91] part of the MSSM parameter space where heavy Higgs bosons would be inaccessible at the LHC (see below). High level b-tagging capabilities would be required for this purpose, since the QED cross section for light $q\bar{q}$ pair production is 34 $(= \sum_q e_q^4/e_b^4)$ times that for $b\bar{b}$ production. For circularly polarized photons, the production of H and A would contribute about equally to the signal in consequence of the relation $m_H \simeq m_A$ in the decoupling limit. Indeed, it will then be difficult to disentangle the two contributions[43]. Smaller values of $\tan\beta$ would be less favorable, since the Higgs branching ratios into $b\bar{b}$ will then be quite small (in many cases), and the background in the $t\bar{t}$ channel would get enhanced by a factor of 16. Large branching ratios into charginos and neutralinos would be advantageous in this case [15.91]. The chargino pair signal would in fact get overwhelmed by the large continuum background. However, a substantial Higgs decay branching ratio into charginos would imply a similarly large branching ratio into neutralinos $\tilde{\chi}_l^0 \tilde{\chi}_n^0$ $(l, n \neq 1, 1)$. Some of these channels, e.g. $\tilde{\chi}_1^0 \tilde{\chi}_n^0$ with $\tilde{\chi}_n^0 \to \tilde{\chi}_1^0 f\bar{f}$, should be fairly background free even in the presence of chargino pair production. No existing laser system as yet satisfies the specifications for both the single pulse intensity and the repetition rate required for the construction of such a $\gamma\gamma$ collider via back scattering from $e^\pm$ beams. However, one is hopeful for such a system to emerge some day.

---

[42] At the tree level the cross section for the process $\gamma\gamma \to q\bar{q}$, with β_q being the final state CM velocity, is proportional to the factor $1 - \beta_q^4$ which goes as m_q^2/s if the two photon helicities satisfy $\lambda_1 = \lambda_2 = \pm1$. However, this helicity suppression can be spoilt by the emission of real gluons. A careful treatment of higher order corrections is therefore essential when evaluating the background, say for the $b\bar{b}$ channel [15.91].

[43] If the photons are linearly polarized, the production of A (H) can be completely suppressed when the two polarization vectors are parallel (orthogonal) [15.92]. However, the continuum $b\bar{b}$ background will then no longer be helicity-suppressed.

Hadron colliders

The mechanism, yielding the largest cross section for producing an SM-like neutral Higgs particle at a hadron collider, is gluon fusion $gg \to \Phi$, Φ being h, H or A. To the leading order, the inclusive total cross section is given by [15.86]

$$\sigma(pp \to \Phi X) = \frac{\Gamma(\Phi \to gg)}{m_\Phi} \frac{\pi^2}{8s} \int_\tau^1 dx\, x^{-1} g(x, \mu_f) g(\tau/x, \mu_f). \tag{15.37}$$

In (15.37) $\tau = m_\Phi^2/s$ and the gluon distribution function g has been taken[44] at a factorization scale μ_f of the order of m_Φ. The total cross section for Φ production in gg fusion is thus proportional to the partial width for the decay $\Phi \to gg$. The latter can only occur through quark or squark loops, as shown in Fig. 15.24a,b respectively. For a CP even Higgs boson $\phi = h$ or H, this partial width is given by [15.86]

$$\Gamma(\phi \to gg) = \frac{G_F \alpha_s^2 m_\phi^3}{64\sqrt{2}\pi^3} \left| -\sum_q X_\phi T_q(\tau_q) + \sum_{\tilde q} c_{\phi\tilde q\tilde q} M_W^2 m_{\tilde q}^{-2} T_{\tilde q}(\tau_{\tilde q}) \right|^2. \tag{15.38}$$

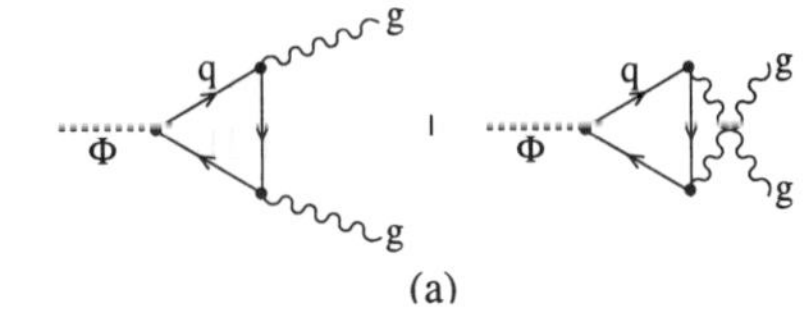

(a)

(b)

Fig. 15.24. Feynman diagrams contributing to the decays of neutral Higgs particles to two gluons. The squark loop contributions (b) only exist for the CP even Higgs particles h and H.

Here $\tau_i = m_\phi^2/(4m_i^2)$ for $i = q, \tilde q$, X_ϕ is the $\phi q\bar q$ coupling strength in units of $g_2 m_q/(2M_W)$, as listed alongside the topmost diagram of Fig. 10.3, and the dimensionless quantities $c_{\phi\tilde q\tilde q}$ are the couplings $C[\phi, \tilde f', \tilde f]$ of Fig. 10.7 divided by $g_2 M_W$ and with $\tilde f' = \tilde f = \tilde q$. Finally, the reduced quark and squark loop amplitudes T_q and $T_{\tilde q}$ are given respectively by

$$T_q(\tau) = -2\tau^{-2}[\tau + (\tau - 1)f(\tau)], \tag{15.39a}$$

$$T_{\tilde q}(\tau) = \tau^{-2}[\tau - f(\tau)]. \tag{15.39b}$$

---

[44]For $\mu_f = m_\Phi$, NLO QCD corrections increase the cross section (15.37) by about a factor of two, cf. Balázs et al. [15.93]. Note that (15.37) also applies to $p\bar p$ collisions, since the gluon density is the same in a proton and an antiproton.

In (15.39) the loop function $f(\tau)$ is defined as

$$f(\tau) \equiv \begin{cases} \arcsin^2(\sqrt{\tau}), & \tau \leq 1, \\[2ex] -\frac{1}{4}[\ln\{(1+\sqrt{1-\tau^{-1}})/(1-\sqrt{1-\tau^{-1}})\} - i\pi]^2, & \tau > 1. \end{cases} \qquad (15.40)$$

The function $f(\tau)$ goes as $\tau + \tau^2/3 + \cdots$ in the limit $\tau \to 0$. When large masses circulate in the loop, (15.39) yield the limiting values $T_q \to -4/3$ and $T_{\tilde{q}} \to -1/3$. The contribution of a heavy quark to the decay width (15.38) thus approaches a constant if $4m_q^2 \gg m_\phi^2$. However, the factor $M_W^2/m_{\tilde{q}}^2$ in the $\tilde{q}$ loop contribution makes a heavy squark decouple if $c_{\phi\tilde{q}\tilde{q}}$ is kept fixed. This underscores the difference between the couplings of Higgs bosons to SM particles and to MSSM sparticles, as mentioned at the beginning of this section. Finally, the partial width $\Gamma(A \to gg)$ is given [15.94] by (15.38) with the replacements $m_\phi \to m_A$, $c_{A\tilde{q}\tilde{q}} \to 0$, $X_\phi \to X_A$ and $T_q \to \tilde{T}_q = 2f(\tau)/\tau \to 2$, where the limit is again for $4m_q^2 \gg m_A^2$.

Recall that the MSSM neutral Higgs particles decay predominantly into third generation quarks or squarks. Unfortunately, these decays by and large do not lead to detectable signals for Higgs production from gluon fusion. That is since the QCD backgrounds are far too high. The branching ratio into $\tau^+\tau^-$ is also often sizable. The existence of at least two neutrinos, originating from τ decays in the final state, unfortunately makes it impossible to reconstruct the mass of the Higgs in this channel unless the latter happens to possess a large transverse momentum, see below. Hence the τ pair final state cannot serve as a signal for inclusive Higgs production, either. The most robust signal for the process $gg \to \phi$, in fact, often comes from the utilization of the rare radiative decay mode $\phi \to \gamma\gamma$. The corresponding partial width for CP even Higgs particles is [15.86]:

$$\Gamma(\phi \to \gamma\gamma) = \frac{\alpha_{em}^2 G_F m_\phi^3}{128\sqrt{2}\pi^3}\left| c_{ZZ\phi}T_W(\tau_W) - \sum_f N_{c,f}e_f^2 X_\phi T_q(\tau_f) \right.$$

$$\left. + \sum_{\tilde{f}} N_{c,\tilde{f}}e_{\tilde{f}}^2 c_{\phi\tilde{f}\tilde{f}}M_W^2 m_{\tilde{f}}^{-2}T_{\tilde{q}}(\tau_{\tilde{f}}) + \sum_{i=1}^{2} c_{\phi\tilde{\chi}_i^+\tilde{\chi}_i^-}T_q(\tau_{\tilde{\chi}_i^\pm}) \right|^2, \qquad (15.41)$$

with

$$T_W(\tau) \equiv 2 + 3\tau^{-1}\left[1 + \left(2 - \tau^{-1}\right)f(\tau)\right]. \qquad (15.42)$$

Here $N_{c,f} = 3$ (1) for $f = q$ (ℓ) and similarly for $N_{c,\tilde{f}}$. The coefficient $c_{ZZ\phi}$ has already been introduced in (15.34). Finally, $c_{\phi\tilde{\chi}_i^+\tilde{\chi}_i^-}$ is the $\phi\tilde{\chi}_i^+\tilde{\chi}_i^-$ coupling strength $c_L = c_R$ (assuming CP conservation) of Fig. 10.6 with $M_{\tilde{\chi}_i^\pm}/(2M_W)$ factored out. The factoring has been done to give the SM fermion and chargino loop contributions a similar form; however, $c_{\phi\tilde{\chi}_i^+\tilde{\chi}_i^-}$ vanishes as $M_{\tilde{\chi}_i^\pm} \to \infty$. Since $T_W(\tau) \to 7$ as $\tau \to 0$, the potentially biggest contribution to (15.41) comes from W loops. Thanks to this contribution, the $\gamma\gamma$ branching ratio of an SM-like Higgs boson can reach a value $\sim 2 \times 10^{-3}$ if the Higgs mass is around 130 GeV. This branching ratio falls at smaller Higgs masses due to the m_ϕ^3 factor in (15.41), and also at larger masses due to the rapid increase of the partial width into (virtual) W bosons. Current experiment bounds imply 114 GeV $\leq m_h$ in the decoupling limit and $m_h < 132$ GeV theoretically. Therefore, the process $gg \to h \to \gamma\gamma$ should be visible [15.6] as a peak over a smooth background in the diphoton invariant mass distribution, provided sparticle loop contributions to (15.38) and (15.41) are small[45]. However, most events

---

[45]The probability of misidentifying a jet as a single photon has been assumed here to be less than 10^{-4}; otherwise the huge dijet background will overwhelm the signal. This defines one of the design specifications of LHC detectors.

would be from the background even in the peak region. A much better signal to noise ratio could be secured by requiring the Higgs to be produced in association with a leptonically decaying W boson (this includes $t\bar{t}h$ production), i.e. by looking for $\ell^{\pm}\gamma\gamma$ events with a large missing p_T. The signal cross section would nonetheless be reduced by several orders of magnitude. Even then it implies a statistical significance of the signal which is comparable to that in the gluon fusion channel [15.6].

Somewhat paradoxically, the $\gamma\gamma$ signal from Higgs production at the LHC becomes less viable at smaller values of m_A. The sum rule $c_{ZZh}^2 + c_{ZZH}^2 = 1$ implies that the partial widths for the decys $h \to \gamma\gamma$ and $H \to \gamma\gamma$ cannot be suppressed simultaneously to values much below the SM prediction. However, if m_A is small and $\tan\beta \gg 1$, the partial widths of both h and H into $b\bar{b}$ and $\tau^+\tau^-$ would get enhanced, thereby reducing their branching ratios into the $\gamma\gamma$ final state. As a result, the $\gamma\gamma$ signal ceases to be measurable [15.6] if $m_A < 200$ GeV. Fortunately, it has been demonstrated [15.6] that, with about 100 fb^{-1} of data, this region of parameter space can be explored through $t\bar{t}h$ production, followed by the decay $h \to b\bar{b}$. One should look for events with exactly four hard b jets as well as a hard lepton (for triggering). The background could be kept at a managable level only if signal events are reconstructed *completely*. Not only good b-tagging, but also good calorimetry, enabling the reconstruction of the $t, \bar{t}$ and h masses[46], would now be required. Note that the $t\bar{t}h \to t\bar{t}b\bar{b}$ signal, which remains viable at a large value of m_A, relies entirely on couplings extant at the tree level. This therefore is much less sensitive to sparticle loop contributions than the (inclusive) $\gamma\gamma$ signal (see below).

We have thus established that, once nearly 100 fb^{-1} of data get accumulated, at least the light CP even Higgs particle h should be detectable at the LHC with a significance greater than the 5σ level. The data accumulation should not be a problem if the LHC performs as planned, with an instantaneous luminosity $\mathcal{L} = 10^{34}$ cm^{-2} s^{-1} during the second "high luminosity" phase of the collider[47]. However, for $m_A > 300$ GeV or so, the $h \to \gamma\gamma$ signal might be essentially indistinguishable from that of the SM Higgs particle. Precise measurements of Higgs couplings would be very difficult at the LHC, since all calculated cross sections for processes there suffer from large theoretical uncertainties on account of unknown higher order QCD effects. Moreover, even if detailed analyses of the production and decay of h at the LHC or elsewhere show significant deviations from SM predictions, one would still like to see direct experimental evidence for the production of the heavy Higgs bosons of the MSSM.

The branching ratios of H and A into two photons are unfortunately too small to be useful. The W loop contribution to (15.41) is either absent (for A) or strongly suppressed (for H if $m_A > 200$ GeV). The contribution from top loops is also suppressed by the factor $\cot\beta$. Their couplings to b and τ are enhanced by $\tan\beta$, but the corresponding loop functions are very small: $T_q(\tau) \to 1/(2\tau)(\ln 4\tau)^2$, $\tilde{T}_q(\tau) \to -1/(2\tau)(\ln 4\tau)^2$ for $\tau \gg 1$. At the same time the partial widths into $b\bar{b}$ and $\tau^+\tau^-$ are enhanced by the $\tan^2\beta$ factor, suppressing the $\gamma\gamma$ branching ratios even further. The smallness of the b loop functions as well as the $\cot\beta$ behavior of $c_{(A,H)tt}$ also suppress the

---

[46]If only one top quark decays semileptonically, the neutrino four momentum p_ν can be determined. This is since the neutrino comes from a $W \to \ell\nu$ decay with $(p_\nu + p_\ell)^2 = M_W^2$. The two transverse components of the three momentum $\vec{p}_\nu$ can be identified with the missing transverse momentum. These relations, together with the masslessness of the neutrino $p_\nu^2 = 0$, determine p_ν completely.

[47]It has recently been claimed [15.95] that, with about 100 fb^{-1} of data, the entire currently allowed part of the m_A–$\tan\beta$ plane can also be covered via the production of h and/or H in W^+W^- and ZZ fusion. The Higgs particle needs to decay into a $\tau^+\tau^-$ pair, though in the "pathological" situation $\sin\alpha = 0$ (i.e. $c_{h\tau\tau} = 0$), the decay $h \to \gamma\gamma$ can be used instead. However, this result from a parton level analysis has not yet been reproduced by a full experimental Monte Carlo simulation.

cross sections for H and A production via gluon fusion to values well below the corresponding SM ones. The $\tan\beta$ enhancement of $c_{(A,H)bb}$ does nonetheless lead to an enhancement of the *total* A and H production cross sections for large $\tan\beta$ with dominant contributions now coming from associated $b\bar{b}(A,H)$ production [15.6]. Of course, this increase in the production rates is insufficient to compensate for the tiny branching ratios into $\gamma\gamma$. However, the branching ratios into the even cleaner $\mu^+\mu^-$ final state, while also very small (typically of order $m_\mu^2/3m_b^2 \simeq 3 \times 10^{-4}$), would be sufficient [15.6] to lead to a signal greater than the 5σ level with 300 fb^{-1} of data if $\tan\beta = 10$ (30) for $m_A = 150$ (500) GeV. Such a signal would become unobservable again for $m_A < 110$ GeV since the Drell-Yan background increases very quickly as the invariant mass $M_{\mu^+\mu^-}$ approaches M_Z.

The A, H branching ratios into $\tau^+\tau^-$ exceed those into $\mu^+\mu^-$ by a factor $(m_\tau/m_\mu)^2 \simeq 300$. We had stated earlier that the presence of at least two neutrinos in this final state makes it impossible to reconstruct the $\tau^+\tau^-$ invariant mass. That is, in fact, correct only if the τ^+ and τ^- are essentially back-to-back in the transverse plane, i.e. if the parent Higgs particle has a small transverse momentum. The ultrarelativistic nature of each τ can be utilized to identify the direction of its three momentum with those of its decay products. This leaves only two unknowns, the energies (or absolute values of the three momenta) of the $\tau^\pm$. On the other hand, the assumption that the only $\not{p}_T$ in the event originates from τ decays[48] leads to two kinematic constraints from the conservation of momentum in the two directions transverse to the beam axis. Note that these two constraints are independent only if the two τ leptons are *not* back-to-back in the transverse plane. Naively, the reaction $gg \to A, H$ might be expected to produce either Higgs particle with a vanishing transverse momentum. This is no longer true once additional gluon radiation (which occurs in just about every real event) has been included. Moreover, in $gg \to b\bar{b}$ (H, A) events, which dominate the total A, H cross sections for $\tan\beta > 5$, the Higgs bosons can already have nonzero p_T even in the absence of additional radiation[49]. Hence the procedure, described above, allows the reconstruction of the Higgs mass with an accuracy of $\sim 10\%$ if the transverse opening angle between the τ^+ and τ^- decay products satisfies the constraint [15.6] $1.8 < \phi_{\tau\tau} < 2.9$ or $3.4 < \phi_{\tau\tau} < 4.5$. However, H and A production cannot be distinguished in this way since they contribute to the same $\tau^+\tau^-$ invariant mass bin. In order to provide a trigger for the event, one has to require that one of the τ leptons decays leptonically, with $p_T(\ell^\pm) > 20$ GeV. The best sensitivity is then obtained if the other τ decays semileptonically. In order to suppress QCD backgrounds, one has to further require that the hadrons from the second τ decay have p_T of at least 40 GeV; QCD jets with such high p_T would usually have much higher charged particle multiplicities and can thus be discarded. Finally, in order to suppress the background even more, one can focus on the $b\bar{b}$ (H, A) production channel by requiring the event to contain at least one tagged b-jet. With about 300 fb^{-1} of data, a greater than 5σ signal for combined $H + A$ production should be found if $\tan\beta > 6$ (16) for $m_A = 140$ (500) GeV. Note that the sensitivity again gets worse for $m_A < 140$ GeV [15.6].

The part of the m_A–$\tan\beta$ plane where one can directly detect one of the heavy MSSM Higgs particles can be extended further by searching for $gg \to \bar{b}tH^-$ production (as well as its charge

---

[48]This assumption is obviously violated in sparticle production events, where much of the missing $\not{p}_T$ is carried away by LSPs. Therefore, the algorithm described here cannot be used to determine $M_{\tau\tau}$ in, say $\tilde{\chi}_2^0 \to \tilde{\chi}_1^0 \tau^+\tau^-$ events.

[49]Current event generator programs predict [15.6] the p_T spectrum of Higgs particles produced in gg fusion to be *harder* than that from associated $b\bar{b}$ (H, A) production. The NLO QCD corrections to these latter processes, which are needed to predict the p_T spectrum reliably, have become available only recently [15.96]. Note that there are three distinct mass scales in this problem, namely m_b, p_T and the Higgs mass, out of which the latter two are much bigger than the former.

conjugate reaction)[50]. The $t\bar{b}H^-$ vertex is proportional to $m_t \cot\beta \pm m_b \tan\beta$, where the upper (lower) sign refers to the scalar (pseudoscalar) coupling. The absolute value of each attains a minimum at $\tan\beta = \sqrt{m_t(m_{H^\pm})/m_b(m_{H^\pm})} \simeq 8$. The best signature for this process comes from semileptonic top decay $t \to b\ell^+\nu_\ell$. For $m_{H^\pm} > m_t + m_b$, the leptonic branching ratio of $H^+ \to \bar{\tau}\nu_\tau$ is sizable only at large $\tan\beta$; it saturates at $m_\tau^2/3m_b^2(m_{H^\pm}) \simeq 0.1$ for $\tan\beta \gtrsim 20$. In order to suppress the background from the process $gg \to b\bar{t}W^+$, it is crucial to determine the chirality ($\simeq$ helicity, for $E_\tau \gg m_\tau$) of the produced τ^+: left handed in signal events, right handed in the background. As mentioned earlier, the τ helicity could be determined from the energy distribution of the visible τ decay products [15.90]. Altogether, a signal at a $\geq 5\sigma$ level should be visible [15.6] with an integrated luminosity of 300 fb^{-1} of data for $\tan\beta > 7$ (12) at $m_{H^\pm} = 200$ (500) GeV.

Prospects for MSSM Higgs searches at the LHC are summarized in Fig. 15.25 [15.8]. This figure is for the "maximal mixing" scenario, which is more difficult to probe than the "no mixing" case. Moreover, sparticle loop contributions to the production and decay of any Higgs particle have been taken to be negligible here. We see that the discovery of h is guaranteed in the entire

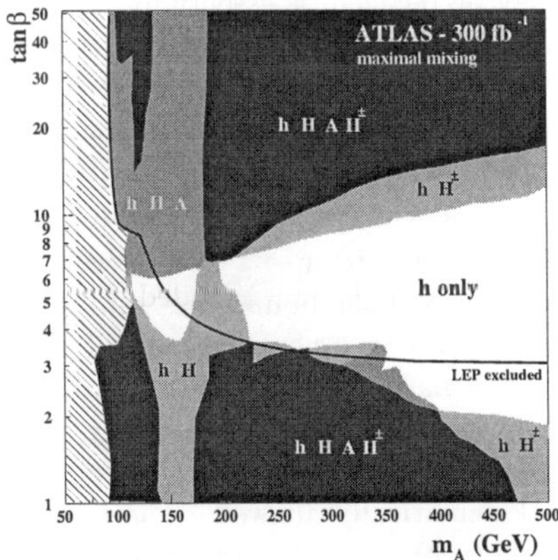

Fig. 15.25. MSSM Higgs search prospects in the ATLAS (LHC) experiment, adapted from Ref. [15.8], shown in the m_A–$\tan\beta$ plane with details described in the text. Only h will be found in the white region, while darker shadings correspond to the discovery regions for more Higgs particles. The LEP-excluded region is to the left and below the solid line.

currently allowed region of the parameter space. Were $\tan\beta$ sufficiently large, (some of) the heavy Higgs bosons could also be discovered. The lowest value of $\tan\beta$ making this feasible, in fact, increases with m_A. That is simply because all heavy Higgs production cross sections decrease rapidly as m_A goes up. Note that, for a significant region of the parameter space, it will be difficult to distinguish between the SM and the MSSM just using Higgs signals at the LHC. Of course, as described earlier in this chapter, the discovery of sparticles at the LHC is guaranteed if sparticle masses are "natural". Nevertheless, one will eventually want to experimentally study the entire MSSM Higgs sector as well.

A few comments on the possible role of sparticles are in order before this discussion of Higgs searches at the LHC is concluded. Given present sparticle mass bounds and the assumption (12.25)

---

[50]This is often called "$gb \to tH^-$", i.e. the virtual b quark in $gg \to \bar{b}tH^-$ is considered to be an (almost) onshell parton "in" the proton. Such a simplification allows one to partly resum large logarithms of μ_f/m_b, where the natural choice of the factorization scale would be $\mathcal{O}(m_{H^\pm})$. A caveat is that a full NLO calculation, which alone can decide the best description of this process, is nonexistent as yet.

of universal high scale gaugino masses, most supersymmetric loop contributions cannot change the partial width for the decay $H \to \gamma\gamma$ by more than $\sim 20\%$ [15.93]; ultimately, they are unable to compete with large contributions from the W loops. However, if high scale gaugino masses are nonuniversal and $|M_1/M_2| < 1/2$ at the EW scale, one could still have in the region of small $|\mu|$, a nearly 60–70% branching ratio for the invisible tree level decay $h \to \tilde{\chi}_1^0 \tilde{\chi}_1^0$. Then both BR $(h \to \gamma\gamma)$ and BR $(h \to b\bar{b})$ could be reduced to 30–40% of their SM values [15.87], thereby compromising the observability of the h at the LHC. In such a situation, the production channels with WW/ZZ fusion [15.95] or Zh [15.97] or $t\bar{t}h$ [15.6] would offer the discovery possibility for h. On the other hand, for large values of $\tilde{A}^t \equiv A^t + \mu \cot\beta$ (cf. 9.62c), $\tilde{t}$ loop contributions *can* reduce the partial width $\Gamma(h \to gg)$, and hence the production cross section $\sigma(gg \to h)$, by an order of magnitude [15.93]. In fact, if $m_{\tilde{t}_L} \simeq m_{\tilde{t}_R}$ and m_A is large, the contribution proportional to $\tilde{A}^t$ to the $h\tilde{t}_1\bar{\tilde{t}}_1$ coupling always reduces $\Gamma(h \to gg)$, independently of the sign of $\tilde{A}^t$. The reason is that the sign of $\tilde{A}^t$ also determines the sign of the $\tilde{t}_L$-$\tilde{t}_R$ mixing angle and this contribution to the $\tilde{t}_1\bar{\tilde{t}}_1$ coupling goes as $\sin 2\theta_{\tilde{t}}$. The signal for $gg \to h \to \gamma\gamma$ might become unobservably small in such a scenario, while that for $gg \to t\bar{t}h \to t\bar{t}b\bar{b}$, it may be recalled, is essentially independent of $\tilde{A}^t$ (except for the mild influence from m_h). Moreover, a very large $h\tilde{t}_1\bar{\tilde{t}}_1$ coupling and a light $\tilde{t}_1$ are needed to significantly reduce $\Gamma(h \to gg)$. In this case the cross section for $gg \to \tilde{t}_1\bar{\tilde{t}}_1 h$ production can be even larger than that for the process $gg \to t\bar{t}h$ [15.98], possibly leading to new h discovery channels.

We have seen earlier that the heavy MSSM Higgs particles could have substantial branching ratios into charginos and neutralinos when $\tan\beta$ is not very large. In that case, the lower bound on $\tan\beta$, where signatures for the decays $H, A \to \tau^+\tau^-$ as well as $H^- \to \tau^-\bar{\nu}_\tau$ become observable, have to be pushed up somewhat beyond the boundaries shown in Fig. 15.25. On the other hand, the decays $H, A \to \tilde{\chi}_2^0 \tilde{\chi}_2^0$ can also lead to a new signature if both neutralinos decay leptonically. This novel 4ℓ signature should be visible [15.6] at small and intermediate values of $\tan\beta$ if the corresponding branching ratio is not too small. Conversely, Higgs particles could be produced in sparticle decays. In particular, the branching ratio for the decay $\tilde{\chi}_2^0 \to h\tilde{\chi}_1^0$ would usually be significant when this decay is kinematically allowed. The latter would dominate $\tilde{\chi}_2^0$ decays if in addition all slepton masses lie above $M_{\tilde{\chi}_2^0}$ (e.g. for the region with $m_0 \gtrsim M_{1/2}$ in mSUGRA). This decay is most easily detectable if the $\tilde{\chi}_2^0$ in turn comes from the decay of a squark or gluino. In this case one could use generic cuts on jets and $\not{\!E}_T$ to isolate a sample of "SUSY events" with little SM background. The h signal[51] could then be seen in the invariant mass distribution of two tagged b-jets and should be [15.6] at a greater than 5σ signal in an mSUGRA context if the unified gaugino mass lies in the range 300 GeV $< M_{1/2} < 2m_0$ so that $m_{\tilde{\ell}_L} > M_{\tilde{\chi}_2^0} > M_{\tilde{\chi}_1^0} + m_h$. Analogous results should hold in other supersymmetry breaking models.

If the machine and the detectors perform more or less as planned, a discovery of h at the LHC is guaranteed. In fact, a single channel should provide a $\gtrsim 5\sigma$ significance. Indeed, in many cases, several independent channels will each provide a signal of such significance. Unfortunately, the prospects at the Tevatron look far less promising. This is even if an integrated luminosity of tens of fb^{-1} gets accumulated there before the commencement of the first LHC run. The cross section for the process $gg \to h \to \gamma\gamma$ is too small at the Tevatron to be useful[52]. The main discovery channels for an SM-like Higgs boson with a mass $\lesssim 130$ GeV are instead Wh and Zh production. These two production channels lead to several different final states, depending on how the gauge and Higgs

---

[51] Once again, in regions where the h might become invisible on account of nonuniversal high scale gaugino masses, the cross sections for the production of h in the decays of heavier neutralinos might become nonnegligible [15.97].

[52] That is true except in "pathological" situations with $|\sin\alpha| \ll 1$ where $BR(h \to \gamma\gamma)$ is strongly enhanced.

bosons decay. The $\ell\nu b\bar{b}$ final state (from Wh production) offers the best significance, as measured in terms of the ratio of the signal to the square root of the background. However, an SM-like Higgs particle with a mass exceeding 114 GeV (the LEP lower bound) would be discoverable with the anticipated integrated luminosity of ≤ 30 fb^{-1} only after adding results from both the CDF and the DØ experiments and from several final states, the next most useful ones being $\nu\bar{\nu}b\bar{b}$ and $\ell^+\ell^- b\bar{b}$ (both from Zh production). Since all these channels exploit $h \to b\bar{b}$ decays and look for a bump in the $b\bar{b}$ invariant mass distribution, good b-tagging as well as a good $b\bar{b}$ invariant mass resolution are crucial to reduce the background. Even after adding these three final states and summing results from both Tevatron experiments, one would need [15.99] an integrated luminosity of 2 (4) fb^{-1} to exclude an SM-like Higgs particle with mass 115 (130) GeV at 95% c.l. A discovery at the 5σ level would require 15 (30) fb^{-1} of data. These results apply to the light CP even state of the MSSM if $m_A^2 \gg M_Z^2$. For smaller m_A, the total $V\phi$ signal ($V = Z/W, \phi = h, H$) would be distributed over Vh and VH production, so that the signal in any one channel would become correspondingly weaker. Finally, the detection of any other Higgs particle would probably be possible only in very restricted regions of the parameter space [15.99]. With 20 fb^{-1} of accumulated data, the decay $t \to bH^+$ should be detectable for $m_{H\pm} = 100$ (140) GeV provided $\tan\beta > 15$ (30), there being no sensitivity to this channel if $m_{H\pm} > 160$ GeV. Moreover, with 30 fb^{-1} of data, a greater than 5σ discovery (per detector) of $b\bar{b}$ (H, A) production should be possible[53] in the $4b$ channel for $\tan\beta \geq 15$ (50) if $m_A = 100$ (160) GeV.

Muon colliders

In the last few years, muon colliders [15.10] have been discussed as serious alternatives to or complements of future e^+e^- and pp colliders. On account of negligible energy loss to synchroton radiation, a muon collider could be built as a storage ring with a beam spread that is possibly smaller than that at LEP [15.100]. The luminosity would be significantly less than at a linear e^+e^- collider, though. Since the Yukawa coupling strength of the muon exceeds that of the electron by a factor of 206, s channel resonant production of neutral Higgs bosons would be the dominant process if the muon collider could be operated at a CM energy $\sqrt{s} \simeq m_\Phi$ in the "Higgs factory" mode[54].

A small beam spread and a large muon-Higgs coupling are both crucial for the study of h production, in the context of the MSSM, on account of the very small width of this state; the latter is typically ~ 3 MeV in the decoupling limit. The spread $\sigma_{\sqrt{s}}$ in $\sqrt{s}$ would then at best be comparable to the intrinsic width [15.100] of the h:

$$\sigma_{\sqrt{s}} = (2 \text{ MeV}) \frac{R}{3 \times 10^{-5}} \frac{\sqrt{s}}{100 \text{ GeV}}, \tag{15.43}$$

where

$$R \equiv \delta E_{\text{beam}}/E_{\text{beam}}. \tag{15.44}$$

Allowing for this spread, the Breit-Wigner expression for the resonant cross section into the final

---

[53] These results are based on a lowest order QCD analysis; the cross sections for both the signal and the background are therefore quite uncertain. The main physics background $gg, q\bar{q} \to 4b$ occurs only at $\mathcal{O}(\alpha_s^4)$, implying a strong dependence on the renormalization scale (the scale in α_s).

[54] Proponents of e^+e^- colliders point out that the total production rate of an SM-like Higgs particle with mass ~ 120 GeV would be higher at the planned 350 GeV to 500 GeV linear colliders than at a $\mu^+\mu^-$ collider, since the former machine boasts a 500 times larger luminosity. The designation "Higgs factory" for muon colliders is thus controversial.

state X, with $\sqrt{s}$ centered at m_Φ becomes [15.100]

$$\sigma_{\mathrm{res}}(\mu^+\mu^- \to \Phi \to X) \simeq \frac{4\pi}{m_\Phi^2}\,\frac{B(\Phi \to \mu^+\mu^-)B(\Phi \to X)}{[1 + 8\pi^{-1}(\sigma_{\sqrt{s}}/\Gamma_\Phi)^2]^{1/2}}. \tag{15.45}$$

For $\Phi = h$, the most important final state is $X = b\bar{b}$. Hopefully, before experiments at a muon collider start, m_h would be known with an error of ~ 50 MeV, e.g. from studies at an e^+e^- collider, as discussed earlier. A muon collider could reduce this error to the MeV (one part in 10^6) level by scanning through the resonance region in $\sqrt{s}$ steps of one or two MeV. However, for $m_h \geq 115$ GeV, the direct measurement of Γ_h at a muon collider, being based on the analysis of the resonance lineshape, would not have better precision than what could be achieved at an e^+e^- collider with about 500 fb^{-1} of data. Once the resonance peak has been located, a muon collider with $R = 3 \times 10^{-5}$ could be expected to collect about 10^4 signal events in the $b\bar{b}$ mode per year, with a comparable $b\bar{b}$ continuum background. The measurement of the cross section $\sigma(\mu^+\mu^- \to h \to b\bar{b})$ would then allow one to see a $\geq 5\sigma$ deviation from the SM prediction[55] if $m_A \leq 500$ GeV. It may be recalled here that experiments at an e^+e^- collider would be able to detect a nearly two standard deviation departure from the SM only if $m_A \simeq 500$ GeV.

A muon collider could really come into its own in the study of the heavy neutral MSSM Higgs particles H and A. Since the latter are much broader the lightest scalar than h, a beam energy spread $R \simeq 10^{-3}$ (with higher luminosity) would be preferable to a smaller value of R (and smaller $\mathcal{L}$). One would first run the muon collider at the maximum energy and with maximum luminosity. The (average) mass of H and A would then appear as peaks in the $b\bar{b}$ invariant mass distribution, using the subsample of events with hard initial state radiation[56]. Having determined the Higgs mass(es), one would then run the machine at $\sqrt{s} \simeq m_A$. An energy scan should reveal two well-separated peaks in the cross section if $\tan\beta \lesssim 5$. However, for $\tan\beta \gtrsim 10$, the mass difference $|m_A - m_H|$ is usually smaller than the widths Γ_A, Γ_H, so that both states would overlap, leading to a broad resonance in the cross section [15.100]. Such a beam energy scan would allow the determination of m_A and m_H, each with an error of a few tens of MeV [15.100]. Moreover, measurements of branching ratios would permit new precision tests of the model. For example, if $m_A > 2m_t$ and $3.5 \lesssim \tan\beta \lesssim 15$, the branching ratios into both $b\bar{b}$ and $t\bar{t}$ should be measurable, allowing a direct determination of the mixing angles (cf. Ch.10) α and/or β. The ratio of branching ratios into $b\bar{b}$ and $\tau^+\tau^-$ would also probe (potentially sizable) vertex corrections. Last but not the least, decays into sparticles, if kinematically allowed, would give direct access to soft supersymmetry breaking parameters that are very difficult to determine otherwise (e.g., A_τ) [15.100].

15.6   Collider signals in the presence of R_p violation

The presence of $\not{R}_p$ couplings can substantially change sparticle search strategies and phenomenological studies of supersymmetry at colliders. This is since the type, number and energies of final state particles, consequent upon sparticle production and decay, can now be quite different from

---

[55]This is partly due to a fortuitous cancellation. For $\Phi = h$ and $X = b\bar{b}$, (15.45) is almost independent of the b mass (within the current errors on this quantity). Increasing m_b reduces $BR(h \to \mu^+\mu^-)$ and hence the numerator in (15.43); but it also increases Γ_h, which reduces the denominator as well. In this regard the finite beam energy spread is, in fact, an advantage.

[56]The probability of emitting a hard photon from the initial state goes as $\ln(\sqrt{s}/m_\mu)$ and is thus reduced by less than a factor of two, as compared to an e^+e^- collider operating at the same $\sqrt{s}$.

those in the situation with R_p conservation. No matter how weak the R_p couplings are, they will cause the decay of the LSP. The consequent effects, however, do fall into three separate categories depending on the strengths of those couplings. (i) If the magnitudes of all those coupling strengths are $\lesssim 5 \times 10^{-7}$, an LSP, produced in a collider, will decay outside the detectors, cf. §14.3. In such a case, if the LSP is neutral, the phenomenology of sparticle search at colliders will remain the same as discussed so far. This can be called the case of 'superweak' R_p couplings about which we have nothing more to say. However, the cosmological argument forbidding charged LSP's, to be presented in Ch.16, no longer applies if the LSP is unstable. Scenarios, where e.g. $\tilde{\tau}_1$ is the LSP, therefore become feasible now. In the given case of superweak R_p violation, the resulting phenomenology would resemble that of a GMSB model with a long-lived $\tilde{\tau}_1$ NLSP. (ii) In case the R_p coupling strengths are only 'weak', i.e. $5 \times 10^{-7} \lesssim |\lambda_{R_p}| \lesssim 10^{-2}$, the LSP decays within the detectors, thereby qualitatively changing the character of the finally detected signals of sparticle production at colliders. However, if $\tilde{\chi}_1^0$ is the LSP, its detectable decay is the only new effect that one will have here. (iii) With still larger magnitudes, these couplings will begin to compete with gauge couplings and affect the decays of heavier produced sparticles. Then the relative branching fractions of R_p decays will be decided by the sparticle mass spectrum and the magnitudes of the R_p and RPC coupling strengths under consideration. Moreover, in this (relatively) strong R_p regime, new channels with *single* sparticle production will open up with sizable rates. In addition, R_p contributions to processes, mediated by exchanged sparticles but involving external SM particles, can lead to interesting interference effects.

Collider signals in models of R_p violation depend not only on the sizes of the 27 possible new couplings but also on sparticle masses and their relative ordering, resulting in a really huge parameter space. For the sake of brevity, we shall make here the (usual) assumption that $\tilde{\chi}_1^0$ is the LSP. Even so our subsequent discussion will have to remain a bit sketchy. The present bounds on the magnitudes of the R_p coupling strengths were shown in §14.4 to depend sensitively on their generation structure. Trilinear couplings with two indices in the first generation could be probed most easily through sfermion production at particle colliders and strong constraints already exist on their magnitudes. On the other hand, bounds on couplings involving the second or the third generation are often quite weak. Thus those couplings can yet play significant roles in the decays not only of the LSP, but also of heavier neutralinos and charginos, if the RPC decays of the latter are only into three body final states. So these couplings will not be easily accessible. R_p couplings, with only one index referring to the first generation, can also affect the production of second or third generation fermions at colliders via virtual sfermion exchange. All these indicate that collider searches for R_p violation in final states containing second or third generation sfermions might be especially promising. However, Fig. 14.9 shows that the bounds on $|\lambda_{i33}|$ and $|\lambda'_{i33}|$ are among the tightest of all bounds on R_p couplings due to the strong upper bounds on neutrino masses to which these couplings contribute at the one loop level.

'Weak' R_p case

While sparticle production and decay occur through RPC couplings here, the $\tilde{\chi}_1^0$ pair, emerging at the end of the usual supersymmetric decay chains, decay via R_p couplings into detectable final states. If only trilinear R_p couplings were nonzero, $\tilde{\chi}_1^0$ has only three body decays (Fig. 14.3). Complete lowest order expressions for these partial widths are given in the second and third articles cited in Ref. [14.43]. The relative contributions of the different diagrams in the said figure, for a given final state, depend not only on the concerned coupling strengths but also on the masses of the exchanged sparticles. Another important factor is the gaugino/higgsino content of $\tilde{\chi}_1^0$. A higgsinolike $\tilde{\chi}_1^0$ will have suppressed couplings to the first and second generations while third generation sfermions will have sizable couplings to both higgsinolike and gauginolike states. Significant phase space effects can occur if any of the final state SM fermions is heavy. If the dominant R_p coupling strength is λ'_{13k}, these effects suppress the $\ell_i t \bar{d}_k$ final state relative to the $\nu_i b \bar{d}_k$ one. Moreover, if the only R_p coupling present is λ''_{3jk}, the top produced in $\tilde{\chi}_1^0$ decay may have to be quite off-shell, forcing the $\tilde{\chi}_1^0$ to decay into four (five) body final states if $M_{\tilde{\chi}_1^0} > (<) M_W$. As a result, with $M_{\tilde{\chi}_1^0} \sim 100$ GeV and all squark masses $\gtrsim 600$ GeV, the $\tilde{\chi}_1^0$ can become sufficiently long-lived so as to decay outside the detector even with a dominant $|\lambda''_{3jk}| \sim 10^{-2}$ [15.101]. For dominant R_p couplings involving the first or second generation or λ, λ' couplings involving the third generation, $\tilde{\chi}_1^0$ decay is almost always prompt.

The following three distinct decay patterns emerge for $\tilde{\chi}_1^0$ in the case of weak R_p violation.

- $\tilde{\chi}_1^0 \to \nu_j \ell_i^+ \ell_k^-$ for a nonzero λ_{ijk} coupling strength, yielding two charged leptons and some $\not{E}_T$ carried by the neutrino. For $k = 3$, at least one of the charged leptons is a τ and the final state contains its decay products. The antisymmetry of the λ's in the first two indices implies that, at least half the time, one of the two charged leptons is *not* a τ – enabling an easy detection of this decay even at a hadron collider.

- $\tilde{\chi}_1^0 \to u_j \bar{d}_k \ell_i,\, d_j \bar{d}_k \nu_i$ plus the charge conjugate channels for a nonzero λ'_{ijk} coupling, providing two jets and one charged lepton or $\not{E}_T$ carried by a neutrino. The branching fractions of these two modes may be very similar unless $j = 3$, as remarked above. Some of the jets will be b jets and there might be additional jets/leptons if ℓ_i is a τ.

- $\tilde{\chi}_1^0 \to u_i d_j d_k$ plus the charge conjugate channel for a nonzero λ''_{ijk} coupling leading to three jets per decaying $\tilde{\chi}_1^0$. If one of the final state quarks/antiquarks carries the bottom or charm flavor, the corresponding jet may contain a lepton coming from its semileptonic decay.

Since $\tilde{\chi}_1^0$ is a Majorana particle, charge conjugate final states are produced with equal branching ratios. The possibility of utilizing displaced vertices to obtain additional information on the decays of a relatively long-lived $\tilde{\chi}_1^0$ has not been fully explored so far. What have been considered are characteristic final states with multiple leptons, jets and neutrinos carrying $\not{E}_T$. Clearly, the larger number of leptons in the final state makes search procedures easier for the L violating λ, λ' couplings rather than for the B nonconserving λ'' couplings which mostly yield leptonless events. One of the motivations for introducing R_p violation has been the possibility (cf. §14.4) of generating small neutrino masses, as required by experiment. Different R_p neutrino mass generation mechanisms lead to different predictions for the flavor structure of the λ, λ' couplings and hence for the relative branching fractions of the $\tilde{\chi}_1^0$ decaying into different L violating channels. Many of these can be probed in collider experiments [15.102].

Even in models with bilinear R_p terms in the superpotential, trilinear R_p terms get induced (cf. §14.5), so that the above decays of the $\tilde{\chi}_1^0$ can take place. However, now additional modes open up via ν-$\tilde{\chi}_1^0$ mixing, namely $\tilde{\chi}_1^0 \to \ell^\pm W^\mp (W^{\star\mp})$, $\nu Z(Z^\star)$, $\nu\gamma$, $\nu h(h^\star)$, the virtual gauge/Higgs boson leading to a fermion-antifermion pair. For the range of $M_{\tilde{\chi}_1^0}$, that is of interest to LEP and/or TEVATRON, the gauge/Higgs boson is virtual so that effectively $\tilde{\chi}_1^0$ will have three body decays into final states containing at least one charged lepton or a neutrino. For the fraction of events where the $Z/Z^\star$ transits into a $\nu\bar{\nu}$ pair, the decay of the $\tilde{\chi}_1^0$ will become invisible so that these will look like events from an RPC scenario. If in the minimal bilinear R_p model, considered within an mSUGRA picture, the different R_p coupling strengths in the effective Lagrangian are fixed to explain the neutrino data from the solar, atmospheric and reactor experiments, a light $\tilde{\chi}_1^0$

will almost always have [14.60] a decay length > 1 cm. However, a more prompt $\tilde{\chi}_1^0$ decay becomes possible [15.102] if one takes the same model in a GMSB framework.

As per our previous practice, we first discuss the visible effects of R_p violation at e^+e^- colliders. Let us focus initially on signals of lepton number violation caused by nonzero $\lambda, \lambda', \epsilon_i$ terms. Figs. 15.3 and 15.4 show that the cross section for $\tilde{\chi}_1^0$ pair production, though less than that for $\tilde{\chi}_1^+\tilde{\chi}_1^-$ production, is not insignificant in e^+e^- collision. This process, followed by $\not{R}_p$ decays of the $\tilde{\chi}_1^0$'s, will yield [14.43] four charged leptons with $\not{E}_T$ for the λ_{ijk} case and at least four jets plus two charged or neutral leptons for the λ'_{ijk} case. For the latter case $\ell, \ell\nu$ and $\nu\nu$ will be produced with the relative abundances 1:2:1 unless $j = 3$, as remarked earlier. Since each of the $\tilde{\chi}_1^0$'s can independently decay into a lepton of either positive or negative charge, half of the $\ell\ell$ events would have same sign (SS) dileptons. If the event contains no neutrinos, lepton flavor violation would be indicated. Such events with SS ditaus are spectacular $\not{R}_p$ signatures, cf. Godbole, Roy and Tata [14.48]. The flavor content of the final state would evidently depend on the relative sizes of the different $\not{R}_p$ coupling strengths. Most studies have been performed under the assumption that a single coupling dominates LSP decays[57]. If $M_{\tilde{\chi}_1^0} > m_t$, so that the $\tilde{\chi}_1^0$ is producable only at future colliders and its decay final states could contain the top quark, the signal analysis would become more complicated owing to possible additional hard leptons in the final state coming from semileptonic top decays. So long as no top quark can be produced, $\not{R}_p$MSSM events should have little SM background despite the loss of $\not{E}_T$ as a major characteristic of the signal. Other sparticle pairs, produced in e^+e^- collision, will always decay into a pair of $\tilde{\chi}_1^0$'s plus additional visible/invisible particles. For example, a produced $\tilde{\chi}_1^\pm$ pair will yield four charged leptons (in the nonzero λ case) – all decay products of a pair of χ_1^0's – *in addition* to the usual energetic leptons/jets from the 'canonical' MSSM decays of the chargino pair. A detailed analysis [15.103] shows that signals from pair production and decays of EW bosinos in an $\not{R}_p$ scenario can be easily picked up above the SM background.

Similar statements can be made about the pair production and decays of sparticles at $pp/p\bar{p}$ colliders. Once more, the final states will be characterized by the $\not{R}_p$ decay products of the inevitable $\tilde{\chi}_1^0$ pair, rather than by large $\not{E}_T$, though some $\not{E}_T$ may be carried off by neutrinos. Same sign dileptons with little $\not{E}_T$ can again be used [15.104] as a $\not{R}_p$ signal. Most of these studies actually show that, for nonzero λ, λ' couplings, the reach[58] in the mSUGRA parameter space from different sparticles searches is at least as good as in the RPC case. This is largely on account of multiple leptons in the final state which help in suppressing backgrounds and providing triggers. For instance, a $q\bar{q}'$ annihilation subprocess can now create a $\tilde{\chi}_1^\pm\tilde{\chi}_1^0$ pair giving rise to hadronically quiet events with upto five charged leptons, the SM background in this channel being negligible. When $\tilde{\chi}_1^0$ decays hadronically via λ'' couplings, multijet events are expected. If the concerned coupling is λ''_{3jk}, displaced vertices[59] might help to distinguish the signal from QCD backgrounds. Otherwise, $\tilde{\chi}_1^0$ decays are difficult to disentangle from QCD radiation unless one finds an excess of final state b, c quarks which could be tagged to reduce the background. Another possibility is to use additional leptons in the signal coming from earlier cascade decay chains which produce $\tilde{\chi}_2^0, \tilde{\chi}_1^\pm$. Such an approach holds promise [15.6, 15.105, 15.106] in tackling the challenging cases of $\lambda''_{212}, \lambda''_{112}$ couplings effecting $\tilde{\chi}_1^0$ decay. It needs to be mentioned also that such hadronic decays of the $\tilde{\chi}_1^0$ can spoil the observability of the lightest Higgs h (produced from the decay $\tilde{\chi}_2^0 \to \tilde{\chi}_1^0 h$) in the $h \to b\bar{b}$ channel.

---

[57]Since the branching fractions scale like squares of the appropriate coupling strengths, the actual hierarchy between coupling strengths need not be very large for one kind of final state to be dominant.

[58]This statement is true for both TEVATRON and the LHC.

[59]The use of such displaced vertices might offer some chance of measuring the strengths of the concerned $\not{R}_p$ couplings. No detailed investigation of this issue has yet been made, though.

We end this subsection with a final comment. No search strategy, that has been proposed to date, has specifically shown how to discriminate between RPC and $\not{R}_p$ scenarios, should both lead to similar detectable final state particles such as the SS dilepton signal mentioned above. A preponderance of multileptons as well as the occurrence of nonstandard lepton flavor combinations and low jet multiplicity will strongly indicate the existence of L_i violating couplings. In contrast, events containing leptons and yet with high jet multiplicity in excess of known backgrounds might suggest the presence of baryon nonconserving λ'' couplings.

Larger $\not{R}_p$ coupling strengths

With stronger $\not{R}_p$ coupling strengths, the signal sensitivity to their precise magnitudes increases. As a result, one may have a chance to measure these couplings in a collider environment once $\not{R}_p$ signals are detected. Such opportunities arise in resonant or nonresonant single sparticle production, in decays of sparticles and particles caused by $\not{R}_p$ couplings as well as in virtual sparticle effects in scattering processes. Let us discuss these situations separately.

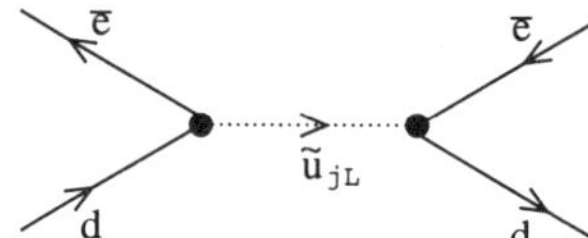

Fig. 15.26. One of the diagrams for resonant production and decay of a squark in e^+p collision.

<u>Production of a single sparticle</u>

Consider, as an example, the resonant electroproduction of a squark in ep collisions at HERA, occuring via an s channel exchange process with a λ' coupling, as shown in Fig. 15.26. With an e^+ beam, the dominant subprocess will be $e^+d \to \tilde{u}_{jL}$ via the λ'_{1j1} coupling. The produced squark can decay too by the same coupling. That will give rise to [15.107] a 'leptoquark signature', i.e. a hard ℓ^+ balanced by a jet. Two body decay kinematics leads to a peak in the $\ell + $ jet invariant mass on top of a smooth background of deep inelastic scattering events. The squark, however, can also decay by the modes of Table 15.4, ending in a $\tilde{\chi}_1^0$ which then decays via a $\not{R}_p$ coupling, cf. Fig. 14.3. Indeed, the $\not{R}_p$ and RPC decays of the squark will compete with one another. Similar resonant production [15.108] of a sneutrino is possible at an e^+e^- collider via the λ_{j11} coupling. In case one particular coupling strength dominates, the sneutrino $\tilde{\nu}_j$ must decay by the same coupling or through the RPC decays of (15.18), eventually yielding a $\tilde{\chi}_1^0$ which then decays via the coupling λ_{j11}. The RPC decay of $\tilde{\nu}_j$ can now yield a single $\tilde{\chi}_1^\pm$. Single $\tilde{\chi}_1^\pm$ can also come from e^+e^- collision via nonresonant production of $\tilde{\nu}_j$ as well as t channel $\tilde{\nu}_j$ exchange in [15.109] $e^+e^- \to \tilde{\chi}_1^\mp \ell^\pm$.

A single $\tilde{\chi}_1^\pm$ may emerge in a similar fashion at a $pp/p\bar{p}$ collider by [15.110–15.112] resonant or nonresonant $\tilde{\ell}_{iL}/\tilde{\nu}_i$ production via a λ'_{ijk} coupling. Turning to λ'' couplings, λ''_{3jk} can lead to resonant production [15.113] of a single stop $\tilde{t}$: $d_k + d_j \to \overline{\tilde{t}_a}$. Even a $|\lambda''_{3jk}|$ as small as 0.1 will lead to a substantial cross section for this process. Of course, the $\lambda''_{(1,2)jk}$ couplings lead to analogous single production of $\tilde{u}_R$ and $\tilde{c}_R$ squarks respectively as well as to $u_i + d_j \to \overline{\tilde{d}_k}$. However, in many models these are significantly heavier than $\tilde{t}_1$; moreover, the coupling strengths λ''_{ijk} are already tightly constrained, cf. §14.4. One can also study single resonant slepton production in pp ($p\bar{p}$) collisions via the λ'_{i11} coupling [15.114]. The potentially larger coupling λ'_{i32} can give rise to associated single slepton production in $s\bar{t}\ell_j^+ + $ c.c. final states [15.114]. The analogous process with $s \to b$ [15.115] is unlikely to be detectable, given the tight limits on $|\lambda'_{i33}|$, cf. Fig. 14.9.

R-parity violating decays of particles and sparticles

The decays of $\tilde{\chi}_1^{\pm}$ and $\tilde{\chi}_1^0$ due to R_p couplings have already been discussed. Such couplings can also cause unusual decays of other sparticles and of SM particles like the top quark. For instance, λ''_{3ji} (λ'_{j3i}) couplings can cause top decays like $t \to \bar{d}_i \tilde{d}_j$ ($d_i \tilde{\ell}_j$), clearly affecting the top signal at the TEVATRON [15.115, 15.116]. The absence of such unusual top decays has been used [15.115, 15.117] to put upper bounds on $|\lambda'_{j3k}|$ and $|\lambda''_{3jk}|$ in scenarios with light sfermions. Three body decays of the top via R_p couplings to states containing a single $\tilde{\chi}_1^0$ or $\tilde{\chi}_1^+$ [15.115, 15.118] can produce a single neutralino/chargino in pp ($p\bar{p}$) $\to t\bar{t}X$, if the other top decays into Wb. This situation is complementary to that of the slepton/squark production process, discussed earlier. In the latter case the production occurs through R_p couplings and the decays via RPC ones. For even larger magnitudes of the R_p coupling strengths, with two of the $|\lambda'_{ijk}|$ simultaneously being large, the top can decay into three SM fermions with a partial width proportional to $|\lambda'_{i3k}\lambda'_{jlk}|$, $i, j, l \neq 3$.

We have already mentioned sfermion decays via λ, λ' couplings in the context of leptoquark signals at ep colliders. In $pp/p\bar{p}$ colliders the effects of R_p couplings on strongly interacting sparticles can be tested in their decays, provided they are light enough to be abundantly pair-produced by SQCD processes. For instance, a $\tilde{u}_{jL}$ – produced by a strong interaction mechanism – can decay into $d_k \ell_i^+$ via the coupling λ'_{ijk} or equivalently via the chargino-tau mixing in a bilinear R_p theory, cf. §14.5. For $|\lambda'_{ijk}| \sim 0.1$–$0.2$, two body R_p decays can compete with RPC ones. They might even dominate if the only kinematically accessible RPC decays involve higgsinolike $\tilde{\chi}$ states.

Virtual Effects

The amplitude for Drell-Yan dilepton production at a pp ($p\bar{p}$) collider has a t channel squark exchange diagram with λ' couplings [15.119]. Similarly, in $t\bar{t}$ production the amplitude for the subprocess $q\bar{q} \to t\bar{t}$ contains the t channel exchange of a slepton/squark with λ'/λ'' couplings. Given the current lower bounds on sfermion masses and the upper bounds on the concerned R_p coupling strengths, these effects on total rates are not too large. It is thus not easy to extract information on the sizes of these couplings from those rates. This is particularly so because of inherent normalization uncertainties in the calculated SM cross sections. Nevertheless, R_p effects can be looked for at the LHC by studying distributions in different kinematic variables of the dilepton pair in the first case [15.120], and in the polarization asymmetry [15.121] between the top and the antitop in the second case, showing up in the angular and energy distributions of the decay leptons. The latter could yield useful information on $\lambda'_{i31}, \lambda''_{311}$ at the TEVATRON where the $q\bar{q}$ annihilation component of $t\bar{t}$ production is relatively more important. Similarly, a two fermion final state, produced in a lepton collider, can have [15.122] t channel sfermion exchanges (via λ, λ' couplings) whose effects can manifest themselves in distributions of cleverly combined kinematic variables.

Current constraints and future prospects

Anomalous $\mu\gamma\not{E}_T$ events, which may have been seen at CDF [15.123] can be explained [15.124] in the context of R_pGMSB models in terms of resonant smuon production in $q\bar{q}'$ annihilation via the coupling λ'_{311} and the consequent decay chain $\tilde{\mu} \to \mu\tilde{\chi}_1^0 \to \mu\tilde{G}\gamma$, a possibility to be tested in TEV II. Existing TEVATRON searches [15.125] for events with dileptons and two hard jets, intended to study leptoquark production, can actually be used to constrain SQCD pair production of squarks, each of which decays via λ' couplings into a lepton and a light quark. A similar TEVATRON search for multileptons accompanied by jets in general has looked for decays through λ' couplings of a

$\tilde{\chi}_1^0$ pair created from the decays of a squark or gluino pair produced by SQCD processes. Most of these direct searches [15.126] have put limits on combinations of λ and λ' coupling strengths and squark and gluino masses. However, since only the charged leptons from the first two generations are easily detectable, the constrained couplings involve just those. Nevertheless, ditau final states from a decaying squark pair have also been looked for. All these searches have been negative, ruling out any stop upto a mass of 100 GeV irrespectively of the size of the λ' couplings. Analyses of resonant slepton production [15.111, 15.112] have begun at the time of writing and preliminary results put bounds on the magnitudes of some of the λ' couplings in the ballpark of 10^{-2}.

No squark resonance has been observed at HERA. In consequence, a first generation squark upto a mass of 260 GeV has been ruled out for $|\lambda'| \sim \sqrt{4\pi\alpha_{\rm em}}$ by an analysis [15.127] performed both in the mSUGRA framework and in unconstrained MSSM with assumed high scale gaugino mass unification. LEP has established limits [15.128, 15.129] on sparticle masses, assuming the sparticles to decay directly via $\not{R}_p$ processes with λ, λ' and λ'' couplings as well as from an inclusively produced $\tilde{\chi}_1^0$ which then decays. It has been possible to translate these limits to exclusion zones in the M_2–μ parameter space. The nonobservation of any new s channel resonance, as also the lack of any evidence for a t channel sfermion exchange, have yielded upper bounds on $|\lambda_{121}|, |\lambda_{131}|$ and $|\lambda_{232}|$, independently of the supersymmetry breaking parameters in the chargino/neutralino sector. These bounds[60] are quite stringent for low values of sneutrino masses, viz. 100 GeV $< m_{\tilde{\nu}} <$ 200 GeV, but increase to nearly gauge coupling strength values for higher $m_{\tilde{\nu}}$.

Choosing λ'_{211} for definiteness, we note the following. A study of resonant slepton production at [15.111] TEV-II could provide a reach in the $M_{1/2}$–m_0 plane of mSUGRA upto $M_{1/2} \simeq$ 800–900 GeV and $m_0 \simeq$ 500–600 GeV, even with $|\lambda'_{211}|$ as low as 0.05. Corresponding studies at the LHC should be sensitive to $|\lambda'_{211}| >$ 0.01 (0.05) for $m_{\tilde{\nu}} =$ 350 (900) GeV [15.112]. It should also be possible at the LHC to use DY dimuon production as a means of measuring λ'_{2jk} coupling strengths, thereby improving significantly on TEVATRON limits [15.120]. The decay of any $\tilde{\chi}_1^0$, produced at the LHC, into visible fermions (especially leptons) should facilitate its mass determination by event reconstruction. Given the much smaller amoung of $\not{E}_T$ in these events, as compared with the RPC case, a good measure of the supersymmetry breaking scale M_s could now be provided by $M_{eff} = \not{E}_T + \sum_{jets,\ \ell's} E_T$. However, in case $\tilde{\chi}_1^0$ decay proceeds only via λ'' couplings, the goal of measuring its mass would seem achievable [15.6, 15.105] at the LHC only for $M_{\tilde{\chi}_1^0} \lesssim$ 100 GeV.

References

[15.1] M. Schmitt in *Supersymmetric Particle Searches, Review of Particle Physics*, Particle Data Group, *loc. cit., Bibl.*

[15.2] M. Brhlik and G.L. Kane, Phys. Lett. **B437** (1998) 331.

[15.3] E. Accomando *et al.*, Phys. Rept. **299** (1998) 1.

[15.4] G. Arnison *et al.*, Phys. Lett. **B139** (1984) 115.

[15.5] V.D. Barger *et al.*, *Report of the SUGRA working group for Run II of the Tevatron*, hep-ph/0003154. S. Ambrosanio et al, *Report of the Beyond the MSSM subgroup for the Tevatron*

---

[60]There is a caveat here. Many of these analyses, conducted within the mSUGRA framework, have not accounted for nonnegligible effects of relatively large $\not{R}_p$ coupling strengths on the sparticle mass spectra used.

Run II SUSY/Higgs workshop, hep-ph/0006162. H. Baer *et al.*, *Report of the low scale and gauge mediated supersymmetry working group at Tevatron Run II*, hep-ph/0008070.

[15.6] The ATLAS collaboration, *loc. cit.*, *Bibl.*

[15.7] CMS Technical Proposal, CERN/LHCC/94-38 (1994). S. Abdullin *et al.*, J. Phys. **G28** (2002) 469.

[15.8] J.A. Aguilar-Saavedra *et al.*, *loc. cit. Bibl.* T. Abe et al., *loc. cit. Bibl.*, K. Abe *et al.*, *loc. cit. Bibl.*

[15.9] B. Badelek *et al.* (2001), *TESLA Technical Design Report, Part VI, Ch.1*: "Photon collider at TESLA", hep-ex/0108012.

[15.10] B. Autin, A. Blondel and J. Ellis, CERN 99-02, ECFA 99-197. M.S. Berger, in *Proc. APS/DPF/DPB Summer Study on Future of Particle Physics* (Snowmass, 2001), hep-ph/0110390.

[15.11] A. Bartl, H. Fraas and W. Majerotto, Z. Phys. **C30** (1986) 441; Nucl. Phys. **B278** (1986) 1. S.Y. Choi, A. Djouadi, H.S. Song and P.M. Zerwas, Eur. Phys. J. **C8** (1999) 669. G. Moortgat-Pick, H. Fraas, A. Bartl and W. Majerotto, *ibid* **C9** (1999) 521. S.Y. Choi, J. Kalinowski, G. Moortgat-Pick and P.M. Zerwas, *ibid.* **C22** (2001) 563. A. Djouadi, M. Drees and J.L. Kneur, JHEP **0108** (2001) 055.

[15.12] M.A. Diaz, S.F. King and D.A. Ross, Nucl. Phys. B**529** (1998) 12.

[15.13] S. Dawson, E. Eichten and C. Quigg, Phys. Rev. **D31** (1985) 1581.

[15.14] R.D. Field, *op. cit.*, *Bibl.* G. Altarelli, *op. cit.*, *Bibl.*

[15.15] V.D. Barger and R.J.N. Phillips, *op. cit.*, *Bibl.*

[15.16] S.Y. Choi, M. Guchait, H.S. Song and W.Y. Song, hep–ph/0007276.

[15.17] H. Baer and X. Tata, Phys. Rev. **D47** (1993) 2739. H. Baer, C-H. Chen, M. Drees, F. Paige and X. Tata, *ibid.* **D58** (1998) 075008.

[15.18] A. Bartl, W. Majerotto and W. Porod, Phys. Lett. **B465** (1999) 187.

[15.19] A. Djouadi, Y. Mambrini and M. Mühlleitner, Eur. Phys. J. **C20** (2001) 563.

[15.20] D. Pierce and A. Papadopoulos, Nucl. Phys. **B430** (1994) 278.

[15.21] C.H. Chen, M. Drees and J.F. Gunion, Phys. Rev. **D55** (1997) 330; Errtm. *ibid.* **D60** (1999) 039901.

[15.22] H.E. Haber and D. Wyler, Nucl. Phys. **B323** (1989) 267.

[15.23] http://lepsusy.web.cern.ch/lepsusy/.

[15.24] P. Chen, Phys. Rev. **D46** (1992) 1186. T. Behnke *et al.*, in *TESLA Technical Design Report*, Ch. 4; http://tesla.desy.de/tdr/

[15.25] K. Abe *et al.*, *loc. cit.*, *Bibl.*, pp. 87–138 and 384–442.

[15.26] S.Y. Choi, A. Djouadi, M. Guchait, J. Kalinowski, H.S. Song, and P.M. Zerwas, Eur. Phys. J C **14** (2000) 535; S.Y. Choi, A. Djouadi, H. Dreiner, J. Kalinowski and P.M. Zerwas, *ibid.* **C7** (1999) 123; S.Y. Choi, M. Guchait, J. Kalinowski and P.M. Zerwas, Phys. Lett. **B479** (2000) 235.

[15.27] G. Bélanger, F. Boudjema, A. Pukov and S. Rosier-Lee in *Hamburg 2002 : Supersymmetry and unification of fundamental interactions*, vol. 2, p 919, hep-ph/0212227.

[15.28] A. Djouadi, M. Drees and J.L. Kneur, *loc. cit.* [15.11].

[15.29] S. Ambrosanio and G.A. Blair, Eur. Phys. J **C12** (2000) 287.

[15.30] S. Ambrosanio *et al.*, *loc. cit.* [15.5].

[15.31] H. Baer *et al.*, *loc. cit.* [15.5].

[15.32] H. Baer, C-H. Chen, F. Paige and X. Tata., Phys. Rev. **D 50** (1994) 4508.

[15.33] M.M. Nojiri and Y. Yamada, Phys. Rev. **D60** (1999) 015006.

[15.34] H. Baer, A. Bartl, D. Karatas, W. Majerotto and X. Tata, Int. J. Mod. Phys. **A4** (1989) 4111.

[15.35] M.M. Nojiri, Phys. Rev. **D51** (1995) 6281.

[15.36] H. Baer, R. Munroe and X. Tata, Phys. Rev. **D54** (1996) 6735, errtm. *ibid.* **D56** (1997) 4424. A. Djouadi, M. Drees and J.L. Kneur, *loc. cit.* [15.11].

[15.37] M. Peskin, Int. J. Mod. Phys. **A13** (1998) 2299; Prog. Theor. Phys. Suppl. **123** (1996) 507.

[15.38] J.L. Feng and M.E. Peskin, Phys. Rev. **D64** (2001) 115002.

[15.39] A. Bartl, H. Eberl, S. Kraml, W. Majerotto and W. Porod, Eur. Phys. J. **C6** (2000) 1.

[15.40] M. Bisset, S. Raychaudhuri and P. Roy, hep-ph/9602430.

[15.41] H. Baer, C-H. Chen, F. Paige and X. Tata, Phys. Rev. **D49** (1994) 3283.

[15.42] H. Baer, C-H. Chen, R. Munroe, F.E. Paige and X. Tata, Phys. Rev. **D51** (1995) 1046.

[15.43] H. Baer, C-H. Chen, F.E. Paige and X. Tata, Phys. Rev. **D53** (1996) 6241.

[15.44] H. Baer, P.G. Mercadante, X. Tata and Y. Wang, Phys. Rev. **D62** (2000) 095007.

[15.45] S. Ambrosanio, B. Mele, A. Nisati, S. Petrarca, G. Polesello, A. Rimoldi and G. Salvini, hep-ph/0012192. S. Ambrosanio, B. Mele, S. Petrarca, G. Polesello and A. Rimoldi, JHEP **0101** (2001) 014.

[15.46] M.M. Nojiri, K. Fujii and T. Tsukamoto, Phys. Rev. **D54** (1996) 6756.

[15.47] J.K. Mizukoshi, H. Baer, A.S. Belyaev and X. Tata, Phys. Rev. **D64** (2001) 115017.

[15.48] M. Peskin, H. Murayama, J.L. Feng and X. Tata, Phys. Rev. **D52** (1995) 1418. M.M. Nojiri, D.M. Pierce and Y. Yamada, Phys. Rev. **D57** (1998) 1539.

[15.49] H.C. Cheng, J.L. Feng and N. Polonsky, Phys. Rev. **D57** (1998) 152. E. Katz, L. Randall and S. Su, Nucl. Phys. **B536** (1998) 3.

[15.50] D.K. Ghosh, P. Roy and S. Roy, JHEP **0008** (2000) 015. D.K. Ghosh, A. Kundu, P. Roy and S. Roy, Phys. Rev. **D64** (2001) 115001.

[15.51] P.R. Harrison and C.H. Llewellyn Smith, Nucl. Phys. **B213** (1983) 223; errtm. *ibid.* **223** (1983) 542.

[15.52] H. Baer, M. Drees, R.M. Godbole, J.F. Gunion and X. Tata, Phys. Rev. **D44** (1991) 725.

[15.53] W. Beenakker, R. Höpker, M. Spira and P.M. Zerwas, Nucl. Phys. **B492** (1997) 51.

[15.54] W. Beenakker, M. Krämer, T. Plehn, M. Spira and P.M. Zerwas, Nucl. Phys. **B515** (1998) 3.

[15.55] M. Carena, R. Culbertson, S. Eno, H.J. Frisch and S. Mrenna, *loc. cit.*, *Bibl.*

[15.56] H. Baer, C-H. Chen, F. Paige and X. Tata, Phys. Rev. **D52** (1995) 2746.

[15.57] T. Affolder *et al.*, Phys. Rev. Lett. **88** (2002) 041801.

[15.58] J. Hisano, K. Kawagoe, R. Kitano and M.M. Nojiri, Phys. Rev. **D66** (2002) 115004.

[15.59] H. Eberl, S. Kraml and W. Majerotto, JHEP **9905** (1999) 016.

[15.60] M. Drees and K. Hikasa, Phys. Lett. **B252** (1990) 127.

[15.61] W. Beenakker, R. Höpker and P.M. Zerwas, Phys. Lett. **B349** (1995) 403.

[15.62] A. Bartl, H. Eberl, W. Majoretto, Nucl. Phys. **B472** (1996) 481.

[15.63] A. Djouadi and Y. Mambrini, Phys. Lett. **B493** (2000) 120. A. Djouadi, Y. Mambrini and M. Mühlleitner, *loc. cit.* [15.19].

[15.64] H. Baer, X. Tata and J. Woodside, Phys. Rev. **D42** (1990) 1568.

[15.65] A. Bartl, W. Majerotto, B. Mösslacher, N. Oshimo and S. Stippel, Phys. Rev. **D43** (1991) 2214.

[15.66] A. Bartl, W. Majerotto and W. Porod, Z. Phys. **C64** (1995) 499; errtm. *ibid.* **68** (1995) 518.

[15.67] J. Guasch, W. Hollik and J. Sola, Phys. Lett. **B437** (1998) 88.

[15.68] A. Arhrib, A. Djouadi, W. Hollik and C. Jünger, Phys. Rev. **D57** (1998) 5860. A. Bartl, H. Eberl, K. Hidaka, S. Kraml, W. Majerotto, W. Porod and Y. Yamada, *ibid.* **D59** (1999) 115007.

[15.69] K. Hikasa and M. Kobayashi, Phys. Rev. **D36** (1987) 724.

[15.70] C. Boehm, A. Djouadi and Y. Mambrini, Phys. Rev. **D61** (2000) 095006. A. Djouadi and Y. Mambrini, *ibid.* **D63** (2001) 115005.

[15.71] M. Drees and O.J. Eboli, Eur. Phys. J **C10** (1999) 337.

[15.72] A.K. Datta, A. Djouadi, M. Guchait and Y. Mambrini, Phys. Rev. **D65** (2001) 015007-1.

[15.73] D. Acosta *et al.*, Phys. Rev. Lett. **90** (2003) 251801. V.M. Abazov *et al.*, *ibid.* **88** (2002) 171802. T. Affolder *et al.*, *ibid.* **84** (2000) 5704.

[15.74] H. Baer, X. Tata and J. Woodside, Phys. Rev. **D45** (1992) 142. R.M. Barnett, J.F. Gunion and H.E. Haber, Phys. Lett. **B315** (1993) 349.

[15.75] H. Baer, A. Belyaev, T. Krupovnickas and X. Tata, Phys. Rev. **D65** (2002) 075024.

[15.76] A. Djouadi, M. Guchait and Y. Mambrini, Phys. Rev. **D64** (2001) 095014.

[15.77] H. Baer, P.G. Mercadante and X. Tata, Phys. Rev. **D59** (1999) 015010.

[15.78] H. Baer, C. Balazs, A. Belyaev, T. Krupovnickas and X. Tata, JHEP 0306 (2003) 054. See also D.R. Tovey, EPJ Direct, Sec. A-E:4 (2002) no. CN4, p 1.

[15.79] I. Hinchliffe, F.E. Paige, M.D. Shapiro, J. Sodderqvist and W. Yao, Phys. Rev. **D55** (1997) 5520.

[15.80] H. Baer, P.G. Mercadante, F. Paige, X. Tata and Y. Wang, Phys. Lett. **B435** (1998) 109. H. Baer, P.G. Mercadante, X. Tata and Y. Wang, Phys. Rev. **D62** (2000) 095007.

[15.81] I. Hinchliffe and F.E. Paige, Phys. Rev. **D60** (1999) 095002. S. Ambrosanio et al., *loc. cit.* [15.45].

[15.82] H. Baer, J.K. Mizukoshi and X. Tata, Phys. Lett. **B488** (2000) 367.

[15.83] G.A. Blair, W. Porod and P.M. Zerwas, Phys. Rev. **D63** (2001) 017703; Eur. Phys. J. **C27** (2003) 263. B.C. Allanach, G.A. Blair, S. Kraml, H.-U. Martyn, G. Polesello, W. Porod and P.M. Zerwas, hep-ph/0403133.

[15.84] A. Pilaftsis and C.E.M. Wagner, Nucl. Phys. **B553** (1999) 3; S.Y. Choi, K. Hagiwara and J.S. Lee, Phys. Rev. **D64** (2001) 032004.

[15.85] H. Eberl, K. Hidaka, S. Kraml, W. Majerotto and Y. Yamada, Phys. Rev. **D62** (2000) 055006.

[15.86] J.F. Gunion, H.E. Haber, G.L. Kane and S. Dawson, *op. cit.*, *Bibl.* J.F. Gunion, H.E. Haber and R. van Kooten, hep-ph/0301023.

[15.87] G. Bélanger, F. Boudjema, A. Cottrant, R.M. Godbole and A. Semenov, Phys. Lett. **B519** (2001) 93. G. Bélanger, F. Boudjema, F. Donato, R.M. Godbole and S. Rosier-Lees, Nucl. Phys. **B581** (2000) 3.

[15.88] The LEP collaborations' joint report, hep-ex/0107030.

[15.89] M. Carena et. al., Nucl. Phys. **B580** (2000) 29.

[15.90] B.K. Bullock, K. Hagiwara and A.D. Martin, Phys. Rev. Lett. **67** (1991) 3055; Nucl. Phys. **B395** (1993) 499.

[15.91] M.M. Mühlleitner, M. Krämer, M. Spira and P.M. Zerwas, Phys. Lett. **B508** (2001) 311.

[15.92] B. Badelek et al., *loc. cit.* Ref. [15.9].

[15.93] A. Djouadi, Phys. Lett. **B435** (1998) 201. C. Balázs, A. Djouadi, V. Ilyin and M. Spira, in Proc. *Workshop on Physics at Future TeV Colliders* (Les Houches, 1999).

[15.94] J.F. Gunion and H.E. Haber, Nucl. Phys. **B278** (1986) 449.

[15.95] T. Plehn, D. Rainwater and D. Zeppenfeld, Phys. Lett. **B454** (1999) 297.

[15.96] S. Dittmaier, M. Krämer and M. Spira, hep-ph/0309204.

[15.97] M. Guchait, R.M. Godbole, K. Mazumdar, S. Moretti and D.P. Roy, Phys. Lett. **B571** (2003) 184.

[15.98] G. Bélanger, F. Boudjema and K. Sridhar, Nucl. Phys. **B568** (2000) 3. A. Djouadi, J.L. Kneur and G. Moultaka, Nucl. Phys. **B569** (2000) 53.

[15.99] M. Carena et al., Report of the Higgs Working Group to the *TEV2000 Workshop*, hep-ph/0010338.

[15.100] V.D. Barger, M.S. Berger, J.F. Gunion and T. Han, in *Proc. APS/DPF/DPB Summer Study on the Future of Particle Physics* (Snowmass, 2001).

[15.101] B.C. Allanach, A.J. Barr, M.A. Parker, P. Richardson and B.R. Webber, JHEP **09** (2001) 021.

[15.102] e.g. E.J. Chun, D-W. Jung, A.K. Kang and J.D. Park, Phys. Rev. **D66** (2002) 073003.

[15.103] D.K. Ghosh, R.M. Godbole and S. Raychaudhuri, Z. Phys. **C75** (1997) 375; LC-TH-2000-051:http://www.desy.de/lcnotes/notes.html.

[15.104] H.K. Dreiner and G.G. Ross, Nucl. Phys. **B365** (1991) 597; D.P. Roy, Phys. Lett. **B128** (1992) 270.

[15.105] B.C. Allanach et al., JHEP **0103** (2001) 048.

[15.106] See, for example, H. Baer et al., in (B.C. Allanach et al.) *Searching for R-parity violation at Run-II of the Tevatron*, hep-ph/9906224.

[15.107] See, for example, D. Choudhury and S. Raychaudhuri, Phys. Lett. **B401** (1997) 54.

[15.108] J. Erler, J.L. Feng and N. Polonsky, Phys. Rev. Lett. **78** (1997) 3063.

[15.109] G. Moreau, LC-TH-2000-040, http://www.desy.de/lcnotes/notes.html.

[15.110] H.K. Dreiner, P. Richardson and M.H. Seymour, Phys. Rev. **D63** (2001) 055008.

[15.111] F. Déliot, G. Moreau and C. Royon, Eur. Phys. J. **C19** (2001) 155.

[15.112] G. Moreau, E. Perez and G. Polesello, Nucl. Phys. **B604** (2001) 3.

[15.113] E.L. Berger, B.W. Harris and Z. Sullivan, Phys. Rev. Lett. **83** (1999) 4472; Phys. Rev. **D63** (2000) 115001.

[15.114] F. Borzumati, J.L. Kneur and N. Polonsky, Phys. Rev. **D60** (1999) 115011.

[15.115] T. Han and M.B. Magro, Phys. Lett. **B476** (2000) 79.

[15.116] D.K. Ghosh, S. Raychaudhuri and K. Sridhar, Phys. Lett. **B396** (1997) 177.

[15.117] K.J. Abraham, K. Whisnant, J.M. Yang and B.L. Young, Phys. Rev. **D63** (2001) 034011; Phys. Lett. **B514** (2001) 72.

[15.118] A. Belyaev, J.R. Ellis and S. Lola, Phys. Lett. **B484** (2000) 79.

[15.119] G. Bhattacharyya, D. Choudhury and K. Sridhar, Phys. Lett. **B349** (1995) 118; J. Kalinowski, R. Rückl, H. Spiesberger and P.M. Zerwas, *ibid.* **B414** (1997) 297; J.L. Hewett and T.G. Rizzo, Phys. Rev. **D56** (1997) 5709.

[15.120] D. Choudhury, R.M. Godbole and G. Polesello, JHEP **08** (2002) 004.

[15.121] K. Hikasa, J.M. Yang and B. Young, Phys. Rev. **D60** (1999) 114041.

[15.122] J. Kalinowski, R. Rückl, H. Spiesberger and P.M. Zerwas, Phys. Lett. **B406** (1997) 314. T.G. Rizzo in Trieste 1999: *Perspectives in hadronic physics*, p467, hep-ph/9907344.

[15.123] D. Acosta *et al.*, Phys. Rev. **D66** (2002) 012004.

[15.124] B.C. Allanach, S. Lola and K. Sridhar, Phys. Rev. Lett. **89** (2002) 011801.

[15.125] B. Abbott *et al.*, Phys. Rev. Lett. **79** (1997) 4321; F. Abe *et al.*, *ibid.* **79** (1997) 4327; B. Abbott *et al.*, *ibid.* **80** (1998) 2051.

[15.126] B. Abbott *et al.*, Phys. Rev. Lett. **84** (2000) 2088; Phys. Rev. **D62** (2000) 071701; F. Abe *et al.*, Phys. Rev. Lett. **83** (1999) 2133.

[15.127] C. Adloff *et al.*, Eur. Phys. J. **C20** (2001) 639.

[15.128] M. Acciarri *et al.*, Phys. Lett. **B414** (1997) 373; G. Abbiendi *et al.*, Eur. Phys. J. **C13** (2000) 553; R. Barate *et al.*, *ibid.* **C19** (2001) 415.

[15.129] G. Abbiendi *et al.*, Eur. Phys. J. **C11** (1999) 619 and *ibid.* **C12** (2000) 1; R. Barate *et al.*, *ibid.* **C13** (2000) 29; P. Abreu *et al.*, Phys. Lett. **B487** (2000) 36 and *ibid.* **B500** (2001) 22; M. Acciarri *et al.*, Eur. Phys. J. **C19** (2001) 397.

Chapter 16

SUPERSYMMETRIC COSMOLOGY

16.1   Introductory Comments

There were brief allusions in several previous chapters to cosmological considerations that constrain supersymmetric models. We elaborate on those in the present one. The starting assumption is that there was a Big Bang (BB), i.e. that the Universe was once much hotter (and denser) than it is today. Before describing the specifically supersymmetric aspects of the physics of the very early Universe, we first give a brief summary[1] of the standard BB cosmology in §16.2. We shall see, in particular, that the superpartners of Standard Model particles were likely to have been in thermal equilibrium once. This has a number of interesting consequences. If the LSP is stable, supersymmetric relics from the BB era should still be around today. That would be in the same way that the well-known cosmic microwave background photons are and the (as yet undetected) relic neutrinos are supposed to be. The consequences of this simple fact will be analyzed in §16.3, under the assumption that the LSP resides in the observable sector of the supersymmetric theory. Efforts to detect these relic LSPs will also be described there. As we have seen, any supersymmetric field theory, that includes gravity must contain the gravitino. If the LSP resides in the visible sector, the gravitino will decay, but with a rather long lifetime. Conversely, if the gravitino is the LSP, the lightest sparticle of the observable sector will be quite long-lived. Furthermore, if the gravitino is the LSP, additional constraints can be derived from the requirement of a sufficiently small gravitino relic density in order not to overclose the universe. These aspects of the "gravitino problem" will be highlighted in §16.4. Finally, baryogenesis is one of the open questions in BB cosmology: why is there much more matter than antimatter in the Universe? Possible answers to this question within the framework of supersymmetric field theories will be discussed in §16.5.

A word on the interpretation of the above issues is in order. The "final theory" will eventually have to provide solutions to the problems discussed in this chapter. These questions are thus of considerable interest to model builders. On the other hand, we shall see that many of these problems can be solved, or at least circumvented, by adjusting those parameters of the supersymmetric theory that have no bearing on collider phenomenology. An example is the gravitino mass. If $m_{3/2}$ exceeds 1 keV, the decays of the lightest visible sparticle with the gravitino as a product become undetectable in the laboratory. Another such parameter is the "reheating" temperature after inflation (see below). As a rule, cosmological constraints can therefore be largely ignored by

---

[1]Through a somewhat elementary introduction to the subject, we hope to provide sufficient background to enable the reader, without prior knowledge of this area, to follow the gist of the arguments given in the later sections. For reasons of space, however, we omit many derivations. These can be found elsewhere [16.1].

experimenters engaged in collider searches of sparticles. In specific models, though, allowed regions in the parameter space, derived from negative collider searches, can sometimes be tightened further with cosmological inputs.

16.2   Standard Big Bang Cosmology

This scenario rests on three observational "pillars":

i) Hubble's law: Spectral lines, emitted by distant objects, are observed to be redshifted. Suppose we describe the redshift, which is known to be independent of the wavelength, by multiplying all wavelengths by the factor $1 + z$, z being a positive quantity specific to that source. The distance d of an object is given by $\sqrt{(L/4\pi F)}$, L being its absolute luminosity (energy emitted per unit time) and F being the flux (energy per unit time per unit area) measured by the detector. The most fundamental aspect of the observable Universe is Hubble's law that this d grows approximately linearly [16.1] with z:

$$H_0 d = z + \mathcal{O}(z^2) \, . \tag{16.1}$$

H_0 in (16.1) is the Hubble constant, usually expressed in terms of a dimensionless parameter h, as

$$H_0 = \frac{100 \text{ km}}{\text{sec} \cdot \text{Mpc}} h \, . \tag{16.2}$$

Recent determinations of the Hubble constant, many of which make use of the appropriately named Hubble Space Telescope, yield the dimensionless constant h to be

$$h = 0.72 \pm 0.1 \, , \tag{16.3}$$

the error being mostly systematic. This observed redshift is most easily explained through the Doppler effect. The implication is that all distant objects are receding from us, with velocities that are approximately proportional to their distance, i.e. the universe is currently *expanding*[2]. An immediate consequence is that the universe was denser, and hence hotter, in the past.

ii) The cosmic microwave background (CMB): First identified in the 1960's, it has a nearly perfect black body spectrum, with a temperature $T \simeq 2.73$ K. This thermal bath effectively decoupled from matter in the Universe when photon-electron scattering reactions "froze out" (see below). Detailed calculations show that this happened at a temperature $T \simeq 3,500$ K $\simeq 0.3$ eV. The CMB is *isotropic* to better than 1 part in 10^4. In other words the CMB temperature, measured in different patches in the sky, varies[3] by less than one part in 10^4. This is the strongest evidence for the isotropy of the (observable part of the) Universe. Deviations from isotropy in the CMB, at the level of a few times 10^{-5}, have been measured in recent satellite and balloon borne as well as polar experiments and constitute a legacy of the structure formed in the early Universe.

---

[2] A real object in the Universe, in general, has a peculiar velocity in addition to this universal Hubble velocity. On an average, the peculiar velocity should be more or less independent of the distance, while the Hubble velocity grows proportionally to the distance. The latter will therefore always be more important at larger distances. Moreover, peculiar velocities should average out when one measures from a large sample of objects in different regions of the sky.

[3] This is after subtracting a dipole distortion of the CMB. The dipole anisotropy can be explained in terms of the Doppler shift of the Earth relative to the Universal rest frame, which will be introduced below.

iii) Big Bang nucleosynthesis (BBN): Most (luminous) baryons in the Universe occur in the form of (atomic or molecular) hydrogen. However, roughly 25% by weight is in ^{4}He, while some small but nonvanishing fractions occur in the form of other light elements such as ^{2}H, ^{3}He and ^{7}Li. The Universe is not nearly old enough to have produced this much ^{4}He in stellar burning. Most stars, in fact, tend to destroy the other light isotopes (with the possible exception of ^{3}He). On the other hand, if we project the observed expansion of the Universe sufficiently far into the past, we eventually reach densities and temperatures where nuclear reactions occur. The detailed outcome of these reactions includes predicted abundances of the light isotopes. These depend basically on only two quantities: (1) the expansion rate of the Universe at that time, which can be calculated in terms of the relativistic degrees of freedom available; (2) the baryon to photon number density η. The former is predicted by the Standard Model of particle interactions which dictates that the only sufficiently light particles are photons, neutrinos and, during the early part of BBN, electrons and positrons. The abundances of the four light elements, enumerated earlier, can then be computed as a function of η [16.1]. Good agreement with observation is achieved for [16.2][4]

$$1.3 \times 10^{-10} \le \eta \le 4.1 \times 10^{-10}. \tag{16.4}$$

BBN commenced at a temperature $T \sim 1$ MeV. The good agreement between theory and observation means that the Universe must once have been at least this hot.

The quantitative study of modern cosmology rests on the *Friedmann–Robertson–Walker (FRW)* metric. The latter can be read off from the following four dimensional line element[5]:

$$ds^2 = dt^2 - R^2(t)\left[\frac{dr^2}{1 - kr^2} + r^2 d\theta^2 + r^2 \sin^2\theta d\phi^2\right], \tag{16.5}$$

where spherical coordinates (r, θ, ϕ) have been employed. Note that r is dimensionless here and so it can be normalized such that $k = -1$, 0 or $+1$. These three values of k correspond to an open (negative curvature), flat (no curvature) or closed (positive curvature) Universe. The time coordinate t in eq.(16.5) is called "universal time" and, by convention, the BB occurred at $t = 0$. Finally $R(t)$ is the "**scale factor**" and, for $k = +1$, R describes the size of the Universe. Physically, (16.5) follows from the assumptions of isotropy and homogeneity as motivated e.g. by the observed isotropy of the CMB. Hubble's law (16.1) follows directly from (16.5), by Taylor expansion of all quantities around their values at present $(t = t_0)$, with

$$H(t) = \frac{\dot{R}(t)}{R(t)} \tag{16.6}$$

being the (time–dependent!) Hubble parameter and $H_0 \equiv H(t_0)$. The redshift of light emitted at time $t < t_0$ is then

$$1 + z = \frac{R(t)}{R(t_0)}. \tag{16.7}$$

---

[4]There is an ongoing controversy between different observations of deuterium in certain distant locations in the Universe which yield ^{2}H abundances differing by about one order of magnitude. For this reason a significantly broader band for η is sometimes advocated [16.2], e.g. $0.4 \times 10^{-10} \le \eta \le 9 \times 10^{-10}$, as the "reliable range". Eq.(16.4) is the "reasonable range" given in Ref.[16.2].

[5]Recall that our flat spacetime Minkowski metric $\eta_{\mu\nu}$ is $\eta_{oo} = 1$, $\eta_{oi} = 0 = \eta_{io}$ and $\eta_{ij} = -\delta_{ij}$, for $i, j = 1, 2, 3$.

The time dependence of the scale factor can be determined from the Einstein equations which read[6]:

$$\mathcal{R}_{\mu\nu} - \frac{1}{2}g_{\mu\nu}\mathcal{R} = 8\pi G_N T_{\mu\nu} \, . \tag{16.8}$$

In (16.8) $\mathcal{R}_{\mu\nu}$ and $\mathcal{R}$ are the Ricci tensor and the curvature scalar respectively, being computable [16.3] from the metric $g_{\mu\nu}$. Furthermore, $G_N = 1/(8\pi M_{Pl}^2)$ is Newton's constant and $T_{\mu\nu}$ is the energy-momentum tensor of all "matter" fields (i.e. all fields except the metric itself). In the context of cosmological considerations, all "matter" can be treated as a perfect fluid. In cartesian coordintes, the corresponding energy-momentum tensor is

$$T_\mu^\nu = \mathrm{diag}(\rho, -p, -p, -p) \, , \tag{16.9}$$

ρ being the energy density and p the pressure. These two quantities are related by the equation of state. Three cases relevant to us are:

$$\text{nonrelativistic matter}: \quad p \ll \rho \quad (p = 0 \text{ in practice}) \, ; \tag{16.10a}$$

$$\text{relativistic matter} \equiv \text{radiation}: \quad p = \frac{\rho}{3} \, ; \tag{16.10b}$$

$$\text{cosmological constant} \equiv \text{vacuum energy}: \quad p = -\rho \, . \tag{16.10c}$$

All three cases are described by $p = w\rho$, with $w = 0$, $+1/3$ and -1, respectively. These three possible contributions to $T_{\mu\nu}$ also differ in the dependence of ρ on the scale factor R:

$$\text{nonrelativistic matter} \quad : \quad \rho(R) \propto R^{-3} \, ; \tag{16.11a}$$

$$\text{radiation} \quad : \quad \rho(R) \propto R^{-4} \, ; \tag{16.11b}$$

$$\text{vacuum energy} \quad : \quad \rho(R) = \text{const} \, . \tag{16.11c}$$

Clearly, the number density of noninteracting particles scales like R^{-3}. Then (16.11a) follows since the density of nonrelativistic matter is simply the product of the (R-independent) rest mass and the number density. However, the three–momentum of a freely propagating particle also scales like R^{-1}. This can be seen, for example, by writing the absolute value of the three–momentum as the inverse of the de Broglie wavelength and by using the fact that *all* wavelengths get redshifted, cf.(16.7). This explains the stronger R-dependence of the energy density of radiation. Eqs.(16.11) can again be written in a unified form as

$$\rho \propto R^{-3(1 + w)} \, .$$

Using the explicit form (16.5) of the FRW metric, the (00) component of the Einstein equations becomes

$$\frac{\dot{R}^2}{R^2} + \frac{k^2}{R^2} = \frac{8\pi G_N}{3}\rho \, . \tag{16.12}$$

Another useful relation can be derived by combining the (00) and (*ii*) components of the Einstein equations, namely

$$\frac{\ddot{R}}{R} = -\frac{4\pi G_N}{3}(\rho + 3p) \, . \tag{16.13}$$

The combination of the measurement of the temperature of the CMB and the determination of the baryon to photon ratio η shows that at present the Universe is matter dominated (leaving aside for

---

[6]We include a possible cosmological constant Λ as a 'vacuum energy term' $\Lambda g_{\mu\nu}$ in $T_{\mu\nu}$

the moment the contribution from a cosmological constant Λ). However, the stronger dependence of the energy density of radiation on R^{-1} makes it clear that at sufficiently early times the Universe must have been radiation dominated.

We can make this last point more quantitative. Note that the use of (16.6) enables one to rewrite (16.12) in the form

$$1 + \frac{k}{H^2 R^2} = \frac{\rho}{\rho_c} \equiv \Omega \,, \tag{16.14}$$

with $\rho_c = 3H^2/(8\pi G_N)$. Clearly, an open, flat or closed Universe corresponds to $\rho < \rho_c$, $\rho = \rho_c$ and $\rho > \rho_c$ respectively; ρ_c is therefore called the **critical density**. Note that ρ_c depends on t, the current value being about $2 \times 10^{-29} h^2 \text{g/cm}^3$. Studies of the large scale structure of the universe had indicated earlier [16.1] that the current value of Ω, namely Ω_0, should fall in the range $0.2 \lesssim \Omega_0 \lesssim 2$. However, our knowledge of the value of Ω_0 has become much sharper in the wake of later observations on the CMB power spectrum. In particular, the recent satellite born observation [16.4] (WMAP) of the location of the first acoustic peak, expected at the angular scale $\ell \sim 200/\sqrt{\Omega_0}$ and measured at $\ell = (220.1 \pm 0.8)$, has fixed Ω_0 at

$$\Omega_0 = 1.02 \pm 0.02 \tag{16.15}$$

in the 'standard' ΛCDM (cosmological constant plus cold dark matter) model. Though more complicated models could allow somewhat different values of Ω_0, the range 0.2 to 2 is very conservative. The Universe is thus quite flat. The "curvature term" k/R^2 in (16.12) can therefore to first approximation be ignored. One then finds that

$$R(t) \propto t^{2/(3+3w)} \,. \tag{16.16}$$

Recall from (16.10b) that $w = 1/3$ for a Universe dominated by radiation, while w vanishes if it is dominated by matter. Clearly, (16.16) is not applicable if 70% of the critical density, as suggested by several current observations, is due to **Dark Energy** with negative pressure, the simplest example of which is a cosmological constant with $w = -1$.

Let us consider the important issue of the thermal evolution of the Universe, The time dependence of the temperature T follows from the simple fact that most, though not all, processes of relevance in the early Universe occurred (essentially) in thermal equilibrium. In other words, most events in the thermal history of the Universe were *adiabatic*. This means that the entropy per unit comoving volume remained constant, i.e. the entropy density s (per physical volume) scaled like $R(t)^{-3}$. Unlike the energy density, s has always been dominated by relativistic particles. It can thus be written as [16.1]:

$$s = \frac{2\pi^2}{45} g_{*s} T^3 \,, \tag{16.17}$$

where the effective number g_{*s} of the degrees of freedom contributing to the entropy density (as specified by the subscript s) is given by summing over all relativistic particle species in existence:

$$g_{*s} = \sum_{\text{bosons}} g_i \left(\frac{T_i}{T}\right)^3 + \frac{7}{8} \sum_{\text{fermions}} g_i \left(\frac{T_i}{T}\right)^3 \,. \tag{16.18}$$

Here g_i denotes the number of intrinsic degrees of freedom for a particle species i (e.g. due to spin and color), and the factor 7/8 arises from the difference between the Bose-Einstein and Fermi-Dirac distributions.

In (16.18) we have allowed different species of particles to have different temperatures T_i. Such a situation can arise when some species of particles decouples from the thermal bath of the other particles. This is called the **freeze-out** process for that species. For example, the neutrinos did "freeze out" before e^+e^- pairs could. Therefore, while the comoving entropy density of neutrinos remained constant during this process, all the entropy of the electromagnetic plasma before the freeze-out of e^+e^- pairs went into photons. As a result, the CMB temperature exceeds that of massless relic neutrinos by a factor $(11/4)^{1/3} \simeq 1.40$. If all the three types of neutrinos were effectively massless at freeze-out (i.e., each has a mass < 1 MeV), the current value of g_{*s} becomes 3.9. For $T \gtrsim 300$ GeV, where the electroweak gauge symmetry is (essentially) restored, $g_{*s} = 106.75$ in the SM. Finally, for temperatures large compared to the mass of the heaviest visible sector sparticle, $g_{*s} = 228.75$ in the MSSM. The total number increases by a factor which is slightly more than two. This is since we are adding more bosons than fermions to the SM. Also, a second Higgs doublet superfield has to be included.

In the MSSM g_{*s} changed "only" by a factor of 55 or so over the history of the Universe (at least for temperatures below the GUT scale). This implies that g_{*s} is a relatively slowly varying function of time. The adiabaticity condition s $\propto R^{-3}$, together with (16.17), thus implies approximately that

$$T \propto R(t)^{-1} = \left[\frac{R(t_0)}{1+z} \right]^{-1} , \tag{16.19}$$

where (16.7) has been used. This relation (16.19), together with (16.16), allows one to date those events in the history of the universe of interest to particle physics. Going back in time from the present, one has [16.1]:

- Decoupling of the CMB from matter at $T \simeq 0.3$ eV, $z \simeq 1,100$, $t \simeq 3.7 \times 10^5$ yrs. This more or less coincides with the formation of atoms, and with the beginning of the formation of large scale baryonic structures, i.e. galaxies etc. Prior to that time, nuclei, electrons and photons formed an essentially homogeneous plasma. No higher redshifts can be measured (optically).

- Equilibrium between the energy densities of matter and radiation at $T \simeq 5.5\Omega_m h^2$ eV, $t_{eq.} \simeq 1.4 \times 10^3 \left(\Omega_m h^2 \right)^{-2}$ yrs, Ω_m standing for the matter contribution to Ω. At this time the expansion rate of the Universe changed from $R \propto t^{1/2}$ ($t < t_{eq.}$) to $R \propto t^{2/3}$ ($t > t_{eq.}$).

- Commencement of BBN at $T \simeq 1$ MeV, $t \simeq 1$ sec, and its termination at $T \simeq 0.1$ MeV, $t \simeq 3$ minutes. The freeze-out (annihilation) of e^+e^- pairs occurred in the same era, at $T \simeq m_e/3$, as did the freeze-out of (light $SU(2)$ doublet) neutrinos at $T \simeq 1$ MeV.

- Occurrence of the phase transition(s), associated with the restoration of chiral symmetry and deconfinement, at $T \simeq 200$ MeV, $t \simeq 10^{-5}$ sec.

- Occurrence of the phase transition (or cross-over) associated with the (approximate) restoration of $SU(2)_L \times U(1)_Y$ gauge invariance at $T \simeq 200$ GeV, $t \simeq 10^{-11}$ sec.

As hinted earlier, at present we have direct observational evidence only for the first three of these events. In fact, (16.16) and (16.19) can probably not be used at arbitrarily early times. The reason is that the standard cosmology, as outlined so far, suffers from severe fine tuning difficulties, which are usually referred to as the "flatness" and "horizon" problems. The solution of these problems probably involves large deviations from (16.16) and (16.19) during some very early epoch, as is certainly true for the currently most widely discussed solution, namely inflation.

The **flatness problem** follows from the fact that the quantity $\Omega_0 \equiv \Omega(t_0)$ is of order unity. Recall that Ω is not constant (unless it equals 1 exactly). Its evolution follows [16.1] from (16.14), (16.6), (16.16) and (16.19):

$$|\Omega - 1| \simeq |\Omega_0 - 1| \cdot \begin{cases} (1+z)^{-1}, & \text{matter dominated}, \\ 10^4(1+z)^{-2}, & \text{radiation dominated}. \end{cases} \tag{16.20}$$

For example, at the onset of BBN, $|\Omega - 1| \sim \mathcal{O}(10^{-16})$! At early times, Ω must have been "tuned" to unity to very high precision[7] recollapsed long ago (for $\Omega > 1$), or would currently have a far lower expansion rate, i.e. a much smaller Hubble constant (for $\Omega < 1$). Next, the **horizon problem** can be described as follows. The distance to the horizon $d_H(t)$ is given by the distance traveled by a light ray emitted at time $t = 0$ [16.1]:

$$d_H(t) = R(t) \int_0^t \frac{dt'}{R(t')} . \tag{16.21}$$

Eq. (16.21) follows directly from the FRW metric (16.5). The use of (16.16) then implies $d_H(t) \propto t$. In other words, the fraction of the currently visible Universe, that was causally connected at earlier times, scales like

$$\left[\frac{d_H(t)}{R(t)}\right]^3 \propto t$$

in the matter dominated era, while

$$\left[\frac{d_H(t)}{R(t)}\right]^3 \propto t^{3/2}$$

in the radiation dominated era. For example, at the time of the CMBR freeze-out, the Universe consisted of $\sim 10^5$ regions that (according to the standard cosmology as described so far) had never been in causal contact. It is then a great mystery why all these regions should have the same temperature, to better than one part in 10^4!

As mentioned earlier, the currently most popular solution of both of these problems invokes the idea of **inflation**. This amounts to assuming that at some early times the total energy density of the universe was dominated by vacuum energy, so that $\rho(R) = const.$, cf.(16.11c). Recall that vacuum energy is characterized by the equation of state $p = -\rho$. Eq. (16.13) then immediately implies that

$$R(t) \propto e^{Ht} , \tag{16.22}$$

with a constant Hubble parameter

$$H = \frac{1}{M_{Pl}} \sqrt{\frac{\rho}{3}} . \tag{16.23}$$

Eq. (16.22) represents a rapid expansion, allowing one to begin with a small, causally connected "patch" of the universe and inflate it to a sufficiently large size which contains the entire presently visible Universe. The horizon problem is thereby solved. So is the flatness problem, since (16.14) implies $|\Omega(t) - 1| \propto \exp(-2Ht)$ during the inflationary era. A solution to the horizon problem requires that inflation lasted for at least 60 e-folds, i.e. for $t_I \gtrsim 60H^{-1}$, or so. This means that inflation reduced the pre–inflationary value of $|\Omega - 1|$ by at least a factor $e^{120} \simeq 10^{52}$. Simple models of inflation therefore predict that $\Omega_0 = 1$ to very high accuracy[8] today.

---

[7]This observation also shows that the 'curvature term' k/R^2 is indeed completely negligible at early times.
[8]See, however, Ref. [16.5].

There is one problem with this scenario. Inflation also lowers the temperature by at least of a factor $|\Omega - 1| \simeq 10^{26}$, cf. (16.19) and (16.20). Even if $T \sim M_{Pl} \sim 10^{18}$ GeV before inflation, the Universe would have been far too cold afterwards to allow BBN[9]. The simplest solution then is to postulate that the "exit" from inflation was nonadiabatic. As a result, the Universe got reheated since its entropy increased by a large factor. Such an occurrence is in fact natural in most models of inflation. These postulate the existence of a (not necessarily elementary) scalar "inflaton" field ϕ. The temperature dependent potential $V(\phi, T)$ is chosen in many inflationary models in such a way that, at a high temperature, ϕ is close to the origin. One could then write $V \simeq V(0, T)$. At lower temperatures, ϕ starts to "roll" towards the minimum ϕ_0 of the zero temperature potential $V(\phi, 0)$. The potential has to be chosen to be **very flat** in those models. This flatness needs to ensure that, for quite a "long" time t_I (recall $t_I \gtrsim 60H^{-1}$), one still has $\rho = V \simeq V(0, 0)$ – leading to inflation. Once close to the minimum ϕ_0, the inflaton field starts coherent oscillations around the minimum. The detailed description of what happens next is quite complicated. At least part of the vacuum energy gets transferred (through out of equilibrium decays of ϕ) to the thermal bath, leading to reheating. It has recently been pointed out [16.6] that the inflaton field could transfer energy to other modes through nonperturbative processes. Such a possibility allows at the end of inflation the production of particles whose masses exceed the **reheating temperature** T_R.

For our purpose of describing a supersymmetric cosmology, it will usually be sufficient to assume the Universe to have been nearly flat ($\Omega \simeq 1$) and homogeneous just after inflation and treat the reheating temperature T_R as a free parameter. Recall that direct observation puts only the relatively mild constraint[10] $T_R > 1$ MeV. Suppose we also make the economical and attractive assumption that the formation of large scale structures like galaxies in the Universe was initiated by quantum fluctuations of the inflaton field. The natural scale of inflation then would be [16.1] $\rho^{1/4} \sim 10^{14}$ GeV. It is thus plausible to take $T_R \gg 1$ TeV at least, in which case all visible sector MSSM particles and sparticles were in thermal equilibrium after reheating. We shall proceed with this assumption in most of this chapter. However, other attempts to explain structure formation also exist [16.7]. Note, moreover, that the gravitino problem in relation to reheating, cf. §16.4, most likely requires $T_R \ll 10^{14}$ GeV. So one should keep in mind that the proposition $T_R \gg 1$ TeV might be incorrect.

16.3   Dark Matter as a Supersymmetric BB relic

We now investigate the cosmological consequences of assuming R-parity to be exact. As emphasized in Ch.4, this implies the stability of the lightest supersymmetric particle (LSP). At the end of the previous section, we made it plausible that the LSP was in thermal equilibrium with the SM particles immediately after the end of the inflationary era. A significant number of supersymmetric BB relics should then still be around today, joining CMB photons and relic neutrinos as witnesses of the infancy of the Universe. In the first subsection we discuss the calculation of the current mass density of these relic LSPs relative to the critical density ρ_c. The requirement of that ratio not exceeding observational limits on Ω_0 leads to significant constraints on the MSSM parameter space. We shall see that relic LSPs make a strong candidate for the **missing mass** or **dark matter** in the Universe. In the second subsection we describe currently ongoing attempts to test this hypothesis experimentally by detecting some of these supersymmetric relics.

---

[9]BBN could not have occurred during inflation, since the Hubble parameter was far too large then, i.e. time evolution was far too rapid.

[10]The requirement of baryogenesis, discussed later, tightens this constraint.

The LSP relic density

Let n_χ be the number density of a stable particle χ of mass m_χ in the early Universe. Its evolution is described by the Boltzmann equation. For the cases of interest to us, the latter can be written in the form [16.1]

$$\frac{dn_\chi}{dt} + 3Hn_\chi = -\langle\sigma_{\mathrm{ann}}v\rangle\left(n_\chi^2 - n_{\chi,eq.}^2\right),\tag{16.24}$$

$n_{\chi,eq.}$ being the equilibrium value of n_χ. Let us make some explanatory remarks on (16.24). The second LHS term describes the effect of the expansion of the Universe. After freeze-out, the total number of χ particles would be constant. So n_χ would be proportional to R^{-3}, i.e. its time derivative $\dot{n}_\chi$ would behave as

$$\dot{n}_\chi \propto -3\frac{\dot{R}}{R^4} \propto -3Hn_\chi\,,$$

cf. (16.6). The RHS of (16.24) describes the change of n_χ due to $\chi\bar{\chi}$ annihilation into and production from lighter particles. Note that only SM particles (including Higgs bosons) can appear in the final state, since we are interested in the density of the lightest sparticles. The prefactor in the RHS of (16.24) is the thermal average of the product of the $\chi\bar{\chi}$ annihilation cross section and the relative velocity v between the members of the $\chi\bar{\chi}$ pair. Of course, the squared matrix element for the process $\chi\bar{\chi} \to f\bar{f}$ is the same as that for $f\bar{f} \to \chi\bar{\chi}$. Hence the annihilation and creation of $\chi\bar{\chi}$ pairs will occur with equal rates so long as χ and $\bar{\chi}$ are in complete thermal equilibrium with SM particles, i e so long as n_χ equals the equilibrium density $n_{\chi,eq.}$. Moreover, n_χ must appear quadratically in the RHS of (16.24) as a consequence of this being a two-body process. This is since, by assumption, "χ number" cannot be destroyed in collisions between a single χ and any number of SM particles. These two considerations heuristically explain the form of (16.24); a careful derivation can be found in Ref. [16.8].

In general, the thermally averaged product appearing in (16.24) is given by [16.1]

$$\langle\sigma_{\mathrm{ann}}v\rangle = \frac{g_\chi^2}{(n_{\chi,eq.})^2}\sum_{f,\bar{f}}\int\frac{d^3p}{(2\pi)^3 2E}\frac{d^3\bar{p}}{(2\pi)^3 2\bar{E}}g_f g_{\bar{f}}e^{-(E+\bar{E})/T}\frac{d^3q}{(2\pi)^3 2q^0}\frac{d^3\bar{q}}{(2\pi)^3 2\bar{q}^0}$$
$$\cdot\left[1 \pm f_f(q,T)\right]\left[1 \pm f_{\bar{f}}(\bar{q},T)\right]\tag{16.25}$$
$$\cdot(2\pi)^4\delta^{(4)}(p+\bar{p}-q-\bar{q})\left|\mathcal{M}[\chi(p)\bar{\chi}(\bar{p}) \to f(q)\bar{f}(\bar{q})]\right|^2,$$

with

$$n_{\chi,eq.}(T) = g_\chi\int\frac{d^3p}{(2\pi)^3}f_{\chi,eq.}(p,T)\,.\tag{16.26}$$

In (16.25) we have used the symbol $g_\chi, g_f, g_{\bar{f}}$ for the number of degrees of freedom of a single $\chi, f, \bar{f}$ particle (including color degrees of freedom) respectively. Furthermore, f_i denotes the phase space density of particle type i, which is related to its number density through the analogue of (16.26). The factor $[1 \pm f_f][1 \pm f_{\bar{f}}]$ in the RHS of (16.25) takes care of the Bose enhancement or Fermi suppression for final state particles f and $\bar{f}$, depending on whether they are bosons or fermions. For the cases of interest to us, these factors can be set to 1, since freeze-out occurs at temperatures well below the LSP mass M_χ, so that

$$f_f \sim \exp(-q^0/T) \sim \exp(-E_\chi/T) \ll 1\,,$$

and similarly for $f_{\bar f}$. For the same reason, f_χ can be approximated by the Maxwell-Boltzmann distribution

$$f_\chi = \exp[-(E_\chi + \mu_\chi)/T] \; ,$$

where μ_χ is the chemical potential. The latter vanishes if χ is in thermal equilibrium. Eq.(16.26) then becomes [16.9]

$$n_{\chi,eq.}(T) = g_\chi \frac{m_\chi^2 T}{2\pi^2} K_2 \left(\frac{M_\chi}{T}\right) , \tag{16.27}$$

where K_2 is the modified Bessel function of the second kind (MacDonald function) of degree two. Moreover, (16.25) reduces to an integral over s, the squared CM energy of the $\chi\bar\chi$ pair, namely

$$\langle \sigma_{\rm ann} v \rangle = \frac{g_\chi^2}{[n_{\chi,eq.}(T)]^2} \frac{T}{8\pi^4} \int_{4M_\chi^2}^\infty ds \; |\vec{p}|^2 \sqrt{s} K_1 \left(\frac{\sqrt{s}}{T}\right) \sigma_{\rm ann}(s) \; , \tag{16.28}$$

with K_1 as the modified Bessel function of the second kind of degree one. Moreover,

$$\sigma_{\rm ann}(s) = \sum_{f,\bar f} \sigma(\chi\bar\chi \to f\bar f)(s) \tag{16.29}$$

is the energy dependent total $\chi\bar\chi$ annihilation cross section, which can be evaluated using standard particle physics methods. Finally, $|\vec{p}| = \sqrt{s - 4M_\chi^2}/2$ is the absolute value of the three-momentum of the χ in the $\chi\bar\chi$ CM frame.

Further simplifications occur if the freeze-out temperature is so low that a nonrelativistic expansion becomes possible. Indeed, this is a good approximation for most LSP candidates since they are heavy. By convention, the said expansion is written in terms of v so that

$$|\vec{p}| \simeq \frac{1}{2}M_\chi v; \quad s \simeq M_\chi^2(4 + v^2); \quad \sqrt{s} \simeq 2M_\chi \left(1 + \frac{v^2}{8}\right) . \tag{16.30}$$

For $M_\chi \gg T$, i.e. $v^2 \ll 1$, the arguments of the modified Bessel functions in (16.27) and (16.28) become quite large. Using $K_n(x) \simeq \sqrt{\pi/(2x)}e^{-x}$ for $x \gg 1$, we have

$$n_{\chi,eq.} \simeq g_\chi \left(\frac{M_\chi T}{2\pi}\right)^{3/2} e^{-M_\chi/T} \tag{16.31}$$

and

$$\langle \sigma_{\rm ann} v \rangle \simeq \frac{x^{3/2}}{2\sqrt{\pi}} \int_0^\infty dv v^2 (\sigma_{\rm ann} v) e^{-xv^2/4} \; , \tag{16.32}$$

where we have introduced the inverse scaled temperature

$$x \equiv \frac{M_\chi}{T} \; . \tag{16.33}$$

Note that (16.31) and (16.32) are valid only for $x \gg 1$. Finally, $\sigma_{\rm ann} v$ can be expanded in many, though not all, applications as

$$\sigma_{\rm ann} v = a + bv^2 + \mathcal{O}(v^4) \; , \tag{16.34}$$

a, b being appropriate constants. On substituting the expansion, the thermal average (16.32) develops the form

$$\langle \sigma_{\rm ann} v \rangle = a + \frac{6b}{x} + \mathcal{O}(x^{-2}) \; . \tag{16.35}$$

Let us comment on (16.35). It is not identical to an expansion of (16.28) in inverse powers of x. That is since we have omitted terms $\mathcal{O}(v^4)$ from the integrand of (16.32). An expression for the complete expansion of (16.28) can indeed be found in Ref. [16.10]. Eq. (16.35) is nonetheless sufficient for most purposes if errors of 5–10% can be ignored[11]. The reason is that the second term in the RHS of (16.34) is important only if $b \gg a$. However, the latter does happen in some important cases, where a is suppressed by some symmetry. Terms of order a/x, which have been omitted in (16.35), are then practically negligible. However, there are also situations [16.11] where the expansion (16.34) fails altogether, even for $x \gg 1$. The most important example of such a situation is resonant $\chi\bar{\chi}$ annihilation through the s channel exchange of a massive particle V with a mass $m_V \simeq 2M_\chi$. In this case the expansion of the V propagator in powers of v^2 does not converge and one has to go back to (16.32) to compute the thermal average of this contribution to the annihilation cross section.

Let us now try to solve the Boltzmann equation (16.24). On using (16.17) and (16.19), we are led to the equation

$$\frac{ds}{dt} = s\left(-3H + \frac{\dot{g}_{*s}}{g_{*s}}\right). \tag{16.36}$$

The second term in the RHS of (16.36) can be neglected[12]. Moreover, the time derivative can be replaced by a derivative w.r.t. the scaled inverse temperature x of (16.33), by utilizing

$$\frac{\dot{T}}{T} = -\frac{\dot{R}}{R} = -H = -\frac{\pi T^2}{M_{Pl}}\sqrt{\frac{g_{*s}}{90}}, \tag{16.37}$$

where (16.19) has again been used. The expression for the Hubble parameter is valid for the radiation dominated epoch: that follows, provided the curvature term is ignored i.e. k is set equal to zero, from (16.12), (16.10b) and (16.17) together with the identity[13] $s = (p + \rho)/T$. Suppose we normalize n_χ to the entropy density s by introducing $Y_\chi = n_\chi/s$. The Boltzmann equation (16.24) can then be written as

$$\begin{aligned}
\frac{dY_\chi}{dx} &= -\frac{xM_{Pl}}{\pi M_\chi^2}\sqrt{\frac{90}{g_{*s}}}s\langle\sigma_{\mathrm{ann}}v\rangle\left(Y_\chi^2 - Y_{\chi,eq.}^2\right) \\
&= -1.32\, M_\chi M_{Pl}\sqrt{g_{*s}}\langle\sigma_{\mathrm{ann}}v\rangle x^{-2}\left(Y_\chi^2 - Y_{\chi,eq.}^2\right),
\end{aligned} \tag{16.38}$$

where we have used (16.17) in the second step.

Note that both Y_χ and x are dimensionless, while $\langle\sigma_{\mathrm{ann}}v\rangle$ has the dimension GeV^{-2} in natural units. If this cross section is very roughly of the order of a typical electroweak cross section $\sim \alpha^2/M_Z^2$, the factor multiplying $x^{-2}(Y_\chi^2 - Y_{\chi,eq.}^2)$ in (16.38) becomes very large owing to the occurrence of M_{Pl}. This implies that $Y_\chi \simeq Y_{\chi,eq.}$ unless x is very large. In other words, until one has $x \gg 1$, χ will remain locked in thermal equilibrium, where $Y_{\chi,eq.}$ is exponentially suppressed as proportional to e^{-x}. The factor of M_{Pl} arises since the RHS of (16.38) is inversely proportional to the Hubble parameter which, in turn, is proportional to $1/M_{Pl}$. The physical interpretation of

---

[11]This kind of accuracy is generally adequate in such cosmological calculations.

[12]That is provided T is not close to the QCD phase transition temperature $T_{QCD} \simeq 200$ MeV, where the number of degrees of freedom changes quickly. Such a condition is always satisfied for the LSP freeze-out, given current experimental constraints on LSP masses in most supersymmetric models.

[13]In general, the effective number g_* of degrees of freedom contributing to the radiation energy density, which appears in the Friedmann equation (16.12), can differ [16.1] from the corresponding quantity g_{*s} appearing in the entropy density. However, this difference is very small until the decoupling of the SM neutrinos, something which occurs long after the freeze-out of the LSP.

this result is clear: a reaction remains in equilibrium so long as it occurs faster than the Hubble expansion. This puts our qualitative discussion of freeze-out in the previous section onto a more quantitative footing.

No general analytical solution to (16.38) exists. It can of course be easily integrated numerically. For our purposes, though, it is sufficient to use a fairly accurate semi-analytical approximation to the exact solution. We refer to Refs. [16.1,16.11] for a heuristic derivation of this solution. One begins by defining the χ particle **freeze-out temperature** T_F through

$$x_F = \ln \frac{0.382 \; cM_\chi M_{Pl} g_\chi \langle \sigma_{\mathrm{ann}} v \rangle (x_F)}{\sqrt{x_F g_{*s}}} \; . \tag{16.39}$$

In (16.39) c is a numerical constant; the choice $c = \sqrt{2} - 1$ has been shown to give a good approximation for the final χ relic density. Eq. (16.39) has been formulated by requiring an approximate solution to (16.38), valid for high temperatures ($x < x_F$), to deviate significantly from $Y_{\chi,eq.}$ at $x = x_F$.

$$Y_\chi(x = x_F) = (c + 1)Y_{\chi,eq.} \; .$$

For $x > x_F$ one then uses a low temperature approximation to (16.38). The latter leads to

$$Y_{\chi,0} \equiv Y_\chi(x = \infty) = \frac{1}{1.32 \; M_\chi M_{Pl} \sqrt{g_{*s}} J(x_F)} \; , \tag{16.40}$$

where the "annihilation integral" $J(x_F)$ is defined as

$$J(x_F) \equiv \int_{x_F}^{\infty} dx \; x^{-2} \langle \sigma_{\mathrm{ann}} v \rangle (x) \; . \tag{16.41}$$

On account of the dependence of $\langle \sigma_{\mathrm{ann}} v \rangle$ on x, (16.39) has to be solved iteratively, a good first guess being $x_F = 20$. If the expansion in (16.34–16.35) can be used, the integral in (16.41) can easily be evaluated to be

$$J(x_F) = \frac{a}{x_F} + \frac{3b}{x_F^2} + \mathcal{O}(x_F^{-3}) \; . \tag{16.42}$$

One can see, however, from (16.32) that, in general, the evaluation of $J(x_F)$ entails a double integration. Finally, one computes the contribution of the χ particles to the present mass density of the Universe, in units of the critical density ρ_c, defined in (16.14), to be

$$\Omega_\chi h^2 = M_\chi s_0 Y_{\chi,0} \frac{h^2}{\rho_c} = \frac{2.09 \times 10^8 \; \mathrm{GeV}^{-1}}{M_{Pl} \sqrt{g_{*s}(x_F)} J(x_F)} \; , \tag{16.43}$$

where we have used [16.1] $s_0 \simeq 2.9 \times 10^3 \; \mathrm{cm}^{-3}$.

We intend to apply (16.43) to numerically estimate the LSP relic density. But it is a good idea to first briefly recapitulate the assumptions that went into the derivation of this result.

- The relic particle χ must be sufficiently stable, i.e. its lifetime must exceed the age of the Universe: $\tau_\chi > \tau_U \gtrsim 10^{10}$ yrs.

- The relic particles must have been in thermal equilibrium after the inflationary era, if there was one. This means that the Universe must once have been sufficiently hot. For LSP candidates from the observable sector, a reheating temperature $T_R \gtrsim M_\chi/10$ is sufficient.

- We have assumed equal densities for the χ and the $\bar{\chi}$. This condition is trivially fulfilled for the most promising stable LSP candidate, the lightest neutralino, due to its Majorana nature. However, this condition is violated for baryons, as will be shown in §16.4.

- The Universe must have evolved adiabatically after χ decoupled. If there was a period of entropy production, e.g. due to the out–of–equilibrium decay of another massive particle, only a small fraction of today's CMB photons would originate from the SM plasma at T_F which is the freeze-out temperature. Only this fraction should be used in computing $n_{\chi,0}$ from $s_0 Y_{\chi,0}$ and the prediction (16.43) would be diluted correspondingly. An example where this happens, with a dilution factor $\gtrsim 10^3$, is described in Ref.[16.12]. However, this mechanism requires the existence of particles beyond those contained in the MSSM.

In the remainder of this section we will assume that the conditions, stipulated above, are satisfied – as is most likely the case in the MSSM with conserved R-parity.

A first, very important consequence of (16.43) is that *the LSP, if stable, must be neutral.* In fact, we have already used this result in earlier chapters. In order to reach this conclusion, observe first that the relic density scales inversely proportional to the annihilation cross section. Sparticles with the largest annihilation cross section would therefore have the smallest relic density. They would thus be least constrained by observations of the present Universe. Let us therefore consider an LSP candidate that carries color since, if this can be excluded, so can charged color singlet candidates. Its annihilation cross section would be roughly of order $\pi(\alpha_s/M_\chi)^2$, α_s being the QCD fine structure coupling. The present χ number density, normalized to the baryon density Ω_b, would then be

$$\frac{n_\chi}{n_b} = \frac{M_p \Omega_\chi}{M_\chi \Omega_b} \sim 7 \times 10^{-6} \frac{M_\chi}{100 \text{ GeV}} \,, \tag{16.44}$$

where M_p is the proton mass and we have used $\Omega_b \sim 0.05$ from analyses of BBN. In arriving at (16.44), we have also conservatively assumed a relatively high freeze-out temperature, $x_F \sim 20$. Negatively charged χ (or $\bar{\chi}$) particles, or (most likely) those carrying color, would bind into nuclei to form "exotic isotopes", with very large mass to charge ratios. The best bounds on the existence of such isotopes in various materials limits their abundance to values below 10^{-15} to 10^{-25} per ordinary nucleon [16.13], many orders of magnitude below the prediction (16.44). This conclusion can be evaded only if the LSP does not bind to nuclei, i.e. if it is neutral.

Since neutral particles do not get bound to ordinary matter, most neutral relics will still be distributed throughout the Universe. Since this kind of matter does not interact with photons, it is all but invisible. It is therefore usually called **dark matter**[14]. In fact, there is substantial evidence [16.1] for the existence of such matter in the Universe. Most of this evidence comes from the observation of luminous matter that is subject to gravitational forces. The gravitational field strength can then be inferred from the velocity with which these observed objects move. One finds that there is not enough visible matter in the vicinity of these objects to produce this field strength[15]. The simplest example is the observation of globular clusters, gas clouds etc. that are orbiting galaxies. These studies typically show that at least 90% of the mass of a galaxy is dark. The observation of larger structures (clusters and superclusters of galaxies) indicates an even larger total matter density:

$$\Omega_\text{m} \gtrsim 0.2 \,. \tag{16.45}$$

---

[14]This expression is somewhat misleading, since this kind of matter is not "dark" in the colloquial sense; it does not occlude light sources.

[15]Purists therefore refer to a "missing force" problem, rather than a missing mass or dark matter problem.

In comparison, the total contribution from luminous matter (stars, gas, dust) amounts to $\Omega_{\text{luminous}} \simeq 0.007$. Moreover, analyses of BBN limit the total baryon mass density in the universe. From (16.4) and the current value of the CMB photon density, one has

$$0.0047 \lesssim \Omega_{\text{baryon}} h^2 \lesssim 0.015 \ . \tag{16.46}$$

On using $h > 0.4$, one is led to the conservative upper limit $\Omega_{\text{baryon}} < 0.1$. At least half the dark matter, and possibly a much larger fraction, should therefore be *nonbaryonic*. It is tempting to see whether supersymmetric relics can provide (part of) this dark matter. To do so, their relic density should fall in the broad range

$$0.02 \lesssim \Omega_{\text{LSP}} h^2 < 1.0 \ , \tag{16.47}$$

where the lower bound comes from analyses of galactic rotation curves[16], and the upper bound is due [16.1] to the conservative lower limit on the age of the universe $\tau_U > 10^{10}$ yrs. Recall also that simple inflationary models predict $\Omega = 1$ for the total density of the Universe. However, recent analyses of high redshift supernovae indicate [16.15] that, if $\Omega = 1$, about 2/3 of this density is due to a 'Dark Energy' contribution with negative pressure, e.g. a cosmological constant[17]. In view of this result, the upper limit in (16.47) should be reduced by a factor of three or so. Recent studies [16.4] of the CMB anisotropy by the WMAP satellite have led, within the ΛCDM model, to the determination of the nonbaryonic matter density to be $\Omega_{DM} h^2 = 0.113 \pm 0.009$. The same studies also yield $\Omega_\Lambda \simeq 0.73$ for the 'Dark Energy' contribution. Finally, it should be stressed that the two limits in (16.47) have different character: models with $\Omega_{\text{LSP}} h^2 > 1.0$ are strictly excluded, while models with $\Omega_{\text{LSP}} h^2 \lesssim 0.02$ still have the possibility that something else provides the required dark matter, e.g. primordial black holes or axions [16.1].

In the MSSM there are two neutral LSP candidates in the observable sector, the lightest neutralino $\tilde{\chi}_1^0$ and a sneutrino $\tilde{\nu}$. We saw in Ch.12 that most models predict $m_{\tilde{\nu}} > M_{\tilde{\chi}_1^0}$, so that the neutralino emerges as most plausible LSP candidate. Let us nevertheless briefly discuss the possibility of a sneutrino LSP. In this case the most important contributions to the annihilation cross section come from (1) $\tilde{\nu}\tilde{\bar{\nu}} \to f\bar{f}$ through the exchange of a Z boson in the s channel, where f is an SM fermion with a mass $m_f \leq m_{\tilde{\nu}}$, (2) from $\tilde{\nu}\tilde{\nu} \to ll$ through the exchange of a neutralino or chargino in the t channel and (3) for $m_{\tilde{\nu}} \geq M_V$, from $\tilde{\nu}\tilde{\bar{\nu}} \to V\bar{V}$ ($V = W^\pm, Z$). The relevant couplings are electroweak gauge couplings, unsuppressed by any mixing angles. Moreover, no particle with a mass much larger than $m_{\tilde{\nu}}$ needs to be exchanged (except possibly for the ll final state). Recall the implication of LEP 1 data that $m_{\tilde{\nu}} \gtrsim M_Z/2$, so that the relevant Z propagator factor $|4m_{\tilde{\nu}}^2 - M_Z^2|$ is less than $4m_{\tilde{\nu}}^2$. This results in a large $\tilde{\nu}\tilde{\bar{\nu}}$ annihilation cross section and consequently a small $\tilde{\nu}$ relic density, unless $m_{\tilde{\nu}}$ is quite large. The most complete calculation of the sneutrino relic density [16.16] finds that the upper bound in (16.47) only excludes sneutrinos with a mass above 2.3 TeV, well above the range indicated by fine tuning arguments. Moreover, $\Omega_{\tilde{\nu}} h^2 < 0.02$ for $m_{\tilde{\nu}} \leq 250$ GeV, which is already somewhat heavy for the lightest sparticle. Worse, as we shall see in the next subsection, direct dark matter searches exclude the possibility that sneutrinos form a major part of the dark halo of our galaxy. In the remainder of this subsection we therefore focus on the lightest neutralino as LSP [16.17].

If the soft supersymmetry breaking gaugino masses unify at a GUT scale, cf. (12.25), then in most regions of the MSSM parameter space $\tilde{\chi}_1^0$ is either mostly a bino, the superpartner of the

---

[16]These assume that dark baryons contribute at most 50% of the galactic dark matter. Such an assumption is compatible with the most recent analysis of the observation of MACHOs [16.14].

[17]Contrast this with the fact that nonrelativistic matter has essentially vanishing pressure, while relativistic matter has a positive pressure.

$U(1)_Y$ gauge boson, or a higgsino. The former occurs for $|M_1| < |\mu|$, and the latter for $|M_1| > |\mu|$. Strong bino-higgsino mixing occurs only for $|M_1| \simeq |\mu|$. Let us consider the somewhat simpler situation of a higgsinolike LSP first. In such a case the mass scale for the light chargino and the two lightest neutralinos is set by $|\mu|$. Sparticle search limits at LEP then imply $M_{\tilde{\chi}_1^0} \gtrsim m_W$. LSP pairs can therefore annihilate into W^+W^- pairs and, in most of the allowed parameter space, also into Z pairs. These reactions proceed predominantly through t and u channel diagrams, where another higgsinolike state is exchanged between the two LSPs. The relevant $W^\pm \tilde{\chi}_1^\mp \tilde{\chi}_1^0$ and $Z\tilde{\chi}_2^0 \tilde{\chi}_1^0$ couplings have full $SU(2)_L$ gauge strength. The $Z\tilde{\chi}_1^0 \tilde{\chi}_1^0$ coupling is small, so that annihilation into $f\bar{f}$ pairs is suppressed (with the possible exception of $t\bar{t}$ and – for large $\tan\beta$ – $b\bar{b}$ pairs, if the corresponding squarks are not too heavy).

However, since the $\tilde{\chi}_2^0$-$\tilde{\chi}_1^0$ and $\tilde{\chi}_1^\pm$-$\tilde{\chi}_1^0$ mass differences are quite small, one also has to include "coannihilation" [16.11] of the LSP with these slightly heavier sparticles. It is noteworthy that the cross sections for $\tilde{\chi}_1^0 \tilde{\chi}_2^0 \to f\bar{f}$ and $\tilde{\chi}_1^0 \tilde{\chi}_1^\pm \to f\bar{f}'$ are not pulled down by any small mixing angle. Let us recall that the LSP freeze-out occurs at the temperature $T_F \simeq M_{\tilde{\chi}_1^0}/20$. The LSP number density at freeze-out is therefore suppressed by a Boltzmann factor $\sim e^{-20}$ compared to that of light SM fields. At $T \simeq T_F$, reactions of the type $\tilde{\chi}_1^0 f \leftrightarrow \tilde{\chi}_2^0 f, \tilde{\chi}_1^\pm f'$ therefore occur about e^{20} times more frequently than reactions that change the total number of sparticles. As a result, the number densities of $\tilde{\chi}_1^0, \tilde{\chi}_2^0$ and $\tilde{\chi}_1^\pm$ remain in *relative* thermal equilibrium with each other for quite some time after sparticles have decoupled from the SM plasma. Of course, eventually all heavier sparticles decay into the LSP (and some SM particles). This enables one to treat the coannihilation process using (16.39)–(16.43), with the replacements [16.11]

$$g_\chi \to g_{\text{eff}} = \sum_i g_i \left(1 + \Delta_i\right)^{3/2} e^{-x\Delta_i} \,, \tag{16.48a}$$

$$\sigma_{\text{ann}} \to \sigma_{\text{eff}} = \sum_{i,j} \frac{g_i g_j}{g_{\text{eff}}^2} \left(1 + \Delta_i\right)^{3/2} \left(1 + \Delta_j\right)^{3/2} e^{-x(\Delta_i+\Delta_j)} \sigma_{\text{ann}}(\tilde{\chi}_i \tilde{\chi}_j) \,, \tag{16.48b}$$

where the sum runs over each sparticles species $\tilde{\chi}_i$ with a mass close to that of the LSP. Here g_i is the number of degrees of freedom of a $\tilde{\chi}_i$ particle and $\Delta_i = m_{\tilde{\chi}_i}/m_{\tilde{\chi}_1^0} - 1$. Note that the contribution of heavy sparticles is suppressed by exponential Boltzmann factors. Coannihilation is therefore only important for $\Delta_i \lesssim 0.2$.

The annihilation cross section of higgsinolike LSPs into W and Z pairs is already quite large [16.17]. Coannihilation increases the effective cross section further by another factor of 3 to 5 [16.9]. The resulting LSP relic density is shown in Fig. 16.1. We have chosen a common sfermion mass $m_0 = 1$ TeV at the GUT scale $M_U = 2 \times 10^{16}$ GeV. However, we have allowed the soft supersymmetry breaking Higgs scalar masses to differ from m_0, so that μ, m_A and $\tan\beta$ are all treated as free parameters. We have further taken a very large value of m_A here, much above 1 TeV. All sfermions are also much heavier than the LSP due partly to the Renormalization Group Evolution from the GUT scale to the weak scale. In this situation the relic density of higgsinolike LSPs depends essentially only on the LSP mass and is hence maximized. We see nonetheless that the upper bound in (16.47) allows LSP masses in excess of 2 TeV. Indeed, higgsinos start contributing significantly to the matter density of the Universe only if their mass exceeds 300 GeV and the WMAP data [16.4] favors $M_{\tilde{\chi}} \sim 1$ TeV. A higgsinolike LSP therefore makes a good dark matter candidate only if sparticles are somewhat heavier than preferred by fine tuning arguments (cf. Ch.1

and later discussion in Ch.17). This is sometimes given as a reason to disfavor a higgsinolike LSP.

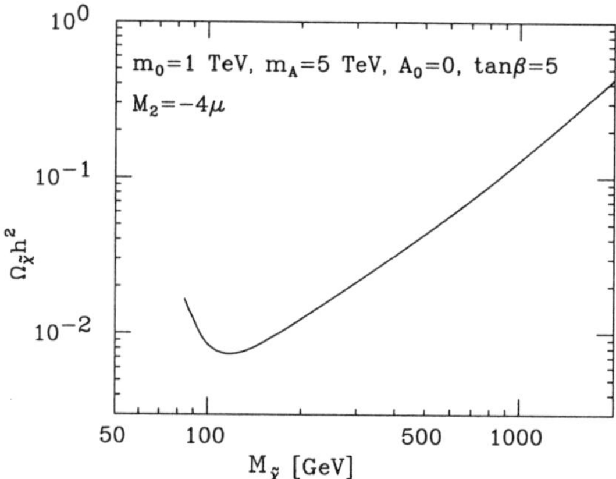

Fig. 16.1. The relic density of a higgsinolike LSP as a function of its mass; the labels are explained in the text.

The situation is quite different for a binolike LSP. The latter has only $U(1)_Y$ gauge interactions. As a result, the annihilation cross section is suppressed by a factor $\tan^4\theta_W \simeq 1/9$, as compared with the case of a higgsinolike $SU(2)_L$ doublet LSP. Moreover, the only annihilation diagrams, with couplings unsuppressed by a small bino-higgsino mixing, involve the exchange of heavier sparticles in the t or u channel. Examples of such diagrams are higgsino exchange contributions to the production of Higgs or longitudinal gauge bosons and, most importantly, sfermion exchange contributions to the production of $f\bar{f}$ pairs. The latter cross section scales like the fourth power of the hypercharge of the exchanged sfermion. The usually most important contribution therefore comes from the exchange of $SU(2)_L$ singlet sleptons, which have the highest hypercharge of all sfermions ($Y_{\tilde{l}_R} = 2$). Recall also that in mSUGRA these are usually the lightest sfermions, since their masses are increased only mildly by RGE effects. These contributions, therefore, generally dominate. This is despite the constraint that, in the absence of CP violation, two identical Majorana fermions can annihilate into massless (chiral) $f\bar{f}$ final states only from a $P-$wave initial state[18], which increases the LSP relic density by another factor of ~ 7 for $x_F \simeq 20$, cf. (16.42).

Thus the relic density of a "generic" binolike LSP, which annihilates predominantly through slepton exchange, depends on the masses of the LSP and of $SU(2)_L$ singlet sleptons. An approximate expression for the case of three mass degenerate sleptons is

$$\Omega_{\tilde{\chi}}h^2 \simeq \frac{\Sigma^2}{(1\text{ TeV})^2 M_{\tilde{\chi}}^2}\frac{1}{\left(1 - M_{\tilde{\chi}}^2/\Sigma\right)^2 + M_{\tilde{\chi}}^4/\Sigma^2}\,, \tag{16.49}$$

where $\Sigma = M_{\tilde{\chi}}^2 + M_{\tilde{l}_R}^2 \geq 2M_{\tilde{\chi}}^2$, $M_{\tilde{\chi}}$ being the mass of the LSP. Under the stated assumptions this simple approximation reproduces the exact numerical result to $\sim 10\%$. It shows that, for a given $M_{\tilde{\chi}}$, the relic density is minimal if $m_{\tilde{l}_R}$ is at its lower bound, i.e. $m_{\tilde{l}_R} = M_{\tilde{\chi}}$. The constraint $\Omega_{\tilde{\chi}}h^2 < 1$ then implies

$$M_{\tilde{\chi}} \leq m_{\tilde{l}_R} \lesssim 300 \text{ GeV}. \tag{16.50}$$

---

[18]Fermi statistics permits two identical fermions to reside in an $S-$wave state only if their spins are anti-parallel, i.e. $J = 0$ and $CP = -1$. This does not match with the CP transformation properties of a massless $f\bar{f}$ pair.

Assuming high scale gaugino mass unification, cf. (12.25), this corresponds to upper bounds of about 0.6 and 1.8 TeV on the masses of the lighter chargino and gluino respectively. Instead, if the much stronger WMAP [16.4] upper bound on $\Omega_{\tilde{\chi}}h^2$ is used, the *upper* bound on $m_{\tilde{\ell}_R}$ has to be reduced by a factor of $\sqrt{3}$ – bringing it close to the *lower* bound on it from negative LEP results.

However, these bounds hold *only if* the annihilation cross section is dominated by slepton exchange. As shown in Fig. 16.2, such need not be the case even if $\tilde{\chi}_1^0$ is binolike. One important exception occurs if $M_{\tilde{\chi}_1^0} \simeq m_A/2$. Even though the $A\tilde{\chi}_1^0\tilde{\chi}_1^0$ coupling is suppressed by a small higgsino-bino mixing angle, the resonance enhancement factor, proportional to $(m_A/\Gamma_A)^2$, suffices to increase the total annihilation cross section. Correspondingly, the relic density is reduced by more than two orders of magnitude. Note that the CP odd Higgs boson A can be produced through LSP annihilation from an $S-$wave initial state, while the CP even Higgs bosons h and H can only be produced from a $P-$wave initial state. The much shallower H-pole is therefore not separately visible in Fig. 16.2. A second comment on the figure is that the solid line terminates at $M_{\tilde{\chi}} \simeq 300$ GeV since for larger gaugino masses the lightest neutralino becomes heavier than $SU(2)$ singlet sleptons, leading to a charged LSP. Such scenarios are excluded, as discussed earlier in this chapter. Another deviation from the prediction (16.49) can be seen in the dashed curve in Fig. 16.2 for $M_{\tilde{\chi}_1^0} \geq 300$ GeV. Here one of the heavy Higgs bosons $(H, A, H^\pm)$ can be produced together with a W or Z boson or the light Higgs boson h, leading to a slight increase in the total annihilation cross section. Finally, as mentioned above, the bounds (16.50) are saturated if $m_{\tilde{\ell}_R} = M_{\tilde{\chi}_1^0}$. In this case $\tilde{\chi}_1^0$-$\tilde{\ell}_R$ coannihilation has to be taken into account and that increases [16.18] the upper bound on $M_{\tilde{\chi}_1^0}$ by about a factor of 3 to 4, allowing gluino masses even beyond the reach of the LHC.

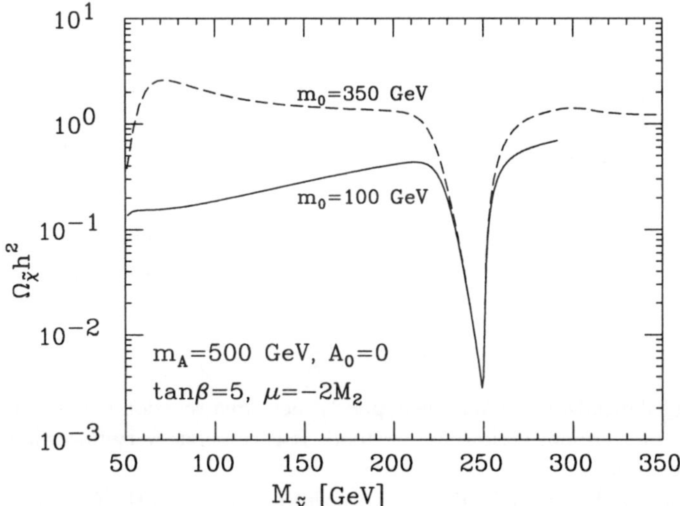

Fig. 16.2. The relic density of a binolike LSP as a function of its mass.

These loopholes reduce the "bound" (16.50) to a status similar to that of considerations of fine tuning, cf. Chs. 1 and 17. It can be evaded, but only if two (s)particle masses fulfill an "accidental" relationship, namely $M_{\tilde{\chi}_1^0} \simeq m_A/2$ or $M_{\tilde{\chi}_1^0} \simeq m_{\tilde{\ell}_R}$. One can nonetheless draw two important lessons from (16.49) and Fig. 16.2. First, if the LSP is stable and binolike, the upper bound (16.47) on the LSP relic density would exclude most scenarios with $m_{\tilde{\ell}_R} \gg M_{\tilde{\chi}_1^0}$. Second, the choice of soft supersymmetry breaking parameters in the 100 to 300 GeV range usually leads to just about the right total dark matter density, as derived from cosmological observations. Unlike in the case of higgsinolike LSPs, this "cosmologically favored" mass range does not imply excessive fine-tuning. Recall also that, in models with radiative EW symmetry breaking (e.g. mSUGRA, cf. Ch.12), one usually has $|\mu| > |M_1|$, i.e. the LSP is indeed binolike. A binolike LSP is therefore a theoretically well motivated dark matter candidate.

It should be evident by now that, in general, the LSP relic density depends quite sensitively on many details of the sparticle and Higgs mass spectra. Let us briefly discuss two nonminimal models to further illustrate this point. If gaugino masses do not unify at some GUT scale, one can have $|M_2| < |M_1|, |\mu|$ in which case the lightest neutralino would be mostly a neutral wino, i.e. an $SU(2)_L$ gaugino. This leads to an even smaller LSP relic density, for a given LSP mass, than in the case of a higgsinolike LSP, chiefly owing to very strong $\tilde{\chi}_1^0 \tilde{\chi}_1^\pm$ coannihilation [16.19]. These two states, namely the lightest neutral and charged winos, are closely mass degenerate in such a scenario. Moreover, they are $SU(2)_L$ triplets, which couple more strongly to W and Z bosons than do doublets (higgsinos). Conversely, if one introduces an extra $SU(2)_L \times U(1)_Y$ singlet Higgs superfield (cf. §14.2), the LSP could have a large "singlino" component. This tends to decrease its annihilation cross section, i.e. to increase its relic density [16.20]. In this case the LSP can be a good dark matter candidate even if sparticle masses lie in the range that is currently being probed by experiments at the TEVATRON. Therefore, which region of parameter space is excluded by the requirement $\Omega_{\mathrm{LSP}} h^2 < 1$ and which region is "cosmologically preferred" (on account of the prediction $\Omega_{\mathrm{LSP}} h^2 \sim 0.1$) are questions that have to be investigated for each model separately.

Search for Supersymmetric Dark Matter

All the currently most promising search experiments seek supersymmetric dark matter in the form of relic LSPs inside our own solar system. Their local mass density can be inferred from a study of the dynamics of our galaxy. This entails (1) a determination of the total mass density through a fit of the measured rotation curve (rotation velocity as a function of distance from the center) and (2) the measurement of "peculiar velocities" of stars in the solar neighborhood, i.e. velocities relative to the overall rotation. This mass density is then compared to an estimate of the mass in stars, gas, dust, as well as the Massive Astronomical Compact Objects (MACHOs) responsible for the microlensing of light from stars in the Large Magellanic Cloud. The result in particle physics units is [16.17]

$$\rho_{\mathrm{DM}}^{\mathrm{local}} \simeq 0.3 \frac{\mathrm{GeV}}{\mathrm{cm}^3} \ . \tag{16.51}$$

However, this value depends on the relatively poorly measured rotation curve and on the details of galactic modelling that are not entirely understood. It is therefore very uncertain – to perhaps as much as a factor of two.

In order to compute the local relic LSP flux, we also need to know the LSP velocity distribution. In principle, that can be derived from the mass distribution in our neighborhood – at least if one assumes[19] that the halo of our galaxy is essentially static. The simplest viable model, the "isothermal sphere", leads to an essentially Maxwellian velocity distribution [16.17]:

$$\frac{d^3 n_{\mathrm{LSP}}}{dv^3} = \frac{1}{\pi^{3/2} v_0^3} e^{-v^2/v_0^2} \ . \tag{16.52}$$

---

[19] This assumption cannot be strictly valid since dark matter should still be falling into our galaxy.

Here v_0 is identical to the rotational velocity of our galaxy at large radii which approaches a constant value in this model. One usually assumes that this saturation value has already been reached[20], so that v_0 is equal to the velocity with which the solar system orbits the center of our galaxy:

$$v_0 \simeq v_\odot \simeq 220 \text{ km/sec} . \tag{16.53}$$

Note that (16.52) has been written in the galactic rest frame. The LSP velocity distribution in the solar system can then be obtained by boosting into the rest frame of the sun or the earth. One is then led to an average relic LSP velocity of about $10^{-3}c$. Together with (16.51), this yields a local flux

$$\Phi_{\text{DM}}^{\text{local}} \simeq 10^5 \left(\frac{100 \text{ GeV}}{M_{\text{LSP}}} \right) \text{ cm}^2\text{sec}^{-1} . \tag{16.54}$$

The local flux of an LSP of mass 100 GeV is thus nearly nine orders of magnitude larger than that of cosmic muons at the sea level. We shall see nonetheless that, despite the high flux, these supersymmetric relics are much more difficult to detect.

The most direct, and hence conceptually simplest, way to search for these objects is based on the energy deposited by an LSP, while scattering off a nucleus in a detector. Since ambient LSPs are nonrelativistic, the scattering off electrons deposits too little energy. Moreover, owing to the limited phase space, the process has too small a cross section to be useful. Further, the kinematic limit on the momentum transfer is so small that most LSP-nucleus scattering events will be elastic. One then tries to detect the energy of the recoiling nucleus. The most promising experiments of this kind, that are currently underway, use germanium, silicon or sodium iodide detectors. The strength of the signal in such a direct relic LSP search experiment is thus proportional to the elastic LSP-nucleus scattering cross section. The corresponding scattering amplitude in general receives two kinds of contributions: (1) those, proportional to the product of the spins of the LSP and the target nucleus, due to pseudoscalar and axial vector currents and (2) spin-independent contributions from scalar, vector and tensor currents. Spin-independent interactions are coherent, since the relevant currents couple to the entire nucleus. In contrast, spin–dependent interactions are incoherent, since the spin of a nucleus does not grow with its mass. In fact, in most cases the entire spin is effectively carried by a single nucleon.

The two types of interactions, mentioned above, do not interfere. We can thus write [16.17] the LSP nucleus scattering cross section as an incoherent sum:

$$\sigma(\tilde{\chi}\mathcal{N} \to \tilde{\chi}\mathcal{N}) = \frac{4M_{\tilde{\chi}}^2 M_{\mathcal{N}}^2}{\pi(M_{\tilde{\chi}} + M_{\mathcal{N}})^2} \Bigg\{ [Z_{\mathcal{N}} f_p + (A_{\mathcal{N}} - Z_{\mathcal{N}}) f_n]^2$$

$$+ 4\lambda_{\mathcal{N}}^2 J_{\mathcal{N}} (J_{\mathcal{N}} + 1) \left(\sum_{q=u,d,s} d_q \Delta q \right)^2 \Bigg\} . \tag{16.55}$$

Here f_p and f_n are the respective coefficients of the effective $\tilde{\chi}\tilde{\chi}\bar{\mathcal{N}}\mathcal{N}$ interaction with $\mathcal{N} = n$ and p respectively.. Furthermore, $Z_{\mathcal{N}}, A_{\mathcal{N}}$ and $J_{\mathcal{N}}$ are respectively the number of protons, the total number of nucleons, and the total spin of the nucleus $\mathcal{N}$, while $\lambda_{\mathcal{N}}$ is a coefficient that basically describes the fraction of $J_{\mathcal{N}}$ that is due to the spin (rather than angular momentum) of nucleons. Here we have made use of the fact that products of vector or tensor currents, giving rise to spin

---

[20]Observations show a more or less constant rotation velocity from the location of the solar system outwards. However, these observations only extend to about twice the distance of the solar system to the galactic center.

independent interactions, can be written in terms of a product of scalar currents through the equations of motion. Finally, the d_q are coefficients of the spin dependent LSP-quark effective interaction, which can be written as a product of two axial vector currents. The spin dependent LSP-nucleon interaction can then be expressed in terms of the first moments of the polarized parton densities Δq, which are by now fairly well known from experiment[21]. One point to be underlined is that (16.55) holds for a nonrelativistic LSP and a pointlike nucleus. For heavier nuclei ($A_N \gtrsim 20$), the momentum transfer can become sufficiently large to make form factor ("decoherence") effects significant. These depend through the momentum transfer on the velocity distribution of the LSPs. Moreover, for practical applications, one often needs the cross section differentiated with respect to the energy transfer, or the latter integrated from a minimal (nonvanishing) threshold value of the energy transfer which is determined by experimental conditions. We refer the interested reader to Ref. [16.17] for a detailed discussion of these issues. The numerical results presented below are for total cross sections, using a simple Gaussian ansatz for the relevant form factors.

The remaining task is the calculation of the coefficients f_p, f_n and d_q in (16.55). The latter directly describes LSP-quark interactions. Of course, it vanishes for a scalar LSP, but can be significant for the more usual case where the spin half $\tilde{\chi}_1^0$ is the LSP. In general, it receives contributions from Z exchange in the t channel, as well as squark exchange in the s or u channel[22]. We saw in the previous section that for most regions of the MSSM parameter space the $Z\tilde{\chi}_1^0\tilde{\chi}_1^0$ coupling is quite small. On the other hand, we already know (cf. p xviii) that the first and second generation squarks are quite heavy, $m_{\tilde{q}} \geq 250$ GeV, suppressing the corresponding propagators. The spin dependent interaction can nonetheless be important for light nuclei with nonvanishing spin, the proton included. For heavier nuclei, the factors A_N and $A_N - Z_N$ appearing in (16.55) imply that the spin independent contributions nearly always dominate. This is trivially true for a sneutrino LSP. The latter has unsuppressed vector couplings to the Z boson and hence a large scattering cross section exceeding that of Dirac neutrinos with the same mass by a factor of four. In fact, this cross section is so large that direct searches exclude MSSM sneutrinos as a major component of the dark halo of our galaxy [16.17].

Spin independent interactions between a neutralino and a nucleon are mediated either by the exchange of CP even Higgs bosons in the t channel or again by squark exchange in the s or u channel. Squark exchange can give rise to spin independent contributions only if chirality is violated somewhere in the amplitude. The latter can occur through the interference of gauge and Yukawa couplings (which requires gaugino-higgsino mixing in the neutralino sector) or else through $\tilde{q}_L$-$\tilde{q}_R$ mixing. The amplitude is proportional to the quark mass in both cases, just as for the Higgs exchange diagram[23]. Since (most) squarks are now known to be heavier than the lighter CP even Higgs boson in the MSSM, squark exchange contributions are usually subdominant. This is always true in models with gaugino mass unification at the GUT scale, where first and second generation squarks are at least five times heavier than the LSP.

As emphasized above, both spin independent contributions are proportional to matrix elements of the type $m_q\langle N|\bar{q}q|N\rangle$, where N is a nucleon. These can be calculated perturbatively for heavy

---

[21]Note that the Δq appear in the amplitude here, rather than only in the cross section as in the case of deep inelastic scattering. The reason is that the optical theorem relates the inclusive inelastic cross section to the forward amplitude for elastic scattering, i.e. that at vanishing momentum transfer. The latter is an excellent approximation to the kinematic situation with which we are dealing in LSP-nucleus scattering.

[22]The pseudoscalar current $\bar{\tilde{\chi}}\gamma_5\tilde{\chi}$ being proportional to the LSP velocity and the latter being nonrelativistic, the effect of the exchange of the CP odd Higgs boson is negligible.

[23]There is also a contribution where chirality violation occurs on the neutralino line. However, that is suppressed by additional powers of the inverse of the mass of the exchanged squark.

(c, b, t) quarks, giving a quark mass independent result of about 70 MeV [16.17]. The contribution from u and d quarks can be estimated using chiral perturbation theory, and is essentially negligible. The consequent implication is that $f_p \simeq f_n$ in (16.55), so that to a good approximation the spin independent contribution to the LSP-nucleus scattering cross section can be taken as proportion to $A_\mathcal{N}^2$. Unfortunately, the strange quark is too light to be treated perturbatively, but too heavy to be treated reliably in the framework of chiral perturbation theory. Different model calculations for $m_s \langle N|\bar{s}s|N \rangle$ give quite different results, from as low as 70 MeV to as high as 400 MeV. This is currently the biggest theoretical uncertainty in the calculation of the LSP-nucleus scattering cross section (for a given MSSM spectrum). In the numerical results presented below we use the "standard" value [16.17] $m_s \langle N|\bar{s}s|N \rangle = 130$ MeV.

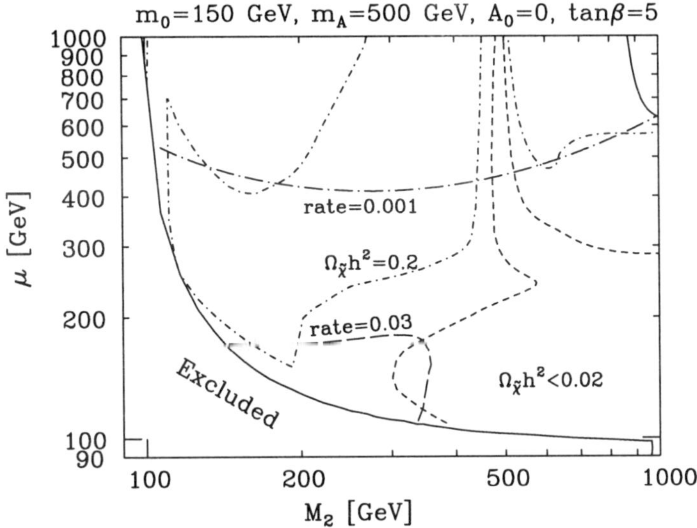

Fig. 16.3. Contours of constant scaled LSP relic density (short dashed: $\Omega_{\tilde{\chi}} h^2 = 0.02$; dot-short dashed: $\Omega_{\tilde{\chi}} h^2 = 0.2$), as well as of constant LSP scattering rate in an isotopically pure ^{76}Ge detector [long dashed: rate = 0.03 events/(kg·day), dot-long dashed: rate = 0.001 events/(kg·day)]. Excluded regions are explained in the text.

In Fig. 16.3 we show contours of constant scaled $\tilde{\chi}_1^0$ relic density (short dashed and dot–short dashed curves) and of the total relic neutralino counting rate in a ^{76}Ge detector[24] (long dashed and dot–long dashed lines) in the M–μ plane. As in Figs. 16.1 and 16.2 we have assumed a universal sfermion mass m_0 at the GUT scale, but allowed the soft supersymmetry breaking Higgs boson masses, or equivalently m_A and μ, to be chosen independently. The region to the left and below the lower solid line is excluded by experimental searches at colliders (mostly by the lower bound on the lightest chargino mass). On the other hand, the lightest neutralino is heavier than the lighter $\tilde{\tau}$ mass eigenstate in the small region delineated by the solid line in the upper right corner. As expected from Fig. 16.1, the LSP relic density is very small for $|M_2| \gtrsim 2|\mu|$, where the LSP is predominantly a higgsino. It is not even sufficient for the known dark halos of galaxies. This region of small relic density extends to large values of $|\mu|$ for $|M_2| \simeq 500$ GeV, where $M_{\tilde{\chi}_1^0} \simeq |M_1| \simeq |M_2|/2 \simeq m_A/2$, so that A-exchange diagrams become (nearly) resonant. In contrast, for much of the region with

---

[24]Whereas ^{76}Ge is a spinless nucleus, the isotope ^{73}Ge carries spin.

$|M_2| \lesssim |\mu|$, where the LSP is binolike, the relic density falls in the region favored by observations, as indicated by the contours for $\Omega_{\tilde{\chi}} h^2 = 0.2$. For the given modest value of m_0, the scaled relic density never exceeds 0.42 (0.62) for $|M_2| < m_A$ ($|M_2| > m_A$). The upper bound in (16.47) is therefore respected everywhere.

In calculating the rate for LSP-^{76}Ge scattering, the local LSP density has been assumed to be given by (16.51), independent of the actual prediction for $\Omega_{\tilde{\chi}} h^2$ at that point in parameter space. This is presumably not realistic for $\Omega_{\tilde{\chi}} h^2 \leq 0.02$. Even under this optimistic assumption, the scattering rate turns out to be at least two orders of magnitude below the current experimental bound of ~ 10 events/(kg·day) [16.13]. A counting rate above 0.03 events/(kg·day) can be achieved only below and to the left of the long dashed curve. Here the $U(1)_Y$ gaugino mass parameter M_1 and the higgsino mass parameter μ are comparable to each other, leading to substantial bino-higgsino mixing. The latter maximizes the LSP coupling to Higgs bosons and hence the scattering cross section. Moreover, the relative lightness of the LSP in this region also enhances the counting rate which is proportional to the ambient LSP number density which, in turn, varies as $1/M_{\tilde{\chi}_1^0}$ for a fixed mass density. This also explains why the contour for counting rate $= 10^{-3}$ events/(kg·day) nearly coincides with a line of constant μ. A reduction of $|M_2|$, for a fixed μ, reduces the LSP mass, increasing the relic LSP flux. It also lowers the bino-higgsino mixing angle, and hence the LSP-^{76}Ge scattering cross section. Note that an increase in the sensitivity of present direct relic LSP search experiments by even four orders of magnitude will still leave unexplored a substantial part of the parameter space with a light binolike LSP.

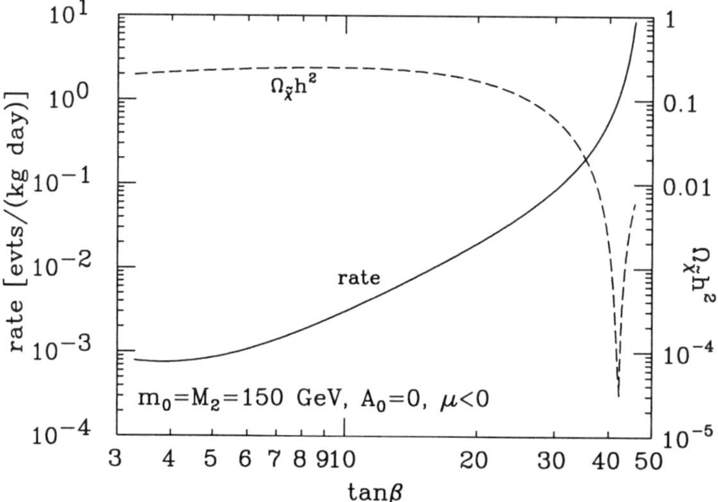

Fig. 16.4. The relic LSP detection rate in a ^{76}Ge detector (solid line, referring to the scale at the left) and the scaled relic density (dashed curve, referring to the scale on the right) as a function of $\tan\beta$ in an mSUGRA scenario.

The next round of experiments is unlikely to achieve much better sensitivity than 0.1 events/(kg·day). One might conclude from Fig. 16.3 that these searches for relic neutralinos would also be fruitless. Such a pessimistic conclusion need not be true, though. In Fig. 16.3 we have taken a relatively large value for the mass of the CP odd Higgs boson A, and a modest value of $\tan\beta$.

This is the 'decoupling limit' of Ch.10 in which the couplings of the light CP even Higgs scalar h are essentially the same as those of the single Higgs of the Standard Model. Moreover, the heavier CP even Higgs boson H is too massive to contribute significantly to the LSP-nucleus scattering amplitude. One can, however, find higher scattering rates than shown in Fig. 16.3 if m_A is reduced and/or some Higgs couplings are enhanced by increasing $\tan\beta$. These two changes of parameters actually tend to go together in models with radiative electroweak symmetry breaking, as illustrated in Fig. 12.6 for the case of mSUGRA. Fig. 16.4 shows the LSP scattering rate on ^{76}Ge (solid curve, referring to the scale on the left) as well as the scaled LSP relic density (dashed curve, referring to the scale at the right) as a function of $\tan\beta$, as predicted in mSUGRA. For the given choice of the soft supersymmetry breaking parameters $m_0, M_{1/2}(\simeq M_2/0.81)$ and A_0, $\tilde{t}$ squarks are not very heavy and $\tilde{t}_L$-$\tilde{t}_R$ mixing is small. Unsuccessful searches for neutral Higgs bosons at LEP then imply $\tan\beta \geq 3.3$. Increasing $\tan\beta$ from this experimental lower bound at first reduces the LSP scattering rate slightly, since m_h is increased. However, soon this is overcompensated by the reduction of m_H and the increase of the $H\bar{s}s$ and $H\bar{b}b$ couplings, which essentially grow proportional to $\tan\beta$. For $\tan\beta > 40$, even counting rates in excess of 1 event / (kg·day) seem possible, in easy reach of next generation experiments.

Two caveats have to be made, though, in relation to the above remarks. First, the reduction of m_A and the increase of its coupling to b quarks also leads to a decrease in the scaled LSP relic density, owing to the enhancement of s channel A exchange diagrams which actually become resonant at $\tan\beta \simeq 42$ for the given choice of mass parameters. If we insist on $\Omega_{\tilde{\chi}} \geq 0.02$, cf. (16.47), the counting rate does not exceed 0.2 events/(kg·day), which is nonetheless well above the maximum found in Fig. 16.3. Note further that the relic density can be increased by increasing m_0: larger values of $\tan\beta$ (and hence larger predicted counting rates for fixed local LSP density) are then allowed while keeping the overall scaled relic density within the range of (16.47). The second problem in scenarios with a high $\tan\beta$ and not very high sparticle masses consists of large $\tilde{t}$-$\tilde{\chi}^\pm$ loop contributions to the amplitude for $b \to s\gamma$ decay [16.21] on account of the sizable bottom Yukawa coupling. These can be made to respect experimental measurements [16.13] by a judicious choice of the trilinear soft breaking parameter A^t (or A_0 in case of mSUGRA). They can also be cancelled by other terms from sparticle loops if small deviations are taken from flavor universal scalar soft breaking masses at the GUT scale itself. However, both solutions require some amount of fine tuning. Rates of LSP-^{76}Ge scattering, in excess of 0.1 events/(kg·day), thus seem somewhat unlikely at present, though they cannot be strictly excluded.

We conclude this section with a brief description of "indirect" searches for relic LSPs. The most promising method of this kind makes use of the fact that these relics should accumulate at the centers of celestial bodies like the sun and the earth. When an LSP scatters off a nucleus in one of these bodies, it may lose sufficient energy to become gravitationally bound. The orbit of such an LSP will then take it repeatedly through this celestial body itself. It then loses more and more of its energy in additional interactions and eventually falls into the center. After some time the LSP density at the center of the earth or of the sun should then become so large that LSP pairs would begin to annihilate there with a significant rate. Eventually LSP capture and LSP annihilation at the centers of these celestial bodies would reach an equilibrium, i.e. the annihilation rate will become half the capture rate[25]. This equilibrium should have been reached quite some time ago in case of the sun, but not necessarily in the earth in case the LSP mass exceeds m_W. During the pre-equilibrium phase the annihilation signal is correspondingly suppressed [16.17]. The only LSP annihilation products with any chance of escaping from the center of the sun or of the earth are

---

[25]Note that each annihilation event destroys two LSPs.

neutrinos and, among those, muon (anti)neutrinos are the most easily detectable. They can produce a muon through a charged current (CC) reaction in the material surrounding an underground (or underwater) detector. This muon can be detected quite easily, either through a track in a conventional particle detector, or through Čerenkov radiation, observable by photomultiplier tubes, in detectors that use (liquid or frozen) water as the active medium. Examples for the former kind of detector are the Baksan and MACRO experiments in Russia and Italy, respectively, while those for the latter kind are SuperKamiokande in Japan and AMANDA as well as ICECUBE in the Antarctic ice cap.

Once equilibrium between capture and annihilation is reached, the strength of this "indirect" (neutrino) signal will again be proportional to the LSP-nucleus scattering cross section, weighted by the appropriate isotopic abundances in the sun or the earth. This indirect signal therefore correlates fairly well with the signal in direct LSP-nucleus scattering experiments. However, three effects have to be considered. First, while current direct search experiments, as well as the indirect signal from the center of the earth, are mostly sensitive to spin independent interactions, the capture rate in the sun does depend quite strongly on the spin dependent contribution to the LSP-nucleon scattering amplitude, owing to the large abundance of protons in the sun. Spin dependent and spin independent interactions should therefore be treated separately when comparing direct and indirect relic LSP search experiments. Second, two factors favor the indirect signal for higher values of LSP masses. Both the $\nu_\mu \to \mu$ charged current cross section and the range of the produced muon (and hence the effective target mass) increase essentially linearly with the neutrino energy, and hence with the LSP mass, at least for masses up to a few TeV. On the other hand, the fraction of LSPs losing sufficient energy in the first scattering reaction to become gravitationally bound decreases with an increasing LSP mass once $M_{\rm LSP}$ exceeds the mass of the heaviest relevant nucleus. A third and final factor is that the spectrum of neutrinos resulting from LSP annihilation depends on the dominant annihilation channel. Recall[19] that Majorana neutralinos at rest cannot annihilate into massless $f\bar{f}$ pairs. Their annihilation into $b\bar{b}$ or $c\bar{c}$ pairs yields a fairly soft neutrino spectrum, since the quarks hadronize and possibly interact with the surrounding medium before they undergo three body decay. The annihilation into $\tau^+\tau^-$ pairs gives rise to a harder spectrum and the processes of annihilation into W and Z pairs, which undergo two body decays, lead to the hardest predicted neutrino spectra. These effects introduce a further model dependence in any comparison of direct and indirect relic LSP searches.

One can nonetheless establish rough relations between the strengths of those two kinds of signals[26]. For purely spin independent interactions one has

$$\frac{\text{rate in } {}^{76}\text{Ge}}{\text{indirect rate}} \simeq 4 \times 10^5 \frac{\text{m}^2}{\text{kg}} \cdot \frac{100 \text{ GeV}}{M_{\rm LSP}} , \tag{16.56}$$

while purely spin dependent couplings lead to

$$\frac{\text{rate in hydrogen}}{\text{indirect rate}} \simeq 3 \times 10^3 \frac{\text{m}^2}{\text{kg}} \cdot \frac{100 \text{ GeV}}{M_{\rm LSP}} . \tag{16.57}$$

Here "indirect" means the sum of the signals from the earth[27] and the sun. These signals are very roughly equal for spin independent interactions, unless the capture rate is so small that equilibrium

---

[26]Note that, while the direct detection rate is measured in events/(kg · day), the indirect rate comes in events/ $(m^2$ day).

[27]It has recently been pointed out that the path of an LSP, that scatters once in the sun, can subsequently get perturbed by the gravitational influence of the planets (chiefly Jupiter) in such a way that it ends up in a stable ("planetary") orbit around the sun. This process can lead to a new population of low velocity

has not yet been established. Such estimates reproduce exact numerical results [16.17] to within a factor of 3 for $M_{\mathrm{LSP}} \leq 1$ TeV. They indicate that direct search experiments are more promising if spin independent interactions dominate. In this case a km^2 "neutrino telescope" has no better sensitivity than a detector weighing five pounds, for $M_{\mathrm{LSP}} \lesssim 100$ GeV. In contrast, the neutrino signal appears to be more promising if spin dependent interactions dominate in which case the signal from the center of the earth is virtually eliminated. The coefficient in (16.57) is about two orders of magnitude larger than that in (16.56), largely due to the lack of any coherence enhancement of spin dependent interactions. Moreover, constructing a massive hydrogen target for direct detection experiments, as assumed in (16.57), is practically difficult. Yet, heavier nuclei, e.g. ^{73}Ge, which may be more practicable as target material, yield much smaller counting rates per detector mass if spin dependent interactions dominate.

Current limits from direct and indirect searches for relic LSPs are competitive, though in either case only rather extreme scenarios can be probed at present, as seen in the discussion of Figs. 16.3 and 16.4. Of course, the sun is a source of energetic neutrinos even in the absence of LSP annihilation. These are produced, when cosmic rays interact in the solar atmosphere, analogously to atmospheric neutrinos in the earth. The size of this background is quite low [16.23], but eventually it might become the limiting factor in the sensitivity of searches for LSP annihilation in the sun. Though direct searches also face numerous backgrounds, most of those can at least in principle be removed by a careful preparation of the experiment. Lastly, without going into any detail we mention that indirect signals from the annihilation of relic LSPs in the halo of our galaxy have also been considered. Final states containing photons, positrons or antiprotons seem to be most promising here. Unfortunately, a detectable signal can be expected for only small regions of the MSSM parameter space, *and* if the distribution of LSPs in the halo is favorable, i.e. strongly peaked towards the center of the galaxy, and/or "lumpy" throughout the halo. The reader is directed to Ref. [16.17] for further details.

16.4　Cosmology of the Gravitino

We have seen speculations in Chs. 12 and 13 that the gravitino mass $m_{3/2}$ can lie anywhere between a fraction of an eV and well above a TeV. On account of its feeble interactions with matter, lower bounds on $m_{3/2}$ are difficult to derive from collider experiments. Stronger unitarity bounds (cf. §12.2), implying that the gravitino cannot be much lighter than the gluino, require $m_{3/2}$ to have been generated at a very high scale and are therefore not applicable to scenarios (such as GMSB, cf. Ch.13) where such is not the case. Upper bounds from fine tuning arguments are very model dependent, since the gravitino does not contribute to renormalizable radiative corrections to scalar masses. Not surprisingly, cosmological constraints differ, depending on the mass range considered and on whether the gravitino is stable or unstable. In later subsections we shall discuss each mass range in turn, starting with a very light gravitino. In particular, for a gravitino, lying in mass between 1 keV and 30 TeV, we shall derive a stringent upper bound on the temperature that the Universe had after the last major period of entropy production. Within the framework of simple inflationary scenarios, sketched at the end of §16.2, this corresponds to an ($m_{3/2}$-dependent) upper limit on the reheating temperature T_R. In this context one statement needs to be made

---

LSPs in the vicinity of the earth. For LSP masses below ~ 150 GeV, the expected strength of the indirect signal from the center of the earth can then increase by two orders of magnitude [16.22]. However, the size of this new contribution depends very sensitively on the precise form of the LSP velocity distribution in these stable orbits: an effect not included in (16.56).

first about the calculation of any amplitude with the gravitino. Since gravitino interactions are not renormalizable, all calculations can presently be performed only at the tree level. If loop corrections can be estimated by cutting off divergent integrals over loop momenta at some scale $\mathcal{O}(M_{Pl})$, they should be suppressed by a loop factor (inverse powers of 4π), in which case they will not substantially change the limits derived here.

A very light gravitino

The interaction strength of a "longitudinal" (spin $s_z = \pm 1/2$) gravitino is enhanced by a factor $M_s/m_{3/2}$ if the gravitino is much lighter than the scale M_s which is roughly of the order of the masses of sparticles in the observable sector. In this case the $s_z = \pm 3/2$ components of the gravitino can safely be ignored, i.e. the gravitino acts like the two component (Majorana) goldstino (cf. §7.3). Were the gravitinos in the early Universe ever in thermal equilibrium, they decoupled at a temperature $T_{3/2} \gg m_{3/2}$, i.e. while they were still relativistic. The gravitino number density at decoupling, denoted by $n_{3/2}(T_{3/2})$, is thus simply the fraction 3/4th of the photon number density at the same temperature $n_\gamma(T_{3/2})$. The subsequent decoupling of massive particles and neutrinos will contribute to n_γ, but not to $n_{3/2}$. Requiring that the total entropy density s remains constant and recalling that the photonic contribution to the entropy density s_γ goes linearly with n_γ, the ratio of gravitino to photon number density is found to scale like

$$\frac{n_{3/2}(T)}{n_\gamma(T)} = \frac{3}{4}\frac{g_{*s}(T)}{g_{*s}(T_{3/2})} \quad , \quad (T \leq T_{3/2} < T_R) . \tag{16.58}$$

In (16.58) g_{*s} is the number of degrees of freedom, cf. (16.18). The inequality in parentheses means that (16.58) is only applicable if gravitinos have once been in thermal equilibrium, i.e. $T_{3/2}$ is less than the highest reheating temperature T_R (cf. §16.1) of the post-inflationary universe.

The contribution of a very light ($m_{3/2} < 1$ eV) gravitino to the total mass density of the Universe is negligible. However, such gravitinos would still have been relativistic at the time of nucleosynthesis, and would thus have increased the Hubble parameter during that epoch, cf. (16.37). This would have led to an earlier freeze-out of reactions that change protons into neutrons and vice versa. The net result would have been an increase in the neutron to proton ratio at the onset of BBN, and hence an increase of the produced ^{4}He fraction. The resulting constraint on additional relativistic particles (besides photons and neutrinos) can be expressed in units of extra generations of SM neutrinos, namely ΔN_ν. The Hubble parameter depends on the energy density ρ, which is proportional to T^4, while the entropy or number density is proportional to T^3. Therefore, (16.58) implies that the contribution of light gravitinos to ΔN_ν is given by

$$\Delta N_\nu^{(3/2)} = \left[\frac{g_{*s}(T_\nu)}{g_{*s}(T_{3/2})}\right]^{4/3} \quad , \quad (T_\nu < T_{3/2} < T_R). \tag{16.59}$$

Here $T_\nu \simeq 1$ MeV is the neutrino freeze-out temperature in (16.59) and $g_{*s}(T_\nu) = 43/4$. Unfortunately, the current upper bound on ΔN_ν is quite uncertain [16.2]. Values as low as 0.04 and even -0.4 (excluding the SM with three light neutrinos!) have been quoted in the literature; but a more conservative and hence reliable upper bound is

$$\Delta N_\nu \lesssim 0.75 . \tag{16.60}$$

Eq. (16.60) indicates that light gravitinos should be cosmologically safe if they decoupled before muons did. This implies that $g_{*s}(T_{3/2}) \geq 16$, noting that muons — being Dirac particles — have four degrees of freedom, i.e. $\Delta N_\nu^{(3/2)} \leq 0.59$. We should thus require that

$$T_{3/2} > T_\mu \sim 100 \text{ MeV} .\tag{16.61}$$

As described in §16.2, freeze-out temperatures can be (approximately) computed by equating the appropriate reaction rate with the Hubble expansion rate. For the case at hand, the relevant reaction is gravitino pair annihilation into two photons or two light SM fermions. Moreover, (16.61) implies that we need to know these cross sections only for $m_{3/2} \ll \sqrt{s} \ll M_s$, $\sqrt{s}$ being the CM energy. In this limit, one has [16.24]

$$\sigma(\widetilde{G}\widetilde{G} \to \gamma\gamma) = \frac{1}{576\pi} \frac{M_{\tilde\gamma}^2 s^2}{(m_{3/2} M_{Pl})^4} ,\tag{16.62a}$$

$$\sigma(\widetilde{G}\widetilde{G} \to f\bar{f}) = \frac{g_f}{720\pi} \frac{s^3}{(m_{3/2} M_{Pl})^4} ,\tag{16.62b}$$

where g_f is the number of degrees of freedom associated with the fermion f ($g_e = 4, g_\nu = 2$) and $M_{\tilde\gamma}$ is an effective photino mass, defined to absorb neutralino mixing angles.

Since $s \ll M_{\tilde\gamma}^2$ by assumption, the contribution (16.62a) clearly dominates the total annihilation cross section and hence determines the freeze-out temperature $T_{3/2}$. It is determined by

$$n_{3/2}\langle\sigma(\widetilde{G}\widetilde{G} \to \gamma\gamma)v\rangle = H(T_{3/2}) .\tag{16.63}$$

In the present situation, thermal averaging amounts to the replacement [16.24] $s^2 \to 1.3 \times 10^3 T^4$, while $n_{3/2}(T) \simeq 0.18 T^3$ for $T \geq T_{3/2}$. Eqs.(16.37), (16.62a) and (16.63) thus lead to

$$T_{3/2} \simeq \frac{5}{3} \left[g_{*s}(T_{3/2})\right]^{1/10} \left(\frac{m_{3/2}^4 M_{Pl}^3}{M_{\tilde\gamma}^2}\right)^{1/5} .\tag{16.64}$$

Requiring $T_{3/2} \geq 100$ MeV therefore implies that

$$m_{3/2} \geq 1.4 \times 10^{-5} \text{ eV} \cdot \sqrt{\frac{M_{\tilde\gamma}}{50 \text{ GeV}}} .\tag{16.65}$$

This bound is not very restrictive from the standpoint of model building. In particular, it is well below the range of gravitino masses favored by models of gauge mediated supersymmetry breaking, cf. §12.2.

A light gravitino

Suppose $m_{3/2}$ falls between the lower bound of (16.65) and the scale M_s roughly of observable sector sparticle masses. This is the range of interest for GMSB models, cf. §13.2. According to (16.64), the gravitino decoupling temperature $T_{3/2}$ exceeds the weak scale (~ 100 GeV) if $m_{3/2} \geq 0.02$ eV. The gravitino relic density is still essentially negligible for this mass:

$$\Omega_{3/2}h^2 = \frac{m_{3/2}n_{3/2}(T_0)}{\rho_c}h^2 \simeq \frac{m_{3/2}}{0.85 \text{ keV}} \frac{100}{g_{*s}(T_{3/2})} , \quad (\text{for } T_{3/2} < T_R) .\tag{16.66}$$

In (16.66) we have used (16.58) together with $n_\gamma(T) = 0.24 T^3$, $T_0 = 2.3 \times 10^{-12}$ GeV and $\rho_c/h^2 = 8.1 \cdot 10^{-47}$ GeV^4. Eq. (16.64), which is based on (16.24), is no longer valid for $T_{3/2} \gtrsim 100$ GeV $\sim M_s$.

At such high temperatures, the density of observable sector sparticles is not suppressed. This allows the creation or annihilation of gravitinos through processes involving only one gravitino. Since a goldstino interacts with gauge superfields via dimension five operators and with chiral superfields through dimension four ones, its interactions with gauge superfields dominate at high energies or high temperatures [16.25]. Examples of the latter are $V + \tilde{\lambda} \leftrightarrow V + \tilde{G}$ and $V + V \leftrightarrow \tilde{\lambda} + \tilde{G}$, where V is a gauge boson and $\tilde{\lambda}$ the corresponding gaugino. The zero temperature cross section for the first of these reactions has a singularity from the exchange of gauge bosons in the t or u channel (for nonabelian group factors). This is usually regularized by introducing a lower limit on $|t|$ and $|u|$ of order T^2. For $s \gg M_s^2$, one is led to [16.25] the following expression for the said cross section:

$$\Sigma_{\rm tot}^{(3/2)} \equiv \frac{1}{2} \sum_{x,y,z} \eta_x \eta_y \sigma(x+y \to z+\tilde{G}) \simeq \frac{1}{24\pi(m_{3/2}M_{Pl})^2} \left(2.4 g_1^2 M_1^2 + 9.2 g_2^2 M_2^2 + 26 g_s^2 M_3^2\right). \quad (16.67)$$

Here M_α ($\alpha = 1,2,3$) are the MSSM gaugino masses in (16.67) and the statistics factor $\eta_{x,y}$ equals 1 (3/4) for an incident boson (fermion).

Note that this cross section is independent of the CM energy. The thermal averaging is thus trivial, and the freeze-out condition $n\langle\sigma v\rangle = H$ yields

$$T_{3/2} \simeq 0.62 \frac{m_{3/2}^2 M_{Pl}\sqrt{g_{*s}}}{\alpha_s M_3^2}, \quad (16.68)$$

where we have again used (16.37) and omitted the subdominant electroweak contributions to the cross section (16.67). Substituting $m_{3/2} = 2$ keV, which according to (16.66) is close to the maximal allowed mass of gravitinos that were in thermal equilibrium with $g_{*s} = 228.75$ (the MSSM value), yields $T_{3/2} \simeq 1.4$ TeV $(1\ {\rm TeV}/M_3)^2$. The latter is well below the reheating temperature predicted by most inflationary models [16.1,16.2]. This is the origin of the bound[28] $m_{3/2} \lesssim 1$ keV, used in §12.2.

Heavier gravitinos must never have been in thermal equilibrium (after inflation). One then has to recompute the relic gravitino mass density [16.25]. The starting point of this calculation is a version of the Boltzmann equation (16.38) that ignores reactions destroying gravitinos (since the rate for these reactions is suppressed by the small gravitino density), but adds a contribution from the decay of heavier sparticles into gravitinos[29]:

$$\frac{d\tilde{Y}_{3/2}}{dT} = -\frac{n_\gamma\langle\Sigma_{\rm tot}^{(3/2)} v\rangle}{4H(T)T} - \sum_i \frac{\tilde{Y}_i\langle\Gamma_i\rangle}{2H(T)T}, \quad (16.69)$$

with $\tilde{Y}_i$ defined as n_i/n_γ and differing from $Y_i \equiv n_i/s$ by a constant. The relevant partial decay widths Γ_i are [16.25]:

$$\Gamma(\tilde{\lambda} \to \tilde{G} + V) = \frac{1}{48\pi}\frac{M_{\tilde{\lambda}}^5}{(m_{3/2}M_{Pl})^2}\left[1 - \left(\frac{m_{3/2}}{M_{\tilde{\lambda}}}\right)^2\right]^3, \quad (16.70a)$$

---

[28]The tighter WMAP [16.4] bound on the total dark matter density would imply a correspondingly stronger bound on $m_{3/2}$. Moreover, the WMAP data imply early re-ionization of the Universe, i.e. early star formation. This is difficult to achieve [16.4] if the dark matter is mostly "warm" as is the case with keV gravitinos.

[29]Recall that the difference between bosonic and fermionic initial states has already been included in the definition of $\Sigma_{\rm tot}^{(3/2)}$. Hence the particle density appearing in the Boltzmann equation can be written as n_γ for all relativistic species.

$$\Gamma(\tilde{f} \to \tilde{G} + f) \;=\; \frac{1}{48\pi} \frac{m_{\tilde{f}}^5}{(m_{3/2} M_{Pl})^2} \left[1 - \left(\frac{m_{3/2}}{m_{\tilde{f}}} \right)^2 \right]^2 . \tag{16.70b}$$

In the RHS expressions of (16.70b), the fermion mass m_f has been set equal to zero.

The Boltzmann equation (16.69) has to be solved starting from the boundary condition $\tilde{Y}_{3/2}(T_R) = 0$. Note that the first term in the RHS of (16.69) becomes independent of T for $T > M_s$. We saw earlier that $\Sigma_{\text{tot}}^{(3/2)}$ is independent of the scattering energy at high CM energies, so that its thermal average is independent of T, while n_γ goes as T^3 and H as T^2. For $T > M_s$, this "scattering" contribution to $\tilde{Y}_{3/2}$ can therefore be written as

$$\tilde{Y}_{3/2}^{\text{scatt}}(T) = \frac{n_\gamma(T_R)\Sigma_{\text{tot}}^{(3/2)}}{4H(T_R)T_R}(T_R - T) \longrightarrow \frac{n_\gamma(T_R)\Sigma_{\text{tot}}^{(3/2)}}{4H(T_R)} , \tag{16.71}$$

where the limit is for $T_R \gg T$. In this limit, this contribution is proportional to T_R, but becomes independent of T. Since the scattering contribution drops quickly for $T < M_s$, where the density of observable sector sparticles is exponentially suppressed, (16.71) remains valid at lower temperatures as well, except for a factor $g_{*s}(T)/g_{*s}(T_R)$ that accounts for the relative increase of n_γ due to the subsequent decoupling of heavy particles.

The sparticle decay contribution is given by [16.25]

$$\tilde{Y}_{3/2}^{\text{decay}} = \int_T^{T_R} dT' \sum_i \frac{\tilde{Y}_i(T')\langle \Gamma_i \rangle}{2H(T')T'} . \tag{16.72}$$

For $T \gtrsim M_s$, this integral receives its dominant contribution from the lower end of the integration region, i.e. it is almost independent of T_R for $T_R \gg T$. Now $\langle \Gamma_i \rangle$ depends very weakly on the temperature (through statistics factors), while $H(T')T'$ goes as T'^3. However, contributions from $T < M_s$ will again be suppressed exponentially since the $\tilde{Y}_i$ become very small. The combination of (16.70) and (16.72) shows that this contribution to the gravitino relic (mass) density scales roughly like[30] $M_s^3/m_{3/2}$ if $T_R > M_s$. For 2 keV $\lesssim m_{3/2} \lesssim$ 50 keV, this contribution by itself is then found [16.25] to lead to too large a gravitino relic density for all $T_R \gtrsim M_s$.

For $m_{3/2} > 100$ keV, the sparticle to gravitino decay widths become too small to produce a significant number of gravitinos and the scattering contribution (16.71) takes over. The latter leads to

$$\Omega_{3/2} h^2 \simeq 10^7 \frac{T_R}{1 \text{ TeV}} \frac{M_3^2}{M_{Pl}m_{3/2}} , \tag{16.73}$$

where we have taken $g_{*s}(T_R) = 228.75$ and $\alpha_s = 0.1$, and have again ignored the electroweak contribution to the scattering cross section. The requirement $\Omega_{3/2} h^2 < 1$ then leads to

$$T_R \;<\; 2 \times 10^8 \text{ GeV} \times \frac{m_{3/2}}{1 \text{ GeV}} \left(\frac{1 \text{ TeV}}{M_3} \right)^2 . \tag{16.74}$$

As stated above, for a gravitino in this mass range, sparticle decays contribute negligibly to the gravitino relic density. This is since the lifetimes for these decays are so long that most of them only

____

[30]Recall that we have ignored reactions destroying gravitinos when writing (16.71) and (16.72). This approximation will certainly break down if gravitinos reach thermal equilibrium, i.e. for $m_{3/2} \lesssim 1$ keV and $T_R \gtrsim M_s$.

occur well after observable sector sparticles have decoupled. The lightest observable sparticle, which we refer to as the NLSP (cf. §13.1) will nonetheless decay eventually into gravitinos. Problems can arise [16.2] in case this happens during or after nucleosynthesis. Photons, produced from these decays (after an electromagnetic cascade in the surrounding plasma), will break up ^{4}He nuclei into ^{3}He or ^{2}H ones. Such processes will lead to an overproduction of those light elements if the relevant lifetime exceeds $\sim 10^6$ sec. For lifetimes between $\sim 10^4$ and $\sim 10^6$ sec, the most dangerous process is instead the photodissociation of ^{2}H, leading to too small a primordial deuterium abundance.

The resulting constraint depends on the abundance of NLSP's, which is usually expressed as the contribution $\widetilde{\Omega}_{\rm NLSP} h^2$ that these particles would make to the relic density if they were stable. A detailed numerical analysis [16.2] finds the constraint $\tau_{\rm NLSP} \leq 5 \times 10^6$, $2.5 \times 10^6, 10^5, 2 \times 10^4, 10^4$ sec for $\widetilde{\Omega}_{\rm NLSP} h^2 = 10^{-3}$, $0.01, 0.1, 1$ and 10 respectively. We saw in §14.2 that, at least if the NLSP is the lightest neutralino, it is difficult to get $\widetilde{\Omega}_{\rm NLSP} h^2 < 10^{-3}$. For a given gravitino mass and NLSP (would be) relic density one can then derive a *lower* bound on the NLSP mass from (16.70). For example, if the NLSP is a binolike neutralino with $\widetilde{\Omega}_{\rm NLSP} h^2 = 0.1$, its mass should exceed $3.6, 90, 225, 566$ GeV if $m_{3/2} = 0.1, 1, 10, 100$ GeV respectively. Let us now recall that the would-be relic density tends to increase with the NLSP mass[31]. Therefore, we see that scenarios, with the gravitino as the LSP, become quite constrained if the gravitino mass exceeds a few GeV. It may be noted that this nucleosynthesis constraint only assumes that the LVSP was in thermal equilibrium, i.e. it remains valid for all $T_R \gtrsim M_{\rm NLSP}/10$.

A heavy gravitino

If the mass of the gravitino is comparable to, or larger than, the masses of observable sector sparticles, the interaction strength of its spin $1/2$ component is no longer enhanced compared to that of its spin $3/2$ component. In fact, at high energies, the gravitino production cross section is dominated by the transverse (spin $3/2$) components. The most important single gravitino production processes [16.26] are again those $2 \to 2$ reactions where at least one of the three other external particles (in either the initial or final state) is a member of a gauge supermultiplet. Some of the corresponding cross sections again have t or u channel singularities, which are regularized in the same way as for light gravitinos. The total weighted gravitino production cross section then becomes in the high energy limit [16.26]

$$\Sigma_{\rm tot}^{(3/2)} = \frac{1}{M_{Pl}^2} \left(2.5g_1^2 + 5.0g_2^2 + 11.8g_s^2\right).$$ (16.75)

This cross section is again independent of energy if the running of the gauge couplings is ignored. Thus the solution of the Boltzmann equation (16.69) proceeds as before. Of course, the contribution from the decay of observable sector sparticles into gravitinos is now completely negligible. Taking the gauge couplings at an average scale of 10^{10} GeV, one has the approximate relation

$$\Omega_{3/2} h^2 \simeq \frac{m_{3/2}}{100 \text{ GeV}} \frac{T_R}{10^{11} \text{ GeV}}.$$ (16.76)

A slightly more complicated expression for $\widetilde{Y}_{3/2}(T_0)$, which includes the logarithmic running of the gauge couplings when solving the Boltzmann equation, can be found in Ref. [16.26]. Requiring $\Omega_{3/2} h^2 < 1$, one obtains from (16.76) that

$$T_R \lesssim 10^{11} \text{ GeV} \times \frac{100 \text{ GeV}}{m_{3/2}} , \qquad (m_{3/2} \lesssim M_s).$$ (16.77)

---

[31]Unitarity arguments imply that its annihilation cross section should decrease like the inverse square of its mass for large masses.

In fact, the bound of (16.77) is hardly ever the relevant one. We have assumed here that the gravitino is stable. However, then the lightest observable sector sparticle will be too long-lived, leading to problems with nucleosynthesis, as discussed at the end of the previous subsection, unless it is significantly heavier than the gravitino. In other words, BBN requires a stable gravitino to be significantly lighter than observable sector sparticles and such a light gravitino has already been discussed above. However, (16.76) remains valid even if the gravitino is not the LSP, provided that $\Omega_{3/2}$ is replaced by the would-be relic density $\widetilde{\Omega}_{3/2}$. In this case it is the gravitino itself whose decay can lead to the photodissociation of light elements. We thus have to require that the gravitino is sufficiently shortlived and decays before the onset of BBN. Alternatively, we can demand that the reheating temperature was so low that even late decays of the relatively few produced gravitinos can be tolerated. As an example of the latter situation, let us assume that the only decay mode of the gravitino is into a neutralino and a photon, with [16.26]

$$\Gamma(\widetilde{G} \to \tilde{\chi}_1^0 + \gamma) = |U|^2 \frac{m_{3/2}^3}{32\pi M_{Pl}^2} \left[1 - \left(\frac{M_{\tilde{\chi}}}{m_{3/2}}\right)^2\right]^3 \left[1 + \frac{1}{3}\left(\frac{M_{\tilde{\chi}}}{m_{3/2}}\right)^2\right], \tag{16.78}$$

where U is the same combination of neutralino mixing angles as appears in the first row of Table 13.1. One obtains a corresponding lifetime

$$\tau_{3/2} \geq 4 \times 10^8 \left(\frac{100 \text{ GeV}}{m_{3/2}}\right)^3 \text{ sec.}, \tag{16.79}$$

where the bound is saturated for $M_{\tilde{\chi}}^2 \ll m_{3/2}^2$. The maximal allowed value of $\widetilde{\Omega}_{3/2}h^2$ for $\tau_{3/2} \gtrsim 5\times10^6$ sec is [16.2] about 10^{-4}. Then (16.76) leads to the bound

$$T_R \lesssim 10^7 \text{ GeV} \frac{100 \text{ GeV}}{m_{3/2}}. \tag{16.80}$$

for $\tilde{G} \to \tilde{\chi}_1^0 + \gamma$ only, provided $m_{3/2} \lesssim 500$ GeV.

Let us now consider the opposite limit, where the gravitino is heavier than all observable sector sparticles. The lifetime estimate (16.79) should then be reduced by about a factor 100 on account of the additional final states accessible to gravitino decay. Evidently, such a scenario requires a gravitino mass of at least several hundred GeV. The gravitino lifetime will then be less than 10^4 sec. Moreover, most gravitino decays will lead to hadronic jets, rather than electromagnetic cascades. This also affects nucleosynthesis, but the resulting bounds on the gravitino abundance are weaker [16.2]. One only has to require $\widetilde{\Omega}_{3/2}h^2 \leq 4 \times 10^{-3}$ (0.4) for a lifetime $\tau_{3/2} = 10^4$ (10^2) sec. One is then led to the constraint $T_R \leq 5\times10^7$ (10^9) GeV for the corresponding value $m_{3/2} = 0.7$ (3.4) TeV. The bound on T_R then again decreases in proportionality with $1/m_{3/2}$ for $m_{3/2} \leq 15$ TeV ($\tau_{3/2} \geq 1$ sec). Gravitinos with even shorter lifetimes decay before nucleosynthesis gets underway. The bound on T_R then rapidly weakens and essentially disappears for $m_{3/2} > 30$ TeV or so. However, such a large value of $m_{3/2}$ is difficult to reconcile with fine tuning arguments in generic supergravity models, where the gravitino is not much heavier than observable sector sparticles, as discussed in §12.2. Note, however, that in AMSB scenarios, cf. §12.6, the gravitino mass exceeds the masses of observable sector sparticles by an inverse loop factor. It is possible in such models to arrange $m_{3/2} > 30$ TeV without undue fine tuning.

16.5 Baryogenesis

There is overwhelming evidence for a nonvanishing baryon asymmetry in the Universe. In the vicinity of our solar system there are certainly more baryons than antibaryons. If other parts of the Universe had instead an excess of antibaryons, annihilation reactions would take place on the boundaries between baryon-rich and antibaryon-rich domains. The annihilation products (gamma or X–rays) would have been detected, unless the radius of our baryon-rich domain exceeds the Hubble radius. One can thus safely conclude that there are many more baryons than antibaryons in the (observable) universe. If n_B ($n_{\bar{B}}$) is the baryonic (antibaryonic) number density, $n_B/n_{\bar{B}} \gg 1$.

In order to get a dimensionless ratio, it would be natural to divide the number density of excess baryons, i.e. $n_B - n_{\bar{B}}$, by that of the most predominant particles in the universe, namely photons. However, such a ratio is very small. The best determination, from analyses of nucleosynthesis [16.2], yields $\eta \equiv (n_B - n_{\bar{B}})/n_\gamma =$ (a few) $\times 10^{-10}$, cf. (16.4). In principle, such a tiny baryon asymmetry could simply be imposed as a boundary condition on the initial state of the Universe. However, this "solution" is quite unappealing, since it amounts to a severe fine tuning. A much more elegant explanation would start from a Universe with vanishing net baryon density, and create the asymmetry dynamically later on. As was pointed out by Sakharov [16.27], three conditions would have to be satisfied for such a solution to be possible:

- The theory must contain (sufficiently strong) baryon number violating interactions. This is reasonably evident, since otherwise the net baryon charge of the Universe would retain its initial value which is zero by fiat.

- The theory must violate both C and CP. If either C or CP were conserved, reactions creating some B excess would occur with rates equal to those of C or CP conjugate reactions that create an equal $\bar{B}$ excess, leading to a net baryon number that vanishes. This can be seen from the simple observation that the baryon charge operator is odd under both C and CP, i.e. its expectation value would have to vanish if C or CP were good symmetries of Nature.

- These B, C and CP violating interactions must have been operating during a period when the Universe was *not* in thermal equilibrium. In equilibrium, a reaction (e.g., $X_1 + X_2 \to Y_1 + Y_2$) and its inverse ($Y_1 + Y_2 \to X_1 + X_2$) would occur with equal rate in the soup that was the early Universe. Any baryon number, created through some reaction, would then immediately be erased by the inverse reaction, again leading to a zero net baryon asymmetry.

The top two conditions constrain the Lagrangian, while the bottom done constrains the thermal evolution of the Universe.

The first two of the above conditions are satisfied already in the Standard Model. $SU(2)_L$ gauge interactions violate C maximally. Furthermore, the presence of CP violation is well documented in the neutral kaon and b-flavored meson sectors and is parametrized by the phase δ in the CKM matrix. The occurrence of baryon number violating interactions in the SM is less obvious and takes place in the following way. Above the electroweak scale, when $SU(2)_L$ is unbroken, baryon number becomes a chiral charge in the $SU(2)_L$ gauge theory. Though conserved at the tree level (i.e. classically), this chiral charge suffers a loop induced nonconservation due to the 't Hooft quantum anomaly, with the amount controlled by the topological charge of the gauge field configuration involved in the process. These configurations have distinct, degenerate ground states or vacua – each characterized by a topological index. A process with baryon number violation corresponds to a transition from one vacuum to another, the change in baryon number being proportional to the change of that index. Two neighboring degenerate vacua are actually separated by an energy barrier. The lowest point on this barrier is a (saddle point) solution of the equations of motion known as a *sphaleron*. Transitions from one vacuum to the next are therefore called sphaleron transitions. In the SM they can be envisioned as creating out of the vacuum a special state with vanishing electromagnetic charge, namely

$$\prod_{i=1}^{N_g} |u_{L_i} d_{L_i} d_{L_i} \nu_i\rangle \tag{16.81}$$

or its CP conjugate, where N_g is the number of chiral generations. In case there are other chiral $SU(2)_L$ nonsinglet fermions, they should also be included in the expression (16.81), keeping the state with zero electromagnetic charge. In the SM a sphaleron transition therefore creates nine $SU(2)_L$ doublet quarks and three neutrinos (of different flavors), thereby increasing baryon number B and lepton number L by three units each. Notice that the difference $B - L$ remains unchanged, as do the differences [16.28] $L_i - L_j$ of different lepton type numbers, i and j being type indices.

At zero temperature the sphaleron transition is a tunnelling process with a rate that is negligibly small, being suppressed by a factor[32] $e^{-4\pi/\alpha_W}$, where $\alpha_W = g^2/(4\pi)$. However, at higher temperatures, thermal fluctuations make it much easier to get across the energy barrier. In addition, the barrier becomes lower with increase in temperature. For $T > T_{EW} \sim 200$ GeV it vanishes completely, and gauge symmetry is restored[33]. $B + L$ violating electroweak interactions are therefore in equilibrium for 100 GeV $\lesssim T \lesssim \alpha_W^4 M_{Pl} \sim 10^{12}$ GeV. Any $B+L$ asymmetry, that is produced at $T > 100$ GeV, will subsequently be washed out by these interactions. There are two ways to avoid this "sphaleron erasure". (1) One can create a $B - L$ asymmetry (more precisely [16.28], an asymmetry in $B/3 - L_i$ for at least one effectively conserved lepton type number L_i), or (2) one can create a $B + L$ asymmetry at a temperature below 100 GeV. In the following subsections we shall discuss proposals of how to satisfy the Sakharov conditions and this additional constraint from sphaleron erasure. In order of decreasing energy scale, these proposals are: GUT scale baryogenesis, baryon asymmetry via leptogenesis, the Affleck-Dine mechanism and electroweak baryogenesis.

GUT Baryogenesis

The oldest proposal [16.1] for baryogenesis in the early universe is based on the decay of superheavy Higgs bosons associated with a Grand Unified Theory (GUT). Such bosons can be produced through gauge interactions. They should therefore have been in thermal equilibrium at temperatures $T \gtrsim M_U/10$, cf. §16.2. Moreover, they could be comparatively long lived, if their decays were only through small Yukawa couplings. As a result, they might have decayed well after decoupling from the thermal bath. In this way Sakharov's third condition is satisfied. The other two conditions are also easily met: GUTs generally violate baryon number and Yukawa couplings violate CP even in the SM. Such a scenario may naively seem more attractive in supersymmetric theories. This is since in such theories the disparity between the GUT and weak scales is stable against radiative corrections and moreover one-step unification of all gauge interactions is phenomenologically allowed, cf. Fig. 11.2. However, we saw in the previous section that – for gravitino masses in the broad range between 1 keV and about 30 TeV – the reheating temperature after inflation must

---

[32]Note that the expression for this factor cannot be expanded around $\alpha_W = 0$ showing that sphaleron transitions are intrinsically nonperturbative.

[33]That is unless there is an asymmetry in some charge, e.g. some lepton type number, of at least 10% or so at $T = T_{EW}$ [16.29], at the level of at least eight orders of magnitude above the current baryon number asymmetry (16.4).

have been orders of magnitude smaller than M_U. In that case superheavy GUT fields would never have been in thermal equilibrium after inflation.

For many years the last mentioned difficulty was thought to rule out GUT scale baryogenesis[34]. However, it has recently been realized that superheavy bosons (or fermions) can be produced at the end of inflation, while the inflaton field oscillates coherently around the minimum of its potential. If the superheavy field couples to the inflaton, these oscillations lead to a time-dependence of its mass triggering particle production through a "parametric resonance" [16.6]. The number of superheavy particles produced in this manner can be sufficiently large, even if their masses significantly exceed that of the inflaton itself. Athermal gravitino production through the same mechanism is not dangerous [16.30]. Problems remain, though. For one thing, the simplest GUT models, based on the gauge group $SU(5)$, conserve $B - L$. Any baryon number generated in the decay of superheavy fields in such models would then immediately be erased by electroweak processes, as discussed in the previous Subsection. Moreover, the CP violation contained in the corresponding Yukawa couplings is not sufficient [16.1]. Note that it affects decays of superheavy particles only through the interference of the tree level and the one loop diagrams[35]. Successful GUT scale baryogenesis would thus have to be based on the gauge group $SO(10)$ (or an even larger group). It would have to contain new, strong sources of CP violation, and would impose tight constraints on the couplings of the inflaton(s). Because of these complications, and because the conclusion is not qualitatively changed by the introduction of sparticles, we shall not discuss this mechanism any further.

Baryon asymmetry via leptogenesis

As discussed already, nonperturbative electroweak processes will wipe out any baryon asymmetry created at temperatures $T \gtrsim 100$ GeV if $\langle B - L \rangle = 0$. On the other hand, if $\langle B - L \rangle \neq 0$, the same processes will distribute the total $B - L$ excess roughly equally into a baryon and an antilepton excess. This can be used to partially convert an initial lepton asymmetry into a baryon asymmetry via $\langle L \rangle = -\langle B - L \rangle$. If mass effects can be neglected, the final baryon excess can be computed from the sum rules for the relevant chemical potentials. One finds [16.27] that

$$\langle B_{\text{final}} \rangle \simeq -\frac{8N_g + 4N_H}{22N_g + 13N_H} \langle L_{\text{initial}} \rangle = -\frac{8}{23} \langle L_{\text{initial}} \rangle \,, \qquad (16.82)$$

N_g being the number of generations of matter superfields and N_H being the number of Higgs doublet superfields. The second equality in (16.82) is for the MSSM (with $N_g=3$ and $N_H=2$)[36]. This opens the possibility to generate an excess of baryons via a perturbatively created lepton asymmetry, specifically, an excess of antileptons. The latter is known as "leptogenesis" [16.27] and is usually achieved through the presence of a heavy Majorana neutrino contributing a lepton number violating mass term to the Lagrangian density.

---

[34]As discussed in Ch.13, models with gauge mediated supersymmetry breaking do allow for quite light gravitinos, circumventing the constraint on T_R. However, investigations of the large scale anisotropy of the universe, most notably of the cosmic microwave background, have fixed the inflaton mass to be in the vicinity of 10^{13} GeV. Since T_R cannot plausibly exceed the inflation mass [16.27], this would still mean that particles with mass near $M_U \simeq 2 \times 10^{16}$ GeV did not reach thermal equilibrium after inflation.

[35]Since (partial) decay widths are even under naive time reversal, they can be affected by CP violation only if the CP violating phases act together with a CP conserving (dispersive) phase, which can only be created in higher orders in perturbation theory.

[36]Since sphaleron processes remain in equilibrium down to $T \sim 100$ GeV, mass effect may become important, in particular from Higgs masses [16.28]. In most models this only changes the relation (16.82) by a factor $\mathcal{O}(1)$. However, it also allows baryogenesis from a lepton flavor asymmetry, where $\langle L_i - L_j \rangle \neq 0$ but the total $\langle L \rangle = 0$. We shall not pursue this somewhat contrived possibility.

This mechanism has attracted a fair amount of interest in recent years, since it can be tied in with "see-saw" models of neutrino masses (cf. §14.2). One introduces a set of $SU(2)_L \times U(1)_Y$ singlet superfields $\overline{N}_i$ and adds the following terms to the superpotential of the MSSM [16.31]

$$\mathcal{W}_N = \frac{1}{2} \sum_i \overline{N}_i M_i \overline{N}_i + \sum_{i,j} f^\nu_{ij} \overline{N}_i H_2 \cdot L_j \; . \tag{16.83}$$

The first RHS term in (16.82) produces (large) Majorana masses for the fermions in $\overline{N}_i$ and makes their superpartners heavy, while the second term gives rise to Dirac neutrino masses. We have written (16.82) in a basis where the Majorana mass matrix is diagonal with real entries M_i. The heavy neutrinos in $\overline{N}_i$ can decay into both $h_2 + l_j$ and its charge conjugate channel and there are supersymmetry related modes as well. CP violation comes into the picture through corresponding partial widths differing at the one loop level [16.31], provided the Yukawa coupling matrix f^ν contains a nontrivial phase ψ_ν. This typically leads to a lepton asymmetry [16.32] of $10^{-6} \sin \psi_\nu$ per decay, but significantly larger asymmetries are possible if two of the heavy $\overline{N}_i$ are close in mass to each other [16.33]. However, these neutrinos must not be exactly degenerate; their mass splitting must be at least as large as their total decay widths, since otherwise $\psi_\nu \to 0$. If the $\overline{N}_1$ neutrinos were still in thermal equilibrium when those in $\overline{N}_2$ and $\overline{N}_3$ decay, the final lepton asymmetry is essentially only due to the decays of the lightest Majorana neutrino $\overline{n_1}$ and its superpartner $\widetilde{\overline{n}}_1$.

In the standard approach one assumes that the interactions of these fields are sufficiently strong so that they were (almost) in thermal equilibrium for a temperature $T \gtrsim M_1$. However, at lower energies (or temperatures), $\overline{N}_i$ exchange also gives rise to an effective lepton number violating term in the superpotential:

$$\mathcal{W}_{\Delta L} \sim \frac{f^\nu f^{\nu\dagger}}{M} H_2 \cdot L H_2 \cdot L \tag{16.84}$$

The resulting interactions, which violate lepton number by two units, must *not* have been in thermal equilibrium at $T \lesssim M_1$, since otherwise they would have erased any lepton asymmetry created earlier. Note that the coefficient appearing in (16.84) has the same form as the mass matrix of the light neutrinos; a similar coefficient also appears in the calculation of scattering cross sections and (inverse) decay widths involving $\overline{n_1}$ and $\widetilde{\overline{n}}_1$. The requirement, that these fields once were in thermal equilibrium but the interactions (16.84) were not, thus translates into an allowed range for the "average" light neutrino mass [16.32]:

$$10^{-5} \text{ eV} \lesssim \overline{m_\nu} \lesssim 5 \times 10^{-3} \text{ eV}. \tag{16.85}$$

A slightly larger range is allowed if the lepton asymmetry per n_1 decay is enhanced because $M_2 \simeq M_1$. The precise definition [16.32] of $\overline{m_\nu}$ is model dependent. The range (16.85) is compatible with current models [16.34] that explain the solar and atmospheric neutrino deficits in terms of neutrino oscillations, if $m_{\nu_1} \ll m_{\nu_3}$. The existence of superparticles has a significant impact on the allowed range (16.85). As noted earlier, both $\overline{n_1}$ and its superpartner $\widetilde{\overline{n}}_1$ contribute to the lepton asymmetry. Moreover, the addition of (16.83) to the superpotential of the MSSM leads to a variety of interactions. For example, there are quartic scalar couplings of the form $f^\nu_{ij} f^{u*}_{kl} \tilde{n}_i \tilde{l}_{Lj} \tilde{q}^\dagger_{Lk} \tilde{u}_{Rl}$. In nonsupersymmetric models one often has to introduce new interactions (beyond those needed to generate neutrino masses) if n_1 is to attain thermal equilibrium, but this is unnecessary [16.32] in the supersymmetric version.

Current models of neutrino masses, based on the see-saw mechanism, favor $M_1 \gtrsim 10^{10}$ GeV [16.34]. This can again lead to problems with the overproduction of massive gravitons, as discussed in §16.3. There are several possible solutions:

- The gravitino mass could be less than 1 keV, e.g. in models with gauge mediated supersymmetry breaking (see Ch.13)[37].

- The gravitino could be the LSP, and hence stable. One then has to worry about $\tilde{\chi}_1^0 \to \tilde{G} + X$ decays ($X = \gamma, Z, H, \ldots$) distorting nucleosynthesis. However, as discussed in §16.3, the neutralino relic density is small even prior to its decay if it is higgsinolike. As shown in Ref. [16.35], one can satisfy the BBN constraint, and even "naturally" obtain $\Omega_{3/2}h^2 \sim 0.1$, if $T_R \sim 10^{10}$ GeV and $m_{3/2} \sim 50$ GeV. In this manner one can connect neutrino oscillations, baryogenesis via leptogenesis and dark matter (gravitinos). Unfortunately, this last connection is not amenable to experimental tests, since relic gravitinos are essentially impossible to detect. So their mass $m_{3/2}$, which enters the calculation (cf. §16.4) of their relic density, cannot be measured.

- If the heavy (s)neutrinos couple to the inflaton, they can be produced in perturbative, but athermal, inflaton decays [16.36], without having to appeal to the "parametric resonance" mentioned in the previous subsection.

- Finally, scalar fields with masses well below that of the inflaton and small or vanishing self–couplings are expected to attain large average values through quantum fluctuations during inflation. At least the lighter of the heavy $\overline{n}_i$ often satisfy these conditions, since they are $SU(2)_L \times U(1)_Y$ singlets. The coherence length of these fluctuations can be so large that at the end of inflation, $\langle \widetilde{n_1} \rangle$ is effectively constant over the observable Universe. The decay of this "condensate" can then create a lepton asymmetry [16.37]. This mechanism, which only works for scalar fields, is quite closely related to the "Affleck-Dine" mechanism, which is the topic of the following subsection.

The Affleck-Dine Mechanism

This [16.27] is based on the observation that certain combinations of scalar fields in the MSSM can acquire large values during inflation. Such a possibility arises only along directions in field space that are "D- and F-flat", i.e. where $D^\alpha = F_i = 0 \ \forall \alpha, i$, cf. (8.34), (8.35). In the MSSM many such directions exist [16.38]. These can be characterized by gauge invariant products of chiral superfields like $L_i H_2$, $\overline{U_i D_j D_k}$ etc., such that the flat direction corresponds to equal VEVs for each factor in the group (for appropriately chosen gauge indices). Composites of such "primitive" flat directions are also possible. The dynamics can then be described in terms of an "Affleck–Dine" field ϕ_{AD}, which is associated with the scalar component of the product superfield. Thus, in the $\overline{U_i D_j D_k}$ flat direction, $\phi_{AD} = \left(\tilde{u}_{Ri}^\star \tilde{d}_{Rj}^\star \tilde{d}_{Rk}^\star \right)^{1/3}$. Here we are interested in AD fields that carry nonvanishing baryon number B. Alternatively, if $T_R \gtrsim 100$ GeV, so that electroweak sphaleron processes would be in equilibrium, they need to carry nonvanishing $B - L$.

By definition, the potential $V(\phi_{AD})$ does not receive any contributions that are both supersymmetric and renormalizable. However, in general one expects the F-flatness to be lifted by nonrenormalizable contributions to the superpotential. The most relevant operator is the one with the lowest mass dimension d that generates a nonvanishing F-term along a given "flat" direction. In the MSSM one finds [16.38] values between $d = 4$ and $d = 9$, depending on the direction, if all operators, allowed by gauge invariance and R-parity conservation, are included. In addition,

---

[37]Note that the required T_R is now well below the inflaton mass, making this solution viable, unlike for GUT scale baryogenesis.

the potential receives supersymmetry breaking contributions. One ought to note that supersymmetry must get badly broken during inflation. This is since at that time the (effective) potential develops a large, positive value $\langle V \rangle = 3H^2 M_{Pl}^2$, cf. (16.23). This can be due to either F-term or D-term contributions to the (inflaton) potential. Indeed, the early stages of AD baryogenesis work somewhat differently in these two cases.

F-term inflation

In this case [16.39] all scalars in the theory, including the AD field, will generically receive masses of order of the Hubble parameter H. In addition, A-terms of the same order are often generated. This can be shown by an analysis of the scalar potential in supergravity models, as given by (12.12) or (12.15). Another way to see this fact is that the supergravity Lagrangian density will contain terms of the kind

$$\mathcal{L}_m = \frac{c}{M_{Pl}^2} \int d^4\theta \; \Phi\Phi^\dagger \chi\chi^\dagger \,, \tag{16.86}$$

where χ is the inflaton superfield and Φ some chiral superfield. By definition, in F-term inflation $\langle V \rangle = \langle \int d^4\theta \chi\chi^\dagger \rangle = 3H^2 M_{Pl}^2$, so that the operator of (16.86) generates[38] a scalar mass $m_\phi^2 = -3cM_{Pl}^2$. Note that this contribution might be negative if $c > 0$. An A-term of $\mathcal{O}(H)$ can also be generated by a similar operator that is linear in χ, but such terms might be forbidden by some symmetry under which χ transforms nontrivially. The complete potential of the AD field can thus be written as

$$V(\phi_{AD}) = \left(m_\phi^2 - c_1 H^2\right) |\phi_{AD}|^2 + \left[\frac{(A_\phi - c_2 H)\phi_{AD}^d}{M^{d-3}} + h.c.\right] + \frac{\lambda^2 |\phi_{AD}|^{2d-2}}{M^{2d-6}} \,, \tag{16.87}$$

where m_ϕ and A_ϕ are typical soft supersymmetry breaking parameters $\lesssim 1$ TeV, and $M \lesssim M_{Pl}$ is some "new physics" scale.

During inflation, the Hubble parameter is essentially constant at the value H_I, say. One can then show [16.39] that there is enough time for $|\phi_{AD}|$ to approach the minimum of the potential,

$$\langle |\phi_{AD}| \rangle \simeq |\phi_0| \sim \left(\frac{H_I M^{d-3}}{\lambda}\right)^{1/(d-2)} \,, \tag{16.88}$$

where c_1 has been taken to be positive and $\mathcal{O}(1)$. If $|c_2| \sim \mathcal{O}(1)$ as well, the phase of ϕ_{AD} will be fixed by the phase of c_2; otherwise it will be arbitrary, i.e. can be expected to be $\mathcal{O}(1)$. Either way, at the end of inflation, it will be essentially constant over the observable Universe. Once inflation ends, the inflaton field will start to oscillate coherently around the minimum of its potential, thereby transforming the vacuum energy into a Bose condensate of physical inflaton particles. Note that the average kinetic energy of the inflaton field also contributes to (16.86), and hence to the effective mass of the AD field, which is now time dependent. One then expects the AD field to basically follow its instantaneous minimum [16.39], which is still given by (16.88) with H_I replaced by $H(t)$. Most of the energy of the AD field is thus still stored as vacuum energy. By assumption, the latter must be smaller than the energy stored in inflaton particles, as further discussed later.

Only when $H \sim m_\phi$, will the AD field start oscillating around its current minimum, i.e. $\langle \phi_{AD} \rangle_{now} = 0$, and begin to form a condensate of physical scalar fields (squarks, sleptons, and/or

---

[38] An analogous argument would also create a mass $\mathcal{O}(H)$ for the inflaton itself, thereby spoiling the required flatness of the inflaton potential. This is one of the most serious current problems of F-term inflation. One has to find mechanisms to make the coefficient of this contribution much less than unity [16.40].

Higgs fields, depending on the flat direction). Up to this time the three terms in (16.87) have contributed roughly equally. The second (A-)term violates B or $B - L$ badly, since by assumption the AD field carries nonvanishing B or $B - L$. The first Sakharov condition is thus satisfied. One also has to assume a nonvanishing phase between A_ϕ and c_2 (or, if $|c_2| \ll 1$, between A_ϕ and the random phase of ϕ_{AD}) at the end of inflation. This is the source of CP violation required by the second Sakharov condition. However, once $H < m_\phi$, the B or $B - L$ violating interactions decouple, and the baryon number stored in the AD field, now a Bose condensate of physical particles, remains essentially constant. The third Sakharov condition is also satisfied since the transition from an essentially constant $\langle \phi_{AD} \rangle$ to a Bose condensate of physical particles is a nonequilibrium process.

If the phase difference between A_ϕ and c_2 (or A_ϕ and the random initial phase of ϕ_{AD}) is $\mathcal{O}(1)$, the net baryon charge density of the ϕ_{AD} condensate would be of the same order as the total ϕ_{AD} number density n_ϕ [16.39]. The produced baryon to entropy ratio after reheating (i.e., after the decay of the inflaton particles), will then be given by

$$\frac{n_B}{s} \sim \frac{n_B T_R}{\rho_{\text{tot}}} = \frac{n_B}{n_\phi} \frac{T_R}{m_\phi} \frac{\rho_\phi}{\rho_{\text{tot}}} \, , \tag{16.89}$$

where we have used $\rho_{\text{tot}} \sim T_R s$ in the radiation dominated era after most inflatons have decayed, and ρ_ϕ is $n_\phi m_\phi$. We just argued that n_B/n_ϕ should be $\mathcal{O}(1)$. Since $m_\phi \lesssim 1$ TeV and T_R could be as large as 10^8 GeV without overproducing gravitinos, the second factor could be as large as 10^6 or so. The third factor has to be much less than unity by definition, since the decay products of the inflaton are supposed to dominate the total energy density during and just after inflation; this quantity is not changed by the decay of the inflaton. Using (16.87) and (16.88) at $H \simeq m_\phi$ (i.e. $\rho_{\text{tot}} \simeq 3 m_\phi^2 M_{Pl}^2$), we have

$$\frac{\rho_\phi}{\rho_{\text{tot}}} \sim \left(\frac{m_\phi M^{d-3}}{\lambda M_{Pl}^{d-2}} \right)^{2/(d-2)} . \tag{16.90}$$

If $M \simeq M_{Pl}$, the ratio of (16.89) increases with increasing d. Recall that d is the mass dimension of the lowest dimensional operator generating the nonzero F-term along the flat direction. For $d = 4$, $n_B/s \sim 10^{-10}$ requires $T_R \sim \lambda \times 10^8$ GeV, not far below the upper bound from gravitino production (unless $\lambda \ll 1$). However, for $d = 6$, this reduces to $T_R \sim \sqrt{\lambda} \times 1$ GeV, taking $\sqrt{m_\phi M_{Pl}} \sim 10^{10}$ GeV. In other words, for $d \geq 5$, the Affleck-Dine mechanism produces *too large* a baryon asymmetry, unless the phase of the relevant A-parameter is very small and/or the reheating temperature is quite low[39].

This mechanism can only work if the AD condensate is allowed to oscillate *coherently* around its minimum. Otherwise, the phase of this field will differ for different parts of the Universe, leading to a net vanishing baryon number (since the third Sakharov condition would be violated). This requires that the AD condensate decays before most inflatons do. The latter is weaker than the gravitino production limit, since inflaton decay at $H \gtrsim m_\phi$ would mean $T_R \sim \sqrt{H M_{Pl}} \gtrsim 10^{10}$ GeV. However, some inflaton decay will start as soon as physical inflaton particles are produced at the end of inflation. Even though the universe continues to be matter dominated until most inflatons have decayed, it also contains a hot plasma with temperature $T \sim \left(T_R^2 H M_{Pl} \right)^{1/4}$ [16.1]. On the other hand, any particle that couples to the AD field, and thus could potentially thermalize it, also acquires a mass if $\langle \phi_{AD} \rangle \neq 0$. The AD condensate will thus be protected from thermalization if for

---

[39]Another possibility is to reduce the baryon to entropy ratio later on, by introducing another source of entropy production (besides the decay of the inflaton). This may be advantageous also for other reasons; e.g., it would alleviate the "moduli problem" of cosmological models based on superstring theory [16.41].

all $H \gtrsim m_\phi$, $g\langle\phi_{AD}(H)\rangle > T(H)$, where g is the relevant coupling constant, and $\langle\phi_{AD}(H)\rangle$ is given by (16.88). This is easily satisfied for $d \geq 5$ if T_R is chosen to give $n_B/s \sim 10^{-10}$, but reduces the upper bound on T_R to $\sim 10^6$ GeV for $d = 4$ [16.39].

A particularly interesting set of flat directions, lifted by $d = 4$ operators, are those characterized by $L_i \cdot H_2 L_i \cdot H_2$ (with $\sum_i |\tilde{l}_{L,i}|^2 = |h_2|^2$). It is lifted by the superpotential operator $\lambda_{ij} L_i \cdot H_2 H_2 \cdot L_j/M$, which is nothing but the light neutrino mass term (16.84). Recall that the net baryon number scales like $1/\lambda$ for $d = 4$, cf. (16.88), (16.89). In the given case it is thus related to the mass of the *lightest* neutrino [16.39] ν_1:

$$\left.\frac{n_B}{s}\right|_{LH_2} \sim 10^{-10} \frac{T_R}{10^6 \text{ GeV}} \frac{10^{-8} \text{ eV}}{m_{\nu_1}} \,. \tag{16.91}$$

Thus the AD mechanism along some $L{\cdot}H_2$ flat direction requires a very small mass for ν_1, leading to a hierarchical neutrino mass pattern (since the atmospheric neutrino anomaly seems to require $m_{\nu_3} \gtrsim 0.03$ eV). This particular flat direction is attractive since the required negative $\mathcal{O}(H^2)$ contribution to the squared mass of the AD field can be explained dynamically. Even if this contribution is positive at large scales $\phi_{AD} \sim M_{Pl}$, it can become negative at $\phi_{AD} \sim \sqrt{m_\phi M_{Pl}}$ due to radiative corrections involving the Yukawa coupling of the top quark, in complete analogy with the mechanism of radiative gauge symmetry breaking described in §11.4. Note that this term will indeed by positive at Planckian energy scales if the Kähler metric (12.14) of ϕ_{AD} is flat, i.e. if ϕ_{AD} has a minimal kinetic energy term in the supergravity Lagrangian [16.39].

<u>D term inflation</u>

The AD mechanism proceeds slightly differently here [16.41]. To begin with, in this case by definition $F_\chi = 0$ and $\langle\chi\rangle \sim$ const. during inflation. Therefore, no large $\mathcal{O}(H^2)$ contribution to the squared mass of the AD field gets generated from the operator in (16.86) during inflation. Recall the discussion at the end of the previous subsection of the case of $SU(2)_L$ singlet sneutrinos which provide a very simple example of a flat direction, described by a single field. As in that case, a large and effectively constant $\langle\phi_{AD}\rangle$ will then be generated by quantum fluctuations. Once the inflaton starts to oscillate around its minimum after inflation, large contributions to the operator in (16.86) do appear. On the other hand, the magnitude of the A-term remains much smaller than that of H. Hence the phase of ϕ_{AD} will keep its random, generally $\mathcal{O}(1)$, value acquired during inflation. The AD potential is then given by (16.87) with $|c_1| \sim \mathcal{O}(1)$ and $|c_2| \ll 1$.

Another modification arises because the inflaton sector in D-term inflation contains several fields. A simple example [16.41] is based on the superpotential $\mathcal{W}_I = \lambda\Psi_+\Psi_-\chi$. Here the chiral fields Ψ_+ and Ψ_- carry charges of a $U(1)$ gauge group that supports a large Fayet–Illiopoulos term ξ, while the superfield χ of the "inflaton proper" is neutral under that group. If $\langle\chi\rangle$ is sufficiently large, $\langle\psi_+\rangle = \langle\psi_-\rangle = 0$ and $\langle V\rangle = g^2\xi^2/2$, g being the $U(1)$ gauge coupling. The absolute minimum of the potential is at $\langle\chi\rangle = \langle\psi_+\rangle = 0$ and $\langle|\psi_-|^2\rangle = \xi^2$. Therefore, at the end of inflation, *both* χ and ψ_- start to oscillate. The lifetimes of the particles corresponding to these fields are often quite different. In particular, ψ_- is expected [16.41] to decay quite early. Even though the Universe remains matter dominated (by nonrelativistic χ particles), the plasma that results from ψ_- decay is too hot to allow AD baryogenesis along flat directions lifted by $d = 4$ operators. Directions only lifted at $d \geq 5$ still work [16.41].

Let us remember that a larger d corresponds to a lower T_R if n_B/s is kept fixed. In particular, for $d \geq 6$, $T_R \lesssim 1$ GeV is too low for neutralinos to have been in thermal equilibrium after inflation[40],

---

[40]This is more exactly true after the decay of most inflatons; later, the entropy density per comoving volume remained essentially constant.

thereby excluding thermal LSPs from being a candidate dark matter. However, the low reheating temperature can lead to a different athermal source of relic neutralinos [16.42]. We have assumed so far that the AD condensate releases its energy into free (s)particles (squarks, sleptons and/or Higgs bosons). However, it is quite possible that some of these (s)particles instead get trapped in charged solitons called Q-balls [16.27]. These will be at least metastable if the ratio $V(\phi_{AD})/|\phi_{AD}|^2$ has a minimum away from the origin. Such is often the case if the running mass of ϕ_{AD} decreases with scale, e.g. for the $\overline{U_i D_j D_k}$ flat directions but not for the $L_i H_2$ flat directions, as noted above. The requirement that the Q-balls do not thermalize (in which case essentially the standard AD mechanism would result) leads to $T_R \lesssim 10^4$ GeV [16.42], which is easily satisfied for $d \geq 6$. In models with gauge mediated supersymmetry breaking (cf. Ch.13), the Q-balls can be absolutely stable. This is since there the soft breaking mass of ϕ_{AD} vanishes for values of ϕ_{AD} that exceed the messenger scale. In this case the Q-balls could themselves be the dark matter. Their experimental signatures are similar to those of heavy monopoles, i.e. they are relatively easy to detect [16.43].

In models with gravity mediated supersymmetry breaking (cf. Ch.12), the Q-balls will eventually decay, typically well after inflatons do. For $\overline{U_i D_j D_k}$ Q-balls, one unit of baryon number will release three LSPs when the squarks in the Q-ball finally decay. By then the temperature has reached $T \sim 100$ MeV or so, i.e. it is well below the LSP freeze-out temperature, cf. §16.3. Thus the baryon to LSP ratio will remain constant afterwards:

$$n_{LSP} = 3x_b n_B \ , \tag{16.92}$$

x_b being the fraction of AD particles that get trapped in Q-balls[41]. The authors of Ref. [16.42] estimate that $x_b \gtrsim 0.1$, though a full dynamical calculation is still lacking. For example, the value $x_b = 0.1$ would give $\Omega_{\rm LSP} \sim 1$ for $\Omega_b \sim 0.05$ and $m_{\rm LSP} \sim 60$ GeV, the last being a very reasonable number for the mass of the lightest neutralino. Q-ball baryogenesis thus leads to a (still approximate) prediction for the LSP relic number density which is testable in principle.

In contrast to the above, experimentally testable predictions cannot be derived from the postulate of successful AD baryogenesis except for the $L_i \cdot H_2$ flat directions, cf. (16.91). The specific problem is that the new CP violating phase(s) required (as in all known mechanisms of baryogenesis) are associated here with nonrenormalizable operators, e.g. the A-term in the potential (16.87). The influence of these on lower laboratory energy physics is in most cases negligible (the $L_i \cdot H_2$ directions, which are associated with neutrino masses, again being a possible exception). In this regard, the situation is quite different for electroweak baryogenesis to which we turn next.

Electroweak Baryogenesis

As mentioned at the beginning of this Section, the three Sakharov conditions are satisfied even in the Standard Model during the electroweak phase transition. To recapitulate, the first two conditions obviously hold: charge conjugation symmetry is violated maximally by $SU(2)_L$ gauge interactions, while CP is violated by the CKM phase in the Yukawa couplings and the 't Hooft anomaly, enhanced at a finite temperature by sphaleron processes, leads to baryon nonconservation. In order to understand how the third Sakharov condition of nonequilibration is satisfied, one needs to study the behavior of the Higgs potential at a finite temperature. Focusing on the combination

---

[41]The decay of free AD particles leads to "hot" LSPs which will thermalize. Their contribution to the LSP relic density of today will be greatly diluted by the entropy released during inflaton decay. As noted earlier, the latter occurs well after the decay of the free AD fields. The total relic LSP density is therefore essentially proportional to x_b. In contrast, the final baryon to entropy ratio is always given by (16.89), (16.90), independent of x_b.

ϕ of Higgs fields that is responsible for symmetry breaking, e.g., $\phi = \cos\beta\,\Re eh_1^0 + \sin\beta\,\Re eh_2^0$ in the MSSM, this effective potential can be schematically written as [16.44]

$$V_{\text{eff}}(\phi, T) \simeq \frac{\gamma T^2 - m^2}{2}\phi^2 - ET\phi^3 + \frac{\lambda}{2}\phi^4 \, , \tag{16.93}$$

γ and E being positive dimensionless functions of the parameters of the theory. One has $\gamma > E^2/\lambda$ in realistic models. In this case the effective potential is positive semidefinite for a temperature $T \geq T_c \equiv m/\sqrt{\gamma - E^2/\lambda}$. This is despite the potential having a (relative) minimum away from the origin for some range of temperatures $T > T_c$. At $T = T_c$ this second minimum occurs at

$$\langle \phi(T_c) \rangle \equiv \phi_c = \frac{ET_c}{\lambda} \, , \tag{16.94}$$

satisfying $V(\phi_c, T_c) = 0$. For $T < T_c$ this second minimum is dynamically favored, i.e. occurs at negative values of the effective potential. A positive potential barrier between $\phi = 0$ and the absolute minimum persists nonetheless for some temperatures $T < T_c$. This behavior of the effective potential is illustrated in Fig. 16.5. The existence of a potential barrier means that the Universe

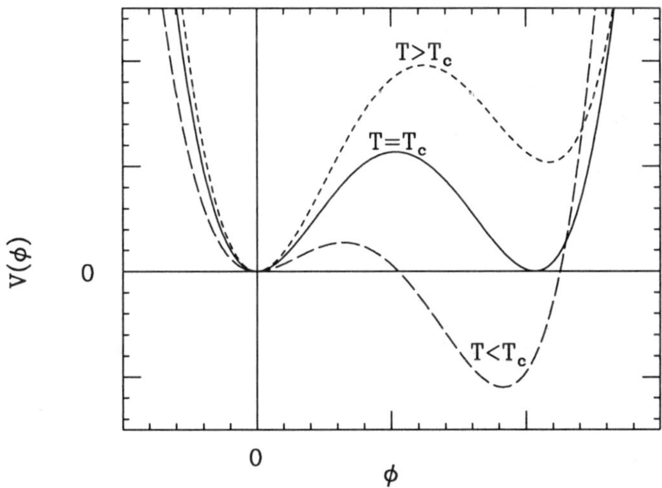

Fig. 16.5. Schematic behavior of the effective potential (16.93) for three different temperatures slightly above, at, and slightly below the critical temperature T_c.

has to tunnel from the vacuum in the unbroken phase ($\langle\phi\rangle = 0$) where the $SU(2)_L \times U(1)_Y$ symmetry is exact to the vacuum in the broken phase ($\langle\phi\rangle \neq 0$) where this symmetry is broken spontaneously. This tunnelling process, which will be described in more detail below, is irreversible, i.e. occurs out of thermal equilibrium.

However, even if a baryon asymmetry is created during the phase transition, it could still be washed out just afterwards in case electroweak sphaleron processes were still in thermal equilibrium in the broken phase. The height of the sphaleron barrier is proportional to the VEV of the Higgs field. The higher this barrier, the more strongly suppressed are the sphaleron processes. Detailed numerical analyses [16.27] show that erasure of a previously created baryon asymmetry can only be avoided if

$$\phi_c \geq T_c \ .$$

(16.95)

If the condition (16.95) is satisfied, the electroweak phase transition is said to have been "sufficiently strongly first order", since the tunneling from $\langle\phi\rangle = 0$ to $\langle\phi\rangle \simeq \phi_c$ entails a large discontinuity in the order parameter $\langle\phi\rangle$.

It is clear from (16.94) that the condition (16.95) is more easily satisfied for a small Higgs self coupling λ, i.e. a small Higgs mass, and a large coefficient E of the trilinear term of the effective potential (16.93). This coefficient receives contributions only from bosons coupling to the Higgs field. The combination of the lower bound on λ from Higgs searches at LEP, and the absence of bosons with strong couplings to the Higgs boson, means that (16.95) is badly violated in the SM[42]. Therefore, baryogenesis during the electroweak phase transition is just not possible in the SM [16.27].

The situation is somewhat more advantageous in the MSSM, mostly due to the existence of additional scalar fields. Stops are of the greatest importance here, owing to their strong coupling to the Higgs field h_2. They contribute terms proportional to $Tm_{\tilde{t}_i}^3(T,\phi)$ to the effective potential. Here

$$m_{\tilde{t}_i}^2(T,\phi) = \widetilde{m}_i^2 + c_{i,1}\phi^2 + c_{i,2}T^2$$

(16.96)

is the field dependent and temperature dependent mass of $\tilde{t}_i$ ($i = 1, 2$). Their contribution to the effective potential reduces to the form $E\phi^3 T$ shown in the simplified expression (16.93) only if $\widetilde{m}_i^2 + c_{i,2}T^2 \ll c_{i,1}\phi^2$. Since we are interested in temperatures $T \sim T_c \sim \mathcal{O}(100)$ GeV, only light stops contribute significantly. In fact, the soft supersymmetry breaking term $\widetilde{m}_i^2$ should be negative in order to make the (effective) coefficient E large enough, given the current experimental lower bounds on MSSM Higgs boson masses. On the other hand, the squared soft supersymmetry breaking masses of left handed stops (and sbottoms) must be positive and fairly large in order to satisfy the constraints from the precisely measured electroweak ρ parameter. If a light stop is to increase the strength of the electroweak phase transition, it must therefore be an almost pure $SU(2)_L$ singlet state, $\tilde{t}_1 \simeq \tilde{t}_R$. The contributions to its mass appearing in (16.96) are then given [16.45], ignoring electroweak gauge interactions, by

$$\widetilde{m}_1^2 \simeq m_{\tilde{t}}^2 \ ,$$

(16.97a)

$$c_{1,1} \simeq f_t^2 \left(1 - \frac{|A^t + \mu^* \cot\beta|^2}{m_{q_3}^2}\right) \ ,$$

(16.97b)

$$c_{1,2} \simeq \frac{4}{9}g_3^2 + \frac{f_t^2}{6}\left(2 - \frac{|A^t + \mu^* \cot\beta|^2}{m_{q_3}^2}\right) \ ,$$

(16.97c)

where f_t is the top Yukawa coupling strength, and $m_{\tilde{t}}^2$, $m_{q_3}^2$ and A^t are soft supersymmetry breaking parameters, all introduced in §9.4. Notice, in particular, that one needs not only that $m_{\tilde{t}}^2 < 0$, but also a small value of $|A^t + \mu^* \cot\beta|$, the latter requirement following from the proportionality between E and $c_{1,1}^{3/2}$.

The above conditions are not easy to satisfy in "high scale" models of supersymmetry breaking discussed in Chs. 12 and 13. In particular, a negative $m_{\tilde{t}}^2$ cannot be realized in models with GMSB. Both constraints can be satisfied in mSUGRA and in a narrow sliver of parameter space

---

[42]Indeed, for experimentally allowed Higgs masses, there is no proper electroweak phase transition at all in the SM, only a continuous crossover from the unbroken to the broken phase.

one can also simultaneously have the LSP as a good cold dark matter candidate [16.44]. However, this comes at the price of a very large trilinear GUT-scale parameter $|A_0|$. The latter generally leads to a breakdown of charge and/or color symmetry at the absolute minimum of the scalar potential. A more natural implementation of electroweak baryogenesis is thus in models without squark degeneracy at the scale where supersymmetry breaking is transmitted from the hidden to the observable sector. It is therefore preferable to analyze the constraints on the spectrum in a model independent way, in terms of weak scale parameters only. The numerical analysis of these constraints, implied by the condition (16.95), is quite involved. After including two loop contributions to the effective potential, which tend to strengthen the phase transition, one is obliged to have [16.27]

$$m_h \lesssim 105 \text{ GeV} , \tag{16.98a}$$

$$m_{\tilde{t}_1} \lesssim m_t , \tag{16.98b}$$

if $m_{\tilde{t}_2} \lesssim 1$ TeV. All of the range (16.98a) has been excluded by Higgs searches at LEP, forcing one to consider a soft supersymmetry breaking $\tilde{t}_2$ mass of 2 TeV or higher [16.46]. Even then $m_h \leq 117$ GeV is required. Stop searches at future runs of the TEVATRON will cover a large part of the region (16.98b) which is not sensitive to the value of $m_{\tilde{t}_2}$. It is fair to say that the MSSM gets extremely constrained if it tries to incorporate electroweak baryogenesis.

We have so far discussed the methods of avoiding the erasure of a primordial baryon asymmetry just after the electroweak phase transition. In order to understand how this asymmetry can be produced in the first place, we have to describe the phase transition in somewhat more detail. As mentioned earlier, this transition occurs when the Universe tunnels from $\langle \phi \rangle = 0$ to $\langle \phi \rangle \simeq \phi_c$ at a temperature slightly below T_c. This does not happen everywhere at once. Rather, in the way water starts boiling, bubbles of the broken phase form spontaneously and start to grow, eventually filling the entire Universe. Baryon number violating processes remain in equilibrium outside the bubbles even if the condition (16.95) is satisfied, since the latter only ensures that they are out of equilibrium inside the bubbles.

Baryogenesis can now occur if some charge can be injected into the unbroken phase for conversion into baryon number by sphaleron transitions. This injection in turn is due to an asymmetry between reflection from and transmission through the bubble wall, which can be created by CP-violating interactions between the wall and some MSSM fields[43]. It turns out that the potentially largest contributions come from charginos and neutralinos[44]. CP-violating interactions, caused by the relative phase between the gaugino mass parameters M_1, M_2 and the higgsino mass parameter μ, lead to the buildup of a chiral charge (an asymmetry between lefthanded and righthanded fermion fields) in the unbroken phase. This charge is then partly converted into baryon number by sphaleron transitions, the process occurring in the region just in front of the moving bubble walls. Once this region is swallowed up by the expanding bubbles, baryon number violating interactions freeze out. One is then left with a net baryon number, as well as a net chiral charge (opposite in sign to that created outside the bubbles). However, this chiral charge does not survive the decay of the charginos and neutralinos.

The details of the above process are quite complicated [16.27]. Various estimates of the final baryon asymmetry for a given set of input parameters differ by almost one order of magnitude.

---

[43]CP-violating and baryon number violating processes occur in distinct regions of space here, in the bubble walls and in the unbroken phase respectively. This is called "nonlocal baryogenesis".

[44]Stops cannot contribute sufficiently [16.45], partly due to the short coherence length of stop currents caused by their strong interactions with the surrounding plasma and partly because the requirement of small off-diagonal entries of the stop mass matrix, cf. (16.97b), suppresses CP violation in the stop sector.

There is agreement that a larger asymmetry tends to arise if $|M_2| \simeq |\mu|$. This is not surprising, since the phases of *both* M_2 and μ could be rotated away if higgsinos and gauginos did not mix. The choice $|M_2| \simeq |\mu|$ maximizes this mixing and hence the effective CP violation in the chargino and neutralino sectors. A conservative set of requirements for electroweak baryogenesis to work in the MSSM, in addition to (16.98), is [16.27]:

$$|M_2|, \ |\mu| \ \lesssim \ 100 \text{ GeV} , \qquad (16.99\text{a})$$

$$|\sin(\psi_\mu + \psi_{M_2})| \ \gtrsim \ 10^{-3} . \qquad (16.99\text{b})$$

Eq. (16.99a) simply means that the density of charginos and neutralinos is not suppressed by the Boltzmann factor at $T \simeq T_c$. This condition can be mildly violated if the second condition (16.99b) on the CP violating phase difference is "overfulfilled", i.e. a larger phase difference allows somewhat heavier charginos and neutralinos. The entire allowed region can nonetheless be easily covered by chargino and neutralino searches at a lepton collider operating at $\sqrt{s} \simeq 500$ GeV. Experiments at such a collider would then also have a good chance to detect CP violation in the production and decay of sparticles [16.47].

References

[16.1] E.W. Kolb and M.S. Turner, *op. cit.*, *Bibl.*

[16.2] For a review, see S. Sarkar, Rept. Prog. Phys. **59** (1996) 1493.

[16.3] See e.g. S. Weinberg #1, *op. cit.*, *Bibl.*

[16.4] D.N. Spergel et al.. WMAP collaboration, *loc. cit.* [14.68].

[16.5] A.D. Linde, Phys. Rev. **D59** (1999) 023503.

[16.6] E.W. Kolb, A.D. Linde and A. Riotto, Phys. Rev. Lett. **77** (1996) 4290.

[16.7] R.H. Brandenberger, in *Field Theoretical Methods in Fundamental Physics*, ed. C.K. Lee, Min Eun Sa, Seoul (1997).

[16.8] J. Bernstein, L.S. Brown and G. Feinberg, Phys. Rev. **D32** (1988) 3261.

[16.9] J. Edsjö and P. Gondolo, Phys. Rev. **D56** (1997) 1879.

[16.10] M. Srednicki, R. Watkins and K. Olive, Nucl. Phys. **B310** (1988) 693.

[16.11] K. Griest and D. Seckel, Phys. Rev. **D43** (1991) 3191.

[16.12] J.E. Kim, Phys. Rev. Lett. **67** (1991) 3465.

[16.13] Particle Data Group, *loc. cit.* , *Bibl.*.

[16.14] E.I. Gates, G. Gyuk, G.P. Holder and M.S. Turner, Astrophys. J **500L** (1998) 145.

[16.15] S.J. Perlmutter et al., Nature **391** (1998) 51; Ap. J. **517** (1999) 565.

[16.16] T. Falk, K. Olive and M. Srednicki, Phys. Lett. **B339** (1994) 248.

[16.17] G. Jungman, M. Kamionkowski and K. Griest, Phys. *loc. cit.*, *Bibl.* J.R. Primack, D. Seckel and B. Sadoulet, *loc. cit.*, *Bibl.* J. Rich, D.M. Owen and M. Spiro, *loc. cit.*, *Bibl.*

[16.18] J.R. Ellis, T. Falk and K.A. Olive, Phys. Lett. **B444** (1998) 367.

[16.19] S. Mizuta, D. Ng and M. Yamaguchi, Phys. Lett. **B300** (1993) 96; C.–H. Chen, M. Drees and J.F. Gunion, Phys. Rev. **D55** (1997) 330.

[16.20] S.A. Abel, S. Sarkar and I.B. Whittingham, Nucl. Phys. **B392** (1993) 83; A. Stephan, Phys. Rev. **D58** (1998) 035011.

[16.21] F.M. Borzumati, M. Drees and M.M. Nojiri, Phys. Rev. **D51** (1995) 341.

[16.22] L. Bergström, T. Damour, J. Edsjö, L.M. Krauss and P. Ullio, JHEP **9908** (1999) 010.

[16.23] L. Bergström, J. Edsjö and P. Gondolo, Phys. Rev. **D58** (1998) 103519.

[16.24] T. Gherghetta, Nucl. Phys. **B485** (1997) 25.

[16.25] T. Moroi, H. Murayama and M. Yamaguchi, Phys. Lett. **B303** (1993) 289.

[16.26] M. Kawasaki and T. Moroi, Prog. Theor. Phys. **93** (1995) 879.

[16.27] A. Riotto and M. Trodden, *loc. cit.*, *Bibl.*

[16.28] H.K. Dreiner and G.G. Ross, Nucl. Phys. **B410** (1993) 188.

[16.29] A.D. Linde, Rep. Prog. Phys. **42** (1979) 389.

[16.30] R. Allahverdi, M. Bastero-Gil and A. Mazumdar, Phys. Rev. **D64** (2001) 023516

[16.31] M. Flanz, E.A. Paschos and U. Sarkar, Phys. Lett. **B345** (1995) 248; errtm. **B382** (1996) 447.

[16.32] M. Plümacher, Nucl. Phys. **B530** (1998) 207.

[16.33] M. Flanz, E.A. Paschos, U. Sarkar and J. Weiss, Phys. Lett. **B389** (1996) 693. Also, for a review, see A. Pilaftsis, Int. J. Mod. Phys. **A14** (1999) 1811.

[16.34] J. Ellis, S. Lola and D.V. Nanopoulos, Phys. Lett. **B452** (1999) 87; W. Buchmüller and T. Yanagida, Phys. Lett. **B445** (1999) 399.

[16.35] M. Bolz, W. Buchmüller and M. Plümacher, Phys. Lett. **B443** (1998) 209.

[16.36] G. Lazarides, R.K. Schaefer and Q. Shafi, Phys. Rev. **D56** (1997) 1324.

[16.37] H. Murayama and T. Yanagida, Phys. Lett. **B322** (1994) 349.

[16.38] T. Ghergetta, C. Kolda and S.P. Martin, Nucl. Phys. **B468** (1996) 37.

[16.39] M. Dine, L. Randall and S. Thomas, Nucl. Phys. **B458** (1996) 291.

[16.40] For a review, see D. Lyth and A. Riotto, Phys. Rep. **314** (1999) 1.

[16.41] C. Kolda and J. March-Russel, Phys. Rev. **D60** (1999) 023504.

[16.42] K. Enqvist and J. McDonald, Nucl. Phys. **B538** (1999) 321.

[16.43] A. Kusenko, M. Shaposhnikov, P.G. Tinyakov and I.G. Tkachev, Phys. Rev. Lett. **80** (1998) 3185.

[16.44] S. Davidson, T. Falk and M. Losada, Phys. Lett. **B463** (1999) 214.

[16.45] M. Carena, M. Quirós, A. Riotto, I. Vilja and C.E.M. Wagner, Nucl. Phys. **B503** (1997) 387.

[16.46] M. Quiros, Nucl. Phys. Proc. Suppl. **101** (2001) 401.

[16.47] S.Y. Choi, H.S. Song and W.Y. Song, Phys. Rev. **D61** (2000) 075004.

Chapter 17

CONCLUSION: WISH LIST, ROADMAP AND FINE TUNING

We have, in this book, discussed many aspects and facets of softly broken $N=1$ supersymmetry expected to be realized in high energy physics not far above the weak scale. The minimal theoretical scheme MSSM has been shown to have four important virtues. (1) It contains a radiatively stable weak scale with a natural solution to the gauge hierarchy problem[1]. (2) It decouples[2] efficiently at low energies, consistent with precision tests of the SM. (3) It incorporates the unification of gauge coupling strengths at a high scale. (4) It predicts a suitable WIMP candidate for cold dark matter. It also predicts the occurrence of sparticles with spectacular signatures like large $\not{E}_T$, usually accompanied by leptons and jets from cascades, in forthcoming laboratory experiments. The most urgent task is to discover some of them. Of course, once sparticles are discovered, there would be other tests. The verification, cf. §15.3, of the equality between certain vertex coupling strengths (such as that between the triple gauge and the gauge-gaugino-gaugino vertices), predicted to be the same by supersymmetry, would be an important confirmatory requirement.

As mentioned earlier, the specific manifestation of supersymmetry in experimental data will depend sensitively on the precise mechanism of supersymmetry breaking. Only soft (mass dimension < 4) supersymmetry breaking terms are allowed in the weak scale effective Lagrangian by the required taming (cf. §1.3) of the quadratic divergence in the radiative correction to the SM Higgs mass. These are characterized by the mass scale M_s. But they are presumed to arise from some sort of spontaneous (dynamical) breakdown of supersymmetry at a high scale Λ_s that is transmitted by specified messengers to our low energy world. It was discussed in Chs. 12 and 13 how different phenomenologically viable supersymmetry models are distinguished not only by the mechanism of this transmission but also by the values of Λ_s and the messenger mass scale[3] $M_M \sim \Lambda_s^2/M_s$. Various theoretical and experimental considerations restrict these two scales to the range

$$10 \text{ TeV} < \Lambda_s \leq M_M \leq M_{Pl} \ .$$

---

[1] Admittedly, it does not explain the hierarchy itself. But the idea of a radiative EW symmetry breakdown, cf. §11.4, does suggest an exponential relation between the electroweak and the high unifying scales.

[2] The quantities, which need to tend to infinity in the decoupling limit, are the set of soft supersymmetry breaking parameters, characterized by M_s, as well as the CP odd Higgs mass m_A and the higgsino mass parameter μ respectively setting the scales in the Higgs and higgsino sectors. They would not all be independent owing to constraints required by the mechanism of supersymmetry breaking.

[3] Recall (cf. Ch.12) that in gravity mediated scenarios M_M approaches M_{Pl}.

501

Weak scale values of soft supersymmetry breaking parameters, controlling the sparticle masses[4], get determined by RGE from their high scale values which are fixed by the mechanism of supersymmetry breaking. Once one measures these low energy values of the parameters describing sparticle properties, one can in principle get a pointer to the high scale physics responsible for supersymmetry breaking. Our wish list of such measurements will be the following.

- Discovery of sparticles and determination of their quantum numbers.

- Quantitative verification of coupling equalities implied by supersymmetry.

- Measurement of the masses of scalars (including Higgs) as well as gauginos.

- Determination of the gaugino-higgsino mixing parameters.

- Study of the properties of third generation sfermions including L-R mixing.

Gravity mediated (including the mSUGRA, $\tilde{\text{C}}$MSSM and AMSB models) and gauge mediated (GMSB) supersymmetry breaking were discussed in Chs. 12 and 13 respectively. Their spectral characteristics are summarized in Table 17.1. In mSUGRA one expects sleptons to be lighter

Mediation mechanism	Model	Gravitino mass $m_{3/2}$	Gaugino mass M_α	Sfermion mass squared m_i^2
Gravity mediated	mSUGRA	$\leq \mathcal{O}\,(\text{TeV})$	$(g_\alpha/g_2)^2 M_2$	$m_0^2 + G_i M_{1/2}^2 + \text{D-terms}$
	$\tilde{\text{C}}$MSSM	$\leq \mathcal{O}\,(\text{TeV})$	$(g_\alpha/g_2)^2 M_2$	$\tilde{q}\ell:\, m_{\tilde{q}_L}^2 + G_{\tilde{q}_L} M_{1/2}^2 + \text{D-terms}$ $\tilde{l}_L:\, m_{\tilde{\ell}}^2 + G_{\tilde{\ell}} M_{1/2}^2 + \text{D-terms}$ $\tilde{e}_R:\, m_{\tilde{e}_R}^2 + G_{\tilde{e}_R} M_{1/2}^2 + \text{D-terms}$ $\tilde{u}_R:\, m_{\tilde{u}_R}^2 + G_{\tilde{u}_R} M_{1/2}^2 + \text{D-terms}$ $\tilde{d}_R:\, m_{\tilde{d}_R}^2 + G_{\tilde{d}_R} M_{1/2}^2 + \text{D-terms}$
	AMSB	20–100 TeV	$(g_\alpha b_\alpha/g_2 b_2)^2 M_2$	$m_0^2 + C_i (16\pi^2)^{-2} m_{3/2}^2$
Gauge mediated	mGMSB	10^{-5} eV – 1 keV	$(g_\alpha/g_2)^2 M_2$	$\tilde{q}_L:\, M_3^2\ (500\ \text{GeV})\ G'_{\tilde{q}_L} + \text{D-terms}$ $\tilde{l}_L:\, M_2^2\ (100\ \text{GeV})\ G'_{\tilde{\ell}_L} + \text{D-terms}$ $\tilde{e}_R:\, M_2^2\ (100\ \text{GeV})\ G'_{\tilde{e}_R} + \text{D-terms}$ $\tilde{u}_R:\, M_3^2\ (500\ \text{GeV})\ G'_{\tilde{u}_R} + \text{D-terms}$ $\tilde{d}_R:\, M_3^2\ (500\ \text{GeV})\ \tilde{G}'_{\tilde{d}_R} + \text{D-terms}$

Table 17.1. Mass spectra in different models of supersymmetry breaking. The coefficients G_i, C_i and G'_i are given in (12.29), Table 12.1 and (13.20) respectively while the other symbols are as defined earlier.

than squarks of the first two generations and right sleptons to be lighter than left sleptons, while the LSP is anticipated to be mostly a bino[5]. Whereas the last anticipation is natural in $\tilde{\text{C}}$MSSM, no basis exists there for the first two expectations. On the other hand, in AMSB the left and

---

[4]In cases where mixing occurs, even sparticle couplings can depend on the supersymmetry breaking mechanism.

[5]In a limited region with $m_0^2 \gg M_{1/2}^2$, the LSP can also have a sizable higgsino component.

right sleptons are expected to be nearly degenerate; so are the lightest chargino and the neutralino LSP – both being very nearly winos. The gravitino is the LSP in GMSB models and is expected to be very light while $\tilde{\chi}_1^0, \tilde{t}_1$ or $\tilde{\tau}_1$ could be the NLSP, depending on model parameters. Sparticle phenomenology will of course be drastically different from one model to another. Detailed analyses show that experiments at the LHC should be able to *distinguish between these high scale models*; they should also succeed in providing sufficient information to determine the few free parameters of these models by using some sort of a global fit. That is likely to be the roadmap of the routes to the true broken supersymmetric model which describes Nature at a high scale.

The above procedure would be rather similar to global fits of the SM, which became possible once LEP and SLC started operations. Less model dependent "bottom-up" analyses would need precise measurements of a rather large number of sparticle masses as inputs. Such a feat might only be possible by including results from an e^+e^- linear collider. For example, one would need both scalar and gaugino masses in order to extrapolate running sfermion masses to higher energy scales. Even then a true bottom-up analysis, with errors comparable to those of current EW precision fits, would be difficult to perform. The reason is that one would first need to extract the weak scale values of $\overline{\text{DR}}$ parameters from the physical sparticle mass spectrum. In the case of EW gauginos, the corresponding one loop correction factors, which need to be taken into account for percent level precision tests, would depend on essentially the entire sparticle and Higgs mass spectra. Similarly, the extraction of the values of the soft supersymmetry breaking parameters from measured cross sections and branching ratios would become very hard once radiative corrections are taken into account. This is since an experimentally measured quantity would then again depend on a great many parameters. A good deal more theoretical work would need to precede the carrying out of such analyses even if the required data did exist.

However, we are getting ahead of ourselves. Let us return to the central step required to establish supersymmetry, namely – discovering sparticles. Early expectations of the existence of sparticles with masses close to those of the weak gauge bosons have been belied, nor have their virtual effects on low energy observables been identified unambiguously so far. As lower bounds on their masses creep up, an inevitable question arises. If searches continue to yield negative results, at what stage would softly broken supersymmetry cease to be an attractive option as the stabilizer of the weak scale? This is the issue that we shall deal with in the rest of this concluding chapter. As already stated, a primary motivation for the supersymmetric extension of the Standard Model is that it stabilizes the hierarchy $M_W \ll M_U$ or M_{Pl} and renders it natural as compared with the nonsupersymmetric version. Setting the ratio $M_W M_{U/Pl}^{-1}$ to be a small number by hand does not work since radiative corrections shift it by a large amount. We have argued in Ch.1 that the quadratic divergence in the one loop Higgs self energy generates a term

$$\delta M_W^2 = \mathcal{O}\left(\frac{\alpha}{\pi}\right) \Lambda^2 \ . \tag{17.1}$$

In the RHS of (17.1) Λ is a cutoff scale representing new physics which could be the unification scale $M_U \sim 10^{16}$ GeV or the Planck scale $M_{Pl} \sim 10^{19}$ GeV making $\delta M_W^2 \gg M_W^2$. In order to avoid this, i.e. to keep the weak scale small, one needs to invoke strong cancellations. In general, this requires fine tuning the bare input parameters to the level of one in $M_{U/Pl}^2 M_W^{-2}$. Supersymmetry makes this cancellation natural by removing the quadratic divergence. If $m_{B,F}$ is the mass of the boson, fermion circulating in the Higgs self-energy loop, (17.1) is now replaced by

$$\delta M_W^2 = \mathcal{O}\left(\frac{\alpha}{\pi}\right) (m_B^2 - m_F^2) = \mathcal{O}\left(\frac{\alpha}{\pi}\right) M_s^2 \ , \tag{17.2}$$

where M_s characterizes the intra-supermultiplet mass difference between a particle and the corresponding sparticle. This was demonstrated in Ch.1 in a toy model, cf. (1.16). The fine tuning disease [17.1] is therefore cured [17.2] by supersymmetry.

The remedy mentioned above, however, works only upto a point. Evidently, M_s is the key dimensional parameter. We already have some idea about M_s from the current experimental lower bounds on sparticle masses. For instance, the electron-selectron mass difference is experimentally known to be greater than about 100 GeV. It is therefore quite likely that M_s exceeds 100 GeV by a significant amount and let us suppose that such is the case. The much touted absence of fine tuning in the supersymmetric stabilization of the weak scale M_W depends delicately on the relative values of M_s and M_W. If we keep increasing M_s, a point would be reached when the supersymmetric Standard Model would no longer be able to provide a natural explanation of the stability of the light weak scale. Where does this point lie? Present folklore is that one should not have M_s more than a few TeV, but it would be desirable to have a quantitative measure of fine tuning and to see how, in the present context, it relates to M_s and M_W.

A system is natural if its observable properties are stable against small variations of the fundamental parameters. In a system that lacks naturalness, say a nonsupersymmetric field theory with a scalar sector such as the SM, any agreement with the measured value of a typical quantity r requires these parameters to conspire by cancelling or adding in an unusually precise fashion. In such a case, we say that r has been fine tuned. The lack of naturalness is then reflected in the extremely strong dependence that r would have on the basic parameters. When we consider the same quantity r in a natural system, e.g. a supersymmetric extension of the previous field theory such as the MSSM, this kind of a strong dependence would be absent. Nevertheless, if supersymmetry is broken, r would depend on M_s through the supersymmetry breaking parameters. What we need to look for in the MSSM is the sensitivity in r with respect to variations in those parameters.

Suppose we write $r \equiv r(a_i)$, i.e. as a function of the most general parameters $\{a_i\}$ of the theory. The sensitivity of r with respect to variations in the input parameters can be measured in terms of a set of $\Delta_i(r, a_k)$, defined [17.3] via logarithmic derivatives as

$$\Delta_i(r, a_k) \equiv \left| \frac{a_i}{r} \frac{\partial r}{\partial a_i} \right| , \tag{17.3}$$

where i is not summed on the RHS. Rescaling the derivative of r with respect to a_i by $a_i r^{-1}$ removes the dependence on the overall scale of a_i and r. For any r, consider now the largest of the Δ_i's, which we call $\Delta(r)$, i.e.

$$\Delta(r) \equiv \max_i \Delta_i(r, a_k) , \tag{17.4}$$

as the measure of the fine tuning in r. If the theory is natural, $\Delta(r)$ should not be large.

Let us return to the gauge hierarchy problem in a nonsupersymmetric field theory with fundamental scalars, as discussed in §1.2. From (1.3), we can rewrite the one loop renormalized scalar mass squared in the notation of (1.5) as

$$m_S^2(\lambda_f) = \lambda_f^2 \Lambda^2 + m_{S,0}^2 . \tag{17.5}$$

For the effective theory, we have introduced an ultraviolet cut off Λ in (17.5). This cutoff has been defined by absorbing the other multiplicative factors. Furthermore, the bare scalar mass $m_{S,0}$ has been chosen so as to keep the renormalized scalar mass small. The fine tuning needed for this purpose is measured by the 'natural' choice $r = m_S^2$ and $a_i = \lambda_f$ in (17.1). By using (17.4), we then have

$$\Delta(m_S^2, \lambda_f) = \left| \frac{\lambda_f}{m_S^2} \frac{\partial m_S^2}{\partial \lambda_f} \right| = 2 \frac{\lambda_f^2 \Lambda^2}{m_S^2(\lambda_f)} . \tag{17.6}$$

Eq. (17.6) dramatically demonstrates the gauge hierarchy problem when $m_S(\lambda_f)$ is put less than 1 TeV while Λ is taken[6] $\sim 10^{16}$ GeV.

We now apply similar considerations to the supersymmetric toy model of §1.3. Identifying δm_S^2 as $i\Pi_{hh}^f(0) + i\Pi_{hh}^{\tilde{f}}(0)$, we have in place of (17.5) the expression

$$m_S^2 = \frac{\lambda_f^2 N_c(f)}{16\pi^2}\left|4\delta^2 + (2\delta^2 + |A_f|^2)\ln\frac{m_f^2}{\mu^2}\right| + \mathcal{O}(\delta^4, A_f^2\delta^2) + m_{S,0}^2 \,, \tag{17.7}$$

where $N_c(f)$ is the number of colors f can have and μ is a renormalization scale where the 'fundamental' parameters are defined. Consequently, the fine tuning measure becomes

$$\Delta(m_S^2, \lambda_f) = \frac{\lambda_f^2 N_c(f)}{8\pi^2 m_S^2}\left|4\delta^2 + (2\delta^2 + |A_f|^2)\ln\frac{m_f^2}{\mu^2}\right| + \mathcal{O}(\delta^4, |A_f|^2\delta^2) \,. \tag{17.8}$$

Thus, so long as $\delta^2 = m_{\tilde{f}}^2 - m_f^2$ and $|A_f|^2$ are comparable to m_S^2, $\Delta(m_S^2, \lambda_f)$ will be $\mathcal{O}(1)$ or less[7]. This illustrates the statement that supersymmetry, with a small amount of breaking, alleviates the fine tuning problem. On the other hand, if $\delta^2, |A_f|^2 \gg m_S^2$, i.e. supersymmetry is badly broken, the fine tuning will increase. Note, moreover, that we can consider here other measures of fine tuning such as

$$\Delta(m_S^2, \delta^2) = \lambda_f^2 \delta^2 (8\pi^2 m_S^2)^{-1} N_c(f)\left|2 + \ln\frac{m_f^2}{\mu^2}\right| \tag{17.9}$$

and draw conclusions similar to those inferred from (17.8).

Turning to presumably realistic models like the MSSM, it is convenient to choose [17.3] $r = M_Z^2$, where M_Z, the pole-value of the Z-mass, is one of the most precisely known electroweak quantities today, to wit [17.4]

$$M_Z = 91.1822 \pm 0.0022 \text{ GeV} \,.$$

There have been quantitative investigations of fine tuning in various versions of the MSSM, e.g. in the one with supergravity boundary conditions at M_U (cf. Ch.12). Various Δ_i's have been constructed[8], as per

$$\Delta_i = \left|\frac{a_i}{M_Z^2}\frac{\partial M_Z^2}{\partial a_i}\right| \,, \tag{17.10}$$

$\{a_i\}$ being the basic supersymmetry breaking parameters of the model. The overall measure of fine tuning is then taken to be

$$\Delta = \text{max.}\{\Delta_i\} \,. \tag{17.11}$$

---

[6]The reader needs to be warned against making these arguments too precise. For example, either of the choices $r = m_S$ or $a_i = \lambda_f^2$, would change the result by a factor of 2. On the other hand, a choice like $r = m_S^{1000}$ would be 'unnatural'. Half an order of magnitude multiplicative ambiguities are typical in fine tuning measures.

[7]Note that this allows first or second generation sfermion masses in the tens of TeV range since their Yukawa coupling strengths λ_f are very small in magnitude. Such a scenario has been discussed in §9.5 as a possible solution to the flavor problem in supersymmetry.

[8]Another proposal [17.5] has been to use the neutralino relic density fine tuning measure $\Delta^\Omega \equiv \sqrt{\Sigma_i(\partial\ln(\Omega_{\tilde{\chi}}h^2)/\partial\ln a_i)^2}$, where the sum runs over input parameters, which might include (relatively) poorly known SM quantities such as $m_{t,b}$ as well as model dependent CMSSM parameters like $m_0, M_{1/2}$ of mSUGRA.

According to a phenomenological investigation [17.6] in the mSUGRA model, with input experimental lower bounds on sparticle masses and other data, Δ varies from 20 to 180, depending on the region of the parameter space. Though weakly interacting sparticles may lie not very far in mass from the W, Z bosons, Δ can be somewhat larger than unity in this scenario. This is since the fundamental parameters $\{a_i\}$ are defined at very high unification or Planck energies. The renormalization in their evolution down to weak interaction energies, together with color factors, can easily generate[9] a multiplicative factor $\mathcal{O}(100)$ in Δ since $\ln(M_{Pl}/M_Z) \simeq 35$. On the other hand, the above numerical values can be decreased by tinkering with the boundary conditions at M_U. Nevertheless, if the lower bounds on the masses of all sparticles and extra Higgs bosons[10], move up and go much beyond 1–2 TeV, Δ would become considerably larger than ~ 1000 [17.7] and the naturalness of the supersymmetric stabilization of the weak scale would be lost. Similar conclusions hold [17.8] about other models that describe the sparticle spectrum in terms of a few parameters defined at some high scale M_M. The experimentalist's sword of Damocles hangs over the head of supersymmetry as a stabilizer of the weak scale, as incorporated in the Standard Model of particle interactions.

References

[17.1] L. Maiani, *Proc. 1979 Gif-sur-Yvette Summer School on Particle Physics*, p 1. G.'t Hooft, in *Recent Developments in Gauge Theories* (Cargése 1979, eds. G.'t Hooft et al., Plenum Press, NY 1980).

[17.2] E. Witten, Nucl. Phys. **B188** (1981) 513. S. Dimopoulos and H. Georgi, Nucl. Phys. **B193** (1981) 150. N. Sakai, Z. Phys. **C11** (1981) 153. R. Kaul, Phys. Lett. **B109** (1982) 19.

[17.3] R. Barbieri and G.F. Giudice, Nucl. Phys. **B306** (1988) 63.

[17.4] PDG group, *loc. cit., Bibl.*

[17.5] J.R. Ellis and K.A. Olive, Phys. Lett. **B514** (2001) 114.

[17.6] P.H. Chankowski, J.R. Ellis and S. Pokorski, Phys. Lett. **B423** (1998) 327.

[17.7] J.R. Ellis, K.A. Olive and Y. Santoso, New Jour. Phys. **4** (2002) 32.

[17.8] G. Bhattacharyya and A. Romanino, Phys. Rev. **D55** (1997) 7015.

---

[9]We give an example. With $\mu = M_{Pl}$, $|\lambda_f| \sim 1$ and $N_c(f) = 3$, as appropriate for the top quark, (17.8) yields $\Delta(m_s^2, \lambda_f) \sim 5\delta^2 m_s^{-2}$. A choice $r = M_Z$ increase the measure of fine tuning even further owing to the nonlinear dependence of M_Z on the parameters of the weak scale Higgs potential, cf. (10.18).

[10]There are, of course, supersymmetry models in which one set of sparticles are significantly heavier than 1 TeV, while the others are in the sub-TeV range. For instance, in focus point scenarios, cf. §12.4, sfermion masses can be in the region of several TeV, while chargino and neutralino masses are restricted to the few hundred GeV range.

Appendix A

Sparticle Vertices in the MSSM

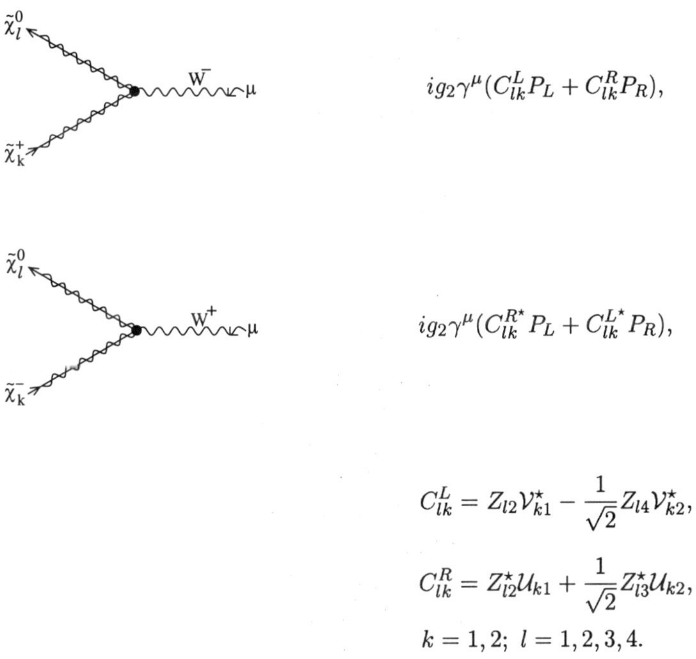

$$ig_2\gamma^\mu(C^L_{lk}P_L + C^R_{lk}P_R),$$

$$ig_2\gamma^\mu(C^{R\star}_{lk}P_L + C^{L\star}_{lk}P_R),$$

$$C^L_{lk} = Z_{l2}\mathcal{V}^\star_{k1} - \frac{1}{\sqrt{2}}Z_{l4}\mathcal{V}^\star_{k2},$$

$$C^R_{lk} = Z^\star_{l2}\mathcal{U}_{k1} + \frac{1}{\sqrt{2}}Z^\star_{l3}\mathcal{U}_{k2},$$

$$k = 1,2; \quad l = 1,2,3,4.$$

Legend

Charged gauge boson

Chargino or neutralino

Fig. 9.3. Vertex with a charged weak boson, a chargino and a neutralino. The $\mathcal{U}$'s, $\mathcal{V}$'s and Z's are as per (9.8) and (9.27). The corresponding vertices for outgoing charginos are complex conjugates of the above.

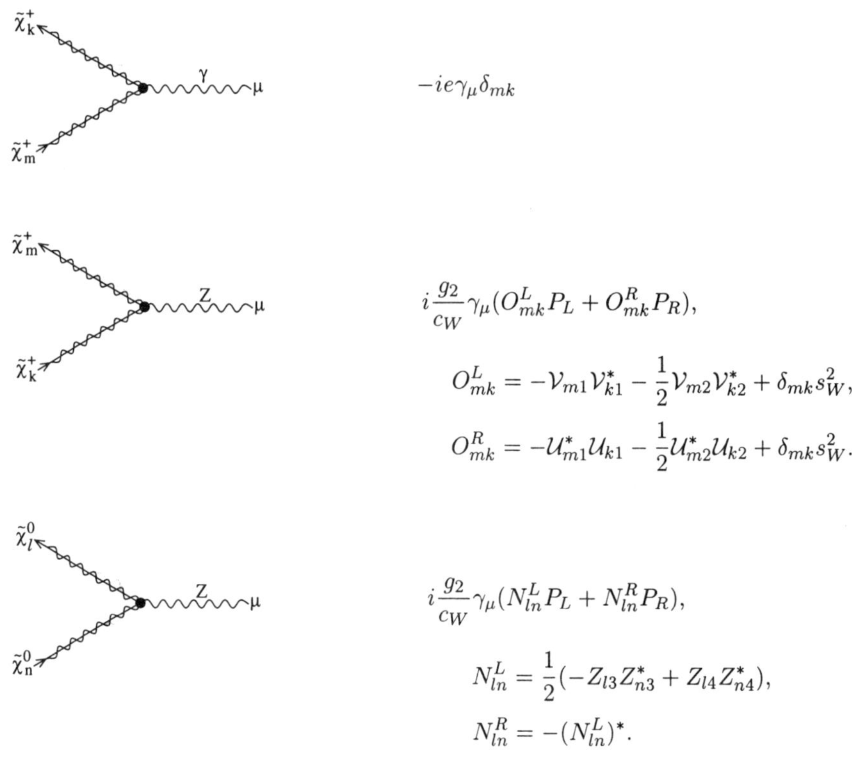

$$-ie\gamma_\mu \delta_{mk}$$

$$i\frac{g_2}{c_W}\gamma_\mu(O^L_{mk}P_L + O^R_{mk}P_R),$$

$$O^L_{mk} = -\mathcal{V}_{m1}\mathcal{V}^*_{k1} - \frac{1}{2}\mathcal{V}_{m2}\mathcal{V}^*_{k2} + \delta_{mk}s^2_W,$$

$$O^R_{mk} = -\mathcal{U}^*_{m1}\mathcal{U}_{k1} - \frac{1}{2}\mathcal{U}^*_{m2}\mathcal{U}_{k2} + \delta_{mk}s^2_W.$$

$$i\frac{g_2}{c_W}\gamma_\mu(N^L_{ln}P_L + N^R_{ln}P_R),$$

$$N^L_{ln} = \frac{1}{2}(-Z_{l3}Z^*_{n3} + Z_{l4}Z^*_{n4}),$$

$$N^R_{ln} = -(N^L_{ln})^*.$$

Legend

Neutral gauge boson

Chargino or neutralino

Fig. 9.4. Vertices of neutral EW bosons with chargino and neutralino pairs. The $\mathcal{U}$'s, $\mathcal{V}$'s and Z's are as per (9.8) and (9.27).

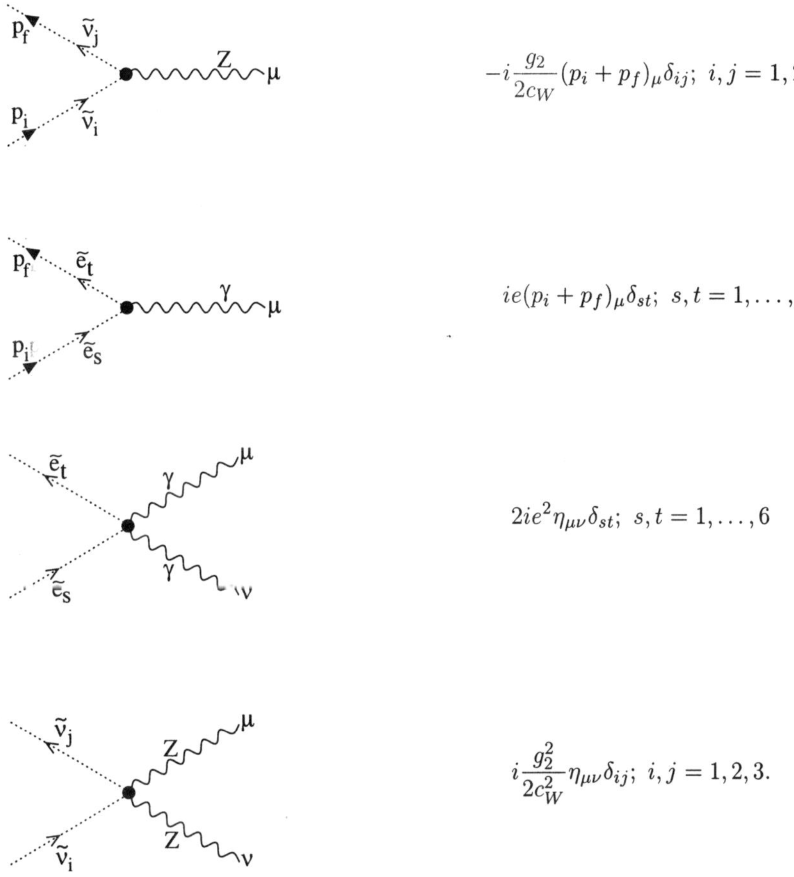

Legend

neutral gauge boson

charged slepton $\tilde{e}_s$ or sneutrino $\tilde{\nu}_i$

Fig. 9.6. Slepton-gauge boson vertices unaffected by any mixing.

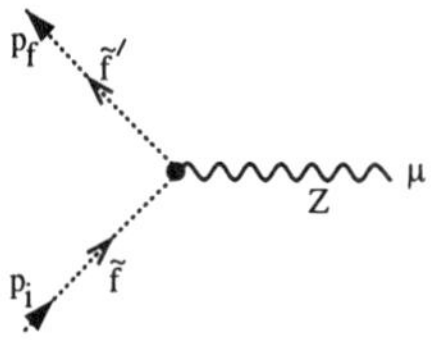

$$i\frac{g_2}{2c_W}(p_i + p_f)_\mu \left(\sum_{i=1}^{3} W_{is}^{\tilde{e}*} W_{it}^{\tilde{e}} - 2s_W^2 \delta_{st} \right);$$

$$\tilde{f} = \tilde{e}_t, \tilde{f}' = \tilde{e}_s; \ \ s,t = 1,\dots,6.$$

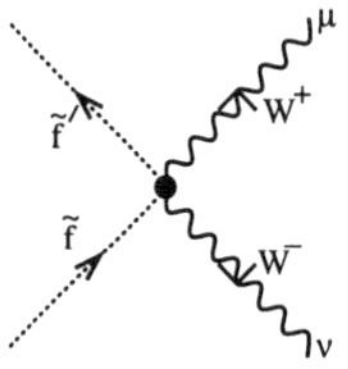

$$i\frac{g_2^2}{2}\eta_{\mu\nu}\delta_{ij}; \ \ \tilde{f} = \tilde{\nu}_j, \ \tilde{f}' = \tilde{\nu}_i; \ \ i,j = 1,2,3$$

$$i\frac{g_2^2}{2}\eta_{\mu\nu} \sum_{i=1}^{3} W_{is}^{\tilde{e}*} W_{it}^{\tilde{e}}; \ \ \tilde{f} = \tilde{e}_t, \tilde{f}' = \tilde{e}_s;$$

$$s,t = 1,\dots,6 \ .$$

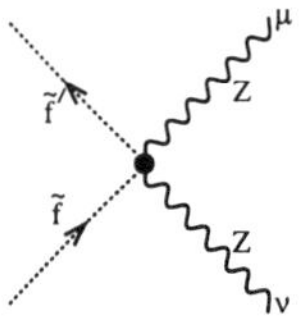

$$i\frac{g_2^2}{2c_W^2}\eta_{\mu\nu} \left[\sum_{i=1}^{3} \left(1 - 4s_W^2\right) W_{is}^{\tilde{e}*} W_{it}^{\tilde{e}} + 4s_W^4 \delta_{st} \right]$$

$$\tilde{f} = \tilde{e}_t, \tilde{f}' = \tilde{e}_s; \ \ s,t = 1,\dots,6$$

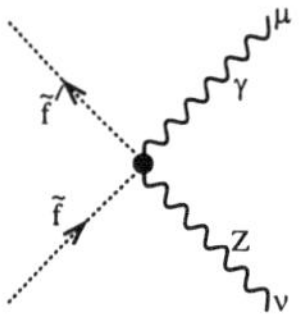

$$i\frac{eg_2}{c_W}\eta_{\mu\nu} \left(\sum_{i=1}^{3} W_{is}^{\tilde{e}*} W_{it}^{\tilde{e}} - 2s_W^2 \delta_{st} \right)$$

$$\tilde{f} = \tilde{e}_t, \tilde{f}' = \tilde{e}_s; \ \ s,t = 1,\dots,6$$

Legend

charged vector boson

neutral vector boson

charged slepton $\tilde{e}_s$ or sneutrino $\tilde{\nu}_i$

Fig. 9.7. Slepton-gauge boson vertices that are affected by generation mixing only in the presence of left-right mixing. $W_{is}^{\tilde{e}}$ and $\tilde{e}_s$ are as in (9.53).

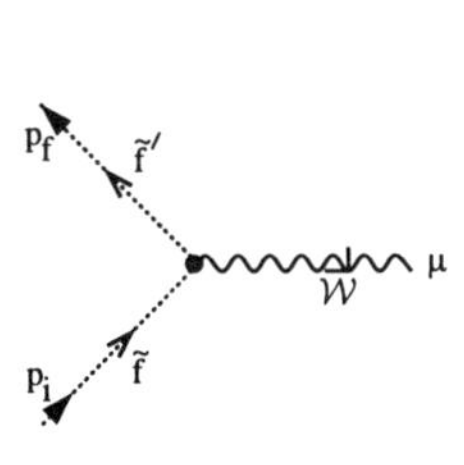

$$-i\frac{g_2}{\sqrt{2}}(p_i+p_f)_\mu \sum_{i=1}^{3} U_{ij}^{\tilde{\nu}*} W_{is}^{\tilde{e}};$$
$$\mathcal{W}=W^-, \ \tilde{f}=\tilde{e}_s, \ \tilde{f}'=\tilde{\nu}_j; \ j=1,2,3, \ s=1,\ldots,6.$$

$$-i\frac{g_2}{\sqrt{2}}(p_i+p_f)_\mu \sum_{i=1}^{3} W_{is}^{\tilde{e}*} U_{ij}^{\tilde{\nu}};$$
$$\mathcal{W}=W^+, \ \tilde{f}=\tilde{\nu}_j, \ \tilde{f}'=\tilde{e}_s; \ j=1,2,3, \ s=1,\ldots,6.$$

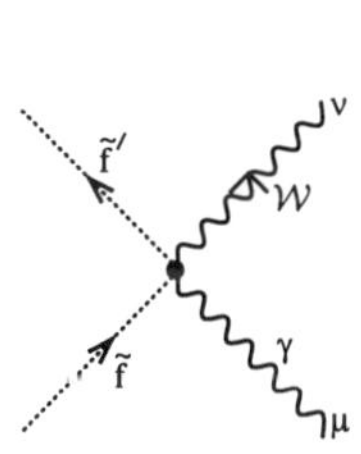

$$-i\frac{eg_2}{\sqrt{2}}\eta_{\mu\nu} \sum_{i=1}^{3} U_{ij}^{\tilde{\nu}*} W_{is}^{\tilde{e}};$$
$$\mathcal{W}=W^-, \ \tilde{f}=\tilde{e}_s, \ \tilde{f}'=\tilde{\nu}_j; \ j=1,2,3, \ s=1,\ldots,6.$$

$$-i\frac{eg_2}{\sqrt{2}}\eta_{\mu\nu} \sum_{i=1}^{3} W_{is}^{\tilde{e}*} U_{ij}^{\tilde{\nu}},$$
$$\mathcal{W}=W^+, \ \tilde{f}=\tilde{\nu}_j, \ \tilde{f}'=\tilde{e}_s; \ j=1,2,3, \ s=1,\ldots,6.$$

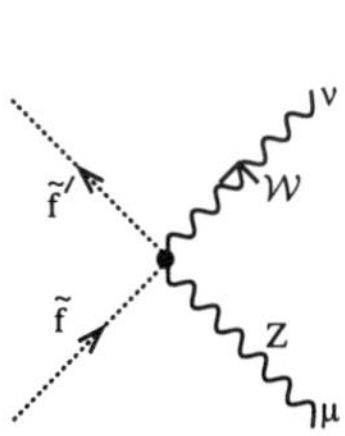

$$i\frac{g_2^2 s_W^2}{\sqrt{2}c_W}\eta_{\mu\nu} \sum_{i=1}^{3} U_{ij}^{\tilde{\nu}*} W_{is}^{\tilde{e}};$$
$$\mathcal{W}=W^-, \ \tilde{f}=\tilde{e}_s, \tilde{f}'=\tilde{\nu}_j; \ j=1,2,3, \ s=1,\ldots,6.$$

$$i\frac{g_2^2 s_W^2}{\sqrt{2}c_W}\eta_{\mu\nu} \sum_{i=1}^{3} W_{is}^{\tilde{e}*} U_{ij}^{\tilde{\nu}};$$
$$\mathcal{W}=W^+, \ \tilde{f}=\tilde{\nu}_j, \ \tilde{f}'=\tilde{e}_s; \ j=1,2,3, \ s=1,\ldots,6.$$

Legend

charged vector boson

slepton

Fig. 9.8. Slepton-gauge boson vertices with generation **and** left-right mixing. $W_{is}^{\tilde{e}}$ and $\tilde{e}_s$ are as in (9.53) and $U^{\tilde{\nu}}$ is as in (9.58a).

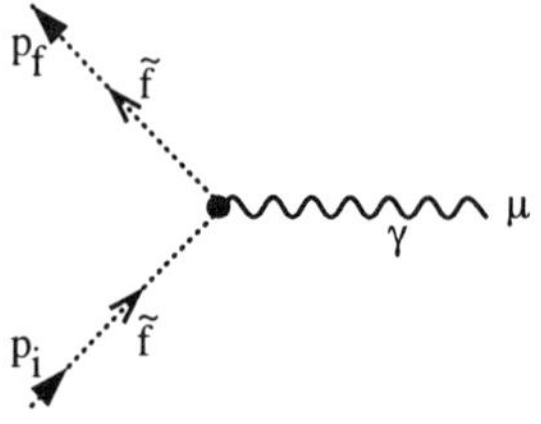

$$-ieQ_{\tilde{f}}(p_i + p_f)_\mu \delta_{st}; \quad \tilde{f} = \tilde{q}_s, \ \tilde{f}' = \tilde{q}_t; \ q = u, d;$$
$$s, t = 1, \ldots, 6$$

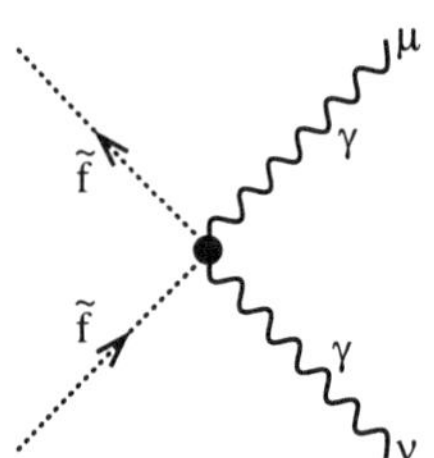

$$2ie^2 Q_{\tilde{f}}^2 \eta_{\mu\nu} \delta_{st}; \quad \tilde{f} = \tilde{q}_s, \ \tilde{f}' = \tilde{q}_t; \ q = u, d;$$
$$s, t = 1, \ldots, 6$$

Legend

squark $\quad\cdots\!\!\rightarrow\!\!\cdots$

photon $\quad\sim\!\!\sim\!\!\sim$

Fig. 9.9. Squark-photon vertices unaffected by any mixing.

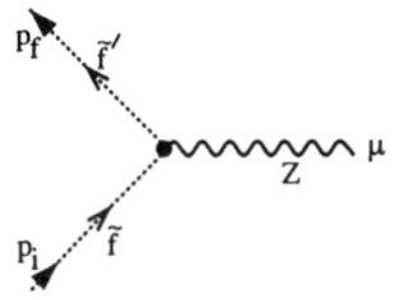

$$-i\frac{g_2}{c_W}(p_i+p_f)_\mu\left(\sum_{i=1}^{3}T^{\tilde{q}}_{3L}W^{\tilde{q}*}_{is}W^{\tilde{q}}_{it}-s^2_W Q_{\tilde{q}}\delta_{st}\right);$$
$$\tilde{f}=\tilde{q}_t,\ \tilde{f}'=\tilde{q}_s;\ q=u,d;\ s,t=1,\ldots,6$$

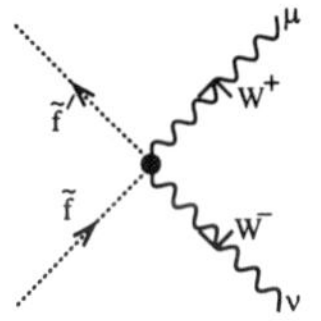

$$i\frac{g_2^2}{2}\eta_{\mu\nu}\sum_{i=1}^{3}W^{\tilde{q}*}_{is}W^{\tilde{q}}_{it};$$
$$\tilde{f}=\tilde{q}_t,\ \tilde{f}'=\tilde{q}_s;\ q=u,d;\ s,t=1,\ldots,6$$

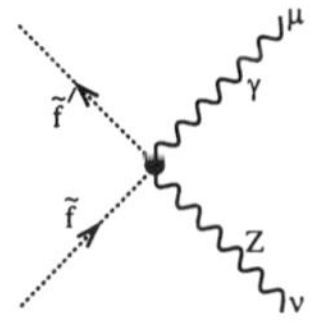

$$2i\frac{g_2 e}{c_W}\eta_{\mu\nu}\left(\sum_{i=1}^{3}Q_{\tilde{q}}T^{\tilde{q}}_{3L}W^{\tilde{q}*}_{is}W^{\tilde{q}}_{it}-Q_{\tilde{q}}^2 s^2_W\delta_{st}\right);$$
$$\tilde{f}=\tilde{q}_t,\ \tilde{f}'=\tilde{q}_s;\ q=u,d;\ s,t=1,\ldots,6$$

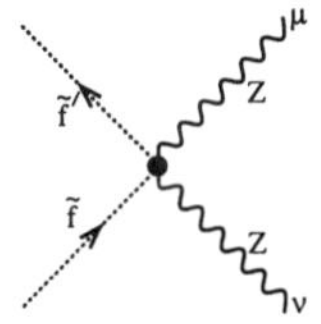

$$i\frac{g_2^2}{2c_W^2}\eta_{\mu\nu}\left[\sum_{i=1}^{3}\left(1-8T^{\tilde{q}}_{3L}Q_{\tilde{q}}s^2_W\right)W^{\tilde{q}*}_{is}W^{\tilde{q}}_{it}+4Q_{\tilde{q}}^2 s^4_W\delta_{st}\right];$$
$$\tilde{f}=\tilde{q}_t,\ \tilde{f}'=\tilde{q}_s;\ q=u,d;\ s,t=1,\ldots,6$$

Legend

squark

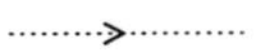

W

Z

Fig. 9.10. Squark-gauge boson vertices with nontrivial generation mixing only in the presence of left-right mixing. $W^{\tilde{q}}_{is}$ and $\tilde{q}_s$ are as in (9.53).

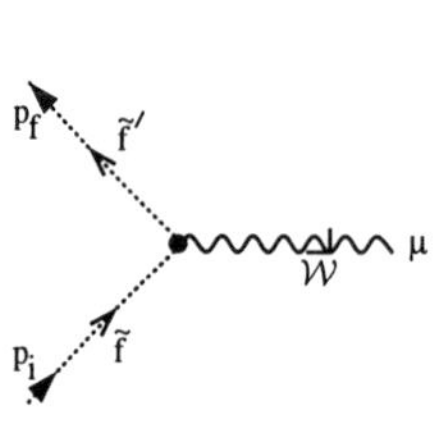

$$-i\frac{g_2}{\sqrt{2}}(p_i+p_f)_\mu \sum_{i=1}^{3} W_{is}^{\tilde{u}*}W_{it}^{\tilde{d}};$$
$$\mathcal{W}=W^-,\ \tilde{f}=\tilde{d}_t,\ \tilde{f}'=\tilde{u}_s;\ s,t=1,\ldots,6.$$

$$-i\frac{g_2}{\sqrt{2}}(p_i+p_f)_\mu \sum_{i=1}^{3} W_{is}^{\tilde{d}*}W_{it}^{\tilde{u}};$$
$$\mathcal{W}=W^+,\ \tilde{f}=\tilde{u}_t,\ \tilde{f}'=\tilde{d}_s;\ s,t=1,\ldots,6.$$

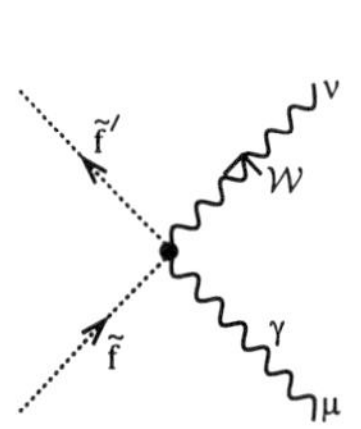

$$i\frac{eg_2}{3\sqrt{2}}\eta_{\mu\nu} \sum_{i=1}^{3} W_{is}^{\tilde{u}*}W_{it}^{\tilde{d}};$$
$$\mathcal{W}=W^-,\ \tilde{f}=\tilde{d}_t,\ \tilde{f}'=\tilde{u}_s;\ s,t=1,\ldots,6.$$

$$i\frac{eg_2}{3\sqrt{2}}\eta_{\mu\nu} \sum_{i=1}^{3} W_{is}^{\tilde{d}*}W_{it}^{\tilde{u}};$$
$$\mathcal{W}=W^+,\ \tilde{f}=\tilde{u}_t,\ \tilde{f}'=\tilde{d}_s;\ s,t=1,\ldots,6.$$

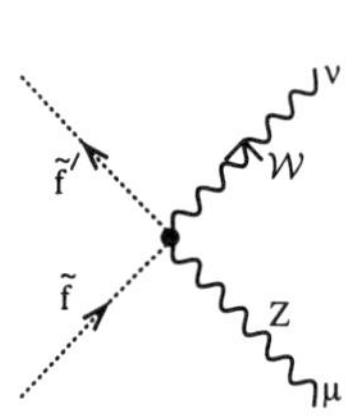

$$-i\frac{g_2^2 s_W^2}{3\sqrt{2}c_W}\eta_{\mu\nu} \sum_{i=1}^{3} W_{is}^{\tilde{u}*}W_{it}^{\tilde{d}};$$
$$\mathcal{W}=W^-,\ \tilde{f}=\tilde{d}_t,\ \tilde{f}'=\tilde{u}_s;\ s,t=1,\ldots,6.$$

$$-i\frac{g_2^2 s_W^2}{3\sqrt{2}c_W}\eta_{\mu\nu} \sum_{i=1}^{3} W_{is}^{\tilde{d}*}W_{it}^{\tilde{u}};$$
$$\mathcal{W}=W^+,\ \tilde{f}=\tilde{u}_t,\ \tilde{f}'=\tilde{d}_s;\ s,t=1,\ldots,6.$$

Legend

neutral vector boson

charged vector boson

squark

Fig. 9.11. Squark-gauge boson vertices with generation **and** left-right mixing. $W_{is}^{\tilde{q}}$ and $\tilde{q}_s$ are as in (9.53).

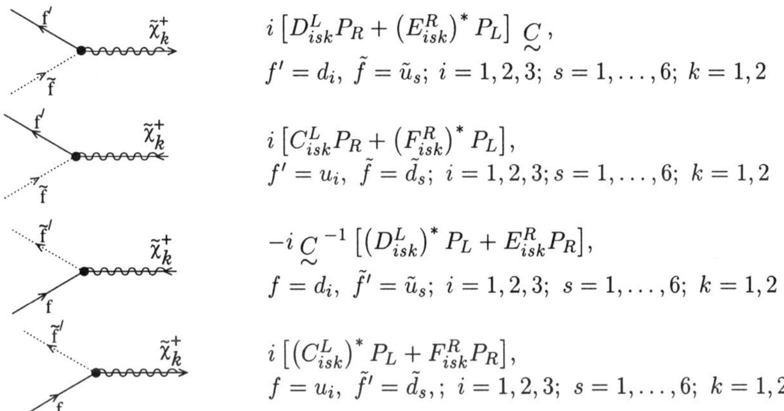

$$i\left[D^L_{isk}P_R + \left(E^R_{isk}\right)^* P_L\right]\underset{\sim}{C},$$
$$f' = d_i,\ \tilde{f} = \tilde{u}_s;\ i = 1,2,3;\ s = 1,\ldots,6;\ k = 1,2$$

$$i\left[C^L_{isk}P_R + \left(F^R_{isk}\right)^* P_L\right],$$
$$f' = u_i,\ \tilde{f} = \tilde{d}_s;\ i = 1,2,3;\ s = 1,\ldots,6;\ k = 1,2$$

$$-i\,\underset{\sim}{C}^{-1}\left[\left(D^L_{isk}\right)^* P_L + E^R_{isk}P_R\right],$$
$$f = d_i,\ \tilde{f}' = \tilde{u}_s;\ i = 1,2,3;\ s = 1,\ldots,6;\ k = 1,2$$

$$i\left[\left(C^L_{isk}\right)^* P_L + F^R_{isk}P_R\right],$$
$$f = u_i,\ \tilde{f}' = \tilde{d}_s,;\ i = 1,2,3;\ s = 1,\ldots,6;\ k = 1,2$$

Legend

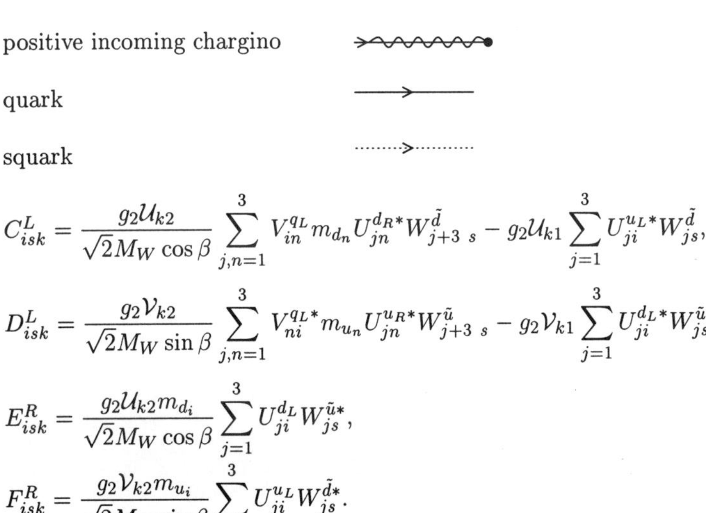

positive outgoing chargino

positive incoming chargino

quark

squark

$$C^L_{isk} = \frac{g_2 \mathcal{U}_{k2}}{\sqrt{2}M_W\cos\beta}\sum_{j,n=1}^{3} V^{q_L}_{in} m_{d_n} U^{d_R *}_{jn} W^{\tilde{d}}_{j+3\ s} - g_2\mathcal{U}_{k1}\sum_{j=1}^{3} U^{u_L *}_{ji} W^{\tilde{d}}_{js},$$

$$D^L_{isk} = \frac{g_2 \mathcal{V}_{k2}}{\sqrt{2}M_W\sin\beta}\sum_{j,n=1}^{3} V^{q_L *}_{ni} m_{u_n} U^{u_R *}_{jn} W^{\tilde{u}}_{j+3\ s} - g_2\mathcal{V}_{k1}\sum_{j=1}^{3} U^{d_L *}_{ji} W^{\tilde{u}}_{js},$$

$$E^R_{isk} = \frac{g_2 \mathcal{U}_{k2} m_{d_i}}{\sqrt{2}M_W\cos\beta}\sum_{j=1}^{3} U^{d_L}_{ji} W^{\tilde{u}*}_{js},$$

$$F^R_{isk} = \frac{g_2 \mathcal{V}_{k2} m_{u_i}}{\sqrt{2}M_W\sin\beta}\sum_{j=1}^{3} U^{u_L}_{ji} W^{\tilde{d}*}_{js}.$$

Fig. 9.12. Quark-squark-chargino vertices, cf. (9.68a). For an external final (initial) chargino, $\underset{\sim}{C}$ acts on $\bar{u}^T_{\tilde{\chi}^+}(\bar{v}^T_{\tilde{\chi}^+})$ to the right giving $v_{\tilde{\chi}^+}(u_{\tilde{\chi}^+})$, while $\underset{\sim}{C}^{-1}$ acts on $v^T_{\tilde{\chi}^+}(u^T_{\tilde{\chi}^+})$ to the left giving $-\bar{u}_{\tilde{\chi}^+}(-\bar{v}_{\tilde{\chi}^+})$. For a chargino line inside a loop, the $\underset{\sim}{C}$ and $\underset{\sim}{C}^{-1}$ factors can be made to cancel.

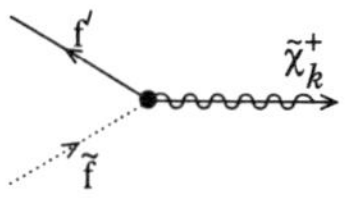

$$i\left[d^L_{ijk}P_R + \left(e^R_{ijk}\right)^* P_L\right]\underset{\sim}{C}$$
$$f' = e_i,\ \ \tilde{f} = \tilde{\nu}_j;\ \ i,j = 1,2,3;\ \ k = 1,2$$

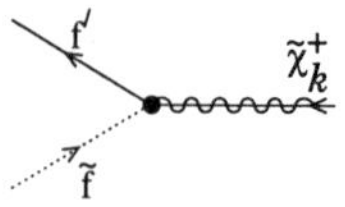

$$ic^L_{isk}P_R,\ \ \ f' = \nu_i,\ \tilde{f} = \tilde{e}_s;\ \ i = 1,2,3;$$
$$s = 1,\ldots,6;\ \ k = 1,2$$

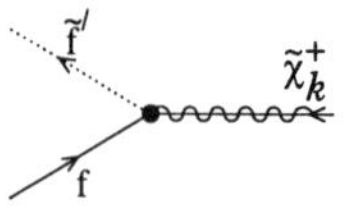

$$-i\underset{\sim}{C}^{-1}\left[\left(d^L_{ijk}\right)^* P_L + e^R_{ijk}P_R\right],$$
$$f = e_i,\ \tilde{f}' = \tilde{\nu}_j;\ \ i,j = 1,2,3;\ \ k = 1,2$$

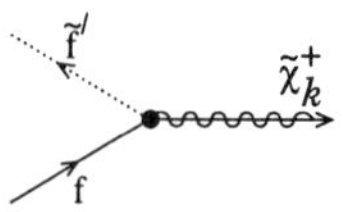

$$i\left(c^L_{isk}\right)^* P_L,\ \ \ f = \nu_i,\ \tilde{f}' = \tilde{e}_s;\ \ i = 1,2,3;$$
$$s = 1,\ldots,6;\ \ k = 1,2$$

Legend

positive outgoing chargino

positive incoming chargino

lepton

slepton

$$c^L_{isk} = -g_2\mathcal{U}_{k1}W^{\tilde{e}}_{is} + \frac{g_2 m_{e_i}}{\sqrt{2}M_W\cos\beta}\mathcal{U}_{k2}W^{\tilde{e}}_{i+3\ s}$$

$$d^L_{ijk} = -g_2 U^{\tilde{\nu}}_{ij}\mathcal{V}_{k1},$$

$$e^R_{ijk} = \frac{g_2 m_{e_i}}{\sqrt{2}M_W\cos\beta}\mathcal{U}_{k2}U^{\tilde{\nu}*}_{ij}$$

Fig. 9.13. Lepton-slepton-chargino vertices, cf. (9.70). The same remark, as made in the caption of Fig. 9.12, applies.

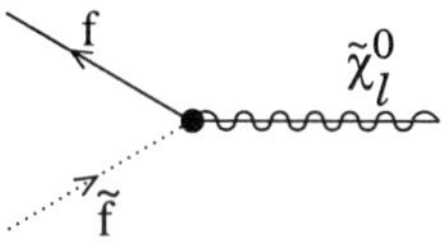

$$i\left[\left(G^{qL}_{isl}\right)^{*}P_R + \left(G^{qR}_{isl}\right)^{*}P_L\right]$$
$$f' = q_i,\ \tilde{f} = \tilde{q}_s;\ q = u, d;\ i = 1,2,3;$$
$$s = 1,\ldots,6;\ l = 1,2,3,4$$

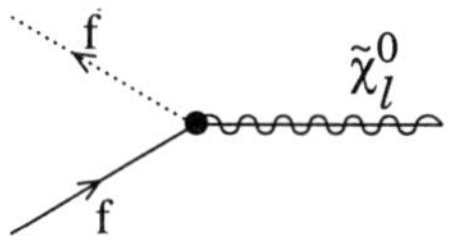

$$i\left(G^{qL}_{isl}P_L + G^{qR}_{isl}P_R\right),$$
$$f = q_i,\ \tilde{f}' = \tilde{q}_s;\ q = u, d;\ i = 1,2,3;$$
$$s = 1,\ldots,6;\ l = 1,2,3,4$$

Legend

neutralino

quark

squark

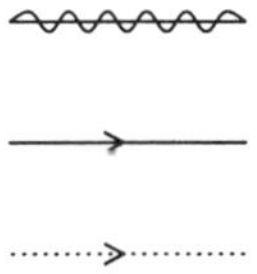

$$G^{u_L}_{isl} = -\sqrt{2}g_2\left(\frac{1}{2}Z^{*}_{l2} + \frac{1}{6}\tan\theta_W Z^{*}_{l1}\right)\sum_{j=1}^{3}W^{\tilde{u}*}_{js}U^{u_L}_{ji} - \frac{g_2}{\sqrt{2}M_W\sin\beta}m_{u_i}Z^{*}_{l4}\sum_{j=1}^{3}W^{\tilde{u}*}_{j+3}\,_s U^{u_R}_{ji},$$

$$G^{u_R}_{isl} = \frac{2\sqrt{2}}{3}g_2\tan\theta_W Z_{l1}\sum_{j=1}^{3}W^{\tilde{u}*}_{j+3}\,_s U^{u_R}_{ji} - \frac{g_2}{\sqrt{2}M_W\sin\beta}m_{u_i}Z_{l4}\sum_{j=1}^{3}W^{\tilde{u}*}_{js}U^{u_L}_{ji},$$

$$G^{d_L}_{isl} = \sqrt{2}g_2\left(\frac{1}{2}Z^{*}_{l2} - \frac{1}{6}\tan\theta_W Z^{*}_{l1}\right)\sum_{j=1}^{3}W^{\tilde{d}*}_{js}U^{d_L}_{ji} - \frac{g_2}{\sqrt{2}M_W\cos\beta}m_{d_i}Z^{*}_{l3}\sum_{j=1}^{3}W^{\tilde{d}*}_{j+3}\,_s U^{d_R}_{ji},$$

$$G^{d_R}_{isl} = -\frac{\sqrt{2}}{3}g_2\tan\theta_W Z_{l1}\sum_{j=1}^{3}W^{\tilde{d}*}_{j+3}\,_s U^{d_R}_{ji} - \frac{g_2}{\sqrt{2}M_W\cos\beta}m_{d_i}Z_{l3}\sum_{j=1}^{3}W^{\tilde{d}*}_{js}U^{d_L}_{ji}.$$

Fig. 9.14. Quark-squark-neutralino vertices, cf. (9.75).

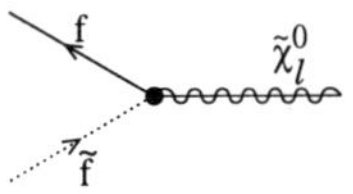

$$i\left[\left(G^{e_L}_{isl}\right)^* P_R + \left(G^{e_R}_{isl}\right)^* P_L\right],$$
$$f' = e_i,\ \ \tilde{f} = \tilde{e}_s;\ \ i = 1,2,3;\ \ s = 1,\ldots,6;$$
$$l = 1,2,3,4$$

$$i\left(G^{\nu}_{ijl}\right)^* P_R,\ \ \ f' = \nu_i,\ \ \tilde{f} = \tilde{\nu}_j;\ \ i,j = 1,2,3;$$
$$l = 1,2,3,4$$

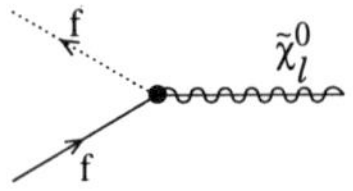

$$i\left(G^{e_L}_{isl} P_L + G^{e_R}_{isl} P_R\right),$$
$$f = e_i,\ \ \tilde{f}' = \tilde{e}_s;\ \ i = 1,2,3;\ \ s = 1,\ldots,6;$$
$$l = 1,2,3,4$$

$$iG^{\nu}_{ijl} P_L,\ \ \ f = \nu_i,\ \ \tilde{f}' = \tilde{\nu}_j;\ \ i,j = 1,2,3;$$
$$l = 1,2,3,4$$

Legend

neutralino

lepton

slepton

$$G^{\nu}_{ijl} = -\frac{1}{\sqrt{2}} g_2 \left(Z^*_{l2} - \tan\theta_W Z^*_{l1}\right) U^{\tilde{\nu}*}_{ij},$$

$$G^{e_L}_{isl} = \frac{1}{\sqrt{2}} g_2 \left(Z^*_{l2} + \tan\theta_W Z^*_{l1}\right) W^{\tilde{e}*}_{is} - \frac{g_2}{\sqrt{2} M_W \cos\beta} m_{e_i} Z^*_{l3} W^{\tilde{e}*}_{i+3\ s},$$

$$G^{e_R}_{isl} = -\sqrt{2} g_2 \tan\theta_W Z_{l1} W^{\tilde{e}*}_{i+3\ s} - \frac{g_2}{\sqrt{2} M_W \cos\beta} m_{e_i} Z_{l3} W^{\tilde{e}*}_{is}.$$

Fig. 9.15. Lepton-slepton neutralino vertices, cf. (9.76).

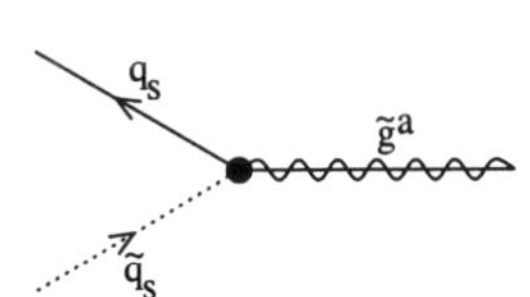

$$-i\sqrt{2}g_s T^a \sum_{j=1}^{3} \left(U_{ji}^{q_L *} W_{js}^{\tilde{q}} P_R - U_{ji}^{q_R *} W_{j+3\ s}^{\tilde{q}} P_L \right),$$

$$q = u, d; \ i = 1, 2, 3; \ s = 1, \dots, 6;$$
$$a = 1, \dots, 8$$

$$-i\sqrt{2}g_s T^a \sum_{j=1}^{3} \left(U_{ji}^{q_L} W_{js}^{\tilde{q}*} P_L - U_{ji}^{q_R} W_{j+3\ s}^{\tilde{q}*} P_R \right),$$

$$q = u, d; \ i = 1, 2, 3; \ s = 1, \dots, 6;$$
$$a = 1, \dots, 8$$

Legend

squark   ········>··········

quark   ⟶

gluino   ∿∿∿∿∿

Fig. 9.16. Quark-squark-gluino vertices, cf. (9.78).

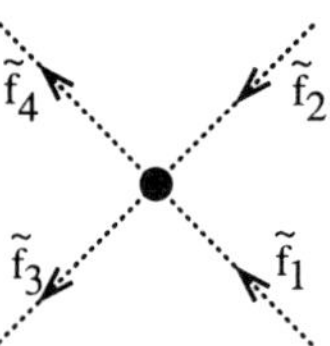

(Fig.9.17)

$$-i\Big\{Y[\tilde{q}_s^{a*},\tilde{q}_t^a,\tilde{q}_u^{b*},\tilde{q}_v^b] + Y[\tilde{q}_u^{b*},\tilde{q}_v^b,\tilde{q}_s^{a*},\tilde{q}_t^a] + \delta^{ab}\Big(Y[\tilde{q}_s^{a*},\tilde{q}_v^a,\tilde{q}_u^{b*},\tilde{q}_t^b]$$
$$+ Y[\tilde{q}_u^{a*},\tilde{q}_t^a,\tilde{q}_s^{b*},\tilde{q}_v^b]\Big)\Big\},$$

$\tilde{f}_1 = \tilde{q}_t^a,\; \tilde{f}_2 = \tilde{q}_v^b,\; \tilde{f}_3 = \tilde{q}_s^a,\; \tilde{f}_4 = \tilde{q}_u^b\;;q = u,d;s,t,u,v = 1,\ldots,6;$
$a,b = 1,2,3$

$$-i\Big(Y[\tilde{u}_s^{a*},\tilde{u}_t^a,\tilde{d}_u^{b*},\tilde{d}_v^b] + \delta^{ab}Y[\tilde{u}_s^{a*},\tilde{d}_v^a,\tilde{d}_u^{b*},\tilde{u}_t^b]\Big),$$

$\tilde{f}_1 = \tilde{u}_t^a,\; \tilde{f}_2 = \tilde{d}_v^b,\; \tilde{f}_3 = \tilde{u}_s^a,\; \tilde{f}_4 = \tilde{d}_u^b;\; s,t,u,v = 1,\ldots,6;\; a,b = 1,2,3$

$$-i\Big(Y[\tilde{u}_s^{b*},\tilde{d}_v^b,\tilde{d}_u^{a*},\tilde{u}_t^a] + \delta^{ab}Y[\tilde{u}_s^{a*},\tilde{u}_t^a,\tilde{d}_u^{b*},\tilde{d}_v^b]\Big),$$

$\tilde{f}_1 = \tilde{u}_t^a,\; \tilde{f}_2 = \tilde{d}_v^b,\; \tilde{f}_3 = \tilde{u}_s^b,\; \tilde{f}_4 = \tilde{d}_u^a;\; s,t,u,v = 1,\ldots,6;\; a,b = 1,2,3$

$$-iY[\tilde{q}_s^{a*},\tilde{q}_t^a,\tilde{e}_u^*,\tilde{e}_v]$$

$\tilde{f}_1 = \tilde{q}_t^a,\; \tilde{f}_2 = \tilde{e}_v,\; \tilde{f}_3 = \tilde{q}_s^a,\; \tilde{f}_4 = \tilde{e}_u;q = u,d;s,t,u,v = 1,\ldots,6;a = 1,2,3$

$$-iY[\tilde{q}_s^{a*},\tilde{q}_t^a,\tilde{\nu}_i^*,\tilde{\nu}_i]$$

$\tilde{f}_1 = \tilde{q}_t^a,\; \tilde{f}_2 = \tilde{\nu}_i,\; \tilde{f}_3 = \tilde{q}_s^a,\; \tilde{f}_4 = \tilde{\nu}_i;\; q = u,d;\; s,t = 1,\ldots,6;\; i,a = 1,2,3$

$$-iY[\tilde{u}_s^{a*},\tilde{d}_t^a,\tilde{e}_u^*,\tilde{\nu}_i]$$

$\tilde{f}_1 = \tilde{d}_t^a,\; \tilde{f}_2 = \tilde{\nu}_i,\; \tilde{f}_3 = \tilde{u}_s^a,\; \tilde{f}_4 = \tilde{e}_u;\; s,t,u = 1,\ldots,6;\; i,a = 1,2,3$

(Fig.9.17 contd.)

$$-iY[\tilde{d}_s^{a*}, \tilde{u}_t^a, \tilde{\nu}_i^*, \tilde{e}_u]$$
$$\tilde{f}_1 = \tilde{u}_t^a,\ \tilde{f}_2 = \tilde{e}_u,\ \tilde{f}_3 = \tilde{d}_s^a,\ \tilde{f}_4 = \tilde{\nu}_i;\ s,t,u = 1,\dots,6;\ i,a = 1,2,3$$

$$-i\left(Y[\tilde{e}_s^*, \tilde{e}_t, \tilde{e}_u^*, \tilde{e}_v] + Y[\tilde{e}_u^*, \tilde{e}_t, \tilde{e}_s^*, \tilde{e}_v] + Y[\tilde{e}_s^*, \tilde{e}_v, \tilde{e}_u^*, \tilde{e}_t] + Y[\tilde{e}_u^*, \tilde{e}_v, \tilde{e}_t^*, \tilde{e}_t]\right),$$
$$\tilde{f}_1 = \tilde{e}_t,\ \tilde{f}_2 = \tilde{e}_v,\ \tilde{f}_3 = \tilde{e}_s,\ \tilde{f}_4 = \tilde{e}_u;\ s,t,u,v = 1,\dots,6$$

$$-iY[\tilde{e}_s^*, \tilde{e}_t, \tilde{\nu}_i^*, \tilde{\nu}_j],\quad \tilde{f}_1 = \tilde{e}_t,\ \tilde{f}_2 = \tilde{\nu}_j,\ \tilde{f}_3 = \tilde{e}_s,\ \tilde{f}_4 = \tilde{\nu}_i;\ s,t = 1,\dots,6;$$
$$i,j = 1,2,3$$

$$-i\frac{g_2^2}{4}\left(1 + \tan^2\theta_W\right)\left(\delta_{ik}\delta_{jl} + \delta_{il}\delta_{jk}\right),\quad \tilde{f}_1 = \tilde{\nu}_i, \tilde{f}_2 = \tilde{\nu}_j, \tilde{f}_3 = \tilde{\nu}_k, \tilde{f}_4 = \tilde{\nu}_l;$$
$$i,j,k,l = 1,2,3$$

Legend

sfermion ....>..........

Fig. 9.17. Four–sfermion vertices, cf. (9.80). The vertices for other combinations of incoming and outgoing sfermions, including different combinations of color indices a, b, vanish.

Appendix B

Higgs Vertices in the MSSM

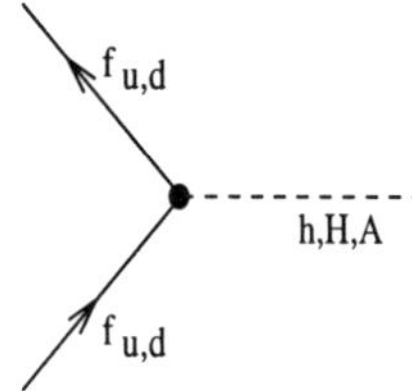

(Fig.10.3)

$$-ig_2\frac{m_f}{2M_W}X_{h,H,A}$$

	u	d
X_h	$\cos\alpha/\sin\beta$	$-\sin\alpha/\cos\beta$
X_H	$\sin\alpha/\sin\beta$	$\cos\alpha/\cos\beta$
X_A	$-i\gamma_5\cot\beta$	$-i\gamma_5\tan\beta$

c.f. $X_h \to 1$ in the SM

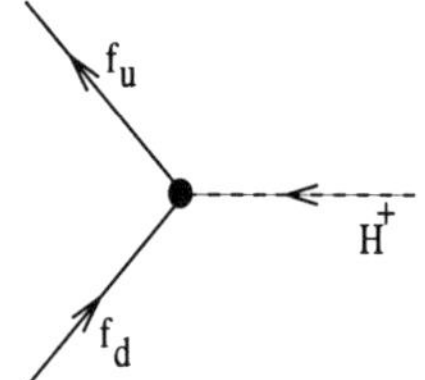

$$ig_2(m_u\cot\beta\ P_L + m_d\tan\beta\ P_R)/(\sqrt{2}M_W)$$

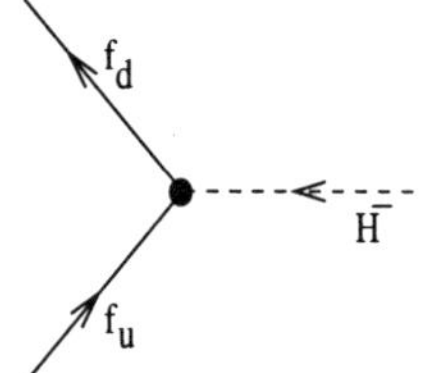

$$ig_2(m_d\tan\beta\ P_L + m_u\cot\beta\ P_R)/(\sqrt{2}M_W)$$

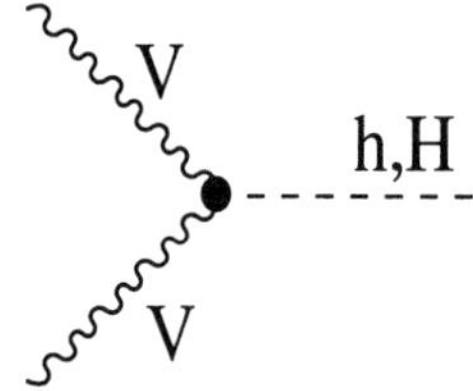

$$ig_2 m_V \eta^{\mu\nu} Y_{h,H}$$

VV	WW	ZZ
Y_h	$\sin(\beta-\alpha)$	$\sin(\beta-\alpha)/\cos\theta_W$
Y_H	$\cos(\beta-\alpha)$	$\cos(\beta-\alpha)/\cos\theta_W$

c.f. in the SM, $Y_h \to 1$ and $(\cos\theta_W)^{-1}$
for WW and ZZ respectively.

(Fig.10.3 cont.)

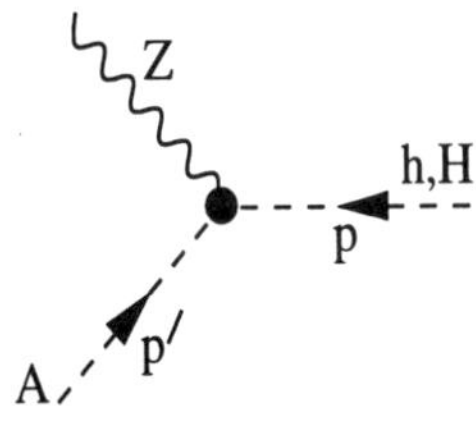

$$\frac{g_2 (p - p')^\mu}{2 \cos \theta_W} Y_{h,H}$$

Y_h	$\cos(\beta - \alpha)$
Y_H	$-\sin(\beta - \alpha)$

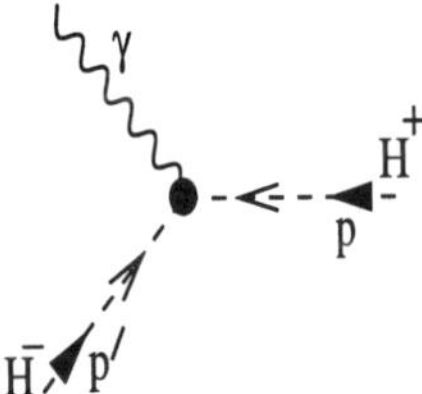

$$- ie(p - p')^\mu$$

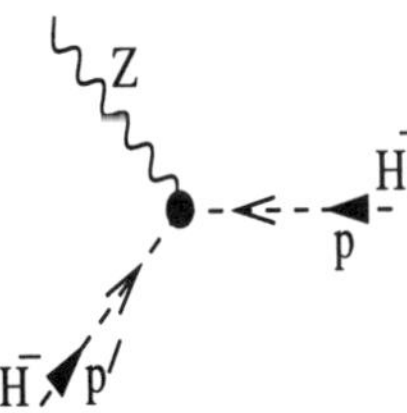

$$- \frac{ig_2 \cos 2\theta_W}{2 \cos \theta_W}(p - p')^\mu$$

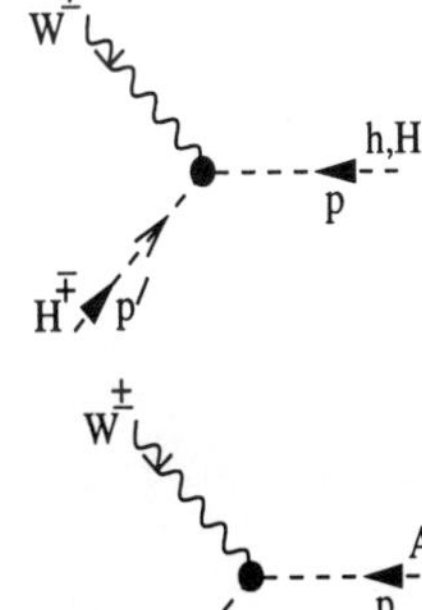

$$\mp \frac{ig_2 (p - p')^\mu}{2} Y_{h,H}$$

Y_h	$\cos(\beta - \alpha)$
Y_H	$-\sin(\beta - \alpha)$

$$\frac{g_2}{2}(p - p')^\mu$$

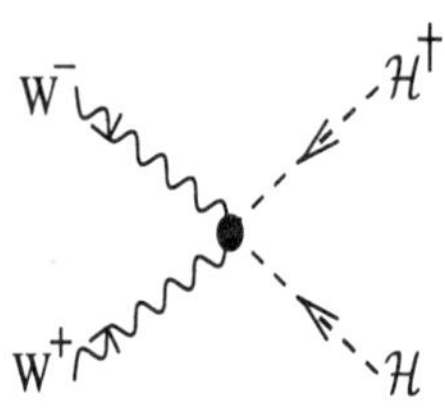

(Fig.10.3 contd.)

$$i\frac{g_2^2}{2}\eta^{\mu\nu}, \quad \begin{aligned} &\mathcal{H} = h, H, A, H^+\\ &\mathcal{H}^\dagger = h, H, A, H^- \end{aligned}$$

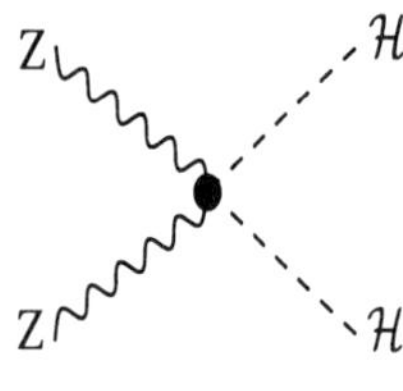

$$i\frac{g_2^2}{2\cos^2\theta_W}\eta^{\mu\nu}, \quad \mathcal{H} = h, H, A$$

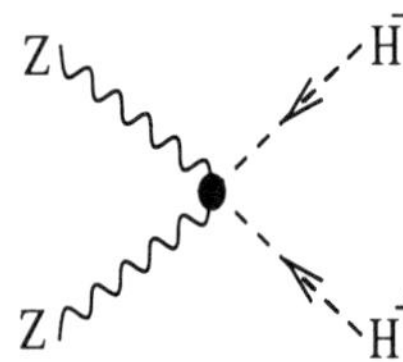

$$i\frac{g_2^2\cos^2 2\theta_W}{2\cos^2\theta_W}\eta^{\mu\nu}$$

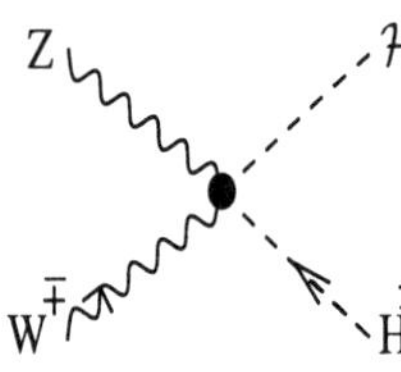

$$i\frac{g_2^2\sin^2\theta_W}{2\cos\theta_W}\eta^{\mu\nu}K_{\mathcal{H}},$$

	K_h	K_H	K_A
	$-\cos(\beta-\alpha)$	$\sin(\beta-\alpha)$	$\mp i$

$$\mathcal{H} = h, H, A$$

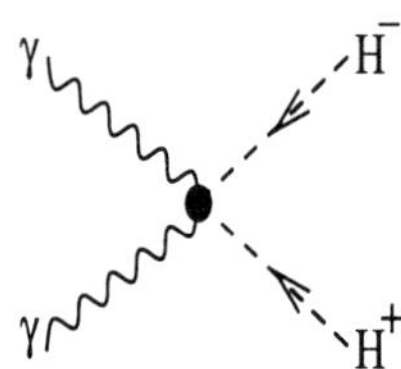

$$2ie^2\eta^{\mu\nu}$$

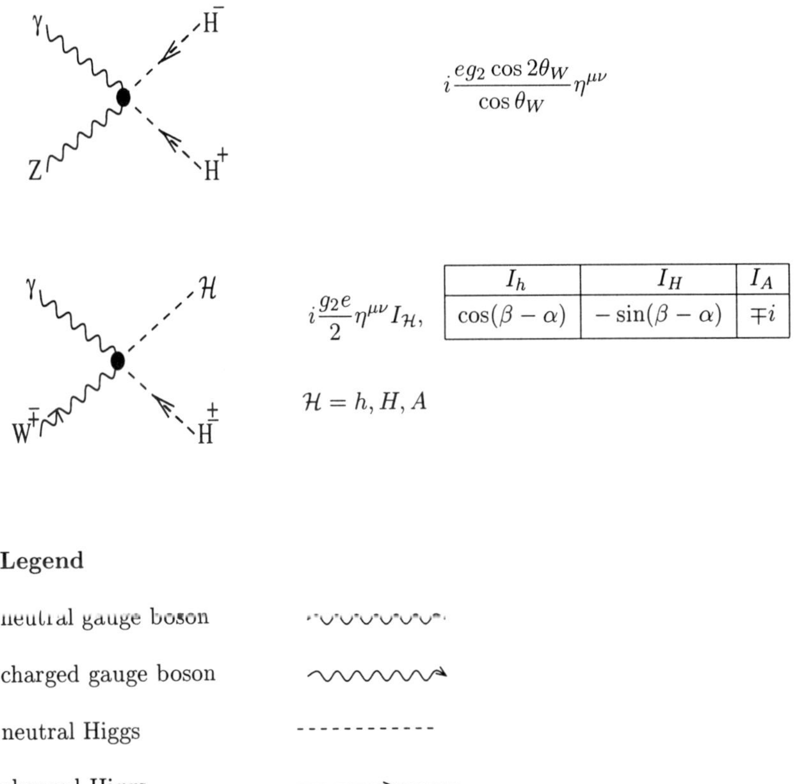

$$i\frac{eg_2\cos 2\theta_W}{\cos\theta_W}\eta^{\mu\nu}$$

$$i\frac{g_2 e}{2}\eta^{\mu\nu}I_{\mathcal{H}},$$

	I_h	I_H	I_A
	$\cos(\beta-\alpha)$	$-\sin(\beta-\alpha)$	$\mp i$

$$\mathcal{H}=h,H,A$$

Legend

neutral gauge boson	
charged gauge boson	
neutral Higgs	
charged Higgs	

Fig. 10.3: Vertex couplings of Higgs bosons with other particles

(Fig.10.4)

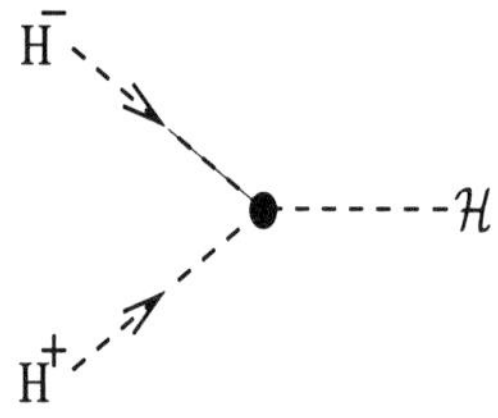

$$-ig_2 A_{\mathcal{H}}$$

A_h	$M_W \sin(\beta - \alpha) + \dfrac{M_Z}{2\cos\theta_W}\cos 2\beta \sin(\beta + \alpha)$
A_H	$M_W \cos(\beta - \alpha) - \dfrac{M_Z}{2\cos\theta_W}\cos 2\beta \cos(\beta + \alpha)$

$$\mathcal{H} = h, H$$

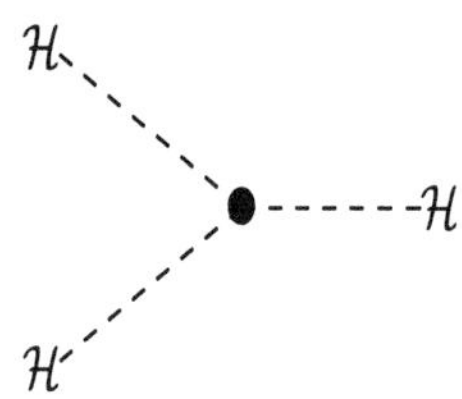

$$-3i\,\frac{g_2 M_Z}{2\cos\theta_W}\,B_{\mathcal{H}}$$

B_h	$\cos 2\alpha \sin(\beta + \alpha)$
B_H	$\cos 2\alpha \cos(\beta + \alpha)$

$$\mathcal{H} = h, H$$

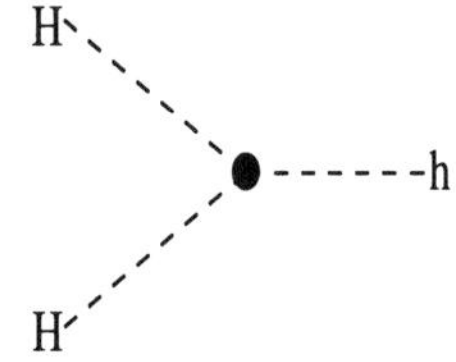

$$i\,\frac{g_2 M_Z}{2\cos\theta_W}\{2\sin 2\alpha \cos(\beta + \alpha) + \sin(\beta + \alpha)\cos 2\alpha\}$$

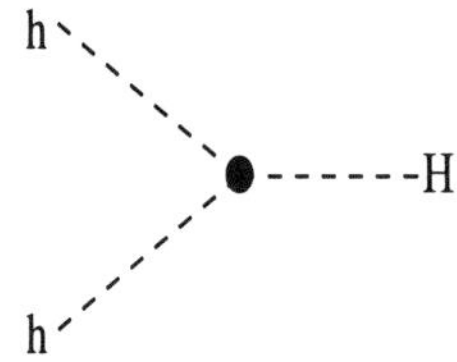

$$-\,i\,\frac{g_2 M_Z}{2\cos\theta_W}\{2\sin 2\alpha \sin(\beta + \alpha) - \cos(\beta + \alpha)\cos 2\alpha\}$$

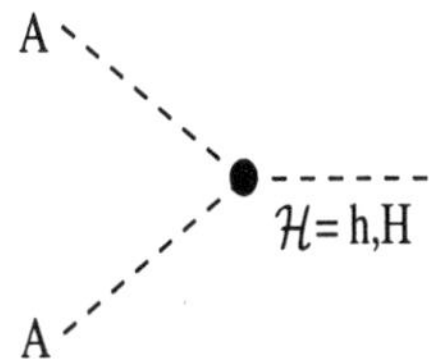

$$-i\frac{g_2 M_Z}{2\cos\theta_W}\cos 2\beta\, D_{\mathcal{H}}$$

D_h	$\sin(\beta+\alpha)$
D_H	$-\cos(\beta+\alpha)$

$$\mathcal{H}=h,H$$

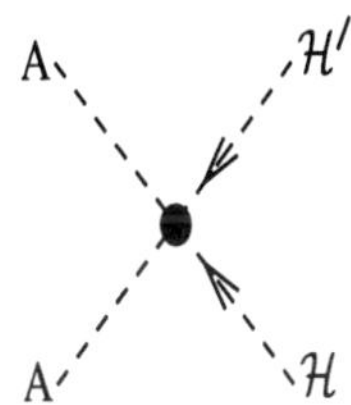

$$-i\frac{g_2^2}{4\cos^2\theta_W}\cos 2\beta\, A_{\mathcal{H}\mathcal{H}'}$$

$A_{H^+H^-}$	$\cos 2\beta$
A_{hh}	$\cos 2\alpha$
A_{hH}	$\sin 2\alpha$
A_{HH}	$-\cos 2\alpha$
A_{AA}	$3\cos 2\beta$

$$\mathcal{H}\mathcal{H}'=H^+H^-,hh,hH,HH,AA$$

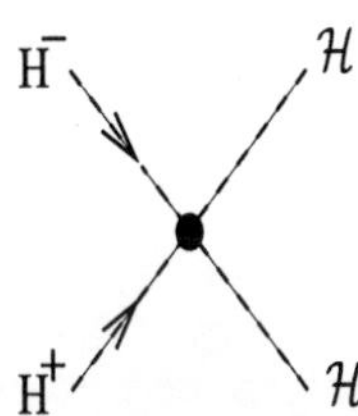

$$-i\frac{g_2^2}{2\cos^2\theta_W}\cos^2 2\beta$$

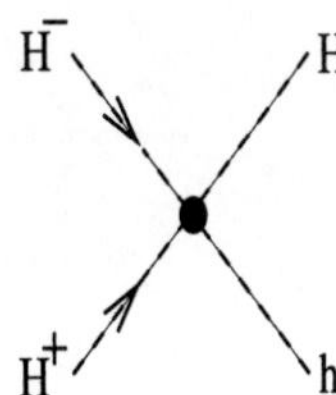

$$-i\frac{g_2^2}{4}\left(1\mp\sin 2\beta\sin 2\alpha\pm\tan^2\theta_W\cos 2\beta\cos 2\alpha\right)$$

$$\mathcal{H}=\begin{matrix}h\\H\end{matrix}$$

$$-i\frac{g_2^2}{4}\left(\sin 2\beta\cos 2\alpha+\tan^2\theta_W\cos 2\beta\sin 2\alpha\right)$$

(Fig.10.4 contd.)

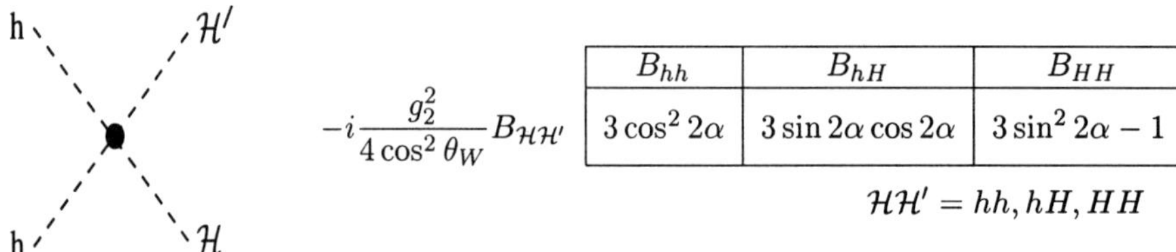

$$-i\frac{g_2^2}{4\cos^2\theta_W}B_{\mathcal{H}\mathcal{H}'}$$

B_{hh}	B_{hH}	B_{HH}
$3\cos^2 2\alpha$	$3\sin 2\alpha\cos 2\alpha$	$3\sin^2 2\alpha - 1$

$$\mathcal{H}\mathcal{H}' = hh, hH, HH$$

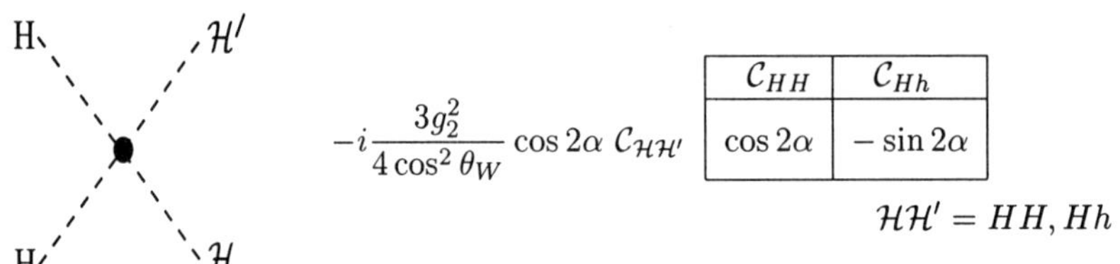

$$-i\frac{3g_2^2}{4\cos^2\theta_W}\cos 2\alpha\ \mathcal{C}_{\mathcal{H}\mathcal{H}'}$$

$\mathcal{C}_{HH}$	$\mathcal{C}_{Hh}$
$\cos 2\alpha$	$-\sin 2\alpha$

$$\mathcal{H}\mathcal{H}' = HH, Hh$$

Legend

neutral Higgs   - - - - - - - - - - - -

charged Higgs   - - - - - - ->- - - - -

Fig. 10.4: Vertices for Higgs self-couplings

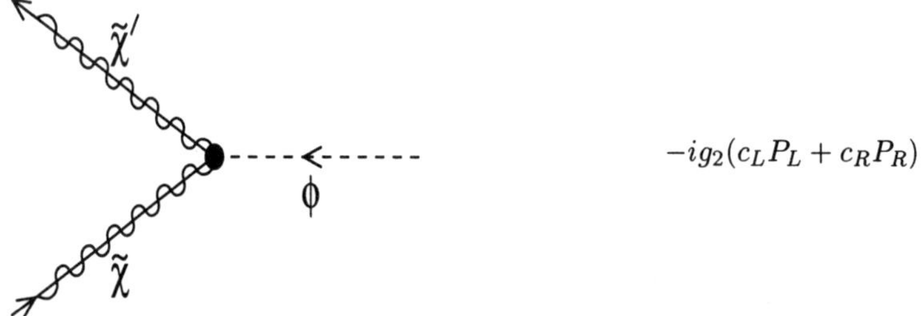

$$-ig_2(c_L P_L + c_R P_R)$$

ϕ	$\tilde\chi$	$\tilde\chi'$	c_L	c_R
H^-	$\tilde\chi_k^+,\, k=1,2$	$\tilde\chi_l^0,\, l=1,\cdots,4$	$Q_{lk}'^L$	$Q_{lk}'^R$
H^+	$\tilde\chi_l^0,\, l=1,\cdots,4$	$\tilde\chi_k^+,\,=1,2$	$Q_{lk}'^{R\star}$	$Q_{lk}'^{L\star}$
H	$\tilde\chi_m^+,\, m=1,2$	$\tilde\chi_k^+,\, k=1,2$	$Q_{mk}^\star \cos\alpha + S_{mk}^\star \sin\alpha$	$Q_{km}\cos\alpha + S_{km}\sin\alpha$
h	$\tilde\chi_m^+,\, m=1,2$	$\tilde\chi_k^+,\, k=1,2$	$-Q_{mk}^\star \sin\alpha + S_{mk}^\star \cos\alpha$	$-Q_{km}\sin\alpha + S_{km}\cos\alpha$
A	$\tilde\chi_m^+,\, m=1,2$	$\tilde\chi_k^+,\, k=1,2$	$-iQ_{mk}^\star \sin\beta - iS_{mk}^\star \cos\beta$	$iQ_{km}\sin\beta + iS_{km}\cos\beta$
H	$\tilde\chi_l^0,\, l=1,\cdots,4$	$\tilde\chi_n^0,\, n=1,\cdots,4$	$Q_{ln}''^\star \cos\alpha - S_{ln}''^\star \sin\alpha$	$Q_{nl}''\cos\alpha - S_{nl}''\sin\alpha$
h	$\tilde\chi_l^0,\, l=1,\cdots,4$	$\tilde\chi_n^0,\, n=1,\cdots,4$	$-Q_{ln}''^\star \sin\alpha - S_{ln}''^\star \cos\alpha$	$-Q_{nl}''\sin\alpha - S_{nl}''\cos\alpha$
A	$\tilde\chi_l^0,\, l=1,\cdots,4$	$\tilde\chi_n^0,\, n=1,\cdots,4$	$-iQ_{ln}''^\star \sin\beta + iS_{ln}''^\star \cos\beta$	$iQ_{nl}''\sin\beta - iS_{nl}''\cos\beta$

Legend

Higgs boson	$------\!\!>-----$
chargino or neutralino	$\sim\!\sim\!\sim\!\sim\!\sim\!\sim\!\!>$

Fig. 10.6. Higgs interactions with charginos and neutralinos. All non-vanishing coefficients c_L and c_R are listed in the above Table. The coupling factors Q, S, Q'^L, Q'^R, Q''^L, Q''^R and S'' have been defined in (10.36). Note that the extra factor of 2 for the case of two Majorana neutralinos has been taken into account.

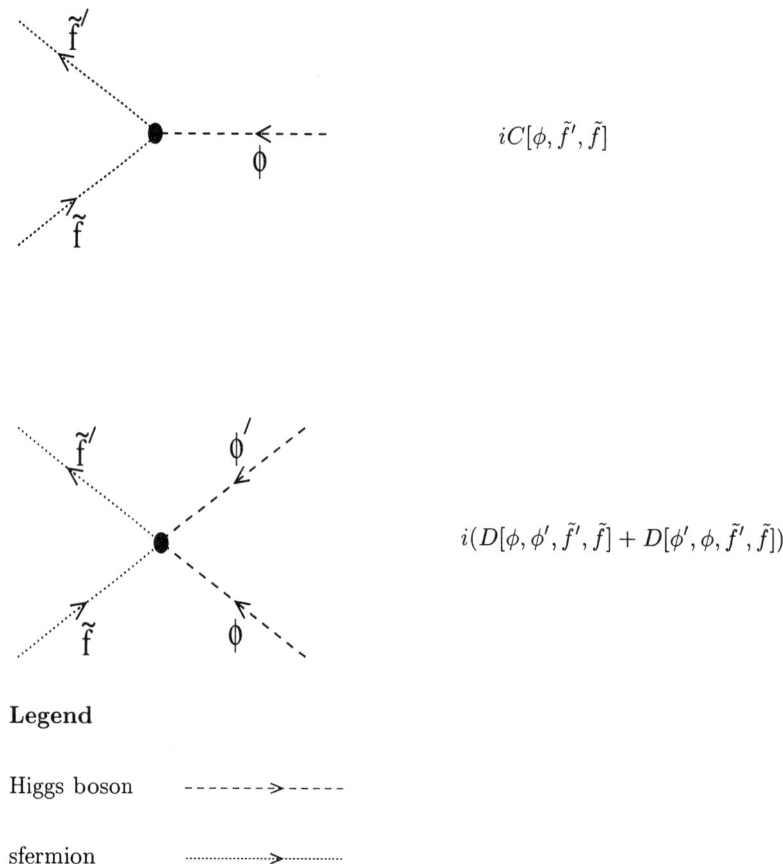

$$iC[\phi, \tilde{f}', \tilde{f}]$$

$$i(D[\phi, \phi', \tilde{f}', \tilde{f}] + D[\phi', \phi, \tilde{f}', \tilde{f}])$$

Legend

Higgs boson

sfermion

Fig. 10.7. Higgs interactions with sfermions. The coupling factors $C[\phi, \tilde{f}', \tilde{f}]$ have been defined in (10.42) and (10.46) for sleptons and squarks respectively. The corresponding quartic couplings $D[\phi, \phi', \tilde{f}', \tilde{f}]$ are given respectively in (10.44) and (10.48). For $\phi \neq \phi'$, only one of the two contributions to the quartic Feynman rule is nonzero, whereas for $\phi = \phi'$ both contributions are equal. This automatically takes into account the factor of two appearing in Feynman rules for identical particles.

Bibliography

A

S. Abdullin *et al.* in *Proc. 1996 DPF/DPB Studies on New Directions for High Energy Physics*, http://www.slac.stanford.edu/pubs/snowmass96/.

K. Abe *et al.*, (2001), ACFA LC Working Group Collaboration, *Particle Physics Experiments at JLC*, hep-ph/0109166.

T. Abe *et al.*, (American Linear Collider Working Group), *Linear Collider Resource Book for Snowmass* 2001, hep-ex/0106055–058.

J.A. Aguilar-Saavedra *et al.* (2001), *TESLA Technical Design Report, Part III, Physics in an e^+e^- Linear Collider*, hep-ph/0106315.

A. Ali and D. London, Phys. Rept. **320** (1999) 79.

B. Allanach *et al.*, *The Snowmass Points and Slopes: Benchmarks for SUSY Searches*, Eur. Phys. J. **C25** (2002) 113.

G. Altarelli, *The Development of Perturbative QCD*, World Scientific (Singapore, 1994).

J. Amundson *et al.*, *Report of the Supersymmetry Theory Subgroup*, in S. Abdullin *et al.*, *loc. cit.*, *Bibl.*, p655, hep-ph/9609374.

R. Arnowitt, A.H. Chamseddine and P. Nath, *Applied $N = 1$ Supergravity*, World Scientific (Singapore, 1984).

R. Arnowitt and P. Nath, *Proc. VII J.E. Swieca Summer School*, ed. E. Eboli (World Scientific, Singapore 1994).

The ATLAS Collaboration, *Detector and Physics Performance Technical Design Report*, Vols. 1 and 2 (CERN, Geneva), CERN/LHCC/99-14, 99-15 (1999), Ch.14, *Supersymmetry*.

B

H. Baer *et. al. Low Energy Supersymmetry Phenomenology*, hep-ph/9503479.

J.A. Bagger, in *QCD and Beyond*, TASI 95 lectures, University of Colorado, Boulder, 1995, hep-ph/9604232.

D. Bailin and A. Love, *Supersymmetric Gauge Field Theory and String Theory*, Institute of Physics Publishing (Bristol, 1994).

R. Barbieri, Riv. Nuov. Cim. **11** (1998) 1.

V.D. Barger and R.N. Phillips, *Collider Physics*, Addison-Wesley (Redwood City, CA, 1987).

F.A. Berezin, *The Theory of Second Quantization*, Academic Press (New York, 1996); *Introduction to Superanalysis*, D. Reidel (Dordrecht, 1987).

A. Bilal, *Introduction to Supersymmetry*, hep-th/0101055;

J.D. Bjorken and S.D. Drell, *Relativistic Quantum Mechanics*; *Relativistic Quantum Fields*, McGraw Hill (New York, 1964).

J.L. Buchbinder and S.M. Kuzenko, *Ideas and Methods of Supersymmetry and Supergravity or a walk through superspace*, Institute of Physics (Bristol, 1995).

W. Buchmüller and M. Plumacher, Phys. Rept. **320** (1999) 329.

C

M. Carena, R.L. Culbertson, S. Eno, H.J. Frish and S. Mrenna, Rev. Mod. Phys. **71** (1999) 937.

M. Carena and H.E. Haber, Prog. Part. Nucl. Sci. **50** (2003) 63.

L. Castellini, R.D. Auria and P. Fré, *Supergravity and Superstrings*, Vol. 2, World Scientific (Singapore, 1991).

T.P. Cheng and L.-F. Li, *Gauge Theory of Elementary Particle Physics*, Clarendon Press (Oxford, 1984).

M. Chen, C. Dionisi, M. Martinez and X. Tata, Phys. Rept. **159** (1988) 201.

S. Coleman, *Aspects of Symmetry*, Cambridge University Press (Cambridge, 1985).

F. Cooper, A.V. Khare and U.P. Sukhatme, *Supersymmetry and Quantum Mechanics*, World Scientific (Singapore, 2001).

C. Csaki, Mod. Phys. Lett. **A11** (1996) 549.

M. Cvetic and P. Langacker (eds.), *SUSY 97*, Nucl. Phys. B (Proc. Suppl.) **62** (1998).

D

S. Dawson, in *Supersymmetry, Supergravity and Supercolliders*, TASI lectures, Univ. Colorado, Boulder 1997, ed. J.A. Bagger, World Scientific (Singapore, 1999).

B. De Witt, *Supermanifolds*, Cambridge University Press (Cambridge, 1984).

J.-P. Derendinger and C. Lucchesi (ed.), Proc. Conf. *Quantum Aspects of Gauge Theories, Supersymmetry and Unification*, Fortsch. Phys. **47** (1999) 1.

M. Dine and A. Kusenko, *The Origin of the Matter-Antimatter Asymmetry*, hep-ph/0303065.

K.R. Dienes, Phys. Rept. **287** (1997) 447.

J.F. Donoghue, E. Golowicz and B.R. Holstein, *Dynamics of the Standard Model*, Cambridge University Press (Cambridge, 1992).

N. Dragon, U. Ellwanger and M.G. Schmidt, Prog. Part. Nucl. Sci. **18** (1987) 1.

M. Drees, in *Field Theoretical Methods in Fundamental Physics* (ed. C.K. Lee, Min Eun Sa, Seoul 1997).

S. Duplij, W. Siegel and J. Bagger (eds.), *Concise Encyclopedia of Supersymmetry and Non-commutative Structures in Mathematics and Physics*, Kluwer (Dordecht, 2003).

E

E. Eichten, I. Hinchliffe, K.D. Lane and C. Quigg, Rev. Mod. Phys. **56** (1984) 579; *ibid.* **58** (1986) 1065.

J.R. Ellis #1, in *Lake Louise Winter Institute* 1986: 0225.

J.R. Ellis #2, in *Beatenberg 2001, High Energy Physics*, hep-ph/0203114.

J.R. Ellis #3, in *AIP Conf. Proc.*, **562** (2001) 9.

G. Eigen, R. Gaitskell, G.D. Kribs and K.T. Matchev in *DPF Summer Study on the Future of Particle Physics* (Snowmass 2001), http://www.snowmass2001.org/, p342, hep-ph/0112312..

F

L.D. Faddeev in *Les Houches 1975 Proceedings, Methods in Field Theory*, eds. R. Balian and J. Zinn-Justin, North Holland (Amsterdam, 1976).

P. Fayet, Phys. Rept. **105** (1984) 461.

P. Fayet and S. Ferrara, Phys. Rept. **105** (1984) 5.

J.M. Figueora-O'Farrill, *Busstep Lectures on Supersymmetry*, hep-th/0109172.

J.L. Feng, *Supersymmetry and Cosmology*, hep-ph/0405215.

J.L. Feng and M.M. Nojiri, hep-ph/0210390, to appear as a chapter in *Supersymmetry and the Linear Collider*, (eds. D. Miller, K. Fujii and A. Soni), World Scientific (Singapore).

S. Ferrara #1, Phys. Rept. **105** (1984) 5.

S. Ferrara (ed.) #2, *Supersymmetry*, North Holland/World Scientific (Amsterdam/ Singapore, 1987), Vols. 1 and 2.

R.D. Field, *Applications of Perturbative QCD*, Addision-Wesley (Redwood City, CA, 1989).

P.G.O. Freund, *Introduction to Supersymmetry*, Cambridge University Press (Cambridge, 1986).

G

A.S. Galperin, E.A. Ivanov, V.I. Ogievetsky and E.S. Sokatchev, *Harmonic Superspaces*, Cambridge University Press (New York, 2001).

S.J. Gates, M.T. Grisaru, M. Rocek and W. Siegel, *Superspace or One Thousand And One Lessons in Supersymmetry*, Benjamin-Cummings (Reading, Mass. 1983).

H. Georgi, *Weak Interactions and Modern Particle Theory*, Benjamin-Cummings (Menlo Park, CA., 1984).

G.F. Giudice and R. Rattazzi, Phys. Rept. **322** (1999) 419.

A. Giveon and D. Kutasov, Rev. Mod. Phys. **71** (1999) 983.

R.M. Godbole in *Batavia 2000, Physics and experiments with future linear e^+e^- colliders* (Fermilab), p136, hep-ph/0102191.

K. Griest and M. Kamionkowski, Phys. Rept. **333** (2000) 309.

J.F. Gunion, H.E. Haber, G.L. Kane and S. Dawson, *The Higgs Hunter's Guide*, Addision-Wesley (Redwood City, CA, 1990).

J.F. Gunion in *Santa Barbara 1996, Future High Energy Colliders*, hep-ph/9704349.

H

H.E. Haber and G.L. Kane, Phys. Rept. **117** (1985) 75.

H.E. Haber, Nucl. Phys. Proc. Suppl. **101** (2001) 217.

W. Heitler, *Quantum Theory of Radiation*, Clarendon Press (Oxford, 1954).

I. Hinchliffe, Ann. Rev. Nucl. Part. Sci. **36** (1986) 505.

I

C. Itzykson and J-B. Zuber, *Quantum Field Theory*, McGraw Hill, International Edition (Singapore, 1985).

J

G. Jungman, M. Kamionkowski and K. Greist, Phys. Rept. **267** (1996) 195.

K

J. Kalinowski, Acta Phys. Polon. **B33** (2002) 3869.

G.L. Kane (ed.), *Perspectives on Higgs Physics*, World Scientific (Singapore, 1993).

G.L. Kane (ed.), *Perspectives on Supersymmetry*, World Scientific (Singapore, 1998).

G.L. Kane and M.A. Shifman (ed.), *The Supersymmetric World: The Beginning of the Theory*, World Scientific (River Edge, NJ, 2000).

G.L. Kane, *Weak Scale Supersymmetry*, TASI 2001 (Boulder) lectures, hep-ph/0202185.

D.I. Kazakov, Phys. Rept. **344** (2001) 309.

S.F. King, Rept. Prog. Phys. **67** (2004) 107.

E.W. Kolb and M.S. Turner, *The Early Universe*, Addison-Wesley (Redwood City, CA., 1990).

C.F. Kolda, Nucl. Phys. Proc. Suppl. **62** (1998) 266.

L

A.B. Lahanas and D.V. Nanopoulos, Phys. Rept. **145** (1987) 1.

L.D. Landau and I.M. Liefshitz #1, *The Classical Theory of Fields*, Pergamon Press (Oxford, 1987).

L.D. Landau and I.M. Liefshitz #2, *Quantum Mechanics*, Pergamon Press, (Oxford, 1977).

P. Langacker (ed.), *Precision Tests of the Standard Model*, World Scientific (Singapore, 1995).

E. Leader and E. Predazzi, *An Introduction to Gauge Theories and Modern Particle Physics*, Vols. 1 and 2, Cambridge University Press (Cambridge, 1990).

J. Louis, I. Brunner and S.J. Huber, *The Supersymmetric Standard Model*, hep-ph/9811341.

J. Łopuszhanski, *Introduction to Symmetry and Supersymmetry in Quantum Field Theory*, World Scientific (Singapore, 1990).

J.D. Lykken in Boulder 1996, *Fields, strings and duality*, p85, hep-th/9612114.

M

S.P. Martin, *A Supersymmetry Primer* in (G.L. Kane, ed.) *Perspectives on Supersymmetry*, *loc. cit.*

W. Miller, *Symmetry Groups and Their Applications*, Academic Press (New York, 1972).

S.P. Misra, *Introduction to Supersymmetry and Supergravity*, Wiley Eastern (New Delhi, 1992).

R.N. Mohapatra, *Unification and Supersymmetry*, Springer Verlag (New York, 1986).

R.N. Mohapatra and A. Rasin (eds.), *Supersymmetry 96: Theoretical Perspectives and Experimental Outlook*, Nucl. Phys. B (Proc. Suppl.) **52A** (1997).

H.W. Müller-Kirsten and A. Wiedemann, *Supersymmetry: an Introduction with Conceptual and Calculational Details*, World Scientific (Singapore, 1987).

H. Murayama and M.E. Peskin, Ann. Rev. Nucl. Part. Sci. **46** (1996) 533.

N

P. Nath and P. Zerwas (eds.), *Supersymmetry and Unification of Fundamental Interactions*, Proc. 10th Intl. Conf. SUSY02 (Hamburg, DESY, 2002).

H.P. Nilles, Phys. Rept. **110** (1984) 1.

O

K.A. Olive, S. Rudaz and M.A. Shifman (eds.), *SUSY 30*, Nucl. Phys. B (Proc. Suppl.) **101** (2001).

K.A. Olive, *Introduction to supersymmetry: astrophysical and phenomenological constraints*, Course 5, Les Houches LXXI, p221 (Springer-Verlag, Berlin 2000).

P

Particle Data Group, *Review of Particle Physics*, Eur. J. Phys. **C15** (2000) 1.

M.E. Peskin and D.V. Schroeder, *An Introduction to Quantum Field Theory*, Addison-Wesley (Redwood City, CA., 1995).

M.E. Peskin #1, in *Fields, Strings and Duality*, TASI lectures, Univ. Colorado, Boulder 1996, eds. C. Efthimion and B. Greene, World Scientific (Singapore, 1997).

M.E. Peskin #2, *Supersymmetry: the Next Spectroscopy*, hep-ph/0212204.

S. Pokorski, *Gauge Field Theories*, Cambridge University Press (Cambridge, 1987).

N. Polonsky, *Supersymmetry: Structure and Phenomena*, Springer-Verlag (Berlin, 2002).

E.R. Poppitz and S. Trivedi, Ann. Rev. Nucl. Part. Sci. **48** (1998) 307.

J.R. Primack, B. Sadoulet and D. Seckel, Ann. Rev. Nucl. Part. Sci. **38** (1988) 751.

Q

C. Quigg, *Gauge Theories of the Strong, Weak and Electromagnetic Interactions*, Benjamin-Cummings (Reading, Mass. 1983).

R

S. Raby, *Desperately seeking supersymmetry [SUSY]*, hep-ph/0401155.

R. Rajaraman, *Solitons and Instantons: An Introduction to Solitons and Instantons in Quantum Field Theory*, North-Holland (Amsterdam, 1982).

P. Ramond, *Journeys Beyond the Standard Model*, Perseus Books (Reading Mass., 1999).

J. Rich, D.L. Owen and M. Spiro, Phys. Rept. **151** (1987) 239.

A. Riotto and M. Trodden, Ann. Rev. Nucl. Part. Sci. **49** (1999) 35.

G.G. Ross, *Grand Unified Theories*, Benjamin-Cummings (Menlo Park, CA., 1984).

P. Roy and V. Singh (eds.), *Supersymmetry and Supergravity: Nonperturbative QCD*, Springer Lecture Notes in Physics **208**, Springer-Verlag (Berlin, 1984).

S

J.J. Sakurai, *Advanced Quantum Mechanics*, Addison-Wesley (Reading, Mass., 1967).

A. Salam and E. Sezgin (eds.), *Supergravities in diverse dimensions*, North-Holland/World Scientific (Amsterdam/Singapore, 1989).

A. Salam and J. Strathdee, Fortschr. Phys. **26** (1978) 57.

M.A. Shifman (ed.), *The Many Faces of the Superworld*, World Scientific (Singapore, 2000).

M.F. Sohnius, Phys. Rept. **128** (1985) 31.

P.P. Srivastava, *Supersymmetry, Superfields and Supergravity*, Adam Hilger (Bristol, 1986).

T

X. Tata #1, *Supersymmetry : Where It Is and How To Find It*, TASI 1995 (Boulder) lectures, hep-ph/9510287.

X. Tata #2, *What Is Supersymmetry and How Do We Find It*, 1997 Andre Swieca Summer School (Sao Paulo) lectures, hep-ph/9706307.

S.B. Treiman, R. Jackiw and D.J. Gross, *Current Algebra and Its Applications*, Princeton University Press (Princeton, 1977).

V

P. van Nieuwenhuizen, Phys. Rept. **68** (1981) 189.

W

R.M. Wald, *General Relativity*, University of Chicago Press (Chicago, 1984).

S. Weinberg #1, *Gravitation and Cosmology*, John Wiley (New York, 1972).

S. Weinberg #2, *The Quantum Theory of Fields*, Vols. 1 and 2, Cambridge University Press (Cambridge, 1995).

S. Weinberg #3, *The Quantum Theory of Fields*, Vol. 3: *Supersymmetry*, Cambridge University Press (Cambridge, 2000).

J. Wess and V.P. Akulov (eds.) *D. Volkov Memorial Volume*, (Springer-Verlag, Berlin, 1998).

J. Wess and J.A. Bagger, *Supersymmetry and Supergravity*, Princeton University Press, 2nd edition (Princeton, 1993).

P. West (ed.) #1, *Supersymmetry: A Decade of Development*, Adam Hilger (Bristol, 1986).

P. West #2, *Introduction to Supersymmetry and Supergravity*, World Scientific (Singapore, 1990).

Z

A. Zee, *Quantum Field Theory in a Nutshell*, Princeton University Press (Princeton and Oxford, 2003).

B. Zumino, Phys. Rept. **104** (1984) 113.

INDEX

H